Lecture Notes in Computer Scie

T0238678

Commenced Publication in 1973
Founding and Former Series Editors:
Gerhard Goos, Juris Hartmanis, and Jan van Leeuwen

Ignacio Rojas   Gonzalo Joya
Joan Gabestany (Eds.)

Advances in Computational Intelligence

12th International Work-Conference
on Artificial Neural Networks, IWANN 2013
Puerto de la Cruz, Tenerife, Spain, June 12-14, 2013
Proceedings, Part I

 Springer

Volume Editors

Ignacio Rojas
University of Granada
Department of Computer Architecture
and Computer Technology
18071 Granada, Spain
E-mail: irojas@ugr.es

Gonzalo Joya
University of Malaga
Department of Electronics Technology
29071 Malaga, Spain
E-mail: gjoya@uma.es

Joan Gabestany
Universitat Politecnica de Catalunya
Department of Electronics Engineering
08034 Barcelona, Spain
E-mail: cabestany@aha-dee.upc.edu

ISSN 0302-9743   e-ISSN 1611-3349
ISBN 978-3-642-38678-7   e-ISBN 978-3-642-38679-4
DOI 10.1007/978-3-642-38679-4
Springer Heidelberg Dordrecht London New York

Library of Congress Control Number: 2013938983

CR Subject Classification (1998): J.3, I.2, I.5, C.2.4, H.3.4, D.1, D.2

LNCS Sublibrary: SL 1 – Theoretical Computer Science and General Issues

Typesetting: Camera-ready by author, data conversion by Scientific Publishing Services, Chennai, India

Printed on acid-free paper

Springer is part of Springer Science+Business Media (www.springer.com)

Preface

We are proud to present the set of final accepted papers of the 12th edition of the IWANN conference—International Work-Conference on Artificial Neural Networks—held in Puerto de la Cruz, Tenerife (Spain), during June 12–14, 2013.

IWANN is a biennial conference that seeks to provide a discussion forum for scientists, engineers, educators, and students on the latest ideas and realizations in the foundations, theory, models, and applications of hybrid systems inspired on nature (neural networks, fuzzy logic and evolutionary systems) as well as in emerging areas related to the above items. It also aims to create a friendly environment that could lead to the establishment of scientific collaborations and exchanges among attendees. The proceedings include all the presented communications at the conference. The publication of an extended version of selected papers in a special issue of several specialized journals (such as *Neurocomputing, Soft Computing* and *Neural Proccesing Letters*) is also foreseen. Since the first edition in Granada (LNCS 540, 1991), the conference has evolved and matured. The list of topics in the successive Call for Papers has also evolved, resulting in the following list for the present edition:

1. **Mathematical and theoretical methods in computational intelligence**. Mathematics for neural networks. RBF structures. Self-organizing networks and methods. Support vector machines and kernel methods. Fuzzy logic. Evolutionary and genetic algorithms.
2. **Neurocomputational formulations**. Single-neuron modelling. Perceptual modelling. System-level neural modelling. Spiking neurons. Models of biological learning.
3. **Learning and adaptation**. Adaptive systems. Imitation learning. Reconfigurable systems. Supervised, non-supervised, reinforcement and statistical algorithms.
4. **Emulation of cognitive functions**. Decision making. Multi-agent systems. Sensor mesh. Natural language. Pattern recognition. Perceptual and motor functions (visual, auditory, tactile, virtual reality, etc.). Robotics. Planning motor control.
5. **Bio-inspired systems and neuro-engineering**. Embedded intelligent systems. Evolvable computing. Evolving hardware. Microelectronics for neural, fuzzy and bioinspired systems. Neural prostheses. Retinomorphic systems. Brain-computer interfaces (BCI). Nanosystems. Nanocognitive systems.
6. **Advanced topics in computational intelligence**. Intelligent networks. Knowledge-intensive problem-solving techniques. Multi-sensor data fusion using computational intelligence. Search and meta-heuristics. Soft computing. Neuro-fuzzy systems. Neuro-evolutionary systems. Neuro-swarm. Hybridization with novel computing paradigms.

7. **Applications**. Expert Systems. Image and Signal Processing. Ambient intelligence. Biomimetic applications. System identification, process control, and manufacturing. Computational Biology and bioinformatics. Parallel and distributed computing. Human-computer Interaction, Internet modeling, communication and networking. Intelligent systems in education. Human-robot interaction. Multi-agent Systems. Time series analysis and prediction. Data mining and knowledge discovery.

At the end of the submission process, and after a careful peer review and evaluation process (each submission was reviewed by at least 2, and on average 2.9, Program Committee members or additional reviewers), 116 papers were accepted for oral or poster presentation, according to the recommendations of reviewers and the authors' preferences.

It is important to note, that for the sake of consistency and readability of the book, the presented papers are not organized as they were presented in the IWANN 2013 sessions, but are classified into 16 chapters. The organization of the papers is in two volumes arranged following the topics list included in the call for papers. The first volume (LNCS 7902), entitled *Advances in Computational Intelligence. Part I* is divided into nine main parts and includes the contributions on:

1. Invited Talks to IWANN 2013
2. Applications of Computational Intelligence
3. Hybrid Intelligent Systems
4. Kernel Methods and SVM
5. Learning and Adaptation
6. Mathematical and Theorical Methods in Computational Intelligence
7. Data Mining with Evolutionary Computation and ANN
8. Self-Organizing Network
9. Advances in Computational intelligence

In the second volume (LNCS 7903), entitled *Advances in Computational Intelligence. Part II* is divided into seven main parts and includes the contributions on:

1. Metaheuristics
2. Bioinformatics/Biomedicine in Computational Intelligence
3. Fuzzy Logic and Soft Computing Applications
4. Artificial Intelligence and Games
5. Biological and Bio-inspired Dynamical Systems for Computational Intelligence
6. Brain-Computer Interfaces and Neurotechnologies
7. Video and Image Processing

The 12th edition of the IWANN conference was organized by the University of Granada, University of Malaga, Polytechnical University of Catalonia, and University of La Laguna, together with the Spanish Chapter of the IEEE Computational Intelligence Society. We wish to thank to the Spanish Ministerio de

Ciencia e Innovacion and the University of La Laguna for their support and grants.

We would also like to express our gratitude to the members of the different committees for their support, collaboration, and good work. We especially thank the Local Committe, Program Committe, the reviewers, and special session organizers. Finally, we want to thank Springer, and especially Alfred Hoffman and Anna Kramer for their continuous support and cooperation.

June 2013

Ignacio Rojas
Gonzalo Joya
Joan Cabestany

Organization

Program Committee

Leopoldo Acosta
Vanessa Aguiar-Pulido   University of Coruña
Arnulfo Alanis Garza   Instituto Tecnologico de Tijuana
Ali Fuat Alkaya
Amparo Alonso-Betanzos   University of A Coruña
Juan Antonio Alvarez-García   University of Seville
Jhon Edgar Amaya   University of Tachira (UNET)
Gabriela Andrejkova
Cesar Andres   Universidad Complutense de Madrid
Miguel Angel Lopez
Anastassia Angelopoulou   University of Westminster
Plamen Angelov   Lancaster University
Davide Anguita   University of Genoa
Cecilio Angulo   Universitat Politcnica de Catalunya
Angelo Arleo   CNRS - University Pierre and Marie Curie Paris VI
Corneliu Arsene   SC IPA SA
Miguel Atencia
Jorge Azorín-López   University of Alicante
Davide Bacciu   University of Pisa
Javier Bajo   Universidad Politécnica de Madrid
Juan Pedro Bandera Rubio   ISIS Group, University of Malaga
Cristian Barrué   Technical University of Catalunya
Andrzej Bartoszewicz   Technical University of Lodz
Bruno Baruque   University of Burgos
David Becerra Alonso   University of the West of Scotland
Lluís Belanche   UPC
Sergio Bermejo   UPC
Julio Brito
Joan Cabestany   Universitat Politecnica de Catalunya
Inma P. Cabrera   University of Malaga
Tomasa Calvo   Universidad de Alcala
Jose Luis Calvo Rolle   Universidad de A Corunha
Francesco Camastra   University of Naples Parthenope
Carlos Carrascosa   GTI-IA DSIC Universidad Politecnica de Valencia
Luis Castedo   Universidad de A Coruña
Pedro Castillo   UGR

Pablo García Sánchez   University of Granada
Maribel García-Arenas   Universidad de Granada
Patrick Garda   Université Pierre et Marie Curie - Paris 6
Peter Gloesekoetter   Muenster University of Applied Sciences
Juan Gomez Romero   Universidad Carlos III de Madrid
Juan Gorriz   University of Granada
Karl Goser   Technical University Dortmund
Bernard Gosselin   University of Mons
Manuel Grana   University of Basque Country
Bertha Guijarro-Berdiñas   University of A Coruña
Nicolás Guil Mata   University of Málaga
Alberto Guillen   University of Granada
Barbara Hammer   Barbara Hammer
Yadira Hernandez
Francisco Herrera   University of Granada
Álvaro Herrero   University of Burgos
Cesar Hervas
Tom Heskes   Radboud University Nijmegen
Wei-Chiang Hong   Oriental Institute of Technology
Pedro Isasi   University Carlos III of Madrid
Jose M. Jerez   Universidad de Málaga
M. Dolores Jimenez-Lopez   Rovira i Virgili University
Juan Luis Jiménez Laredo   University of Granada
Gonzalo Joya
Vicente Julian   GTI-IA DSIC UPV
Christian Jutten   University of Grenoble
Paul Keether
Fernando L. Pelayo   University of Castilla - La Mancha
Alberto Labarga   University of Granada
Raul Lara Cabrera
Nuno Lau   Universidade de Aveiro
Amaury Lendasse   Aalto University
Miguel Lopez   University of Granada
Otoniel Lopez Granado   Miguel Hernandez University
Rafael Marcos Luque Baena   University of Málaga, Spain
Ezequiel López-Rubio   University of Málaga
Kurosh Madani   LISSI / Université PARIS-EST Creteil (UPEC)
Mario Martin   Universitat Politecnica de Catalunya
Bonifacio Martin Del Brio   University of Zaragoza
Jose D. Martin-Guerrero   University of Valencia
Luis Martí   Universidad Carlos III de Madrid
Francisco Martínez Estudillo   ETEA
José Luis Martínez Martínez   Universty of Castilla-La Mancha
José Fco. Martínez-Trinidad   INAOE

Francesco Masulli   University of Genova
Montserrat Mateos   Universidad Pontificia de Salamanca
Jesús Medina-Moreno   University of Cadiz
Maria Belen Melian Batista
Mercedes Merayo   Universidad Complutense de Madrid
Jj Merelo   Universidad de Granada
Gustavo Meschino   Universidad Nacional de Mar del Plata
Jose M. Molina   Universidad Carlos III de Madrid
Jose Muñoz   University of Malaga
Augusto Montisci   University of Cagliari
Antonio Mora   University of Granada
Angel Mora Bonilla   University of Malaga
Claudio Moraga   European Centre for Soft Computing
Gines Moreno   University of Castilla-La Mancha
Jose Andres Moreno
Juan Moreno Garcia   Universidad de Castilla-La Mancha
J. Marcos Moreno Vega
Susana Muñoz Hernández   Technical University of Madrid
Nadia Nedjah   State University of Rio de Janeiro
Manuel Nuñez   UCM
Erkk Oja   Aalto University
Manuel Ojeda-Aciego   University of Malaga
Sorin Olaru   Suplec
Iván Olier   The University of Manchester
Madalina Olteanu   SAMM, Université Paris 1
Julio Ortega   Universidad de Granada
Alfonso Ortega de La Puente
Emilio Ortiz-García   Universidad de Alcala
Osvaldo Pacheco   Universidade de Aveiro
Esteban José Palomo   University of Málaga, Spain
Diego Pardo   Barcelona Tech
Miguel Angel Patricio   Universidad Carlos III de Madrid
Alejandro Pazos Sierra
Francisco J. Pelayo   University of Granada
Jose Manuel Perez Lorenzo   Universidad de Jaen
Vincenzo Piuri   University of Milan
Hector Pomares   University of Granada
Alberto Prieto   Universidad de Granada
Alexandra Psarrou   University of Westminster
Francisco A. Pujol   University of Alicante
Pablo Rabanal   Universidad Complutense de Madrid
Juan Rabuñal   University of A Coruña
Vladimir Rasvan
Leonardo Reyneri   Politecnico di Torino

Ismael Rodriguez — Universidad Complutense de Madrid
Juan A. Rodriguez — Universidad de Malaga
Sara Rodríguez — University of Salamanca
Ignacio Rojas — University of Granada
Samuel Romero-Garcia — University of Granada
Ricardo Ron-Angevin — University of Málaga
Eduardo Ros
Fabrice Rossi — SAMM - Université Paris 1
Fernando Rubio — Universidad Complutense de Madrid
Ulrich Rueckert — University of Paderborn
Addisson Salazar — Universidad Politecnica Valencia
Sancho Salcedo-Sanz — Universidad de Alcalá
Albert Samà — Universitat Politècnica de Catalunya
Francisco Sandoval — Universidad de Málaga
Jose Santos — University of A Coruña
Jose A. Seoane — University of Bristol
Eduardo Serrano — UAM
Olli Simula — Helsinki University of Technology
Evgeny Skvortsov
Jordi Solé-Casals — Universitat de Vic
Shiliang Sun
Carmen Paz Suárez Araujo
Peter Szolgay — Pazmany Peter Catholic University
Javier Sánchez-Monedero
Ricardo Tellez — Pal Robotics
Ana Maria Tome — Universidade Aveiro
Carme Torras — IRI (CSIC-UPC)
Claude Touzet — University of Provence
Olga Valenzuela — University of Granada
Miguel Ángel Veganzones — Universidad del País Vasco (UPV/EHU)
Francisco Velasco-Alvarez — Universidad de Málaga
Sergio Velastin — Kingston University
Marley Vellasco — PUC-Rio
Alfredo Vellido — Universitat Politecnica de Catalunya
Francisco J Veredas — Universidad de Málaga
Michel Verleysen — Universite catholique de Louvain
Thomas Villman — University of Applied Sciences Mittweida
Changjiu Zhou — Singapore Polytechnic
Ahmed Zobaa — University of Exeter
Pedro Zufiria — Universidad Politécnica de Madrid

Additional Reviewers

Acosta, Leopoldo
Affenzeller, Michael
Alonso, Concepcion
Angulo, Cecilio
Azorín-López, Jorge
Calabuig, Daniel
Cazorla, Miguel
Chaczko, Zenon
Comas, Diego Sebastián
Contreras, Roxana
Danciu, Daniela
Estévez, José Ignacio
Gabadinho, Alexis
Garcia-Rodriguez, Jose
Luque Baena, Rafael Marcos

López-Rubio, Ezequiel
Marichal, Graciliano Nicolas
Moreno, Jose Andres
Orts Escolano, Sergio
Palomo, Esteban José
Perez-Suay, Adrian
Prochazka, Ales
Ruiz de Angulo, Vicente
Selisteanu, Dan
Stoica, Cristina
Takac, Boris
Toledo, Pedro A.
Veredas, Francisco J.
Zhou, Yang

Table of Contents – Part I

Hybrid Intelligent Systems

Kernel Methods and SVM

Learning and Adaptation

Mathematical and Theorical Methods in Computational Intelligence

Data Mining with Evolutionary Computation and ANN

Self Organizing Network

Advances in Computational Intelligence

Table of Contents – Part II

Metaheuristics

Bioinformatics/Biomedicine in Computational Intelligence

Fuzzy Logic and Soft Computing Applications

Artificial Intelligence and Games

Biological and Bio-inspired Dynamical Systems for Computational Intelligence

Brain-Computer Interfaces and Neurotechnologies

Video and Image Processing

It's as Easy as ABC

Introducing Anthropology-Based Computing

John N.A. Brown[1,2]

Erasmus Mundus Joint Doctoral Programme in Interactive and Cognitive Environments
[1] Alpen-Adria Universität Klagenfurt,
Universitätsstraße 65-67, 9020 Klagenfurt, Austria
[2] Universität Politècnica de Catalunya
Neàpolis Building, Rbla. Exposició 59-69, 08800 Vilanova i la Geltrú, Spain
jna.brown@gmail.com

Abstract. The evolution and adaptation of humans is intractably intertwined with the evolution and adaptation of our technology. This was true when we added wooden handles to stone adzes, and it is true today. Weiser and Brown warned that ubiquitous computing would require the development of Calm Technology, a total change to the way in which we interact with computers, so that the entire process could become more suitable to human perceptual abilities and limitations. Our failure to do so is responsible for a daily onslaught of injury and death, from Carpal Tunnel Syndrome to plane crashes. We propose a solution based on one of the underlying concepts of Artificial Neural Networks. For decades, attempts have been made to recreate the basic physiological step of human information processing. It is time to go one step further and consider the basic human parameters of input and output, as proposed by Weiser and Brown. Their term Calm Technology has been modified and re-defined over the past twenty years and their true intent has been lost. In order to avoid the territorial battles that surround the term, and in an attempt to assist engineers and human factors specialists in their efforts to preserve health and save lives, we introduce the concept of Anthropology-Based Computing (ABC). We define ABC as any input and output design parameters based on the basic physiological, psychological and social requirements of the human animal in our natural habitat.

Keywords: Ubiquitous Computing, Calm Technology, Anthropology-Based Computing, Human Factors, Anthropology, Human-Computer Interaction, Cross-Generational Habit.

My purpose here today is to introduce a concept that is – theoretically – very simple to understand and yet – in practical terms – a little hard to apply in the real world; a concept I call *Anthropology-Based Computing* (ABC). Once introductions have been made, I'm going to challenge you to bring ABC to life in your field, and I hope by then that you'll see why you should try, for your sake and for the sake of your children.

I. Rojas, G. Joya, and J. Cabestany (Eds.): IWANN 2013, Part I, LNCS 7902, pp. 1–16, 2013.

Why "*Anthropology-Based Computing*"? Why not "Human-Centered Computing" or "Intuitive Computing" or "Naturalistic Computing"? Why not "Affective Computing" or "Ubiquitous Computing" "Pervasive Computing" or even "Calm Computing"? The reason is quite simple: I call it *Anthropology-Based Computing* because it is a basic grasp of the fundamentals of Anthropology that can remake traditional Human-Computer Interaction into science that is truly based on humans, instead of the motley series of brilliant innovations, glorified mistakes and obscure *Cross-Generational Habits* that we practice today.

I know that's a harsh statement. I'll provide evidence to support those allegations, and I will make more harsh statements, too. There will be more of them, but there will also be proposals for a solution.

1   Ubiquitous Computing and the Need for "Calm"

Approximately 22 years ago, Mark Weiser published a monograph in which he described the changing nature of the basic relationship between humans and computers in three stages, and suggested that the changes to date could be described numerically. In the first stage, the age of Mainframe Computing, many people had shared access to a single computer, so the ratio there was lopsided: many people in a relationship with a single computer. At the time that Weiser was writing his monograph it was becoming common for individuals (or families) to have private computers, and he called this the age of Personal Computing. Here, the ratio was balanced with a single computer serving a single human... or, at least, serving a single group of humans, defined by shared work or living space. Based on increasing miniaturisation and the proliferation of embedded networks, Weiser predicted that the near future would see individuals using a great many computers and, as a name for the coming age, coined the term Ubiquitous Computing. This glorious future, this age of technological ubiquity, has come to pass, and our man/machine ratio is again imbalanced, with a single human using dozens of computers, or maybe hundreds or even thousands, if you are willing to count the machines that are used individually by many individuals – machines like the servers run by Google or by Wikipedia, the website that gives you your local weather, or the one that supports the television weather forecaster in their nightly performance. This is, of course, in addition to the computers or computerised systems you share at work, at home and during your transition between the two, to say nothing of whichever personal systems you are using either deliberately or – as is much more likely – without any conscious awareness at all.

Please don't take that as an insult. Being unaware – or rather – consciously unaware – of a pervasive technology is in fact a sign of being very well-adjusted. I promise you that your grandparents were the same way, as were their grandparents before them. All that changes from generation to generation is the technology we learn to ignore.

To explain this perspective and in an attempt to establish a historical foundation for the theory I am presenting, let's look back in time just a little ways, back to when some of our earliest hominoid ancestors were standing around in the tall grass at the base of a tree on a piece of land that will eventually migrate halfway around the globe and become known as North America.

1.1   The Proto-Prosimian and the Workstation, Part 1

Imagine that you are a prosimian about 45 million years ago. You and your cousins are a handsome bunch, skinny fingers on hands and feet, long bushy tails, and pleasantly pointed faces that are not as pointed as the faces of some of your more distant relations. You and your cousins stand taller than those others and you can look at a single object with both eyes at once, giving you a good sense of distance. You can hear the others all around you, but you are trying to focus on a piece of ripe fruit that you are holding in both hands. The problem is that your cousins would also like to focus on that particular piece of fruit and they are prepared to take it from you. You switch your attention back and forth between the environment around you and the task at hand. If one of them touches you, you will jump reflexively and that may mean dropping the fruit. What should you do?

Let's look for a similar case a little more recently, about 30 million years later, that is, about 14 or 15 million years ago. This time you don't have a tail at all and the fingers on your feet are a little shorter than those on your hands. Your face is flatter, too, and your family is spending time out of the trees. You and your cousins can often be found now, sitting on your haunches in the tall grass of the savannah of Southern Europe. Like your great-great-great (x 10^6) grandfather in the previous example, your hands and fingers seem perfectly made for holding a piece of ripe fruit in front of your nose and mouth. Like his cousins, yours are interested in the fruit you are holding. You could run away into the savannah, or you could run back into the forest and climb a tree, or you could sit right where you are, and try to eat the fruit quickly. Your cerebral neocortex is more developed than that of your ancestor, probably allowing you to better compare possible outcomes, but what will you do?

Let's add one more hungry hominoid to our history by skipping forward another 6 or 7 million years, and look in on the flower of another branch of our family tree. In the meantime, our ancestors have followed the receding warmth and migrated into Africa, so we find this fellow, the biggest so far, with a much bigger head and a much bigger brain, facing the same small problem of trying to decide whether to run from his cousins or eat the piece of fruit held in his delicate, precise, and very powerful fingers. He is as related to you and me as he is to chimpanzees and bonobos. You are in his shoes now, figuratively-speaking, so what do you do?

What do you do?

The truth is that there are many possible answers to that question. The more so, because I haven't clearly defined all of the parameters, but that's because the specific answer isn't the least bit important to me. What should we care about the choice made by three fictional characters from millions of years ago? What is important here is what they would not, what they could not possibly have done.

Not one of our remote ancestors, from this series of examples or from any other, could possibly have approached the problem facing them by thinking like a computer.

If we are going to consider the issue of Human-Computer Interaction in any kind of a meaningful way, then we must remember that humans come from stock that spent millions of years not thinking like computers.

I believe that this is the basic fact that Weiser was trying to communicate when he stressed that the ubiquitous presence of computers in our lives would make it absolutely necessary to change the way they work. Please read the original papers, not my impression or someone else's. Weiser said:

> "Calmness is a new challenge that UC brings to computing. When computers are used behind closed doors by experts, calmness is relevant to only a few. Computers for personal use have focused on the excitement of interaction. But when computers are all around, so that we want to compute while doing something else and have more time to be more fully human, we must radically rethink the goals, context and technology of the computer and all the other technology crowding into our lives. Calmness is a fundamental challenge for all technological design of the next fifty years."[1]

If that quotation is not relevant to you for some reason that I cannot imagine, then please apply your own logic to the concept of life with ubiquitous computers. As an aide, I offer another intellectual exercise.

1.2   The Smartphone as a Source of Constant Interruption

Most all of us now use smart phones. No one I know uses all of the technology available on their phone and most, in fact, use only the technology that reflects tool use with which they were already familiar before the smart phone entered their lives. If you played games before, you are likely playing games on your phone now…
…possibly right now. If you were already texting or taking photos or using a calendar to track your schedule, or using alarms to make sure you pay attention to your calendar, then you probably switched all or some of your previous attention in these areas to your smart phone. We can all agree that some undefined subset of the technology we carry around in our most personal computer is very useful. Maybe we can also all agree that some other subset of the phone's technology is not at all useful, so long as these subsets remain undefined, allowing each of us to preserve our personal likes and dislikes.

So, we have established that there is some unspecified probability that we all use smartphones and consider them useful. How do we feel when our phone rings? Is it different than how we feel when someone else's phone rings? When our phone rings because a dear old friend is calling with welcome news, do we feel differently than when it is an unsolicited robocall offering an unwanted opportunity?

These events, which should inspire very different reactions and emotions, are often indistinguishable when the phone rings. It's true that this is probably no worse than it

---

[1] Weiser, M., & Brown, J. S. (1998). The coming age of calm technology (Denning, P. J. & Metcalfe, R. M., Eds). In *Beyond calculation: The next fifty years of computing* (pp. 75-85). New York, Springer.

was when we used landlines, but so what? I'm not concerned with whether or not things were better in the past; I'm concerned with the fact that we could be making them better now. What's more, I believe that we can make things vastly better now, just by applying some of the underused technology that is currently available on our computers and smartphones in combination with some well-understood psychological principles that have never been fully applied to human-computer interaction. We'll discuss that in detail a little later.

The question I want to ask you now is how your feelings about smartphones change when you hear someone else's phone ring and ring and ring again? Do you think it depends on the music they're using as a ringtone? Think again! I promise you that a collocutor who gets too many calls, or gets even a few calls at exactly the wrong time can make you hate your favourite song and literally inspire you to cringe every time you hear it. But that's also not the point I'm trying to make about ubiquitous computing. I believe that there's an easy solution to that problem, too, and that it also makes use of technology that already co-exists with all but the most basic phones. That's another topic we'll get back to later.

The question I want to ask you about ubiquitous computing, as illustrated by smartphones, is this. If alarms and alerts annoy you when they come from a smartphone at a nearby workstation, park bench or seat in a movie theatre, how much more would they annoy you if you were living inside the phone? The smart environment of the near future will surround us with computerised recommender systems, ambient information systems and distributed interfaces and displays. No one expected that the widespread dissemination of electronic mail systems would lead to incessant interruption or that text messaging functions on portable phones would mean that teenagers would be in constant low-fidelity communication throughout their waking day. How will incessant communication expand when every wall, window and door of our homes is automated? How will our "time to be more fully human" diminish when every device in and around our lives is always on the verge of demanding that we stop everything and reply to the 21st century equivalent of "Error 404".

2   Inhuman-Computer Interaction

Many research teams around the world are working on the technological side of that very issue – finding ways to make networked and embedded systems with which humans might surround themselves; systems that can anticipate human requirements and enrich our lives. But they are doing so with the same perverse idea of Human-Computer Interaction that dominates other parts of the industry. The flaw in their reasoning is obvious, but most of us are simply choosing not to consider it.

Modifying a device so that it becomes less harmful to the user is a vital step in the early evolution of any tool. This is one of the reasons that our ancestors added stone handles to stone adzes. It was a technological improvement, in that it increased the length of the lever arm and made possible a series of adaptations that led to further tool specialisation, but another major part of the improvement is that a handle made it less likely for the tool user to hurt herself. There has always been an accepted trade-off between the danger and the value of using a tool. If a tool is too dangerous, but must be used, society tends to restrict its availability. Eventually the tool is improved, given up, or moved entirely into the realm of specialist use. These matters take time, but the amount of time is dependent upon societal and governmental perception of the dangers involved. Governments and societies disagree about the danger/value ratio of handguns, but all seem to agree on restricting access to fissionable materials. The multi-purpose axe, on the other hand, has changed little in the past five thousand years. There is a version of the axe that is common to Ancient Egypt, and Medieval Europe, and modern hardware stores, but commercial and professional pressures make a wide range of axes available.

2.1   Computer-Centered Computing

Of similar value are the surface modifications of the computer mouse, the desk chair, or the computer keyboard. These small "ergonomic" adjustments do not address the fact that none of these tools are really designed based on human needs and abilities. In fact, these tools are all examples of machines that force humans to enter their world. Please allow me to explain.

The computer keyboard is based on the typewriter of the mid-twentieth century, a tool that has a rich history of bad design – at least in terms of human use. Consider that the arrangement of keys was specifically developed to slow down the rate at which people could physically type (so that the typewriter's arms would have time to fall back out of each other's way). If that is not bad enough, then please consider a syndrome, or series of related medical conditions, that used to be called "Secretary's Disease". It was very rare when first described and remained rare until the proliferation of computers for personal and professional work. The name has changed to Carpal Tunnel Syndrome, and it provides clear evidence that keyboard use is injurious. Splitting the keypad in order to relax forearm and wrist rotation is helpful, as is changing the declination and height of the typing surface, but the keyboard demands a strenuous posture, and further demands rapid, repetitive movements of the fingers while in this strenuous posture. The keyboard is not human-centered.

The desk chair is not much better. A tremendous amount of time and money is spent in the attempt to make sitting at a desk less strenuous for office workers, and this industry has followed the computer from the office to the home. There is a lot of expertise out there, but the truth is that humans evolved as walking creatures. It puts our bodies under strain to assume a seated position and that strain is greatly increased as the posture is maintained. Sitting upright for an entire working day is not human-centered.

The first computer mouse was a small block of wood, with a hollowed-out bottom, a red button and a single wire. The hollowed-out bottom was filled with 2 perpendicular wheels and the means to measure their rotation. When the block was rubbed across a level surface, the relative rotation count of the two wheels captured the direction in which the block was moving. The button nose and wire tail earned the device its name and, though it took almost twenty years for it to find a market, it is now one of the most ubiquitous devices in the world. But moving a mouse on a desk requires constant interruption of anything else one might be doing with that hand, and puts the wrist and forearm under stress that is very similar to that caused by the keyboard. Furthermore, it requires a peculiar set of very controlled motions based on the position and orientation of the desk, table, tray or other surface one is using. This means that one of two strategies is possible:

a) one holds one's arm in the air, pushing the mouse around in a state of constant dynamic tension and using the large muscles of the shoulder and elbow joints to execute precise movements, or,

b) one rests one's arm on an available surface and performs the steering action with the distal segments of the upper limb. If the elbow is resting, then mousing is accomplished by rotating the forearm, wrist and fingers. If the forearm is resting, then it is the wrist and the fingers which do all of the work. If the wrist is resting, then all of the work must be done with the fingers.

In all of these cases, muscular contraction or resting weight puts pressure on the nervous, circulatory, and musculo-skeletal systems of the limb, and precise and repetitive movements while under pressure causes repetitive strain. It can cause discomfort, numbness or outright pain to people who are using the device properly. Mousing, especially when it is done for both work and play, is not human-centered.

If the tools by which we sit at a computer and input information are not human-centered, then how can changing a software interface be considered an improvement to Human-Computer Interaction? This question is equally valid the colour of a casing, the size of a monitor, and the speed of a processor. There is another fundamental aspect of the common computer that is as pervasive as the three discussed above, and equally non-human-centered. It will be discussed in depth a little later.

2.2   Cross-Generational Habit

It is also of interest to note that, in all of the years since the mouse appeared, and despite all of the surface re-designs driven by commercial, legal or therapeutic motivations, very few have ever asked just why it is assumed that the best way to interact with a computer is to rub a block of wood on a desk. This, I believe, is due to another aspect of human behavior, our tendency to maintain with unconscious fervor the behaviors of our grandparents. This is what I call *Cross-Generational Habit*.

We hang lights or stand them, where our grandparents hung gas lights and stood lanterns or candles, as their grandparents had done before them. At a glance, this seems only practical. The first embedded electric wires ran through the spaces in walls where gas tubes had run before, so of course the same sconce would be used...
...but do you work in an office that was converted from gas to electricity? Maybe you think it is simpler than that: that we simply put the light where we need it. A desk lamp puts light on your desk, right? An overhead light brightens the whole room, doesn't it? Well, no, it doesn't. An overhead light casts shadows into corners and under desks and it stings your eyes anytime you glance upwards. It is no wonder that modern, city-raised people look up so rarely; the behavior is usually punished with stinging brightness. We build dark rooms and illuminate them with electric lights,

rather than building rooms to take advantage of distributed and diffuse lighting systems. These have been available for decades (some of the technology for centuries), but "that's not how we do it around here".

Wrapping one's foot in leather and tying it in place with laces dates back to the last Ice Age and, despite massive technological innovation in the intervening time, we still accept that as the generic model of a shoe. Have you ever wondered why we steer a car through curves on one plane (the ground), using a circular device set in a different plane? Have you ever wondered why our showers rain water down on us, when we would be better served by a shower that pushes water upwards? Why computers promote the illusion that we are typing onto a sheet of paper visible in our monitor?

We are more comfortable when surrounded by small details that are familiar. It seems to me that this is part of an evolutionary coincidence; a side effect of being mutants with big brains. You see, big brains require big skulls, and big skulls change everything.

3   Big Skulls, Brain Development, Culture, and Conformity

Genetically modern humans (GMH) have unusually big and complex upper brains, and so require unusually big skulls. There are a number of interesting theories about the way in which our brains were pressured to grow. I don't have time to discuss those today, but I would like to mention two of my favorites. Some say that our increased brain size is due to the proliferation of broad-leafed trees and other vegetation during the late Mesozoic Era, which increased the available amount of oxygen in the atmosphere. Others theorize that extra thickness in the cerebral cortex was an evolutionary response to our need for insulation from heatstroke when we moved into a diurnal life on the African savannah.

Whatever caused our swollen craniums, the fully-developed, adult-sized version would just be too big to be borne by our mothers or, if you'll pardon the pun, to be born at all. A skull big enough to house a fully-developed GMH brain cannot fit through GMH pelvic openings. We are born with a brain that grows and develops over years. GMH are born with a brain only 25% of adult size. Our closest cousins in the modern world, bonobos and chimpanzees are born with brains roughly 35% of adult size. By comparison, the brain of a capuchin or a rhesus monkey is already close to half adult size at birth and their skulls have stopped growing. A GMH brain takes about eight years to reach full size and then spends about another eight years (or more) maturing. The result of this is that our young need to be protected for at least eight years while they develop their mental abilities. I believe that this period of learning how to think while learning what to think is responsible to some degree for the richness of human culture and language. Children learn much more quickly and deeply than most people realize. You know it if a child has ever lectured you about dinosaurs or a collectable card game. If our brains learn to think while exposed to a particular home environment, wouldn't we naturally develop an understanding of the world, of how things should and must be, based on that source? If the familiar gives us comfort, couldn't some of that comfort come from the fact that a familiar environment makes it easier for the unusual (and possibly dangerous) to stand out.

To see how this might affect (and might be affected by) the ways in which we take in information, let's go way back in time again, twenty-million years further back than before. Imagine a cat-sized proto-prosimian sitting in a tree, surrounded by a large extended family. All of their pointy little snouts are sniffing at the food being held by our protagonist. This is before the era in which flowering plants and succulents will spread widely around the world, so it is unlikely that our hero is holding a piece of fruit. Let's assume, in its place, a nice juicy insect. Each cousin and sibling of our hero has their head cocked so that one eye can focus on the lovely snack. They twitch their long tails, and drool past sharply-pointed teeth.

The proto-prosimian is dealing with the situation in a very human way, processing information in a manner that is probably very similar to the way that his descendants in our earlier examples would have done it.

3.1   The Proto-Prosimian at the Workstation, Part 2

The little proto-prosimian, closer in shape to a modern mouse than to a modern man, was using the natural abilities of brain and body to deal with the problem. The cerebral neocortex, the part of the brain that lays like a wrinkled blanket overtop of all the rest, would likely have been very small in the skull of our proto-prosimian protagonist. It would be small, that is, in comparison to modern humans, but vastly bigger in comparison to earlier creatures, if it had existed in their skulls at all. Using two separate parts of the brain simultaneously, the creature is processing input in two different ways. Let's try to imagine it in more detail.

You are a proto-prosimian sitting in the crook of a branch about sixty-five million years ago. Your hands have fingers that curve only inwards, and your arms reach only about as far backwards as your peripheral vision can see. At rest, your arms fall to your sides with bent elbows and your hands overlap in front of your chest. I believe that this is your region of focus. Your precise little fingers overlap here and hold things where you can best smell and taste them. It is also the area in which you can most easily focus your eyes on near objects. Of course, this is where you want to hold your food, so that you can really focus on it. At the same time, though, you are aware of your surroundings. Your ears point outwards, shells cupping and adding directional information. Your hair detects wind movement and you sweep your tail back and forth, to add to your chances of early detection of shifting air currents. Our senses work together to form perceptual units out of these data, grouping them according to characteristics like spatial continuity, chronological coincidence and symmetry. Every now and then you pause, looking away or tipping your head to one side. These interruptions of your routine happen when you have detected something on the periphery of your attention, something that doesn't seem to fit an anticipated pattern and so might become important; something that you might need to consider more deeply. You weigh the importance of interrupting your meal and you either return to eating or drop your hands a little and stop chewing so that you can divert more of your cognitive resources to processing the information.

This is how we feel comfortable, surrounded by large slow streams of perceptual data, most of which we feel safe to ignore. Though we do not focus our attention on all of these currents of information, we feel their comforting presence around us and we believe that we can access them at any time, shifting our attention, and reassuring ourselves that all is well.

We have processed information this way for the last sixty-five million years. There have, of course, been situations where something demands immediate attention. Such situations, if they derive naturally, and if we have the opportunity to influence our own chances of survival, must trigger responses as quickly as possible. It is a survival characteristic to be able to respond quickly to stimuli that demand attention, just in case it turns out to be a matter of life or death. Similarly, it is a corresponding survival mechanism to avoid false alarms that might needlessly reduce our resources for dealing with real threats or might even desensitise us to stimuli that will be important later.

3.2 Homo Sapiens Sapiens, or Homo Sapiens Reagens

Here is another story for you; another attempt to illustrate the way that humans process information. Imagine that you are taking a refreshing walk in the woods. The sun's bright light shines warmly on the footpath, speckled in places with cool shadows cast by the leaves and branches overhead. The autumn air is cool, but warms occasionally with a breeze that smells of late summer. At times the path is narrow and the trees crowd in above you. Other times the view opens up to your left and you can see that the path has taken you to the edge of a ridge; a cliff looking down on the dappled greens and oranges of the forest far below. As you round a turn, a snake is on the path before you. The visual stimuli are sorted by your limbic system into simultaneous messages for two parts of your brain that react very differently. Your cerebral neocortex begins immediately to compare the movement, size, shape and color pattern of the snake to your database of known movements, sizes, shapes and color patterns. If your neocortex were to talk to itself, it might sound like this:

"Woah! A live, moving snake... ...Now let's see, there are red, yellow and black stripes, which means either a harmless milk snake or possibly a deadly coral snake. Which pattern is which? Is it 'red and yellow will kill a fellow', is that right? Or is it 'black and red will kill you dead'?"

At the same time, back in your limbic system, an older part of your brain would also be reacting to the same visual stimuli, without the delays of re-routing and comparisons. Its much more primitive reaction might be something like this:

"Snake! Aaugh!" Whereupon you would leap blindly away from the snake and over the edge of the cliff, starting the fall to your death before your neocortex could make a decision about the actual danger posed by the snake.

Now, if you had seen a snake on the trail before, or if you were more comfortable about the stresses inherent in walking in the woods, then you probably wouldn't have jumped to your death. Even when scared, our ingrained response can be trained to better suit a given set of environmental circumstances. A trained driver reacts to the shock of going into a skid by driving in what seems to be the wrong direction and a trained martial artist will react to a physical threat without any conscious intent at all... ...or at least, without any intent at the time. The intent of the driver and of the martial artist was impressed upon them through repetitive training at an earlier time, specifically in anticipation of the possibility that it might one day be needed.

The truth is that every day, each one of us is training to respond, or not to respond, to selected environmental stimuli. We don't even know we are doing it most of the time.

Have you learned to ignore an alarm on your smart phone? Maybe the first alarm that rings in the morning? Have you learned to ignore the train that passes near your window several times a day? We live in an environment full of stimuli we have chosen to ignore. Did you know that there is a blind spot in the middle of the image sent to your brain from each of your eyes? There are no light detectors in the spot at the back of your eyeball where the optic nerve is rooted, so there is a small spot of blindness in both of your eyes. Your brain unconsciously fills in the little blind spot by matching it with whichever colour is nearby. It's the same with your hearing. You only hear some of the noise around you and your brain fills in the blanks. This is why it is possible to sing along with a song you know, and suddenly find that you are halfway through the wrong verse. You were actually singing along with your memory of the song. It is strange to think that our senses have these limitations, and it is even stranger to think that our brain is lying to us, filling in the blanks in a way that lets us believe that we are more aware of our surroundings than we really are. But now, combine these ideas with what we have been discussing about the way that humans perceive and process information.

3.3   Sapio, Sapis, Sapit, Sapimus, Sapitis, Sapiunt

Remember those sensory streams I mentioned; those groupings of perceptual data in which we are forever floating? There's another funny thing about those patterns. If the flow of data stops while we are busily focussed on something else, our perception of the identifiable pattern in the background continues for four more seconds. That's right, for four seconds we are drifting along supported only by our own false certainty. Now it makes sense that such a system should have a buffer, otherwise we might be startled into reflexive action by even the slightest interruption in the stream. If four seconds seems like a long time to you, I assure you that I feel the same way. Fortunately, or perhaps out of simple necessity, if a pattern changes or is over-ridden by a different pattern, we respond to that instantly.

It seems to me that all of these weaknesses in our sensory systems, all of the false data we accept, must be a threat to our survival. If this seems to be so, then the logic of evolutionary forces tells us that some compensatory force must exist in our natural environment, in our culture or in our behaviour. I believe that compensatory force is teamwork. We are social animals and we watch each other's backs. While I am immersed in one sensory stream, you are immersed in another and, if all goes well, the two streams will not dry up within four seconds of each other. The more of us who participate, the better our chance of detecting an important change will be. In this way, the group has a robustness of perception that might be too demanding of an individual. This might offer an explanation of our persistent self-delusion and over-confidence. But how would this compensation function in environments where people are expected to work alone? What effect would it have if we were suddenly entrusted to work with machines, in a machine-like manner that depends on our constant awareness of minute details, rather than with colleagues who would be helping us to stay aware of these details through generalised pattern recognition and cooperation?

When this happens with the advance of personal technology, it seems to me that we have to choose between two common reactions. Either we immerse ourselves in the new technology with some degree of (possibly delusional) self-confidence, or we hide from it with some degree of (possibly delusional) certainty that we will never learn how to use it. Those of us whose brains are still learning how to think are more likely to accept new technology and try to tie it into the existing fabric of the world as they see it. The rest of us have to learn how to think differently, or hope that the way we currently think shares some common ground with the way we should think to be able to use the new technology. While we are adapting to the new technology, and trying to learn how to adapt it to better suit our use, we run the risk of making more than our usual share of mistakes. These mistakes are much more striking when we consider new technology in the workplace. In some workplaces, they can be disastrous.

4   The Human Factors

It seems that the majority of workplace accidents these days are attributed to human error. I agree with that judgement in most cases, but I can't help but wonder why everyone always restricts the meaning of that phrase to the last human to touch the part or to execute the process that failed. The truth, as I see it, is that many of the human errors responsible for accidents are errors of design; design of the system for executing actions, design of the system for monitoring performance and design of the system for maintaining equipment. It is a truism that badly-designed systems lead to badly-performed tasks. In light of what we have been discussing, how often are complex systems truly designed well?

Individual human beings tend to assume that data shows a pattern - even when it doesn't. Worse than that, we are prone to delude ourselves with false confidence about how perceptive we are. These tendencies are well-understood and it is becoming increasingly accepted that people make bad supervisors of machine-based systems, yet we persist in designing automated or semi-automated systems that require humans to process data streams rapidly and accurately.

Consider the two Human Factors interventions that have greatly reduced aviation accidents:

1) Open Channels of Communication – where Captains are required to listen to the opinions of other members of the cabin crew, even if they believe that they are fully aware of all aspects of the situation, and

2) Checklists – a written record of everything that must be done and of everything that has been done, removing false-confidence and poor memory from the equation.

This safety net hasn't prevented all accidents, but it now takes an exceptional series of unlikely circumstances or bad choices to create a hole that is big enough to fly a plane through.

4.1   Getting Lost in One's Work

On October 22nd, 2009, in the skies over the United States of America, a Northwest Airlines Airbus A320 was flying from San Diego to Minneapolis with 144 passengers on board. Flight 188 was out of radio contact for 77 minutes and flew about 160 kilometers past the airport at which they were supposed to land. Captain Tim Cheney and First Officer Richard Cole failed to respond to radio calls from more than a dozen air traffic controllers and other pilots in the area, and only realized their problem when a flight attendant called the cockpit to ask how soon they would be landing. How did it happen? Based on the findings of the investigation, I believe that it was due to a combination of bad design and entirely human responses to computer-centered computing.

To start with, there is a radio channel dedicated to constant contact with any flight in the air. This channel is supposed to be open at all times, in case there is some reason that the plane must be contacted. It is intended solely for important communication, but that is not how it is used. It is an open secret of the Airline industry and their monitoring agencies in North America that, since this channel receives a lot of unnecessary chatter, and since that chatter is distracting, many pilots turn the volume so far down as to make it silent. Of course, it would be against aviation regulations to actually turn off the channel, but lowering the volume situationally is allowed. Most pilots would tell you that if they reduce the volume during chatter, they will certainly increase it again as soon as they notice a need for it. This, of course, assumes that they will notice the need. No one has said that these two pilots had the emergency contact channel turned too low, but it would explain why they didn't hear any of the attempts to contact them. But that's not the only instance of an inappropriate but very human response to bad design in this case.

You see, the story is that Northwestern had just been purchased by Delta and that the new owners had introduced new software for booking time off work. Once they had taken off and programmed the flight into the Airbus's computer, the two pilots took out their laptops and worked together to figure out how to arrange time off so that one of them could attend an important family event. They immersed themselves in trying to figure out the software with the same depth and intensity that most of us use when trying to learn new software. Unlike our ancient ancestors, we computer users have access to all kinds of currents of sensory data that have nothing at all to do with the world around us. A computer directed to work deeply on a problem, while regularly checking for pattern changes in an outside data stream would be able to do so in two ways: true multitasking if more than one processor were available, or task-switching. Task-switching is when a computer goes back and forth between multiple tasks, often at a speed that gives the impression of multitasking. A human directed to do the same thing would have a much harder time of it.

We can only multi-task if each task uses a different type of processor – like walking and chewing gum at the same time, or talking while knitting. Real multi-tasking, say reading a book while carrying on a conversation, is beyond our normal range of capabilities. We can fake it, but we would really be task-switching, which is what we've all experienced when talking with someone who is also reading, composing or sending a text message at the same time. Their divided attention results in divided resources and that means sub-standard performance at one or both tasks. Just like the two pilots, we might believe that we can multi-task but, if so, we are lying to ourselves and running the risk of getting lost.

Any one of us who has tried to make a phone call, send an SMS, apply make-up, eat a sandwich or carry on a deep and meaningful conversation while driving has made the same sort of mistake as these two pilots. They felt safe while doing something that was clearly unsafe. This is what comes of being enveloped in a steady stream of unimportant data – especially if that stream provides constant rewards, feelings of success, and unconscious reassurances that this success could continue. That cycle of "reward-promise of further reward" is actually built into computer games, and it is very similar to how we feel when we are "in the zone". As rewarding as it seems to be, though, it is also dangerous. That rewarding data stream slips into the space in our sensory system that should be filled by important data and we fly blissfully past our targets with the volume of our internal alarm systems turned low enough to keep the screaming of our companions from disturbing us. All the while, we reassure ourselves that we are doing just fine.

4.2   On the Other Side of the Window

Marie Sklowdowska-Curie discovered polonium and radium, coined the term radioactivity and won Nobel Prizes in both Physics and Chemistry. We all remember that she carried test tubes of radium in her pockets and that the notebooks she left behind are so radioactive they must be stored in lead. We have forgotten the immediate social and commercial success of radium-based products. People bathed in Radium Hot Springs and drank Radium water for their health, and it killed a great many of them. The effects of radium, beyond the immediate suffusion of visible energy, were completely unknown. The same is true about the use of our ubiquitous computers.

Studies over the last ten years have confirmed that playing video games affects the blood flow and electrochemical activity of certain regions of the brain. More than that, many tests have shown that the size and density of the brain matter itself can be altered, changing one's ability to solve abstract problems and even improving practical, "real-world" skills. In fact, our increasing understanding of Brain Plasticity, the ability of the brain to change itself in response to deliberate or coincidental training, has led to the development of regular, regimented training programs shown to have a long term effect on the brain and on the individual's ability to use it.

Given that this is true, does it not seem strange to you that no one has measured the effect of intensive, long-term exposure to a Graphic User Interface (GUI)? We are not receiving a small dose. Most of us spend hours every day performing a wide variety of formerly diverse tasks through a GUI. Many of us spend most of our waking hours using a GUI for both work and play and, in nearly all cases, we are trying to varying degrees of success to pretend that it is not there. How is that affecting us? Are we blissfully unaware of the death we are carrying in our pockets? Have we learned to ignore the sound of history passing by our window, as our brains adapt to accidental stimuli and change in ways that we are not even trying to imagine?

I would like to propose to you that we should begin to imagine these changes; that we should begin to study them and that, while waiting for our results, we should begin to take control of the unknown experiment in which we are currently participating.

5   Anthropology-Based Computing

I am a great fan of the field of Artificial Neural Networks. I was introduced to it through Science Fiction and I have spent the bulk of my life with a layman's interest in the process of digitally emulating the manner in which information is stored and processed in the human brain. People in this field see a distinction between biological and mechanical information technology that is invisible or unimportant to many other engineers and computer scientists. That said, I believe that it is not enough to change the elemental means by which information is processed, it is necessary and, I believe, increasingly urgent, to change the way that we interact with it.

5.1   As We May Think

In 1945, the Atlantic Monthly published a monograph entitled "As We May Think". In this article, Vannevar Bush described the Memex, a special and entirely theoretical desk of the future. This desk would be able to receive, store, display, edit and delete information from around the world. His description may well be the reason that the developers of the first commercial GUIs used the analogy of the desktop. That analogy is central to our most-used computerized devices – the laptop and the smart-phone. The thing is, when Bush proposed this possible future technology he wasn't doing so just for the sake of talking about technology. Bush was saying that this access to huge amounts of data would affect our cognitive processing, changing us "as we may think". We have not followed his example, but have slipped into the pattern of developing new technology without worrying at all about how the use of it might affect human thought.

Consider the amount of time that an office worker or a researcher spends at a computer. What did we do with our days to fill the hours when they were free of emails and web-surfing and on-line videos and pictures of cats? It's hard to imagine a world in which writing to someone usually meant waiting days, weeks or even months for a response, and talking with someone usually meant being in close proximity. That was how we communicated for thousands of years, and our communicative tools evolved and adapted with us over that time. Then, in middle of the last century, everything changed and we have become faced with technology that changes faster than we can keep track of it. It is our further misfortune that the ubiquitous manifestations of that technology are tools that focus our attention on a computer interface, regardless of the nature of the task we are trying to perform.

5.2   Once and Future Thinking

In my ongoing work with Professor Martin Hitz and his Interactive Systems Research Group (ISYS) at the Alpen-Adria Universität Klagenfurt, we are developing tools to avoid the pitfalls discussed above. We are not trying to make the computer disappear, but rather to make the interface more human-centered. I'd like to tell you about two of these tools, one that is still in development and another that is already in use.

As I type this, an image appears on my monitor. It is an email notice, but not a pop-up summary of whichever email happens to have landed in my inbox. I have those turned off. Too often, they appear directly over the space at the bottom right-hand corner of the monitor where I am typing the last words on a line. I don't mind not being able to see what I'm typing, but it is well-understood that a written interruption that appears in line-of-sight during a writing task has a catastrophic effect on productivity. It can take up to twenty minutes to regain deep focus after only an unconscious moment of reading the interrupting message. On my computer, the email alert appears as an almost entirely transparent version of the icon my colleague uses to represent herself in our email exchanges. Her ghostly visage fades up and fades out again. When I'm busy, I hardly notice her – unless I am specifically waiting to hear from her. Then, a visual version of the *cocktail party effect* keeps me unconsciously alert to her possible appearance. Today I am happy to watch the image fade up and then fade out again. I want to finish these examples and conclude this paper.

As I continue, my phone rings. By this point, you would be surprised if I were to describe a common ring tone or say that I use an excerpt from a once-popular song. The ring tone is, in fact, the soft but happy voice of the same colleague. Her gentle laughter is discernible as she pronounces "dzing, dzing, dzing, dzing" - her spoken version of the sound of a phone ringing in her native Russia. Another colleague, working at the next desk, can barely hear her because the volume is set so low. Even if he does hear her, there is no chance that he might think it is his own phone ringing. To me, however, the ring tone tells me who it is, and, as with the email alert, the personal nature of the message is enough to attract my attention even at low volume.

These are just two of my own attempts at improving the computer's ability to communicate with humans in a manner less similar to two machines exchanging spools of data and more similar to the rich interaction between our proto-prosimian, prosimian and simian protagonists and their environments. Using these systems, I detect rich information from the periphery of my awareness and decide whether or not to focus on it. The system, you see, has put the information there in a manner that allows me this discerning behavior; this freedom to interact at my own discretion. I invite you to do the same thing in your own laboratories, offices and homes. Keep advancing technology, of course, but please put some thought into making the interaction more human-centered and less machine-centered. It will require some planning, and it will require the acquisition of new, truly human-centered perspective. That, I believe, is the key to what Weiser called Calm Technology. As far as I'm concerned, it's as easy as ABC.

Extreme Learning Machine:
A Robust Modeling Technique? Yes!

Amaury Lendasse[1,2,3], Anton Akusok[1], Olli Simula[1], Francesco Corona[1],
Mark van Heeswijk[1], Emil Eirola[1], and Yoan Miche[1]

[1] Information and Computer Science Department
Aalto School of Science and Technology
FI-00076 Aalto, Finland
[2] IKERBASQUE, Basque Foundation for Science
48011 Bilbao, Spain
[3] Computational Intelligence Group, Computer Science Faculty,
University of the Basque Country
Donostia/San Sebastian, Spain

Abstract. In this paper is described the original (basic) Extreme Learning Machine (ELM). Properties like robustness and sensitivity to variable selection are studied. Several extensions of the original ELM are then presented and compared. Firstly, Tikhonov-Regularized Optimally-Pruned Extreme Learning Machine (TROP-ELM) is summarized as an improvement of the Optimally-Pruned Extreme Learning Machine (OP-ELM) in the form of a L_2 regularization penalty applied within the OP-ELM. Secondly, a Methodology to Linearly Ensemble ELM (ΜЛ3-ELM) is presented in order to improve the performance of the original ELM. These methodologies (TROP-ELM and ΜЛ3-ELM) are tested against state of the art methods such as Support Vector Machines or Gaussian Processes and the original ELM and OP-ELM, on ten different data sets. A specific experiment to test the sensitivity of these methodologies to variable selection is also presented.

1   Introduction

Data sets in Machine Learning and Statistical Modeling are becoming larger: thanks to improvements in acquisition processes it becomes possible to obtain large amounts of information about a studied phenomenon, with data to analyze more abundant, in terms of number of variables and samples. While it is usually desirable to have a large data set —as opposed to a small one from which very few information is available—, it raises various problems. First of, the increase in the number of variables is likely to introduce new relevant data regarding the phenomenon at hand, but causes an accordingly high increase in the number of required samples, to avoid ill-posed problems. Irrelevant variables are also likely to appear, creating a new difficulty for the model building. The increase in the number of samples can also become problematic, for it leads to increased computational times for model building.

The Extreme Learning Machine (ELM) as presented by Huang *et al.* in [1] by its very design is fast enough to accommodate such large data sets, where other traditional

I. Rojas, G. Joya, and J. Cabestany (Eds.): IWANN 2013, Part I, LNCS 7902, pp. 17–35, 2013.

machine learning techniques have very large computational times. The main idea lies in the random initialization of the weights of a Single Hidden Layer Feedfoward Neural Network (SLFN), instead of the traditional —much more time-consuming— learning of these weights through back-propagation [2], for example. In addition to its speed, which takes the computational time down by several orders of magnitude, the ELM is usually capable to compare with state of the art machine learning algorithms in terms of performance [1].

It has however been remarked in [3] that the ELM tends to suffer from the presence of irrelevant variables in the data set, as is likely to happen when dealing with real-world data. In order to reduce the effect of such variables on the ELM model, Miche *et al.* proposed in [3,4,5] a wrapper methodology around the original ELM, which includes a neuron ranking step (via a L_1 regularization known as Lasso [6]), along with a criterion used to prune out the most irrelevant neurons of the model (regarding this criterion): the Optimally-Pruned Extreme Learning Machine (OP-ELM). Section 2 gives a short introduction to the original ELM and fixes the notations for the following presentation of the OP-ELM as proposed in [3,4,5].

Section 2 then elaborates on one problem encountered by the original OP-ELM, in the computation of the pruning criterion. The Leave-One-Out criterion is originally used in the OP-ELM for the pruning, which can be a computationally costly choice. Thanks to the use of a closed form formula (Allen's PRESS statistic [7]), its computation is nevertheless very fast, but raises numerical problems which possibly "disturb" the pruning strategy.

This proposed solution to this situation is by the use of L_2 regularization in the OP-ELM. The concept of regularization —using L_1, L_2 or other norms-based penalties on the regression weights— for regression problems has been studied extensively (see for example [8,9,10,11,12,13,6,14,15,16]) and some of the most widely used methods are presented in section 3: Lasso [6], Tikhonov regularization [14,10], but also hybrid penalties such as the Elastic Net [16].

While these penalties are either of only one kind —L_1 or L_2, traditionally—, or a hybrid using both simultaneously (see Owen's hybrid [11] for example), an approach that could be described as in cascade is used in this paper, for the TROP-ELM. Indeed, a L_1 penalty is first used to rank the neurons, followed sequentially by a L_2 penalty to prune the network accordingly. Section 4 details the approach used, by a modification of Allen's PRESS statistic [7]. This improvement of the OP-ELM is denoted as the Tikhonov-Regularized Optimally-Pruned Extreme Learning Machine (TROP-ELM) and is first introduced in [17].

The second methodology which is described is based on an ensemble of ELM. This methodology is denoted: Methodology to Linearly Ensemble ELM (MꓶE-ELM) [18,19]. Several ensemble techniques have been proposed, out of which two kinds can be distinguished: the variable weights approach and the average ones. Traditionally, average weights ensemble techniques are used and simply take an average of all the built models. While this obviously has the advantage of having immediately the weights of all models, it yields suboptimal results. The variable weights ensemble techniques try to optimize the weight of each model in the ensemble according to a criterion. Techniques such as the Genetic Algorithm [20] have been recently used for such optimization but are very time consuming. This presented methodology (MЈᕬ-ELM) proposes the use of a Leave-One-Out (LOO) output for each model and a Non-Negative constrained Least-Squares problem solving algorithm, leading to an efficient solution coupled with a short computation time. Section 5.1 details this methodology, with Section 5.2 giving a proof on the applicability of the methodology under some hypotheses.

This TROP-ELM and the MЈᕬ-ELM are tested in section 6 against three state of the art machine learning techniques (Gaussian Processes, Support Vector Machines and Multi-Layer Perceptron) but also against the original ELM and OP-ELM. The experiments are carried out using ten publicly available regression data sets and report the performances and timings for all methods. Finally, in section 7, a specific experiment to test the sensitivity of these methodologies to variable selection is also presented.

2 The Optimally-Pruned Extreme Learning Machine

2.1 The Extreme Learning Machine

The Extreme Learning Machine (ELM) algorithm is proposed by Huang *et al.* in [1] as an original way of building a Single Hidden Layer Feedforward Neural Network (SLFN). The main concept behind the ELM is the random initialization of the SLFN internal weights and biases, therefore bypassing a costly training usually performed by time-consuming algorithms (Levenberg-Marquardt [21], back-propagation [2]...).

In [1] is proposed a theorem — on which lies the efficiency of the ELM — stating that with a random initialization of the input weights and biases for the SLFN, and under the condition that the activation function is infinitely differentiable, the hidden-layer output matrix can be determined and will provide an approximation of the target values as good as wished (nonzero).

Under the conditions detailed in [22] — that is, randomly generated hidden nodes weights and bounded non-constant piecewise continuous activation function — the ELM is a universal function approximator [23,24]. It is worth noting that several possible activation functions have been investigated for the ELM nodes, for example thresholds [25], complex [26] and Radial Basis Functions [27].

In this paper, the case of single-output regression is considered, but the ELM, OP-ELM and the proposed approach in section 4 can be modified to solve multi-output regression and classification problems.

Consider a set of n distinct samples $(\mathbf{x}_i, y_i)$, $1 \le i \le n$, with $\mathbf{x}_i \in \mathbb{R}^p$ and $y_i \in \mathbb{R}$. A SLFN with m hidden neurons in the hidden layer can be expressed by the following sum

$$\sum_{i=1}^{m} \beta_i f\left(\mathbf{w}_i \mathbf{x}_j + b_i\right), \ 1 \le j \le n, \tag{1}$$

with β_i the output weights, f an activation function, $\mathbf{w}_i$ the input weights and b_i the biases. Denoting by $\hat{y}_i$ the outputs estimated by the SLFN, in the hypothetical case where the SLFN perfectly approximates the actual outputs y_i, the relation is

$$\sum_{i=1}^{m} \beta_i f\left(\mathbf{w}_i \mathbf{x}_j + b_i\right) = y_j, \ 1 \le j \le m, \tag{2}$$

which is written in matrix form as $\mathbf{H}\boldsymbol{\beta} = \mathbf{y}$, with

$$\mathbf{H} = \begin{pmatrix} f\left(\mathbf{w}_1 \mathbf{x}_1 + b_1\right) \cdots f\left(\mathbf{w}_m \mathbf{x}_1 + b_m\right) \\ \vdots \qquad \ddots \qquad \vdots \\ f\left(\mathbf{w}_1 \mathbf{x}_n + b_1\right) \cdots f\left(\mathbf{w}_m \mathbf{x}_n + b_m\right) \end{pmatrix}, \tag{3}$$

$\boldsymbol{\beta} = (\beta_1, \ldots, \beta_m)^T$ and $\mathbf{y} = (y_1, \ldots, y_n)^T$. The ELM approach is thus to initialize randomly the $\mathbf{w}_i$ and b_i and compute the output weights $\boldsymbol{\beta} = \mathbf{H}^\dagger \mathbf{y}$ by a Moore-Penrose pseudo-inverse [28] (which is identical to the Ordinary Least Squares solution for a regression problem, see section 3) of $\mathbf{H}$, $\mathbf{H}^\dagger$.

There have been recent advances based on the ELM algorithm, to improve its robustness (OP-ELM [4], CS-ELM [29]), or make it a batch algorithm, improving at each iteration (EM-ELM [30], EEM-ELM [29]). Here the case of the OP-ELM is studied, and specifically an approach aimed at regularizing the output layer determination and pruning.

2.2   The OP-ELM

The Optimally-Pruned Extreme Learning Machine (OP-ELM) is proposed in [4,31,3,5] in an attempt to solve the problem that ELM faces with irrelevant (or highly correlated) variables present in the data set that can "corrupt" some of the neurons. As described at more length in [4,3,5], it can be illustrated on a toy example as in Figure 1: the plots give the ELM fit in light blue dots over the training points in black crosses. On the leftmost part of the figure, the fit by the ELM is good, but when a pure random noise variable is added, on the rightmost figure (the added noise variable is not pictured on the figure), the fit becomes loose and spread.

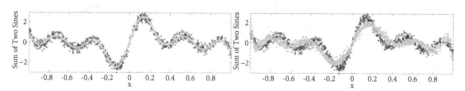

Fig. 1. Illustration of the ELM model fit (light blue dots) on a toy example (sum of sines, black crosses), for the normal data (leftmost part) and for the same data augmented with a random noise variable (not displayed), on the rightmost part. Due to the irrelevant additional variable, the fit of the ELM model is less accurate. From [4].

Indeed, the ELM is not designed to cope with such variables irrelevant to the problem at hand. In this spirit, the OP-ELM proposes a three-steps methodology, shortly described here, to address this issue.

Data → SLFN Construction using ELM → Ranking of the best neurons by LARS → Selection of the optimal number of neurons by LOO → Model

Fig. 2. Illustration of the three OP-ELM steps: the SLFN is first built using the ELM approach (random initialization of internal weights and biases); then a LARS algorithm is used to rank the neurons of the hidden layer; finally the selection of the optimal number of neurons for the OP-ELM model is performed using a Leave-One-Out criterion

The idea is to build a wrapper around the original ELM, with a neuron pruning strategy. For this matter, as can be seen on Figure 2, the construction of the SLFN by the ELM is retained, and two steps are added afterwards. First comes a ranking of the neurons by a Least Angle Regression (LARS [8]; in practice the MRSR [12] implementation of LARS is used for it also applies to multi-output cases), which sorts them by their usefulness regarding the output. And then a Leave-One-Out criterion is used to determine how many of the —sorted— neurons should be kept for the final OP-ELM model structure.

The LARS algorithm is not detailed here since it is described and discussed at length (or more precisely the idea it implements, Lasso) in section 3, but it has the property of providing an exact ranking of the hidden layer neurons in the case of the OP-ELM, since the relation between the neurons and the output is linear (by design of the OP-ELM).

The Leave-One-Out (LOO) method is usually a costly approach to optimize a parameter since it requires to train the model on the whole data set but one sample, and evaluate on this sample, repeatedly for all the samples of the data set. In the OP-ELM structure though, the situation is linear (between the hidden layer and the output one), and the LOO error has a closed matrix form, given by Allen's Prediction Sum of Squares (PRESS) [7] (details of the computation of the PRESS LOO error are given in section 4). This closed form allows for fast computation of the Mean Square Error and hence of the output weights β, making the OP-ELM still computationally efficient and more robust than the original ELM to irrelevant/ correlated variables.

Hence, the OP-ELM can be seen as a "regularized" ELM, by the use of a LARS approach, which is a L_1 penalty on a regression problem, here.

Meanwhile, the decision over the final number of neurons to retain (by a LOO criterion) has shown potential instabilities (numerically), due to the nature of the matrix operations performed in the PRESS formula (see section 4 for these calculations). The proposed solution in this paper is to use regularization in the calculations of the PRESS formula. In the following are reviewed the most well-known algorithms used to perform regularization, using a L_1 and L_2 (and jointly L_1 and L_2) penalty on the regression problem. The proposed approach in section 4 combines both L_1 and L_2 penalties in the OP-ELM, to regularize the network.

3 The Problem of Regularization

Here are presented some of the most widely used methods to regularize a regression problem (which is the situation between the hidden layer and the output layer of the OP-ELM).

In the following, matrices are denoted by boldface and $\mathbf{A}$ is a $n \times p$ matrix with $\mathbf{A} = \left(\mathbf{a}_1^T, \ldots, \mathbf{a}_n^T\right)^T$, $\mathbf{a}_i \in \mathbb{R}^p$. Also $\mathbf{A}$ can be referred by $\mathbf{A} = \left(a_{i,j}\right)_{1 \leq j \leq p}^{1 \leq i \leq n}$. Capital boldface $\mathbf{A}$ are used for matrices and low-case boldface $\mathbf{b}$ for vectors.

3.1   General Case

For the general setup, assume a single-output regression problem of the form

$$\mathbf{y} = \mathbf{X}\mathbf{w} + \boldsymbol{\varepsilon}, \tag{4}$$

with $\mathbf{X} = \left(\mathbf{x}_1^T, \ldots, \mathbf{x}_n^T\right)^T$ the inputs of the problem (data set), $\mathbf{y} = (y_1, \ldots, y_n)^T$ the actual output, $\mathbf{w} = \left(w_1, \ldots, w_p\right)^T$ the regression weights and $\boldsymbol{\varepsilon} = (\varepsilon_1, \ldots, \varepsilon_n)^T$ the residuals.

Traditionally, the Ordinary Least Squares (OLS) solution (a.k.a. Gauss-Markov solution) is a possible approach to solve this problem. The problem can be formulated as a minimization of the Mean Square Error as

$$\min_{\hat{\mathbf{w}}} (\mathbf{y} - \mathbf{X}\hat{\mathbf{w}})^T (\mathbf{y} - \mathbf{X}\hat{\mathbf{w}}), \tag{5}$$

or in a non matrix form

$$\min_{\hat{\mathbf{w}}} \sum_{i=1}^{n} (y_i - \mathbf{x}_i\hat{\mathbf{w}})^2, \tag{6}$$

with $\hat{\mathbf{w}} = (\hat{w}_1, \ldots, \hat{w}_n)^T$ the estimated regression weights.

The solution of Eq. 5 is then obtained by a classical pseudo-inverse (Moore-Penrose [28]) as

$$\hat{\mathbf{w}}^{\text{OLS}} = \left(\mathbf{X}^T\mathbf{X}\right)^{-1} \mathbf{X}^T\mathbf{y}, \tag{7}$$

assuming that $\mathbf{X}$ is full rank.

This way of computing the solution involves matrix inversion (for the computation of the inverse covariance matrix $\left(\mathbf{X}^T\mathbf{X}\right)^{-1}$) which tends to pose numerical problems in practice, since $\mathbf{X}$ is sometimes not full rank (there might very well be irrelevant or linear combinations of samples and/ or variables in the data set). A numerically more stable solution is to use the Singular Value Decomposition (SVD) of $\mathbf{X}$ to compute the pseudo-inverse. The proposed approach presented in section 4 makes use of the SVD for faster computations and numerical stability.

Two classical critiques of the OLS solution relate to the two main aspects that one expects from a model. First, the OLS is likely to perform poorly on real data (for example for the numerical reasons invoked before), while it is expected that the model should perform reasonably well on the training data. Second, it is usually desirable to have a sparse model which makes interpretation possible, regarding the relationships between variables and the output. Again, the OLS is not designed in this sense and does not provide sparse models at all.

Also, it has been shown (e.g. in [32,13]) that there exists solutions which achieve lower Mean Square Error (MSE) than the OLS one—for numerical instability reasons, in practice—, for example by the use of regularization factors, which can be seen as penalties added to the minimization problem in Eq. 5. In addition, regarding the generalization error, the OLS solution found in training is possibly not the best one (in terms of generalization MSE).

Here are detailed two different approaches to regularization, using either a L_1 or L_2 penalty term.

3.2 The L_1 Penalty: LASSO

Let us first consider the case of the L_1 penalty. In this setup, the minimization problem of Eq. 6 becomes

$$\min_{\lambda,\hat{\mathbf{w}}} \left[\sum_{i=1}^{n} (y_i - \mathbf{x}_i\hat{\mathbf{w}})^2 + \lambda \sum_{j=1}^{p} |\hat{w}_j| \right], \tag{8}$$

again with $\hat{\mathbf{w}} = (\hat{w}_1, \ldots, \hat{w}_n)^T$. An instance of this very problem is studied by Tibshirani in [6] and is commonly known as the LASSO (Least Absolute Shrinkage and Selection Operator). Due to the nature of the minimization problem (L_1 penalty on the regression coefficients), the Lasso produces solutions that exhibit sparsity, making interpretability possible. Control over this sparsity of the final model is obtained by modifying the λ value: the smaller is λ, the more $\hat{w}_j$ coefficients are non-zero and hence the more variables are retained in the final solution.

Generally, the computation of the solution to Eq. 8 is a quadratic programming problem with linearity constraint which can be intensive. In [8], a computationally more efficient algorithm is presented as the LARS algorithm (Least Angle Regression), of which the Lasso is a specific instance. LARS actually generalizes both the Lasso and the Forward Stagewise regression strategy (see [33] for example): the algorithm starts similarly to Forward selection, with all coefficients equal to zero and finds the variable most correlated with the output. The direction of this first selected variable is followed until another variable has as much correlation with the output. LARS then follows the direction of the equi-angular between first and second selected variables, until a third variable as much correlated with the output is found. The set of selected variables grows until none remain to be chosen (please refer to the original paper [8] for the computationally efficient implementation proposed by the authors).

By enforcing a restriction on the sign of the weights (which has to be the same as that of the current direction of the correlation), the LARS algorithm thus implements Lasso effectively. The authors claim an order of magnitude greater speed than the classical quadratic programming problem, using their algorithm.

Meanwhile, as noted by Zou and Hastie in [16] for example, the Lasso presents some drawbacks:

- If $p > n$, i.e. there are more variables than samples, the Lasso selects at most n variables [8];

– For classical situations where $n > p$, and if the variables are correlated, it seems (from experiments in [6]) that the Tikhonov regularization (in the following subsection 3.3) outperforms the Lasso.

A common drawback of the L_1 penalty and therefore of the Lasso approach is that it tends to be too sparse in some cases, i.e. there are many j such that $\hat{w}_j = 0$. In addition, the control over the sparsity by the parameter λ can be challenging to tune.

3.3 The L_2 Penalty: Tikhonov Regularization

Another possible approach to find a solution which deems a lower MSE than the OLS one is to use regularization in the form of Tikhonov regularization proposed in [14] (a.k.a. Ridge Regression [10]).

This time, the minimization problem involves a penalty using the square of the regression coefficients

$$\min_{\lambda,\hat{\mathbf{w}}} \left[\sum_{i=1}^{n} (y_i - \mathbf{x}_i \hat{\mathbf{w}})^2 + \lambda \sum_{j=1}^{p} \hat{w}_j^2 \right]. \tag{9}$$

Thanks to a bias–variance tradeoff, the Tikhonov regularization achieves better prediction performance than the traditional OLS solution. And as mentioned in the previous subsection 3.2, it outperforms the Lasso solution in cases were the variables are correlated. One famous advantage of the Tikhonov regularization is that it tends to identify/ isolate groups of variables, enabling further interpretability (this grouping can be very desirable for some data sets, as mentioned in [16]).

The major drawback of this regularization method is similar to one mentioned for the OLS: it does not give any parsimonious solution, since all variables are retained, due to the L_2 penalty. Therefore, contrary to the Lasso which actually performs variable selection "internally"—given that λ is large enough to set some coefficients to zero—, the Tikhonov regularization does not select variables directly.

The Elastic Net. Zhou and Hastie in [16] propose to alleviate the problems encountered by the Tikhonov regularization (lack of sparsity) while keeping its good performance thanks to the L_2 penalty. This is done using a composite of the Lasso and Tikhonov regularization, by combining the two penalties L_1 and L_2 in the form of a weighted penalty

$$\lambda_1 \sum |\hat{w}_j| + \lambda_2 \sum \hat{w}_j^2, \tag{10}$$

with λ_1 and λ_2 positive (controlling the sparsity of the model). In practice, the algorithm is implemented as a modification of the LARS algorithm (the LARS-EN) since once λ_2 is fixed, the computations are similar to that of a Lasso. While the LARS-EN is a very efficient way of implementing the elastic net approach, it remains that two parameters need optimizing: λ_1 and λ_2. Usually, this is done by the use of classical Cross-Validation (CV) which is unfortunately costly for it requires a two-dimensional search, which is hardly feasible if one wants to keep the ELM speed property. This is why a cascade method is proposed in the next Sections.

4   Regularized ELM

Recently, Deng *et al.* in [34] proposed a Regularized Extreme Learning Machine algorithm, which is essentially a L_2 penalized ELM, with a possibility to weight the sum of squares in order to address outliers interference. Using the notations from the previous section, the minimization problem is here

$$\min_{\lambda,\mathbf{d},\hat{\mathbf{w}}} \left[\lambda \sum_{i=1}^{n} (d_i (y_i - \mathbf{x}_i \hat{\mathbf{w}}))^2 + \sum_{j=1}^{p} \hat{w}_j^2 \right], \tag{11}$$

where the d_i are the weights meant to address the outliers.

This extension of the ELM clearly (from the results in [34]) brings a very good robustness to outliers to the original ELM. Unfortunately, it suffers from the problems related to L_2 penalties, that is the lack of sparsity for example.

As described before, the original OP-ELM already implements a L_1 penalty on the output weights, by performing a LARS between the hidden and output layer.

It is here proposed to modify the original PRESS LOO criterion for the selection of the optimal number of neurons by adding a Tikhonov regularization factor in the PRESS, therefore making the modified PRESS LOO a L_2 penalty applied on the L_1 penalized result from the LARS.

In the following are used matrix operations such as $\frac{\mathbf{A}}{\mathbf{B}}$ to refer to the matrix $\mathbf{C}$ such that $\left(c_{i,j} \right) = \frac{a_{i,j}}{b_{i,j}}$. Also the diag $(\cdot)$ operator is used to extract the diagonal of a matrix, diag $(\mathbf{A}) = \left(a_{1,1}, \ldots, a_{n,n} \right)^T$.

4.1 L_1 and L_2 Regularized OP-ELM

Allen's PRESS. The original PRESS formula used in the OP-ELM was proposed by Allen in [7]. The original PRESS formula can be expressed as

$$\mathrm{MSE}^{\mathrm{PRESS}} = \sum_{i=1}^{n} \left(\frac{y_i - \mathbf{x}_i \left(\mathbf{X}^T \mathbf{X} \right)^{-1} \mathbf{x}_i^T y_i}{1 - \mathbf{x}_i \left(\mathbf{X}^T \mathbf{X} \right)^{-1} \mathbf{x}_i^T} \right)^2, \tag{12}$$

which means that each observation is "predicted" using the other $n-1$ observations and the residuals are finally squared and summed up. Algorithm 1 proposes to implement this formula in an efficient way, by matrix computations.

The main drawback of this approach lies in the use of a pseudo-inverse in the calculation (in the Moore-Penrose sense), which can lead to numerical instabilities if the data set $\mathbf{X}$ is not full rank. This is unfortunately very often the case, with real-world data sets. The following approach proposes two improvements on the computation of the original PRESS: regularization and fast matrix calculations.

Tikhonov-Regularized PRESS (TR-PRESS). In [9], Golub *et al.* note that the Singular Value Decomposition (SVD) approach to compute the PRESS statistic is preferable to the traditional pseudo-inverse mentioned above, for numerical reasons. In this very

---

Algorithm 1. Allen's PRESS algorithm, in a fast matrix form

1: Compute the utility matrix $\mathbf{C} = \left(\mathbf{X}^T\mathbf{X}\right)^{-1}$
2: And $\mathbf{P} = \mathbf{XC}$;
3: Compute the pseudo-inverse $\mathbf{w} = \mathbf{CX}^T\mathbf{y}$;
4: Compute the denominator of the PRESS $\mathbf{D} = 1 - \text{diag}\left(\mathbf{PX}^T\right)$;
5: And finally the PRESS error $\varepsilon = \frac{\mathbf{y}-\mathbf{Xw}}{\mathbf{D}}$;
6: Reduced to a MSE, $\text{MSE}^{\text{PRESS}} = \frac{1}{n}\sum_{i=1}^{n}\varepsilon_i^2$.

---

same paper is proposed a generalization of Allen's PRESS, as the Generalized Cross-Validation (GCV) method, which is technically superior to the original PRESS, for it can handle cases were the data is extremely badly defined —for example if all $\mathbf{X}$ entries are 0 except the diagonal ones.

In practice, from our experiments, while the GCV is supposably superior, it leads to identical solutions with an increased computational time, compared to the original PRESS and the Tikhonov-Regularized version of PRESS presented below.

Algorithm 2 gives the computational steps used, in matrix form, to determine the $\text{MSE}^{\text{TR-PRESS}}(\lambda)$ from

$$\text{MSE}^{\text{TR-PRESS}}(\lambda) = \sum_{i=1}^{n}\left(\frac{y_i - \mathbf{x}_i\left(\mathbf{X}^T\mathbf{X} + \lambda\mathbf{I}\right)^{-1}\mathbf{x}_i^T y_i}{1 - \mathbf{x}_i\left(\mathbf{X}^T\mathbf{X} + \lambda\mathbf{I}\right)^{-1}\mathbf{x}_i^T}\right)^2, \tag{13}$$

which is the regularized version of Eq. 12.

Algorithm 2. Tikhonov-Regularized PRESS. In practice, the REPEAT part of this algorithm (convergence for λ) is solved by a Nelder-Mead approach [35], a.k.a. downhill simplex.

1: Decompose $\mathbf{X}$ by SVD: $\mathbf{X} = \mathbf{USV}^T$;
2: Compute the products (used later): $\mathbf{A} = \mathbf{XV}$ and $\mathbf{B} = \mathbf{U}^T\mathbf{y}$;
3: **repeat**

4:     Using the SVD of $\mathbf{X}$, compute the $\mathbf{C}$ matrix by: $\mathbf{C} = \mathbf{A} \times \begin{pmatrix} \frac{S_{11}}{S_{11}^2+\lambda} & \cdots & \frac{S_{nn}}{S_{nn}^2+\lambda} \\ \vdots & \vdots & \vdots \\ \frac{S_{11}}{S_{11}^2+\lambda} & \cdots & \frac{S_{nn}}{S_{nn}^2+\lambda} \end{pmatrix}$;

5:     Compute the $\mathbf{P}$ matrix by: $\mathbf{P} = \mathbf{CB}$;
6:     Compute $\mathbf{D}$ by: $\mathbf{D} = 1 - \text{diag}\left(\mathbf{CU}^T\right)$;
7:     Evaluate $\varepsilon = \frac{\mathbf{y}-\mathbf{P}}{\mathbf{D}}$ and the actual MSE by $\text{MSE}^{\text{TR-PRESS}} = \frac{1}{n}\sum_{i=1}^{n}\varepsilon_i^2$;
8: **until** convergence on λ is achieved
9: Keep the best $\text{MSE}^{\text{TR-PRESS}}$ and the λ value associated.

Globally, the algorithm uses the SVD of $\mathbf{X}$ to avoid computational issues, and introduces the Tikhonov regularization parameter in the calculation of the pseudo-inverse by the SVD. This specific implementation happens to run very quickly, thanks to the pre-calculation of utility matrices ($\mathbf{A}$, $\mathbf{B}$ and $\mathbf{C}$) before the optimization of λ.

In practice, the optimization of λ in this algorithm is performed by a Nelder-Mead [35] minimization approach, which happens to converge very quickly on this problem (fminsearch function in Matlab).

Through the use of this modified version of PRESS, the OP-ELM has an L_2 penalty on the regression weights (regression between the hidden and output layer), for which the neurons have already been ranked using an L_1 penalty. Figure 3 is a modified version of Figure 2 illustrating the TROP-ELM approach.

Fig. 3. The proposed regularized OP-ELM (TROP-ELM) as a modification of Figure 2

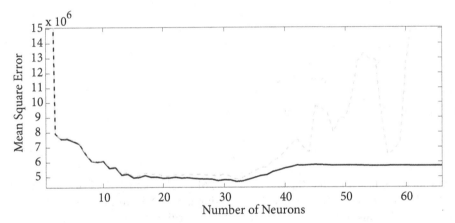

Fig. 4. Comparison of the MSE for the original OP-ELM (grey dashed line) and the proposed TROP-ELM (solid black line) for one data set (Auto Price, see section 6) for a varying amount of neurons (in the order ranked by the LARS). The regularization enables here to have a more stable MSE along the increase of the number of neurons.

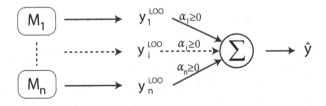

Fig. 5. Illustrative scheme of the Ensemble of models using LOO outputs

Figure 4 illustrates the effect of the regularization factor introduced in the TR-PRESS: the Mean Square Error is more stable regarding the increase of the number of neurons following the ranking provided by LARS (L_1 penalty). The introduction of the L_2 penalty has a very visible regularization effect here (the situation is similar for the other datasets), avoiding numerical instabilities, for example.

5 Ensemble of ELM: A Methodology to Linearly Ensemble ELM

In this section, the second methodology which is presented is based on an ensemble of ELM. Several ensemble techniques have been proposed, out of which two kinds can be distinguished: the variable weights approach and the average ones. Traditionally, average weights ensemble techniques are used and simply take an average of all the built models. While this obviously has the advantage of having immediately the weights of all models, it yields suboptimal results. The variable weights ensemble techniques try to optimize the weight of each model in the ensemble according to a criterion. The proposed ensemble methodology is presented below.

5.1 General Method

In order to build an ensemble of models as a linear combination of them, it is proposed to use their Leave-One-out output and solve the linear system it constitutes in a least-squares sense, under positivity constraint.

The global aim is to find the optimal weights α_i for a given set of models M_i (giving the prediction $\hat{y}_i$) to form an ensemble as a linear combination:

$$\hat{y} = \sum_{i=1}^{n} \alpha_i \hat{y}_i. \tag{14}$$

The solution is based on the Leave-One-Out [36] (LOO) output y_i^{LOO} of each of the models M_i, and determining the coefficients using a Non-Negative constrained Least-Squares (NNLS) algorithm. The classical NNLS algorithm presented in [37] is used to compute this solution. The overall idea is depicted in Fig. 5.

For each model $M_i, 1 \leq i \leq n$, the LOO output y_i^{LOO} is computed by omitting the considered point from the training, and evaluating the model on that specific point. Hence, a set of $\left\{y_i^{\text{LOO}}\right\}_{1 \leq i \leq n}$ outputs are computed, one for each model. The coefficients α_i are then solved from the constrained linear system:

$$\arg\min_{\alpha} \left\| y - \sum_{i=1}^{n} \alpha_i y_i^{\text{LOO}} \right\|^2 \quad \text{s.t. } \alpha_i \geq 0. \tag{15}$$

One advantage of this method is its low computational time, in terms of finding an optimal linear combination. The NNLS algorithm has been widely used and is known to converge in $\frac{1}{2}n$ steps, as noticed in [37]. In the idea of keeping the computational time low, for the whole method to be fast, the class of models used should be such that the Leave-One-Out output is rapidly computed or eventually approximated. In section 5.2, theoretical results concerning the weights α_i are proposed, under some assumptions on the models.

5.2 Theory of the Ensemble Methodology

While the NNLS algorithm mentioned previously solves the equation 15 in practice, there exists an analytical solution to this problem, under some assumptions on the noise and models.

Assuming there are n independent, unbiased models for y – each of the form $y_i = y + \varepsilon_i$, where the additive noise has zero mean and variance σ_i^2 – it is possible to directly derive the optimal weights α_i. The aim is to minimise the (expected) mean squared error $E\left[(y - \sum_i \alpha_i y_i)^2\right]$, which can be separated by exploiting the independence of the noise:

$$E\left[\left(y - \sum_i \alpha_i y_i\right)^2\right] = E\left[\left(y - \sum_i \alpha_i y - \sum_i \alpha_i \epsilon_i\right)^2\right]$$
$$= E\left[\left(y - \sum_i \alpha_i y\right)^2\right] + E\left[\left(\sum_i \alpha_i \epsilon_i\right)^2\right] = \left(1 - \sum_i \alpha_i\right)^2 s^2 + \sum_i \alpha_i^2 \sigma_i^2$$

where $s^2 = E\left[y^2\right]$. Differentiating w.r.t. α_k, and setting to zero:

$$\frac{d}{d\alpha_k}\left[\left(1 - \sum_i \alpha_i\right)^2 s^2 + \sum_i \alpha_i^2 \sigma_i^2\right] = -2\left(1 - \sum_i \alpha_i\right) s^2 + 2\alpha_k \sigma_k^2 = 0$$

This leads to the equation

$$\alpha_k \sigma_k^2 = \left(1 - \sum_i \alpha_i\right) s^2 \tag{16}$$

Here the right side (call it c), while still dependent on the parameter α_k, is independent of k. Hence it holds that $\alpha_k \propto \sigma_k^{-2}$, with the proportionality coefficient c. Substituting $\alpha_i = c\sigma_i^{-2}$ into Equation 16, we can solve for c:

$$\implies c = \frac{1}{s^{-2} + \sum_i \sigma_i^{-2}}$$

Finally, the optimal weights are

$$\alpha_k = \frac{\sigma_k^{-2}}{s^{-2} + \sum_i \sigma_i^{-2}}, \tag{17}$$

and the resulting MSE with these weights can be calculated to be $\frac{1}{s^{-2} + \sum_i \sigma_i^{-2}}$. This is lower than $\min_k \sigma_k^2$, meaning that the ensemble is more performant than any single constituent model. The error is also lower than the error achieved by the naïve average weighting $\alpha_k = \frac{1}{n}$, which is $\frac{1}{n^2} \sum_i \sigma_i^2$.

There is a trade-off between bias and variance here: minimising the variance introduces a slight bias to the ensemble model. This can be rectified by restricting the weights to $\sum_i \alpha_i = 1$. In this case, the optimal weights are

$$\alpha_k = \frac{\sigma_k^{-2}}{\sum_i \sigma_i^{-2}}, \tag{18}$$

and the resulting MSE $\frac{1}{\sum_i \sigma_i^{-2}}$. The difference is, however, practically insignificant. If the collections of models includes even a few reasonably accurate ones – that is, $\exists k$, s.t., $\sigma_k \ll s$ – the term s^{-2} is insignificant compared to the sum in the denominator in Equation 17 for α_k, and the weights (and resulting MSEs) calculated by formula 17 or 18 are essentially equivalent.

Some observations concerning the formulas 17 and 18 can be made. First, the weight of a model is inversely proportional to the variance of the error of that model, resulting in good models having large weight, and, correspondingly, poor models low weight. Second, all the weights are strictly positive. As the assumptions specified that the models are independent, even the poor models can still contribute with a slight bit of predictive power.

The exposition in this section assumes that all the models are entirely independent. In practice, this scenario is unattainable. If the particular dependencies are unknown, it is difficult to derive an exact expression for the optimal weights of the ensemble, but in any reasonable case there certainly *exists* a set of weights such that the resulting MSE is significantly lower than that of any single model.

The alternative is to solve the weights from the linear system 15, as this way, the dependencies between models can be accounted for. Solving the system is a very aggressive method of fitting and runs the risk of over-fitting. To counteract this, the leave-one-out output of the models is used, and the α_i are restricted to be non-negative. Having negative weights for some models essentially corresponds to "over-correcting" by using the errors of those models to balance the errors in other models. This quickly leads to fitting the models to the noise, that is: over-fitting. Solving the constrained linear system naturally results in higher weight for more accurate models, and low weight for poor models. This inverse relationship between the weight and MSE is in correspondance with formula 17. As the ELM models used in the experimental section are decidedly not independent, the weights are determined by solving system 15 instead of using formula 17. In order to obtain a set of model that are as independent as possible, each model is an ELM with a different number of hidden neurons. For example, in the following experiments, the maximum number of neurons being P and the total number of ELM being E, each model M_i has $\frac{iP}{E}$ neurons.

6 Experiments

In order to compare the proposed TROP-ELM and MJƎ-ELM with the original ELM and other typical machine learning algorithms, ten data sets from UCI Machine Learning Repository [38] have been used. They are chosen for their heterogeneity in terms of problem, number of variables, and sizes.

Table 1 summarizes the details of each data set.

The data sets have all been processed in the same way: for each data set, ten different random permutations are taken without replacement; for each permutation, two thirds are taken for the training set, and the remaining third for the test set (see Table 1). Training sets are then normalized (zero-mean and unit variance) and test sets are also normalized using the very same normalization factors than for the corresponding training set. The results presented in the following are hence the average of the ten repetitions

Table 1. Details of the data sets used and the proportions for training and testing sets for each (two thirds of the whole set for training and one third for testing), along with the number of variables

	Abalone	Ailerons	Elevators	Computer	Auto P.	CPU	Servo	Bank	Stocks	Boston
# of Variables	8	5	6	12	15	6	4	8	9	13
Training	2784	4752	6344	5461	106	139	111	2999	633	337
Test	1393	2377	3173	2731	53	70	56	1500	317	169

for each data set. This also enables to obtain an estimate of the standard deviation of the results presented (see Table 4).

It should be noted that most of the results presented in Tables 4 and 3 are from [4,17] and are reproduced here for comparison purposes.

As mentioned in the original paper [4], experiments are performed using the online available versions of the methodologies, unaltered. All experiments have been run on the same $x86\_64$ Linux machine with at least 4 GB of memory (no swapping for any of the experiments) and 2+ GHz processor. Also, even though some methodologies implementations are taking advantage of parallelization, computational times are reported considering single-threaded execution on one single core, for the sake of comparisons.

The SVM is performed using the SVM toolbox [39]; MLP [21] is using a neural network toolbox, part of the Matlab software from the MathWorks, Inc; the GPML toolbox for Matlab from Rasmussen and Williams [40] is used for the GP; finally, the OP-ELM was used with all possible kernels, linear, sigmoid, and Gaussian, using a maximum number of 100 neurons and similarly for the TROP-ELM. For more details on the parameters used for each toolbox, please refer to [4]. For the MJE-ELM, the total number of ELM being E is always 100.

First are reported the Mean Square Errors (and standard deviations) for the six algorithms tested. It can be seen that the proposed TROP-ELM is always at least as good as the original OP-ELM, with an improvement on the standard deviation of the results, over the ten repetitions for each data set (only for the Boston Housing case is the standard deviation larger for the TROP-ELM than the OP-ELM): over the ten data sets, the

Table 2. Mean Square Error results in boldface (standard deviations in regular) for all six methodologies for regression data sets. "Auto P." stands for Auto Price dataset.

	Abalone	Ailerons	Elevators	Computer	Auto P.	CPU	Servo	Bank	Stocks	Boston
SVM	**4.5**	**1.3e-7**	**6.2e-6**	**1.2e+2**	**2.8e+7**	**6.5e+3**	**6.9e-1**	**2.7e-2**	**5.1e-1**	**3.4e+1**
	2.7e-1	2.6e-8	6.8e-7	8.1e+1	8.4e+7	5.1e+3	3.3e-1	8.0e-4	9.0e-2	3.1e+1
MLP	**4.6**	**2.7e-7**	**2.6e-6**	**9.8**	**2.2e+7**	**1.4e+4**	**2.2e-1**	**9.1e-4**	**8.8e-1**	**2.2e+1**
	5.8e-1	4.4e-9	9.0e-8	1.1	9.8e+6	1.8e+4	8.1e-2	4.2e-5	2.1e-1	8.8
GP	**4.5**	**2.7e-8**	**2.0e-6**	**7.7**	**2.0e+7**	**6.7e+3**	**4.8e-1**	**8.7e-4**	**4.4e-1**	**1.1e+1**
	2.4e-1	1.9e-9	5.0e-8	2.9e-1	1.0e+7	6.6e+3	3.5e-1	5.1e-5	5.0e-2	3.5
ELM	**8.3**	**3.3e-8**	**2.2e-6**	**4.9e+2**	**7.9e+9**	**4.7e+4**	**7.1**	**6.7e-3**	**3.4e+1**	**1.2e+2**
	7.5e-1	2.5e-9	7.0e-8	6.2e+1	7.2e+9	2.5e+4	5.5	7.0e-4	9.35	2.1e+1
OP-ELM	**4.9**	**2.8e-7**	**2.0e-6**	**3.1e+1**	**9.5e+7**	**5.3e+3**	**8.0e-1**	**1.1e-1**	**9.8e-1**	**1.9e+1**
	6.6e-1	1.5e-9	5.4e-8	7.4	4.0e+6	5.2e+3	3.3e-1	1.0e-6	1.1e-1	7.4
TROP-ELM	**4.8**	**2.7e-7**	**2.0e-6**	**2.4e+1**	**7.0e+6**	**4.1e+3**	**6.1e-1**	**1.1e-1**	**8.4e-1**	**1.9e+1**
	4.2e-1	1.5e-9	5.2e-8	6.2	2.2e+6	2.9e+3	2.2e-1	3.4e-5	5.8e-2	4.4
MJE-ELM	**4.6**	**2.6e-8**	**2.0e-6**	**4.3e+1**	**1.9e+7**	**2.5e+3**	**6.9e-1**	**1.7e-1**	**1.2e+1**	**2.1e+1**
	1.7e-1	1.7e-9	5.3e-8	3.6	3.8e+6	1.1e+3	2.6e-1	1.2e-4	2.6e-1	6.9

TROP-ELM performs on average 27% better than the original OP-ELM and gives a standard deviation of the results 52% lower than that of the OP-ELM (also on average over the ten data sets).

Also, the TROP-ELM is clearly as good (or better) as the GP in six out of the ten data sets —Ailerons, Elevators, Auto Price, Bank and Boston— in which cases it has a similar (or lower) standard deviation of the results. This with a computational time usually two or three orders of magnitude lower than the GP.

Table 3 gives the computational times for each algorithm and each data set (average of the ten repetitions).

Table 3. Computational times (in seconds) for all five methodologies on the regression data sets. "Auto P." stands for Auto Price dataset.

	Abalone	Ailerons	Elevators	Computer	Auto P.	CPU	Servo	Bank	Stocks	Boston
SVM	6.6e+4	4.2e+2	5.8e+2	3.2e+5	2.6e+2	3.2e+2	1.3e+2	1.6e+3	2.3e+3	8.5e+2
MLP	2.1e+3	3.5e+3	3.5e+3	8.2e+3	7.3e+2	5.8e+2	5.2e+2	2.7e+3	1.2e+3	8.2e+2
GP	9.5e+2	2.9e+3	6.5e+3	6.3e+3	2.9	3.2	2.2	1.7e+3	4.1e+1	8.5
ELM	4.0e-1	9.0e-1	1.6	1.2	3.8e-2	4.2e-2	3.9e-2	4.7e-1	1.1e-1	7.4e-2
OP-ELM	5.7	16.8	29.8	26.2	2.7e-1	2.0e-1	2.1e-1	8.03	1.54	7.0e-1
TROP-ELM	12.2	14.6	44.3	13.9	4.8e-1	1.2	8.4e-1	4.4	1.1	1.5
M∃-ELM	20	35	51	43	2.7e-1	1.6e-1	2.6e-1	23	13	2.9

It can be seen that the TROP-ELM keeps computational times of the same order as that of the OP-ELM (although higher on average), and remains several orders of magnitudes faster than the GP, MLP or SVM. Of course, as for the OP-ELM, the computational times remain one to two orders of magnitude above the original ELM.

The results obtained with the M∃-ELM are better than with the TROP-ELM for 3 datasets and similar for 4 other datasets. For the 3 datasets for which the M∃-ELM is not as good as TROP-ELM, the performances are anyway better than with SVM or MLP. The computational time of the M∃-ELM is in general larger (2 or 3 times slower); but it should be noticed that the M∃-ELM was not parallelized for these experiments. In fact, the M∃-ELM can intrinsically be parallelized and the computational time can be approximatively divided by the number of available cores. A number of cores equal to the number of models that are assembled is probably optimal.

7 Sensitivity to Variable Selection: A Simple Test

In this section, a simple test to verify and test the robustness of ELM techniques is introduced. The abalone data set is used and in order to add artificially some irrelevant but dependent variables, a subset of the aileron dataset is concatenated to the abalone dataset. The new dataset has then the same number of samples than the original dataset but the number of variables is now 13 instead of 8. Obviously, the new 5 variables cannot help building any regression model. Furthermore, these extra variables may pollute the hidden neurons of the ELM techniques since they are bringing information by means of the random projection.

In the next table, the results of the ELM, OPELM, TROP-ELM and M∃-ELM are presented.

Table 4. Mean Square Error results in boldface (standard deviations in regular) for all four methodologies for robustness test datasets

	Abalone	Concatenated Dataset
ELM	**8.3**	**15.2**
OP-ELM	**4.9**	**5.0**
TROP-ELM	**4.8**	**4.9**
MꓕƎ-ELM	**4.6**	**6.4**

It can be notice that the basic ELM is very sensitive to irrelevant additional variables. This fact was already illustrated in Figure 1. The OP-ELM and the TROP-ELM are nearly insensitive to those additional variables. This property is due to the regulations that are pruning or decreasing the importance of the neurons that are "polluted" by the additional variables.

8   Conclusions and Future Work

In this paper is compared a modification of the original Optimally-Pruned Extreme Learning Machine (OP-ELM) with Methodology to Linearly Ensemble ELM (MꓕƎ-ELM).

The OP-ELM was proposed in the first place as a wrapper around ELM to improve its robustness by adding a neuron pruning strategy based on LARS (L_1 penalty) and Leave-One-Out (LOO). Here the LOO criterion is modified to add a L_2 penalty (Tikhonov regularization) to the estimate, in order to regularize the matrix computations and hence make the MSE computation more reliable. The modified OP-ELM (TROP-ELM) therefore uses "in cascade" L_1 and L_2 penalties, avoiding the large computational times problems commonly encountered when attempting to intertwine the two penalties (as in the Elastic Net).

The TROP-ELM shows better performance than the original OP-ELM, with an average of 27% better MSE for the considered data sets (and improvements between 0 and 96% over the data sets used). Also notable is the decrease of the standard deviation of the results over the multiple repetitions for each data set, illustrating that the regularization introduced has a visible effect. In the end, the TROP-ELM performs rather similarly to the Gaussian Processes on more than half the data sets tested, for a computational time which remains two to three orders of magnitude below —and less than an order of magnitude slower than the OP-ELM, in the worst case among the data sets used.

The MꓕƎ-ELM is providing results that are equivalent or better than the results obtained by the OP-ELM and the TROP-ELM. The computational time of the MꓕƎ-ELM is slightly increased but it can be reduced easily and naturally using a large number of cores to build the ensemble of ELM. Nevertheless, the MꓕƎ-ELM is not as robust as the TROP-ELM. In case of a very large dataset with a large number of sample, the MꓕƎ-ELM should be preferred. In case of a large number of eventually irrelevant variables, preference should be given to the TROP-ELM.

In the future, ensemble of TROP-ELM should be investigated, in order to keep both the robustness property of the TROP-ELM and the scalability of the MꓕƎ-ELM. Furthermore, ensemble of ELM and regularized ELM should be combined in the future with incremental ELM (see e.g. [23,24]).

References

1. Huang, G.B., Zhu, Q.Y., Siew, C.K.: Extreme learning machine: Theory and applications. Neurocomputing 70(1-3), 489–501 (2006)
2. Haykin, S.: Neural Networks: A Comprehensive Foundation, 2nd edn. Prentice Hall (July 1998)
3. Miche, Y., Bas, P., Jutten, C., Simula, O., Lendasse, A.: A methodology for building regression models using extreme learning machine: OP-ELM. In: European Symposium on Artificial Neural Networks, ESANN 2008, Bruges, Belgium, April 23-25 (2008)
4. Miche, Y., Sorjamaa, A., Bas, P., Simula, O., Jutten, C., Lendasse, A.: OP-ELM: Optimally-pruned extreme learning machine. IEEE Transactions on Neural Networks 21(1), 158–162 (2010)
5. Miche, Y., Sorjamaa, A., Lendasse, A.: OP-ELM: Theory, experiments and a toolbox. In: Kůrková, V., Neruda, R., Koutník, J. (eds.) ICANN 2008, Part I. LNCS, vol. 5163, pp. 145–154. Springer, Heidelberg (2008)
6. Tibshirani, R.: Regression shrinkage and selection via the lasso. Journal of the Royal Statistical Society, Series B 58, 267–288 (1994)
7. Allen, D.M.: The relationship between variable selection and data agumentation and a method for prediction. Technometrics 16(1), 125–127 (1974)
8. Efron, B., Hastie, T., Johnstone, I., Tibshirani, R.: Least angle regression. Annals of Statistics 32, 407–499 (2004)
9. Golub, G.H., Heath, M., Wahba, G.: Generalized cross-validation as a method for choosing a good ridge parameter. Technometrics 21(2), 215–223 (1979)
10. Hoerl, A.E.: Application of ridge analysis to regression problems. Chemical Engineering Progress 58, 54–59 (1962)
11. Owen, A.B.: A robust hybrid of lasso and ridge regression. Technical report, Stanford University (2006)
12. Similä, T., Tikka, J.: Multiresponse sparse regression with application to multidimensional scaling. In: Duch, W., Kacprzyk, J., Oja, E., Zadrożny, S. (eds.) ICANN 2005. LNCS, vol. 3697, pp. 97–102. Springer, Heidelberg (2005)
13. Thisted, R.A.: Ridge regression, minimax estimation, and empirical bayes methods. Technical Report 28, Division of Biostatistics, Stanford University (1976)
14. Tychonoff, A.N.: Solution of incorrectly formulated problems and the regularization method. Soviet Mathematics 4, 1035–1038 (1963)
15. Zhao, P., Rocha, G.V., Yu, B.: Grouped and hierarchical model selection through composite absolute penalties. Annals of Statistics 37(6A), 3468–3497 (2009)
16. Zou, H., Hastie, T.: Regularization and variable selection via the elastic net. Journal of the Royal Statistical Society Series B 67(2), 301–320 (2005)
17. Miche, Y., van Heeswijk, M., Bas, P., Simula, O., Lendasse, A.: Trop-elm: A double-regularized elm using lars and tikhonov regularization. Neurocomputing 74(16), 2413–2421 (2011)
18. Miche, Y., Eirola, E., Bas, P., Simula, O., Jutten, C., Lendasse, A., Verleysen, M.: Ensemble modeling with a constrained linear system of leave-one-out outputs. In: Verleysen, M. (ed.) ESANN 2010: 18th European Symposium on Artificial Neural Networks, Computational Intelligence and Machine Learning, April 28-30, pp. 19–24. d-side Publications, Bruges (2010)
19. van Heeswijk, M., Miche, Y., Oja, E., Lendasse, A.: GPU-accelerated and parallelized ELM ensembles for large-scale regression. Neurocomputing 74(16), 2430–2437 (2011)
20. Hua Zhou, Z., Wu, J., Tang, W.: Ensembling neural networks: Many could be better than all. Artif. Intell. 137(1-2), 239–263 (2002)

21. Bishop, C.M.: Neural Networks for Pattern Recognition. Oxford University Press, USA (1996)
22. Huang, G.B., Chen, L., Siew, C.K.: Universal approximation using incremental constructive feedforward networks with random hidden nodes. IEEE Transactions on Neural Networks 17, 879–892 (2005)
23. Huang, G.B., Chen, L.: Enhanced random search based incremental extreme learning machine. Neurocomputing 71(16-18), 3460–3468 (2008)
24. Huang, G.B., Chen, L.: Convex incremental extreme learning machine. Neurocomputing 70(16-18), 3056–3062 (2007)
25. Huang, G.B., Zhu, Q.Y., Mao, K., Siew, C.K., Saratchandran, P., Sundararajan, N.: Can threshold networks be trained directly? IEEE Transactions on Circuits and Systems II: Express Briefs 53(3), 187–191 (2006)
26. Li, M.B., Huang, G.B., Saratchandran, P., Sundararajan, N.: Fully complex extreme learning machine. Neurocomputing 68, 306–314 (2005)
27. Huang, G.B., Siew, C.K.: Extreme learning machine with randomly assigned rbf kernels. International Journal of Information Technology 11(1), 16–24 (2005)
28. Rao, C.R., Mitra, S.K.: Generalized Inverse of Matrices and Its Applications. John Wiley & Sons Inc. (1971)
29. Yuan, L., Chai, S.Y., Huang, G.B.: Random search enhancement of error minimized extreme learning machine. In: Verleysen, M. (ed.) European Symposium on Artificial Neural Networks, ESANN 2010, April 28-30, pp. 327–332. d-side Publications, Bruges (2010)
30. Feng, G., Huang, G.B., Lin, Q., Gay, R.: Error minimized extreme learning machine with growth of hidden nodes and incremental learning. IEEE Transactions on Neural Networks 20(8), 1352–1357 (2009)
31. Group, E.: The op-elm toolbox (2009),
    http://www.cis.hut.fi/projects/eiml/research/downloads/op-elm-toolbox
32. Berger, J.: Minimax estimation of a multivariate normal mean under arbitrary quadratic loss. Journal of Multivariate Analysis 6(2), 256–264 (1976)
33. Hastie, T., Tibshirani, R., Friedman, J.: The Elements of Statistical Learning: Data Mining, Inference, and Prediction, 2nd edn. Springer (2009)
34. Deng, W., Zheng, Q., Chen, L.: Regularized extreme learning machine. In: IEEE Symposium on Computational Intelligence and Data Mining, CIDM 2009, March 30-April 2, pp. 389–395 (2009)
35. Nelder, J.A., Mead, R.: A simplex method for function minimization. The Computer Journal 7(4), 308–313 (1965)
36. Lendasse, A., Wertz, V., Verleysen, M.: Model selection with cross-validations and bootstraps - application to time series prediction with RBFN models. In: Kaynak, O., Alpaydin, E., Oja, E., Xu, L. (eds.) ICANN/ICONIP 2003. LNCS, vol. 2714, pp. 573–580. Springer, Heidelberg (2003)
37. Lawson, C.L., Hanson, R.J.: Solving least squares problems, 3rd edn. SIAM Classics in Applied Mathematics (1995)
38. Frank, A., Asuncion, A.: UCI machine learning repository (2010),
    http://archive.ics.uci.edu/ml
39. Chang, C.C., Lin, C.J.: LIBSVM: a library for support vector machines (2001), Software available at http://www.csie.ntu.edu.tw/~cjlin/libsvm
40. Rasmussen, C.E., Williams, C.K.I.: Gaussian Processes for Machine Learning. The MIT Press (2006)

A Novel Framework to Design Fuzzy Rule-Based Ensembles Using Diversity Induction and Evolutionary Algorithms-Based Classifier Selection and Fusion

Oscar Cordón[1,2] and Krzysztof Trawiński[1]

[1] European Centre for Soft Computing, Edificio Científico-Tecnológico, planta 3,
C. Gonzalo Gutiérrez Quirós s/n, 33600 Mieres (Asturias), Spain
{oscar.cordon,krzysztof.trawinski}@softcomputing.es
[2] Dept. of Computer Science and Artificial Intelligence (DECSAI) and the Research
Center on Information and Communication Technologies (CITIC-UGR),
University of Granada, 18071 Granada, Spain
ocordon@decsai.ugr.es

Abstract. Fuzzy rule-based systems have shown a high capability of knowledge extraction and representation when modeling complex, non-linear classification problems. However, they suffer from the so-called curse of dimensionality when applied to high dimensional datasets, which consist of a large number of variables and/or examples. Multiclassification systems have shown to be a good approach to deal with this kind of problems. In this contribution, we propose an multiclassification system-based global framework allowing fuzzy rule-based systems to deal with high dimensional datasets avoiding the curse of dimensionality. Having this goal in mind, the proposed framework will incorporate several multi-classification system methodologies as well as evolutionary algorithms to design fuzzy rule-based multiclassification systems. The proposed framework follows a two-stage structure: 1) fuzzy rule-based multiclassification system design from classical and advanced multiclassification system design approaches, and 2) novel designs of evolutionary component classifier combination. By using our methodology, different fuzzy rule-based multiclassification systems can be designed dealing with several aspects such as improvement of the performance in terms of accuracy, and obtaining a good accuracy-complexity trade-off.

1 Introduction

Multiclassification systems (MCSs), also called classifier ensembles, are machine learning tools capable to obtain better performance than a single classifier when dealing with complex classification problems. They are especially useful when the number of dimensions or the size of the data are really large [1]. The most common base classifiers are decision trees [2] and neural networks [3]. More recently, the use of fuzzy classifiers has also been considered [4–6].

I. Rojas, G. Joya, and J. Cabestany (Eds.): IWANN 2013, Part I, LNCS 7902, pp. 36–58, 2013.
© Springer-Verlag Berlin Heidelberg 2013

On the other hand, fuzzy rule-based classification systems (FRBCSs) have shown a high capability of knowledge extraction and representation when modeling complex, non-linear classification problems. They consider soft boundaries obtained through the use of a collection of fuzzy rules that could be understood by a human being [1, 7]. Interpretability of fuzzy systems is a characteristic that definitely favors this type of models, as it is often a need to understand the behavior of the given model [8, 9].

FRBCSs, however, have one significant drawback. The main difficulty appears when it comes to deal with a problem consisting of a high number of variables and/or examples. In such a case the FRBCS suffers from the so-called *curse of dimensionality* [7]. It occurs due to the exponential increase of the number of rules and the number of antecedents within a rule with the growth of the number of inputs in the FRBCS. This issue also causes a scalability problem in terms of the run time and the memory consumption.

This paper aims to propose an MCS-based global framework allowing FRBCSs to deal with high dimensional datasets avoiding the curse of dimensionality. With this aim, this framework will incorporate several MCS methodologies taken from the machine learning field as well as evolutionary algorithms to design fuzzy rule-based multiclassification systems (FRBMCSs). The proposed framework follows a two-stage structure: 1) component fuzzy classifier design from classical and advanced MCS design approaches, and 2) novel designs of evolutionary component classifier combination. This methodology will allow us to design different FRBMCSs dealing with several aspects such as improvement of the performance in terms of accuracy and obtaining a good accuracy-complexity trade-off.

This manuscript is organized as follows. In the next section, the preliminaries required to understand our work are reviewed. Section 3 briefly presents the proposed framework. Then, Section 4 introduces the proposed FRBMCS design methods, while Section 5 describes evolutionary the classifier combination designs. Each subsection in the latter section will introduce different approaches, referring the author to the corresponding publication, as well as reporting a brief performance analysis considering wide experimentations developed on a large number of UCI datasets. Finally, Section 6 concludes this contribution, suggesting also some future research lines.

2   State of the Art

This section reports a state of the art about MCSs and fuzzy MCSs. We also review FURIA, a novel and good performing fuzzy rule-based classifier, which will be used as the component base classifier. Finally, we briefly describe genetic fuzzy systems, which is a fundamental tool for development of the component fuzzy classifier combination method presented in the current contribution.

2.1   Multiclassification Systems

MCS design is mainly based on two stages [10]: the learning of the component classifiers and the combination mechanism for the individual decisions provided by them into the global MCS output. Since a MCS is the result of the combination of the outputs of a group of individually trained classifiers, the accuracy of the finally derived MCS relies on the performance and the proper integration of these two tasks. The best possible situation for an ensemble is that where the individual classifiers are both accurate and fully complementary, in the sense that they make their errors on different parts of the problem space [3]. Hence, MCSs rely for their effectiveness on the "instability" of the base learning algorithm.

On the one hand, the correct definition of the set of base classifiers is fundamental to the overall performance of MCSs. Different approaches have been thus proposed to succeed on generating diverse component classifiers with uncorrelated errors such as data resampling techniques (mainly, bagging [11] and boosting [12]), specific diversity induction mechanisms (feature selection [2], diversity measures [13], use of different parameterizations of the learning algorithm, use of different learning models, etc.), or combinations between the latter two families, as the well known random forests approach [14].

On the other hand, the research area of combination methods is also very active due to the influential role of this MCS component. It does not only consider the issue of aggregating the results provided by all the initial set of component classifiers derived from the first learning stage to compute the final output (what is usually called *classifier fusion* [15, 16]). It also involves either locally selecting the best single classifier which will be taken into account to provide a decision for each specific input pattern (static or dynamic classifier selection [17]) or globally selecting the subgroup of classifiers which will be considered for every input pattern (overproduce-and-choose strategy [18]). Besides, hybrid strategies between the two groups have also been introduced [1]. In any case, the determination of the optimal size of the ensemble is an important issue for obtaining both the best possible accuracy in the test data set without overfitting it, and a good accuracy-complexity trade-off [19].

2.2   FURIA

Fuzzy Unordered Rules Induction Algorithm (FURIA) [20] is an extension of the state-of-the-art rule learning algorithm called RIPPER [21], considering the derivation of simple and comprehensible fuzzy rule bases, and introducing some new features. FURIA provides three different extensions of RIPPER:

- It takes an advantage of fuzzy rules instead of crisp ones. Fuzzy rules of FURIA are composed of a class C_j and a certainty degree CD_j in the consequent. The final form of a rule is the following:

$$\text{Rule } R_j : \text{If } x_1 \text{ is } A_{j1} \text{ and } \ldots \text{and } x_n \text{ is } A_{jn}$$
$$\text{then Class } C_j \text{ with } CD_j; \quad j = 1, 2, ..., N.$$

The certainty degree of a given example x is defined as follows:

$$CD_j = \frac{2\frac{D_T^{C_j}}{D_T} + \sum_{x \in D_T^{C_j}} \mu_r^{C_j}(x)}{2 + \sum_{x \in D_T} \mu_r^{C_j}(x)} \tag{1}$$

where D_T and $D_T^{C_j}$ stands for the training set and a subset of the training set belonging to the class C_j respectively. In this approach, each fuzzy rule makes a vote for its consequent class. The vote strength of the rule is calculated as the product of the firing degree $\mu_r^{C_j}(x)$ and the certainty degree CD_j. Hence, the fuzzy reasoning method used is the so-called voting-based method [22, 23].

- It uses unordered rule sets instead of rule lists. This change omits a bias caused by the default class rule, which is applied whenever there is an uncovered example detected.
- It proposes a novel rule stretching method in order to manage uncovered examples. The unordered rule set introduces one crucial drawback, there might appear a case when a given example is not covered. Then, to deal with such situation, one rule is generalized by removing its antecedents. The information measure is proposed to verify which rule to "stretch".

The interested reader is referred to [20] for a full description of FURIA.

2.3   Related Work on Fuzzy Multiclassification Systems

Focusing on fuzzy MCSs, only a few contributions for bagging fuzzy classifiers have been proposed considering fuzzy neural networks (together with feature selection) [24], neuro-fuzzy systems [4], and fuzzy decision trees [25, 26] as component classifier structures.

Especially worth mentioning is the contribution of Bonissone et al. [25]. This approach hybridizes Breiman's idea of random forests [14] with fuzzy decision trees [27]. Such resulting fuzzy random forest combines characteristics of MCSs with randomness and fuzzy logic in order to obtain a high quality system joining robustness, diversity, and flexibility to not only deal with traditional classification problems but also with imperfect and noisy datasets. The results show that this approach obtains good performance in terms of accuracy for all the latter problem kinds.

Some advanced GFS-based contributions should also be remarked. On the one hand, an FRBCS ensemble design technique is proposed in [28] considering some niching genetic algorithm (GA) [29] based feature selection methods to generate the diverse component classifiers, and another GA for classifier fusion by learning the combination weights. On the other hand, another interval and fuzzy rule-based ensemble design method using a single- and multiobjective genetic selection process is introduced in [30, 31]. In this case, the coding scheme allows an initial set of either interval or fuzzy rules, considering the use of different features in their antecedents, to be distributed among different component classifiers trying to make them as diverse as possible by means of two accuracy and one entropy measures. Besides, the same authors presented a previous proposal in [32], where an evolutionary multiobjective (EMO) algorithm generated a Pareto set of FRBCSs with different accuracy-complexity trade-offs to be combined into an ensemble.

2.4   Genetic Fuzzy Systems

Fuzzy systems, which are based on fuzzy logic, became popular in the research community, since they have ability to deal with complex, non-linear problems being too difficult for the classical methods [33]. Besides, its capability of knowledge extraction and representation allowed them to become human-comprehensible to some extent (more than classical black-box models) [8, 9].

The lack of the automatic extraction of fuzzy systems have attracted the attention of the computational intelligence community to incorporate learning capabilities to these kinds of systems. In consequence, a hybridization of fuzzy systems and GAs has become one of the most popular approaches in this field [34–37]. In general, genetic fuzzy systems (GFSs) are fuzzy systems enhanced by a learning procedure coming from evolutionary computation, i.e. considering any evolutionary algorithm (EA).

Fuzzy rule-based systems (FRBSs), which are based on fuzzy "IF-THEN" rules, constitute one of the most important areas of fuzzy logic applications. Designing FRBSs might be seen as a search problem in a solution space of different candidate models by encoding the model into the chromosome, as GAs are well known optimization algorithms capable of searching among large spaces with the aim of finding optimal (usually nearly optimal) solutions.

The generic coding of GAs provides them with a large flexibility to define which parameters/components of FRBS are to be designed [36]. For example, the simplest case would be a parameter optimization of the fuzzy membership functions. The complete rule base can also be learned. This capability allowed the field of GFSs to grow over two decades and to still be one of the most important topics in computational intelligence.

In the current contribution, we will relay on the GFS paradigm to define some of the proposed FRBMCS designs.

3   Proposal of the Framework

The main objective of this paper is to enable FRBCSs to deal with high dimensional datasets by means of different MCS approaches. Thus, we sketched a global framework containing several FRBMCSs designs. This framework is composed of two stages (see Fig. 1). The first one, called "component fuzzy classifier design from classical ML approaches", includes the use of FURIA to derive the component classifiers considering the classical MCS design approaches such as:

– *Static* approaches. From this family we incorporate classical MCS approaches to obtain accurate FRBMCSs such as bagging, feature selection, and the combination of bagging and feature selection. Thanks to the intrinsic parallelism of bagging they will also be time efficient.
– *Dynamic* approaches. From this family we employ the combination of bagging and random oracles (ROs) [38, 39], since ROs induce an additional diversity to the base classifiers, the accuracy of the final FRBMCSs is thus improved.

In [19], a study to determine the size of a parallel ensemble (e.g. bagging) by estimating the minimum number of classifiers that are required to obtain stable aggregate predictions was shown. The conclusion drawn was that the optimal ensemble size is very sensitive to the particular classification problem considered. Thus, the second stage of our framework, called "Evolutionary component classifier combination", is related to post-processing of the generated ensemble by means of EAs to perform component classifier combination. All the approaches used consider classifier selection and some of them also combine it with classifier fusion.

Of course, the second stage follows the approaches from the first stage. This is indicated by a red arrow in the figure, showing exactly which approach is used for the FRBMCS design (Stage 1) together with its corresponding evolutionary post-processing (Stage 2). A dashed red arrow points out a proposal that was not developed and is left for the future works.

The second stage includes the following evolutionary component classifier selection designs:

- *Classifier Selection*. Within this family, we opted for a EMO overproduce-and-choose strategy (OCS) [18] (also known as test-and-select methodology [40]) strategy, using the state-of-the-art NSGA-II algorithm [41], in order to obtain a good accuracy-complexity trade-off.
- *Classifier Selection and Fusion*. As a combination method joining both families, classifier selection and classifier fusion, we proposed the use of a GFS, which allows us to benefit from the key advantage of fuzzy systems, i.e., their interpretability.

4   Component Fuzzy Classifier Design Methods

4.1   Static Approaches: Bagging, Feature Selection, and Bagging with Feature Selection

In [42, 43] it was shown that a combination between bagging and feature selection composed a general design procedure usually leading to good MCS designs, regardless the classifier structure considered. Hence, we decided to follow that approach by integrating FURIA into a framework of that kind. Our aim was to combine the diversity induced by the MCS design methods and the robustness of the FURIA method in order to derive good performance FURIA-based FRBMCSs for high dimensional problems [44]. We also tried a combination of FURIA with bagging and feature selection separately in order to analyze which is the best setting for the design of FURIA-based FRBMCSs.

We considered three different types of feature selection algorithms: random subspace [2], mutual information-based feature selection (MIFS) [45], and the random-greedy feature selection based on MIFS and the GRASP approach [46].

The term *bagging* is an acronym of bootstrap aggregation and refers to the first successful method to generate MCSs proposed in the literature [11].

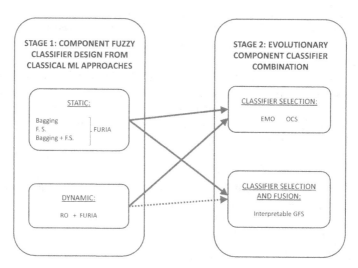

Fig. 1. The proposed framework is composed of several FRBMCSs design methodologies embedded into two stages: 1) FRBMCS design from classical ML approaches and 2) evolutionary component classifier combination

This approach was originally designed for decision tree-based classifiers, however it can be applied to any type of model for classification and regression problems. Bagging is based on bootstrap and consists of reducing the variance of the classification by averaging many classifiers that have been individually tuned to random samples that follow the sample distribution of the training set. The final output of the model is the most frequent value, called voting, of the learners considered. Bagging is more effective when dealing with unstable classifiers (the so-called "weak learners"), what means a small change in the training set can cause a significant change in the final model. In addition, it is recommended when the given dataset is composed of small amount of examples. Furthermore, bagging enables a parallel and independent learning of the learners in the ensemble.

Random subspace is a method in which a subset of features is randomly selected from the original dataset. Alternatively, the greedy Battiti's MIFS method is based on a forward greedy search using the mutual information measure [47], with regard to the class. This method orders a given set S of features by the information they bring to classify the output class considering the already selected features. The mutual information $I(C, F)$ for a given feature F is defined as:

$$I(C, F) = \sum_{c,f} P(c, f) \log \frac{P(c, f)}{P(c)P(f)} \qquad (2)$$

where $P(c)$, $P(f)$ and $P(c, f)$ are respectively the values of the density function for the class, the feature variables, and the joint probability density. In the MIFS method, a first feature f is selected as the one that maximizes $I(C, f)$, and then the features f that maximize $Q(f) = I(C, f) - \beta \sum_{s \in S} I(f, s)$ are sequentially chosen until S reaches the desired size. β is a coefficient to reduce the influence of the information brought by the already selected features.

The random-greedy variant is an approach where the feature subset is generated by iteratively adding features randomly chosen from a restricted candidate list (RCL) composed of the best τ percent features according to the Q measure at each selection step. Parameter τ is used to control the amount of randomness injected in the MIFS selection. With $\tau = 0$, we get the original MIFS method, while with $\tau = 1$, we get the random subspace method.

FURIA-based FRBMCSs are designed as follows. A normalized dataset is split into two parts, a training set and a test set. The training set is submitted to an instance selection and a feature selection procedures in order to provide individual training sets (the so-called *bags*) to train FURIA classifiers. Let us emphasize that FURIA already incorporates an internal feature selection algorithm, being one of the features inherently owned from the RIPPER algorithm.

An exhaustive study was developed comparing all the variants proposed. We selected 21 datasets from the UCI machine learning repository [48] with different characteristics concerning the number of examples, features, and classes. For validation we used Dietterichs 5×2-fold cross-validation (5×2-cv) [49]. Three different feature subsets of different sizes (Small "S", Medium "M", and Large "L") were tested for the FURIA-based fuzzy MCSs using the three different feature selection algorithms. A small number of component fuzzy classifiers (up to 10) was considered in this study. Finally, the best choices of FURIA-based FRBMCSs were compared to two state-of-the-art MCS algorithms such as bagging decision trees and random forests, as well as with the use of the same methodology combined with a different fuzzy classifier generation method, Ishibuchi-based fuzzy MCS [7].

We show Table 4 presenting this final comparison, as the most representative results we have obtained. It consists of 5×2-cv training and test error values. For each algorithm, we only show the best obtained result in terms of accuracy for each dataset and highlight the best values in boldface. Random subspace and random-greedy feature selection are denoted as "R" and "RG", respectively.

The main conclusions obtained in [44] are as follows:

- A MCS framework based on a quick and accurate fuzzy classification rule learning algorithm, namely FURIA, can be competitive if not better than two state-of-the-art machine learning classifier ensembles such as random forests and C4.5 decision tree [50] MCSs generated from bagging [51].
- The proposed FURIA-based FRBMCSs are *accurate* and can be directly applied to high dimensional datasets, high in terms of large number of attributes, number of instances, and/or number of classes, thanks to the fact we use FURIA as a component classifier learning method.
- FURIA-based FRBMCSs with bagging clearly outperform FURIA-based FRBMCSs with feature selection and FURIA-based FRBMCSs with bagging and feature selection. Thus, it is the recommended MCSs combination method.

Table 1. A comparison of the best choice for different approaches for FURIA-based fuzzy MCSs against the best choice of bagging C4.5 MCSs, random forests, and Ishibuchi-based fuzzy MCSs

	aba	bre	gla	hea	ion	let	mag	opt	pbl	pen	pho	pim	sat	seg	son	spa	tex	veh	wav	win	yea
FURIA-based MCSs																					
test err.	0.753	**0.037**	0.313	**0.178**	0.134	0.091	0.136	**0.628**	0.028	0.015	0.136	**0.235**	0.105	0.035	**0.198**	0.061	**0.036**	0.276	**0.156**	**0.036**	0.408
feat sel.	G	R	-	-	RG	-	-	RG	R	R	R	RG	-	-	R	-	-	-	-	RG	-
feat.	L	L	-	-	S	-	-	L	L	L	L	L	-	-	L	-	-	-	-	M	-
sub. size																					
nr of cl.	10	10	7	7	7	10	7	10	10	10	10	10	10	10	10	10	10	10	10	10	10
C4.5 ensembles with bagging																					
test err.	0.772	0.043	0.306	0.194	0.149	0.103	**0.134**	0.697	0.030	0.028	0.131	0.253	0.112	0.042	0.247	0.067	0.051	0.289	0.193	0.097	0.415
nr of cl.	10	7	10	10	10	10	10	10	10	10	10	10	10	10	10	10	10	10	10	10	10
random forests																					
test err.	0.777	0.041	**0.282**	0.211	0.140	**0.080**	**0.134**	0.695	0.031	0.016	0.119	0.264	**0.104**	0.034	0.239	**0.060**	0.040	**0.269**	0.185	0.048	0.438
nr of cl.	7	7	10	10	10	10	10	10	10	10	10	10	10	10	10	10	10	10	10	10	10
Ishibuchi-based fuzzy MCSs																					
test err.	**0.751**	0.056	0.379	0.213	**0.129**	0.420	0.202	0.629	0.075	0.062	0.208	0.238	0.175	0.166	0.245	0.223	0.256	0.398	0.181	0.056	0.482
nr of cl.	3	7	7	10	7	10	7	3	7	10	3	7	7	10	0	10	7	3	7	10	7
feat. sel.	R	R	G	R	RG	RG	R	R	RG	R	G	G	RG	RG	RG	G	RG	RG	RG	G	G

The interested reader is referred to [44] for a deeper explanation of the presented approach.

4.2 Dynamic Approach: Bagging with Random Oracles

This section introduces the use of random oracles (ROs) [38, 39] within the bagging MCS framework to derive FURIA-based FRBMCSs. Our idea is that, thanks to the additional diversity introduced by ROs into the base classifiers, the obtained FRBMCSs are able to achieve an outstanding performance in terms of accuracy [52].

An RO is a structured classifier, also defined as a "mini-ensemble", encapsulating the base classifier of the MCS. It is composed of two subclassifiers and an oracle that decides which one to use in each case. Basically, the oracle is a random function whose objective is to randomly split the dataset into two subsets by dividing the feature space into two regions. Each of the two generated regions (together with the corresponding data subset) is assigned to one classifier. Any shape for the decision surface of the function can be applied as far as it divides the training set into two subsets at random.

Let us emphasize that during the classification phase, the oracle commits an internal dynamic classifier selection, that is to say it decides which subclassifier makes the final decision for the given example to be further used at the ensemble level (classifier fusion). Thus, this MCS method belongs to the *dynamic* family [17, 53].

The RO approach owns several interesting features, making it quite unique among the existing MCS solutions:

- It is a generic approach composing a framework in which ROs embed only the base classifier. Thus, it allows a design choice at two different levels: i) any MCS strategy can be applied; ii) any classifier learning algorithm can be used. Apart from that, it can be used as the MCSs generation method on its own.

- It induces an additional diversity through the randomness coming from the nature of ROs. Generating a set of diverse base classifiers was shown to be fundamental for the MCSs overall performance [3, 54]. Let us emphasize that ROs are applied separately to each of the base classifiers and no training of the oracle is recommended, as it will strongly diminish the desired diversity.
- It embeds the two most common and complementary MCS combination methods, i.e. *classifier fusion* and *(dynamic) classifier selection.*
- A wide study has been carried out over several MCS generation approaches [38, 39] in order to analyse the influence of ROs on these methods. C4.5 [50] (in [38]) and Naïve Bayes [55] (in [39]) were the base classifiers used. All the MCS approaches took an advantage of the ROs, outperforming the original MCSs in terms of accuracy. Especially, the highest accuracy improvement was obtained by random subspace and bagging according to [38].

In particular, we considered two versions of ROs: random linear oracle (RLO) [38, 39] and random spherical oracle (RSO) [39]. The former uses a randomly generated hyperplane to divide the feature space, while the latter does so using a hypersphere.

We selected 29 datasets with different characteristics concerning a high number of examples, features, and classes from the UCI machine learning [48] and KEEL [56] repositories. For validation, 5×2-cv was used. We studied the performance of both RO-based bagging FRBMCSs in comparison with bagging FRBMCSs considering both accuracy and complexity. Then, the best performing FRBMCSs were compared against state-of-the-art RO-based bagging MCSs. By doing so, we wanted to show that RO-based bagging FRBMCSs are competitive against the state-of-the-art RO-based bagging MCSs using C4.5 [38, 39] and Naïve Bayes [39] as the base classifiers, when dealing with high dimensional datasets, thanks to the use of the FURIA algorithm. Finally, we presented some kappa-error diagrams [57] to graphically illustrate the relationship between the diversity and the individual accuracy of the base classifiers among FRBMCSs.

For an illustrative purpose, we include Table 2 in the current contribution, reporting the test results achieved by RSO-based bagging FRBMCSs and RSO-based bagging MCS using C4.5 and NB over the 29 selected datasets.

We highlight the main conclusions drawn from the study developed in [52] as follows:

- Both RO-based bagging FRBMCSs show significant differences in comparison to bagging FRBMCSs considering accuracy, as well as complexity in terms of overall average number of rules. This happens due to the additional diversity induced by the ROs, which was clearly seen in the Kappa-error diagrams [57].
- RSO-based bagging FRBMCSs not only outperform classical RSO-based bagging MCSs using C4.5 and NB, but they also show a lower complexity in comparison to RSO-based bagging MCSs using C4.5. FURIA again turned out to be robust and accurate algorithm, belonging to the fuzzy rule-based classifier family, which obtains an outstanding performance in combination with classical MCS techniques.

Table 2. A comparison of RSO-based bagging MCSs using FURIA, C4.5, and NB in terms of accuracy

Dataset	FURIA Test err.	C4.5 Test err.	NB Test err.
abalone	**0.7472**	0.7696	0.7624
bioassay_688red	**0.0090**	**0.0090**	0.0153
coil2000	**0.0601**	0.0616	0.1820
gas_sensor	**0.0081**	0.0094	0.3003
isolet	**0.0727**	0.0813	0.1253
letter	0.0760	**0.0658**	0.2926
magic	0.1304	**0.1268**	0.2366
marketing	**0.6690**	0.6745	0.6875
mfeat_fac	**0.0461**	0.0501	0.0655
mfeat_fou	**0.1924**	0.1948	0.2205
mfeat_kar	0.0737	0.0867	**0.0597**
mfeat_zer	**0.2220**	0.2294	0.2473
musk2	0.0321	**0.0283**	0.1121
optdigits	**0.0289**	0.0297	0.0717
pblocks	0.0341	**0.0330**	0.0705
pendigits	**0.0136**	0.0161	0.0861
ring_norm	0.0326	0.0397	**0.0202**
sat	0.1007	**0.0967**	0.1731
segment	**0.0296**	0.0326	0.1198
sensor_read_24	**0.0231**	0.0232	0.3703
shuttle	**0.0009**	**0.0009**	0.0157
spambase	**0.0640**	0.0658	0.1777
steel_faults	0.2379	**0.2286**	0.3429
texture	**0.0280**	0.0351	0.1426
thyroid	0.0218	**0.0215**	0.0393
two_norm	0.0288	0.0327	**0.0222**
waveform	**0.1482**	0.1698	0.1672
waveform1	**0.1459**	0.1654	0.1541
wquality_white	0.3825	**0.3737**	0.5216
Avg.	**0.1312**	0.1357	0.2068
Std. Dev.	0.1819	0.1856	0.1892

5 Evolutionary Component Classifier Combination

5.1 Evolutionary Multiobjective Overproduce-and-Choose Static Classifier Selection

In this section, we describe our proposal of an EMO method defining an OCS strategy for the component classifier selection [58]. Our goal is to obtain a good accuracy-complexity trade-off in the FURIA-based FRBMCSs when dealing with high dimensional problems. That is, we aim to obtain FRBMCSs with a low number of base classifiers, which jointly keep a good accuracy. Thus, we have selected the state-of-the-art NSGA-II EMO algorithm [41] in order to generate good quality Pareto set approximations.

NSGA-II is based on a Pareto dominance depth approach, where the population is divided into several fronts and the depth of each front shows to which front an individual belongs to. A pseudo-dominance rank being assigned to each individual, which is equal to the front number, is the metric used for the selection of an individual.

We have used a standard binary coding in such a way that a binary digit/gene is assigned to each classifier. When the variable takes value 1, it means that the current component classifier belongs to the final ensemble, while when the variable is equal to 0, that classifier is discarded. This approach provides a low operation cost, which leads to a high speed of the algorithm.

Five different biobjective fitness functions combining the three existing kinds of optimization criteria (accuracy, complexity, and diversity) are proposed in order to study the best setting. We use the following measures: the training error (accuracy), the number of classifiers (complexity), and the difficulty measure θ and the double fault δ (diversity). Table 3 presents the five combinations proposed.

Table 3. The five fitness function proposed

1st obj.	2nd obj.
TE	Complx
TE	θ
TE	δ
θ	Complx
δ	Complx

The initial fuzzy classifier ensembles are based on applying a bagging approach with the FURIA method as described in Section 4.1. Each FRBMCS so generated is composed of 50 weak learners.

We carried out an experiment comparing all five biobjective fitness functions. We have selected 20 datasets from the UCI machine learning repository with different characteristics concerning the number of examples, features, and classes. To compare the Pareto front approximations of the global learning objectives (i.e. MCS test accuracy and complexity) we considered two of the usual kinds of multiobjective metrics, namely hypervolume ratio (HVR) [59] and C-measure [60], respectively. We also analyzed single solutions extracted from the obtained Pareto front approximations.

In Table 4, we show a representative comparison for this study. FURIA-based fuzzy MCSs are comprised by 7 or 10 classifiers, the small ensemble sizes providing the best results in our previous contribution [44] (see Section 4.1), and with 50 classifiers, the initial structure of the EMO-selected fuzzy MCSs. We also compare them with two state-of-the-art algorithms, random forests [14] and bagging C4.5 MCSs [50], comprised by 7 or 10 classifiers [44]. Besides, for illustration purposes, the aggregated Pareto fronts are represented graphically for the magic and waveform datasets in Figure 2, which allows an easy visual comparison of the performance of the different EMO OCS-based FRBMCSs variants.

The main conclusions drawn from the study developed are as follows [58]:

— Comparing Pareto Fronts using the HVR metric, the fitness function composed of training error (accuracy) and variance (diversity) clearly reported the best performance, while combining variance (diversity) with the number of classifiers (complexity) and double fault (diversity) with the number of classifiers (complexity) turned out to be deceptive combinations. To make a fair comparison, the reference Pareto Fronts, that is to say those based on test error and the number of classifiers, were considered.
— NSGA-II bagging FURIA-based FRBMCSs turned out to be competitive with the static bagging FURIA-based FRBMCSs and classical MCSs such as random forests and bagging C4.5 decision trees in terms of accuracy.

Table 4. A comparison of the NSGA-II FURIA-based fuzzy MCSs against static FURIA-based MCS

NSGA-II combined with FURIA-based MCSs.

	aba	bre	gla	hea	ion	mag	opt	pbl	pen	pho	pim	sat	seg	son	spa	tex	veh	wav	win	yea
test err.	0.741	0.037	0.283	0.170	0.126	0.132	0.625	0.027	0.014	0.125	0.231	0.101	0.027	0.188	0.056	0.028	0.255	0.146	0.018	0.396
fit. func.	2b	2b	2c	2b	2c	2a	2b	2c	2c	2e	2b	2c	2e	2b	2c	2b	2c	2c	2b	
# cl.	18.6	2.7	5.5	2	18.7	5.6	26	4.8	21.8	9	2	14.6	17.6	2	6.8	23.2	7.5	18.7	18.7	7.1

FURIA-based MCSs algorithms Small ensemble sizes.

	aba	bre	gla	hea	ion	mag	opt	pbl	pen	pho	pim	sat	seg	son	spa	tex	veh	wav	win	yea
test err.	0.753	0.037	0.313	0.178	0.134	0.136	0.628	0.028	0.015	0.136	0.235	0.105	0.035	0.198	0.061	0.036	0.276	0.156	0.036	0.408
# cl.	10	10	7	7	7	7	10	10	10	10	10	10	10	10	10	10	10	10	10	10

FURIA-based MCSs algorithms. Ensemble size 50.

	aba	bre	gla	hea	ion	mag	opt	pbl	pen	pho	pim	sat	seg	son	spa	tex	veh	wav	win	yea
test err.	0.748	0.041	0.287	0.182	0.145	0.135	0.630	0.028	0.016	0.135	0.241	0.102	0.034	0.226	0.059	0.031	0.275	0.149	0.035	0.400

C4.5 ensembles with bagging. Small ensemble sizes.

	aba	bre	gla	hea	ion	mag	opt	pbl	pen	pho	pim	sat	seg	son	spa	tex	veh	wav	win	yea
test err.	0.772	0.043	0.306	0.194	0.149	0.134	0.697	0.03	0.028	0.131	0.253	0.112	0.042	0.247	0.067	0.051	0.289	0.193	0.097	0.415
# cl.	10	7	10	10	10	10	10	10	10	10	10	10	10	10	10	10	10	10	10	10

Random forests. Small ensemble sizes.

	aba	bre	gla	hea	ion	mag	opt	pbl	pen	pho	pim	sat	seg	son	spa	tex	veh	wav	win	yea
test err.	0.777	0.041	0.282	0.211	0.14	0.134	0.695	0.031	0.016	0.119	0.264	0.104	0.034	0.239	0.06	0.04	0.269	0.185	0.048	0.438
# cl.	7	7	10	10	10	10	10	10	10	10	10	10	10	10	10	10	10	10	10	10

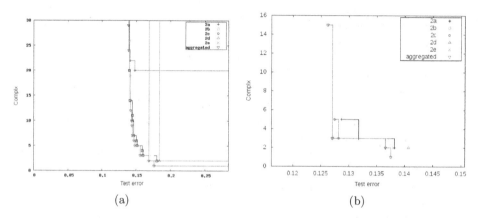

(a)   (b)

Fig. 2. The Pareto front approximations obtained for two datasets using the five fitness functions: (a) waveform and (b) magic. Objective 1 stands for test error and objective 2 for complexity. The pseudo-optimal Pareto front is also drawn for reference.

– NSGA-II combined with FURIA-based FRBMCSs is a good approach to obtain high quality, well performing ensembles with a good accuracy-complexity trade-off, when dealing with high dimensional datasets.

5.2 Joint Classifier Selection and Fusion via an Interpretable Genetic Fuzzy System

The aim of the current section is to present a fuzzy linguistic rule-based classification system playing the role of MCS combination method (a FRBCS-CM) [61]. Our design fulfills several requirements, namely: i) showing a human-understandable structure; ii) being able to deal with high dimensional problems avoiding the curse of dimensionality; iii) having the chance to be automatically learned from training data; and iv) being able to perform both classifier fusion and selection in order to derive low complexity fuzzy classifier ensembles with a good accuracy-complexity trade-off [1].

Using the novel FRBCS-CM together with a fuzzy classifier ensemble, we have the additional advantage of handling a two-level hierarchical structure composed of the individual classifiers in the first level and the FRBCS-CM in the second. These kinds of hierarchical structures [62–65] are well known in the area as they allow fuzzy systems to properly deal with high-dimensional problems while maintaining their descriptive power, especially when considering the single-winner rule fuzzy reasoning method in the component fuzzy classifiers as done in our case.

One step further, using it in combination with a bagging fuzzy classifier ensemble strategy as done in this proposal, we can also benefit from some collateral advantages for the overall design of the FRBMCS: a) the simplicity of the implicit parallelism of bagging, which allows for an easy parallel implementation; and b) the problem partitioning due to the internal feature selection at the component classifier level and the classifier selection capability of the fuzzy linguistic combination method, resulting in a tractable dimension for learning fuzzy rules for each individual classifier and for achieving a compact fuzzy classifier ensemble. These characteristics make the fuzzy ensemble using the FRBCS-CM specially able to deal with the curse of dimensionality.

Our approach might thus be assigned to the stacking (or stacked generalization) group [66], which after bagging and boosting is probably the most popular approach in the literature. Its basis lay in the definition of the meta-learner, playing a role of (advanced) MCS combination method, giving a hierarchical structure of the ensemble. Its task is to gain knowledge of whether training data have been properly learned and to be able to correct badly trained base classifiers. The FRBCS-CM proposed acts as the meta-learner, by discarding the rule subsets in the base fuzzy classifiers providing incorrect decisions at individual class level and promoting the ones leading to a correct classification.

Moreover, fuzzy classification rules with a class and a certainty degree in the consequent used in FRBCS-CM allows the user to get an understandable insight to the MCS. This means that this approach allows interpretability (to some extent) of such complicated system.

The proposed FRBCS-CM is built under the GFS approach (in particular, being an interpretable GFS). A specific GA, which uses a sparse matrix to codify features and linguistic terms in the antecedent parts of the rules and a fitness function based on three accuracy components performs both classifier fusion and classifier selection at class level. The complexity of the final ensemble, defined by the number of terms in the sparse matrix different than zero ("nonzero value"), which is a designed parameter provided by the user.

To evaluate the performance of the FRBCS-CM in the ensembles generated, 20 popular datasets from the UCI machine learning repository have been selected with a number of features varying from a small value (i.e., 5) to a large one

---

[1] We should remind that the proposed combination method can be applied to any multiclassification system with the only restriction that the component classifiers must additionally provide certainty degrees associated to each class in the dataset.

(i.e., 64), while the number of examples scales from 208 to 19 020. In order to compare the accuracy of the considered classifiers, we used 5×2-cv. This study was carried in a three-fold manner. Firstly, we compared bagging FRBMCSs combined with our interpretable GFS performing classifier selection and fusion over bagging FRBMCSs with the full ensemble using standard majority voting (MV). Secondly, we compared the novel interpretable GFS with state-of-the-art crisp and fuzzy multiclassification combination methods, as well as with a hybrid method based on GA considering both classifier selection and classifier fusion [67]. Finally, we showed some interpretability aspects of the proposed fuzzy linguistic combination method.

For the comparison, apart from the standard MV, we select average (AVG) [1] and decision templates (DT) [68] based on Euclidean distance, as crisp and fuzzy fusion methods respectively, being the best methods of each group according to Kuncheva [69]. Since the proposed FRBCS-CM includes classifier selection and classifier fusion, we also apply classifier selection with the mentioned classifier fusion methods in order to make a fair comparison. To select classifiers we will use two standard greedy approaches, Greedy Forward Selection (FS) and Greedy Backward Selection (BS) [70], which will use the abovementioned classifier fusion methods (these methods are also used to guide the search of the greedy algorithms). The hybrid method based on GA proposed in [67] (GA-Dimililer) embeds both classifier selection and classifier fusion, thus we directly apply it without any modifications.

For illustrative purpose, Tables 5 and 6 present a comparison between FRBCS-CM (interpretable GFS) and the other MCS combination methods in terms of accuracy and complexity, respectively. Table 5 shows the test error obtained for MV (operating on the full original ensemble), FRBCS-CM (nonzero values: 10%, 25%, 50%, 75%, and 90%), Greedy FS with MV, AVG, and DT, Greedy BS with MV, AVG, and DT, and GA-Dimililer. Then, Table 6 reports the total number of rules in the ensembles considering the same approaches. The comparison was conducted with respect to the complexity of the obtained FRBMCSs. For example, FRBCS-CM with nonzero values 10% and 25% were compared to Greedy FS with MV, AVG, and DT.

The experiments conducted in this study allowed us to obtain the following conclusions [61]:

- Bagging FRBMCSs combined with the interpretable GFS obtain good results in comparison with bagging FRBMCSs with the full ensemble using standard MV. Apart from obtaining good performance in terms of accuracy, it is also very competitive in terms of complexity reduction, after the selection of the component classifiers. We notice that, the final results highly depends on the parameter defining the complexity of the FRBCS-CM, which leads to different accuracy-complexity trade-offs.

- Our approach turned out to be competitive with the algorithms compared in terms of accuracy, while showing low complexity of the FRBMCSs obtained. Notice that, we aimed to propose a MCS combination method providing a good accuracy-complexity trade-off.

Table 5. Accuracy of the fuzzy MCSs, FRBCS-CM, and the other MCS combination methods in terms of test error

Dataset	fuzzy MCSs	FRBCS-CM					Greedy FS			Greedy BS			GA
		10%	25%	50%	75%	90%	MV	AVG	DT	MV	AVG	DT	Dimil.
Low dim.:													
abalone	0.7458	0.7581	0.7537	0.7493	0.7470	0.7461	0.7524	0.7582	0.7610	0.7484	0.7524	0.7511	0.7494
breast	0.0409	0.0472	0.0469	0.0452	0.0438	0.0432	0.0455	0.0418	0.0398	0.0412	0.0386	0.0372	0.0409
glass	0.2822	0.3159	0.2879	0.2832	0.2692	0.2710	0.2981	0.3271	0.3000	0.2832	0.2720	0.2776	0.3131
heart	0.1822	0.1785	0.1733	0.1719	0.1696	0.1696	0.1859	0.2015	0.1874	0.1778	0.1770	0.1674	0.1726
magic	0.1346	0.1340	0.1314	0.1309	0.1302	0.1300	0.1329	0.1328	0.1323	0.1338	0.1326	0.1298	0.1336
pblocks	0.0288	0.0285	0.0265	0.0271	0.0268	0.0261	0.0282	0.0302	0.0296	0.0286	0.0269	0.0263	0.0402
phoneme	0.1332	0.1277	0.1252	0.1261	0.1256	0.1264	0.1260	0.1232	0.1258	0.1291	0.1271	0.1248	0.1301
pima	0.2385	0.2492	0.2484	0.2411	0.2432	0.2424	0.2503	0.2516	0.2596	0.2385	0.2375	0.2414	0.2398
wine	0.0393	0.0461	0.0382	0.0303	0.0404	0.0393	0.0629	0.0551	0.0607	0.0393	0.0371	0.0360	0.0348
yeast	0.4008	0.4155	0.4054	0.3985	0.4034	0.4013	0.4116	0.4142	0.4189	0.4011	0.3978	0.4018	0.4116
Avg. Low	0.2227	0.2301	0.2237	0.2204	0.2199	0.2196	0.2294	0.2336	0.2315	0.2221	0.2199	0.2193	0.2266
High dim.:													
ionosphere	0.1459	0.1527	0.1413	0.1458	0.1430	0.1430	0.1584	0.1532	0.1646	0.1476	0.1430	0.1413	0.1464
optdigits	0.0329	0.0337	0.0327	0.0327	0.0318	0.0313	0.0367	0.0352	0.0351	0.0329	0.0284	0.0279	0.0721
pendigits	0.0156	0.0174	0.0152	0.0140	0.0140	0.0138	0.0171	0.0150	0.0162	0.0156	0.0129	0.0126	0.0160
sat	0.1021	0.1067	0.1027	0.0997	0.0986	0.1005	0.1044	0.1010	0.1005	0.1022	0.0967	0.0971	0.1040
segment	0.0336	0.0334	0.0319	0.0304	0.0316	0.0302	0.0318	0.0326	0.0336	0.0330	0.0309	0.0306	0.0345
sonar	0.2269	0.2404	0.2183	0.2077	0.2077	0.2058	0.2163	0.2337	0.2452	0.2260	0.2183	0.2163	0.2231
spambase	0.0587	0.0569	0.0559	0.0555	0.0539	0.0546	0.0576	0.0573	0.0574	0.0579	0.0554	0.0549	0.0574
texture	0.0307	0.0343	0.0312	0.0304	0.0291	0.0285	0.0343	0.0330	0.0336	0.0308	0.0268	0.0270	0.0325
vehicle	0.2726	0.2773	0.2664	0.2690	0.2664	0.2674	0.2671	0.2690	0.2693	0.2723	0.2641	0.2600	0.2721
waveform	0.1492	0.1554	0.1490	0.1503	0.1489	0.1479	0.1508	0.1535	0.1533	0.1498	0.1468	0.1472	0.1532
Avg. High	0.1068	0.1108	0.1045	0.1036	0.1025	0.1023	0.1075	0.1084	0.1109	0.1068	0.1023	0.1015	0.1111
Avg. All	0.1647	0.1704	0.1641	0.1620	0.1612	0.1609	0.1684	0.1710	0.1712	0.1644	0.1611	0.1604	0.1689

Table 6. Complexity of the fuzzy MCSs, FRBCS-CM, and the other MCS combination methods in terms of the number of rules

Dataset	fuzzy MCSs	FRBCS-CM					Greedy FS			Greedy BS			GA
		10%	25%	50%	75%	90%	MV	AVG	DT	MV	AVG	DT	Dimil.
Low dim.:													
abalone	3990.9	398.2	995.7	1996.9	2983.6	3578.4	1211.0	1047.6	1037.7	2711.3	3306.9	3398.7	2391.9
breast	435.2	46.1	110.9	217.0	326.2	391	33.0	25.7	24.1	415.9	426.6	427.4	221.1
glass	590.3	57.4	140.6	289.9	434.4	528	88.7	43.6	54.7	560.5	576.4	577.5	173.8
heart	466.0	49.4	120.3	235.3	352.6	421	48.9	35.7	33.4	444.6	455.7	454.7	221.1
magic	3882.1	421.0	968.3	1965.6	2969.9	3475.8	528.2	424.6	417.3	2247.8	3203.6	3319	2123.6
pblocks	1329.4	131.2	328.9	628.1	967.8	1182.2	248.2	108.9	106.1	1259	1288	1297.3	314.1
phoneme	2197.3	241.7	587.8	1132.5	1679.0	2000	493.2	381.1	339.4	1442.8	2046	2049.4	996.9
pima	1050.9	110.9	260.7	530.1	782.4	946	239.3	149.4	118.1	957	1025	1027.7	530
wine	231.4	23.7	57.9	116.4	172.7	208	9.1	6.8	6.2	222.4	226.9	226.9	71.2
yeast	2449.0	260.8	630.9	1198.4	1825.1	2198.4	511.5	389.5	434.9	1901.3	2296.7	2291.9	902.4
Avg. Low	1662.3	174.0	420.2	831.0	1249.4	1492.8	341.1	261.3	257.2	1216.3	1485.2	1507.1	794.6
High dim.:													
ionosphere	367.7	37.8	95.4	211.0	279.8	334	27.0	22.2	24.4	353.3	361.2	360.6	190.3
optdigits	3584.6	359.2	893.5	1787.7	2678.8	3227.2	652.7	428.7	423.7	3398.5	3513.8	3513.1	661.5
pendigits	4395.3	448.8	1098.1	2208.7	3299.9	3964.3	892.1	569.8	470.8	4167.2	4306.4	4307.5	1874.6
sat	4207.2	427.2	1046.9	2107.2	3128.1	3762.8	1214.0	728.7	800.6	3575.4	4006.8	4055	1431.9
segment	1175.3	130.1	290.9	593.4	876.9	1051.4	165.6	109.2	86.7	1100.5	1151.3	1151.4	414.2
sonar	319.3	32.4	80.4	162.0	240.0	288	24.4	22.9	19.8	306.4	312.1	311.9	158.8
spambase	2220.9	229.0	557.2	1115.5	1661.7	2002.6	340.7	286.1	292.8	2135.5	2152.4	2139.8	1026
texture	2912.2	300.1	716.6	1458.8	2175.0	2610.9	433.6	333.8	352.5	2759.8	2852.9	2852.8	1240.4
vehicle	1415.3	154.3	380.4	735.3	1075.3	1283	364.1	173.3	193.4	1304.7	1387.6	1380	425.7
waveform	3484.3	354.0	861.5	1749.8	2601.2	3137.6	1355.9	753.1	727.1	3125.9	3408.3	3381.1	828.9
Avg. High	2408.2	247.3	602.1	1212.9	1801.7	2166.1	547.0	342.8	339.2	2222.7	2345.3	2345.3	825.2
Avg. All	2035.2	210.7	511.1	1022.0	1525.5	1829.4	444.1	302.0	298.2	1719.5	1915.2	1926.2	809.9

– This proposal allows the user to estimate the reduction of the complexity of the final MCS *a priori* by selecting the appropriate non zero parameter value. This high flexibility, an *a priori* choice of how simple will the MCS obtained be, constitutes an advantage over the compared approaches.

– We showed that the proposed fuzzy linguistic combination method provides a good degree of interpretability to the MCS, making the combination method operation mode more transparent for the user. Furthermore, when combined with a FRBMCS, the whole system takes a pure hierarchical structure based on fuzzy classification rules structure (in the sense that the weak learners constitute individual FRBCSs becoming the input to the FRBCS-based combination method). The type of rules with a class and a certainty degree in the consequent used in FRBCS-CM allows the user to get an understandable insight to the MCS, thus allowing interpretability of such complicated system to some extent.

5.3 Evolutionary Multiobjective Overproduce-and-Choose Dynamic Classifier Selection

This section presents an OCS strategy for the classifier selection of our *dynamic* FRBMCSs, the RSO-based bagging FRBMCSs (see Section 4.2). On the one hand, the aim is again to refine the accuracy-complexity trade-off in the RSO-based bagging FRBMCSs when dealing with high dimensional classification problems. On the other hand, an interesting objective is to study whether the additional diversity induced by RSOs is beneficial for the EMO OCS-based FRBMCSs. Thus, we have again chosen the state-of-the-art NSGA-II EMO algorithm in order to generate good quality Pareto set approximations.

In this study [52], we take one step further and use a three-objective fitness function combining the three existing kinds of optimization criteria: accuracy, complexity, and diversity. We use the following measures: the training error (accuracy), the total number of fuzzy rules in the ensemble (complexity), and the difficulty measure θ (diversity). Notice that, in order to make a fair comparison, we consider the final complexity in terms of the total number of rules instead of the total number of classifiers, since RSO-based classifiers produce twice as much classifiers and usually they are less complex than a standard base classifier.

RSO offers a tremendous advantage over a standard component classifier, because each classifier can be independently selected within each pair component. Because of that, our classifier selection is done at the level of the component classifiers and not at the whole pair of classifiers. A specific coding scheme, which permits that none, one, or both FURIA fuzzy subclassifiers can be selected, is introduced. We also develop a reparation operator, whose objective is to correct the unfeasible solutions.

We compared the proposed NSGA-II for RSO-based bagging FRBMCSs classifier selection with the standard NSGA-II using two different approaches from the first stage. Table 7 summarizes the three EMO OCS-based FRBMCSs approaches.

Table 7. The three EMO approaches used for the classifier selection

abbreviation	base classifier	MCS methodology	OCS strategy
2a	FURIA	bagging	standard NSGA-II
2b	RSO (2×FURIA+oracle)	bagging+RSO	standard NSGA-II
2c	RSO (2×FURIA+oracle)	bagging+RSO	proposed NSGA-II

We conducted exhaustive experiments considering 29 datasets with different characteristics concerning a high number of examples, features, and classes from the UCI [48] machine learning and KEEL [56] repositories. For validation we used 5×2-cv. To compare the Pareto front approximations of the global learning objectives (i.e. MCS test accuracy and complexity) we considered the most common multiobjective metric, HVR [59]. We also analyzed single solutions extracted from the obtained Pareto front approximations. We compared the three EMO variants in order to check whether the additional diversity induced by the RSO is beneficial to the performance of the final FRBMCS selected by the NSGA-II.

To give a brief view to the results obtained, Table 8 shows the average and standard deviation values for the four different solutions selected from each Pareto front approximation in the 29 problems. Besides, the aggregated Pareto fronts for the bioassay 688red dataset are represented graphically in Figure 3, which allows an easy visual comparison of the performance of the different EMO OCS-based FRBMCSs variants.

Table 8. A comparison of the averaged performance of the four single solutions selected from the obtained Pareto sets

	Card.	Best train			Best complx			Best trade-off			Best test		
		Tra	Tst	Cmpl	Tra	Tst	Cmpl	Tra	Tst	Cmpl	Tra	Tst	Cmpl
avg. 2a	40.1	0.0512	0.1321	1175	0.0920	0.1628	159	0.0673	0.1367	338	0.0543	0.1298	966
2b	40.3	0.0441	0.1315	1281	0.0920	0.1679	188	0.0612	0.1368	405	0.0480	0.1288	1078
2c	50.0	0.0442	0.1332	931	0.1516	0.2206	104	0.0745	0.1494	270	0.0469	0.1304	853
dev. 2a	43.1	0.1403	0.1829	2180	0.1643	0.1922	166	0.1514	0.1831	533	0.1449	0.1811	1897
2b	42.5	0.1231	0.1826	2164	0.1579	0.1921	188	0.1380	0.1827	574	0.1293	0.1808	2000
2c	32.4	0.1218	0.1841	1497	0.1454	0.1858	109	0.1417	0.1842	427	0.1246	0.1825	1434

From the wide study carried out we concluded that [52]:

- According to the HVR metric, the variant considering the RSO-based bagging FRBMCSs with the NSGA-II method proposed (2c) clearly outperformed the other approaches, mainly due to the low complexity of the final FRBMCSs. To make a fair comparison, the reference Pareto Fronts (based on test error and the number of classifiers) were considered.
- When selecting the best individual FRBMCS design according to the test error, the proposed approach is not significantly worst than the other variants in terms of accuracy, however it obtains a much lower complexity. On the other hand, the best individual FRBMCS design considering the complexity criterion is obtained by our approach, since it provides a solution with the lowest number of rules.

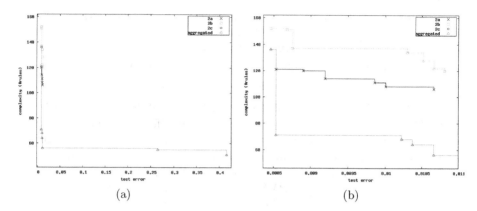

Fig. 3. The Pareto front approximations obtained from the three EMO approaches for three datasets: (a) bioassay_688red, (b) bioassay_688red (zoom). Objective 1 stands for test error and objective 2 for complexity in terms of number of rules. The pseudo-optimal Pareto front is also drawn for reference.

– In general, the additional diversity induced by the RSO have a positive influence on the final FRBMCSs selected by NSGA-II resulting in a strong reduction of complexity, while maintaining a similar accuracy. Thus, the diversity is beneficial for this kind of designs.

6   Conclusions and Future Work

We have proposed a global framework for FRBCS design in order to allow them dealing with high dimensional datasets. Our proposal is composed of different methods for component fuzzy classifier derivation, which consider several MCS methodologies, as well as evolutionary algorithms for classifier selection and fusion. We carried out exhaustive experiments for each component FRBMCS design. The results obtained have shown that we have reached the global goal. Besides, we obtained several sub-goals within the approaches proposed such as improvement of the performance in terms of accuracy and accuracy-complexity trade-off.

The promising results obtained lead to several research lines as future works. Combining bagging RO-based FRBMCSs with interpretable GFS for joint classifier selection and fusion is a future step to take into account. Besides, we will consider a combination of an EMO algorithm with interpretable GFS. Finally, we would like to apply the FRBMCS framework proposed to the real-world applications, consisting of complex and high dimensional classification problems. For instance, a topology-based WiFi indoor localization problem was already solved by one of our FRBMCS designs in [71].

Acknowledgements. This work was supported by the Spanish Ministerio de Economía y Competitividad under Project TIN2012-38525-C02-01, including funding from the European Regional Development Fund.

References

1. Kuncheva, L.I.: Combining Pattern Classifiers: Methods and Algorithms. Wiley (2004)
2. Ho, T.: The random subspace method for constructing decision forests. IEEE Transactions on Pattern Analysis and Machine Intelligence 20(8), 832–844 (1998)
3. Optiz, D., Maclin, R.: Popular ensemble methods: An empirical study. Journal of Artificial Intelligence Research 11, 169–198 (1999)
4. Canul-Reich, J., Shoemaker, L., Hall, L.O.: Ensembles of fuzzy classifiers. In: IEEE International Conference on Fuzzy Systems (FUZZ-IEEE), London, pp. 1–6 (2007)
5. Pedrycz, W., Kwak, K.C.: Boosting of granular models. Fuzzy Sets and Systems 157(22), 2934–2953 (2006)
6. Cordón, O., Quirin, A.: Comparing two genetic overproduce-and-choose strategies for fuzzy rule-based multiclassification systems generated by bagging and mutual information-based feature selection. International Journal of Hybrid Intelligent Systems 7(1), 45–64 (2010)
7. Ishibuchi, H., Nakashima, T., Nii, M.: Classification and Modeling With Linguistic Information Granules. Springer (2005)
8. Casillas, J., Cordon, O., Herrera, F., Magdalena, L.: Interpretability Issues in Fuzzy Modeling. Springer, Heidelberg (2003)
9. Alonso, J.M., Magdalena, L., González-Rodríguez, G.: Looking for a good fuzzy system interpretability index: An experimental approach. International Journal of Approximate Reasoning 51, 115–134 (2009)
10. Dasarathy, B.V., Sheela, B.V.: A composite classifier system design: Concepts and methodology. Proceedings of IEEE 67(5), 708–713 (1979)
11. Breiman, L.: Bagging predictors. Machine Learning 24(2), 123–140 (1996)
12. Schapire, R.: The strength of weak learnability. Machine Learning 5(2), 197–227 (1990)
13. Tsymbal, A., Pechenizkiy, M., Cunningham, P.: Diversity in search strategies for ensemble feature selection. Information Fusion 6(1), 83–98 (2005)
14. Breiman, L.: Random forests. Machine Learning 45(1), 5–32 (2001)
15. Xu, L., Krzyzak, A., Suen, C.Y.: Methods of combining multiple classifiers and their application to handwriting recognition. IEEE Transactions on Systems, Man, and Cybernetics 22(3), 418–435 (1992)
16. Woods, K., Kegelmeyer, W.P., Bowyer, K.: Combination of multiple classifiers using local accuracy estimates. IEEE Transactions on Pattern Analysis and Machine Intelligence 19(4), 405–410 (1997)
17. Giacinto, G., Roli, F.: Dynamic classifier selection based on multiple classifier behaviour. Pattern Recognition 34(9), 1879–1881 (2001)
18. Partridge, D., Yates, W.B.: Engineering multiversion neural-net systems. Neural Computation 8(4), 869–893 (1996)
19. Hernández-Lobato, D., Martínez-Muñoz, G., Suárez, A.: How large should ensembles of classi ers be? Pattern Recognition 46(5), 1323–1336 (2013)
20. Hühn, J.C., Hüllermeier, E.: FURIA: an algorithm for unordered fuzzy rule induction. Data Mining and Knowledge Discovery 19(3), 293–319 (2009)

21. Cohen, W.W.: Fast effective rule induction. In: Proceedings of the Twelfth International Conference on Machine Learning, pp. 115–123. Morgan Kaufmann (1995)
22. Ishibuchi, H., Nakashima, T., Morisawa, T.: Voting in fuzzy rule-based systems for pattern classification problems. Fuzzy Sets and Systems 103(2), 223–238 (1999)
23. Cordón, O., del Jesus, M.J., Herrera, F.: A proposal on reasoning methods in fuzzy rule-based classification systems. International Journal of Approximate Reasoning 20, 21–45 (1999)
24. Takahashi, H., Honda, H.: Lymphoma prognostication from expression profiling using a combination method of boosting and projective adaptive resonance theory. Journal of Chemical Engineering of Japan 39(7), 767–771 (2006)
25. Bonissone, P.P., Cadenas, J.M., Garrido, M.C., Díaz-Valladares, R.A.: A fuzzy random forest. International Journal of Approximate Reasoning 51(7), 729–747 (2010)
26. Marsala, C.: Data mining with ensembles of fuzzy decision trees. In: IEEE Symposium on Computational Intelligence and Data Mining, Nashville, USA, pp. 348–354 (2009)
27. Janikow, C.Z.: Fuzzy decision trees: issues and methods. IEEE Transactions on Systems, Man, and Cybernetics, Part B 28(1), 1–14 (1998)
28. Aguilera, J.J., Chica, M., del Jesus, M.J., Herrera, F.: Niching genetic feature selection algorithms applied to the design of fuzzy rule based classification systems. In: IEEE International Conference on Fuzzy Systems (FUZZ-IEEE), London, pp. 1794–1799 (2007)
29. Goldberg, D.E.: Genetic Algorithms in Search Optimization and Machine Learning. Addison-Wesley (1989)
30. Nojima, Y., Ishibuchi, H.: Designing fuzzy ensemble classifiers by evolutionary multiobjective optimization with an entropy-based diversity criterion. In: International Conference on Hybrid Intelligent Systems and Conference on Neuro-Computing and Evolving Intelligence, CD-ROM, 4 pages (2006)
31. Nojima, Y., Ishibuchi, H.: Genetic rule selection with a multi-classifier coding scheme for ensemble classifier design. International Journal of Hybrid Intelligent Systems 4(3), 157–169 (2007)
32. Ishibuchi, H., Nojima, Y.: Evolutionary multiobjective optimization for the design of fuzzy rule-based ensemble classifiers. International Journal of Hybrid Intelligent Systems 3(3), 129–145 (2006)
33. Yager, R.R., Filev, D.P.: Essentials of fuzzy modeling and control. Wiley-Interscience, New York (1994)
34. Cordón, O., Herrera, F., Hoffmann, F., Magdalena, L.: Genetic Fuzzy Systems. Evolutionary Tuning and Learning of Fuzzy Knowledge Bases. World Scientific (2001)
35. Cordón, O., Gomide, F., Herrera, F., Hoffmann, F., Magdalena, L.: Ten years of genetic fuzzy systems: Current framework and new trends. Fuzzy Sets and Systems 141(1), 5–31 (2004)
36. Herrera, F.: Genetic fuzzy systems: taxonomy, current research trends and prospects. Evolutionary Intelligence 1, 27–46 (2008)
37. Cordón, O.: A historical review of evolutionary learning methods for mamdani-type fuzzy rule-based systems: Designing interpretable genetic fuzzy systems. International Journal of Approximate Reasoning 52(6), 894–913 (2011)
38. Kuncheva, L.I., Rodríguez, J.J.: Classifier ensembles with a random linear oracle. IEEE Transactions on Knowledge and Data Engineering 19(4), 500–508 (2007)

39. Rodríguez, J.J., Kuncheva, L.I.: Naïve bayes ensembles with a random oracle. In: Haindl, M., Kittler, J., Roli, F. (eds.) MCS 2007. LNCS, vol. 4472, pp. 450–458. Springer, Heidelberg (2007)

40. Sharkey, A.J.C., Sharkey, N.E.: The *test and select* approach to ensemble combination. In: International Workshop on Multiclassifier Systems, Cagliari, pp. 30–44 (2000)

41. Deb, K., Pratap, A., Agarwal, S., Meyarivan, T.: A fast and elitist multiobjective genetic algorithm: NSGA-II. IEEE Transactions on Evolutionary Computation 6, 182–197 (2002)

42. Panov, P., Džeroski, S.: Combining bagging and random subspaces to create better ensembles. In: Berthold, M., Shawe-Taylor, J., Lavrač, N. (eds.) IDA 2007. LNCS, vol. 4723, pp. 118–129. Springer, Heidelberg (2007)

43. Stefanowski, J.: An experimental study of methods combining multiple classifiers - diversified both by feature selection and bootstrap sampling. In: Atanassov, K.T., Kacprzyk, J., Krawczak, M., Szmidt, E. (eds.) Issues in the Representation and Processing of Uncertain and Imprecise Information, pp. 337–354. Akademicka Oficyna Wydawnicza EXIT, Warsaw (2005)

44. Trawiński, K., Cordón, O., Quirin, A.: On designing fuzzy rule-based multiclassification systems by combining furia with bagging and feature selection. International Journal of Uncertainty, Fuzziness and Knowledge-Based Systems 19(4), 589–633 (2011)

45. Battiti, R.: Using mutual information for selecting features in supervised neural net learning. IEEE Transactions on Neural Networks 5(4), 537–550 (1994)

46. Feo, T.A., Resende, M.G.C.: Greedy randomized adaptive search procedures. Journal of Global Optimization 6, 109–133 (1995)

47. Shannon, C.E., Weaver, W.: The Mathematical Theory of Communication. University of Illlinois Press (1949)

48. Blake, C.L., Merz, C.J.: UCI repository of machine learning databases (1998), http://archive.ics.uci.edu/ml

49. Dietterich, T.G.: Approximate statistical test for comparing supervised classification learning algorithms. Neural Computation 10(7), 1895–1923 (1998)

50. Quinlan, J.R.: C4.5: programs for machine learning. Morgan Kaufmann Publishers Inc., San Francisco (1993)

51. Dietterich, T.G.: An experimental comparison of three methods for constructing ensembles of decision trees: Bagging, boosting, and randomization. Machine Learning 40(2), 139–157 (2000)

52. Trawiński, K., Cordón, O., Sánchez, L., Quirin, A.: Multiobjective genetic classifier selection for random oracles fuzzy rule-based multiclassifiers: How benefical is the additional diversity? Technical Report AFE 2012-17, European Centre for Soft Computing, Mieres, Spain (2012)

53. Dos Santos, E.M., Sabourin, R., Maupin, P.: A dynamic overproduce-and-choose strategy for the selection of classifier ensembles. Pattern Recognition 41(10), 2993–3009 (2008)

54. Kuncheva, L.I., Whitaker, C.J.: Measures of diversity in classifier ensembles and their relationship with the ensemble accuracy. Machine Learning 51(2), 181–207 (2003)

55. Domingos, P., Pazzani, M.J.: On the optimality of the simple bayesian classifier under zero-one loss. Machine Learning 29(2-3), 103–130 (1997)

56. Alcalá-Fdez, J., Fernández, A., Luengo, J., Derrac, J., García, S.: Keel data-mining software tool: Data set repository, integration of algorithms and experimental analysis framework. Journal of Multiple-Valued Logic and Soft Computing 17(2-3), 255–287 (2011)
57. Margineantu, D.D., Dietterich, T.G.: Pruning adaptive boosting. In: Proceedings of the Fourteenth International Conference on Machine Learning, ICML 19897, pp. 211–218. Morgan Kaufmann Publishers Inc., San Francisco (1997)
58. Trawiński, K., Quirin, A., Cordón, O.: A study on the use of multi-objective genetic algorithms for classifier selection in furia-based fuzzy multiclassifers. International Journal of Computational Intelligence Systems 5(2), 231–253 (2012)
59. Coello, C.A., Lamont, G.B., van Veldhuizen, D.A.: Evolutionary Algorithms for Solving Multi-Objective Problems, 2nd edn. Springer (2007)
60. Zitzler, E., Thiele, L.: Multiobjective evolutionary algorithms: a comparative case study and the strength pareto approach. IEEE Transactions on Evolutionary Computation 3, 257–271 (1999)
61. Trawiński, K., Cordón, O., Sánchez, L., Quirin, A.: A genetic fuzzy linguistic combination method for fuzzy rule-based multiclassifiers. IEEE Transactions on Fuzzy Systems (in press, 2013), doi:10.1109/TFUZZ.2012.2236844.
62. Torra, V.: A review of the construction of hierarchical fuzzy systems. International Journal of Intelligent Systems 17(5), 531–543 (2002)
63. Gegov, A.E., Frank, P.M.: Hierarchical fuzzy control of multivariable systems. Fuzzy Sets and Systems 72, 299–310 (1995)
64. Yager, R.R.: On the construction of hierarchical fuzzy systems model. IEEE Transactions on Systems, Man, and Cybernetics - Part B 28(1), 55–66 (1998)
65. Cordón, O., Herrera, F., Zwir, I.: A hierarchical knowledge-based environment for linguistic modeling: Models and iterative methodology. Fuzzy Sets and Systems 138(2), 307–341 (2003)
66. Wolpert, D.: Stacked generalization. Neural Networks 5(2), 241–259 (1992)
67. Dimililer, N., Varoglu, E., Altincay, H.: Classifier subset selection for biomedical named entity recognition. Applied Intelligence 31, 267–282 (2009)
68. Kuncheva, L.I., Bezdek, J.C., Duin, R.P.W.: Decision templates for multiple classifier fusion: An experimental comparison. Pattern Recognition 34(2), 299–314 (2001)
69. Kuncheva, L.I.: "Fuzzy" versus "nonfuzzy" in combining classifiers designed by boosting. IEEE Transactions on Fuzzy Systems 11(6), 729–741 (2003)
70. Ruta, D., Gabrys, B.: Classifier selection for majority voting. Information Fusion 6(1), 63–81 (2005)
71. Trawiński, K., Alonso, J.M., Hernández, N.: A multiclassifier approach for topology-based wifi indoor localization. Soft Computing (in press, 2013)

Using Nonlinear Dimensionality Reduction to Visualize Classifiers

Alexander Schulz, Andrej Gisbrecht, and Barbara Hammer

University of Bielefeld - CITEC Centre of Excellence, Germany
{aschulz,agisbrec,bhammer}@techfak.uni-bielefeld.de

Abstract. Nonlinear dimensionality reduction (DR) techniques offer the possibility to visually inspect a given finite high-dimensional data set in two dimensions. In this contribution, we address the problem to visualize a trained classifier on top of these projections. We investigate the suitability of popular DR techniques for this purpose and we point out the benefit of integrating auxiliary information as provided by the classifier into the pipeline based on the Fisher information.

Keywords: Visualization of Classifiers, Supervised Dimensionality Reduction, Fisher Information.

1 Introduction

Scalable visual analytics constitutes an emerging field of research which addresses problems occurring when humans interactively interpret large, heterogeneous, high-dimensional data sets, thereby iteratively specifying the learning goals and appropriate data analysis tools based on obtained findings [26]. Besides classical inference tools and classification techniques, interpretability of the models and nonlinear data visualization play a major role in this context [23,15]. Here, the question of how to visualize not only the given data sets, but also classifiers inferred thereof occurs. The possibility to visualize a classifier allows us to extract information beyond the mere classification accuracy such as the questions: are there potential mis-labelings of data which are observable as outliers, are there noisy data regions where the classification is inherently difficult, are there regions where the flexibility of the classifier is not yet sufficient, what is the modality of single classes, etc. A visualization of data together with classification boundaries opens immediate access to this information.

At present, however, the major way to display the result of a classifier and to judge its suitability is by means of the classification accuracy only. Visualization is often restricted to intuitive interfaces to set certain parameters of the classification procedure, such as e.g. ROC curves to set the desired specificity, or more general interfaces to optimize parameters connected to the accuracy [11]. There exists relatively little work to visualize the underlying classifier itself. For the popular support vector machine (SVM), for example, one possibility is to let the user decide an appropriate linear projection by means of tour methods [4].

I. Rojas, G. Joya, and J. Cabestany (Eds.): IWANN 2013, Part I, LNCS 7902, pp. 59–68, 2013.

As an alternative, some techniques rely on the distance of the data points to the class boundary and present this information using e.g. nomograms [12] or by using linear projection techniques on top of this distance [18]. A few nonlinear techniques exist such as SVMV [25], which visualizes the given data by means of a self-organizing map and displays the class boundaries by means of sampling. Summing up, all these techniques constitute either linear approaches or are specific combinations of a given classifier with a given visualization technique.

In this contribution we discuss a general framework which allows to visualize the result of a given classifier and its training set in general, using nonlinear dimensionality reduction techniques. We investigate the benefit of integrating auxiliary information provided by the classifier in the DR method, and we empirically test the suitability of different DR techniques in benchmark scenarios.

2   The General Framework

We assume the following scenario: a finite data set including points $\mathbf{x}_i \in X = \mathbb{R}^n$ and labeling $l_i \in L$ is given. Furthermore, a classifier $f : X \to L$ has been trained on the given training set, such as a SVM or a learning vector quantization network. To evaluate the performance, typically the classification error of the function f on the given training set or a hold out test set is inspected. This gives us an indication whether the classifier is nearly perfect, corresponding to 100% accuracy, or whether errors occur. However, the classification error does not give us a hint about the geometric distribution of the errors (are they equally distributed in the space, or do they accumulate on specific regions), whether errors are unavoidable (due to overlapping regions of the data or outliers), whether the class boundaries are complex (e.g. due to multiple modes in the single classes), etc. A visualization of the given data set and the classifier would offer the possibility to visually inspect the classification result and to answer such questions. We propose a general framework how to create such a visualization.

In recent years, many different nonlinear DR techniques have been proposed to project a given data set onto low dimensions (usually 2D), see e.g. [2,15,22]. These techniques substitute the points $\mathbf{x}_i \in X$ by low-dimensional counterparts $p(\mathbf{x}_i) = \mathbf{y}_i \in Y = \mathbb{R}^2$, such that the structure of the original data points $\mathbf{x}_i$ is preserved by the projections $p(\mathbf{x}_i) = \mathbf{y}_i$ as much as possible. These techniques, however, map a given finite set of data points only. The techniques do neither represent the structure of the data points as concerns a given classifier nor their relation to the classification boundary. Which possibilities exist to extend a given nonlinear DR method such that an underlying classifier is displayed as well?

We assume a classifier f is present. In addition, we assume that the label $f(\mathbf{x})$ is accompanied by a real value $r(\mathbf{x}) \in \mathbb{R}$ which is a monotonic function depending on the minimum distance from the class boundary. Assuming a nonlinear DR method p is given, a naive approach could be to sample the full data space X, classify those samples and to project them down using p, this way visualizing the class to which each region belongs. This simple method, however, fails unless X is low-dimensional because of two reasons: (i) sampling a high-dimensional data space X sufficiently requires an exponential number of points and (ii) it is impossible to map a full high-dimensional data space X faithfully to low dimensions. The problem lies in the fact that this procedure tries to visualize the class boundaries in the full data space X. It would be sufficient to visualize only those parts of the boundaries which are relevant for the given training data $\mathbf{x}_i$, the latter usually lying on a low-dimensional sub-manifold only.

How can this sub-manifold be sampled? We propose the following three steps:

- Project the data $\mathbf{x}_i$ using a DR technique leading to points $p(\mathbf{x}_i) \in Y = \mathbb{R}^2$.
- Sample the projection space Y leading to points $\mathbf{z}_i'$. Determine points $\mathbf{z}_i$ in the data space X which correspond to these projections $p(\mathbf{z}_i) \approx \mathbf{z}_i'$.
- Visualize the training points $\mathbf{x}_i$ together with the contours induced by the sampled function $(\mathbf{z}_i', |r(\mathbf{z}_i)|)$.

Unlike the naive approach, sampling takes place in $\mathbb{R}^2$ only and, thus, it is feasible. Further, only those parts of the space X are considered which correspond to the observed data manifold as represented by the training points $\mathbf{x}_i$, i.e. the class boundaries are displayed only as concerns these training data. Figure 1 shows the application of these steps to a toy data set.

However, two crucial question remain: How can we determine inverse points $\mathbf{z}_i$ for given projections $\mathbf{z}_i'$ which correspond to inverse images in the data manifold? What properties should the DR technique fulfill? (See [20] for prior work.) After discussing the first two questions, we evaluate the suitability of different DR techniques, taking into account *discriminative* DR, in particular.

3   Inverse Nonlinear Dimensionality Reduction

Given a nonlinear projection of points $\mathbf{x}_i \in X$ to $p(\mathbf{x}_i) = \mathbf{y}_i \in \mathbb{R}^2$ and additional data points $\mathbf{z}_i' \in \mathbb{R}^2$, what are points $\mathbf{z}_i$ such that its projections approximate $\mathbf{z}_i' \approx p(\mathbf{z}_i)$ and, in addition, $\mathbf{z}_i$ are contained in the data manifold? There exist a few problems: usually, an explicit mapping p is not given, rather only discrete projections of the data, albeit a few approaches to extend a mapping of points to a mapping of data have recently been proposed for the general case [9,2]. Second, since X is high-dimensional, the projection p is not uniquely invertible.

Here, we propose an interpolation technique similar to the kernel DR mapping as introduced in [9]. We assume the inverse mapping to be of the form

$$p^{-1} : Y \to X, \mathbf{y} \mapsto \frac{\sum_i \alpha_i k_i(\mathbf{y}_i, \mathbf{y})}{\sum_i k_i(\mathbf{y}_i, \mathbf{y})} = \mathbf{A}\mathbf{k}, \tag{1}$$

where $\alpha_i \in X$ are parameters of the mapping and $k_i(\mathbf{y}_i, \mathbf{y}) = \exp(-0.5\|\mathbf{y}_i - \mathbf{y}\|^2/\sigma_i^2)$ constitutes a Gaussian kernel with bandwidth determined by σ_i. The matrix $\mathbf{A}$ contains the α_i in its columns and $\mathbf{k}$ is a vector of normalized kernel values. The sum is either over a subset of the given data projections $\mathbf{y}_i = p(\mathbf{x}_i)$, or over codebooks resulting from a clustering of the $\mathbf{y}_i$. Now this mapping is trained on the points $(\mathbf{x}_i, p(\mathbf{x}_i))$ corresponding to the data manifold X only.

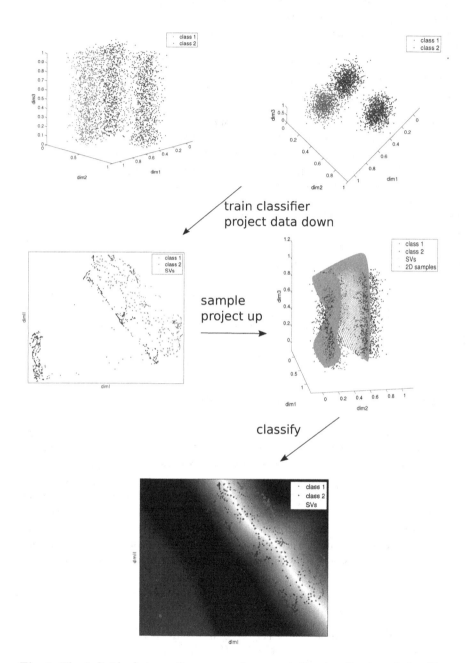

Fig. 1. The individual steps of our general approach for visualization of classifiers

Due to this training set, an inversion of the projection p is emphasized which maps points in Y to inverse points which lie in the original data manifold.

The parameters α_i can be obtained by minimizing the regularized Euclidean error of the projection p^{-1} on the training set $\sum_i \left\| \mathbf{x}_i - p^{-1}(\mathbf{y}_i) \right\|^2 + \lambda \left\| \mathbf{A} \right\|^2$. Although the solution can be computed directly with the Moore-Penrose pseudo inverse, such an error function assumes that all dimensions in X are equally important which is not generally true. Better results can be obtained if dimensions in X, which are locally relevant for the classification, are emphasized. This information is implicitly available in form of the trained classifier: if for two close-by points the distance from the class boundary differs, then the location of these points varies along relevant dimensions. Including this thought, we propose to use the following error function

$$E = \sum_i \left\| r(\mathbf{x}_i) - r\left(p^{-1}(\mathbf{y}_i)\right) \right\|^2 + \lambda \sum_i \left\| \mathbf{x}_i - p^{-1}(\mathbf{y}_i) \right\|^2, \qquad (2)$$

which emphasizes the relevance of errors as concern their distance from the class boundaries. Examples for a suitable choice of $r(\mathbf{x})$ are $(\mathbf{w}^\top \phi(\mathbf{x}) + b)/\sqrt{\mathbf{w}^\top \mathbf{w}}$ for a two class SVM. Minimization of these costs takes place by gradient techniques.

4 Discriminative Nonlinear Visualization

A large variety of DR techniques exists with the aim to map high-dimensional data to low dimensions such that as much structure as possible is preserved. Since many formalizations of 'structure preservation' exist, a variety of different techniques results, as summarized e.g. in [15,3]. Most nonlinear DR methods are non-parametric. Three popular DR methods, which we will use in experiments, are briefly described in the following.

- *t-Distributed Stochastic Neighbor Embedding* (t-SNE) projects high-dimensional data such that the probabilities of data pairs are preserved in the low-dimensional space [22]. A Gaussian distribution is assumed in the high-dimensional space and a Student-t distribution in the low-dimensional space. This addresses the crowding problem.
- *Isomap* [21] is a manifold learner which estimates geodesic distances in the data manifold based on a k nearest neighborhood graph, and maps these distances to two dimensions using classical multi-dimensional scaling.
- *Maximum Variance Unfolding* (MVU) is another manifold learner [27] which estimates a similarity matrix of the data by maximizing the overall variance while keeping the distances of each k nearest neighbors fixed.

DR is an inherently ill-posed problem, and the result of a DR tool largely varies depending on the chosen technology, the parameters, and partially even random aspects for non-deterministic algorithms. Often, the reliability and suitability of the obtained visualization for the task at hand is not clear at all since a DR tool might focus on irrelevant aspects or noise in the data. Discriminative DR, i.e. the integration of auxiliary information by an explicit labeling of data can help to partially overcome these problems: in discriminative DR, the aim is to visualize those aspects of the data which are particularly relevant for the given class information. Thus, the information which is neglected by the DR method is no longer arbitrary but directly linked to its relevance for the given classes. Since, in our setting, auxiliary information is available in terms of the trained classifier, using supervised DR tools seems beneficial.

A variety of different discriminative DR techniques has been proposed, such as the linear techniques Fisher's linear discriminant analysis (LDA), partial least squares regression (PLS), informed projections [6], or global linear transformations of the metric to include auxiliary information [10,3], or kernelization of such approaches [16,1]. A rather general idea to include supervision is to locally modify the metric [17,8] by defining a Riemannian manifold: the information of $\mathbf{x}$ for the class labeling can be incorporated into the distance computation: $d_{\mathbf{J}}^2(\mathbf{x}, \mathbf{x}+d\mathbf{x}) = (d\mathbf{x})^\top \mathbf{J}(\mathbf{x})(d\mathbf{x})$, where $\mathbf{J}(\mathbf{x})$ is the local Fisher information matrix

$$\mathbf{J}(\mathbf{x}) = E_{p(c|\mathbf{x})} \left\{ \left(\frac{\partial}{\partial \mathbf{x}} \log p(c|\mathbf{x}) \right) \left(\frac{\partial}{\partial \mathbf{x}} \log p(c|\mathbf{x}) \right)^\top \right\}. \qquad (3)$$

For practical applications, $\mathbf{J}(\mathbf{x})$ has to be approximated. See [17] for more details.

Obviously, this approach can be integrated in all DR methods which work on distances or similarities, in particular, it can be integrated into the methods described above, resulting in Fisher t-SNE, Fisher Isomap or Fisher MVU.

5   Experiments

We investigate the influence of different DR methods including discriminative techniques based on the Fisher information on our general framework to visualize classifiers. We use the three unsupervised DR methods t-SNE, Isomap and MVU and compare their visualizations of classifiers to those obtained by Fisher t-SNE, Fisher Isomap and Fisher MVU.

We employ three benchmark data sets. Similarly as in [24], we use a randomly chosen subsample of 1500 samples for each data set to save computational time.

- The *letter recognition* data set (referred to as letter in the following) comprises 16 attributes of randomly distorted images of letters in 20 different fonts. The data set contains 26 classes and is available at the UCI Machine Learning Repository [7].
- The *phoneme* data set (denoted phoneme) consists of phoneme samples which are encoded with 20 attributes. 13 classes are available and the data set is taken from LVQ-PAK [13].
- The *U.S. Postal Service* data set (abbreviated via usps) contains 16×16 images of handwritten digits, and hence comprises 10 classes. It can be obtained from [19]. The data set has been preprocessed with PCA by projecting all data samples on the first 30 principal components.

For each of these data sets, we train a SVM provided by the LIBSVM toolbox [5]. It employs an one versus one classification with majority vote for problems with more then two classes. For such an approach, the resulting class boundaries coincide mostly with those of the two class SVMs (which is not the case for the one versus all scheme, see [14]). Therefore, we can choose the accumulated function r as the minimum output taken over all pairwise measures r which include the class of the current data point.

For each data set, we apply the six DR methods to project all data points. We use a 10-fold cross-validation setup to evaluate the mapping p^{-1}: in each fold, we use one of the subsets for evaluation and the remaining to train the mapping. The training is done by, first, performing a clustering in the two-dimensional space. We choose the number of codebooks as 20 times the number of classes and we estimate the bandwidth of the kernel by $\sigma_i^2 = \sigma^2 = c \cdot mean(dist_i)$ where $dist_i$ is the distance of the codebook $\mathbf{y}_i$ to its closest neighboring codebook and $c = 10$. Second, we use 9 subsets to train the mapping p^{-1} by minimizing (2) via gradient descent. We evaluate the mapping on the remaining subset by calculating the accordance of the labels assigned by the SVM and the labels which would be assigned by the low-dimensional visualization of the classifier.

This evaluation measure is averaged over the folds to produce the mean and standard deviation. These values are depicted for all DR mappings and all data sets in figure 2. Obviously, in all cases, integration of the Fisher information is beneficial. For the most cases, this even leads to significantly better results: Only for the phoneme data set, the difference of the means between Isomap and Fisher

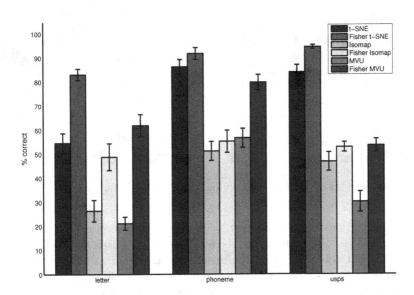

Fig. 2. Accuracy of the visualization of SVM classifiers with three supervised and three unsupervised DR methods

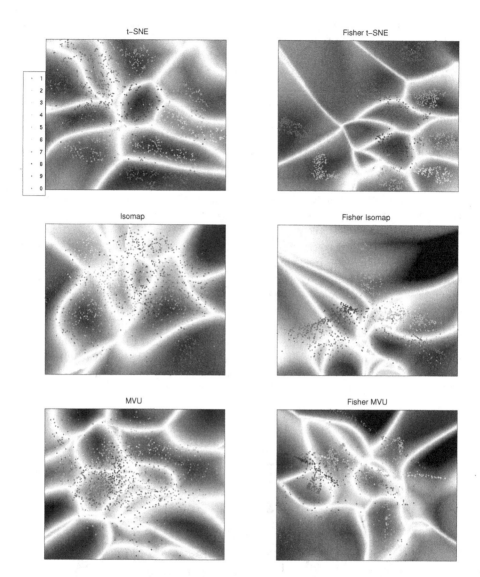

Fig. 3. Visualizations of the usps data set and the SVM with the six used DR methods

Isomap does not show any significant (at a significance level of 0.05). This can also be verified visually by example visualizations of the usps data set shown in figure 3. The left column depicts the unsupervised visualizations and the right the supervised ones. Additionally, these images show also different visualizations of the same SVM classifier. This is encoded by the background: The color of a region indicates the label which would be assigned to it by the classifier while the intensity specifies the certainty of the classifier, i.e. the brighter the color, the more uncertain is the classifier in the corresponding region. The regions of highest uncertainty (i.e. white regions) imply class borders. Furthermore, figure 2 suggests that t-SNE is the most appropriate technique in these clustering scenarios. This suggestion is supported by the visualizations in figure 3 where the t-SNE results show clearer cluster.

6   Conclusions

We investigated the possibility to extend DR techniques for classifier visualization, and we experimentally compared the suitability of different supervised and unsupervised DR techniques in this context. It turns out that supervision of the DR techniques greatly enhances the performance, since it streamlines the aspects which should be visualized by the mapping towards the relevant local directions. In addition, using visualization techniques which are tailored to emphasize cluster structure such as t-SNE are particularly suited in this context. For the visualization of regression tasks (for which a similar framework could be used), alternatives which focus on manifold structure might be better suited. This will be the subject of future work.

Acknowledgements. Funding from DFG under grant number HA2719/7-1 and by the CITEC centre of excellence is gratefully acknowledged.

References

1. Baudat, G., Anouar, F.: Generalized discriminant analysis using a kernel approach. Neural Computation 12, 2385–2404 (2000)
2. Bunte, K., Biehl, M., Hammer, B.: A general framework for dimensionality reducing data visualization mapping. Neural Computation 24(3), 771–804 (2012)
3. Bunte, K., Schneider, P., Hammer, B., Schleif, F.-M., Villmann, T., Biehl, M.: Limited rank matrix learning, discriminative dimension reduction and visualization. Neural Networks 26, 159–173 (2012)
4. Caragea, D., Cook, D., Wickham, H., Honavar, V.G.: Visual methods for examining svm classifiers. In: Simoff, S.J., Böhlen, M.H., Mazeika, A. (eds.) Visual Data Mining. LNCS, vol. 4404, pp. 136–153. Springer, Heidelberg (2008)
5. Chang, C.-C., Lin, C.-J.: LIBSVM: A library for support vector machines. ACM Transactions on Intelligent Systems and Technology 2, 27:1–27:27 (2011), Software available at http://www.csie.ntu.edu.tw/~cjlin/libsvm
6. Cohn, D.: Informed projections. In: Becker, S., Thrun, S., Obermayer, K. (eds.) NIPS, pp. 849–856. MIT Press (2003)
7. Frank, A., Asuncion, A.: UCI machine learning repository (2010)
8. Gisbrecht, A., Hofmann, D., Hammer, B.: Discriminative dimensionality reduction mappings (2012)
9. Gisbrecht, A., Lueks, W., Mokbel, B., Hammer, B.: Out-of-sample kernel extensions for nonparametric dimensionality reduction. In: ESANN 2012, pp. 531–536 (2012)
10. Goldberger, J., Roweis, S., Hinton, G., Salakhutdinov, R.: Neighbourhood components analysis. In: Advances in Neural Information Processing Systems 17, pp. 513–520. MIT Press (2004)

11. Hernandez-Orallo, J., Flach, P., Ferri, C.: Brier curves: a new cost-based visualisation of classifier performance. In: International Conference on Machine Learning (June 2011)
12. Jakulin, A., Možina, M., Demšar, J., Bratko, I., Zupan, B.: Nomograms for visualizing support vector machines. In: Proceedings of the Eleventh ACM SIGKDD International Conference on Knowledge Discovery in Data Mining, KDD 2005, pp. 108–117. ACM, New York (2005)
13. Kohonen, T., Hynninen, J., Kangas, J., Laaksonen, J., Torkkola, K.: LVQ_PAK: The Learning Vector Quantization program package. Report A30, Helsinki University of Technology, Laboratory of Computer and Information Science (January 1996)
14. Kreßel, U.H.-G.: Advances in kernel methods. In: Chapter Pairwise Classification and Support Vector Machines, pp. 255–268. MIT Press, Cambridge (1999)
15. Lee, J.A., Verleysen, M.: Nonlinear dimensionality reduction. Springer (2007)
16. Ma, B., Qu, H., Wong, H.: Kernel clustering-based discriminant analysis. Pattern Recognition 40(1), 324–327 (2007)
17. Peltonen, J., Klami, A., Kaski, S.: Improved learning of riemannian metrics for exploratory analysis. Neural Networks 17, 1087–1100 (2004)
18. Poulet, F.: Visual svm. In: Chen, C.-S., Filipe, J., Seruca, I., Cordeiro, J. (eds.) ICEIS (2), pp. 309–314 (2005)
19. Roweis, S.: Machine learning data sets (2012), http://www.cs.nyu.edu/~roweis/data.html
20. Schulz, A., Gisbrecht, A., Bunte, K., Hammer, B.: How to visualize a classifier? In: New Challenges in Neural Computation, pp. 73–83 (2012)
21. Tenenbaum, J., da Silva, V., Langford, J.: A global geometric framework for nonlinear dimensionality reduction. Science 290, 2319–2323 (2000)
22. van der Maaten, L., Hinton, G.: Visualizing high-dimensional data using t-sne. Journal of Machine Learning Research 9, 2579–2605 (2008)
23. Vellido, A., Martin-Guerroro, J., Lisboa, P.: Making machine learning models interpretable. In: ESANN 2012 (2012)
24. Venna, J., Peltonen, J., Nybo, K., Aidos, H., Kaski, S.: Information retrieval perspective to nonlinear dimensionality reduction for data visualization. Journal of Machine Learning Research 11, 451–490 (2010)
25. Wang, X., Wu, S., Wang, X., Li, Q.: Svmv - a novel algorithm for the visualization of svm classification results. In: Wang, J., Yi, Z., Żurada, J.M., Lu, B.-L., Yin, H. (eds.) ISNN 2006. LNCS, vol. 3971, pp. 968–973. Springer, Heidelberg (2006)
26. Ward, M., Grinstein, G., Keim, D.A.: Interactive Data Visualization: Foundations, Techniques, and Application. A. K. Peters, Ltd. (2010)
27. Weinberger, K.Q., Saul, L.K.: Unsupervised learning of image manifolds by semidefinite programming. Int. J. Comput. Vision 70(1), 77–90 (2006)

Which Dissimilarity Is to Be Used When Extracting Typologies in Sequence Analysis? A Comparative Study

Sébastien Massoni[1], Madalina Olteanu[2], and Nathalie Villa-Vialaneix[2,3]

[1] Centre d'Economie de la Sorbonne, UMR CNRS 8174, Université Paris 1
sebastien.massoni@gmail.com
[2] SAMM, EA 4543, Université Paris 1, Paris, France
madalina.olteanu@univ-paris1.fr
[3] Unité MIAT, INRA de Toulouse, Auzeville, France
nathalie.villa@toulouse.inra.fr

Abstract. Originally developed in bioinformatics, sequence analysis is being increasingly used in social sciences for the study of life-course processes. The methodology generally employed consists in computing dissimilarities between the trajectories and, if typologies are sought, in clustering the trajectories according to their similarities or dissemblances. The choice of an appropriate dissimilarity measure is a major issue when dealing with sequence analysis for life sequences. Several dissimilarities are available in the literature, but neither of them succeeds to become indisputable. In this paper, instead of deciding upon one dissimilarity measure, we propose to use an optimal convex combination of different dissimilarities. The optimality is automatically determined by the clustering procedure and is defined with respect to the within-class variance.

1 Introduction

Originally developed in bioinformatics, sequence analysis is being increasingly used in social sciences for the study of life-course processes. The methodology generally employed consists in computing dissimilarities between the trajectories and, if typologies are sought, in clustering the trajectories according to their similarities or dissemblances. However, measuring dissimilarities or similarities for categorical sequences has always been a challenge in practice. This challenge becomes even harder in social sciences where these measures need some theoretical foundations. Choosing the appropriate dissimilarity or dissimilarity for life-sequence analysis is a key issue which relates to the resulting typologies. The literature on this topic is very rich and still very debated. Each method has its own advantages and drawbacks [1,2].

In this paper, we introduce a different approach. Instead of deciding upon one specific dissimilarity, we propose to use several ones, optimally combined. We consider three main categories of dissimilarities : χ^2-metric [3], optimal matching [2] and non-alignement techniques [1]. Since our final goal is to extract typologies

I. Rojas, G. Joya, and J. Cabestany (Eds.): IWANN 2013, Part I, LNCS 7902, pp. 69–79, 2013.
© Springer-Verlag Berlin Heidelberg 2013

for life sequences, we are looking for the best convex combination of the different dissimilarities which provides the best clusters in terms of homogeneity. The algorithm used for clustering is a self-organizing map (SOM). We use a modified version of the online relational SOM introduced in [4]. In the algorithm proposed here, an additional step is added to each iteration. During this step, the coefficients of the convex combination of dissimilarities are updated according to a gradient-descent principle which aims at minimizing the extended within-class variance.

The rest of the manuscript is organized as follows : Section 2 reviews the different dissimilarities usually used to handle categorical time series. Section 3 describes the online relational SOM for multiple dissimilarities. Section 4 presents a detailed application for sequences related to school-to-work transitions.

2   Dissimilarities for Life Sequences

Three main categories of dissimilarities were addressed in our study. Each of them is briefly described below.

χ^2-distance. Historically, factor analysis was used first to extract typologies from life sequences, [3]. The sequences, which are categorical data, were transformed by running a multiple correspondence analysis (MCA) on the complete disjunctive table. Then, clustering methods adapted to continuous data were applied and the main typologies were extracted. Performing MCA and then computing the Euclidean distance on the resulting vectors is equivalent to computing the χ^2-distance on the rows of the complete disjunctive table. The χ^2-distance is weighting each variable by the inverse of the associated frequency. Hence, the less frequent situations have a larger weight in the distance and the rare events become more important. Also, the χ^2-distance emphasizes the contemporary identical situations, whether these identical moments are contiguous or not. However, the transitions between statuses are not taken into account and input vectors are close only if they share contemporary statuses throughout time.

Optimal-matching Dissimilarities. Optimal matching, also known as "edit distance" or "Levenshtein distance", was first introduced in biology by [5] and used for aligning and comparing sequences. In social sciences, the first applications are due to [6]. The underlying idea of optimal matching is to transform the sequence i into the sequence i' using three possible operations: insertion, deletion and substitution. A cost is associated to each of the three operations. The dissimilarity between i and i' is computed as the cost associated to the smallest number of operations which allows to transform i into i'. The method seems simple and relatively intuitive, but the choice of the costs is a delicate operation in social sciences. This topic is subject to lively debates in the literature [7,8] mostly because of the difficulties to establish an explicit and sound theoretical frame. Among optimal-matching dissimilarities, we selected three dissimilarities: the OM with substitution costs computed from the transition matrix between statuses as proposed in [9], the Hamming dissimilarity (HAM, no insertion or deletion costs and a substitution cost equal to 1) and the Dynamic Hamming dissimilarity (DHD as described in [10]). Obviously, other choices are equally possible and the costs may be adapted, depending whether the user wants to highlight the contemporaneity of situations or the existence of common, possibly not contemporary, sub-sequences.

Non-alignment Techniques. Since the definition of costs represents an important drawback for optimal-matching dissimilarities, several alternatives were proposed in the literature. Here, we considered three different dissimilarities introduced by C. Elzinga [1,11]: the longest common prefix (LCP), the longest common suffix or reversed LCP (RLCP) and the longest common subsequence (LCS). Dissimilarities based on common subsequences are adapted to handle transitions between statuses while they take into account the order in the sequence. They are also able to handle sequences of different lengths.

3 Relational SOM

Extracting typologies from life sequences requires clustering algorithms based on dissimilarity matrices. Generally, hierarchical clustering or K-means are used in the literature, [2]. In this paper, we focus on a different approach, based on a Self-Organizing Map (SOM) algorithm [12]. The interest of using a SOM algorithm adapted to dissimilarity matrices was shown in [13]. Self-organizing maps possess the nice property of projecting the input vectors in a two-dimensional space, while clustering them. In [13], the authors used dissimilarity SOM (DSOM) introduced by [14]. OM with substitution cost defined from the transition matrix was used to measure the dissimilarity between sequences. While DSOM improves clustering by additionally providing a mapping of the typologies, it still has a major drawback: prototypes have to be chosen among the input vectors. Thus, the clustering doesn't allow for empty clusters, which may be quite restrictive in some cases. Moreover, this property of DSOM makes it very sensitive to the initialization. The computation time is also very important, since the research of the prototype is done exhaustively among all input vectors and the algorithm is of batch type.

Online Relational SOM. Inspired by the online kernel version of SOM [15], [4] recently proposed an online version of SOM for dissimilarity matrices, called online relational SOM. Online relational SOM is based on the assumption that prototypes may be written as convex combinations of the input vectors, as previously proposed in [16]. This assumption gives more flexibility to the algorithm, which now allows for empty clusters. Moreover, since the algorithm is online, the dependency on the initialization lessens and the computation time also decreases.

In the online relational SOM, n input data, $x_1, \ldots, x_n$, taking values in an arbitrary input space $\mathcal{G}$, are described by a dissimilarity matrix $\mathbf{\Delta} = (\delta_{ij})_{i,j=1,\ldots,n}$ such that $\mathbf{\Delta}$ is non negative ($\delta_{ij} \geq 0$), symmetric ($\delta_{ij} = \delta_{ji}$) and null on the diagonal ($\delta_{ii} = 0$). The algorithm maps the data into a low dimensional grid composed of U units which are linked together by a neighborhood relationship

$H(u, u')$. A prototype p_u is associated with each unit $u \in \{1, \ldots, U\}$ in the grid. To allow computation of dissimilarities between the prototypes $(p_u)_u$ and the data $(x_i)_i$, the prototypes are symbolically represented by a convex combination of the original data $p_u \sim \sum_i \beta_{ui} x_i$ with $\beta_{ui} \in [0, 1]$ and $\sum_i \beta_{ui} = 1$.

Online Multiple Relational SOM. As explained in the introduction, the choice of a dissimilarity measure in social sciences is a complex issue. When the purpose is to extract typologies, the results of the clustering algorithms are highly dependent on the criterion used for measuring the dissemblance between two sequences of events. A different approach is to bypass the choice of the metric: instead of having to choose one dissimilarity measure among the existing ones, use a combination of them. However, this alternative solution requires an adapted clustering algorithm.

Similarly to the multiple kernel SOM introduced in [17], we propose the multiple relational SOM (MR-SOM). Here, D dissimilarity matrices measured on the input data, $\mathbf{\Delta}^1, \ldots, \mathbf{\Delta}^D$, are supposed to be available. These matrices are combined into a single one, defined as a convex combination: $\mathbf{\Delta}^\alpha = \sum_d \alpha_d \mathbf{\Delta}^d$ where $\alpha_d \geq 0$ and $\sum_{d=1}^D \alpha_d = 1$.

If the (α_d) are given, relational SOM based on the dissimilarity $\mathbf{\Delta}^\alpha$ aims at minimizing over $(\beta_{ui})_{ui}$ and $(\alpha_d)_d$ the following energy function :

$$\mathcal{E}((\beta_{ui})_{ui}, (\alpha_d)_d) = \sum_{i=1}^n \sum_{u=1}^U H\left(f(x_i), u\right) \delta^\alpha \left(x_i, p_u(\beta_u)\right),$$

where $f(x_i)$ is the neuron where x_i is classified[1], $\delta^\alpha \left(x_i, p_u(\beta_u)\right)$ is defined by $\delta^\alpha \left(x_i, p_u(\beta_u)\right) \equiv \mathbf{\Delta}_i^\alpha \beta_u - \frac{1}{2}\beta_u^T \mathbf{\Delta}^\alpha \beta_u$ and $\mathbf{\Delta}_i^\alpha$ is the i-th row of the matrix $\mathbf{\Delta}^\alpha$.

When there is no a-priori on the $(\alpha_d)_d$, we propose to include the optimization of the convex combination into an online algorithm that trains the map. Following an idea similar to that of [18], the SOM is trained by performing, alternatively, the standard steps of the SOM algorithm (i.e., affectation and representation steps) and a gradient descent step for the $(\alpha_i)_i$. To perform the stochastic gradient descent step on the (α_d), the computation of the derivative of $\mathcal{E}|_{x_i} = \sum_{u=1}^M H\left(f(x_i), u\right) \delta^\alpha \left(x_i, p_u(\beta_u)\right)$ (the contribution of the randomly chosen observation $(x_i)_i$ to the energy) with respect to α is needed. But, $\mathcal{D}_{id} = \frac{\partial \mathcal{E}|_{x_i}}{\partial \alpha_d} = \sum_{u=1}^M H\left(f(x_i), u\right) \left(\mathbf{\Delta}_i^d \beta_u - \frac{1}{2}\beta_u^T \mathbf{\Delta}^d \beta_u\right)$, which leads to the algorithm described in Algorithm 1.

4   Application for the Analysis of Life Sequences

Data. For illustrating the proposed methodology and its relevance for categorical time series analysis, we used the data in the survey "Generation 98" from CEREQ, France (http://www.cereq.fr/). According to the French National Institute of Statistics, 22,7% of young people under 25 were unemployed at the end

---

[1] Usually, it is simply the neuron whose prototype is the closest to x_i: see Algorithm 1.

Algorithm 1. Online multiple dissimilarity SOM

1: $\forall u$ and i initialize β_{ui}^0 randomly in $\mathbf{R}$ and $\forall d$, set α_d.
2: **for** t=1,...,T **do**
3:    Randomly choose an input x_i
4:    *Assignment step*: find the unit of the closest prototype

$$f^t(x_i) \leftarrow \arg\min_{u=1,...,M} \delta^{\alpha,t}(x_i, p_u(\beta_u))$$

5:    *Representation step*: update all the prototypes: $\forall u$,

$$\beta_{ul}^t \leftarrow \beta_{ul}^{t-1} + \mu(t)H(f^t(x_i), u)\left(\delta_{il} - \beta_{ul}^{t-1}\right)$$

6:    *Gradient descent step*: update the dissimilarity: $\forall d = 1,...,D$,

$$\alpha_d^t \leftarrow \alpha_d^{t-1} + \nu(t)\mathcal{D}_d^t \qquad \text{and} \qquad \delta^{\alpha,t} \leftarrow \sum_d \alpha_d^t \delta^d.$$

7: **end for**

---

of the first semester 2012.[2] Hence, the question of how is achieved the transition from school to employment or unemployment is crucial in the current economic context. The dataset contains information on 16 040 young people having graduated in 1998 and monitored during 94 months after having left school. The labor-market statuses have nine categories, labeled as follows: permanent-labor contract, fixed-term contract, apprenticeship contract, public temporary-labor contract, on-call contract, unemployed, inactive, military service, education. The following stylized facts are highlighted by a first descriptive analysis of the data as shown in Figure 1:

- permanent-labor contracts represent more than 20% of all statuses after one year and their ratio continues to increase until 50% after three years and almost 75% after seven years;
- the ratio of fixed-terms contracts is more than 20% after one year on the labor market, but it is decreasing to 15% after three years and then seems to converge to 8%;
- almost 30% of the young graduates are unemployed after one year. This ratio is decreasing and becomes constant, 10%, after the fourth year.

In this dataset, all career paths have the same length, the status of the graduate students being observed during 94 months. Hence, we suppose that there are no insertions or deletions and that only the substitution costs have to be defined for OM metrics. This is equivalent to supposing low substitution costs with respect to the insertion-deletion costs. This choice may be considered restrictive,

---

[2] All computations were performed with the free statistical software environment **R** (`http://cran.r-project.org/`, [19]). The dissimilarity matrices (except for the χ^2-distance) and the graphical illustrations were carried out using the `TraMineR` package [20]. The online multiple dissimilarity SOM was implemented by the authors.

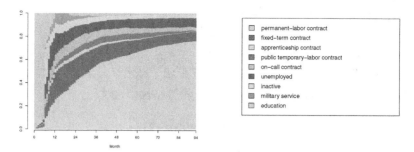

permanent–labor contract
fixed–term contract
apprenticeship contract
public temporary–labor contract
on–call contract
unemployed
inactive
military service
education

Fig. 1. Labor market structure

since in this case the OM metrics will only highlight the contemporaneity of situations. However, Elzinga metrics such as the LCS used in the manuscript are built starting from common, although not contemporary, subsequences and are very similar to OM dissimilarities with insertion-deletion costs lower than the substitution costs.

Seven different dissimilarities were considered: the χ^2-distance, the Hamming dissimilarity (HAM), OM with substitution-cost matrix computed from the transition matrix as shown in Section 2, the dynamic Hamming dissimilarity (DHD) as defined in [10], the longest common prefix (LCP), the longest common suffix or reversed LCP (RLCP), the longest common substring (LCS).

5   Preliminary Study

Since the original data contain more than 16 000 input sequences and since the relational SOM algorithms are based on dissimilarity matrices, the computation time becomes rapidly very important. Training the map on the entire data set requires several hours or days of computing time. Hence, in order to identify the role of the different dissimilarities in extracting typologies, we considered several samples drawn at random from the data. For each of the experiments below, 50 samples containing 1 000 input sequences each were considered. For each sample, the seven dissimilarity matrices listed above were computed and normalized according to the max norm. In order to assess the quality of the maps, three indexes were computed : the quantization error and the dispersion between prototypes for quantifying the quality of the clustering and the topographic error for quantifying the quality of the mapping, [21]. These quality criterai all depend on the dissimilarities used to train the map but the results are made comparable by using normalized dissimilarities.

Optimal-matching Metrics. The first experiment was concerned with the three optimal-matching metrics. The results are listed in Table 1. According to the mean values of the α's, the three dissimilarities contributed to extracting typologies. The Hamming and the dynamical Hamming dissimilarities have similar weights, while the OM with cost-matrix defined from the transition matrix has the largest weight. The mean quantization error computed on the maps trained with the three dissimilarities optimally combined is larger than the quantization error computed on the map trained with the OM metric only. On the other hand, the topographic error is improved in the mixed case. In this case, the joint use of the three dissimilarities provides a trade-off between the quality of the clustering and the quality of the mapping. The results in Table 1 confirm the difficulty to define adequate costs in OM and the fact that the metric has to be chosen according to the aim of the study : building typologies (clustering) or visualizing data (mapping).

Table 1. Preliminary results for three OM metrics

a) Optimally-tuned α

Metric	OM	HAM	DHD
α-Mean	0.43111	0.28459	0.28429
α-Std	0.02912	0.01464	0.01523

b) Quality criteria for the SOM-clustering

Metric	OM	HAM	DHD	Optimally-tuned α
Quantization error	92.93672	121.67305	121.05520	114.84431
Topographic error	0.07390	0.08806	0.08124	0.05268
Prototype dispersion	2096.95282	2255.36631	2180.44264	2158.54172

Elzinga Metrics. When MR-SOM clustering is performed using the three Elzinga metrics only, the results in Table 2 are clearly in favor of the LCS. This result is less intuitive. For example, the LCP metric has been widely used in social sciences and more particularly for studying school-to-work transitions. Indeed, it is obvious that all sequences start with the same status, being in school. Hence, the longer two sequences will be identical, the less different they should be. However, according to our results, it appears that if the purpose of the study is to build homogeneous clusters and identify the main typologies, LCS should be used instead. Thus we can assume that a trajectory is not defined by the first or the final job but rather by the proximity of the transitions during the career-path. As in the previous example, the quality indexes in Table 2 show that the use of an optimally-tuned combination of dissimilarities provides a nice trade-off between clustering (the quantization error) and mapping (the topographic error).

OM, LCS and χ^2 Metrics. Finally, the MR-SOM was run with the three OM metrics, the best Elzinga dissimilarity, LCS, and the χ^2-distance. According to the results in Table 3, the χ^2-distance has the most important weight and it contributes the most to the resulting clustering. The weights of the other dissimilarities are generally below 5%. The clustering and the resulting typologies are then defined by the contemporaneity of their identical situations, rather then by the transitions or the common subsequences. Hence, it appears that the timing and not the duration or the order is important for the clustering procedure. This confirms the importance of the history on the identification of a trajectory.

Table 2. Preliminary results for three Elzinga metrics

a) Optimally-tuned α

Metric	LCP	RLCP	LCS
α-Mean	0.02739	0.00228	0.97032
α-Std	0.02763	0.00585	0.02753

b) Quality criteria for the SOM-clustering

Metric	LCP	RLCP	LCS	Optimally-tuned α
Quantization error	379.77573	239.63652	93.50893	107.1007
Topographic error	0.07788	0.04344	0.07660	0.0495
Prototype dispersion	2693.47676	2593.21763	2094.27678	2080.8514

Some temporal events are crucial on the labor market and a common behavior during these periods is determinant to define a common typology. However, let us remark two things. On the one hand, the quantization error is significantly improved, hence the clustering properties of the mixture of the five dissimilarities are better than for the previous examples. On the other hand, the topographic error becomes very large, hence the mapping properties are degraded. The combination of the five dissimilarities is then particularly adapted for extracting typologies, but is less interesting for visualization purposes.

Table 3. Preliminary results for the five best dissimilarities

a) Optimally-tuned α

Metric	OM	HAM	DHD	LCS	χ^2
α-Mean	0.06612	0.03515	0.03529	0.03602	0.82739
α-Std	0.04632	0.02619	0.02630	0.03150	0.07362

b) Quality criteria for the SOM-clustering

Metric	Optimally-tuned α
Quantization error	75.23233
Topographic error	0.56126
Prototype dispersion	484.00436

5.1   Results on the Whole Data Set

In addition to the statistical indexes computed in the previous section, we can compare different dissimilarities by inspecting the resulting self-organizing maps. Three maps were trained on the whole data set : the first is based on the χ^2-distance, the second on the best performing Elzinga metric in the above section, the length of the longest subsequence (LCS), while the third was obtained by running online multiple-relational SOM on the three optimal-matching dissimilarities (OM, Hamming, DHD). We can note that the three maps provide some common paths: a fast access to permanent contracts (clear blue), a transition through fixed-term contracts before obtaining stable ones (dark and then clear

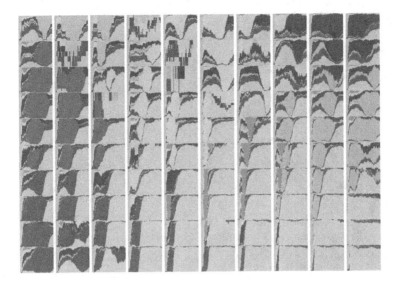

Fig. 2. Final map obtained with the χ^2-distance

Fig. 3. Final map obtained with the LCS-dissimilarity

blue), a holding on precarious jobs (dark blue), a public temporary contract (dark green) or an on-call (pink) contract ending at the end by a stable one, a long period of inactivity (yellow) or unemployment (red) with a gradual return to employment. The maps obtained by LCS and OM dissimilarities are quite similar. A drawback of the OM map is its difficulty to integrate paths characterized by a long return in the educative system (purple). This path is better

Fig. 4. Final map obtained with the OM dissimilarities

integrated in the LCS map. The visual interpretation of the two maps gives
support to the OM map due to a progressive transition on the map between trajectories of exclusion on the west and quick integration on the east. This reading
is less clear on the LCS map. The χ^2 map is a little bit different: we observe
more different trajectories (by example a start by apprenticeship contract (clear
green) ending with a fixed-term or a permanent-term contract). The reading of
the map is easy without any outliers paths and a clear distinction of the trajectories between north (exclusion - education in west, unemployment in east),
middle (specific short-term contracts - public, apprenticeship and on-call from
west to east) and south (integration - long term contracts in east, short term
ones in east). Overall its diversity and its ease to read give support to the χ^2
map against the LCS and OM ones. This confirms that the overweighting of the
χ^2-distance on the five dissimilarities could be attributed to a better fit of this
dissimilarity on our dataset.

6   Conclusion and Future Work

A modified version of online relational SOM, capable of handling several dissimilarity matrices while automatically optimizing a convex combination of them,
was introduced. The algorithm was used for analyzing life sequences for which
the question of selecting an appropriate metric is largely debated. Instead of
one dissimilarity, we used several categories that were automatically mixed in
an optimal combination.

As explained in the previous section, the main drawback of the proposed
relational SOM algorithm is related to the computation time. We are currently
investigating a sparse version which will allow us to handle very large datasets.

References

1. Elzinga, C.H.: Sequence similarity: a nonaligning technique. Sociological Methods & Research 3270(1), 3–29 (2003)
2. Robette, N.: Explorer et décrire les parcours de vie: les typologies de trajectoires. CEPED ("Les Clefs pour"), Université Paris Descartes (2011)
3. Fénelon, J.-P., Grelet, Y., Houzel, Y.: The sequence of steps in the analysis of youth trajectories. European Journal of Economic and Social Systems 14(1), 27–36 (2000)
4. Olteanu, M., Villa-Vialaneix, N., Cottrell, M.: On-line relational SOM for dissimilarity data. In: Estevez, P.A., Principe, J.C., Zegers, P. (eds.) Advances in Self-Organizing Maps. AISC, vol. 198, pp. 13–22. Springer, Heidelberg (2013)
5. Needleman, S., Wunsch, C.: A general method applicable to the search for similarities in the amino acid sequence of two proteins. Journal of Molecular Biology 48(3), 443–453 (1970)
6. Abbott, A., Forrest, J.: Optimal matching methods for historical sequences. Journal of Interdisciplinary History 16, 471–494 (1986)
7. Abbott, A., Tsay, A.: Sequence analysis and optimal matching methods in sociology. Review and prospect. Sociological Methods and Research 29(1), 3–33 (2000)
8. Wu, L.: Some comments on "sequence analysis and optimal matching methods in sociology, review and prospect". Sociological Methods and Research 29(1), 41–64 (2000)
9. Müller, N.S., Gabadinho, A., Ritschard, G., Studer, M.: Extracting knowledge from life courses: Clustering and visualization. In: Song, I.-Y., Eder, J., Nguyen, T.M. (eds.) DaWaK 2008. LNCS, vol. 5182, pp. 176–185. Springer, Heidelberg (2008)
10. Lesnard, L.: Setting cost in optimal matching to uncover contempo-raneous socio-temporal patterns. Sociological Methods et Research 38(3), 389–419 (2010)
11. Elzinga, C.H.: Sequence analysis: metric representations of categorical time series. Sociological Methods and Research (2006)
12. Kohonen, T.: Self-Organizing Maps, 3rd edn., vol. 30. Springer, Heidelberg (2001)
13. Massoni, S., Olteanu, M., Rousset, P.: Career-path analysis using optimal matching and self-organizing maps. In: Príncipe, J.C., Miikkulainen, R. (eds.) WSOM 2009. LNCS, vol. 5629, pp. 154–162. Springer, Heidelberg (2009)
14. Conan-Guez, B., Rossi, F., El Golli, A.: Fast algorithm and implementation of dissimilarity self-organizing maps. Neural Networks 19(6-7), 855–863 (2006)
15. Mac Donald, D., Fyfe, C.: The kernel self organising map. In: Proceedings of 4th International Conference on Knowledge-Based Intelligence Engineering Systems and Applied Technologies, pp. 317–320 (2000)
16. Hammer, B., Hasenfuss, A., Strickert, M., Rossi, F.: Topographic processing of relational data. In: Proceedings of the 6th Workshop on Self-Organizing Maps (WSOM 2007), Bielefeld, Germany (September 2007) (to be published)
17. Olteanu, M., Villa-Vialaneix, N., Cierco-Ayrolles, C.: Multiple kernel self-organizing maps. In: Volume Proceedings of ESANN (2013)
18. Rakotomamonjy, A., Bach, F.R., Canu, S., Grandvalet, Y.: Simplemkl. Journal of Machine Learning Research 9, 2491–2521 (2008)
19. R Development Core Team: R: A Language and Environment for Statistical Computing, Vienna, Austria (2012) ISBN 3-900051-07-0
20. Gabadinho, A., Ritschard, G., Müller, N., Studer, M.: Analyzing and visualizing state sequences in r with traminer. Journal of Statistical Software 40(4), 1–37 (2011)
21. Pölzlbauer, G.: Survey and comparison of quality measures for self-organizing maps. In: Volume Proceedings of the Fifth Workshop on Data Analysis (WDA 2004), pp. 67–82. Elfa Academic Press (2004)

Implementation of the C-Mantec Neural Network Constructive Algorithm in an Arduino Uno Microcontroller

Francisco Ortega-Zamorano[1], José Luis Subirats[1], José Manuel Jerez[1],
Ignacio Molina[2], and Leonardo Franco[1]

[1] Universidad de Málaga, Department of Computer Science, ETSI Informática, Spain
{fortega,jlsubirats,jja,lfranco}@lcc.uma.es
[2] Max Planck Institute, Munich, Germany
imol@uma.es

Abstract. A recently proposed constructive neural network algorithm, named C-Mantec, is fully implemented in a Arduino board. The C-Mantec algorithm generates very compact size neural architectures with good prediction abilities, and thus the board can be potentially used to learn on-site sensed data without needing to transmit information to a central control unit. An analysis of the more difficult steps of the implementation is detailed, and a test is carried out on a set of benchmark functions normally used in circuit design to show the correct functioning of the implementation.

Keywords: Constructive Neural Networks, Microcontroller, Arduino.

1 Introduction

Several technologies like Wireless Sensor Networks [1], Embedded Systems [2] and Real-time Systems [3] are nowadays being extensively used in all kind of industrial applications, most of which use microcontrollers [4] to implement. The recent advances in the computing power of this kind of systems are starting to permitting the use of learning systems, that are able to adjust its functioning as the input data is received, to manage the microcontrollers present in their structure. Neural networks [5] are a kind of flexible and widely used learning systems that are natural candidates for this task as they are very flexible. Nevertheless a disadvantage of neural networks is that learning needs intensive computing power and tends to be prohibitive even for modern systems. In this sense, a recently proposed neural network constructive algorithm has the advantage of being very fast in comparison to standard neural network training and further it creates very compact neural architectures that is useful given the limited memory resources of the microcontrollers.

In this work the C-Mantec[6] algorithm has been fully implemented in a microcontroller, as the training process is part of the software of the controller and it is not carried out externally as it is usually done. We have chosen the Arduino

I. Rojas, G. Joya, and J. Cabestany (Eds.): IWANN 2013, Part I, LNCS 7902, pp. 80–87, 2013.

UNO board [7] as it is a popular, economic and efficient open source single-board microcontroller. C-Mantec is a neural network constructive algorithm designed for supervised classification tasks. One of the critical factors at the time of the implementation of the C-Mantec algorithm is the limited resources of memory of the microcontroller used (32 KB Flash, 2KB RAM & 1KB EPROM memory) and in this sense the implementation has been done with integer arithmetic except for one of the parameters of the algorithm. The paper is structured as follows: we first, briefly describe the C-Mantec algorithm and the Arduino board, secondly we give details about the implementation of the algorithm, to finish with the results and the conclusions.

2   C-Mantec, Constructive Neural Network Algorithm

C-Mantec (Competitive Majority Network Trained by Error Correction) is a novel neural network constructive algorithm that utilizes competition between neurons and a modified perceptron learning rule (termal perceptron) to build compact architectures with good prediction capabilities. The novelty of C-Mantec is that the neurons compete for learning the new incoming data, and this process permits the creation of very compact neural architectures. The activation state (S) of the neurons in the single hidden layer depends on N input signals, ψ_i, and on the actual value of the N synaptic weights (ω_i) and the bias (b) as follows:

$$y = \begin{cases} 1(ON) & \text{if } \phi \geq 0 \\ 0(OFF) & \text{otherwise} \end{cases} \tag{1}$$

where ϕ is the synaptic potential of the neuron defined as:

$$\phi = \sum_{i=1}^{N} \omega_i \, \psi_i - b \tag{2}$$

In the thermal perceptron rule, the modification of the synaptic weights, $\Delta\omega_i$, is done on-line (after the presentation of a single input pattern) according to the following equation:

$$\Delta\omega_i = (t - S) \, \psi_i \, T_{fac} \tag{3}$$

Where t is the target value of the presented input, and ψ represents the value of input unit i connected to the output by weight ω_i. The difference to the standard perceptron learning rule is that the thermal perceptron incorporates the factor T_{fac}. This factor, whose value is computed as shown in Eq. 4, depends on the value of the synaptic potential and on an artificially introduced temperature (T) that is decreased as the learning process advances.

$$T_{fac} = \frac{T}{T_0} e^{-\frac{|\phi|}{T}} \tag{4}$$

C-Mantec, as a CNN algorithm, has in addition the advantage of generating online the topology of the network by adding new neurons during the training phase, resulting in faster training times and more compact architectures. The C-Mantec algorithm has 3 parameters to be set at the time of starting the learning procedure. Several experiments have shown that the algorithm is very robust against changes of the parameter values and thus C-Mantec operates fairly well in a wide range of values. The three parameters of the algorithm to be set are:

- I_{max}: maximum number of iterations allowed for each neuron present in the hidden layer per learning cycle.
- g_{fac}: growing factor that determines when to stop a learning cycle and include a new neuron in the hidden layer.
- F_{itemp}: determines in which case an input example is considered as noise and removed from the training dataset according to the following condition:

$$\forall X \in \{X_1, X_2, ..., X_N\}, delete(X) \mid NTL \geq (\mu + F_{itemp} \cdot \sigma), \qquad (5)$$

where N represents the number of input patterns of the dataset, NTL is the number of times that the pattern X has been presented to the network on the current learning cycle, and the pair $\{\mu, \sigma\}$ corresponds to the mean and variance of the normal distribution that represents the number of times that each pattern of the dataset has been learned during the learning cycle. This learning procedure is essentially based on the idea that patterns are learned by those neurons, the thermal perceptrons in the hidden layer of the neural architecture, whose output differs from the target value (wrongly classified the input) and for which its internal temperature is higher than the set value of g_{fac}. In the case in which more than one thermal perceptron in the hidden layer satisfies these conditions at a given iteration, the perceptron with the highest temperature is the selected candidate to learn the incoming pattern. A new single neuron is added to the network when there is no thermal perceptron that complies with these conditions and a new learning cycle starts.

3   The Arduino UNO Board

The Arduino Uno is a popular open source single-board microcontroller based on the ATmega328 chip [8]. It has 14 digital input/output pins, which can be used as input or outputs, and in addition, has some pins for specialized functions, for example 6 digital pins can be used as PWM outputs. It also has 6 analog inputs, each of which provide 10 bits of resolution, together with a 16 MHz ceramic resonator, USB connection with serial communication, a power jack, an ICSP header, and a reset button. The ATmega328 chip has 32 KB (0.5 KB are used for the bootloader). It also has 2 KB of SRAM and 1 KB of EEPROM. Arduino is a descendant of the open-source *Wiring* platform and is programmed using a Wiring-based language (syntax and libraries); similar to C++ with some slight simplifications and modifications, and a processing-based integrated development environment. Arduino boards can be purchased pre-assembled or do-it-yourself kits, and hardware design information is available. The maximum length and width of the Uno board are 6.8 and 5.3 cm respectively, with the USB connector and power jack extending beyond the former dimension. A picture of the Arduino UNO board is shown in Fig. 1.

Fig. 1. Picture of an Arduino UNO board used for the implementation of the C-Mantec algorithm

4 Implementation of the C-Mantec Algorithm

The C-Mantec algorithm implemented in the wiring code is transferred by USB from the development framework from the PC to the board. The execution of the algorithm comprises two phases or states, because first, the patterns to be learnt have to be loaded into the EEPROM, and then the neural network learning process can begin. The microcontroller state is selected using a digital I/O pin. We explain next, the main technical issues considered for the implementation of the algorithm according to the two phases mentioned before:

4.1 Loading of Patterns

It is necessary to have the patterns stored in the memory board because the learning process work in cycles and use the pattern set repeateadly. The truth (output) value of a given Booelan pattern is stored in the memory position that corresponds to the input. For example, for the case of pattern of 8 inputs, the input pattern "01101001" that corresponds to the decimal number 105 and has a truth value of 0 would be stored by saving a value of 0 in the EEPROM memory position 105. The Arduino Uno EPROM has 1KB of memory, i.e., 8192 bits (2^{13}) and thus this limits the number of Boolean inputs to 13. For the case of using an incomplete truth table, the memory is divided into two parts, a first one to identify the pattern output and a second part to indicate its inclusion or not in the learning set. In this case, of an incomplete truth table, the maximum number of inputs is reduced to 12.

For the case of using real-valued patterns is necessary to know in advance the actual number of bits that will be used to represent each variable. If one byte is used to represent each variable then from the following equation the maximum number of input patterns permitted can be computed:

$$N_P \cdot N_I + N_P/8 \le 1024 \,, \tag{6}$$

where N_I is the number of inputs and N_P is the number of patterns. N_P depends on the number of entries and the number of bits used for each entry.

4.2 Neural Network Learning

C-Mantec is an algorithm which adds neurons as they become necessary, action that is not easily implemented in microcontroller, so we decided to set a value for the maximum number of permitted neurons, that will be stored in the SRAM memory. From this memory, with a capacity of 2 KB, we will employ less than 1 KB for storing the variables of the program; and thus saving at least 1 KB of free memory for saving the following variables related to the neurons:

- T_{fac}: must be a variable of *float* type and occupies 4 bytes.
- Number of iterations: an integer value with a range between 1000 and 100000 iterations, so it must be of type *long*, 4 bytes.
- Synaptic weights: almost all calculatios are based on these variable, so to speed up the computations we choose *integer* types of 2 bytes long.

According to the previous definitions, the maximum number of neurons (N_N) that can be implemented should verify the following constraint:

$$4 \cdot N_N + 2 \cdot N_N + 2 \cdot N_N \cdot (N_I + 1) \leqslant 1024, \qquad (7)$$

where N_I is the number of inputs. For the maximum number of permitted inputs (13), the maximum number of neurons is 30. The computation of T_{fac} is done using a float data type because it requires an exponential operation that can be done only with this type of data, but as its computation involves other data types (integers) , a conversion must be done. To make this change without losing accuracy, we multiply the value of T_{fac} by 1000, leading to values in the range between 0 and 1000. When we convert to integer data type, precision is lost starting from the fourth digital number. Weights are of integer type in the range from -32768 to 32767, and as they are multiplied by the value of T_{fac}, we compensate this change by dividing them by 1000. When any synaptic weight value is greater than 30, or less than -30, all weights are divided by 2. When this change does not affect at all the procedure of the network as neural network are invariant to this type of rescaling. To avoid the overflow of the integer data type, we apply the previous transformation whenever a synaptic weight reach the maximum or minimum permitted values. One very important thing in the implementation is the the execution time needed by the algorithm. In our case, this value depends strongly on the number of neurons actually used, as this time grows exponentially as a function of the number of used neurons. Fig 2 shows the execution time as a function of the neurons used in an architecture generated by the C-Mantec algorithm.

5 Results

We have tested the correct implementation of the C-Mantec algorihtm in the Arduino board by comparing the obtained results, in terms of the number of neurons generated and the generalization accuracy obtained, with those previously observed when using the PC implementation. The test is also carried out

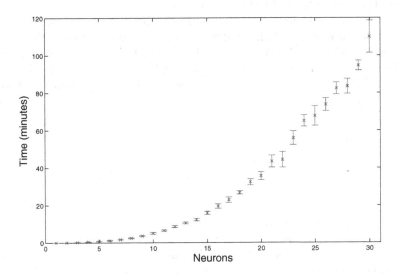

Fig. 2. Mean and standard deviation (indicated by error bars) of the execution time of the learning process as a function of the number of neurons used in a network created by the C-Mantec algorithm. The values shown are averages across 20 samples.

Table 1. Number of neurons and generalization ability obtained for a set of benchmark function for the implementation of the C-Mantec algorithm in an Arduino Uno board. (See text for more details).

Function	# Inputs	# Neurons		Accuracy generalization	
		Theory	Arduino	Theory	Arduino
cm82af	5	3,0±0,0	3,0±0,0	93,3±11,1	87,2±5,3
cm82ag	5	3,0±0,0	3,0±0,0	60.0±37,3	72,5±12,3
cm82ah	5	1,0±0,0	3,0±0,0	100.0±0,0	95,3±4,7
z4ml24	7	3,0±0,0	3,0±0,0	98,3±3,7	97,9±1, 1
z4ml25	7	3,1±0,9	3,1±0,9	90,8±12,3	86,0±0,9
z4ml26	7	3,0±0,0	3,0±0,0	96,7±5,9	94,6±0,4
z4ml27	7	3,0±0,0	3,0±0,0	99,2±2,8	99,9±0,9
9symml	9	3,0±0,0	3,0±0,0	99,4±0,9	97,5±1,2
alu2k	10	11,2±0,9	11,8±1,2	97,4±1,9	95,5±0,9
alu2l	10	18,9±1,5	19,3±1,3	79,2±5,5	70,3±1,3
alu2o	10	11,2±0,9	12,8±0,2	90,2±2,3	85,8±2,2

to analyze the effects of using a limited precision representation for the synaptic weights. A set of 10 single output Boolean functions from the MCNC benchmark were used to test the generalization ability of the C-Mantec algorithm. The C-Mantec algorithm was run with the following parameter values: $g_{fac} = 0.05$ and $I_{max} = 10000$. Table 1 shows the results obtained with the microcontroller for the set of benchmark functions. The first two columns indicate the function reference name and its number of inputs. Third and fourth columns shows the number of neurons obtained by the PC and Arduino implementations, while fifth and last column shows the generalization ability obtained both for the PC and Arduino cases. The averages are computed from 20 samples and the standard deviation is indicated. The generalization ability shown in the table was computed using a ten-fold cross validation procedure.

6   Conclusion

We have successfully implemented the C-Mantec neural network constructive algorithm in an Arduino Uno board. The main issues at the time of the implementation are related to the memory limitations of the board. In this sense, we have analyzed the maximum number of Boolean and Real patterns that can be used for the learning process. For the case of Boolean patterns, we carried out a comparison against published results, showing that the algorithm works almost exact in comparison to the original PC implementation. As the number of inputs of the test functions increases, the Arduino implementation needs just a small extra number of neurons, and also a small degradation in the generalization accuracy is observed. These effects can be related to the limited numerical precision of the synaptic weights. The rounding effects should not in principle degrade the functioning of the algorithm, but affects the number of iterations needed to achieve convergence. Thus, we have also analyzed an important factor as it is the execution time of the algorithm. The results (cf. Figure 2) shows an exponential execution time increase as a function of the number of neurons in the constructed algorithms, and so for networks of approximately 15 neurons the execution time is around 20 minutes, while for 30 neurons this time increases up to two hours.

As a conclusion, and despite the previously mentioned limitations, we believe that the current implementation can be used in several practical applications, and we are planning to incorporate the C-Mantec algorithm in WSN in a near future.

Acknowledgements. The authors acknowledge support from Junta de Andalucía through grants P10-TIC-5770 and P08-TIC-04026, and from CICYT (Spain) through grant TIN2010-16556 (all including FEDER funds).

References

1. Yick, J., Mukherjee, B., Ghosal, D.: Wireless sensor network survey. Comput. Netw. 52(12), 2292–2330 (2008)
2. Marwedel, P.: Embedded System Design. Springer-Verlag New York, Inc., Secaucus (2006)
3. Kopetz, H.: Real-Time Systems: Design Principles for Distributed Embedded Applications, 1st edn. Kluwer Academic Publishers, Norwell (1997)

4. Andersson, A.: An Extensible Microcontroller and Programming Environment. Massachusetts Institute of Technology, Department of Electrical Engineering and Computer Science (2003)
5. Haykin, S.: Neural networks: a comprehensive foundation. Prentice Hall (1994)
6. Subirats, J.L., Franco, L., Jerez, J.M.: C-mantec: A novel constructive neural network algorithm incorporating competition between neurons. Neural Netw. 26, 130–140 (2012)
7. Oxer, J., Blemings, H.: Practical Arduino: Cool Projects for Open Source Hardware. Apress, Berkely (2009)
8. Atmel: Datasheet 328, http://www.atmel.com/Images/doc8161.pdf

A Constructive Neural Network to Predict Pitting Corrosion Status of Stainless Steel

Daniel Urda[1], Rafael Marcos Luque[1], Maria Jesus Jiménez[2], Ignacio Turias[3], Leonardo Franco[1], and José Manuel Jerez[1]

[1] Department of Computer Science, University of Málaga, Málaga, Spain
{durda,rmluque,lfranco,jja}@lcc.uma.es
[2] Department of Civil and Industrial Engineering, University of Cádiz, Cádiz, Spain
mariajesus.come@uca.es
[3] Department of Computer Science, University of Cádiz, Cádiz, Spain
ignacio.turias@uca.es

Abstract. The main consequences of corrosion are the costs derived from both the maintenance tasks as from the public safety protection. In this sense, artificial intelligence models are used to determine pitting corrosion behaviour of stainless steel. This work presents the C-MANTEC constructive neural network algorithm as an automatic system to determine the status pitting corrosion of that alloy. Several classification techniques are compared with our proposal: Linear Discriminant Analysis, k-Nearest Neighbor, Multilayer Perceptron, Support Vector Machines and Naive Bayes. The results obtained show the robustness and higher performance of the C-MANTEC algorithm in comparison to the other artificial intelligence models, corroborating the utility of the constructive neural networks paradigm in the modelling pitting corrosion problem.

Keywords: Constructive neural networks, Austenitic stainless steel, Pitting corrosion.

1 Introduction

Corrosion can be defined as the degradation of the material and its properties due to chemical interactions with the environment. The main consequences of corrosion are important maintenance costs in addition to endangering public safety. The annual cost of corrosion worldwide has been estimated over 3% of the gross world product [1]. Therefore, corrosion has become one of the most relevant engineering problems. This phenomenon occurs so often that it has been necessary to develop models in order to predict corrosion behaviour of materials under specific environmental conditions.

Many authors have applied neural networks models to study corrosion: Kamrunnahar and Urquidi-MacDonald [2] presented a supervised neural network method to study localized and general corrosion on nickel based alloys. Cavanaugh et al. [3] used these models to model pit growth as a function of different environmental factors. Lajevardi et al. [4] applied artificial neural networks to predict the time to failure as a result of stress corrosion cracking in austenitic stainless steel. While, Pidaparti et al. [5] developed computational model based on cellular automata approach to predict the multi-pit corrosion damage initiation and growth in aircraft aluminium.

I. Rojas, G. Joya, and J. Cabestany (Eds.): IWANN 2013, Part I, LNCS 7902, pp. 88–95, 2013.
© Springer-Verlag Berlin Heidelberg 2013

In spite of the numerous researches in corrosion risk of materials, no reliable method to predict pitting corrosion status of grade 316L stainless steel has yet been developed by others authors. Based on our studies about pitting corrosion [6,7], constructive neural networks (CNNs) are proposed in this paper to develop an automatic system to determine pitting corrosion status of stainless steel, with no need to check pits occurrence on surface material by microscopic techniques. Particularly, C-MANTEC model [8] is compared with other different standard classification models such as Linear Discriminant Analysis (LDA), k-Nearest Neighbor (kNN), Multilayer Perceptron (NeuralNet), Support Vector Machines (SVM) and Naive Bayes, in order to check the robustness and reliability of this algorithm on industrial environments. The use of C-MANTEC is motivated in the good performance results previously obtained in other areas [9,10] and due to its relatively small and compact neural network architecture leading to possible hardware implementation on industrial environments.

The remainder of this paper is organized as follows: Section 2.1 and Section 2.2 provides respectively a description of the dataset utilized on the experiments and the use of several classifiers models to be compared with C-MANTEC, and Section 3 shows the experimental results over several classifying algorithms. Finally, Section 4 concludes the article.

2 Material and Methods

2.1 Material

In order to study corrosion behaviour of austenitic stainless steel a European project called "Avoiding catastrophic corrosion failure of stainless steel" CORINOX (RFSR-CT-2006-00022) was partially developed by ACERINOX. In this project, 73 different samples of grade 316L stainless steel were subjected to polarization tests in order to determine pitting potentials values in different environmental conditions: varying ion chloride concentration $(0.0025-0-1M)$, pH values $(3.5-7)$ and temperature $(2-75°C)$ using NaCl as precursor salt.

Pitting potential is one of the most relevant factors used to characterize pitting corrosion [11]. This parameter is defined as the potential at which current density suffers an abrupt increase. It can be determined based on polarization curves as the potential at which current density is $100\mu A/cm^2$ [12].

All the polarization tests were carried out using a Potentiostate PARSAT 273. For each of the 73 sample, the potential and current density values registered during the tests were plotted on semi-logarithmic scale to determine pitting potential values (see Figure 1). After polarization tests, all samples were checked microscopically for evidence of localized corrosion. In this way, all species were characterized by the environmental conditions tested (chloride ion concentration, pH and temperature) in addition to corrosion status: 1 for samples where pits appeared on the material surface and 0 otherwise.

2.2 Methods

In this work, we propose the use of constructive neural networks as classifiers models, in particular C-MANTEC, to predict corrosion behaviour of austenitic stainless steel.

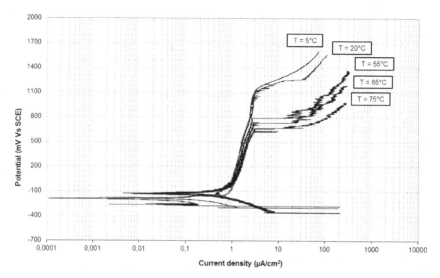

Fig. 1. Polarization curves measured for grade 316L stainless steel using NaCl as precursor salt. The conditions tested were: 0.0025 M (mol/L) chlorides ions, pH = 5.5 and temperature (5°C-75°C).

C-MANTEC (Competitive Majority Network Trained by Error Correction) is a novel neural network constructive algorithm that utilises competition between neurons and a modified perceptron learning rule to build compact architectures with good prediction capabilities. The novelty of C-MANTEC is that the neurons compete for learning the new incoming data, and this process permits the creation of very compact neural architectures. At the single neuronal level, the algorithm uses the thermal perceptron rule, introduced by Marcus Frean in 1992 [13], that improves the convergence of the standard perceptron for non-linearly separable problems. In the thermal perceptron rule, the modification of the synaptic weights, $\Delta\omega_i$, is done on-line (after the presentation of a single input pattern) according to the following equation:

$$\Delta\omega_i = (t - S)\psi_i T_{fac} \tag{1}$$

where t is the target value of the presented input, and ψ represents the value of input unit i connected to the output by weight ω_i. The difference to the standard perceptron learning rule is that the thermal perceptron incorporates the factor T_{fac}. This factor depends on the value of the synaptic potential and on an artificially introduced temperature (T) that is decreased as the learning process advances.

The topology of a C-MANTEC created network consists of a single hidden layer of thermal perceptrons that maps the information to an output neuron that uses a majority function. The choice of the output function as a majority gate is motivated by previous experiments in which very good computational capabilities have been observed for the majority function among the set of linearly separable functions [14]. The results so far

Table 1. Brief pseudo-code summary of the C-MANTEC learning algorithm

C-MANTEC learning algorithm
1 Initialise the parameters of the algorithm;
2
3 **while** (exists patterns to be learned) {
4   input a random pattern;
5   **if** (pattern target value == network output) {
6   remove temporarily the pattern from the dataset;
7   }
8   **else** {
9   the pattern has to be learned by the network;
10   select the wrong neuron with highest temperature;
11   **if** (Tfac >= Gfac) {
12   the neuron will learn the pattern;
13   update its synaptic weights according to the thermal perceptron rule;
14   }
15   **else** {
16   a new neuron is added to the network;
17   this new neuron learns the pattern;
18   iteration counters are reset;
19   noisy patterns are deleted from the training dataset;
20   reset the set of patterns;
21   }
22   }
23 }

obtained with the algorithm [15,8,10] show that it generates very compact neural architectures with state-of-the-art generalization capabilities. It has to be noted that the algorithm incorporates a built-in filtering stage that prevent overfitting of noisy examples.

The C-MANTEC algorithm has 3 parameters to be set at the time of starting the learning procedure. Several experiments have shown that the algorithm is very robust against changes of the parameter values and thus C-MANTEC operates fairly well in a wide range of values. The three parameters of the algorithm to be set are: (i) I_{max} as maximum number of iterations allowed for each neuron present in the hidden layer per learning cycle, (ii) G_{fact} as growing factor that determines when to stop a learning cycle and include a new neuron in the hidden layer, and (iii) Fi_{temp} that determines in which case an input example is considered as noise and removed from the training dataset according to Eq. 2, where N represents the number of input patterns of the dataset, NTL is the number of times that the pattern X has been learned on the current learning cycle, and the pair $\{\mu, \sigma\}$ corresponds to the mean and variance of the normal distribution that represents the number of times that each pattern of the dataset has been learned during the learning cycle.

$$\forall X \in \{X_1, ..., X_N\}, \text{delete}(X) \mid \text{NTL} \geq (\mu + Fi_{temp}\sigma) \tag{2}$$

A summary of the C-MANTEC pseudo-code algorithm is described in Table 1. This learning procedure is essentially based on the idea that patterns are learned by those neurons, the thermal perceptrons in the hidden layer of the neural architecture, whose output differs from the target value (wrongly classified the input) and for which its internal temperature is higher than the set value of G_{fac}. In the case in which more than one thermal perceptron in the hidden layer satisfies these conditions at a given iteration, the perceptron that has the highest temperature is the selected candidate to

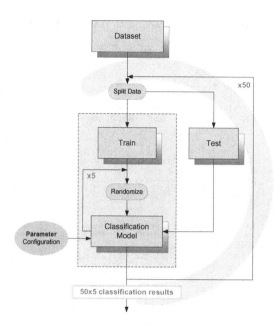

Fig. 2. Bootstrap resampling technique procedure used for each classification model, both for estimating the parameters configuration of each model and to predict pitting corrosion behaviour

learn the incoming pattern. A new single neuron is added to the network when there is no a thermal perceptron that complies with these conditions and a new learning cycle starts. The learning process ends when there are no more patterns to be learned, as all of them are classified correctly or are outside of the initial dataset because are considered noisy by an internal built-in filter.

Several classification models have been used to evaluate our proposal in this paper: LDA, kNN, NeuralNet, SVM and Naive Bayes. As Figure 2 shows, first a bootstrap resampling technique is applied $50x5$ times for each of these models varying the values of their required parameters, including C-MANTEC. Although it is not an honest parameter estimation procedure, it allows us to estimate a parameter configuration set in order to test the robustness of different classification models [16]. Afterwards, bootstrapping is reapplied $200x10$ for each model with the obtained parameters in order to predict pitting corrosion behaviour in terms of accuracy and standard deviation.

3   Experimental Results

It is not easy to determine in advance the appropriate parameters to get a good generalization rate, which requires a tedious empirical assessment of the data to assign these values. In this way, multiple configurations of the comparative techniques are generated by combining the values of the parameters shown in Table 2 in every possible manner, which also includes the final quantitative results in the column "Accuracy". These results are obtained by setting the algorithms parameters as follows: $\{k = 1, d = \text{cosine-similarity}\}$ in kNN; $\{NHidden = 20, \alpha = 0.05, NCycles = 25\}$

Table 2. Parameter settings tested during evaluation of the classification algorithms. The combination of all the values of the parameters generate a set of configurations for each method. The third column shows the quantitative results for the best parameter setting of each algorithm.

Algorithm	Test Parameters	Accuracy
LDA	No parameters	72.560±0.49
kNN	Neighbours, k= {1, 2, 3, 4, 5, 6, 7, 8, 9, 10} Distance type, d= {euclidean, chi-squared, cosine-similarity}	79.867±0.44
NeuralNet	Hidden neurons, $NHidden$= {2, 4, 6, 8, 10, 15, 20} Alpha, α= {0.05, 0.1, 0.2, 0.3, 0.5} Number cycles, $NCycles$= {10, 25, 50}	**87.254±0.47**
SVM	Kernel type, t= {linear, polynomial, radial base function, sigmoid} Cost, C= {1, 3, 5, 7, 9, 10, 12, 15} Degree, d= {1, 2, 3, 4, 5} Gamma, g= {0.001, 0.005, 0.1, 0.15, 0.2, 0.4, 0.6, 0.8, 1, 2, 3, 5} Coef0, r= {0, 1, 2}	85.508±0.50
NaiveBayes	Kernel density, K= {0, 1}	66.882±0.55
C-MANTEC	Max. Iterations, I_{max}= {1000, 10000, 100000} GFac, g_{fac} = {0.01, 0.05, 0.1, 0.2, 0.25, 0.3} Phi, ϕ = {1, 1.5, 2, 2.5, 3, 3.5, 4, 4.5, 5, 5.5, 6}	**89.788±0.56**

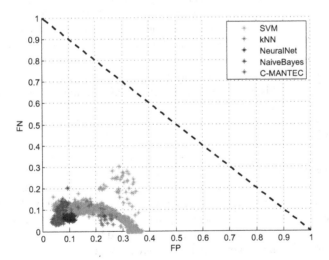

Fig. 3. False Positives (FP) and False Negatives (FN) ratios after applying each method to the dataset with all the parameter configurations. Each coloured point '*' is considered as a different configuration of that method. The closer the points are to the origin, the better the segmentation is. Additionally, the method is less sensible to a parameters' change if the points' cloud keeps compact and grouped.

in NeuralNet; $\{t = \text{polynomial}, C = 15, d = 2, g = 5, r = 0\}$ in SVM; $\{K = 0\}$ in NaiveBayes; and $\{I_{max} = 10000, g_{fac} = 0.3, \phi = 4.5\}$ in C-MANTEC. In concrete, C-MANTEC (89.78% in accuracy) clearly outperforms LDA (72.56%), kNN (79.86%) and NaiveBayes (66.88%) models, and it also improves the NeuralNet and SVM classification accuracies but only in 2 and 4 percentage points respectively.

A thorough analysis is presented in Figure 3, where the influence of the parameter setting for different algorithms is evaluated in the classification accuracy variability. The horizontal axis corresponds to the average percentage of the false positives (FP) on the data, while the vertical axis is associated with the false negatives values (FN). Each point of the plot represents the average FP and FN of a generated configuration when it is applied to the dataset. The closer the points are to the origin, the better the classification process. The optimum performance occurs if $FN = 0$ and $FP = 0$, which implies there is a perfect match between the output of the algorithm and the real output of the problem. The results are always below the diagonal of the plot because we always have $FN + FP <= 1$.

The variability for each classifier depends largely on the analysed dataset, but the robustness of the method has also an influence, i.e. more robust methods yield smaller values. If the configuration cloud is compact, it means that the results do not vary significantly after a change in its parameters. On the other hand, if several configurations are far from each other, it implies that the variation of a parameter causes abrupt changes in the results, which is a very undesirable property for a classification algorithm. As shown in Figure 3, the compactness for kNN and SVM methods is rather poor, while our C-MANTEC approach and NeuralNet model have their configurations very close together. In other words, the performance of the proposed method is not very sensitive to the parameter selection. Additionally, our approach is closer to the zero point than the remaining alternatives, which implies that C-MANTEC provides the best classification result. Since the NaiveBayes classifier do not require many values for its parameters, the cloud of points for this method (i.e. number of configurations) is not relevant.

4   Conclusions

This work presents a novel application of a constructive neural network (the C-MANTEC algorithm) to the prediction of pitting corrosion status of stainless steel as function of environmental conditions. The results demonstrate that C-MANTEC outperforms, in terms of classification accuracy, the other algorithms under study. In addition, the compact neural network architecture generated by the C-MANTEC algorithm makes it suitable to be implemented in a hardware architecture for industrial scope.

The high cost of the polarization tests, in addition to the complexity of its interpretation because of the multiples factor involved, justifies the development of a model to predict pitting corrosion behaviour of austenitic stainless steel without resorting polarization tests. In this sense and as further work, it could be interesting to provide a useful tool to determine the existence of pits on the material surface by automatic technique without need of microscope analysis reducing the cost of experimental tests. Moreover, due to the multiple variables affecting pitting corrosion behaviour of stainless steel, it would also be interesting to test classification models varying some of the environmental factors studied in this paper. A remarkable case of study would be to analyse the influence of precursor salts on classification performance.

Acknowledgements. The authors acknowledge support through grants TIN2010-16556 from MICINN-SPAIN and P08-TIC-04026 (Junta de Andalucía), all of which include FEDER funds.

References

1. Schmitt, G.: Global needs for knowledge dissemination, research, and development in materials deterioration and corrosion control. The World Corrosion Organization (2009)
2. Kamrunnahar, M., Urquidi-Macdonald, M.: Prediction of corrosion behaviour of alloy 22 using neural network as a data mining tool. Corrosion Science 53, 961–967 (2011)
3. Cavanaugh, M., Buchheit, R., Birbilis, N.: Modeling the environmental dependence of pit growth using neural network approaches. Corrosion Science 52, 3070–3077 (2010)
4. Lajevardi, S., Shahrabi, T., Baigi, V., Shafiei, M.A.: Prediction of time to failure in stress corrosion cracking of 304 stainless steel in aqueous chloride solution by artificial neural network. Protection of Metals and Physical Chemistry of Surfaces 45, 610–615 (2009)
5. Pidaparti, R.M., Fang, L., Palakal, M.J.: Computational simulation of multi-pit corrosion process in materials. Computational Materials Science 41, 255–265 (2008)
6. Jiménez-Come, M.J., Muñoz, E., García, R., Matres, V., Martín, M.L., Trujillo, F., Turias, I.: Austenitic stainless steel en 1.4404 corrosion detection using classification techniques. In: Corchado, E., Snášel, V., Sedano, J., Hassanien, A.E., Calvo, J.L., Ślęzak, D. (eds.) SOCO 2011. AISC, vol. 87, pp. 193–201. Springer, Heidelberg (2011)
7. Jiménez-Come, M.J., Muñoz, E., García, R., Matres, V., Martín, M.L., Trujillo, F., Turias, I.: Pitting corrosion behaviour of austenitic stainless steel using artificial intelligence techniques. J. Applied Logic 10, 291–297 (2012)
8. Subirats, J.L., Franco, L., Jerez, J.M.: C-mantec: A novel constructive neural network algorithm incorporating competition between neurons. Neural Networks 26, 130–140 (2012)
9. Urda, D., Cañete, E., Subirats, J., Franco, L., Llopis, L., Jerez, J.: Energy efficient reprogramming in WSN using Constructive Neural Networks. International Journal of Innovative Computing, Information and Control 8, 7561–7578 (2012)
10. Urda, D., Subirats, J.L., Franco, L., Jerez, J.M.: Constructive neural networks to predict breast cancer outcome by using gene expression profiles. In: García-Pedrajas, N., Herrera, F., Fyfe, C., Benítez, J.M., Ali, M. (eds.) IEA/AIE 2010, Part I. LNCS, vol. 6096, pp. 317–326. Springer, Heidelberg (2010)
11. Galvele, J.: Present state of understanding of the breakdown of passivity and repassivation. The Electrochemical Society, 285–326 (1979)
12. Merello, R., Botana, F., Botella, J., Matres, M., Marcos, M.: Influence of chemical composition on the pitting corrosion resistance of non-standard low-Ni high-Mn N duplex stainless steels. Corrosion Science 45, 909–921 (2003)
13. Frean, M.: A "thermal" perceptron learning rule. Neural Comput. 4, 946–957 (1992)
14. Subirats, J.L., Franco, L., Gómez, I., Jerez, J.M.: Computational capabilities of feedforward neural networks the role of the output function. In: Proceedings of the XII CAEPIA 2007, Salamanca, Spain, vol. 2, pp. 231–238 (2008)
15. Subirats, J.L., Jerez, J.M., Franco, L.: A new decomposition algorithm for threshold synthesis and generalization of boolean functions. IEEE Transactions on Circuits and Systems 1, 3188–3196 (2008)
16. Jiang, W., Simon, R.: A comparison of bootstrap methods and an adjusted bootstrap approach for estimating the prediction error in microarray classification. Statistics in Medicine 26, 5320–5334 (2007)

Robust Sensor and Actuator Fault Diagnosis with GMDH Neural Networks

Marcin Witczak, Marcin Mrugalski, and Józef Korbicz

Institute of Control and Computation Engineering,
University of Zielona Góra,
ul. Podgórna 50, 65–246 Zielona Góra, Poland
{M.Witczak,M.Mrugalski,J.Korbicz}@issi.uz.zgora.pl

Abstract. The uncertainty of neural model influences the effectiveness of the neural model-based FDI and FTC systems. The application of the GMDH approach to the state-space neural model structure selection allows reducing the model uncertainty. The state-space representation of the neural model enables to develop a new technique of estimation of the neural model inputs based on the RUIF. This result enables performing robust fault detection and isolation of the actuators.

Keywords: State-space GMDH neural networks, non-linear system identification, robust fault diagnosis.

1 Introduction

The quality of the models of systems and processes determines the effectiveness of the *Fault Detection and Isolation* (FDI) [1–4] and the *Fault-Tolerant Control* (FTC) [5–7]. In the case of non-linear dynamic system identification the *Artificial Neural Networks* (ANNs) are often applied [8]. Unfortunately, the ANNs have disadvantages e.g. they do not usually available the state-space representation of a neural model, only rare approaches ensure the stability and there is a limited number of approaches that can settle the robustness problems regarding neural model uncertainty [9, 10]. These problems is important in the case of model-based FDI systems, where it is assumed that the residual, which is a difference of the system and the nominal model response, is distinguishably different from zero in the faulty case. If this condition is fulfilled, the faults are detected when the residual crosses arbitrary defined constant threshold. The weakness of this approach relies on the corruption of residual by disturbances or/and neural model uncertainty, what results in the undetected faults or false alarms. To solve such a problem, a methodology of dynamic non-linear system identification based on the state-space *Group Method of Data Handling* (GMDH) neural network [11–13] has been proposed. The concept of this approach relies on replacing the complex neural model by the set of the hierarchically connected partial models (neurons) what result in the reduction of the neural model inaccuracy. The application of the *Unscented Kalman Filter* (UKF) [14] to neurons parameters estimation also allows to obtain the output adaptive thresholds and enable to

I. Rojas, G. Joya, and J. Cabestany (Eds.): IWANN 2013, Part I, LNCS 7902, pp. 96–105, 2013.

perform the robust fault detection of the system and sensors [3]. Unfortunately, this approach does not enable to distinguish a faulty system component, which often consists of several actuators and sensors. In order to solve it, a new approach of fault detection and isolation of the actuators faults is proposed. The developed approach is based on the estimation of the GMDH neural model inputs with the application of the *Robust Unknown Input Filter* (RUIF) [15–17]. This method allows obtaining the adaptive thresholds for each input signal of the diagnosed system. In the consequence this approach enables to perform robust fault detection and isolation of the actuators faults.

2   GMDH Neural Network

The GMDH neural model (c.f. Fig. 1) is gradually increasing by adding new layers of neurons according following repeated steps [11, 18, 19, 4]:

- Creation of a neuron layer on the basis of system inputs combinations,
- Parameters estimation of each neuron,
- Calculation of neurons responses and its uncertainty,
- Quality evaluation of neurons,
- Selection of neurons,
- Termination condition testing.

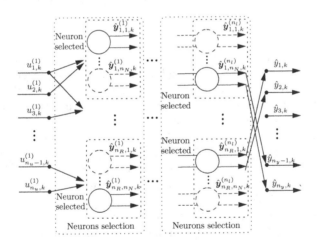

Fig. 1. Synthesis process of the GMDH neural network

In the GMDH network it is assumed that each neuron should reflect the behaviour of the dynamic system being identified. Moreover, the neurons parameters are estimated seriately for each neuron in such a way that its outputs are the best approximation of the real system outputs. The behavior of the proposed state-space neuron is defined by:

$$\hat{x}_{k+1} = A\hat{x}_k + Bu_k, \tag{1}$$

$$\hat{y}_{k+1} = g(C\hat{x}_{k+1}), \tag{2}$$

where $u_k \in \mathbb{R}^{n_u}$ and $y_k \in \mathbb{R}^{n_y}$ are the inputs and outputs of the dynamic neuron created on the combination of systems inputs. $g(\cdot) = [g_1(\cdot), ..., g_{n_y}(\cdot)]^T$ where $g_i(\cdot)$ are a non-linear activation functions. $A \in \mathbb{R}^{n_z \times n_z}$, $B \in \mathbb{R}^{n_z \times n_u}$, $C \in \mathbb{R}^{n_y \times n_z}$, $z_k \in \mathbb{R}^{n_z}$, where n_z represents the order of the dynamics. Additionally, the matrix A has an upper-triangular form, i.e.

$$A = \begin{bmatrix} a_{11} & a_{12} & \cdots & a_{1,n_z} \\ 0 & a_{22} & \cdots & a_{2,n_z} \\ \vdots & & \ddots & \vdots \\ 0 & 0 & \cdots & a_{n_z,n_z} \end{bmatrix}. \tag{3}$$

This mean that the dynamic neuron is asymptotically stable iff:

$$|a_{i,i}| < 1, \quad i = 1, ..., n_z. \tag{4}$$

In order to estimate the parameters of the neurons the UKF [14] can be applied. This task relies on the estimation of x_k which satisfies the constraint:

$$-1 + \delta \le e_i^T x_k \le 1 - \delta, \quad i = 1, ..., n \tag{5}$$

where: $e_i \in \mathbb{R}^{n_p+n}$ whereas $e_1 = [1, 0, ..., 0]^T$, $e_2 = [0, 1, ..., 0]^T$, ect., and δ is a small positive value. These constrains follow directly from the asymptotic stability condition (4), while δ is introduced in order to make the problem tractable. The model has a cascade structure what results from that the neuron outputs constitute the neuron inputs in the subsequent layers. The neural model which is the result of the cascade connection of dynamic neurons is asymptotically stable, when each of neurons is asymptotically stable [20].

The application of the UKF with the procedure of truncation of the probability density function [14] allows to obtain the state estimates as well as the uncertainty of the GMDH model in the form of a matrixes P^{xxt} which can be then applied to the calculation of the system outputs adaptive thresholds. The real system responses in the fault-free mode should be contained in the following output adaptive thresholds:

$$\mathcal{F}_i(c_i\hat{x}_k - t_{n_t-n_p}^{\alpha/2}\hat{\sigma}_i\sqrt{c_i P^{xxt} c_i^T}) \le y_{i,k} \le \mathcal{F}_i(c_i\hat{x}_k + t_{n_t-n_p}^{\alpha/2}\hat{\sigma}_i\sqrt{c_i P^{xxt} c_i^T}). \tag{6}$$

where c_i stands for the i-th row ($i = 1, ..., n_y$) of the matrix C of the output neuron, $\hat{\sigma}_i$ is the standard deviation of the i-th fault-free residual and $t_{n_t-n_p}^{\alpha/2}$ is the t-Student distribution quantile. The faults are signaled when system outputs y_k crosses the output adaptive threshold (6).

3   Estimation of the GMDH Neural Model Inputs

The state-space representation of the GMDH network enables to develop a new RUIF-based approach in order to estimate the input signals of the each neurons and the whole GMDH network (c.f. Fig. 2). The calculation of the input adaptive thresholds for each input signal of the diagnosed system allows to perform the robust fault detection and isolation of the actuators simultaneously (c.f. Fig. 3).

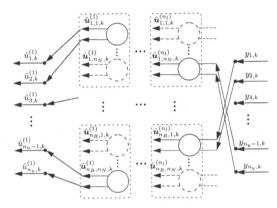

Fig. 2. Estimation of the system inputs via GMDH model and RUIF

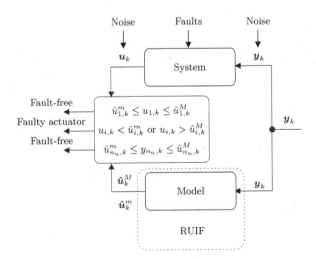

Fig. 3. Application of the GMDH neural model and RUIF to robust fault detection and isolation of actuators

Let us consider a non-linear discrete-time system for neuron model (1-2):

$$\boldsymbol{x}_{k+1} = \boldsymbol{A}\boldsymbol{x}_k + \boldsymbol{B}\boldsymbol{u}_k, \tag{7}$$

$$\boldsymbol{y}_{k+1} = \boldsymbol{g}(\boldsymbol{C}\boldsymbol{x}_{k+1}) + \boldsymbol{w}_{k+1}, \tag{8}$$

where $\boldsymbol{w}_k \in l_2$ is an exogenous disturbance vector, $l_2 = \{\boldsymbol{w} \in \mathbb{R}^n, \|\boldsymbol{w}\|_{l_2} < +\infty,\}$ where $\|\boldsymbol{w}\|_{l_2} = \left(\sum_{k=0}^{\infty} \|\boldsymbol{w}_k\|^2\right)^{\frac{1}{2}}$. The system output can be written as follows:

$$\boldsymbol{g}^{-1}(\boldsymbol{y}_{k+1} - \boldsymbol{w}_{k+1}) = \boldsymbol{C}\boldsymbol{x}_{k+1} = \boldsymbol{g}^{-1}(\boldsymbol{y}_{k+1}) + \boldsymbol{v}_{k+1}, \tag{9}$$

$$\boldsymbol{H}[\boldsymbol{g}^{-1}(\boldsymbol{y}_{k+1}) + \boldsymbol{v}_{k+1}] = \boldsymbol{H}\boldsymbol{C}\boldsymbol{x}_{k+1}, \tag{10}$$

where $\boldsymbol{v}_k \in \mathcal{L}_2$. Substituting (7) into (10):

$$\boldsymbol{H}[\boldsymbol{g}^{-1}(\boldsymbol{y}_{k+1}) + \boldsymbol{v}_{k+1}] = \boldsymbol{H}\boldsymbol{C}\boldsymbol{A}\boldsymbol{x}_k + \boldsymbol{H}\boldsymbol{C}\boldsymbol{B}\boldsymbol{u}_k, \tag{11}$$

and assuming $\boldsymbol{H}\boldsymbol{C}\boldsymbol{B} = \boldsymbol{I}$ which implies that $\text{rank}(\boldsymbol{C}\boldsymbol{B}) = \text{rank}(\boldsymbol{B}) = n_u$, the system input receives the following form:

$$\boldsymbol{u}_k = \boldsymbol{H}\boldsymbol{g}^{-1}(\boldsymbol{y}_{k+1}) + \boldsymbol{H}\boldsymbol{v}_{k+1} - \boldsymbol{H}\boldsymbol{C}\boldsymbol{A}\boldsymbol{x}_k. \tag{12}$$

On the basis of (12) the input estimate can be defined as:

$$\hat{\boldsymbol{u}}_k = \boldsymbol{H}\boldsymbol{g}^{-1}(\boldsymbol{y}_{k+1}) - \boldsymbol{H}\boldsymbol{C}\boldsymbol{A}\hat{\boldsymbol{x}}_k. \tag{13}$$

The state estimation error is given by $\boldsymbol{e}_k = \boldsymbol{x}_k - \hat{\boldsymbol{x}}_k$ and the input estimation error can be defined as follows:

$$\varepsilon_{k,u} = \boldsymbol{u}_k - \hat{\boldsymbol{u}}_k = -\boldsymbol{H}\boldsymbol{C}\boldsymbol{A}\boldsymbol{e}_k + \boldsymbol{H}\boldsymbol{v}_{k+1}. \tag{14}$$

Substituting (12) into (7):

$$\boldsymbol{x}_{k+1} = \boldsymbol{A}\boldsymbol{x}_k + \boldsymbol{B}\boldsymbol{H}\boldsymbol{g}^{-1}(\boldsymbol{y}_{k+1}) + \boldsymbol{B}\boldsymbol{H}\boldsymbol{v}_{k+1} - \boldsymbol{B}\boldsymbol{H}\boldsymbol{C}\boldsymbol{A}\boldsymbol{x}_k, \tag{15}$$

and assuming $\bar{\boldsymbol{A}} = \boldsymbol{A} - \boldsymbol{B}\boldsymbol{H}\boldsymbol{C}\boldsymbol{A}$ and $\bar{\boldsymbol{B}} = \boldsymbol{B}\boldsymbol{H}$, (7) receives the form:

$$\boldsymbol{x}_{k+1} = \bar{\boldsymbol{A}}\boldsymbol{x}_k + \bar{\boldsymbol{B}}\boldsymbol{g}^{-1}(\boldsymbol{y}_{k+1}) + \bar{\boldsymbol{B}}\boldsymbol{v}_{k+1}. \tag{16}$$

The observer structure is:

$$\hat{\boldsymbol{x}}_{k+1} = \bar{\boldsymbol{A}}\hat{\boldsymbol{x}}_k + \bar{\boldsymbol{B}}\boldsymbol{g}^{-1}(\boldsymbol{y}_{k+1}) + \boldsymbol{K}(\boldsymbol{g}^{-1}(\boldsymbol{y}_k) - \boldsymbol{C}\hat{\boldsymbol{x}}_k), \tag{17}$$

while the state estimation error is given by:

$$\begin{aligned}
\boldsymbol{e}_{k+1} &= \boldsymbol{x}_{k+1} - \hat{\boldsymbol{x}}_{k+1} = \bar{\boldsymbol{A}}\boldsymbol{e}_k + \bar{\boldsymbol{B}}\boldsymbol{v}_{k+1} - \boldsymbol{K}(\boldsymbol{g}^{-1}(\boldsymbol{y}_k) - \boldsymbol{C}\hat{\boldsymbol{x}}_k) \\
&= \bar{\boldsymbol{A}}\boldsymbol{e}_k + \bar{\boldsymbol{B}}\boldsymbol{v}_{k+1} - [\boldsymbol{K}(\boldsymbol{C}\boldsymbol{x}_k - \boldsymbol{v}_k - \boldsymbol{C}\hat{\boldsymbol{x}}_k)] = \boldsymbol{A}_1\boldsymbol{e}_k + \bar{\boldsymbol{B}}\boldsymbol{v}_{k+1} + \boldsymbol{K}\boldsymbol{v}_k,
\end{aligned} \tag{18}$$

where: $\boldsymbol{A}_1 = \bar{\boldsymbol{A}} - \boldsymbol{K}\boldsymbol{C}$.

The objective is to design the observer in such a way that the state estimation error is asymptotically convergent and the following upper bound is guaranteed:

$$\|\varepsilon_{k,u}\|_{l_2} \leq \upsilon\|\boldsymbol{v}_k\|_{l_2}, \tag{19}$$

where $\upsilon > 0$ is a prescribed disturbance attenuation level. Thus, μ should be achieved with respect to the input estimation error but not to the state estimation error. Thus, the problem of $\mathcal{H}_\infty$ observer design [17] is to determine the gain matrix $\boldsymbol{K}$ such that: $\lim_{k\to\infty} \boldsymbol{e}_k = \boldsymbol{0}$ for $\boldsymbol{v}_k = \boldsymbol{0}$, and $\|\varepsilon_{k,u}\|_{l_2} \leq \upsilon\|\boldsymbol{v}\|_{l_2}$ for $\boldsymbol{v}_k \neq \boldsymbol{0}$ and $\boldsymbol{e}_0 = \boldsymbol{0}$. In order to settle the above problem it is sufficient to find a Lyapunov function V_k such that:

$$\Delta \boldsymbol{V} + \varepsilon_{u,k}^T \varepsilon_{u,k} - \mu^2 \boldsymbol{v}_{k+1}^T \boldsymbol{v}_{k+1} - \mu^2 \boldsymbol{v}_k^T \boldsymbol{v}_k < 0, \tag{20}$$

where $\Delta V_k = V_{k+1} - V_k$, $\boldsymbol{v}_k = \boldsymbol{e}_k^T \boldsymbol{P} \boldsymbol{e}_k$ and $\mu > 0$. Indeed, if $\boldsymbol{v}_k = \boldsymbol{0}$, ($k = 0, \ldots, \infty$) then (20) boils down to

$$\Delta V_k + \varepsilon_{u,k}^T \varepsilon_{u,k} < 0, \; k = 0, \ldots \infty, \tag{21}$$

and hence $\Delta V_k < 0$, which leads to $\lim_{k\to\infty} \boldsymbol{e}_k = \boldsymbol{0}$ for $\boldsymbol{v}_k = \boldsymbol{0}$. If $\boldsymbol{v}_k \neq \boldsymbol{0}$, ($k = 0, \ldots, \infty$) then inequality (20) yields:

$$J = \sum_{k=0}^{\infty} \left(\Delta V_k + \varepsilon_{u,k}^T \varepsilon_{u,k} - \mu^2 \boldsymbol{v}_k^T \boldsymbol{v}_k - \mu^2 \boldsymbol{v}_{k+1}^T \boldsymbol{v}_{k+1} \right) < 0, \tag{22}$$

which can be written as:

$$J = V_\infty - V_0 + \sum_{k=0}^{\infty} \varepsilon_{u,k}^T \varepsilon_{u,k} - \mu^2 \sum_{k=0}^{\infty} \boldsymbol{v}_k^T \boldsymbol{v}_k - \mu^2 \sum_{k=0}^{\infty} \boldsymbol{v}_{k+1}^T \boldsymbol{v}_{k+1} < 0. \tag{23}$$

Bearing in mind that $\mu^2 \sum_{k=0}^{\infty} \boldsymbol{v}_{k+1}^T \boldsymbol{v}_{k+1} = \mu^2 \sum_{k=0}^{\infty} \boldsymbol{v}_k^T \boldsymbol{v}_k - \mu^2 \boldsymbol{v}_0^T \boldsymbol{v}_0$ inequality (23) can be written as:

$$J = V_\infty - V_0 + \sum_{k=0}^{\infty} \varepsilon_{u,k}^T \varepsilon_{u,k} - 2\mu^2 \sum_{k=0}^{\infty} \boldsymbol{v}_k^T \boldsymbol{v}_k + \mu^2 \boldsymbol{v}_0^T \boldsymbol{v}_0 < 0. \tag{24}$$

Knowing that $V_0 = 0$ for $\boldsymbol{e}_0 = \boldsymbol{0}$ and $V_\infty \geq 0$, (24) leads to $\|\varepsilon_{k,u}\|_{l_2} \leq \upsilon\|\boldsymbol{v}\|_{l_2}$ with $\upsilon = \sqrt{2}\mu$.

In particular the following form of the Lyapunov function is proposed [17]:

$$\Delta V = \boldsymbol{e}_{k+1}^T \boldsymbol{P} \boldsymbol{e}_{k+1} - \boldsymbol{e}_k^T \boldsymbol{P} \boldsymbol{e}_k, \tag{25}$$

Thus, for $\boldsymbol{z}_k = [\boldsymbol{e}_k, \boldsymbol{v}_k, \boldsymbol{v}_{k+1}]^T$ the inequality (20) becomes:

$$\boldsymbol{z}_k^T \boldsymbol{X} \boldsymbol{z}_k < 0, \tag{26}$$

where the matrix $\boldsymbol{X} \prec 0$ has following form:

$$\boldsymbol{X} = \begin{bmatrix} \boldsymbol{A}_1^T \boldsymbol{P} \boldsymbol{A}_1 - \boldsymbol{P} + \boldsymbol{A}^T \boldsymbol{C}^T \boldsymbol{H}^T \boldsymbol{H} \boldsymbol{C} \boldsymbol{A} & \boldsymbol{A}_1^T \boldsymbol{P} \boldsymbol{K} & \boldsymbol{A}_1^T \boldsymbol{P} \bar{\boldsymbol{B}} - \boldsymbol{A}^T \boldsymbol{C}^T \boldsymbol{H}^T \boldsymbol{H} \\ \boldsymbol{K}^T \boldsymbol{P} \boldsymbol{A}_1 & \boldsymbol{K}^T \boldsymbol{P} \boldsymbol{K} - \mu^2 \boldsymbol{I} & \boldsymbol{K}^T \boldsymbol{P} \bar{\boldsymbol{B}} \\ \bar{\boldsymbol{B}} \boldsymbol{P} \boldsymbol{A}_1 - \boldsymbol{H}^T \boldsymbol{H} \boldsymbol{C} \boldsymbol{A} & \bar{\boldsymbol{B}} \boldsymbol{P} \boldsymbol{K} & \bar{\boldsymbol{B}} \boldsymbol{P} \bar{\boldsymbol{B}} + \boldsymbol{H}^T \boldsymbol{H} - \mu^2 \boldsymbol{I} \end{bmatrix} \tag{27}$$

Moreover, by applying the Schur complements, (27) is equivalent to

$$
\begin{bmatrix}
-P + A^T C^T H^T H C A & 0 & -A^T C^T H^T H & A_1^T \\
0 & -\mu^2 I & 0 & K^T \\
-H^T H C A & 0 & H^T H - \mu^2 I & \bar{B} \\
A_1 & K & \bar{B} & -P^{-1}
\end{bmatrix} < 0. \tag{28}
$$

Multiplying (28) from both sites by $\mathrm{diag}(I, I, I, P)$, and then substituting $A_1 = \bar{A} - KC$, $PA_1 = P\bar{A} - PKC = P\bar{A} - NC$, $A_1^T P = \bar{A}^T P - C^T N^T$ and $N = PK$, (28) receives the form:

$$
\begin{bmatrix}
-P + A^T C^T H^T H C A & 0 & -A^T C^T H^T H & \bar{A}^T P - C^T N^T \\
0 & -\mu^2 I & 0 & N^T \\
-H^T H C A & 0 & H^T H - \mu^2 I & \bar{B}^T P \\
P\bar{A} - NC & N & P\bar{B} & -P
\end{bmatrix} < 0. \tag{29}
$$

Note that (29) is a usual Linear Matrix Inequality (LMI), which can be solved, e.g. with MATLAB. As the result for the given disturbance attenuation level μ the observer gain matrix K and the estimate of the inputs $\hat{u}_k$ can be obtained.

The presented above approach allows to obtain estimates of GMDH neural networks inputs. Moreover, on the basis of the (20):

$$
\varepsilon_{u,k}^T \varepsilon_{u,k} \le \mu^2 v_{k+1}^T v_{k+1} + \mu^2 v_k^T v_k. \tag{30}
$$

Assuming that $v_k^T v_k = \|v_k\|_2^2 < \delta$, where $\delta > 0$ is a given bound then

$$
\varepsilon_{u,k}^T \varepsilon_{u,k} \le 2\mu^2 \delta, \tag{31}
$$

the adaptive thresholds for the inputs of the GMDH neural model receive the following form:

$$
\hat{u}_{i,k} - \mu\sqrt{2\delta} \le u_{i,k} \le \hat{u}_{i,k} + \mu\sqrt{2\delta}. \tag{32}
$$

An occurrence of the fault for each i-th actuator is signaled when input $u_{i,k}$ crosses the input adaptive threshold (32).

4   Illustrative Example

For the modeling and fault diagnosis purpose a tunnel furnace was chosen [3]. The considered tunnel furnace (c.f. Fig. 4) is designed to mimic, in the laboratory conditions, the real industrial tunnel furnaces, which can be applied in the food industry or production of ceramics among others. The furnace is equipped in three electric heaters and four temperature sensors. The required temperature of the furnace can be kept by controlling the heaters behaviour. This task can be achieved by the group regulation of the voltage with the application of the controller PACSystems RX3i manufactured by GE Fanuc Intelligent Platforms and semiconductor relays RP6 produced by LUMEL. The temperature of the

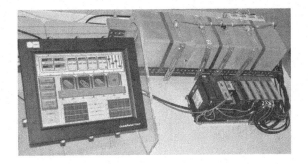

Fig. 4. Tunnel furnace

furnace is measured via IC695ALG600 module with Pt100 Resistive Thermal Devices (RTDs). The modeled furnace is a three-input and four-output system $(t_1, t_2, t_3, t_4) = f(u_1, u_2, u_3)$, where $t_1, \ldots, t_4$ are temperatures from sensors and $u_1, \ldots, u_3$ are input voltages allowing to control the electric heaters. The data used for the identification and validation were collected in two data sets consisting of 2600-th samples. It should be also pointed out that these data sets were scaled for the purpose of neural networks designing. The parameters of the dynamic neurons were estimated with the application of the UKF algorithm [14]. The selection of best performing neurons in terms of their processing accuracy was realized with the application of the soft selection method [12] based on SSE evaluation criterion.

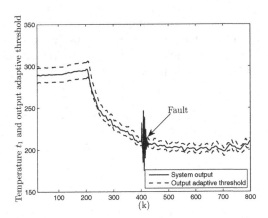

Fig. 5. Detection of faulty temperature sensor via output adaptive threshold

After the synthesis of the GMDH model according to the methodology presented in Sect. 2, it is possible to employ it for robust fault detection. The detection of the faulty sensor for the temperature t_1 (simulated during 10sec.) via output adaptive threshold and the faulty first electric heater of the tunnel furnace via input adaptive threshold are presented in Fig. 5 and Fig. 6, respectively.

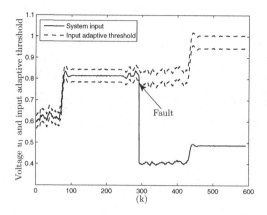

Fig. 6. Detection of faulty electric heater via input adaptive threshold

As it can be seen the faults are detected when the measurements of temperature t_1 and voltage u_1 cross the output (6) and input (32) adaptive thresholds.

5   Conclusion

The objective of this paper was to elaborate a novel robust FDI method on the basis of the state-space GMDH neural model. The application of the UKF allows to obtain the asymptotically stable model and calculate the output adaptive threshold, which can be used in the robust fault detection of the system and sensors. Moreover, the GMDH model inputs estimation using RUIF approach was developed. It allows to calculate the input adaptive thresholds and enables to perform the robust fault detection and isolation of the actuators.

Acknowledgments. The work was supported by the National Science Centre of Poland under grant: 2011-2014

References

1. Ding, S.: Model-based Fault Diagnosis Techniques: Design Schemes, Algorithms, and Tools. Springer, Heidelberg (2008)
2. Korbicz, J., Kościelny, J.: Modeling, Diagnostics and Process Control: Implementation in the DiaSter System. Springer, Berlin (2010)
3. Mrugalski, M.: An unscented kalman filter in designing dynamic gmdh neural networks for robust fault detection. International Journal of Applied Mathematics and Computer Science 23(1), 157–169 (2013)
4. Mrugalski, M., Witczak, M.: State-space gmdh neural networks for actuator robust fault diagnosis. Advances in Electrical and Computer Engin. 12(3), 65–72 (2012)

5. De Oca, S., Puig, V., Witczak, M., Dziekan, L.: Fault-tolerant control strategy for actuator faults using lpv techniques: Application to a two degree of freedom helicopter. International Journal of Applied Mathematics and Computer Science 22(1), 161–171 (2012)
6. Niemann, H.: A model-based approach to fault-tolerant control. International Journal of Applied Mathematics and Computer Science 22(1), 67–86 (2012)
7. Noura, H., Theilliol, D., Ponsart, J., Chamseddine, A.: Fault-tolerant Control Systems: Design and Practical Applications. Springer, London (2009)
8. Haykin, S.: Neural Networks and Learning Machines. Prentice Hall, NY (2009)
9. Mrugalski, M., Witczak, M., Korbicz, J.: Confidence estimation of the multi-layer perceptron and its application in fault detection systems. Engineering Applications of Artificial Intelligence 21(6), 895–906 (2008)
10. Patan, K., Witczak, M., Korbicz, J.: Towards robustness in neural network based fault diagnosis. International Journal of Applied Mathematics and Computer Science 18(4), 443–454 (2008)
11. Ivakhnenko, A., Mueller, J.: Self-organization of nets of active neurons. System Analysis Modelling Simulation 20, 93–106 (1995)
12. Korbicz, J., Mrugalski, M.: Confidence estimation of gmdh neural networks and its application in fault detection system. International Journal of System Science 39(8), 783–800 (2008)
13. Witczak, M., Korbicz, J., Mrugalski, M., Patton, R.: A gmdh neural network based approach to robust fault detection and its application to solve the damadics benchmark problem. Control Engineering Practice 14(6), 671–683 (2006)
14. Teixeira, B., Torres, L., Aguirre, L., Bernstein, D.: On unscented kalman filtering with state interval constraints. Journal of Process Control 20(1), 45–57 (2010)
15. Korbicz, J., Witczak, M., Puig, V.: Lmi-based strategies for designing observers and unknown input observers for non-linear discrete-time systems. Bulletin of the Polish Academy of Sciences: Technical Sciences 55(1), 31–42 (2007)
16. Witczak, M., Pretki, P.: Design of an extended unknown input observer with stochastic robustness techniques and evolutionary algorithms. International Journal of Control 80(5), 749–762 (2007)
17. Zemouche, A., Boutayeb, M., Iulia Bara, G.: Observer for a class of Lipschitz systems with extension to $\mathcal{H}_\infty$ performance analysis. Systems and Control Letters 57(1), 18–27 (2008)
18. Mrugalski, M., Arinton, E., Korbicz, J.: Dynamic gmdh type neural networks. In: Rutkowski, L., Kacprzyk, J. (eds.) Neural Networks and Soft Computing, pp. 698–703. Physica-Verlag, Heidelberg (2003)
19. Mrugalski, M., Korbicz, J.: Least mean square vs. outer bounding ellipsoid algorithm in confidence estimation of the gmdh neural networks. In: Beliczynski, B., Dzielinski, A., Iwanowski, M., Ribeiro, B. (eds.) ICANNGA 2007, Part II. LNCS, vol. 4432, pp. 19–26. Springer, Heidelberg (2007)
20. Lee, T., Jiang, Z.: On uniform global asymptotic stability of nonlinear discrete-time systems with applications. IEEE Trans. Automatic Control 51(10), 1644–1660 (2006)

Diffusion Methods for Wind Power Ramp Detection

Ángela Fernández, Carlos M. Alaíz, Ana M. González,
Julia Díaz, and José R. Dorronsoro

Departamento de Ingeniería Informática and Instituto de Ingeniería del Conocimiento
Universidad Autónoma de Madrid, 28049, Madrid, Spain
{a.fernandez,carlos.alaiz,ana.marcos,julia.diaz,jose.dorronsoro}@uam.es

Abstract. The prediction and management of wind power ramps is currently receiving large attention as it is a crucial issue for both system operators and wind farm managers. However, this is still an issue far from being solved and in this work we will address it as a classification problem working with delay vectors of the wind power time series and applying local Mahalanobis K-NN search with metrics derived from Anisotropic Diffusion methods. The resulting procedures clearly outperform a random baseline method and yield good sensitivity but more work is needed to improve on specificity and, hence, precision.

Keywords: Diffusion Methods, Anisotropic Diffusion, diffusion distance, wind power ramps.

1 Introduction

The growing presence of wind energy is raising many issues in the operation of electrical systems and some of them can be conceivably addressed through the application of Machine Learning (ML) techniques. One important example that we shall deal with in this paper is the prediction of wind ramps, i.e., sudden, large increases or decreases of wind energy production over a limited time period [8]. In fact, algorithms to detect possible ramps and raise alerts about them are of obvious interest to system operators and wind farm managers to support wind farm control, to decide how much energy should be dispatched or to modify generation schedules. However, there are still few methodologies for ramp prediction and even there is not a standard ramp definition yet, making this topic a wide open research area.

From a ML point of view, two approaches to wind ramp detection with different final goals have been proposed in the literature. If we want to determine not only the starting of a ramp but also its magnitude, regression models are the natural choice. This approach has been followed, for example, in [11], that applies multivariate time series prediction models and uses mean absolute prediction error and standard deviation as accuracy measures. In [3] probabilistic numerical weather prediction systems are used to associate uncertainty estimates to wind

I. Rojas, G. Joya, and J. Cabestany (Eds.): IWANN 2013, Part I, LNCS 7902, pp. 106–113, 2013.
© Springer-Verlag Berlin Heidelberg 2013

energy predictions and to provide risk indices to warn about possible large deviations and ramp events. On the other hand, we can pursue a classification-based approach to predict wind ramps through event detection models. Examples of this are [4] or [7], that try to locate ramp presence some hours before or after the real occurrence, given the wind energy forecasts for some time into the future.

In this work we will also consider ramp detection as a classification problem but we will seek to provide for each hour a forecast on whether a ramp is about to start, which differs from the few available state-of-the-art results and makes them non comparable. Our overall approach is to relate the conditions at a given hour to similar conditions in the past and to somehow derive a ramp forecast from what happened in these previous similar situations. More precisely, we can consider for each hour t a certain feature vector X_t that should adequately represent wind energy behavior up to time t and find a subset of K past vectors X_{t_i} close to X_t in an appropriate metric. There is a growing number of options to choose data that characterize X_t but in this work we shall simply consider the wind energy production time series as the only such information and X_t will be a delay vector built from the last D wind energy production values, $X_t = (p_{t-D+1}, \ldots, p_{t-1}, p_t)^T$. This is certainly not an optimal choice, as the well-known chaotic behavior of the weather implies that past behavior of wind energy up to time t has only a weak influence on its behavior afterwards. However, ramps are also local phenomena and, in any case, our approach can easily accommodate the use of more relevant information. For instance, the quality of wind energy prediction is steadily improving and can easily be incorporated to the procedures pursued here.

Going back to our approach, the most relevant issue is the definition of the metric to be used to compare X_t with previous values. To do so, we will work in an Anisotropic Diffusion context. In general, diffusion methods assume that sample values, the D-dimensional delay vectors X_t in our case, lie in a manifold $\mathcal{M}$ whose geometry corresponds to a diffusion distance associated with a Markov process. Then, the relationship between the spectral properties of the Markov chain and the manifold geometry allows the definition of a Diffusion Map into a lower dimensional space in such a way that Euclidean distance in the projected space corresponds to the diffusion metric on $\mathcal{M}$. However, this requires a computationally very costly eigenanalysis of the Markov transition matrix and we will pursue here an alternative Anisotropic Diffusion model which assumes that the sample data points are the result of the application of an unknown map f to the latent variables l_t that govern the X_t data and that follow a particular independent stochastic Itô process. This allows to estimate the Euclidean distance in the inaccessible l space through local Mahalanobis distances in the sample manifold $\mathcal{M}$ *without* having to go through any costly eigenanalysis.

Wind power clearly has a time structure and if we assume weather and, thus, wind power as governed by a latent variable model, the wind ramp detection problem might fit nicely in the Anisotropic Diffusion framework. In this paper we will explore this approach and, as we shall see, our methods clearly improve on a baseline random model and have good sensitivity. However, specificity and, hence, precision must be improved. Moreover, while slightly better, the Maha-lanobis models still give results similar to those achieved using a simple Euclidean metric. Still, there is a clear room from improvement. In fact, it is well known that delay vectors are a too crude representation of the wind power time series and that they cannot be used, for instance, for forecasting future power values. We will briefly discuss this and other related questions at the end of the paper, that is organized as follows. In Sect. 2 the diffusion theory framework is introduced and in Sect. 3 the wind ramp detection problem is presented and our prediction methods proposed. Sect. 4 contains the numerical experiments and, finally, Sect. 5 ends this paper with a brief discussion, some conclusions and hints for further work.

2   Diffusion Methods Review

We give first a simplified review of standard Diffusion Methods (DMs) fol-lowing the notation of [10]. The first step is to build a complete connectivity graph G where the original points are the graph nodes and where the weight distances reflect the local similarity between two points X_i, X_j, i.e., we have $w_{i,j} = W_\sigma(X_i, X_j) = h\left(\rho(X_i, X_j)^2/\sigma\right)$, where h is a function with exponential de-cay, such as a Gaussian kernel, ρ is some metric and σ is a parameter that defines the "locality" of a neighborhood. Weights are then normalized as $K = D^{-1}W$, with $D_{ii} = \sum_j w_{ij}$ the graph degree, being D a diagonal matrix. K is then a Markov matrix that can be iterated to generate a Markov process with transition probabilities $P_t(X_i, X_j)$. This can in turn be used to define the spectral distance

$$\mathcal{D}_t(X_i, X_j) = \|P_t(X_i, \cdot) - P_t(X_j, \cdot)\|_{L_2}^2 \simeq \sqrt{\sum_k |P_t(X_i, X_k) - P_t(X_j, X_k)|^2} \ ,$$

that express the similarity after t steps between two diffusion processes starting from X_i and X_j. While it is rather hard to compute this distance, it turns out that the eigenfunctions $\{\Psi_i\}$ of the operator K coincide with the eigenfunctions $\{\Phi_i\}$ of the graph Laplacian (see [2,5]), which is defined as

$$\mathcal{L} = D^{-\frac{1}{2}}WD^{-\frac{1}{2}} - I = D^{\frac{1}{2}}KD^{-\frac{1}{2}} - I \ .$$

This can be used to show that $\mathcal{D}_t(X_i, X_j)$ coincides with Euclidean distance in the DM space.

The study of DMs has opened a world of possibilities in dimensionality reduc-tion [5], clustering [2] or function approximation [10]. However, the eigenanalysis needed to compute the DMs is still quite costly computationally and, moreover, their application to new patterns is not straighforward and requires the use of some approximation tool such as Nyström's method.

We will focus our attention here on the anisotropic version of these meth-ods [9], which fits nicely to the problem we want to solve. The starting point is to assume that the sample is generated by a non linear function f acting on some d-dimensional parametric features l_t that follow an Itô process

$$dl^j = a^j(l)dt + b^j(l)dw^j, \ j = 1, \ldots, d,$$

where a^j is the drift coefficient, b^j is the noise coefficient and w^j is a Brownian motion. Itô's Lemma ensures that our observable variables $X_t = f(l_t)$ are also Itô processes. Thanks to this fact, and assuming an appropriate feature rescaling, we can locally estimate the distortion in the transformation f through the covariance matrix C of the observable data, namely $C = JJ^T$, where J is the Jacobian of the function f. The important fact now is that the Euclidean distance $\|l_i - l_j\|$ in the latent variable space can be approximated as

$$\|l_i - l_j\|^2 \simeq (X_i - X_j)^T [C^{-1}(X_i) + C^{-1}(X_j)](X_i - X_j). \tag{1}$$

We can now build a diffusion kernel based on this distance whose infinitesimal generator coincides with a backward Fokker–Planck operator. In particular, the original latent features could be recovered by the appropriate eigenanalysis of this operator. However, we do not need this to estimate distances in the inaccessible latent space as they can be approximated directly on the sample manifold $\mathcal{M}$ using (1) without having to go through any costly eigenanalysis.

3   Predicting Wind Ramps

As mentioned before, while the idea of a wind ramp is intuitively clear, there is not a universally accepted characterization. Thus, here we shall discuss first the definition of wind ramps, present then an approach for issuing wind ramp warnings and close this section with the methodology that will be used to evaluate its effectiveness.

As mentioned in [6], an intuitive description of a wind power ramp could be a large change in wind production in a relatively short period of time. To turn this description into a formal definition we need to specify what are a "large change" and a "short time period". Several options are discussed in [3,6] but possibly the simplest one is to consider derivatives or, rather, first order differences, and say that a ramp will happen at time t if in a time period Δt we have

$$|P(t + \Delta t) - P(t)| > \Delta P_{th} \ .$$

Notice that this definition detects equally upward and downward ramps and it requires to determine the values of Δt and the threshold ΔP_{th}. Starting with Δt, if t is given in hours, a low value such as $\Delta t = 1$ leaves no reaction time to the system operator; on the other hand, a larger value will not imply a big impact on the electrical system. Because of these and similar considerations (see [7]), we have settled on the value $\Delta t = 3$. Notice that once Δt is chosen, ΔP_{th} essentially determines how often ramps happen. A low threshold results in many ramp events but most of them will be of little consequence, while large values result very relevant but also very infrequent ramps. We have settled in a ΔP_{th} that marks the top 5% percentile of ramp events. In other words, the probability of a ramp jump $|P(t + \Delta t) - P(t)|$ larger than ΔP_{th} is 0.05.

In order to apply Anisotropic Diffusion to ramp event prediction, we have to assume that extreme power ramps correspond to particular values of the unknown latent variables that determine wind energy production. More precisely, we have to define wind energy patterns X_t that somehow capture the structure of wind production at time t and that are determined by latent variable values l_t. Thus, a possible approach to predict ramps at time t is to identify previous latent vectors l_{t_i} that are close to the current latent vector l_t and to exploit the corresponding previous wind energy patterns X_{t_i} to deduce whether the current pattern X_t is associated to a ramp event. To make this work, we must have an estimate of the distance $\|l_{t_i} - l_t\|$ and is in this context where we can benefit of an Anisotropic Diffusion approach. As explained in Sect. 2, this framework allows to approximate $\|l_{t_i} - l_t\|$ by (1). This estimate requires to compute and invert the covariances $C(X_{t_i})$ at each possible X_{t_i}. To alleviate the possibly large computational cost, we simplify the Mahalanobis distance to $d(X_{t_i}, X_t) = (X_{t_i} - X_t)^T C_t^{-1}(X_{t_i} - X_t)$, with C_t^{-1} the inverse of the local covariance matrix in a cloud of points around X_t.

We shall apply this approach working with D-dimensional energy patterns of the form $X_t = (p_{t-D+1}, \ldots, p_{t-1}, p_t)^T$ that correspond to a delay window of length D, for which we will find a subset $\mathcal{S}_t$ with the K sample patterns X_{t_i} nearest to X_t, with K appropriately selected. This will be done for both the Mahalanobis and the Euclidean (i.e., isotropic) distances. Once $\mathcal{S}_t$ is found, we will classify X_t as a ramp if we have $\nu_t \geq \rho$, with ν_t the number of ramp-associated patterns in $\mathcal{S}_t$ and $1 \leq \rho \leq K$ a threshold value; we will give results only for $\rho = 1$ but larger ρ values would be associated to more confidence in a ramp happening at time t. In the Mahalanobis case we also have to select a pattern cloud $\mathcal{C}_t$ to compute the covariance matrix C_t at time t. The simplest way is just to work with a *time cloud*, i.e., to select $\mathcal{C}_t = \{X_t, X_{t-1}, \ldots, X_{t-M+1}\}$, using the M patterns closest to X_t in time. Alternatively, we shall consider a *cluster cloud* where we fix a larger time cloud with κM patterns, apply κ-means clustering to it and choose the new cloud $\mathcal{C}_t^\kappa$ as the cluster that contains X_t. Besides the parameter ρ, that will affect the confidence on the ramp prediction, performance will of course depend on the concrete selection of the parameters used, namely the number K of patterns closest to X_t, the dimension D of the patterns, and the M and κ used to determine the covariance cloud. The complete method is summarized in Alg. 1.

Since we want to solve what essentially is a supervised classification problem, confusion matrix-related indices seem to be the best way to evaluate algorithm performance. More precisely, we use the sensitivity Sens = $^{\text{TP}}/_{\text{TP + FN}}$ and specificity Spec = $^{\text{TN}}/_{\text{TN + FP}}$ values, as well as precision Prec = $^{\text{TP}}/_{\text{TP + FP}}$, that measures the proportion of correct ramp alerts. In order to select the best K, D, M and κ values we will combine TP, TN, FP and FN in the Matthews correlation coefficient [1]

$$\Phi = \frac{\text{TP} \cdot \text{TN} - \text{FP} \cdot \text{FN}}{\sqrt{(\text{TP} + \text{FP}) \cdot (\text{TN} + \text{FN}) \cdot (\text{TP} + \text{FN}) \cdot (\text{TN} + \text{FP})}}$$

that returns a $[-1, 1]$ value with $\Phi = 1$ if FP = FN = 0, i.e., we have a diagonal confusion matrix and perfect classification, and $\Phi = -1$ if TP = TN = 0.

Algorithm 1. Ramp Events Detection

Input: $P = \{p_1, \ldots, p_s\}$, wind power time series; D, pattern dimension; Δt, ramp
duration; $R = \{r_1, \ldots, r_{s-\Delta t}\}$, $\{0, 1\}$ ramp labels; ρ, ramp threshold.
Output: $\hat{r}_{s+1}$, the ramp prediction at time $s + 1$.
1: Build patterns $X_t = (p_{t-D+1}, \ldots, p_{t-1}, p_t)^T$;
2: Select the cloud $\mathcal{C}_s$;
3: Compute the covariance C_s;
4: $\mathcal{S}_s = NearestNeightbors(X_s, C_s, K)$, the K patterns closest to X_s;
5: $\nu_s = \sum_{\mathcal{S}_s} r_{s_i}$;
6: **if** $\nu_s \geq \rho$ **then**
7:     **return** 1;
8: **else**
9:     **return** 0;
10: **end if**

---

4 Experiments

In this section we will illustrate the application of the previous methods to the
Sotavento wind farm[1], located in the northern Spanish region of Galicia and
that makes its data publicly available. The training data set that we will use is
composed of hourly productions from July 1, 2010, to July 31, 2012. Of these,
each hour starting from August 1 2011 will be used for test purposes, with the
training set formed by one year and one month, up to the hour before. Wind
ramp hours have been defined as those hours h for which the absolute 3-hourly
difference between productions at hours h and $h + 3$ falls in the top 5%. This
means a power rise of at least 4.38MW, which essentially correspond to a 25%
of the nominal power of this wind farm, a value also used in other studies [3,4].
We recall that straight wind ramp prediction is a rather difficult problem for
which there are not reference results in the literature. Thus we will use as a
baseline reference the performance of a random prediction that assigns at each
hour a ramp start with a 0.05 probability. For the test period considered, we
have $N = 8699$ patterns of which 5%, i.e. $N_p = 450$, are ramps and the rest,
$N_n = 8249$, are not. Table 1 shows the expected values of the confusion matrix
of this random model as well as the mean and deviation of sensitivity, specificity
and precision. We will compare these baseline results with the three K-nearest
neighbors (NN) models previously considered, that is, standard Euclidean K-
NN, called NN^E, and Mahalanobis K-NN with either a time cloud covariance,
called NN_T^M, or a cluster cloud, called NN_C^M. We have to appropriately set the
hyper-parameter values, namely pattern dimension D and time cloud size M. To
arrive to some good values of K, D and M we have considered K values in the
set $\{5, 10, 15\}$, D values in $\{4, 8, 12\}$ and M values in $\{10, 20, 50\}$ (we fix $\kappa = 4$
to define clouds in the cluster approximation) and chosen as the best parameters
those giving a largest Matthews coefficient Φ, which are $K = 15$ and $D = 4$ for
all cases, and $M = 50$ and $\kappa M = 200$ for the time and cluster cloud sizes.

---

[1] Sotavento Galicia, http://www.sotaventogalicia.com/index.php.

Table 1. Baseline model confusion matrix

	Pred. +	Pred. −	$\sum$		Mean	Deviation
Real +	181	269	450	Sens	40.13%	2.31%
Real −	3310	4939	8249	Spec	59.87%	0.54%
$\sum$	3491	5208	8699	Prec	5.17%	0.30%

Table 2. K-NN models confusion matrices

Time Cloud (NN_T^M)

	P. +	P. −	$\sum$
R. +	321	129	450
R. −	3048	5201	8249
$\sum$	3369	5330	8699

Sens	71.33%
Spec	63.05%
Prec	9.53%

Cluster Cloud (NN_C^M)

	P. +	P. −	$\sum$
R. +	318	132	450
R. −	3001	5248	8249
$\sum$	3319	5380	8699

Sens	70.67%
Spec	63.62%
Prec	9.58%

Euclidean (NN^E)

	P. +	P. −	$\sum$
R. +	314	136	450
R. −	3034	5215	8249
$\sum$	3348	5351	8699

Sens	69.78%
Spec	63.22%
Prec	9.38%

The results obtained with each optimal model are presented in Table 2. As it can be seen, all K-NN methods clearly outperform the random baseline model, as the sensitivity and specificity of any random predictor always sum 100%. The biggest improvement can be appreciate with respect to sensitivity, that goes from near 40% to about 70%, and precision, that goes from near 5% to about 9.5%. The specificity gain is smaller, about 4%, but still quite larger than the 0.54% standard deviation of the random model. On the other hand, the NN_T^M and NN_C^M models are only slightly better than the purely Euclidean model NN^E and none of the methods can be considered as exploitation-ready models, for while they give a good sensitivity, but specificity and, thus, precision are far from good enough. However, as we discuss next, it is known that delay vectors provide a rather crude information about the wind power time series and adding more information to the X_t is a clear first step toward better wind ramp detection.

5 Discussion and Conclusions

While they are a key problem in wind energy and system operation management, there is still no standard definition of wind power ramps and their detection is therefore a question far from being solved. In this work we have applied an Anisotropic Diffusion approach where we consider wind power delay vectors as visible events derived from latent vectors that follow some Itô processes. This leads naturally to define a covariance based Mahalanobis distance for the delay vectors and, in turn, to apply K-NN methods to detect past vectors close to the current one X_t and to use this information to predict whether or not a ramp is going to start at time t. The resulting methods clearly outperform a baseline random model and show a good sensitivity.

However, specificity must be improved which, in turn, would lead to better precision and, hence, to systems ready to industrial use. A first step to achieve this would be to refine ramp prediction using some weighted combination of the ramp states of the K nearest neighbors of X_t. A second step would be to work with patterns X_t richer than plain delay vectors, adding for instance numerical weather prediction (NWP) information or even short time wind power predictions derived from this NWP information. Finally, we could also exploit the time evolution of previous wind ramp alerts to improve specificity. We are working on these and similar directions.

Acknowledgement. With partial support from Spain's grant TIN2010-21575-C02-01 and the UAM–ADIC Chair for Machine Learning. The first author is also supported by an FPI–UAM grant and kindly thanks the Applied Mathematics Department of Yale University for receiving her during her visits. The second author is supported by the FPU–MEC grant AP2008-00167.

References

1. Baldi, P., Brunak, S., Chauvin, Y., Andersen, C., Nielsen, H.: Assessing the Accuracy of Prediction Algorithms for Classification: An Overview. Bioinformatics 16, 412–424 (2000)
2. Belkin, M., Nyogi, P.: Laplacian Eigenmaps for dimensionality reduction and data representation. Neural Computation 15(6), 1373–1396 (2003)
3. Bossavy, A., Girard, R., Kariniotakis, G.: Forecasting ramps of wind power production with numerical weather prediction ensembles. Wind Energy 16(1), 51–63 (2013)
4. Bradford, K., Carpenter, R., Shaw, B.: Forecasting southern plains wind ramp events using the wrf model at 3-km. In: Proceedings of the AMS Student Conference, Atlanta, Georgia (2010)
5. Coifman, R., Lafon, S.: Diffusion Maps. Applied and Computational Harmonic Analysis 21(1), 5–30 (2006)
6. Ferreira, C., Gama, J., Matias, L., Botterud, A., Wang, J.: A survey on wind power ramp forecasting. Technical report, Argonne National Laboratory (February 2011)
7. Greaves, B., Collins, J., Parkes, J., Tindal, A.: Temporal forecast uncertainty for ramp events. Wind Engineering 33(4), 309–319 (2009)
8. Kamath, C.: Associating weather conditions with ramp events in wind power generation. In: 2011 IEEE/PES Power Systems Conference and Exposition (PSCE), pp. 1–8 (March 2011)
9. Singer, A., Coifman, R.: Non-linear independent component analysis with Diffusion Maps. Applied and Computational Harmonic Analysis 25(2), 226–239 (2008)
10. Szlam, A., Maggioni, M., Coifman, R.: Regularization on graphs with function-adapted diffusion processes. Journal of Machine Learning Research 9, 1711–1739 (2008)
11. Zheng, H., Kusiak, A.: Prediction of wind farm power ramp rates: A data-mining approach. Journal of Solar Energy Engineering 131(3), 031011-1–31011-8 (2009)

Computational Study Based on Supervised Neural Architectures for Fluorescence Detection of Fungicides

Yeray Álvarez Romero[1], Patricio García Báez[2],
and Carmen Paz Suárez Araujo[1]

[1] Instituto Universitario de Ciencias y Tecnologías Cibernéticas, Universidad de Las Palmas de Gran Canaria, Las Palmas de Gran Canaria, Canary Islands, Spain
`yeray.alvarez102@estudiantes.ulpgc.es, cpsuarez@dis.ulpgc.es`
[2] Departamento de Estadística, Investigación Operativa y Computación, Universidad de La Laguna, La Laguna, Canary Islands, Spain
`pgarcia@ull.es`

Abstract. Benzimidazole fungicides (BFs) are a type of pesticide of high environmental interest characterized by a heavy spectral overlap which complicates its detection in mixtures. In this paper we present a computational study based on supervised neural networks for a multi-label classification problem. Specifically, backpropagation (BPN) with data fusion and ensemble schemes is used for the simultaneous resolution of difficult multi-fungicide mixtures. We designed, optimized and compared simple BPNs, BPNs with data fusion and BPN ensembles. The information environment used is made up of synchronous and conventional BF fluorescence spectra. The mixture spectra are not used in the training stage. This study allows the use of supervised neural architectures to be compared to unsupervised ones, which have been developed in previous works, for the identification of BFs in complex multi-fungicide mixtures. The study was carried out using a new software tool, MULLPY, which was developed in Python.

Keywords: Artificial Neural Networks, Neural Ensembles, Benzimidazole Fungicides, Fluorescence Detection, Environment, Multi-label classification.

1 Introduction

Fungicides and their benzimidazole derivatives, (benomyl (BM), carbendazim, (MBC), fuberidazole (FB) and thiabendazole (TBZ)), are an important type of pesticide that are extensively used in agriculture. Even though they protect crops and eliminate fungus [1], they also produce harmful side effects on the environment and the health of its inhabitants. A large number of adverse effects has been detected after decades of its application which has brought about regulation of their use by European agencies [2]. The development of methods with precise and sensitive detection of pesticides are essential, especially if they can

I. Rojas, G. Joya, and J. Cabestany (Eds.): IWANN 2013, Part I, LNCS 7902, pp. 114–123, 2013.

detect and control which compounds are damaging the ecosystem and human health.

The suitability of fluorescent spectroscopic techniques for mixture resolution has been proven [1]. Their main drawback lies in the spectral interferences of the BFs present in the mixture. Thus, the resolution of complex mixtures with a high degree of overlap among its compounds produces challenges in chemical analysis. Traditional instrumental techniques, such as layer chromatography, gas chromatography, and high performance liquid chromatography [1] are used to tackle this difficult problem, however complications such as cost, time, analytical complexity, and the need for substance pre-treatment must also be considered if they are to be used.

A search for alternative techniques and methods would certainly be convenient when studying this problem. A complementary approach to the instrumental and chemometric methods to obtain the resolution of multi-analyte systems is based on neural computation. ANNs capture high-dimensional inputs and generate relationships between the inputs and outputs from a training set. The encoded internal representation also captures the similarity in the input that results in generalizations. Neural network computation tools are a legitimate computational approach to tackle mixture resolution problems, and have also been backed in a wide range of studies [3,4,5,6,7,8]. In these studies where supervised training was performed, mixture spectra were used during the training phase [4,5,6,8]. One aim of this study consists in using only pure substances spectra, because it allows the complexity of the necessary chemical experiments to reduced.

In this paper we present a computational study on intelligent systems for fluorescence detection of BFs, based on supervised neural networks with data and decision fusion schemes. The Backpropagation (BPN) neural architecture is used. We work with conventional and synchronous fluorescence spectra, analyzing the convenience of the data-fusion and that from a neural ensemble approach. This is a very successful technique where the outputs of a set of separately trained neural networks are combined to form one unified prediction [9]. This computational study is an extension of our previous ones [3,7,8], where we used unsupervised ANNs, HUMANN, and also supervised ANNs, where mixture spectra were used in the training set, with only one kind of fluorescence spectra, the synchronous one.

The objective of the paper is to identify which neural learning paradigm (supervised or unsupervised) is the most appropriate for the problem under study. A second objective is to determine which fluorescence spectra provides more and better information. The importance of this study lies in its application to a significant public health concern that has a widespread impact on environmental protection and control.

2   Data Set

All fluorescent systems are generally characterized by an excitation (or absorption) spectra and an emission spectra. These spectra also allow their synchronous spectra to be acquired. BFs are examples of fluorescent systems. Carbendazim 99.7% (methyl (1 H-benzimidazol-2-yl) carbamate), benomyl 99.3% (methyl 1-(butylcarbomayl) benzimidazole-2-yl carbamate), thiabendazole 99.6% (2-(4-thiazol)benzimidazole) and fuberidazole 99.6% (2-(2furanyl)-1 h-benzimidazole) were obtained from Riedel-de Haen (Seelze, Germany). All conventional and synchronous fluorescence spectra in the study were obtained using a Perkin-Elmer LS-50 luminescence spectrophotometer (Beaconsfeld, Buckinghamshire, UK) fitted with a xenon discharge lamp.

The data set was provided by the Environmental Chemical Analysis Group at the ULPGC. An experimental design with chemical and computational requirements was performed [3] to obtain the data set. For each solution we generated eight spectra types for the mean, median and optimal λ or $\Delta\lambda$ values, see Table 1. All spectra were repeated three times to guarantee measurements and define error margins in measurements.

The data set in a supervised scheme must be divided into 3 subsets to avoid overtraining. The training set was made up of spectra from 16 solutions of pure substances to different concentrations plus the spectra from the clean sample. The test set was made up of spectra from the remaining 8 solutions, where the spectra of each substance were distributed by 6 concentrations. One hundred combinations of mixtures, used as the validation set, were generated automatically for each kind of spectra and λ, $\Delta\lambda$ values, with the only conditioning factors being that the compound distributions should be as balanced as possible with respect to the number of compounds present in each mixture, the concentrations of the same and the type of compound used [3]. We labeled each fungicide used as a fungicide class, except for BM and MBC, which belong to the same class. The distinction is based on the spectral characteristics of these compounds given by the correlation matrix [7]. Thus, there are 3 classes of fungicides, MBC-BM, FB, TBZ, and a fourth class which represents a clean sample (CS).

3   Methods

Systems analysis was performed in the pre-processing and processing modules, both based on ANNs. The pre-processing stage includes the fluorescence spectra modeling and the attainment of the characteristics vector. Fluorescence spectra can be modeled by a Gaussian distribution of intensity versus reciprocal wavelength. Spectral representation via Gaussian distribution was carried out using Radial Basis Function networks [3,10]. The results obtained in this stage make up the information environment of the BPN-based systems for the BFs fluorescence identification, which are employed in the processing module.

Three kinds of ANN-based systems as processing modules are proposed. The first is the well-known BPN system [11] with momentum, a hidden layer and an output layer with the same number of neurons as there are BF classes in a complex mixture plus the clean sample (CS) case, where there is no fungicide. The number of neurons in the hidden layer was determined by an iterative process of trial and error by selecting the smallest possible number. All the optimization

Table 1. General characteristics of the data set of benzimidazole fungicides

	Characteristics	Benzimidazole family
	Compounds	4: Benomyl (BM), Carbendazim (MBC), Fuberidazol (FB), Thiabendazol (TBZ)
	Concentrations/compound	C0=absence, C1 to C6:
	BM	Intervale=$250 - 1,500 \mu g/l$, $\Delta c = 250 \mu g/l$
	MBC	Intervale=$250 - 1,500 \mu g/l$, $\Delta c = 250 \mu g/l$
	FB	Intervale=$25 - 150 \mu g/l$, $\Delta c = 25 \mu g/l$
	TBZ	Intervale=$2.5 - 15 \mu g/l$, $\Delta c = 2.5 \mu g/l$
4 synchronous	S1: Mean/TBZ optimum	$\Delta\lambda 1 = 47 nm$, intervale=$200 - 400 nm$
	S2: Median	$\Delta\lambda 2 = 53 nm$, intervale=$200 - 400 nm$
	S3: MBC-BM optimum	$\Delta\lambda 3 = 59 nm$, intervale=$200 - 400 nm$
	S4: FB optimum	$\Delta\lambda 4 = 29 nm$, intervale=$200 - 400 nm$
3 excitation	S5: Mean	$\lambda_{em} 5 = 327 nm$, intervale=$200 - 315 nm$
	S6: Median	$\lambda_{em} 6 = 325 nm$, intervale=$200 - 315 nm$
	S7: FB optimum	$\lambda_{em} 7 = 341 nm$, intervale=$200 - 315 nm$
1 emission	S8: Mean/Median	$\lambda_{ex} 8 = 277 nm$, intervale=$300 - 400 nm$

processes of the RNAs were based on the test subset and following the reduction of the false negatives (FN).

The input in these systems acts as the feature vector of one single type of fluorescence spectra. The second ANN system is BPN-based with a data fusion scheme, Fig. 1(a). The input is a combination of several fluorescence spectra with different spectral features, that is, a multi-fluorescence spectra. We study three different data fusion schemes, one with conventional spectra (emission, excitation) (BPN_{5-8}), one with synchronous spectra (BPN_{1-4}) and the last one with all the spectra (BPN_{1-8}). The third ANN system is characterized by the neural ensemble approach. A neural network ensemble offers several advantages and benefits over a monolithic neural network: It can perform more complex tasks than any of its components. It can make an overall system easier to understand and modify. It is more robust than a monolithic neural network. It can produce a reduction of variance and increase confidence in the decision taken. Finally it can also show graceful performance degradation in situations where only a subset of neural networks in the ensemble are performing correctly [12].

Two strategies are needed to build an ensemble system: diversity strategy and combination strategy. Specifically we use BPN ensembles with n members, concretely some of the BPNs from the first proposed system with diverse character in the input space. The combination strategy used is the very straightforward Simple Majority Voting (SMV) as a collective decision strategy. This collective decision strategy allows us to group the individual ANNs outputs that makes up the ensemble in such a way that the correct decisions are amplified, and incorrect ones are eliminated. Furthermore, SMV offers the possibility of comparing the

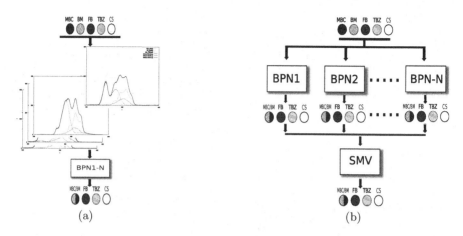

Fig. 1. BPN-based systems for detection of BFs.: (a) BPN Data fusion systemand (b) BPN ensemble system.

diversity based on the same classifier to highlight the significance of the different spectral characteristics. Another simple and effective combination strategy is Weighted Majority Voting (WMV), where the weight is calculated based on the test error. In our study, all the BPNs have a test error equal to zero, therefore the WMV is reduced to an SMW scheme.

We propose three types of BPN ensembles in our study, see Fig. 1(b), where the differences lie in the input space for each one similar to the approach with the proposed data fusion systems.

The simultaneous resolution of complex BF mixtures is a multi-label problem, thus any input pattern will be associated as belonging to as many classes as neurons having been fired in the output layer. This particular problem characteristic is extremely important in the design and implementation of the proposed systems. A new computational tool (MULLPY) has been designed and developed which can efficiently and easily construct the proposed systems. The tool is written in Python (v2.7), a language with high attributes in a scientific setting and proven performance in intensive computational processes despite being a scripting language. All of the architectures as well as the validation and visualization methods have been fully developed in this language using numpy and matplotlib libraries support. The optimization processes of the RNAs were also automated in MULLPY.

MULLPY is a software tool that can integrate different types of classifiers, comprising the learning phase and the presentation of different types of results. It also offers the possibility of generating independent-classifier ensembles. The increasing amount of auxiliary libraries in python and the ability to execute code in other languages, can serve as a channel for creating a unified environment for the study and development of different machine learning paradigms.

4   Results and Discussion

The evaluation of the overall efficiency of the systems used required a mixture error function, which we defined with respect to the class detected in any one mixture. This error function is shown by the equations in (1), where $NCND$ is the number of classes present in the undetected mixture by the system, false negatives; $NCBD$ is the number of classes detected by the system which were not present in the mixture, false positives (FP) and $NCIM$ is the total number of classes in the mixture.

$$E = E_{FP} + E_{FN} \; ; \; E_{FP} = \frac{NCBD}{NCIM} \; ; \; E_{FN} = \frac{NCND}{NCIM} \; . \tag{1}$$

Four analysis blocks were performed to evaluate the suitability of the ensembles: 1) decision fusion with synchronous spectra (SMV_{1-4}), 2) decision fusion with all conventional spectra (SMV_{5-8}), 3) decision fusion of synchronous, emission and excitation spectra (SMV_{1-8}), and 4) decision fusion of all the possible 8 module combinations, 247 different ensembles. Only those ensembles and modules with the lowest error were selected and displayed. Table 2 shows the validation of the results of these experiments in addition to all of the modules ($BPN_1,\ldots,BPN_8$).

Data fusion models with synchronous spectra (BPN_{1-4} and BPN_{1-8}) improved the (SMV_{1-4} and SMV_{1-8}) ensembles, respectively. When conventional spectra are used, SMV_{5-8} not only improves the BPN_{5-8} but is also the initial proposed system that produces lowest E_{FN}. The purpose of reducing FN is extremely

Table 2. Average mixture errors of identification values and their standard deviation between parentheses using the various different systems

Detectors	Mixture error		
	E	E_{FN}	E_{FP}
BPN_1	0.066 (0.139)	0.066 (0.139)	0.000 (0.000)
BPN_2	0.168 (0.202)	0.168 (0.202)	0.000 (0.000)
BPN_3	0.029 (0.099)	0.029 (0.099)	0.000 (0.000)
BPN_4	0.104 (0.185)	0.104 (0.185)	0.000 (0.000)
BPN_5	0.252 (0.239)	0.210 (0.201)	0.042 (0.138)
BPN_6	0.062 (0.158)	0.030 (0.101)	0.032 (0.122)
BPN_7	0.260 (0.297)	0.202 (0.206)	0.058 (0.171)
BPN_8	0.283 (0.250)	0.283 (0.250)	0.000 (0.000)
BPN_{1-4}	0.037 (0.104)	0.037 (0.104)	0.000 (0.000)
BPN_{5-8}	0.133 (0.195)	0.099 (0.165)	0.033 (0.125)
BPN_{1-8}	0.033 (0.100)	0.033 (0.100)	0.000 (0.000)
SMV_{1-4}	0.045 (0.118)	0.045 (0.118)	0.000 (0.000)
SMV_{5-8}	0.053 (0.142)	0.028 (0.098)	0.025 (0.109)
SMV_{1-8}	0.050 (0.126)	0.050 (0.126)	0.000 (0.000)
SMV_{1+3}	0.016 (0.070)	0.016 (0.070)	0.000 (0.000)

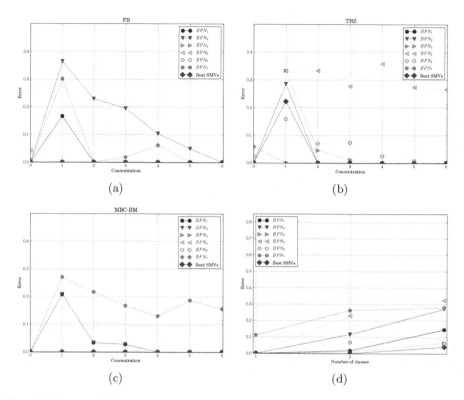

Fig. 2. Influence of the concentration of the analytes in the mixture over the average of the errors: (a) FB, (b) TBZ and (c) MBC-BM. (d) Influence of the number of classes from the mixture in the average of the errors of the mixtures.

important especially in problems concerning diagnosis and/or detection or identification of failures. In general, the use of synchronous spectra results in a better
mixture resolution, while the emission spectra is the only one which worsens.
Combining just the synchronous spectra ($\Delta\lambda 1$ and $\Delta\lambda 3$) leads to an acceptable
decision region for the SMV_{1+3} ensemble. This SMV is one of the best results in
solving the multi-label classification problem addressed in this study, together
with the $SMV_{1+2+3+5+6+7}$ and $SMV_{1+2+3+6}$ ensembles. The three ensembles
have the same validation error. This result demonstrates the importance of diversity among the ensemble modules to reach the right decision, despite the low
accuracy of each of them individually as seen in the two ensembles with a larger
number of modules. This higher number of modules in these ensembles against
SMV_{1+3}, can also help increase the reliability of decision making by the system.
The SMV_{1+3} ensemble has the lowest computational and experimental cost. In
this sense, it could be proposed as good solution for detection of BFs in complex
mixtures.

Disaggregating the error of this ensemble by classes and concentrations, it is
produced by TBZ class. Note in Fig. 2(b) that all errors occur at low concen-

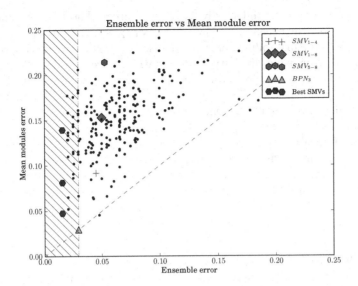

Fig. 3. Ensemble error vs mean module error for all possible ensemble combinations. The shadow region contains twenty-two ensembles that improve the best single neural architecture BPN_3.

trations of this class. Furthermore, this error is mainly produced in mixtures of three classes, Fig. 2(d), The one class mixtures were free of error. In addition, this model properly detected FB and MBC-BM in mixtures which included up to three classes.

Most of the generated combinations improved the results given by the average of its modules, as observed in Fig. 3, which confirms the advantage of using a diversity of information sources in decision making. Despite that, only twenty-two ensembles improve the best module BPN_3 , see region on Fig. 3. There is just one ensemble that does not contain the BPN_3 module but improves it, the $SMV_{1+2+5+6}$, with an error of 0.020. Moreover, the $SMV_{1+2+5+6}$ ensembles have the same FN error as the best combinations found.

5   Conclusions

We have presented a computational study on BF detection, based on supervised neural networks (BPN) with data and decision fusion schemes, which has characteristics similar to those found in a multi-label problem. All analyzed neural architectures were only trained with the fluorescence spectra of each BF. Optimization of figures of merit such as precision, sensitivity and limit of detection was also undertaken. A new software tool (MULLPY) written in the python language was designed and built to carry out the study.

BPNs with data fusion and BPN ensembles provide better performance of simple detector systems and reduce the risk of making a particularly poor selection. This result confirms the advantage of extracting complementary pieces of information from different and/or diverse data sources. The best obtained detector systems were the ensembles SMV_{1+3}, $SMV_{1+2+3+6}$ and $SMV_{1+2+3+5+6+7}$ which can reduce the instabilities of its modules. This fact shows the importance of generating and/or using diversity of information sources in decision making. The more appropriate best detector is SMV_{1+3}. It only requires synchronous spectra and presents satisfactory results in the detection of the BFs mixtures of up to four components or the absence of any pollutant. However it does have difficulties in mixtures with a very low TBZ concentration.

This study indicates that both, the supervised and the unsupervised neural architecture based methods can be appropriate for the simultaneous resolution of difficult multi-fungicide mixtures also according to the results given in [3]. It also aids in the modeling and construction of simple, fast, economic and smart computational tools for environmental monitoring. In conclusion this study contributes valuable results which address environmental and human health challenges.

Acknowledgements. We would like to thank the Environmental Chemical Analysis Group at the University of Las Palmas de Gran Canaria for providing the data set used in this paper. We also appreciate the comments made by the referees, which have improved the quality of this article.

References

1. Suarez Araujo, C.P., García Báez, P., Hernández Trujillo, Y.: 23. In: Carisse, O. (ed.) Neural Computation Methods in the Determination of Fungicides, Fungicides. Intech (2010)
2. CEE: Directive 1107/2009(91/414) (2009)
3. Suárez Araujo, C.P., García Báez, P., Sánchez Rodríguez, A., Santana Rodríguez, J.J.: Humann-based system to identify benzimidazole fungicides using multi-synchronous fluorescence spectra: An ensemble approach. Analytical and Bioanalytical Chemistry 394, 1059–1072 (2009)
4. Almhdi, K.M., Valigi, P., Gulbinas, V., Westphal, R., Reuter, R.: Classification with artificial neural networks and support vector machines: application to oil fluorescence spectra. EARSeL eProceedings 6, 115–129 (2007)
5. Vasilescu, J., Marmureanu, L., Carstea, E.: Analysis of seawater pollution using neural networks and channels relationship algorithms. Romanian Journal of Physics 56, 530–539 (2011)
6. Clarke, C.: Development of an automated identification system for nanocrystal encoded microspheres in flow cytometry. PhD thesis, Cranfield University (2008)
7. García Báez, P., Suárez Araujo, C.P., Sánchez Rodríguez, A., Santana Rodríguez, J.J.: Towards an efficient computational method for fluorescence identification of fungicides using data fusion and neural ensemble techniques. Luminescence 25, 285–287 (2010)
8. García Báez, P., Álvarez Romero, Y., Suárez Araujo, C.P.: A computational study on supervised and unsupervised neural architectures with data fusion for fluorescence detection of fungicides. Luminescence 27, 534–572 (2012)
9. Opitz, D.W., Shavlik, J.: Actively searching for an effective neural network ensemble. Connection Science 8, 337–353 (1996)

10. García Báez, P., Suárez Araujo, C., Fernández López, P.: A parametric study of humann in relation to the noise. appl. to the ident. of comp. of env. interest. Systems Analysis Modelling and Simulation 43(9), 1213–1228 (2003)
11. Werbos, P.: Beyond Regression: New Tools for Prediction and Analysis in the Behavioral Sciences. PhD thesis, Harvard University (1974)
12. Liu, Y., Yao, X., Higuchi, T.: Designing Neural Network Ensembles by Minimising Mutual Information. In: Mohammadian, M., Sarker, R.A., Yao, X. (eds.) Computational Intelligence in Control, pp. 1–21. Hershey: Idea Group Pub., USA & London (2003)

Study of Alternative Strategies to Selection of Peer in P2P Wireless Mesh Networks

Lissette Valdés[1], Alfonso Ariza[2], Sira M. Allende[1], Rubén Parada[2],
and Gonzalo Joya[1,2,⋆]

[1] Universidad de La Habana, Cuba
{lissette.valdes,sira}@matcom.uh.cu
[2] Universidad de Málaga, España
{aarizaq,gjoya}@uma.es, rubenparadamartin@gmail.com

Abstract. In this paper we study the use of various strategies for selecting the server node in a P2P network, especially oriented networks with
limited resources such as wireless mesh networks (WMN) based on WiFi
technology. Three strategies are examined: *Min-Hop*, using the path with
least number of hops, *Min-Hop-Fuzzy*, where in case of several paths with
equal length an additional criterion of selection based on fuzzy logic is
applied; and *Purely Fuzzy*, where the selection is made exclusively from
a fuzzy inference process. These strategies based on resource optimization criteria are an alternative to the currently used, which are based
on information sharing criteria. The study of performance was carried
out using a wireless network simulation tool for discrete event simulation
OMNeT + +. A comparison of the results for the different approaches
is presented.

Keywords: P2P Network, Fuzzy Inference, Min-hop.

1 Introduction

Applications P2P (Peer to Peer) are responsible for a significant percentage of
total traffic generated in the Internet in recent years. Such applications were
initially designed to share information among multiple users, and in them there
is no clear distinction between client and server nodes. Instead, each node can act
as both client and server (peer network)[1]. These networks are highly efficient
in information sharing, allowing quick dissemination of information avoiding
bottlenecks created on dedicated servers. To achieve this, data are divided into
segments, which are distributed in the network in order to maximize the number
of nodes acting as servers of the information sharing process . Figure 1 shows
the evolution of the data distribution in a P2P network in three time instants.
Initially only node A has the four segments that constitute the data and this
node serves a different segment for each of the nodes that request. In the second
moment, node B provides segment 1 and node C provides segment 2. At this

⋆ This work has been partially supported by the AECID, Projects PCI A2/038418/11.

I. Rojas, G. Joya, and J. Cabestany (Eds.): IWANN 2013, Part I, LNCS 7902, pp. 124–132, 2013.

moment node A proceeds to send the segments 3 and 4 to the nodes B and C, respectively as long as they exchange the segments 1 and 2. Finally, nodes B and C exchange segments 3 and 4 so that all nodes receive all the segments. In this process, node A has sent every segment only once.

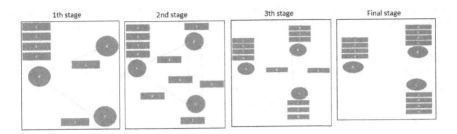

Fig. 1. An illustrative example of the evolution of data distribution in a P2P network

One of the limitations in internet P2P networks is that most of the nodes do not use a permanent IP address; therefore, this they should solve the problem of locating each node. A classification of P2P networks has been established, depending on how this problem is solved:

- *Centralized*, where all transactions are carried out based on a central server that stores and distributes information about the contents of the nodes.
- *Hybrid*, where there is a central server that manages some resources, but the nodes are responsible for maintaining the data.
- *Completely decentralized*, where there is no central server.

One aspect that has been repeatedly discussed in recent years is how to distribute the bandwidth among different users making requests in such networks, since this is a limited resource. Currently, the most commonly used criteria are based on sharing by other users, i.e. the most sharing user is attempted to be prioritized, relegating with low priority users that share less (leeches). This prioritization criterion may be suitable for wired networks where the bandwidth available for infrastructure is large. However, in wireless networks where bandwidth resources are very limited and shared among all users, it is inadvisable to establish criteria of priority based on sharing, since they can lead to a waste of resources. This issue is especially critical in multihop wireless networks, because if a distant client is been served, the information must cross a large number of nodes, increasing the probability of interfering with the transmissions of other nodes. Considering that the probability that an information transmission error occurs along a path P is given by equation 1:

$$B_p = 1 - \prod_{\forall i \in P} (1 - B_i) \qquad (1)$$

Where i is the i-th hop in path P and B_i is the probability of error in the i-th hop. It is obvious that increasing the size of the route increases the chance of losing information. Furthermore, each error B_i increases the total traffic in the network, i.e. the number of transmitting nodes, which in turn, grow while established routes grow.

With this in mind, we conclude that in wireless networks, the choice of a node server must reply to criteria that minimize the probability of error in the information provided, avoiding the need to retransmit damaged packages. These criteria should be imposed over the ones, currently applied, based on awarding the sharing.

There are several metrics that provide information about resources optimization and the warranty of the delivering of packages, which can be used as criteria for selecting the node server. Among them, the simplest is the number of hops. This metric, additive in nature, is used to find the shortest physical path. In homogeneous networks it is the most widely used due to its effectiveness in reducing the resources. However, this metric does not consider the differences in network load or if different hops have different bandwidths available. Other easily implemented metrics are Expected Transmission Count (ETX), which measures the guarantee of deliver successfully a package in one link, and Expected Transmission Time (ETT) which measures the time taken to transmit a package in a link considering the ETX value and binary speed (greater ETX values imply greater number of retransmissions, and therefore a higher value of ETT).

Certainly, the use of these metrics can introduce dynamic instabilities in the network so it is recommended to apply them with caution.

In this paper, we study the implementation of various strategies for choosing the node server using the above mentioned criteria. The first strategy is to minimize the number of hops (Min-Hop), currently the most widely used, especially in homogeneous networks. The second strategy, which we call Min-Hop-Fuzzy, uses as first criterion the number of hops, and in case of equality, applies a fuzzy inference system that takes as inputs ETX and number of nodes connected to the same Access Point (AP) of the destination node. The third strategy involves the application of a purely fuzzy system, wherein inputs are number of hops, ETX and number of neighbors. The first strategy is the simplest to implement, and is very efficient in wireless networks due to its conservative character in the use of resources. However, we believe that a fuzzy strategy can help the distribution of network load more consistently because it acts as a compromise between different factors that increase network traffic. Moreover, we think the fuzzy solution will be less vulnerable to changes in the network, i.e. the value of the metrics used. This advantage will become more visible as the network becomes more complex in structure (e.g., irregular structures, irregular background traffic, node mobility...)

The results presented are a preliminary stage, which has as its primary objective to validate the use of these new strategies against those already used, setting the base for further work in more complex networks. Basically, we present the fuzzy inference system applied in Section 2. In section 3, we describe the experimental environment. The tests carried out and the results obtained are presented in Section 4, and finally Section 5 covers the conclusions.

2   Description of Fuzzy Strategy for P2P Problem

A fuzzy inference system (FIS) is an intelligent system that uses fuzzy set theory to map inputs to outputs according to a set of inference rules, described by experts. Inputs of the system are the data, elements of a (crisp or fuzzy) set. Data and inference rules constitute the knowledge base of the system. For computing the output of the fuzzy inference system, it is necessary:

1. Fuzzyfication of the input.
2. Determining a set of fuzzy rules and adequacy of premises.
3. Fuzzy inference (finding the consequence of each rule).
4. Aggregation.
5. Defuzzification of the output.

In the following we describe the fuzzy system implemented for the application of fuzzy strategy.

2.1   Fuzzification of Input Variables

The input variables in our fuzzy system are:

- Number of hops
- ETX metric
- Number of neighbors

Numerical values of input variables are entered (crisp values), then Fuzzification process is performed and they become linguistic variables, i.e. characterized by the quadruple $\{X, T(X), U, G\}$ where,

X-Name of linguistic variable

$T(X)$-Set of terms (linguistic values) defined in X

U-Real physical domain where the values applied to the linguistic variable are defined

G-Semantic function that gives a "meaning" (interpretation) of the linguistic variable depending on the elements that X represents (membership function). Usually, the most used membership functions are triangular and trapezoidal.

So, for the inputs we have the quadruples:

$\{NHops, T(NHops), U_{NHops}, G_{NHops}\}$ $\{ETX, T(ETX), U_{ETX}, G_{ETX}\}$
    where: where:
        $NHops$ =Number of hops ETX =ETX metric
        $T(NHops) = \{$Low,High$\}$ $T(ETX) = \{$Low,Middle,High$\}$
        $U_{NHops} = [0, 10]$ $U_{ETX} = [1, \infty]$

$\{NNeighbors, T(NNeighbors), U_{NNeighbors}, G_{NNeighbors}\}$
    where:

NNeighbors =Number of Neighbors
$T(NNeighbors) = \{$Low,Middle,High$\}$
$U_{NNeighbors} = [0, 10]$

Semantic functions are for Number of Hops ($\mu_{\text{Low}} = [0, 0, 2, 5]$, $\mu_{\text{High}} = [2, 5, 10,$
$10]$); for ETX ($\mu_{\text{Low}} = [0, 0, 2, 4]$, $\mu_{\text{Middle}} = [2, 4, 6]$, $\mu_{\text{High}} = [4, 6, 50, 50]$) and for
Number of neighbors ($\mu_{\text{Low}} = [0, 0, 3, 5]$, $\mu_{\text{Middle}} = [3, 5, 7]$, $\mu_{\text{High}} = [5, 7, 10, 10]$).
Figure 2 shows the membership functions corresponding to input variables in
the fuzzy system.

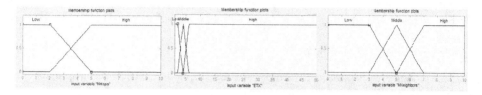

Fig. 2. Membership functions for Number of Hops, ETX metric and Number of Neighbours

3   Rules and Inference System

The output variable is the "Goodness index of server-client path" which is also
a lenguistic variable described as:

$\{GPath, T(GPath), U_{GPath}, G_{GPath}\}$ where
    $GPath =$Goodness index of server-client path
    $T(GPath) = \{$Low,Middle Low,Middle,Middle High,High$\}$
    $U_{GPath} = [0, 1]$

Semantic functions for output variable are $\mu_{\text{Low}} = [0, 0, 0.2, 0.3]$, $\mu_{\text{MidLow}} =$
$[0.2, 0.3, 0.4]$, $\mu_{\text{Middle}} = [0.3, 0.4, 0.6, 0.7]$, $\mu_{\text{MidHigh}} = [0.6, 0.7, 0.8]$ and $\mu_{\text{High}} =$
$[0.7, 0.8, 1, 1]$.

The membership function of output variable in the fuzzy system is shown in
figure 3. We introduced 14 rules based on network performance as determined
by an expert.

1. If NHops is (Low) AND ETX is (Low) AND NNeighbors is (Low) THEN GPath
   is (High)
2. If NHops is (Low) AND ETX is (Low) AND NNeighbors is (Middle) THEN GPath
   is (Middle High)
3. If NHops is (Low) AND ETX is (Low) AND NNeighbors is (High) THEN GPath
   is (Middle)
4. If NHops is (Low) AND ETX is (Middle) AND NNeighbors is (Low) THEN GPath
   is (Middle)
5. If NHops is (Low) AND ETX is (Middle) AND NNeighbors is (Not Low) THEN
   GPath is (Middle Low)
6. If NHops is (Low) AND ETX is (High) THEN GPath is (Low)
7. If NHops is (High)

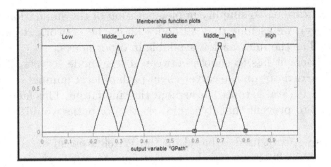

Fig. 3. Membership function for Goodness index of server-client path

Apply the fuzzy operator to the rules to obtain one number that represents the result of the antecedent for each rule. Specifically, in our system we apply:

AND : min operator
 OR : max operator

The fuzzy inference method used is the **Mamdani's method** [2], which is the most commonly used in applications due to its simple structure of 'min-max' operations. We use the minimum operator, based in the sentence $A \Rightarrow B \equiv A \wedge B$.

$$\mu_{A \Rightarrow B}(x, y) = \min(\mu_A(x), \mu_B(y)) \tag{2}$$

3.1   Defuzzification

In defuzzification, we switch the output variable from fuzzy to numerical value. There are several methods for defuzzifing the output variable such as the centroid method, maximum membership methods, center of gravity and others. In our analysis we apply the centroid method, whose formula is shown in equation 3, [3]. This is, perhaps, the most popular defuzzification method, which returns the center of the area under the curve.

$$y_c = \frac{\int y\mu(y)\mathrm{d}y}{\int \mu(y)\mathrm{d}y} \tag{3}$$

4   Experimental Enviroment

The simulation of different strategies has been carried out on a simple model of P2P network, which has been developed using the *framework inetmanet 2.0*, [4], and the simulation of IP networks on OMNET++ (a tool of simulation of a discrete event) [5]. Initially, this source provides a series of datum segments of predetermined size, which have to be distributed to all network nodes. Our simulations end when all nodes have all the segments. These segments may be further divided into smaller units so that they can be introduced into IP packets without fragmentation. To simplify implementation of the simulations (without loss of generality), we assume that all nodes know the network status in every moment, as well as the information available in other nodes.

When a node-client has to choose between several node servers, the analysis of each one is carried out on the server-client path of least number of hops, since this criterion force a stable route throughout the simulation. This factor has been explicitly forced to prevent that dynamic selection of paths would influence our study.

The wireless network used is based on the Wireless mesh extension present in the IEEE 802.11-2012, where the routing and forwarding mechanisms are implemented at the link level [6]. Simulation conditions are shown in table 4.

Table 1. Simulation conditions

Simulation area	1000 x 1000m
Nodes in backbone network	64
Maximum transmission distance	130m
Propagation model	Two ray
Separation between nodes in the backbone network	80m
Simulation time period	Limited to which all nodes have complete information
Interference model	Aditive
WIFI model	802.11g
Bit rate	54Mbit/s
Number of segments to be transmitted	10
Size of the segments	100000B
Maximum packet size	1000B
Number of repetitions with different seeds	5

Figure 4 shows the backbone network used in our simulation. In a simple P2P model, nodes have perfect knowledge of what information is present in each node. When a node has no information, it begins by itself the transfer request. To simplify the system, without loss of generality, nodes have a precise knowledge of the network status. In an actual implementation, OLSR packages themselves are those who propagate this information through the network.

5   Simulation and Results

We have simulated several P2P networks with different number of nodes. Specifically, networks with 15, 30, 60 and 90 nodes have been simulated. In all cases the number of nodes that initially had the required information was 4.

Figure 5 shows the graphics of the average download time and the maximum download time obtained with different numbers of P2P nodes and different strategies used. Each graphic shows the average total value and confidence intervals with a confidence level of 90%. We note that the higher number of nodes,

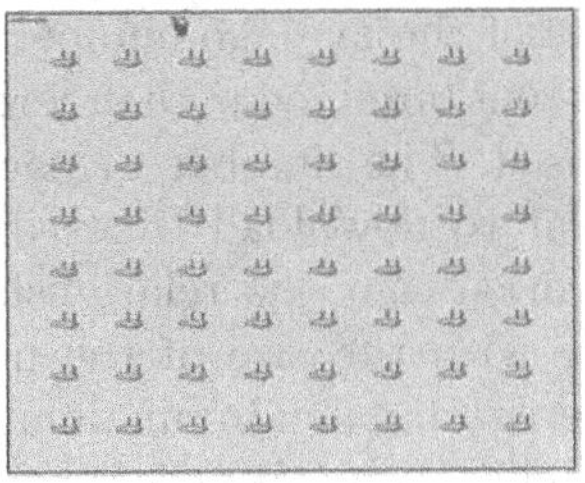

Fig. 4. Backbone network

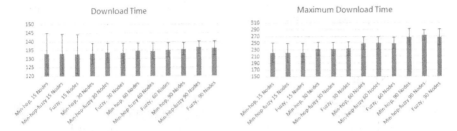

Fig. 5. Average and maximum download time

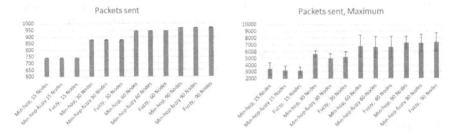

Fig. 6. Average and maximum number of packages sent

the greater the average time regardless of the strategy, due to the fact that we are simulating wireless networks. Also, we believe this time will grow depending on the number of nodes, since in such networks a larger number of nodes mean a larger traffic and interference. It can be seen that the increase in download time is not proportional with the increase in the number of nodes in network. This behavior is reasonable since it responds to the "Epidemic" character of P2P network. This conicidencia with the expected behavior provides a confidence factor for the proposed model. In general there are no relevant differences between the strategies. However, we note that the smallest deviation occurs for the strategy Min-Hop-Fuzzy, and the greater for the strategy Min-Hop. This leads us to believe that the application of fuzzy criteria produces a more balanced download time of different clients.

Figure 6 shows the graphics of the average number of packages sent and the maximum number of packages sent obtained with different numbers of P2P nodes and the different strategies used. The number of packages sent by each node has a growing behavior similar to download times discussed above. However, we note that, for larger networks, that the relationship of the increase in the number of packages per node with the size of the network is not significant. Although differences in the average and maximum number of packages sent aren't relevant for the different strategies, we can observe that their values are lower in the algorithms including a fuzzy module (Min-Hop-Fuzzy, Purely Fuzzy). Also, the deviation in the maximum number of packages is lower in the strategy Purely fuzzy, which means better distribution of the load and more homogeneous consumption in the network. The latter is important in battery-powered systems. This causes a greater distribution of the load on the node servers.

6   Conclusions

In this paper we have proposed three strategies of selection of the node server in P2P networks with limited resources. These are Min-Hop, Min-Hop-Fuzzy and purely fuzzy. We have shown that fuzzy criteria produces a more balanced download time and greater distribution of the load among the clients. We are at a preliminary stage of our work, since we have considered only static nodes, homogeneous P2P traffic and networks with regular structure. For future work, we are developing test scenarios more complex: nonuniform scenarios, obstacles, traffic interferences, etc. Also, we are exploring the use of different metrics for ETX parameter (additive and concave). We believe that concave metric can be effective in reducing the bottle necks and background traffic.

References

1. Buford, J.F., Lua, E.K.: P2P Networking and Applications. Elsevier Inc. (2008), http://www.sciencedirect.com/science/book/9780123742148
2. Lee, K.: First Course on Fuzzy Theory and Applications. Springer (2005) ISBN 3-540-22988-4
3. Martín, B., Sanz, A.: Redes Neuronales y Sistemas Borrosos. RA-MA Editorial (2001) ISBN 84-7897-466-0
4. inet-framework, https://github.com/inet-framework/inet
5. OMNeT++, http://www.omnetpp.org/
6. IEEE Standard for Information technology-Telecommunications and information exchange between systems Local and metropolitan area networks–Specific requirements Part 11: Wireless LAN Medium Access Control (MAC) and Physical Layer (PHY) Specifications. IEEE Std 802.11-2012
7. Calusen, T., Jacquet, P.: Optimized Link State Routing Protocol (OLSR), IETF RFC 3623 (2003)
8. De Couto, D.S.J.: High-Throughput Routing for Multi-Hop Wireless Networks, Ph D. Thesis. MIT (June 2004)

A Cloud-Based Neural Network Simulation Environment

Erich Schikuta and Erwin Mann

University of Vienna, Faculty of Computer Science,
A-1090 Währingerstr. 29, Vienna, Austria
erich.schikuta@univie.ac.at

Abstract. We present N2Sky, a novel Cloud-based neural network simulation environment. The system implements a transparent environment aiming to enable arbitrary and experienced users to do neural network simulations easily and comfortably. The necessary resources, as CPU-cycles, storage space, etc., are provided by using Cloud infrastructure. N2Sky also fosters the exchange of neural network specific knowledge, as neural network paradigms and objects, between users following a virtual organization design blue-print. N2Sky is built using the RAVO reference architecture which allows itself naturally integrating into the Cloud service stack (SaaS, PaaS, and IaaS) of service oriented architectures.

Keywords: Artificial Neural Network Simulation, Cloud computing, SOA/SOI, Virtual Organization.

1 Introduction

We are living in the era of virtual collaborations, where resources are logical and solutions are virtual. Advancements on conceptual and technological level have enhanced the way people communicate. The exchange of information and resources between researchers is one driving stimulus for development. This is just as valid for the neural information processing community as for any other research community. As described by the UK e-Science initiative [1] several goals can be reached by the usage of new stimulating techniques, such as enabling more effective and seamless collaboration of dispersed communities, both scientific and commercial, enabling large-scale applications and transparent access to "high-end" resources from the desktop, providing a uniform "look & feel" to a wide range of resources and location independence of computational resources as well as data.

A Virtual Organisation is a logical orchestration of globally dispersed resources to achieve common goals. It couples a wide variety of geographically distributed computational resources (such as PCs, workstations and supercomputers), storage systems, databases, libraries and special purpose scientific instruments to present them as a unified integrated resource that can be shared transparently by communities.

I. Rojas, G. Joya, and J. Cabestany (Eds.): IWANN 2013, Part I, LNCS 7902, pp. 133–143, 2013.

In the Computational Intelligence community these current developments are not used to the maximum possible extent until now. As an illustration for this we highlight the large number of neural network simulators that have been developed, as for instance SOM-PAK [2] and SNNS [3] to name only a few. Many scientists, scared of existing programs failing to provide an easy-to-use, comprehensive interface, develop systems for their specific neural network applications. This is also because most of these systems lack a generalized framework for handling data sets and neural networks homogeneously. This is why we believe that there is a need for a neural network simulation system that can be accessed from everywhere.

We see a solution to this problem in the N2Sky system. N2Sky is an artificial neural network simulation environment providing basic functions like creating, training and evaluating neural networks. The system is Cloud based in order to allow for a growing virtual user community. The simulator interacts with Cloud data resources (i.e. databases) to store and retrieve all relevant data about the static and dynamic components of neural network objects and with Cloud computing resources to harness free processing cycles for the "power-hungry" neural network simulations. Furthermore, the system allows to be extended by additional neural network paradigms provided by arbitrary users.

The layout of the paper is as follows: In the following section we give the motivation behind the work done. In section 3 we present the design principles behind the N2Sky development. The system deployment within a Cloud environment is described in section 4. The interface of N2Sky is laid out in section 5. The paper closes with a look at future developments and research directions in Section 5.

2   Towards a Cloud-Based ANN Simulator

In the last years the authors developed several neural network simulation systems fostering up-to-date computer science paradigms then.

NeuroWeb [4] is a simulator for neural networks which exploits Internet-based networks as a transparent layer to exchange information (neural network objects, neural network paradigms). NeuroAccess [5] and NeuroOracle [6] identify neuronal network elements as complex objects in the sense of database technology and integrate them conceptually and physically into the object-relational model. This approach supports an object-oriented point of view which enables a natural mapping of neural network objects and their methods to the service-oriented landscape. The N2Cloud system [7] is based on a service oriented architecture (SOA) and is a further evolution step of the N2Grid systems [8]. The original idea behind the N2Grid system was to consider all components of an artificial neural network as data objects that can be serialized and stored at some data site in the Grid, whereas N2Cloud will use the storage services provided by the Cloud environment.

The presented N2Sky environment takes up the technology of N2Cloud to a new dimension using the virtual organisation paradigm. Hereby the RAVO reference architecture is used to allow the easy integration of N2Sky into the Cloud service stack using SaaS, PaaS, and IaaS. Cloud computing is a large scale distributed computing paradigm for utility computing based on virtualized, dynamically scalable pool of resources and services that can be delivered on-demand over the Internet. In the scientific community it is sometimes stated as the neural evolution of Grid computing, which lacks on usability and accountability. Cloud computing therefore became a buzz word after IBM and Google collaborated in this field followed by IBM's "Blue Cloud" [9] launch. Three categories can be identified in the field of Cloud computing:

- **Software as a Service (SaaS):** This type of Cloud delivers configurable software applications offered by third party providers on an on-demand base and made available to geographically distributed users via the Internet. Examples are Salesforce.com, CRM, Google Docs, and so on.
- **Platform as a Service (PaaS):** Acts as a runtime-system and application framework that presents itself as an execution environment and computing platform. It is accessible over the Internet with the sole purpose of acting as a host for application software. This paradigm offers customers to develop new applications by using the available development tools and API's. Examples are Google's App engine and Microsoft's Azure, and so on.
- **Infrastructure as a Service (IaaS):** Traditional computing resources such as servers, storage, and other forms of low level network and physical hardware resources are hereby offered in a virtual, on-demand fashion over the Internet. It provides the ability to provide on-demand resources in specific configurations. Examples include Amazon's EC2 and S3, and so on.

The motivation behind the development of N2Sky is to

- share neural net paradigms, neural net objects and other data and information between researchers, developers and end users worldwide. Provide for an efficient and standardized solution to neural network problems,
- allow for transparent access to "high-end" neural resources stored within the Cloud from desktop or smart phone,
- provide a uniform "look and feel" to neural network resources, and
- foster location independence of computational, storage and network resources.

3 N2Sky Design

Information Technology (IT) has become an essential part of our daily life. Utilization of electronic platforms to solve logical and physical problems is extensive. Grid computing is often related with Virtual Organisations (VOs) when it comes to creation of an E-collaboration. The layered architecture for grid computing has remained ideal for VOs.

However, grid computing paradigm has some limitations. Existing grid environments are categorized as data grid or computational grid. Today, problems being solved using VOs require both data and storage resources simultaneously. Scalability and dynamic nature of the problem solving environment is another serious concern. Grid computing environments are not very flexible to allow the participant entities enter and leave the trust. Cloud computing seems to be a promising solution to these issues. Only, demand driven, scalable and dynamic problem solving environments are target of this newborn approach. Cloud computing is not a deviation concept from the existing technological paradigms, rather it is an upgradation. Cloud computing centers around the concept of XaaS, ranging from hardware/software, infrastructure, platform, applications and even humans are configured as a service. Most popular service types are IaaS, PaaS and SaaS.

Existing paradigms and technology is used to form VOs, but lack of standards remained a critical issue for the last two decades. Our research endeavor focused on developing a Reference Architecture for Virtual Organizations (RAVO) [10]. It is intended as a standard for building Virtual Organizations (VO). It gives a starting point for the developers, organizations and individuals to collaborate electronically for achieving common goals in one or more domains. RAVO consists of two parts,

1. The requirement analysis phase, where boundaries of the VO are defined and components are identified. A gap analysis is also performed in case of evolution (up-gradation) of an existing system to a VO.
2. The blueprint for a layered architecture, which defines mandatory and optional components of the VO.

This approach allows to foster new technologies (specifically the SOA/SOI paradigm realized by Clouds) and the extensibility and changeability of the VO to be developed.

The basic categorization of the N2Sky design depends on the three layers of the Cloud service stack as they are: Infrastructure as a Service (IaaS), Platform as a Service (PaaS) and Software as a Service (SaaS). Figure 1 depicts the components of the N2Sky framework, where white components are mandatory, and the other components are optional.

Infrastructure as a Service (IaaS) basically provides enhanced virtualisation capabilities. Accordingly, different resources may be provided via a service interface. In N2Sky the IaaS layer consists of two sub-layers: a Factory layer and an Infrastructure Enabler Layer. Users need administrative rights for accessing the resources in layer 0 over the resource management services in layer 1.

- Factory Layer (Layer 0): contains physical and logical resources for the N2Sky. Physical resources comprise of hardware devices for storage, computation cycles and network traffic in a distributed manner. Logical resources contain experts knowledge helping solving special problems like the Paradigm Matching.
- Infrastructure Enabler Layer (Layer 1): allows access to the resources provided by the Factory layer. It consists of protocols, procedures and methods to manage the desired resources.

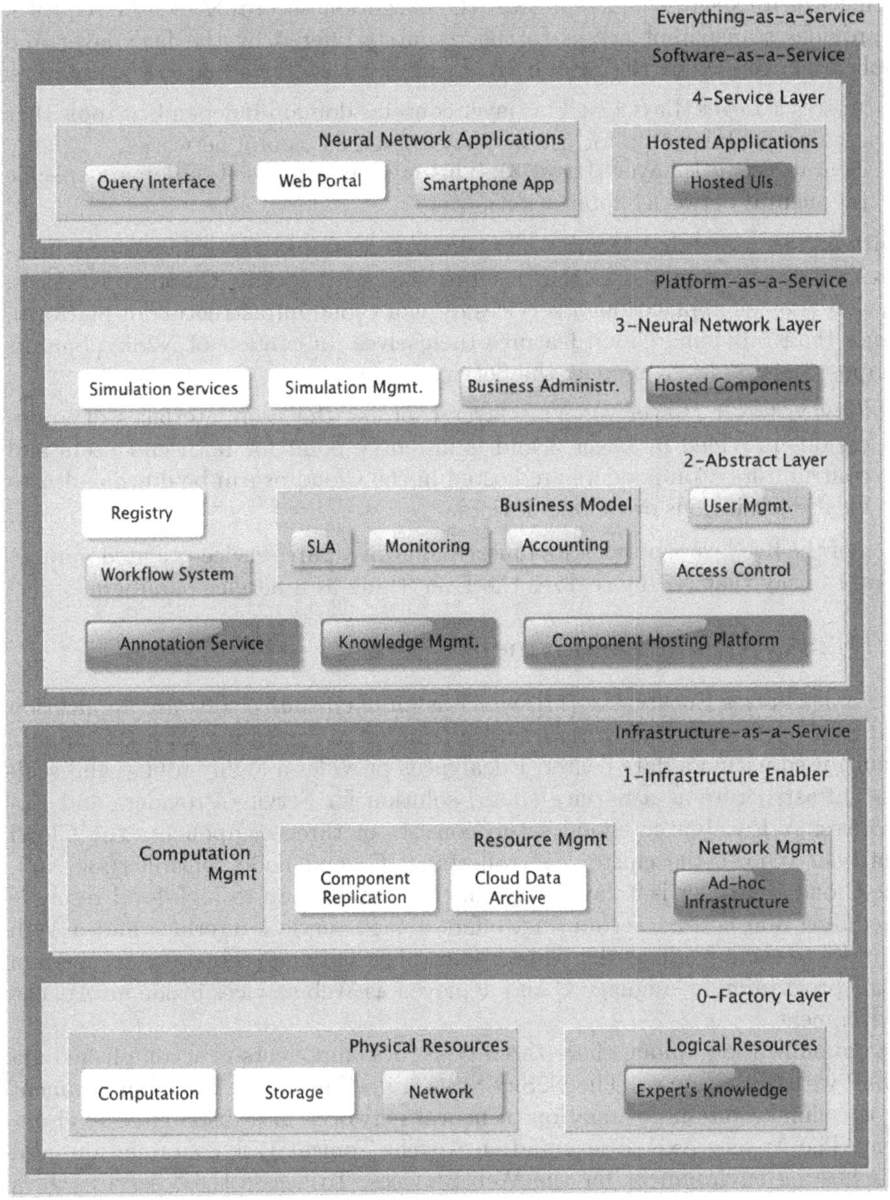

Fig. 1. N2Sky design based on RAVO

Platform as a Service (PaaS) provides computational resources via a platform upon which applications and services can be developed and hosted. PaaS typically makes use of dedicated APIs to control the behaviour of a server hosting engine which executes and replicates the execution according to user requests. It provides transparent access to the resources offered by the IaaS layer and applications offered by the SaaS layer. In N2Sky it is divided into two sublayers:

- Abstract Layer (Layer 2): This layer contains domain-independent tools that are designed not only for use in connection with neural networks.
- Neural Network Layer (Layer 3): This layer is composed of domain-specific (i.e. neural network) applications.

Software as a Service (SaaS) offers implementations of specific business functions and business processes that are provided with specific Cloud capabilities, i.e. they provide applications / services using a Cloud infrastructure or platform, rather than providing Cloud features themselves. In context of N2Sky, SaaS is composed of one layer, namely the Service Layer.

- Service Layer (Layer 4): This layer contains the user interfaces of applications provided in Layer 3 and is an entry point for both end users and contributors. Components are hosted in the Cloud or can be downloaded to local workstations or mobile devices.

Each of the five layers provide its functionality in a pure service-oriented manner so we can say that N2Sky realizes the Everything-as-a-Service paradigm.

4 N2Sky Cloud Deployment

N2Sky facilitates Eucalyptus [11], which is an open source software application that implements a Cloud infrastructure (similar to Amazon's Elastic Compute Cloud) used within a data center. Eucalyptus provides a highly robust and scalable Infrastructure as a Service (IaaS) solution for Service Providers and Enterprises. A Eucalyptus Cloud setup consists of three components the Cloud controller (CLC), the cluster controller(s) (CC) and node controller(s) (NC). The Cloud controller is a Java program that, in addition to high-level resource scheduling and system accounting, offers a Web services interface and a Web interface to the outside world. Cluster controller and node controller are written in the programming language C and deployed as Web services inside an Apache environment.

Communication among these three types of components is accomplished via SOAP with WS-Security. The N2Sky System itself is a Java-based environment for the simulation and evaluation of neural networks in a distributed environment. The Apache Axis library and an Apache Tomcat Web container are used as a hosting environment for the Web Services. To access these services Java Servlets/JSPs have been employed as the Web frontend.

N2Sky system can be deployed on various configurations of the underlying infrastructure. It is even possible to use a federated Cloud approach, by fostering the specific capabilities (affinities) of different Cloud providers (e.g. data Clouds, compute Clouds, etc.). A possible specific deployment is show in Figure 2.

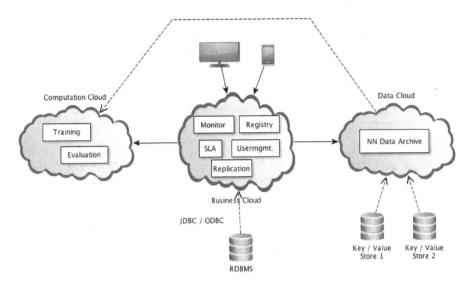

Fig. 2. N2Sky Cloud Deployment

5   N2Sky Interface

The whole system architecture and its components are depicted in Figure 3.

A neural network has to be configured or trained (supervised or unsupervised) so that it may be able to adjust its weights in such a way that the application of a set of inputs produces the desired set of outputs. By using a particular paradigm selected by the user the *N2Sky Simulation Service* allows basically three tasks: **train** (the training of an untrained neural network), **retrain** (training of a previously trained network again in order to increase the training accuracy), **evaluate** (evaluating an already trained network). The *N2Sky Data Archive* is responsible to provide access to data of different objects (respectively paradigms) of neural networks by archiving or retrieving them from a database storage service. It can also publish evaluation data. It provides the method *put* (inserts data into a data source) and *get* (retrieves data from a data source) The main objective of the *N2Sky Database Service* is to facilitate users to benefit from already trained neural networks to solve their problems. So this service archives all the available neural network objects, their instances, or input/output data related to a particular neural network paradigm. This service dynamically updates the database as the user gives new input/output patterns, defines a new paradigm or evaluates the neural network. The *N2Sky Service Monitor* keeps tracks of the available services, publishes these services to the whole system. Initially user interact with it by selecting already published paradigms like Back Propagation, Quick Propagation, Jordan etc. or submit jobs by defining own parameters. This module takes advantage of virtualization and provides a transparent way for the user to interact with the simulation services of the system. It also allows to implement business models by an accounting functionality and restricting access to specific

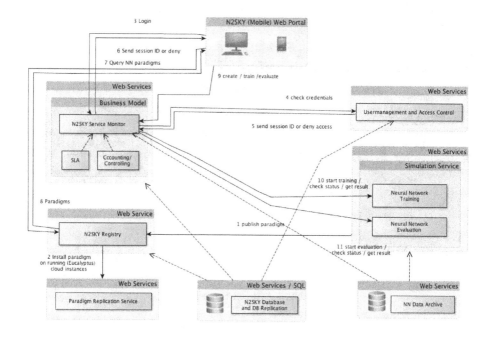

Fig. 3. N2Sky Architecture and Components

paradigms. The *N2Sky Paradigm/Replication Service* contains the paradigm implementation that can be seen as the business logic of a neural network service implementation. The *N2Sky Registry* administrates the stored neural network paradigms. The main purpose of N2Sky system is to provide neural network data and objects to users. Thus the *N2Sky Java Application/Applet* provides a graphical user interface (GUI) to the user. It especially supports experienced users to easily run their simulations by accessing data related neural network objects that has been published by the N2Sky service manager and the N2Sky data service. Moreover the applet provides a facility to end-users to solve their problems by using predefined objects and paradigms. For the purpose of thin clients a simple Web browser, which can execute on a PC or a smart phone, can be used to access the front-end, the *N2Sky (Mobile) Web Portal*. It is relying on the *N2Sky User management Service* which grants access to the system.

The user can choose to work with the N2Sky system via a PC or a smart phone (e.g. an iPhone).

The N2Sky interface provides screen for the classical neural network tasks:

- **Subscription:** Choosing published existing neural network paradigms and instantiating new neural networks based on this paradigm.
- **Training:** Specifying training parameters, starting training and monitoring the training process.
- **Evaluation:** Using trained neural networks for problem solution.

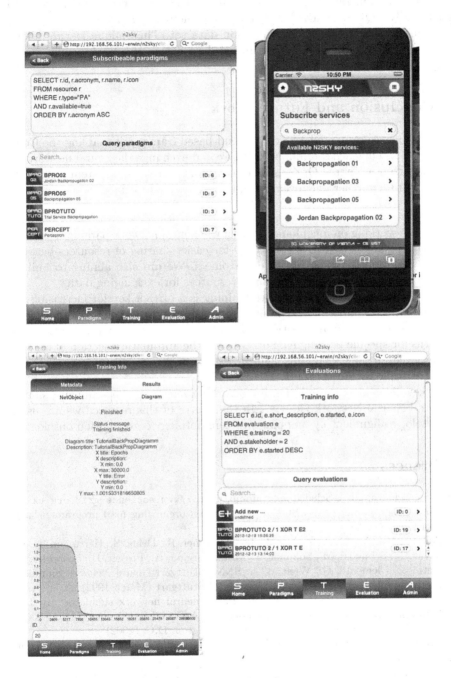

Fig. 4. N2Sky Interface

A specific highlight of the N2Sky system is the use of the standardized and user-friendly SQL language for searching for network paradigms and objects and defining the training and evaluation data set. This unique feature allows for combining globally stored, distributed data within the N2Sky environment easily.

6   Conclusion and Future Work

In this paper we presented N2Sky, a Cloud-based framework enabling the Computational Intelligence community to share and exchange the neural network resources within a Virtual Organisation. N2Sky is a prototype system with quite some room for further enhancement. Ongoing research is done in the following areas:

- We are working on an enhancement of the neural network paradigm description language ViNNSL [12] to allow for easier sharing of resources between the paradigm provider and the customers. We are also aiming to build a generalized semantic description of resources for exchanging data.
- Parallelization of neural network training is a further key for increasing the overall performance. Based on our research on neural network parallelization [13] we envision an automatically definition and usage of parallelization patterns for specific paradigms. Furthermore the automatic selection of capable resources in the Cloud for execution, e.g. multi-core or cluster systems is also a hot topic within this area.
- A key element is to find neural network solvers for given problems, similar to a "Neural Network Google". In the course of this research we are using ontology alignment by mapping problem ontology onto solution ontology.

References

1. UK e-Science: UK e-Science programme, http://www.escience-grid.org.uk
2. SOM Programming Team SOM-PAK: The self-organizing map program pakage, user guide (1992)
3. Zell, A., Mamier, G., Vogt, M., Mache, N., Hbner, R., Dring, S., Herrmann, K.-U., Soyez, T., Schmalzl, M., Sommer, T., Hatzigeorgiou, A., Posselt, D., Schreiner, T., Kett, B., Clemente, G., Wieland, J.: SNNS Stuttgart Neural Network Simulator user manual. Technical report, University of Stuttgart (March 1992)
4. Schikuta, E.: NeuroWeb: an Internet-Based neural network simulator. In: 14th IEEE International Conference on Tools with Artificial Intelligence (ICTAI 2002), pp. 407–414. IEEE Computer Society, Washington, D.C. (2002)
5. Brunner, C., Schulte, C.: NeuroAccess: The Neural Network Data Base System. Master's thasis, University of Vienna, Vienna, Austria (1998)
6. Schikuta, E., Glantschnig, P.: NeuroOracle: Integration of Neural Networks into an Object-Relational Database System. In: Liu, D., Fei, S., Hou, Z., Zhang, H., Sun, C. (eds.) ISNN 2007, Part II. LNCS, vol. 4492, pp. 1115–1124. Springer, Heidelberg (2007)

7. Huqqani, A.A., Xin, L., Beran, P.P., Schikuta, E.: N2Cloud: Cloud based Neural Network Simulation Application. In: The 2010 International Joint Conference on Neural Networks (IJCNN), pp. 1–5 (July 2010)
8. Schikuta, E., Weishäupl, T.: N2Grid: Neural Networks in the Grid. In: Proceedings of the 2004 IEEE International Joint Conference on Neural Networks, pp. 1409–1414 (July 2004)
9. IBM: IBM Blue Cloud, http://www-03.ibm.com/press/us/en/photo/22615.wss (November 2007) (last accessed January 20, 2010)
10. Khalil, W.: Reference Architecture for Virtual Organization. Ph.d. thesis, Faculty of Computer Science, University of Vienna, Austria (2012)
11. Nurmi, D., Wolski, R., Grzegorczyk, C., Obertelli, G., Soman, S., Youseff, L., Zagorodnov, D.: The eucalyptus open-source cloud-computing system. In: 9th IEEE/ACM International Symposium on Cluster Computing and the Grid, CC-GRID 2009, pp. 124–131 (2009)
12. Beran, P.P., Vinek, E., Schikuta, E., Weishäupl, T.: ViNNSL - the Vienna Neural Network Specification Language. In: Proceedings of the International Joint Conference on Neural Networks, IJCNN 2008, part of the IEEE World Congress on Computational Intelligence, WCCI 2008, pp. 1872–1879 (2008)
13. Weishäupl, T., Schikuta, E.: Cellular Neural Network Parallelization Rules. In: CNNA 2004: Proceedings of the 8th IEEE International Biannual Workshop on Cellular Neural Networks and their Applications. IEEE Computer Society, Los Alamitos (2004)

Performance Evaluation over Indoor Channels of an Unsupervised Decision-Aided Method for OSTBC Systems

Paula M. Castro, Ismael Rozas-Ramallal,
José A. García-Naya, and Adriana Dapena

Departament of Electronics and Systems
University of A Coruña
Campus de Elviña s/n, 15071, A Coruña, Spain
{pcastro,ismael.rozas,jagarcia,adriana}@udc.es

Abstract. Unsupervised algorithms can be used in digital communications to estimate the channel at the receiver without using pilot symbols, thus obtaining a considerable improvement in terms of data rate, spectral efficiency, and energy consumption. Unfortunately, the computational load is considerably high since they require to estimate Higher Order Statistics. For addressing this issue, it has been recently presented a decision-aided channel estimation strategy, which implemented a decision rule to determine if a new channel estimate was required or not. If channel estimation is not needed, a previous estimate was used to recover the transmitted signals. Based on this idea, we propose a lower-complexity decision criterion and we evaluate its performance over real-world indoor channels measured using a hardware platform working at the Industrial, Scientific and Medical band at 5 GHz.

1 Introduction

In 1998, S. M. Alamouti proposed a popular Orthogonal Space Time Block Coding (OSTBC) scheme for transmitting in systems with two antennas at the transmitter and a single one at the receiver [1]. This code provides spatial and temporal diversity, while the decoding scheme is very simple because the Maximum Likelihood (ML) criterion is reduced to a matrix-matched filter followed by a symbol-by-symbol detector. Due to such advantages, the Alamouti code has been incorporated in some of the latest wireless communication standards, like IEEE 802.11n [2] or IEEE 802.16 [3].

The decoding procedure performed in the Alamouti scheme requires to estimate a 2×2 channel matrix. For this purpose, current standards define the inclusion of pilot symbols in the data frame. Pilot symbols are used by supervised algorithms to estimate the channel matrix with good precision, but as it is well-known, they reduce the maximum achievable data rate and the spectral efficiency. An alternative is to use unsupervised (also called blind) algorithms to decrease the overhead associated with pilot transmission [4]. Most unsupervised

I. Rojas, G. Joya, and J. Cabestany (Eds.): IWANN 2013, Part I, LNCS 7902, pp. 144–151, 2013.
© Springer-Verlag Berlin Heidelberg 2013

algorithms exploit the statistical independence of the transmitted signals and require to estimate Higher Order Statistics (HOS). For this reason, the computational load of unsupervised decoders is considerably higher than that exhibited by the supervised ones.

In order to reduce the computational load of decoding, different strategies to estimate channel variations have been proposed in [5,6]. The method presented in [6] uses the preambles –included in current digital communication standards– to obtain a coarse channel estimation. Such an estimation is used to decide if the channel has suffered a considerable variation that requires to re-estimate its coefficients. In other case, a previously-estimated channel matrix is used to recover the transmitted signals. The evaluation performed with simulated channels shows a good performance for such a scheme.

In this paper, we present a decision-aided decoding scheme similar to that proposed in [6] but with a different preamble structure. The main contribution of this paper is to present a performance evaluation in realistic scenarios using a hardware testbed. This testbed, configured as a Multiple Input Single Output (MISO) 2×1 system, operates in the Industrial, Scientific and Medical (ISM) band at 5 GHz. Channel coefficients have been acquired in real transmissions performed in indoor scenarios and they are later on plugged in a multilayer software, specifically designed to ease the evaluation of channel estimation algorithms.

This paper is organized as follows. Section 2 presents the Alamouti scheme. Section 3 explains the decision-aided method proposed in this work. Section 4 describes the testbed used for experimental evaluation and also shows the results obtained from the testbed. Finally, Section 5 is devoted to the conclusions.

2   Alamouti Coded Systems

Figure 1 shows the baseband representation of an Alamouti-based system with two antennas at the transmitter and one antenna at the receiver. A digital source in the form of binary data stream, b_i, is mapped into symbols which are split into two substreams, s_1 and s_2. We assume that s_1 and s_2 are independent equiprobable discrete random variables that take values from a finite set of symbols belonging to a real or complex modulation (PAM, PSK, QAM...). The path from the first transmit to the receive antenna is denoted by h_1 and the path from the second transmit antenna to the receive antenna is denoted by h_2. The received signals are given by

$$\texttt{Even instants} \quad r_1 = h_1 s_1 + h_2 s_2 + v_1, \quad (1)$$

$$\texttt{Odd instants} \quad r_2 = h_2 s_1^* - h_1 s_2^* + v_2. \quad (2)$$

The observations are obtained using $x_1 = r_1$ and $x_2 = r_2^*$.

The vector $\mathbf{x} = [x_1 \ x_2]^T = [r_1 \ r_2^*]^T$ of the received signals (observations) can be written as $\mathbf{x} = \mathbf{H}\mathbf{s} + \mathbf{v}$, where $\mathbf{s} = [s_1 \ s_2]^T$ is the source vector, $\mathbf{v} = [v_1 \ v_2]^T$ is the additive white Gaussian noise vector, and the 2×2 channel matrix has the form

$$\mathbf{H} = \begin{bmatrix} h_1 & h_2 \\ h_2^* & -h_1^* \end{bmatrix} \quad (3)$$

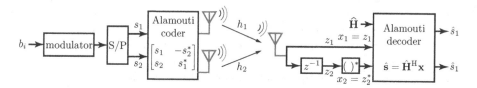

Fig. 1. Alamouti coding scheme

The matrix $\mathbf{H}$ is unitary up to a scalar factor, i.e., $\mathbf{H}\mathbf{H}^{\mathrm{H}} = \mathbf{H}^{\mathrm{H}}\mathbf{H} = \|\mathbf{h}\|^2 \mathbf{I}_2$, where $\|\mathbf{h}\|^2 = |h_1|^2 + |h_2|^2$ is the squared Euclidean norm of the channel vector, $\mathbf{I}_2$ is the 2×2 identity matrix, and $(\cdot)^{\mathrm{H}}$ is the Hermitian operator. It follows that the transmitted symbols can be recovered applying $\hat{\mathbf{s}} = \hat{\mathbf{H}}^{\mathrm{H}}\mathbf{x}$, where $\hat{\mathbf{H}}$ is a suitable channel matrix estimate. As a result, this scheme supports maximum likelihood detection based only on linear processing at the receiver.

3 Decision-Aided Scheme

In static environments, it is common to assume that the channel remains constant during the transmission of several frames (block fading). On the contrary, in mobile environments, channel variations happen faster (for each frame or even within the transmission of a frame). In order to reduce the computational complexity of decoder, in [6] it has been proposed a simple method to detect channel variations from the preambles transmitted before each data frame. The decoder estimates the channel coefficients only when the decision criterion detects a channel variation. In this work we propose a decision-aided scheme similar to that presented in [6], but using a simpler preamble scheme.

We denote by p_1 and p_2 the orthogonal preambles transmitted by each antenna. Unlike the scheme presented in [6], these preambles are not coded with Alamouti. From Equation (1), the receive signal has the form

$$r = h_1\, p_1 + h_2\, p_2 + v. \tag{4}$$

Multiplying each sample of this signal by the preamble samples and summing up over the preamble length P, we obtain

$$c_1[k] = \sum_{n=1}^{P} r[n]p_1[n]^* = h_1 \sum_{n=1}^{P} |p_1[n]|^2 + \sum_{n=1}^{P} v[n]p_1^*[n], \tag{5}$$

$$c_2[k] = \sum_{n=1}^{P} r[n]p_2[n]^* = h_2 \sum_{n=1}^{P} |p_2[n]|^2 + \sum_{n=1}^{P} v[n]p_2^*[n]. \tag{6}$$

Considering that the preamble length is large enough to eliminate the term corresponding to the noise, we have that each result obtained from such a "correlation" is a coarse estimate of each one of the channel coefficients. Comparing the values c_1 and c_2 to a threshold value, the decoder can determine if it is needed to re-estimate the channel matrix or not.

The proposed decision-aided scheme can be summarized as follows:

1: Compute $c_1[k]$ and $c_2[k]$ from the preambles transmitted for the k-th frame.
2: Compute the error
   $\text{Error}_1[k] = |c_1[k] - c_1[k-1]|$ and $\text{Error}_2[k] = |c_2[k] - c_2[k-1]|$.
3: Use the decision criterion

$$(\text{Error}_1[k] > \beta) \text{ OR } (\text{Error}_2[k] > \beta) \rightarrow \text{Channel estimate is required,}$$

where β is a real-valued threshold.

In order to avoid the transmission of pilot symbols, we propose to estimate the channel matrix using an unsupervised algorithm like the Joint Approximate Diagonalization of Eigen-matrices (JADE) algorithm or the Blind Channel Estimation based on Eigenvalue Spread (BCEES) method proposed in [9]. BCEES is a simplification of JADE [7], where the matrix to be diagonalized is selected taking into account the absolute difference between the eigenvalues (eigenvalue spread).

4   Performance Evaluation Based on Measured Indoor Channels

A testbed developed at the University of A Coruña [10] (see Figure 2) was used to extract 2×1 channel matrices corresponding to a realistic indoor scenario in which the transmitter and the receiver were separated approximately 9 m, whereas the antenna spacing at the transmitter was set to 7 cm. In this section we describe the measurement procedure followed to obtain the indoor wireless channel coefficients that are later on plugged in the simulations in order to evaluate the performance of the proposed approaches under real-world indoor channels.

4.1   Measurement Procedure

The testbed is used to estimate the 2×1 MISO channel. For that purpose, we design a frame structure consisting of a preamble sequence (119 symbols) for time and frequency synchronization; a silence (50 symbols) for estimating at the receiver the noise variance; and a long training sequence (4 000 symbols per transmit antenna) for estimating the channel. Note that the preamble length is considerably higher than the preamble introduced in Section 3. The resulting signals are modulated (single carrier) and pulse-shape filtered using a squared root-raised cosine filter with 12 % roll-off, and the resulting signal bandwidth is 1.12 MHz, which leads –according to our tests– to a frequency-flat channel response.

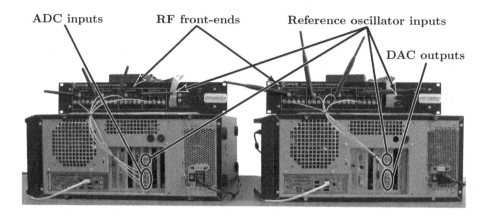

Fig. 2. Picture of the testbed

With the aim of obtaining statistically rich channel realizations, and given that the Lyrtech RF front-end is frequency-agile, we measure at different RF carriers (frequency hopping) in the frequency interval ranging from 5 219 MHz to 5 253 MHz and from 5 483 MHz to 5 703 MHz. Carrier spacing is 4 MHz (greater than the signal bandwidth), which results in 65 different frequencies. Additionally, we repeat the whole measurement procedure for four different positions of the transmitter, giving as a result 260 channel realizations. Note that we have these 260 realizations per each pair of transmit antennas for a given receiver position and therefore, taking into account four receiver locations, a maximum number of 1 040 channel realizations is available for the Alamouti system.

In order to be able to plug the estimated channel coefficients in a simulation, all of them from each of the four sets of 65 channel matrices are normalized, giving as a result unit mean variance channels, but preserving the same statistical distribution as the original channel matrices.

4.2 Experimental Results

A thousand channel realizations have been used to evaluate the performance of the proposed decision-aided scheme. The experiments have been performed using QPSK source symbols coded with the Alamouti scheme. A total of 20 frames consisting of 200 symbols per transmit antenna, i.e. 8 000 QPSK coded symbols (4 000 source symbols), are transmitted in 20 frames. The channel matrix remains constant during the transmission of 5 frames; hence the 20 frames experience 4 different channel realizations.

First, we consider the problem that arises from selecting the threshold value used for the decision criterion. In order to obtain a good estimate of the cross–correlations, the simulations performed in these tests contained a preamble with 100 symbols per antenna. To quantify the difference –in terms of Symbol Error Rate (SER) versus Signal-to-Noise Ratio (SNR)– between the BCEES and the Decision-Aided BCEES scheme (DA-BCEES), the following expression is introduced

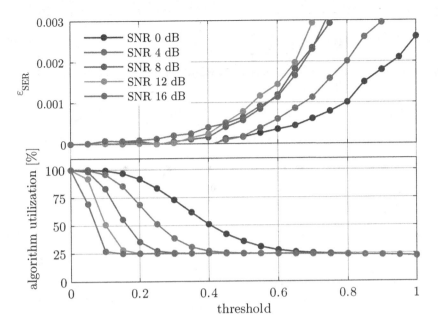

Fig. 3. SER and algorithm utilization for the DA-BCEES approach versus threshold β given several SNR values for measured channels

$$\epsilon_{SER} = \frac{SER_{DA\text{-}BCEES} - SER_{BCEES}}{1 + SER_{DA\text{-}BCEES}}. \qquad (7)$$

Figure 3 plots ϵ_{SER} as well as the percentage of algorithm utilization, defined as the number of frames in which the channel was estimated divided by the total number of frames. We can see that a value of $\beta = 0.6$ gives a good tradeoff between SER and channel estimation, since SER is almost zero and the channel estimation is equal to 25 %, which corresponds to estimate the channel only 5 times per 20 transmitted frames (for the first frame and for each channel variation), which corresponds to the optimum value.

Figure 4 shows the SER and the algorithm utilization percentage for the unsupervised algorithms (JADE and BCEES) when the channel is estimated for all the frames (100 % of algorithm utilization). Observing the curves of the decision-aided schemes (DA-JADE and DA-BCEES), we can conclude that both schemes achieve the same performance as, respectively, JADE and BCEES, in terms of SER versus SNR, but with a considerable reduction of the algorithm utilization percentage. Note also that the SER obtained with DA-BCEES presents an insignificant loss compared to DA-JADE. For comparison reasons, this figure also plots the results obtained with the Least-Squares (LS) algorithm (denoted as *Supervised* in the figure), which estimates the channel using 8 pilots. Note that the curve of SER vs SNR is equal to that obtained with JADE and DA-JADE.

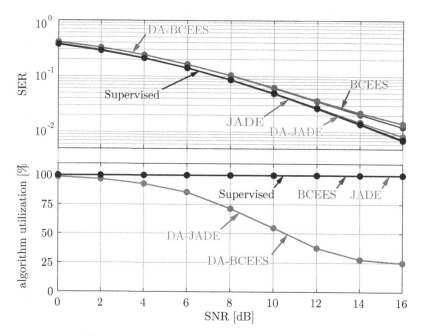

Fig. 4. SER and algorithm utilization versus SNR

5   Conclusions

We present a simple scheme to detect channel variations in Alamouti coded systems. The proposed approach uses information obtained during the synchronization procedure to determine channel variations. When channel variations are significant, the system estimates the channel matrix using an unsupervised method. The experimental results —obtained from real-world channel coefficients measured in indoor scenarios— show that the proposed scheme presents several important benefits: the utilization of an unsupervised algorithm increases the spectral efficiency and the utilization of the proposed decision rule reduces the computational load. Consequently, the unsupervised decision-aided approach arises as a promising method to avoid the transmission of training sequences and thus reducing power consumption in wireless communication devices.

Acknowledgements. This work is supported by the Spanish Ministerio de Ciencia e Innovación under grants TEC2010-19545-C04-01 and CSD2008-00010, and by Xunta de Galicia through 10TIC105003PR, and 10TIC003CT contracts.

References

1. Alamouti, S.M.: A simple transmit diversity technique for wireless communications. IEEE Journal on Selected Areas in Communications 16, 1451–1458 (1998)

2. IEEE, IEEE Standard for Information technology–Telecommunications and information exchange between systems–Local and metropolitan area networks–Specific requirements Part 11: Wireless LAN Medium Access Control (MAC) and Physical Layer (PHY) Specifications Amendment 5: Enhancements for Higher Throughput (October 2009)
3. IEEE, IEEE 802.16-2009: Air interface for fixed broadband wireless access systems (May 2009)
4. Zarzoso, V., Nandi, A.K.: Blind source separation, ch. 4, pp. 167–252. Kluwer Academic Publishers (1999)
5. Castro, P.M., Garca-Naya, J.A., Dapena, A., Iglesia, D.: Channel Estimation Techniques for Linear Precoded Systems: Supervised, Unsupervised and Hybrid Approaches. In: Signal Processing. Elsevier Science, ISSN: 0165-1684, doi:10.1016/j.sigpro.2011.01.001
6. Fernández-Caramés, T.M., Dapena, A., García-Naya, J.A., González-López, M.: A Decision-Aided Strategy for Enhancing Transmissions in Wireless OSTBC-Based Systems. In: Cabestany, J., Rojas, I., Joya, G. (eds.) IWANN 2011, Part II. LNCS, vol. 6692, pp. 500–507. Springer, Heidelberg (2011)
7. Cardoso, J.F.: Blind Signal Separation: Statistical Principles. Proceedings of IEEE 86(10), 2009–2025 (1998)
8. Cardoso, J.-F., Souloumiac, A.: Blind beamforming for non-Gaussian signals. IEE Proceedings F 140(46), 362–370 (1993)
9. Dapena, A., Pérez-Iglesias, H., Zarzoso, V.: Blind Channel Estimation Based on Maximizing the Eigenvalue Spread of Cumulant Matrices in (2 x 1) Alamouti's Coding Schemes. In: Wireless Communications and Mobile Computing (2010) (accepted) (article published online), doi:10.1002/wcm.992
10. García-Naya, J.A., González-López, M., Castedo, L.: Radio Communications, chap. A Distributed Multilayer Software Architecture for MIMO Testbeds. Intech (2010)

A Decision-Making Model for Environmental Behavior in Agent-Based Modeling*

Noelia Sánchez-Maroño[1], Amparo Alonso-Betanzos[1], Óscar Fontenla-Romero[1], Miguel Rodríguez-García[1], Gary Polhill[2], and Tony Craig[2]

[1] University of A Coruña, Department of Computer Science, 15071 A Coruña, Spain
{nsanchez,ofontenla,ciamparo}@udc.es, miguel.bhm@gmail.com
[2] The James Hutton Institute, Aberdeen, United Kingdom
{gary.polhill,tony.craig}@hutton.ac.uk

Abstract. Agent-based modeling (ABM) is an increasingly popular technique for modeling organizations or societies. In this paper, a new approach for modeling decision-making for the environmental decisions of agents in an organization modeled using ABM is devised. The decision-making model has been constructed using data obtained by responses of individuals of the organizations to a questionnaire. As the number of responses is small, while the number of variables measured is relatively high, and obtained decision rules should be explicit, decision trees were selected to generate the model after applying different techniques to properly preprocess the data set. The results obtained for an academic organization are presented.

1 Introduction

Nowadays, big companies and organizations require more and more precise models in order to monitor, inference or simulate their realities in a more detailed way. Agent-Based Modeling (ABM) has been proven as an effective tool for this purpose, allowing the direct modeling of those agents (workers, sections, departments...) participating on its daily life, instead of large and hard-to-understand equation models, which are also harder to develop (extra information has to be gathered in order to obtain the needed equations), justify, perform or even explain.

LOw Carbon At Work (LOCAW, http://www.locaw-fp7.com/) is a FP-7 European Union project, in which seven European research institutions participate with the aim of deepening the knowledge of barriers and drivers for healthy lifestyles concerning carbon, through an integrated investigation of daily practices and behaviors on different organizations, so they can achieve the European Union pollution agreements for the next years, and more specifically in 2050[1]. The project includes case studies of six organizations of different types

---

* This work has been funded in part by the European Commission through Framework Programm 7, grant agreement number 26515, LOCAW:LOw CArbon at Work.

I. Rojas, G. Joya, and J. Cabestany (Eds.): IWANN 2013, Part I, LNCS 7902, pp. 152–160, 2013.
© Springer-Verlag Berlin Heidelberg 2013

and sizes, to be modeled using an ABM approach, that simulates everyday pro-environmental practices of the different kinds of workers, taking into consideration also these barriers and drivers. ABM has become more and more popular as a tool for modeling on social sciences, since it allows the construction of models where individual entities and their interactions are directly represented. Compared with Variable-Based Modeling (using structural equations) or approximations based on systems (using differential equations), ABM offers the possibility of modeling individual heterogeneity, representing explicitly the decision rules for the agents, and locating them on a geography or other kind of space. It allows the modelers to represent on a natural way multiple analysis scales, the emergency of structures at the macro or social level of the individual action and several kinds of adaptation and learning, which are not easy to achieve with other modeling approximations [2]. The potential of ABM is on the direct representation of each of the actors on a social system, and their behaviors, working on their natural environment. Thus, a model for the behavior of the agents is also needed. In this paper, the model for decision-making in environmental responses of the different types of agents involved in an organization is described. The structure of the model is derived based on two main restrictions: (a) the output of the model, that is, the environmental decision of the agent, needs to be explicit, (b) the model should be based on the reported actual behavior of the different individuals of the organization. This behavior is obtained through the responses to a questionnaire elaborated by the sociologists participating in the project.

2   The General Model

The LOCAW project uses ABM as a synthesis tool for representing everyday practices in the workplace pertaining to the use of energy and materials, management and generation of waste, and transport. Different types of organizations were selected as case studies, specifically, two public sector organizations, two private companies which belong to the energy sector and two private companies of the heavy industry sector. Each organization entails different degrees of autonomy for its workers; therefore, the possibilities for making a decision varies from one to another. For example, people involved on the daily activity of one of the public sector organizations (a university), enjoy considerably more autonomy than do factory workers in the private companies. Therefore, the model should be adjusted to these particularities of each organization, but maintaining a core model that facilitates comparative studies between them and to derive polices or guidelines to achieve a more pro-environmental behavior at the workplace. Bearing this in mind, a general ontology [3] and a general schema were developed. The idea is to simulate the behavior of every worker on the organization, according to the tasks they perform and the options available to implement these daily tasks. For instance, an agent has to move from home to the workplace, but there are choices available, such as going by car, bus, walking, etc. Thus, in order to reproduce the behavior of the agents, the ABM model will follow this schedule:

(1) All (or some) agents make their choices; (2) Environmental impact of those choices is computed; (3)All agents who made a choice adjust their choice algorithm according to the inherent feedback from making that choice (i.e. their own personal enjoyment of it); (4)All (or some) agents forming an in connection to the agents who just made a choice reinforce or inhibit that choice; (5)All agents receiving at least one inhibition or reinforcement adjusts their choice selection algorithm accordingly; (6) Any adjustment to the choice set is made according to scenario conditions.

This paper is focused on the first point of this schema, i.e., the decision-making process of the agents. As mentioned before, this model has several restrictions. The first one is related to its output, as the decision of the agent needs to be explicit in order to check if it is theoretically consistent with the knowledge of the experts (psychologists and sociologists). Besides, a comprehensible output may help to its interpretation by the personnel of the organizations involved. These reasons determine the election of *if-then rules* to explain the decision-making process of the agents. The second limitation stipulates that the model should represent the actual behavior of workers, therefore actual data must be collected using a questionnaire. As there are different size organizations, and responding to the questionnaire will be voluntary, it is not expected to obtain a large number of samples, therefore it restricts the validity of the decision-making algorithms applicable. Giving these reduced data set and the need of deriving rules, decision-trees were selected to generate the decision-making process. A large amount of decision-trees are going to be derived, one for each decision with an environmental impact that the agent (worker) has to take under consideration, for example, going to work walking or using some transport, turning on/off the lights when going for lunch,etc. Thus, it is important to design an automatic procedure that help to derive those decision-trees from data. The different techniques applied are explained in the next section.

3   The Decision-Making Model

The LOCAW project is organized on seven work packages (WP) that pursue different objectives regarding the environmental behavior of individuals in organizations. The psychologists and sociologists in this project have discussed different theoretical models to explain human pro-environmental behavior, finally adopting the model presented in the upper section of Figure 1 where behavior is influenced by values, awareness of consequences, outcome efficacy and norms [4]. Values can be seen as abstract concepts or beliefs concerning a person's goals and serve as guiding standards in his or her life. Schwartz identifies 10 human value types [5], however only four different types were considered important for this project: egoistic, hedonic, altruist and biospheric.

Different quantitative and qualitative tools were used in LOCAW project to analyze the different organizations, for example, focus groups, interviews, life story's, etc. Among them, a questionnaire was designed to obtain data regarding individual factors that affects pro-environmental behavior at work. The questionnaire is based in the value-belief-model (VBM) shown in upper part of Figure 1, and therefore it includes three different blocks with questions about: a) values (some of them depicted at figure 2), b)motivations, i.e., efficacy, worldviews and norms, and c) behaviors. Regarding this last block, notice that not only the behavior at work is important for the aim of this project, but also the behavior at home in order to detect if there exists spillover between them. Therefore, 74 questions for behaviors regarding the use of energy and materials, the treatment of waste and the use of transport in both- work and home- were included.

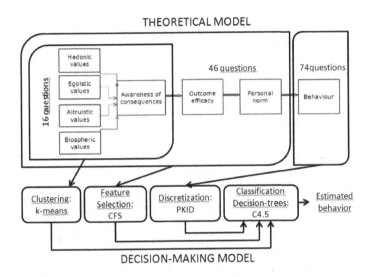

Fig. 1. The decision-making model for the agents in the LOCAW project

The information and data collected are being used to automatically obtain classification decision trees that could explain the agents' pro-environmental decisions when doing a daily task. However, before using decision-trees to determine the possible behaviors of the agents, some techniques have been applied to obtain a representative set of data that maximizes the generalization capability of the resulting decision trees (see Figure 1). The different algorithms are subsequently briefly described, all of them are available in the Weka tool environment [6]. This platform was chosen as it based on Java programming language and the whole project will be developed using this language.

- **Clustering:** Following the VBN model in Fig. 1, the behavior of individuals depends on four different types of values. Therefore, it is expected that workers (and so agents) behave in different ways according to these values, i.e., several profiles can be identified. To generate these profiles and so to represent variability in the model, a clustering technique was used. Since it is an adequate well-known technique, *K-means* [7] was employed in our model.

	Opposed to my values	Not important			Important		Very important	
1. EQUALITY: equal opportunity for all	-1	0	1	2	3	4	5	6	7
2. RESPECTING THE EARTH: harmony with other species	-1	0	1	2	3	4	5	6	7
3. SOCIAL POWER: control over others, dominance	-1	0	1	2	3	4	5	6	7
4. PLEASURE: joy, gratification of desires	-1	0	1	2	3	4	5	6	7
5. UNITY WITH NATURE: fitting into nature	-1	0	1	2	3	4	5	6	7
6. A WORLD AT PEACE: free of war and conflict	-1	0	1	2	3	4	5	6	7
7. WEALTH: material possessions, money	-1	0	1	2	3	4	5	6	7
8. AUTHORITY: the right to lead or command	-1	0	1	2	3	4	5	6	7

Fig. 2. A part of the questionnaire related to values

- **Feature Selection:** Adequate identification of relevant features/variables is fundamental in real world scenarios because it may help to obtain simpler models and to focus experts' attention on the relevant data. In this problem, the ratio samples/features is low, because there are 68 questions (6 personal, 16 on values and 46 for motivations) while the number of responses is expected to be in the order of a few hundred (depending on the size of the organization), so the lack of samples prevents obtaining models that properly generalize in spite of the ability of decision trees to discriminate features. Therefore, feature selection (FS) was applied to determine the relevant features while eliminating the irrelevant or redundant ones [8]. From the different FS methods, a filter was chosen because of its independency of any learning algorithm, specifically, the Correlation-Feature-Selection algorithm (CFS) [9] has been applied to the whole set of data.
- **Discretization:** Most questions in the questionnaire use Likert scales, indicating the degree to which respondents agreed with a proposition, or the frequency with which they performed a behaviour (see Figure 2). Again, as the number of responses is not expected to be high, it could happen that not all the ranges could be equally represented in the final sample. To solve this problem a discretization step was considered as necessary, using the Proportional K-Interval Discretization (PKID) algorithm [10]. This algorithm automatically chooses a number of intervals to divide the sample, taking into account the number of samples obtained in each subinterval.
- **Classification:** Finally, once the data has been preprocessed by the previous steps, decision trees can be constructed to automatically derive rules that will lead to a specific behavior for the agents. For this, the C4.5 algorithm was employed [11] as it is one of the most successful methods for this purpose.

4   Experimenting Results

In this section, we will show the results obtained for one of the organizations to be modelled, the University of A Coruña (UDC). The UDC has a total of 2277 workers, between administration (790) and research/teaching (1487) personnel. The questionnaire has been passed down to workers (that could voluntarily answer it) automatically using the Qualtrics application (`https://www.qualtrics.com/`). The answers of the questionnaire have been preprocessed to clean highly-incomplete (more than 45% blanks) or ambiguous data which could contaminate the model. After that, a total amount of 237 different valid samples have been gathered. The different methods presented in the previous section were subsequently applied to this data set and the results obtained are detailed in the following subsections.

4.1   Step 1: Clustering for UDC

The clustering process was carried out using only those 16 questions pertaining to values included in the questionnaire (some of them shown in Figure 2), leading to a data set size of 237 instances × 16 dimensions. The *k-means* algorithm requires the number of clusters as a parameter, and since four different clusters have been theoretically identified by the experts as adequate for this application study, obtaining four clusters was our first attempt. In general, *k-means* is quite sensitive to how clusters are initially assigned, so different initializations were tested. However, none of the partitions obtained allowed for clearly distinguishing the profiles as indicated by the experts working in the project. Finally, in discussion with the experts, six clusters were identified that drive to adequate separation of the samples and contain hybrid groups. Specifically four "almost-pure" profiles can be identified on clusters zero through three (coinciding with the theoretical ones: egoistic, altruistic, biospheric and hedonic) and two more hybrid groups, that mixed similar profiles (biospheric-altruist and egoistic-hedonic). Columns 3-8 in Table 1 illustrate those clusters, and the parenthesis contain the value of the number of samples they represent. The table details the values of the centroid of each cluster for each value. It can be appreciated that each item is marked with a different symbol (square, triangle, etc.); these shapes are associated with a theoretical profile, so diamond represents biospheric questions, square is used for altruist ones, up-triangle is linked to egoistic items and, finally, down-triangle shows the hedonistic issues. As each dimension value in the centroid represents the mean value for that dimension in the cluster, high values of "up-triangle" dimensions (social power, wealth, authority, influential, ambitious) are expected for the "egoistic" profile, high values of "down-triangle" ones for the hedonic one and so on. Notice that the highest values for each row are in boldface letters. Another important aspect in the clustering with values, is that this section is the only one in the questionnaire that has a column entitled "Opposed to my values" , with a -1 value assigned, than can be checked by individuals answering it. The other sections of the questionnaire have a range between 0 and 7 for the answer. So, not all the ranges in the values part of the questionnaire have the same significance, as only the first column specifies opposing values, while the others specify a continuous range between 0 (Not important) and 7 (Very important). Then, the responses obtained in that column have been weighted with a factor that multiplies by 10 its importance regarding the responses obtained in the other 8 columns. That is the reason why some of the centroid values are negative.

4.2   Steps 2 and 3: Feature Selection and Discretization for UDC

Feature selection allows for determining the relevant features for a giving problem. Actually, this paper copes with 74 problems, one for each election the agent has to consider, i.e., one for each behavior to be modeled. Then, the CFS algorithm has been applied 74 times to determine the relevant inputs (values and motivations) for all the behaviors. Therefore, the final output of this step is a matrix relating behaviors and inputs that has been proven theoretically-consistent by our experts. This matrix shows similarities and differences between behaviors and an extract can be appreciated in Table 2. As explained before, the sample was discretized in order to obtain an adequate representation of the actual intervals obtained in the samples.

Table 1. Clusters obtained for the UDC case. Note that beside the four theoretical clusters initially devised, two more hybrid groups were added.

Attribute	Full set	0(2)	1(62)	2(85)	3(20)	4(56)	5(12)
■Equality	6.37	4.5	6.68	**6.71**	6.55	5.91	4.58
♦Respecting earth	5.71	2.50	**6.42**	6.22	6.20	4.73	2.83
▲Social Power	-3.20	1.00	-10.0	**1.35**	-8.50	-1.40	-0.75
▼Pleasure	4.87	2.5	**5.37**	5.27	4.10	4.07	4.83
♦Unity with nature	5.13	1.00	**6.02**	5.91	5.60	3.63	2.08
■A world at peace	6.41	2.50	**6.79**	**6.79**	6.75	6.02	3.67
▲Wealth	1.65	2.50	2.37	2.54	-7.00	2.30	**2.75**
▲Authority	1.24	**4.50**	-0.27	2.49	-1.15	1.46	2.58
■Social justice	6.37	3.50	6.55	**6.73**	6.45	5.95	5.25
▼Enjoying life	5.20	3.5	**5.84**	5.64	3.40	4.52	5.25
♦Protecting environment	5.75	4.00	**6.40**	6.33	6.15	4.75	2.58
▲Influential	2.25	**4.50**	1.73	3.13	1.05	1.84	2.33
■Helpful	5.39	2.00	5.68	**6.01**	5.85	4.30	4.33
♦Preventing pollution	5.55	3.50	6.15	**6.26**	6.05	4.41	2.33
▼Self-indulgent	4.06	2.00	**4.73**	4.59	2.00	3.36	3.92
▲Ambitious	3.31	**4.50**	4.03	3.84	-0.10	2.96	2.92
		Egoistic▲	Hedonic▼	Altruist■	Biospheric♦	Bio-Altruist	Ego-Hedonic

4.3   Step 4: Classification for UDC

For each behavior, the relevant inputs selected by CFS together with the discretized output provided by the previous step form the data set to be fed to C4.5 algorith for training and testing. In all these cases, 66% of the data has

Table 2. An extract of the results of the feature selection process

Behavior	Sex	Studies Lv	Organiz. Lv	Exempl. role	Equality	Resp.earth	Peace
Total Flights	X	X	X	X			X
Turn lights	X	X	X	X	X	X	

been used for training while the remaining 34% is employed for testing. As 6 different clusters where obtained in Step 1 and 74 different behaviors must be modeled, 74 × 5 = 444 decision-trees were generated, as cluster 0 (column 3 in Table 1) has only 2 samples and thus it was not automatically treated. An example showing one of the trees derived can be seen on Fig. 3.

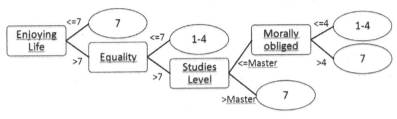

Fig. 3. One example of a tree derived for a behavior related to waste separation, specifically, separating glass from regular garbage at home

5   Conclusions and Future Work

LOCAW project focuses on everyday practices in the workplace and on the interplay of barriers and drivers of sustainable behavior. It will use ABM to study the possible large scale effects of introducing low carbon strategies in the workplace, in different organizations. ABM will include a decision-making algorithm to determine how agents choose between different environmental options in their daily tasks. This paper presents the decision-making algorithm designed based on decision-trees for practical restrictions. This algorithm takes, as input data, the workers' responses to a questionnaire designed by the psychologists in the project. Different methods were employed to make data tractable and, more important, to enhance decision-trees generalization capability. Between the different organizations involved in the project, UDC was selected as starting point because of proximity and familiarity. However, in future stages, this decision-making algorithm has to be adapted to the remaining organizations. Moreover, the decision-making algorithm has to be integrated in the ABM to reflect how the interaction between agents and environment may vary the possible options.

References

1. European Commission: What is EU doing about climate change?
   http://ec.europa.eu/clima/policies/brief/eu/index_en.htm
   (last visited on January 2013)

2. Gilbert, N.: Agent-based models. SAGE Publications, University of Surrey (2007)
3. Polhill, G., Gotts, N., Sánchez-Maroño, N., Pignotti, E., Fontenla-Romero, O., Rodríguez-García, M., Alonso-Betanzos, A., Edwards, P., Craig, T.: An ontology-based design for modelling case studies of everyday proenvironmental behaviour in the workplace. In: Proc. of International Congress on Environmental Modelling and Software Managing Resources of a Limited Planet, Leipzig, Germany (2012)
4. Steg, L., De Groot, J.I.: Environmental values. In: The Oxford Handbook of Environmental and Conservation Psychology, Oxford University Press (2012)
5. Schwartz, S.H.: Universals in the content and structures of values: Theoretical advances and empirical tests in 20 countries. Advances in Experimental Psychology 25, 1–65 (1992)
6. Hall, M., Frank, E., Holmes, G., Pfahringer, B., Reutemann, P., Witten, I.H.: The WEKA data mining software: An update. SIGKDD Explorations 11(1) (2009), http://www.cs.waikato.ac.nz/ml/weka/ (last visited on January 2013)
7. MacQueen, J.B.: Some methods for classification and analysis of multivariate observations. In: Proceedings of 5th Berkeley Symposium on Mathematical Statistics and Probability, pp. 281–297. University of California Press (1967)
8. Guyon, I., Elisseeff, A.: An introduction to variable and feature selection. Journal of Machine Learning Research 3, 1157–1182 (2003)
9. Hall, M.A.: Correlation-based Feature Selection for Machine Learning. PhD thesis, University of Waikato, Hamilton, New Zealand (1999)
10. Yang, Y., Webb, G.I.: Proportional k-interval discretization for naive-bayes classifiers. In: Flach, P.A., De Raedt, L. (eds.) ECML 2001. LNCS (LNAI), vol. 2167, pp. 564–575. Springer, Heidelberg (2001)
11. Quinlan, J.R.: C4.5: Programs for Machine Learning. Morgan Kaufmann Publishers Inc. (1993)

Version of the New SHA Standard Applied to Manage Certificate Revocation in VANETs

Francisco Martín-Fernández and Pino Caballero-Gil

Department of Statistics, Operations Research and Computing
University of La Laguna, Spain
francisco.martin.07@ull.edu.es, pcaballe@ull.es

Abstract. This work describes the application of a new version of the Secure Hash Algorithm SHA-3 that was recently chosen as standard, in order to improve the performance of certificate revocation in Vehicular Ad-hoc NETworks (VANETs), which are interesting self-organizing networks. Specifically, we propose the use of both a duplex construction instead of the sponge one present in the SHA-3 version of the Keccak hash function, and a dynamic authenticated data structure based on B-trees, which allows taking advantage of such a construction.

1 Introduction

Vehicular Ad-hoc NETworks (VANETs) are self-organizing networks built up from moving vehicles that communicate with each other mainly to prevent adverse circumstances on the roads and to achieve a more efficient traffic management. In particular, these networks are considered an emerging research area of mobile communications that offer a wide variety of possible applications, ranging from road safety and transport efficiency, to commercial services, passenger comfort, and infotainment delivery. Furthermore, VANETs can be seen as an extension of Mobile Ad-hoc NETworks (MANETs) where there are mobile nodes, which are On-Board Units (OBUs) in vehicles; and static nodes, which are Road-Side Units (RSUs).

Without security, all network nodes are potentially vulnerable to any misbehaviour of any dishonest user, because this would make that all services provided by the VANET be untrustworthy. Therefore, it is absolutely necessary to have a procedure not only to identify the misbehaving nodes, but also to exclude them from the network. One of the basic solutions to accomplish this task in networks where communications are based on a Public Key Infrastructure (PKI) is the use of certificate revocation. Thus, a critical part in such networks is the management of revoked certificates. Related to this issue, in the bibliography we can find two different types of solutions. On the one hand, a decentralized proposal enables revocation without the intervention of any centralized infrastructure, based on trusting the criteria of network nodes. On the other hand, a centralized approach is based on the existence of a central Certificate Authority (CA), which is the only entity responsible for deciding on the validity of each

I. Rojas, G. Joya, and J. Cabestany (Eds.): IWANN 2013, Part I, LNCS 7902, pp. 161–168, 2013.
© Springer-Verlag Berlin Heidelberg 2013

node certificate, and all nodes trust it. This second approach is usually based on the distribution of the so-called Certificate Revocation Lists (CRLs), which can be seen as blacklists of revoked certificates.

IEEE 1609 is a family of standards based on the IEEE 802.11p, which is an approved amendment to the IEEE 802.11 standard for vehicular communications. Within such a family, 1609.2 deals with the issues related to security services for applications and management messages. In particular, the IEEE 1609.2 standard defines the use of PKIs, CAs and CRLs in VANETs, and implies that in order to revoke a vehicle, a CRL has to be issued by the CA to the RSUs, who are in charge of sending the information to the OBUs. Thus, an efficient management of certificate revocation is crucial for the robust and reliable operation of VANETs.

Once VANETs are implemented in practice on a large scale, as their size grows and the use of multiple temporary certificates or pseudonyms becomes necessary to protect the privacy of the users, it is foreseeable that CRLs will grow up to become very large. Moreover, in this context it is also expected a phenomena known as implosion request, consisting of nodes who synchronously want to download the CRL at the time of its updating, producing serious congestion and overload of the network, what could ultimately lead to a longer latency in the process of validating a certificate.

This paper proposes the use of an Authenticated Data Structure (ADS) known as B-tree, for the management of certificate revocation in VANETs. By using this structure, the process of query on the validity of certificates will be more efficient because OBUs will send queries to RSUs, who will answer them on behalf of the CA. In this way, at the same time the CA will no longer be a bottleneck, and OBUs will not have to download the entire CRL. In particular, the used B-trees are based on the application of a duplex construction of the Secure Hash Algorithm SHA-3 that was recently chosen as standard, because the combination of both structures allows improving efficiency of updating and querying of revoked certificates.

This paper is organized as follows. Section 2 addresses the general problems of the use of certificate revocation lists in VANETs. Then, Section 3 is focused on the explanation of our proposal to improve such problems, which is based on the combination of B-trees and a duplex construction of the Secure Hash Algorithm SHA-3. Finally, Section 4 discusses conclusions and possible future research lines.

2   The CRL Issue in VANETs

In general, when CRLs are used, and a CA has to invalidate a public-key certificate, what it does is to include the corresponding certificate serial number in the CRL. Then, the CA distributes this CRL within the network in order to let users know which nodes are no longer trustworthy. The distribution of this CRL must be done efficiently so that the knowledge about untrustworthy nodes can be spread quickly to the entire network.

In the case of VANETs, previous works assume that the CRL may be distributed by broadcasting it from RSUs directly to the OBUs. However, the large size of VANETs, and consequent large size of the CRLs, makes this approach infeasible due to the overhead it would cause to network communications. This issue is further increased with the use of multiple pseudonyms for the nodes, what has been suggested to protect privacy and anonymity of OBUs.

In particular, knowing that there are almost one thousand million cars in the world, considering the use of pseudonyms the number of revoked certificates might reach soon the same amount, one thousand million. On the other hand, assuming that each certificate takes at least 224 bits, in such a case the CRL size would be 224 Gbits, what means that its management following the traditional approach would not be efficient. Even though regional CAs were used and the CRLs could be reduced to 1 Gbit, by using the 802.11a protocol to communicate with RSUs in range, the maximum download speed of OBUs would be be between 6 and 54 Mbit/s depending on the vehicles speed and the road congestion, so o average an OBU would need more than 30 seconds to download a regional CRL from an RSU.

A direct consequence of this size problem is that a new CRL cannot be issued very often, what would affect the freshness of revocation data. On the other hand, if a known technique for large data transfers were used for CRL distribution as solution for the size problem, it would result in higher latencies, what would also impact in the revocation data validity. Consequently, a solution not requiring the distribution of the full CRL from RSUs to OBUs would be very helpful for the secure and efficient operation of VANETs.

3   Approach Based on B-trees and a Version of SHA-3

In order to improve efficiency of communication and computation in the management of public-key certificates in VANETs, some authors have proposed the use of particular ADSs such as Merkle trees and skip lists [6] [8]. However, to the best of our knowledge no previous work has described the use of B-tree in general as hash tree for the management of certificate revocation.

In general, a hash tree is a tree structure whose nodes contain digests that can be used to verify larger pieces of data. The leaves in a hash tree are hashes of data blocks while nodes further up in the tree are the hashes of their respective children so that the root of the tree is the digest representing the whole structure. Most implemented hash trees require the use of a cryptographic hash function in order to prevent collisions.

The model here proposed is based on the following notation:

- h: Cryptographic hash function used to define the hash tree.
- D (≥ 1): Depth of the hash tree.
- d ($\leq D$): Depth of a node in the hash tree.
- N_{ij} ($i = D - d$ and $j = 0, 1...$): Node of the hash tree obtained by hashing the concatenation of all the digests contained in its children.
- s: Number of revoked certificates.

- RC_j ($j = 1, 2, ..., s$): Serial number of the $j - th$ Revoked Certificate.
- R: Tree-based Repository containing the revoked certificates in the leaves.
- m: Maximum number of children for each internal node in the B-tree.
- f: Basic cryptographic hash function of SHA-3, called Keccak.
- n: Bit size of the digest, which is here assumed to be the lowest possible size of SHA-3 digest, 224.
- b: Bit size of the input to the hash function f, which is here assumed to be one of the possible values of Keccak, 800.
- r: Bit size of input blocks after padding for the hash function h, which is here assumed to be 352.
- c: Difference between b and r, which is here assumed to be as in SHA-3, $2n$, that is 448.
- k: Bit size of revoked certificates, which is here assumed to be around 224.
- l: Bit size of output blocks for the hash function h, which is here assumed to be lower than r.

In this work, the leaves of the hash tree, $N_{00}, N_{01}, ..., N_{0(s-1)}$ contain the digests of the serial number of the s revoked certificates, while each internal node N_{ij} is the digest resulting from the application of the cryptographic hash function h to the concatenation of the digests represented by its children $N_{(i-1)j} : j = 0, 1, ...$

Most hash tree implementations are binary, but this work proposes the use of a more general structure known as B-tree (see Figure 1). A B-tree is a data structure that holds a sorted data set and allows efficient operations to find, delete, insert, and update data. In a B-tree the number of children of each internal node varies between $\frac{m}{2}$ and m [9]. In a B-tree it is required that all leaf nodes are at the same depth, so the depth of the tree will increase slowly as new leaves are inserted in the tree. When leaves are inserted or removed from the tree, in order to maintain the pre-defined range of children between $\frac{m}{2}$ and m, internal nodes may be merged or split. All this means that our proposal is based on a dynamic tree-based data structure that will vary depending on the number of revoked certificates.

The authenticity of the hash tree structure is guaranteed thanks to the CA signature of the root. When an RSU has to respond to an OBU about a query on a certificate, it proceeds in the following way. If it finds the digest of the certificate among the leaves of the tree because it is a revoked certificate RC_j, then the RSU sends to the OBU the route from the root to the corresponding leaf, along with all the siblings of the nodes on this path. After checking all the digests corresponding to the received path and the CA signature of the root, the OBU gets convinced of the validity of the evidence on the revoked certificate received from the RSU.

In our proposal, the used B-tree structure assigns a unique identifier to each revoked certificate represented by its leafs, so that an auxiliar structure linking such identifiers with the corresponding certificate serial number is also stored in the RSU. Thus, when an OBU sends a request about a certificate, the RSU firstly gets the identifier generated by the B-tree structure using the certificate serial number, and then proceeds with the tree search. In this way, thanks to the

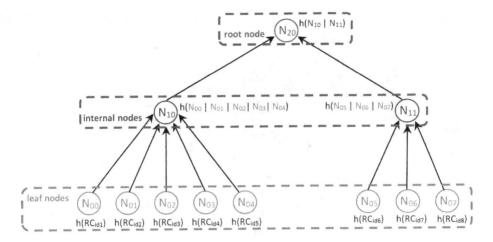

Fig. 1. Hash tree based on a B-tree with $m = 5$

use of B-trees the computational complexity for searching, inserting and deleting any leaf is of O(log n).

Regarding the cryptographic hash function, our proposal is based on the use of a new version of the Secure Hash Algorithm SHA-3 that was chosen as standard in October 2012 [5]. In SHA-3, the input is represented by a 5×5 matrix of 64-bit lanes, but our proposal is based on 32-bit lanes (see Figure 2), the padding of the input is a minimum $10 * 1$ pattern that consists of a 1 bit, zero or more 0 bits (maximum $r - 1$) and a final 1 bit, and the basic cryptographic hash function f called Keccak contains 24 rounds of a basic transformation that involves 5 steps called theta, rho, pi, chi and iota [10].

This work proposes the combination of a duplex version of the sponge structure of SHA-3 [3] and a hash B-tree. On the one hand, like the sponge construction of SHA-3, our proposal based on a duplex construction uses Keccak as fixed-length transformation f, the same padding rule based on the $10 * 1$ pattern, and data bit rate r. On the other hand, unlike a sponge function, the duplex construction output corresponding to an input string may be obtained through the concatenation of the outputs resulting from successive input blocks (see Figure 3).

The use of the duplex construction as hash function in our proposed hash tree allows the insertion of a new revoked certificate as new leaf of the tree by running a new iteration of the duplex construction only on the new revoked certificate. In particular, the RSU can take advantage of all the digests corresponding to the sibling nodes of the new node, which were computed in previous iterations, by simply discarding the same minimum number of the last bits of each one of those digests so that the total size of the resulting digest of all the children remains the same, n. Note that, while the maximum number of children of an internal node has not been reached, the RSU has to store not only all the digests

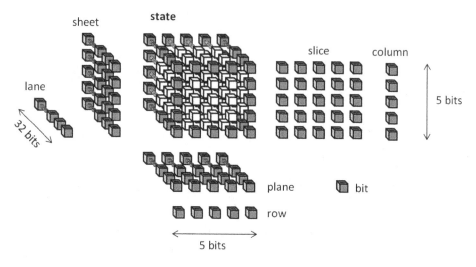

Fig. 2. State of Keccak

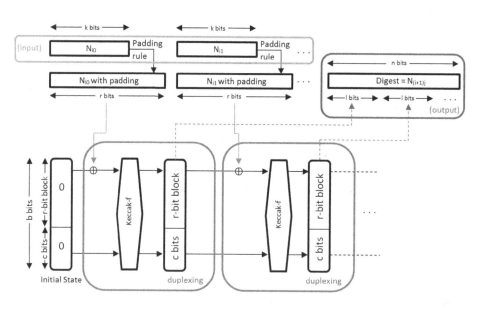

Fig. 3. Duplex Construction

of the tree structure but also the state resulting from the application of Keccak hash function f in the last iteration corresponding to such internal node, in order to use it as input in a next iteration. On the other hand, periodic delete operations of certificates that are in the tree and reach their expiration date, require rebuilding the part of the tree involving the path from those nodes to the root. In particular, in order to maximize our proposal, such tree rebuilding would be linked for example to the fact that all the sibling nodes of some internal node are expired because this would avoid an unnecessary reduction of the system efficiency by having to rebuild the tree very often.

The choice of adequate values for the different parameters in our proposal must be done carefully, taking into account the relationships among them. In particular, the maximum tree size takes the following value:

$$n(1 + m + m^2 + m^3 + \cdots + m^D) = \frac{n(m^{(D+1)} - 1)}{m-1}$$

Thus, since this quantity is upperbounded by the size of available memory in the RSU, and the maximum number of leaves of the B-tree m^D is lowerbounded by the number of revoked certificates s, then both conditions can be used to deduce the optimal value for m.

4   Conclusions and Future Works

One of the most important security issues in VANETs is the problem of certificate revocation management, so an efficient verification of public-key certificates by OBUS is crucial to ensure the safe operation of the network. However, as VANETs grow, certificate revocation lists will also grow, making it impossible their issuance. This paper proposes a more efficient alternative to CRL issuance, which uses an authenticated data structure based on dynamic B-trees. In addition, the proposed mechanism applies the basic hash function of the new SHA-3 standard called Keccak combined with a duplex construction. Thanks to the structure of the used B-tree, the duplex construction allows taking advantage of the digests of previous revoked certificates for calculating the hash of every new revoked certificate, so that its inclusion in the tree can be performed by a single iteration of the hash function. Both the analysis of optimal values for the parameters and the implementation of the proposal are part of work in progress.

Acknowledgements. Research supported by the Ministerio de Economa y Competitividad and the FEDER Fund under Projects TIN2011- 25452 and IPT-2012-0585-370000, and the FPI scholarship BES-2012-051817.

References

1. Andreeva, E., Mennink, B., Preneel, B.: Security reductions of the second round SHA-3 candidates. In: Burmester, M., Tsudik, G., Magliveras, S., Ilić, I. (eds.) ISC 2010. LNCS, vol. 6531, pp. 39–53. Springer, Heidelberg (2011)

2. Bertoni, G., Daemen, J., Peeters, M., Van Assche, G.: Keccak sponge function family main document (2009), http://keccak.noekeon.org/Keccak-main-2.1.pdf
3. Bertoni, G., Daemen, J., Peeters, M., Van Assche, G.: Duplexing the sponge: single-pass authenticated encryption and other applications, Submission to NIST, Round 2 (2010),
http://csrc.nist.gov/groups/ST/hash/sha-3/Round2/
Aug2010/documents/papers/DAEMEN_DuplexSponge.pdf
4. Bertoni, G., Daemen, J., Peeters, M., Van Assche, G.: The Keccak SHA-3 submission, Submission to NIST, Round 3 (2011),
http://keccak.noekeon.org/Keccak-submission-3.pdf
5. Chang, S., Perlner, R., Burr, W., Turan, M., Kelsey, J., Paul, S., Bassham, L.: Third-Round Report of the SHA-3 Cryptographic Hash Algorithm Competition, NIST (2012), http://nvlpubs.nist.gov/nistpubs/ir/2012/NIST.IR.7896.pdf
6. Gañán, C., Muñoz, J.L., Esparza, O., Mata-Díaz, J., Alins, J.: Toward Revocation Data Handling Efficiency in VANETs. In: Vinel, A., Mehmood, R., Berbineau, M., Garcia, C.R., Huang, C.-M., Chilamkurti, N. (eds.) Nets4Cars/Nets4Trains 2012. LNCS, vol. 7266, pp. 80–90. Springer, Heidelberg (2012)
7. Homsirikamol, E., Rogawski, M., Gaj, K.: Comparing Hardware Performance of Fourteen Round two SHA-3 Candidates using FPGAs, Cryptology ePrint Archive, Report 2010/445, 210 (2010), http://eprint.iacr.org/2010/445 (January 15, 2011)
8. Jakobsson, M., Wetzel, S.: Efficient attribute authentication with applications to ad hoc networks. In: Proceedings of the 1st ACM International Workshop on Vehicular Ad Hoc Networks, pp. 38–46 (2004)
9. Knuth, D.: Sorting and Searching, The Art of Computer Programming, 2nd edn., vol. 3, pp. 476–477, pp. 481–491. Addison-Wesley (1998)
10. Martin, F., Caballero, P.: Analysis of the New Standard Hash Function SHA-3. In: Fourteenth International Conference On Computer Aided Systems Theory, Las Palmas de Gran Canaria Spain (2013)

System Identification of High Impact Resistant Structures

Yeesock Kim, K. Sarp Arsava, and Tahar El-Korchi

Department of Civil and Environmental Engineering,
Worcester Polytechnic Institute (WPI) Worcester, MA 01609-2280, USA
{Yeesock,ksarsava,tek}@wpi.edu

Abstract. The main purpose of this paper is to develop numerical models for predicting and analyzing highly nonlinear behavior of integrated structure-control systems subjected to high impact loading. A time-delayed adaptive neuro-fuzzy inference system (TANFIS) is proposed for modeling complex nonlinear behavior of smart structures equipped with magnetorheological (MR) dampers under high impact forces. Experimental studies are performed to generate sets of input and output data for training and validating the TANFIS models. The high impact load and current signals are used as the input disturbance and control signals while the acceleration responses from the structure-MR damper system are used as the output signals. Comparisons of the trained TANFIS models with the experimental results demonstrate that the TANFIS modeling framework is an effective way to capture nonlinear behavior of integrated structure-MR damper systems under high impact loading.

Keywords: adaptive neuro-fuzzy inference system (ANFIS), high impact load, magnetorheological (MR) damper, system identification, and smart structures.

1 Introduction

When a structure is excited by an impact load such as an aircraft or ship collision, key components of the infrastructure can be severely damaged and cause a shutdown of critical life safety systems (Consolazio et al., 2010). One of the most promising strategies to absorb and dissipate the external energy would be to use a smart control mechanism that adjusts the force levels of mechanical devices within the infrastructure in real time. In recent years, with the increase of smart structure applications in many engineering fields, usage of smart control systems in the improvement of the dynamic behavior of complex structural systems has become a topic of major concern (Spencer and Nagarajajah, 2003). In particular, magnetorheological (MR) dampers have received great attention for use in large-scale civil infrastructural systems since they combine the best features of both the passive and active control strategies (Spencer et al., 1997). Many investigators have demonstrated that this technology shows great deal of promise for civil engineering applications in recent years (Dyke et al., 2001; Kim et al., 2009, 2010). However, most of the studies on MR damper technology has focused on nonlinear behavior under low velocity

I. Rojas, G. Joya, and J. Cabestany (Eds.): IWANN 2013, Part I, LNCS 7902, pp. 169–178, 2013.

environments while relatively little research has been carried out on the performance of MR dampers under high impact forces (Wang and Li, 2006; Mao et al., 2007; Ahmadian and Norris, 2008; Hongsheng and Suxiang., 2009). The main focus of these studies was on the behavior of the MR damper itself under impact loads, not specifically a structure equipped with the MR dampers. As of yet, an integrated model to predict nonlinear behavior of smart structures-MR damper systems under high impact loads has not been investigated.

It is quite challenging to develop an accurate mathematical model of the integrated structure-MR control systems due to the complicated nonlinear behavior of integrated systems and uncertainties of high impact forces. For example, when highly nonlinear hysteretic actuators/dampers are installed in structures for efficient energy dissipation, the structure employing the nonlinear control devices behaves nonlinearly although the structure itself is usually assumed to behave linearly (Kim et al., 2011). Moreover, this nonlinear problem becomes more complex with the application of unexpected high impact loads. Hence, the challenge is to develop an appropriate mathematical model for the integrated nonlinear system under high impact loads. Fig. 1 represents the highly nonlinear hysteretic behavior between the high impact force and the structural velocity responses under high impact loads due to nonlinear MR dampers, high speed impact forces and nonlinear contact between structure and MR damper. This issue can be addressed by applying nonlinear system identification (SI) methodologies to a set of input and output data in order to derive a nonlinear input-output mapping function.

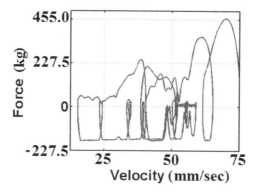

Fig. 1. Nonlinear behavior of the structure equipped with MR dampers under high impact loads

In general, the SI methodologies can be categorized into two parts: parametric and nonparametric SI approaches (Bani-Hani et al., 1999). In the parametric approach, the architecture of the mathematical model is directly dependent on the physical quantities of structural system such as stiffness, damping and mass (Lin et al., 2001; Yang and Lin, 2004). On the other hand, to identify the given system model, the nonparametric SI method trains the input-output map of the structural system (Hung et al., 2003; Kim et al., 2011). The nonparametric SI approaches have been widely used in the field of large civil structures because of their proven usefulness to estimate incomplete and incoherent measurements of large-scale infrastructural systems (Allison and Chase, 1994; Marsri et al., 2000; Hung et al., 2003; Suresh et al., 2008; Kim et al. 2013). However, there is no study on SI for predicting high impact nonlinear behavior of smart structures equipped with highly nonlinear hysteretic devices. With this in mind, a nonlinear SI framework is proposed for estimating complex behavior response of structure-MR control systems under high impact loads in this paper. The approach is developed through the introduction of time-delayed components to adaptive neuro-fuzzy inference system (ANFIS) modeling framework, which is an integrated learning model of fuzzy logic and neural network.

This paper is organized as follows. Section 2 discusses the time-delayed ANFIS (TANFIS). In section 3, the experimental setup and procedures are described. The modeling results, including training and validations are given in section 4. Concluding remarks are given in section 5.

2   Time-Delayed Adaptive Neuro Fuzzy Inference System (TANFIS)

ANFIS can be simply defined as a set of fuzzy 'if-then' rules with appropriate membership functions to generate the stipulated input-output pairs in the solution of uncertain and ill-defined systems (Jang, 1993). The application of ANFIS models in the SI of complex civil engineering structures is a relatively new topic (Mitchell et al., 2012). Although the application of the ANFIS system has been commonly used (Faravelli and Yao, 1996; Alhanafy, 2007; Gopalakrisnan and Khaitan, 2010; Wang, 2010), minimizing the output error to maximize the performance of the SI is still a challenging issue.

ANFIS is a hybrid system that is able to integrate fuzzy inference system and adaptive learning tools from neural networks to get more accurate results (Mitchell et al., 2012). By using a backpropagation neural network learning algorithm, the parameters of the Takagi-Sugeno (TS) fuzzy model are updated until they reach the optimal solution (Tahmasebi and Hezarkhani, 2010). However, it is observed from the simulation that ANFIS predictions are not in aggreement with the actual high impact responses. Only 20% to 40% of the actual acceleration values are predicted correctly by ANFIS. In order to increase the accuracy between the trained and the actual high impact test data, TANFIS, which uses the outputs of the previous steps to predict the features of the following output, is used. The new TANFIS method, which is defined below, increased the accuracy of the trained model significantly.

2.1   Time-delayed ANFIS (TANFIS)

The objective of the method is to estimate the output by using the observations from previous steps. In general, a dynamic input-output mapping (Adeli and Jiang 2006) can be expresed as follows

$$F_j(t) = f(x^{t-d}, f^{t-d}, e^{t-d}) + e(t) \tag{1}$$

where x^t, f^t and e^t represents the input, output and error for time t, respectively. The time delay term is represented by the term d. In this research, impact loading, the electrical current applied to the MR damper and the responses are assigned as input. The fuzzy model is then trained to identify the features of structural responses. In the research, time delay term d is assigned as 1, which means that model uses the observations from previous step (t-1) to estimate the output at time t. By the integration of Eq. (8) and Eq. (9), the proposed TANFIS model is as follows

$$O^5_j = overall \quad output + \sum_j \overline{w}_j \times f_j = \frac{\sum_j w_j \times f_j}{\sum_j w_j}$$

(2)

The architecture of the TANFIS model is depicted in Fig. 2.

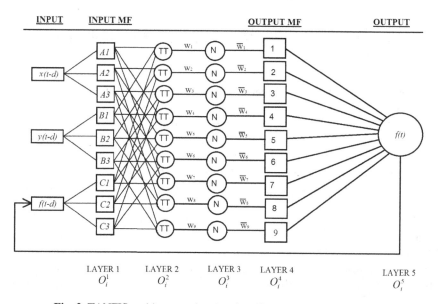

Fig. 2. TANFIS architecture showing three input and one output model

In order to obtain input-output data for training and validating the TANFIS model, experimental studies are performed. Impact load, current on MR damper, and acceleration values are measured and models are trained to predict the noninear behavior of the smart structure. Results are used in the evaluation of the accuracy of TANFIS to predict the actual test data.

3   Experimental Setup

To investigate the effectiveness of smart control systems on the high impact response attenuations of the structure, an experimental test framework is proposed that includes drop tower tests, an aluminum cantilever beam, two MR dampers, data acquisition system, sensors and a high speed camera.

3.1   Drop Tower Test Facility

Drop-tower testing is an effective way of investigating the dynamic response and energy dissipation of structure-MR damper systems under impulse loads.   In this study, the high impact load test facility in Structural Mechanics and Impact Laboratory in the Civil and Environmental Engineering Department at Worcester Polytechnic Institute is used as shown in (Fig. 3). The maximum capacity to apply impulse load of the used mechanism is 22,500 kilogram.   By changing the release heights and drop-masses, the kinetic energy, impact velocity and applied load can be easily adjusted.

Fig. 3. Drop-tower testing facility with a capacity of 22,500 kg

3.2   Aluminum Plate Beam Equipped with MR Dampers

As the structure used to measure the dynamic response, a cantilever aluminum plate beam with dimensions of 615×155×10 mm is used.   The aluminum beam is fixed to the ground to prevent it from shifting during the application of high impact loading. For consistency in each test, the load is applied to the free end of the cantilever beam. The CAD drawing of the beam, placement of the actuators/sensors (MR dampers and accelerometers) and location of the impact load are presented in Fig. 4.

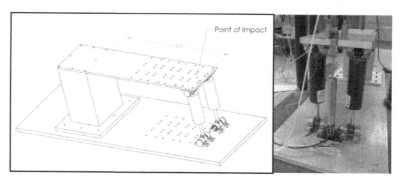

Fig. 4. Details of cantilever aluminum plate beam

The smart control system includes the two MR dampers placed under the cantilever beam and the control system (Fig. 4). The MR dampers consist of the hydraulic cylinders, the magnetic coils and MR fluid. The MR fluid consists of micron sized magnetically polarized particles within an oil-type fluid (Spencer et al. 1997). The feature which makes the MR dampers so attractive is that they can be both operated as passive or active dampers. In the active system, the application of a measured magnetic field to the MR fluid affects its rheological and flow properties which cause it to absorb and dissipate energy effectively. The MR energy dissipation function is adjusted based on feedback of current signals associated with structural response gained through sensors monitoring the structure. In contrast to active systems, MR dampers can still operate as a passive damper if some control feedback component, e.g., wires and sensors, are damaged for some reason (Mitchell et al., 2012).

3.3   Data Acquisition

During the impact tests, acceleration, velocity and impact forces are collected with three sensors connected to a National Instrument (NI) LabView data acquisition system. A 500 g capacity PCB type 302A accelerometer and a 4,500 kg capacity Central HTC-10K type load cells are used in the acceleration and impact force measurements respectively. The sampling rate of the data acquisition system is 10000 data points per second.

The goal of the experimental testing is to measure the dynamic response of the smart structure under different impact loads and different scenarios including with and without the MR dampers. A series of experimental tests are performed by changing the drop release height (25 ~ 80mm) and the current level (0 ~ 1.9A) applied to the MR damper. For each drop release height and current on the MR damper, the drop release test is performed three times to train and validate the proposed models.

A total of 105 impact tests are performed to investigate the structural response under five different force levels without MR damper and with MR damper for six different current levels. To design an optimal control system, a dynamic model to predict the nonlinear behavior of the smart structure needs to be developed. However, as previously discussed, it is challenging to derive an analytical model for describing the nonlinear impact behavior of the time-varying smart structures equipped with highly nonlinear hysteric control devices. To address this issue, the TANFIS model is proposed for predicting nonlinear impact behavior of the smart structure.

4   System Identification Results

4.1   Parameter Setting

To develop the proposed models, sets of input and output data are collected and prepared for training and validation. Fig. 5 shows the input-output data sets for training the TANFIS. In this modeling, acceleration is used as the output while currents and impact loads are the 1st and 2nd input signals.

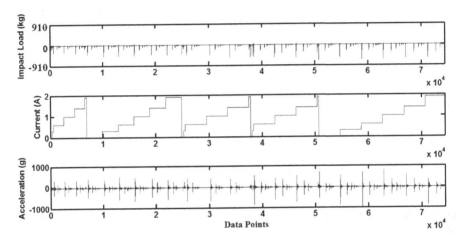

Fig. 5. Input-output data sets to train the model to predict acceleration

In the simulation process, to get the best match, an iterative method is used by changing the training iteration, step size, type and quantity of MFs.

4.2   TANFIS Modeling

Fig. 6 represents the conceptual configuration of the propsed TANFIS model. Each input variable uses two MFs.

Fig. 7 compares the real measured acceleration responses with the estimates from the proposed models for various drop release heights with various current levels. There is a great agreement between the estimates and measurements.

To generalize the trained models, they are validated using different data sets that are not used in the training process. Fig. 8 exhibits the graphs of validated data sets.

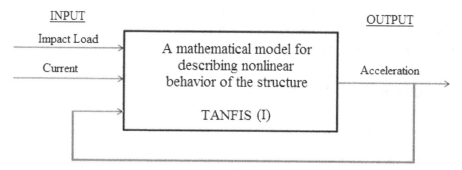

Fig. 6. Configuration of the proposed TANFIS: Impact acceleration prediction

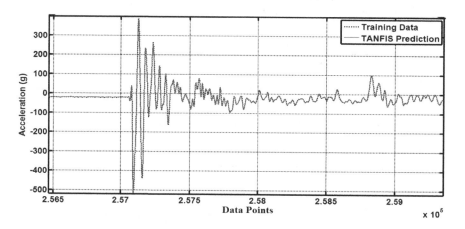

Fig. 7. Comparison of the acceleration measurements with TANFIS model for various currents and drop release heights

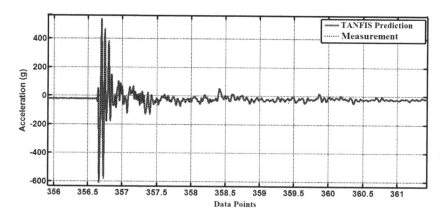

Fig. 8. Validation

5   Conclusion

In this paper, a time-delayed adaptive neuro fuzzy inference system (TANFIS) model is proposed for modeling nonlinear impact responses of smart structures equipped with highly nonlinear hysteretic control devices under high impact loadings. To train the proposed TANFIS models, high impact loads and current signals are used as input signals while the acceleration responses are used as output signals. The proposed TANFIS model is effective in predicting acceleration responses of smart structures. Also, the trained TANFIS models are validated using different data sets that are not used in the training process. It is demonstrated from both the training and validation results that the proposed TANFIS is very effective in estimating nonlinear behaviors of structures equipped with highly nonlinear hysteretic MR damper systems under a variety of high impact loads.

References

1. Adeli, H., Jiang, X.: Dynamic Fuzzy Wavelet Neural Network Model for Structural System Identification. Journal of Structural Engineering 132, 102–111 (2006)
2. Ahmadian, M., Norris, J.A.: Experimental Analysis of Magneto Rheological Dampers when Subjected to Impact and Shock Loading. Communications in Nonlinear Science and Numerical Simulation 13, 1978–1985 (2008)
3. Alhanafy, T.E.: A Systematic Algorithm to Construct Neuro-fuzzy Inference System. In: 16th International Conference on Software Engineering and Data Engineering, vol. 1, pp. 137–142 (2007)
4. Allison, S.H., Chase, J.G.: Identification of Structural System Parameters Using the Cascade-Correlation Neural Network. Journal of Dynamic Systems, Measurement, and Control 116, 790–792 (1994)
5. Bani-Hani, K., Ghaboussi, J., Schneider, S.P.: Experimental Study of Identification and Control of Structures using Neural Network Part 1: Identification. Earthquake Engineering and Structural Dynamics 28, 995–1018 (1999)
6. Consolazio, G.R., Davidson, M.T., Getter, D.J.: Vessel Crushing and Structural Collapse Relationships for Bridge Design, Structures Research Report, Department of Civil and Coastal Engineering, University of Florida (2010)
7. Dyke, S.J., Yi, F., Caicedo, J.M., Carlson, J.D.: Experimental Verification of Multinput Seismic Control Strategies for Smart Dampers. ASCE Journal of Engineering Mechanics 127, 1152–1164 (2001)
8. Faravelli, L., Yao, T.: Use of Adaptive Networks in Fuzzy Control of Civil Structures. Microcomputer in Civil Engineering 12, 67–76 (1996)
9. Gopalakrishnan, K., Khaitan, S.K.: Finite Element Based Adaptive Neuro-Fuzzy Inference Technique for Parameter Identification of Multi-Layered Transportation Structures. Transport 25, 58–65 (2010)
10. Hongsheng, H., Suxiang, Q.: Performance Simulation and Experimental Evaluation for a Magnet-rheological Damper under Impact Load. In: Proceedings of the 2008 IEEE International Conference on Robotics and Biomimetics, pp. 1538–1543 (2009), doi:10.1109
11. Hung, S.L., Huang, C.S., Wen, C.M., Hsu, Y.C.: Nonparametric Identification of a Building Structure from Experimental Data using Wavelet Neural Network. Computer-Aided Civil and Infrastructure Engineering 18, 356–368 (2003)

12. Jang, J.S.R.: ANFIS: Adaptive-Network-Based Fuzzy Inference System. IEEE Transactions on Systems, Man, and Cybernetics 23, 665–685 (1993)
13. Kim, Y., Langari, R., Hurlebaus, S.: Semiactive Nonlinear Control of a Building with a Magnetorheological Damper System. Mechanical Systems and Signal Processing 23, 300–315 (2009)
14. Kim, Y., Langari, R., Hurlebaus, S.: Control of Seismically Exited Benchmark Building using Linear Matrix Inequality-based Semiactive Nonlinear Fuzzy Control. ASCE Journal of Structural Engineering 136, 1023–1026 (2010)
15. Kim, Y., Mallick, R., Bhowmick, S., Chen, B.: Nonlinear system identification of large-scale smart pavement systems. Expert Systems with Applications 40, 3551–3560 (2013)
16. Kim, Y., Langari, R., Hurlebaus, S.: MIMO Fuzzy Identification of Building-MR damper System. International Journal of Intelligent and Fuzzy Systems 22, 185–205 (2011)
17. Mitchell, R., Kim, Y., El-Korchi, T.: System identification of smart structures using a wavelet neuro-fuzzy model. Journal of Smart Materials and Structures 21, 115009 (2012), doi:10.1088/0964-1726/21/11/115009
18. Lin, J.W., Betti, R., Smyth, A.W., Longman, R.W.: On-line Identification of Non-linear Hysteretic Structural Systems using a Variable Trace Approach. Earthquake Engineering and Structural Dynamics 30, 1279–1303 (2001)
19. Mao, M., Hu, W., Wereley, N.M., Browne, A.L., Ulicny, J.: Shock Load Mitigation Using Magnetorheological Energy Absorber with Bifold Valves. In: Proceedings of SPIE, vol. 6527, pp. 652710.1–652710.12 (2007)
20. Masri, S.F., Smyth, A.W., Chassiakos, A.G., Caughey, T.K., Hunter, N.F.: Application of Neural Networks for Detection of Changes in Nonlinear Systems. ASCE Journal of Engineering Mechanics 126, 666–676 (2000)
21. Spencer Jr., B.F., Dyke, S.J., Sain, M.K., Carlson, J.D.: Phenomenological Model for Magnetorheological Dampers. ASCE Journal of Engineering Mechanics 123, 230–238 (1997)
22. Spencer Jr., B.F., Nagarajaiah, S.: State of the Art of Structural Control. ASCE Journal of Structural Engineering 129, 845–856 (2003)
23. Suresh, K., Deb, S.K., Dutta, A.: Parametric System Identification of Multistoreyed Buildings with Non-uniform Mass and Stiffness Distribution. In: Proceedings of 14th WCEE, Paper ID: 05-01-0053 (2008)
24. Tahmasebi, P., Hezarkhani, A.: Application of Adaptive Neuro-Fuzzy Inference System for Grade Estimation; Case Study, Sarcheshmeh Porphyry Copper Deposit, Kerman, Iran. Australian Journal of Basic and Applied Sciences 4, 408–420 (2010)
25. Wang, H.: Hierarchical ANFIS Identification of Magneto-Rheological Dampers. Applied Mechanics and Materials 32, 343–348 (2010)
26. Wang, J., Li, Y.: Dynamic simulation and test verification of MR shock absorber under impact load. Journal of Intelligent Material Systems and Structures 17, 309–314 (2006)
27. Yang, Y.N., Lin, S.: On-line Identification of Non-linear Hysteretic Structures using Adaptive Tracking Technique. International Journal of Non-Linear Mechanics 39, 1481–1491 (2004)

Spikes Monitors for FPGAs, an Experimental Comparative Study[*]

Elena Cerezuela-Escudero, Manuel Jesus Dominguez-Morales,
Angel Jiménez-Fernández, Rafael Paz-Vicente,
Alejandro Linares-Barranco, and Gabriel Jiménez-Moreno

Departamento de Arquitectura y Tecnología de Computadores,
ETS Ingeniería Informática - Universidad de Sevilla,
Av. Reina Mercedes s/n, 41012-Sevilla, Spain
ecerezuela@atc.us.es

Abstract. In this paper we present and analyze two VHDL components for monitoring internal activity of spikes fired by silicon neurons inside FPGAs. These spikes monitors encode each spike according to the Address-Event Representation, sending them through a time multiplexed digital bus as discrete events, using different strategies. In order to study and analyze their behavior we have designed an experimental scenario, where diverse AER systems have been used to stimulate the spikes monitors and collect the output AER events, for later analysis. We have applied a battery of tests on both monitors in order to measure diverse features such as maximum spike load and AER event loss due to collisions.

Keywords: spiking neurons, monitoring spikes, Address-Event Representation, Field Programmable Gate Array, inter-chip communication.

1   Introduction

Neuromorphic systems provide a high level of parallelism, interconnectivity, and scalability; doing complex processing in real time, with a good relation between quality, speed and resource consumption. Neuromorphic engineers work in the study, design and development of neuro-inspired systems, like aVLSI (analog VLSI) chips for sensors [1][2], neuro-inspired processing, filtering or learning [3][4][5][6], neuro-inspired control pattern generators (CPG), neuro-inspired robotics [7][8][11] and so on. Spiking systems are neural models that mimic the neurons layers of the brain for processing purposes. Signals in spikes-domains are composed of short pulses in time, called spikes. Information is carried by spikes, and it is measured in spike frequency or rate [9], following a Pulse Frequency Modulation (PFM) scheme, and also from another point of view, in the inter-spike-time (ISI) [5]. If we have several layers with hundreds or thousands of neurons, it turns very difficult to use a point to multiple-point connection among neurons along the chips that implement different neuronal

[*] This work has been supported by the Spanish government grant project VULCANO (TEC2009-10639-C04-02) and BIOSENSE (TEC2012-37868-C04-02).

I. Rojas, G. Joya, and J. Cabestany (Eds.): IWANN 2013, Part I, LNCS 7902, pp. 179–188, 2013.

layers. This problem is solved thanks to the introduction of the Address-Event Representation (AER), proposed by Mead lab in 1991, facing this problem using a common digital bus multiplexed in time, the AER bus. The idea is to give a digital unique code (address) to each neuron. Whenever a neuron fires a spike a circuit should take note of it, manage the possible collisions with other simultaneously fired spikes and, finally, encode it as an event with its pre-assigned address. This event will be transferred through the AER bus, which uses additional control lines of request (REQ) and acknowledge (ACK), implementing a 4-phase asynchronous hand-shake protocol. In the receiver, neurons will be listening to the bus, looking for the spikes sent to them [10]. Using the AER codification, neurons are virtually connected by streams of spikes.

This work is focused on presenting in a detailed way two spikes monitors, written in VHDL for FPGAs, which encode each spike according to the Address-Event Representation, using different strategies in order to avoid spike collisions in time. Temporal spike collision is known as the situation where two or more spikes have been fired at the same time, and they should be sent using the AER bus. Fig. 1 shows the typical application of spikes monitors, where there are circuits that fire spikes (e.g. a set of spiking neurons), and there is a spikes monitor connected to them , which will encode the spikes and send them to another layer, using the AER bus.

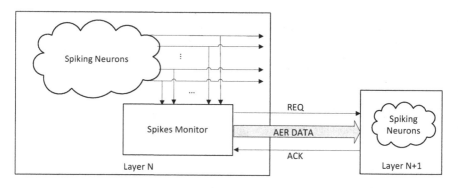

Fig. 1. Typical application of spikes monitor in a multilayer AER system

2   Spikes Monitors Description

We have implemented two spikes monitors, as mentioned before; these monitors implement different strategies to manage the spike collisions in time, being this an important concept, because it is the main difficulty in this kind of system. In an ideal scenario, where spikes are fired neuron by neuron, sequentially, without temporal collisions, spikes encoding as AER events will be automatic using a traditional digital encoder. However, when two or more spikes are fired simultaneously, two or more AER events should be transferred, but the AER bus is unique and multiplexed in time. In consequence, spikes that have been fired in parallel will be transmitted as AER events sequentially. Different strategies can be used to implement this functionality;

the results and monitor behavior will depend directly on the strategy adopted, and both monitors are generics and can be adapted to variable input spike number.

2.1   Massive Spikes Monitor(MSM)

MSM needs three blocks: the first block is used to avoid collisions taking a snapshot of the spikes activity every clock cycle; the second block to encode the spike with its address; and the third block sends the address with the hand-shake protocol. Fig. 2 shows the block diagram. In order to avoid collisions, MSM takes a snapshot of spikes and stores it in a FIFO (Spikes FIFO) if some spike has been fired (Fig. 2 top). If there are many '1's in a single word it means that more than one spike has been fired at the same time. Now we need to encode every spike with its address. We have designed a Finite State Machine (FSM) which, if the Spikes FIFO is not empty, loads a word into a register and looks for a spike bit by bit. If it finds one, it looks for its address in a ROM and writes the address in another FIFO (AER FIFO) to be transferred as an AER event (Fig. 2 middle). The AER FIFO contains the encoded spikes addresses, and they are ready to be sent through the AER port. Finally there is included another FSM for the 4-phase AER handshaking (Fig.2 right) [11].

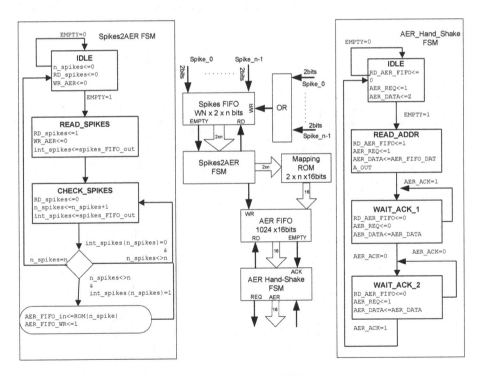

Fig. 2. MSM internal structure and FSM descriptions

The main problem of MSM resides in the fact that all spikes are stored in a single word, and when this word is relatively big MSM demands a high memory quantity for Spikes FIFO in synthesis time and a great number of clock cycles to search for spikes, introducing a high latency in spikes encoding, and consequently losing a high rate of spikes.

2.2   Distributed Spikes Monitor (DSM)

The DSM aims to avoid the problem of MSM, breaking the Spikes FIFO and encoding FSM into several identical sub-circuits, which distributes the task of spikes encoding in different FIFOs and a FSM that can now work in parallel. The DSM distributes spikes in four similar modules; therefore a quarter of the input spikes excite each module, which is shown in Fig. 3. Each module stores its spike portion in a register and looks for a spike bit by bit. If it finds one spike, it works out its partial AER address by the index on the register. Then, the module stores this partial address in a FIFO. Now we need to encode every spike with its complete address. We have designed a FSM which computes the full address from the partial address and empty signals. Finally, the monitor writes the address in the AER FIFO (Fig. 3 bottom). This contains the spikes addresses as events, and they are ready to be sent through the AER port.

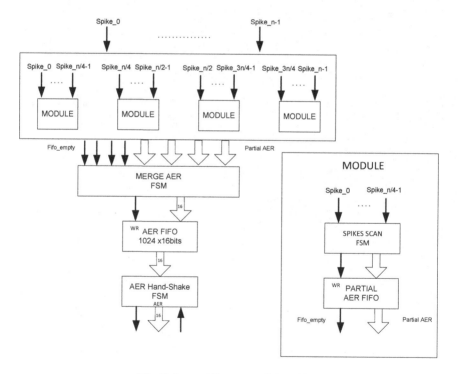

Fig. 3. Internal Structure of the DSM

3   Experimental Setup

In this work we want to study and analyze the monitors' behavior. In order to achieve this, we have designed an experimental scenario, where diverse AER systems have been used to stimulate the spikes monitors and to collect the output AER events, for later analysis. The experiment components are (Fig. 4):

1. First, a PC generates a test battery of spikes using MATLAB.
2. The PC sends the spikes information to an USB-AER board through USB interface. The USB-AER board stores the spikes in its RAM memory. We have implemented a component VHDL to manage RAM memory. These spikes are used to stimulate the spikes monitors.
3. At the end, the USB-AERmini2 receives the monitor outputs and sends them to the PC for later analysis.

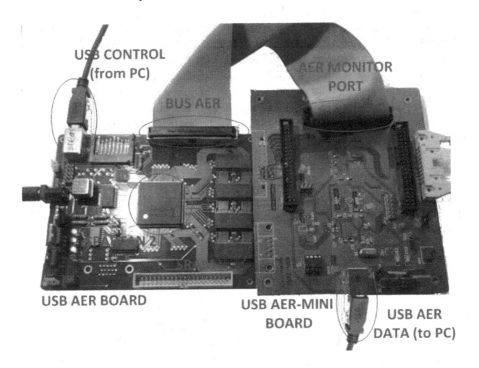

Fig. 4. Experiment Components

3.1   USB-AER and USB-AERmini2 Boards

We used an USB-AER board to load the MSM and DSM with the VHDL stimulus builder component. USB-AER board is based on a Xilinx Spartan II-200 FPGA that can be reconfigured using the USB interface provided by SiliconLabs 8051 microcontroller or SD card. This board includes two AER parallel ports (input and output) and 2 Mbytes of static RAM (SRAM) [12].

We used the USB-AERmini2 board in order to monitor the AER traffic in a PC. This device allows monitoring and sequencing AER events with a time resolution of 200 nanoseconds. The device consists of a Cypress FX2LP microcontroller and a Xilinx Cool runner 2 CPLD. The CPLD is clocked with 30 MHz and achieves a peak monitor rate of 6 Megaevents per second and a sustained rate of 5 Megaevents per second, which is limited by the host computer. This board provides the captured AER events and the time instant at which they have been fired (time stamp) [13].

3.2   Stimulating the Monitors: Generating Spikes and Processing AER Events

The USB-AER board receives the spikes information from the USB interface (Fig. 5-1) and stores them in the SRAM memory, using a component which manages the communication between the USB and the SRAM (Fig.5-2). Then, the system reads SRAM (Fig.5-3) and uses these spikes to stimulate the MSM and DSM (Fig.5-4). The output AER events are sent to the USB-AERmini2 by the AER output parallel port.

We have implemented a MATLAB function which generates random spikes from set input parameters, such as the number of maximum active spikes in time instant, and the probability of this to happen. Being these spikes used to stimulate MSM/DSM inputs, the PC receives the MSM/DSM outputs by USB-AERmini2 and analyzes them.

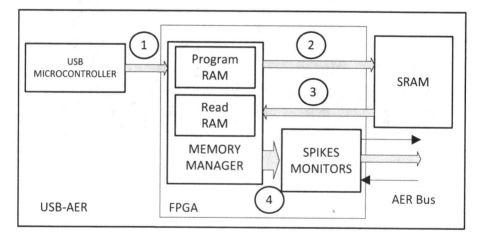

Fig. 5. Experiment Execution

We have designed the battery of tests from equation 1 which calculates the parameters to obtain a particular average spike rate.

4   Experimental Results

In order to characterize the monitors´ behavior we have excited both monitors with diverse stimulus inputs, creating a sweep of stimulus spike rate and changing the number of simultaneously fired spikes using equation 1. The spike rate generated changed from 2 to 20 MSpkes/Sec, and the number of simultaneous spikes from 8 to 16 spikes.

$$Avg.\,Events\,Rate = \frac{Number\,of\,Active\,Spikes * Probability}{Total\,Stimulus\,Time}(Events/_{Sec}) \qquad (1)$$

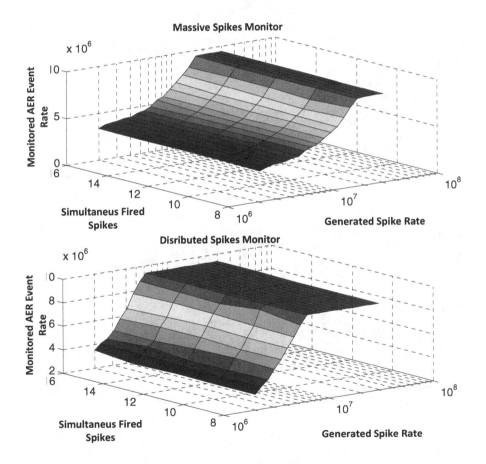

Fig. 6. MSM and DSM output AER events rate

The first measurement done was the average AER events rate monitored for every test case; Fig. 6 shows the results for both MSM (top) and DSM (bottom). Due to the structure of the MSM, many spikes are lost, and for example, when it is excited with 10MSpikes/Sec, it only provides about 6 MEvents/Sec. However, it needs about 8 MSpikes/Sec to reach an AER event rate of 9.8 MEvents/Sec, being this the maximum capacity of AER events monitoring for the USB-AERmini2 board, saturating in consequence the AER bus. Opposite to the MSM, the DSM shows a better behavior, providing at its output an AER event rate very similar to the input spikes, saturating the AER bus when it is excited with 10 MEvents. In both cases, the number of simultaneously fired spikes does not affect significantly the spikes monitors performance.

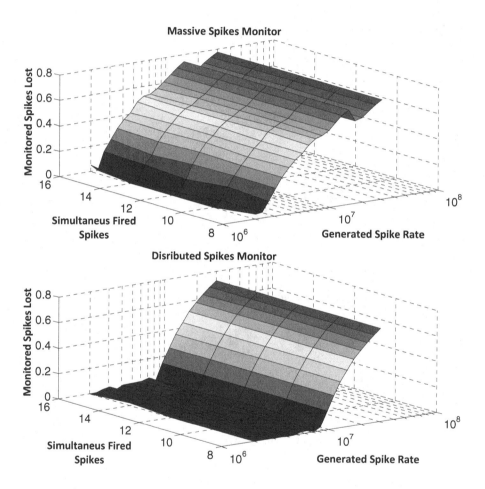

Fig. 7. MSM and DSM spikes loss

Next, we have measured the ratio of spikes lost in the same conditions as in the previous measurement. Fig. 7 contains the spikes loss ratio for the MSM (top) and DSM (bottom). For the MSM, there are a low number of spikes lost for the lower spike rate values, only being discarded a few of them at 2-3MSpikes/Sec. However, when the stimulus spike rate is increased, the MSM Spikes FIFO is full very soon, and many spikes are discarded.

One more time, the DSM presents a better response, losing a very small quantity of spikes, thanks to its FIFOs and FSM distribution, and it only stars losing a considerable amount of spikes when it is excited with more than 10 MSpikes/Sec. However, this is the AER bus maximum reachable event rate using the USB-AERmini2 board, which starts discarding AER events.

After these experiments, the DSM denotes a better behavior than the MSM, being very adequate for this kind of system since it provides a higher bandwidth, in terms of AER events, which can be transferred using the AER bus, being this now the communication bottleneck.

5   Conclusions

In this work we want to study and analyze the behavior of two different spikes monitors' with spiking neurons. We have designed an experimental scenario, where diverse AER systems have been connected together, building a test infrastructure for stimulating MSM and DSM, and collect their AER information for later analysis. Finally, we have analyzed MSM and DSM responses in terms of output AER events rate, and the ratio of spike loss.

The DSM has shown better behavior, providing a higher AER events rate than the MSM, having a low spike loss and reaching a higher AER events rate than the AER bus capacity.

References

1. Lichtsteiner, P., et al.: A 128×128 120dB 15 us Asynchronous Temporal Contrast Vision Sensor. IEEE Journal on Solid-State Circuits 43(2) (2008)
2. Chan, V., et al.: AER EAR: A Matched Silicon Cochlea Pair With Address Event Representation Interface. IEEE T. Circuits and Systems I 54(1) (2007)
3. Serrano-Gotarredona, R., et al.: On Real-Time AER 2-D Convolutions Hardware for Neuromorphic Spike-Based Cortical Processing. IEEE T. Neural Network 19(7) (2008)
4. Oster, M., et al.: Quantifying Input and Output Spike Statistics of a Winner-Take-All Network in a Vision System. In: IEEE International Symposium on Circuits and Systems, ISCAS 2007 (2007)
5. Hafliger, P.: Adaptive WTA with an Analog VLSI Neuromorphic Learning Chip. IEEE T. Neural Networks 18(2) (2007)
6. Indiveri, G., et al.: A VLSI Array of Low-Power Spiking Neurons and Bistables Synapses with Spike-Timig Dependant Plasticity. IEEE T. Neural Networks 17(1) (2006)
7. Linares-Barranco, A., et al.: Using FPGA for visuo-motor control with a silicon retina and a humanoid robot. In: IEEE International Symposium on Circuits and Systemsm, ISCAS 2007 (2007)
8. Jiménez-Fernández, A., et al.: AER and dynamic systems co-simulation over Simulink with Xilinx System Generator. In: IEEE I. Conference on Electronic, Circuits and Systems, ICECS 2008 (2008)

9. Shepherd, G.: The Synaptic Organization of the Brain. Oxford University Press (1990)
10. Boahen, K.: Point-to-Point Connectivity Between Neuromorphic Chips Using Address Events. IEEE T. Circuits and Systems II 47(5) (2000)
11. Jiménez-Fernández, A., et al.: Spike-based control monitoring and analysis with Address Event Representation. In: IEEE Int. Conference on Computer Systems and Applications, AICCSA 2009 (2009)
12. Gómez-Rodríguez, F., et al.: AER tools for communications and debugging. In: Proceedings of IEEE Int. Sym. on Circuits and Systems, ISCAS 2006 (2006)
13. Berner, R., et al.: A 5 Meps $100 USB2.0 Address-Event Monitor-Sequencer Interface. In: IEEE International Symposium on Circuits and Systems, ISCAS 2007 (2007)

On Second Language Tutoring through Womb Grammars

Leonor Becerra Bonache[1], Veronica Dahl[2], and J. Emilio Miralles[2]

[1] Laboratoire Hubert Curien, Jean Monnet University,
18 rue Benoit Lauras, 42100 Saint-Etienne, France
`leonor.becerra@univ-st-etienne.fr`
[2] Simon Fraser University, Burnaby, BC, V5A-1S6, Canada
`veronica@cs.sfu.ca, emiralle@sfu.ca`

Abstract. Womb Grammar Parsing is a novel constraint based paradigm that was devised mainly to induce grammatical structure from the description of its syntactic constraints in a related language. In this paper we explore its uses for second language tutoring, and propose a model that combines automatic proficiency level detection, grammar repair, and automatic training with live training through interaction. Research has shown that live training is indispensable for speedy acquisition, but the parts of the process that can be automated will be of great help, combined with such live training, to increase the learning speed optimally. We believe that freeing the teacher's time from the less creative parts of the live interaction, namely proficiency level detection and grammar repair, will promote a richer experience for both student and teacher.

Keywords: Womb Grammar Parsing, Language Acquisition, Second Language Tutoring, Constraint Based Grammars, Property Grammars, CHRG.

1 Introduction

Womb Grammar Parsing was introduced in [6] as a means to derive the grammar of a language from that of another in automatic fashion. In [7], we tailored and exemplified its uses in terms of a novel application of constraint-based parsing: that of inducing the (incorrect) grammar in use by a person learning a language and detecting the level of proficiency of such a learner. We showed as well how to detect a child's morphological level of proficiency in English. The present paper proposes to use this approach for second language tutoring which proceeds through both automated and interactive stages. Unlike previous work, which focuses on machine learning techniques (e.g. [21]), our contribution to quality assessment of utterances in a language being learned proceeds through pointing out which linguistic constraints are being violated. From these, an accurate (while probably incorrect by academic standards) grammar of the users language proficiency can be produced, as well as a set of exercises targeting his or her progress.

I. Rojas, G. Joya, and J. Cabestany (Eds.): IWANN 2013, Part I, LNCS 7902, pp. 189–197, 2013.
© Springer-Verlag Berlin Heidelberg 2013

After presenting our methodological background in the next section, section 3 presents our second language tutoring model and describes how to use Womb Grammars to detect a second language learner's level of grammatical proficiency. Section 4 presents our concluding remarks.

2 Background

2.1 Womb Grammars

Property Grammars [1] and HPSG [18] are examples of constraint-based linguistic models that represent linguistic information in terms of non-hierarchical constraints. HPSG, or Head-driven phrase structure grammar, is a linguistic formalism developed by Pollard and Sag [19] which simplifies grammar rules through moving much of the complexity of the grammar into the lexicon, which is richly structured. It is highly modular so although thought out as a generative theory, has originated many computer applications. Property Grammars define the acceptance of a phrase in terms of the properties (or constraints) that must be satisfied by a group of categories. In that way, English noun phrases, for instance, can be described through a few constraints: precedence (a determiner must precede a noun), uniqueness (there must be only one determiner), etc. These approaches have several advantages with respect to classical parsing methods; in particular Property Grammars allow characterization of a sentence through the list of the constraints a phrase satisfies and the ones it violates, with the good result that even imperfect phrases will be parsed. In contrast, classical methods of parsing simply fail when the input is incorrect or incomplete.

Womb Grammar Parsing paradigm constitutes a new and original development of constraint-based parsing. It was designed to induce, given a corpus of correct phrases in a target language, the target language's constraints from those of another language called the source. One of the main differences with respect to other approaches that already exists is that, Womb Grammar Parsing focuses on *generating the constraints* that would sanction the input as correct, rather than characterizing the acceptability of a sentence in terms of linguistic constraints. This is because it was conceived for grammar induction rather than only for parsing sentences. Therefore, this paradigm is ideal for grammar correction and grammar induction, not just for flexible parsing.

More concretely: let L^S (the source language) be a human language that has been studied by linguists and for which we have a reliable parser that accepts correct sentences while pointing out, in the case of incorrect ones, what grammatical constraints are being violated. Its syntactic component will be noted L^S_{syntax}, and its lexical component, L^S_{lex}.

Now imagine we come across a dialect or language called the target language, or L^T, which is close to L^S but has not yet been studied, so that we can only have access to its lexicon (L^T_{lex}) but we know its syntax rules overlap significantly with those of L^S. If we can get hold of a sufficiently representative corpus of sentences in L^T that are known to be correct, we can feed these to a hybrid parser consisting of L^S_{syntax} and L^T_{lex}. This will result in some of the sentences being marked as incorrect by the parser. An analysis of the constraints these "incorrect" sentences violate can subsequently reveal how to transform L^S_{syntax} so it accepts as correct the sentences in the corpus of L^T—i.e., how to transform it into L^T_{syntax}. For more information, see [6] and [7].

Language acquisition is a research area where constraint-based approaches, such as Womb Grammars, can make important contributions. Surveys on applications of constraint-based approaches for processing learner language, mostly around error detection, can be found in [10] and [2]. In this paper we propose the application of Womb Grammars for Second Language Acquisition.

2.2 Second Language Acquisition

Second language acquisition (SLA) refers to the process of learning a language which is not your native language. The term *second* is also used to refer to learning third, fourth or subsequent languages. The term *acquisition* was originally used to emphasize the subconscious nature of the learning process, but nowadays, for most SLA researchers, *acquisition* and *learning* are interchangeable.

SLA is a complex phenomenon and despite all research efforts in this domain, there are still many issues unresolved. The study of how learners learn a second language began in the late 60s. Since then, a huge number of SLA theories and hypotheses has been proposed; as Larsen-Freeman and Long stated, "at least forty 'theories' of SLA have been proposed" [13]. However, none of them are accepted as a complete explanation for the phenomenon by all SLA researchers.

Here we are going to briefly summarize some of the theories and hypotheses that have caused a great impact in the field. For more information, see [15].

Interlanguage Theory. In order to explain the systematic errors of second language learners (some of which are not attributed to learner's first language nor age, etc.), the idea of *interlanguage* was developed [20]. Interlanguage is the type of language produced by a learner who is in the process of learning a second language. Therefore, this concept assumes that the learner creates a self-contained linguistic system, different from their first and second language [3]. The concept of interlanguage has been widely extended in SLA research and is often a basic assumption made by SLA researchers.

Universal Grammar-Based Approaches. Linguistic approaches for explaining SLA are mainly based on Chomsky's theory of Universal Grammar [4]. It consists of a set of *principles*, which are shared by all the languages, and *parameters*, which can vary between languages. Hence, learning the grammar of a second language is simply a matter of setting the parameters in the language correctly. For instance, the *pro-dop* (or null-subject) parameter dictates whether or not sentences must have a subject to be grammatically correct. In English, the sentence "he speaks" is grammatically correct, but "speaks" is ungrammatical (i.e., the subject "he" must appear in the sentence). However, in Spanish, the sentence "habla" (i.e., "speaks") is grammatically correct. Therefore, an English speaker learning Spanish would need to deduce that the subject is optional in Spanish, and then set his *pro-dop* parameter accordingly to it.

There exists different views about the role of universal grammar in SLA. Some researchers consider that universal grammar is available or partially available to second language learners, whereas some others argue that it does not have any role in SLA [8].

Monitor Model. This model was developed by S. Krashen in the 70's and 80's [12]. He was influenced by Chomsky's assumptions on language as an innate faculty. The Monitor Model is a group of five interrelated hypotheses. Some of the most well known are: 1) *The input hypothesis*: it states that language acquisition takes place only when learners receive comprehensible input, i.e. input that is just beyond their current level of competence. Comprehensible input is conceptualized as "$i + 1$", where i is the current level of proficiency and $+1$ is the next stage of language acquisition. Krashen believes that processing and understanding of such samples activate the innate language faculty allowing learners to proceed from one stage to another [12]; 2) *The natural order hypothesis*: it states that all learners acquire a language in roughly the same order. This hypothesis is based on morpheme studies that found that certain morphemes were predictably learned before others in SLA [12].

Processability Theory. It was developed by M. Pienemann [17]. This theory states that second language learners restructure their interlanguage knowledge systems in an order of which they are capable at their stage of development. For example, in order to correctly form English questions, learners must transform declarative English sentences, and they do so by a series of stages, consistent across learners.

The application of this theory to language teaching is called the *Teachability Hypothesis*. It assumes that language acquisition can benefit from language instruction as long as this instruction concerns structures for which the interlanguage is developmentally ready. Therefore, according to this hypothesis, instruction can speed up the rate of development in SLA, providing that learners are instructed on one stage beyond their current proficiency level. In fact, Pienemann observed that instruction was most effective when it reflected the stage just beyond the learner's current stage of interlanguage [16].

Rapid Profile is a computer software developed by Pienemann and his collaborators used to assess language learners' level of development. It was empirically tested by Kebler [11]. It can be applied in the classroom to gain quick and valid profiles of second language learners interlanguage development. These profiles tell language teachers and curriculum designers what the learners are ready to acquire at a given point in their process of second language learning.

Interaction Hypothesis. This hypothesis was proposed by Long [14]. It states that language acquisition is strongly facilitated by the use of the target language in interaction. The interaction can play an important role in bringing learners' attention to new structures and can contribute, in this way, to language development.

3 Womb Grammars for Second Language Tutoring

The place of grammar instruction in second language acquisition has been a subject of debate since the origins of SLA research, and many of the debates still remain opened. It has been proved that SLA mirrors to some extent the processes involved in the acquisition of first languages [15], but also that mere exposure to the target language does not guarantee the attainment of high levels of grammatical and discourse competence. Consequently, instruction has been reintroduced into the language classroom and has been widely accepted that grammar teaching should become a vital part of classroom practices [9]. Therefore, second language tutoring seems to play an important role for the linguistic development of second language learners.

Although every theory and hypothesis described in the previous section presents a different view of SLA, there are some aspects of each theory that are complementary. In this paper we are going to take into account mainly 3 aspects:

– As research in SLA has shown, second language learners follow a fairly rigid developmental route. The learner creates a series of interlocking linguistic systems, i.e. interlanguages, which are linguistic systems in their own right, with their own set of rules (ideas extracted from the interlanguage theory and natural order hypothesis).
– If teachers give instruction just exactly at the students' level, there would not be any progress in learner's interlanguage except acquiring some input from their surrounding. We believe that classroom instruction that is a little above the student's level would be obviously more effective for students in learning a second language. However, instruction should not be too much above their level (idea extracted from Krashen's input hypothesis [12] and Pinemann's teachability hypothesis [16]).
– Interaction plays an important role for the linguistic development of the learners (idea extracted from the interaction hypothesis).

Mystkowska-Wiertelak and Pawlak stated in [15] that: "Keeping in mind that the accomplishment of a lower-rank processing procedure enables the learner to reach a higher stage, teachers would have to apply complex diagnostic mechanisms, first, to identify the current level, next, to check if a given stage has been successfully accomplished. It is highly unrealistic that any educational system could afford a teaching programme which would manage to tailor classroom procedures to the needs of every single student". We believe that it would be possible to achieve this task if it is done in an automatic way.

Inspired by all these ideas, we propose a new model based on Womb grammars that would allow us:

– to detect in an automatic way a second language learner's level.
– to provide an automatic and live training to second language learners.

Figure 1 shows its workings: a student's input corpus to the level detection module produces the level the student is at and the corresponding grammar; this grammar is treated by the repair module to get to the next level's grammar, and the automatic training module presents easily mechanizable training exercises that allow the student to advance from level L to level L+1. Next, live training through interaction will complete the job, or advance it as far as it will go in one session, and the level is then tested again, for a further iteration of the whole process.

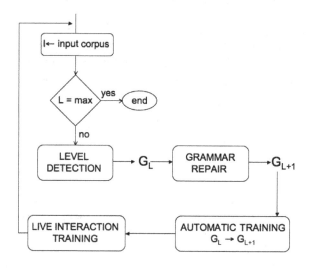

Fig. 1. Second language tutoring model

As stated in [7], the generative power of Womb Grammars can be used to find out the set of linguistic constraints (i.e., the grammar) that a person learning a language is using. In this paper, we propose to detect the level of the second language learner by using a Womb Grammar. We will use, as in [6], a *universal grammar* of our own device, noted as L^U, consisting of all possible grammar properties for the existing constituents (even contradictory ones). Thus, this Universal Womb Grammar will list *all* possible properties or constraints. By running a student's input through this universal grammar and deleting any constraints not manifest in the input, we obtain a characterization of the student's level.

In this paper we propose to apply our approach to the development of English as a second language. SLA research has determined that there is a specific development schedule for English, shown in Figure 2[1]. Based on these studies, we

---

[1] This figure has been taken from
http://kw.uni-paderborn.de/institute-einrichtungen/
institut-fuer-anglistik-und-amerikanistik/personal/
pienemann/rapid-profile/documents/

modify our universal grammar to now include as well the information contained in Figure 2, by specifying its constraints together with the level their knowledge denotes. For instance, for level 4, the requirement of the verb in the first sentence, of the location circumstantial in the second one, and of the preposition in the third would appear marked by the level number– i.e. 4– so that if violated, this requirement would pronounce the student to have achieved a level no higher than 3. It is still possible that requirements in level 2, say, be violated as well. This would result in the student being pronounced to have achieved a level no higher than 1.

Stage	Phenomena	Examples
6	Cancel Aux-2nd	I wonder what he wants.
5	Neg/Aux-2nd-?	Why didn't you tell me? Why can't she come?
	Aux-2nd -?	Why did she eat that? What will you do?
	3sg-s -	Peter likes bananas.
4	Copula S (x)	Is she at home?
	Wh-copula S (x)	Where is she?
	V-Particle	Turn it off!
3	Do-SV(O)-?	Do he live here?
	Aux SV(O)-?	Can I go home?
	Wh-SV(O)-?	Where she went? What you want?
	Adverb-First	Today he stay here.
	Poss (Pronoun)	I show you my garden. This is your pencil.
	Object (Pronoun)	Mary called him.
2	S neg V(O)	Me no live here. / I don't live here.
	SVO	Me live here.
	SVO-Question	You live here?
	-ed	John played.
	-ing	Jane going.
	Plural –s (Noun)	I like cats.
	Poss –s (Noun)	Pat's cat is fat.
1	Words	Hello, Five Dock, Central
	Formulae	How are you? Where is X? What's your name?

Fig. 2. Developmental features for English as a second language

4 Concluding Remarks

We have argued that Womb Grammar Parsing, whose CHRG (Constraint Handling Rule Grammars [5]) implementation is described in [6], is an ideal aid to guide a student through second language acquisition by using our proposed Universal Womb Grammar. We have also complemented this prototype with a component that can detect the level of a second language learner.

We have studied the applicability of Womb Grammars as an aid within a universal model for learning English as a second language, within an application that seeks to speed up the acquisition of some specific linguistic phenomena, as suggested by studies that show that second language learners also progress orderly along the same stages.

Our research has a great potential practical value, in that it not only help a student progress faster through the learning stages, but can also help educators tailor the games, stories, songs, etc. that can aid a second language learner to progress in timely fashion into the next level of proficiency.

To the best of our knowledge, this is the first time the idea of detecting and improving on grammatical performance levels for language acquisition materializes through weeding out constraints from a kind of universal constraint-based grammar fragment. With this preliminary work we hope to have shown that our proposed approach is a promising one, and to stimulate interest in further work along these lines. This is mostly a position paper on a possible approach which we argue would be useful for second language learning. Future work includes an in-depth study of its applicability in that area, and a proper evaluation of its practical results.

References

1. Blache, P.: Property grammars: A fully constraint-based theory. In: Christiansen, H., Skadhauge, P.R., Villadsen, J. (eds.) CSLP 2005. LNCS (LNAI), vol. 3438, pp. 1–16. Springer, Heidelberg (2005)
2. Boyd, A.A.: Detecting and diagnosing grammatical errors for beginning learners of german: From learner corpus annotation to constraint satisfaction problems. Ph.D. thesis, Ohio State University (2012)
3. Brown, H.: Principles of learning and teaching. White Plains, New York (2000)
4. Chomsky, N.: Syntactic Structures. Mouton, The Hague (1957)
5. Christiansen, H.: CHR grammars. TPLP 5(4-5), 467–501 (2005)
6. Dahl, V., Miralles, E.: Womb parsing. In: 9th International Workshop on Constraint Handling Rules, pp. 32–40 (2012)
7. Dahl, V., Miralles, E., Becerra, L.: On language acquisition through womb grammars. In: CSLP, pp. 99–105 (2012)
8. Ellis, N. (ed.): Implicit and explicit language learning. Academic Press, London (1994)
9. Ellis, R.: Current issues in the teaching of grammar: An sla perspective. TESOL Quarterly 40, 83–107 (2006)
10. Heift, T., Schulze, M.: Errors and intelligence in computer-assisted language learning. Parsers and pedagogues. Routledge, New York (2007)
11. Kebler, J.: Assessing efl-development online: A feasibility study of rapid profile. In: Second Language Acquisition Research. Theory Construction and Testing, pp. 111–135 (2007)
12. Krashen, S.: The input hypothesis. Longman, London (1985)
13. Larsen-Freeman, D., Long, M.H.: An introduction to second language acquisition research. Longman, New York (1991)
14. Long, M.: The role of the linguistic environment in second language acquisition. In: Ritchie, W.C., Bhatia, T.K. (eds.) Handbook of Second Language Acquisition, pp. 413–468. Academic Press, San Diego (1996)
15. Mystkowska-Wiertelak, A., Pawlak, M.: Production-oriented and comprehension-based grammar teaching in the foreign language classroom. Springer, Berlin (2012)
16. Pienemann, M.: Is language teachable? Psycholinguistic experiments and hypotheses. Applied Linguistics 10, 52–79 (1989)

17. Pienemann, M.: Language processing and second language development: Process-ability theory. John Benjamin, Amsterdam (1998)
18. Pollard, C., Sag, I.A.: Head-driven Phrase Structure Grammars. Chicago University Press, Chicago (1994)
19. Pollard, C., Sag, I.: Information-based syntax and semantics. CSLI Lecture Notes, Center for the Study of Language and Information (1987)
20. Selinker, L.: Interlanguage. International Review of Applied Linguistics 10, 201–231 (1972)
21. Yannakoudakis, H., Briscoe, T., Medlock, B.: A new dataset and method for automatically grading esol texts. In: ACL: Human Language Technologies, vol. 1, pp. 180–189. Association for Computational Linguistics (2011)

Simulated Annealing for Real-Time Vertical-Handoff in Wireless Networks

María D. Jaraíz-Simon, Juan A. Gómez-Pulido, Miguel A. Vega-Rodríguez, and Juan M. Sánchez-Pérez

Department of Technologies of Computers and Communications, University of Extremadura, Spain
{mdjaraiz,jangomez,mavega,sanperez}@unex.es

Abstract. When a mobile terminal is moving across heterogeneous wireless networks acting as access points, it must decide the best network to connect to, taking into account the values of the quality of service parameters of the networks. Selecting an optimal set of weights for these values in the terminal is an optimization problem that must be solved in real time for embedded microprocessors that manage the Vertical Handoff decision phase in highly dynamic environments. For this purpose, we have developed an adaptive heuristic inspired on the Simulated Annealing algorithm that improves the performance of a former algorithm designed to solve this optimization problem.

Keywords: Simulated Annealing, Embedded Processors, Wireless Networks, Vertical Handoff, Mobile Devices, Quality of Service.

1 Introduction

In our research framework we consider heterogeneous wireless networks (UMTS, WiMax, WLAN, etc) acting as access points for a mobile terminal that must be connected in any time. Each network is characterized by the values of its *Quality-of-Service* (QoS) parameters. In traditional heterogeneous wireless switching processes, only the channel availability and the signal strength were considered as QoS parameters; nowadays, the new generation networks consider other important parameters [1] [2], like service type, monetary cost, bandwidth, response time, latency, packet loss, bit error rate, battery and security levels, etc.

When the mobile terminal discovers new networks (see Figure 1), it could leave the current network and establish a new link to other one depending on their QoS values. This process is named *Vertical Handoff* (VH) [3], and it consists of three phases: discover, decision and execution. The VH decision phase is driven by algorithms and it is where we have centered our efforts.

Sometimes the terminal is moving quickly, so the algorithms that support the VH decision phase must be fast. In order to decide the best network, we need a function or metric able to give us the goodness of each available network. For this purpose, a set or combination of weights assigned to each one of the QoS parameters is used for support a measure of the quality of the network, and it

I. Rojas, G. Joya, and J. Cabestany (Eds.): IWANN 2013, Part I, LNCS 7902, pp. 198–209, 2013.
© Springer-Verlag Berlin Heidelberg 2013

can be based on the user's preferences [4][1]. This measure is given by a function we name *fitness function*, which gives us the quality degree of a determined combination of weights, and it is evaluated in the VH decision phase.

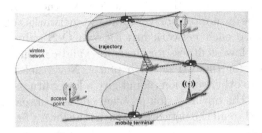

Fig. 1. Vertical Handoff scenery: several heterogeneous wireless networks acting as access points are discovered by a mobile terminal

The fitness function can be formulated as in (1), where n identifies the network, $E^{(n)}$ is an elimination factor, s and i identify the service and the QoS parameter respectively, $w_{s,i}$ is the weight assigned to the QoS parameter i for performing the service s, and N is a normalization function for the cost $p_{s,i}^n$ applied to the parameter i for performing the service s.

$$F^{(n)} = E^{(n)} \sum_s \sum_i w_{s,i} N(p_{s,i}^{(n)}) : \sum_i w_i = 1 \qquad (1)$$

The constraint given in (1) conditions strongly the methodology for solving the optimization problem because not any weight combination can be accepted. Now, we simplify our research considering networks providing only one service, removing the elimination factor and taking as normalization function the natural logarithm [1][4], so the fitness function can be formulated as in (2), where $p_i'^{(n)} = p_i^{(n)}$ if to higher p, higher fitness (in other words, the fitness gets worse as when we consider delay or economic cost), or $p_i'^{(n)} = \frac{1}{p_i^{(n)}}$ if to higher p, lower fitness (the fitness gets better as when we consider bandwidth).

$$F^{(n)} = \sum_i w_i ln(p_i'^{(n)}) : \sum_i w_i = 1 \qquad (2)$$

2   The Optimization Problem

The optimization problem tries to adjust the QoS weights in order to determine the optimal network among the available ones. The search of the best solution is not trivial; it could require a high computational effort. The efficient solving of this optimization problem is very important for wireless networks where many advanced applications need high QoS [6].

The values of the QoS parameters characterize a given network, whereas a combination of weights assigned to them satisfying the constraint (2) gives a measure of the network quality by means of the fitness function. Each weight has a value between 0 and 1, and only the combinations that satisfy that constraint are considered as valid combinations or solutions for the optimization problem. Each solution can be evaluated for the different networks, obtaining different fitness values because each network has its own QoS values. This way, that network offering the lowest fitness for the same valid combination is considered as the best network for the VH decision phase. Nevertheless, there are many valid combinations so other solutions could give a lower fitness for other different networks. Therefore, the optimization problem consists of searching the optimal solution that, applied to all the networks, returns the lowest fitness corresponding to a determined network, which will be chosen for the VH decision phase.

The space of possible solutions is very big, and it depends on the number of QoS parameters $NQoS$ and the precision, hence the need for optimization algorithms for the weights adjustment, like Analytic Hierarchy Process (AHP) [5], Simple Additive Weighting (SAW) [7] and Technique for Order Preference by Similarity to Ideal Solution (TOPSIS) [8], among others. These are good and low complex algorithms that use simple rules to find an optimal solution.

We have used two experimental datasets and three profiles for the user's preferences in order to validate our heuristics.

2.1   Experimental Datasets

The characteristics of the two following datasets are shown in Table 1:

- DS1. This dataset [5] consists of three WLAN and one UMTS networks in an scenery where a terminal moves transferring data files. The interest of this dataset resides in the high number of QoS parameters, because this permits us to supply a high computational effort to the optimization algorithm.
- DS2. This dataset [8] considers two services for conversational and streaming applications (the QoS parameters more important are defined in [9]). The mobile terminal moves in scenery formed by six heterogeneous networks characterized by five QoS parameters. The security level goes from 0 (non-secure) to 5 (high security). The bandwidth values for IEEE802.11b (Wi-Fi), WiMax and UMTS networks are given in [10],[11] and [8] respectively.

2.2   Profiles for User's Preferences

We have defined the following three profiles for the user's preferences:

- Profile P1 (general). A general profile, where the user does not specify any constraint for the QoS parameters, with the following intervals: "Any QoS parameter can have assigned any weight between 0 and 1".
- Profile P2 (conversational). In this profile the most important parameters are delay and cost, because a conversation must be processed in real time and be cheap. The interval for the Delay is: weight between 0.5 and 0.7".

- Profile P3 (streaming). Typical profile for multimedia applications, where delay is lesser important than bandwidth (for which high values permit transmitting many data per second). The intervals for the QoS parameters are: *"Bandwidth: weight between 0.5 and 0.7; Delay: weight between 0.1 and 0.3 ".*

Table 1. Datasets DS1 and DS2, formed by networks characterized by the values of the following QoS parameters: B = Bandwidth (kbps), E = BER (dB), D = Delay (ms), S = Security level, C = Monetary cost (eur/MB), L = Latency (ms), J = Jitter (ms), R = Burst error, A = Average retransmissions/packet, P = Packet loss ratio

Net	Type	B	E	D	S	C	L	J	R	A	P
DS1											
0	UMTS	1,700	0.001	19	8	0.9	9	6	0.5	0.4	0.07
1	WLAN	2,500	10E-5	30	7	0.1	30	10	0.2	0.2	0.05
2	WLAN	2,000	10E-5	45	6.5	0.2	28	10	0.25	0.3	0.04
3	WLAN	2,500	10E-6	50	6	0.5	30	10	0.2	0.2	0.04
DS2 (*)											
0	Wi-Fi	5,100	0.01	70	2	0.2	-	-	-	-	-
1	Wi-Fi	5,100	0.01	65	1	0.2	-	-	-	-	-
2	WiMax	256	0.01	85	3	0.3	-	-	-	-	-
3	Wi-Fi	5,100	0.01	75	3	0.2	-	-	-	-	-
4	Wi-Fi	5,100	0.01	55	3	0.2	-	-	-	-	-
5	UMTS	384	0.03	80	5	0.2	-	-	-	-	-

(*)L, J, R, A are only applied to DS1.

2.3   A Direct Search Algorithm as Basis

Our heuristic proposal starts from an algorithm named *SEFI* (from "Weight Combinations SEarch by Fixed Intervals"), that we have designed to search solutions [12]. SEFI is a non-exhaustive direct search algorithm, where all the possible solutions for a given search precision are found. We have used SEFI for a double purpose: on the one hand, to determine the computation time and the size of the space of solutions (both are related to the search precision and the number of QoS parameters considered); on the other hand, to be integrated in our heuristic proposal that allow us getting better solutions near to real time. We have designed SEFI for this optimization problem, but considering particular sceneries such as a mobile wireless sensor moving along heterogeneous wireless sensor networks [13], that are sceneries of interest nowadays [14].

SEFI explores the space of solutions looking for combinations uniformly distributed according to a given interval h, named *search precision* (with the limit $h > 10^{-9}$). This way, if h decreases, the number of solutions found increases. The uniform search avoids leaving unexplored areas of the space of solutions. SEFI generates all the possible combinations for a given h, analyzes how many of them satisfy the constraint given in (2), computes the fitness of the solutions for the available networks, and finally reports the optimal network, that matches the combination with the minimum fitness found.

We have programmed SEFI in C language using recursive loops for the uniform generation of all the possible combinations. The code has been successfully tested on a custom embedded microprocessor Microblaze [12] based on reconfigurable hardware [15], which has similar features to the current microprocessors in many mobile terminals.

Thanks to SEFI we can perform experiments that inform us about the computing time and the number of generated and valid combinations, in order to get an idea of the computational effort of the optimization problem. For example, for five QoS parameters, $h = 0.01$, DS1 and P1, the results were: 10 seconds, $100, 000, 000$ generated combinations and $8, 000, 000$ solutions. After performing many experiments, we have gotten the following conclusions:

- The computing time comes from the operations made for generating all the possible combinations, evaluating the constraint given in (2) and, for the obtained solutions, calculating the fitness. This computing time increases with $NQoS$ and precision degree. We consider 1 second as the maximum time for obtaining an optimal solution due to the dynamic sceneries. This constraint moves us to restrict the values of h depending on $NQoS$ (for example, $h \geq 0.05$ for six parameters, $h \geq 0.001$ for two parameters).
- The fitness improves always with higher precisions for a given $NQoS$.
- The selection of an optimal network depends on both $NQoS$ and h; in other words, another optimal network could be found if we consider an additional QoS parameter or a higher h. This moves us to consider the need for designing an heuristic that searches efficiently the optimal network. Nevertheless, after analyzing the results of SEFI, we can conclude that it is better to consider more QoS parameters than increasing the search precision, so we can use higher h values without damaging the find of the optimal network.

3   An Adaptive Heuristic Proposal

As we have seen before, the key to find the optimal network is increasing the number of QoS parameters, but this implies to reduce the search precision in order to keep the real-time constraint. The apparent contradiction (low search precision are not good to find optimal solutions) move us to design a heuristic that could find optimal solutions in real time using low search precisions.

We name *SEFISA* the new heuristic proposal that, starting from SEFI, is a Simulated Annealing (SA) adaptation. The SA algorithm [16][17] is inspired in the cooling process of a metal where a final structure of minimum energy is searched and reached after successive stages (we name *generations*) where structures more and more cooled are found. In its original formulation, SA starts looking for an optimal solution within a space of solutions well defined; once found, the following generation reduces this space and centers it in the optimum found before, starting the searching again, this time with higher precision. The amount of the successive reductions is defined by a factor of reduction.

This algorithm is very versatile for many optimization problems. In our case, we use the adaptation aspect for our heuristic. The operation of SEFISA is shown in the pseudo-code given in Algorithm 1 and in Figure 2, and it is as follows: Initially (generation #0), the algorithm starts performing SEFI where the search spaces of sizes D_i for the QoS weights w_i are constrained by the limits u_{min_i} and u_{max_i}. These limits are imposed by the application profile or user's preferences. In this first generation, we establish a constant div, named *division factor*, for determining how many samples the smallest search space D_{min} will be divided. The precision h for SEFI only in this generation is calculated dividing D_{min} (it corresponds to w_1 in Figure 2) by div, so some D_i will have greater or equal number of samples than the division factor for generating combinations. Therefore, h is different in each generation and it depends on D_{min} and div.

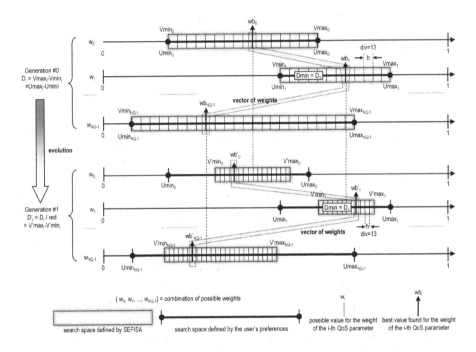

Fig. 2. SEFISA performs SEFI in successive generations, reducing and re-centering the search spaces for the QoS weights

Once the minimum fitness found in the initial generation, we use the corresponding set of weights (optimal solution) in order to center the new and smaller search spaces D_i' on them for the following generation (see generation #1 in Figure 2), where all D_i are reduced by a reduction factor red, usually equal to 2 (successive reductions in half). The new D_i' are used to calculate the newer

Algorithm 1. SEFISA pseudo-code

Select dataset and profile
Determine limits $U_{min_i}, U_{max_i} \Rightarrow D_i \Rightarrow D_{min}$
Select red and $div \Rightarrow h$
$IdGeneration = 0$
while stop criterion not reached **do**
    Run SEFI (h) $\Rightarrow$ obtain optimal wb_i
    $D_i = D_i/2$ and centered in $wb_i \Rightarrow$ determine V_{min_i} and V_{max_i}
    if limits exceeded or other causes **then**
        Take correcting actions on the search spaces
    end if
    Determine $D_{min} \Rightarrow h$
    $IdGeneration++$
end while

search limits v_{min_i} and v_{max_i}, taking into account that, for the first generation, the search spaces were determined by u_{min_i} and u_{max_i} (3).

$$D_i' = \frac{D_i}{red} : (D_i = v_{max_i} - v_{min_i}) \wedge (D_{i,0} = u_{max_i} - u_{min_i}) \qquad (3)$$

The calculation of the newer limits from D_i' must take into account possible special situations, usually when the limits imposed by u_{min_i} and u_{max_i} are exceeded. Let's suppose a generation #J where $D_i^{(J)}$, defined by the interval $\{V_{min_i}^{(J)}, V_{max_i}^{(J)}\}$, is inside the interval $\{U_{min_i}, U_{max_i}\}$ (that does not depend on the generation). The search space is reduced in half in the following generation #J+1, $D_i^{(J+1)} = \frac{D_i^{(J)}}{2}$, so $V_{min_i}^{(J+1)} = wb_i^{(J)} - \frac{D_i^{(J+1)}}{2}$ and $V_{max_i}^{(J+1)} = wb_i^{(J)} + \frac{D_i^{(J+1)}}{2}$, where $wb_i^{(J)}$ is the i-th weight of the best solution found in generation #J. In order to avoid these new limits fall before or after the minimum or maximum possible U_{min_i} and U_{max_i} respectively, we use the following adjustment actions: if $V_{min_i}^{(J+1)} \leq U_{min_i}$ then $V_{min_i}^{(J+1)} = U_{min_i}$, and if $V_{max_i}^{(J+1)} \geq U_{max_i}$ then $V_{max_i}^{(J+1)} = U_{max_i}$. Other casuistries could appear (precision issues), being necessary to take correcting actions.

Once determined the newer search spaces, we run SEFI generating all the possible combinations NC and obtaining the valid solutions NS ($NS \subseteq NC$) that are those satisfying the constraint now formulated in (4), where NQ is the number of QoS parameters.

$$\sum_{i=0}^{i=NQ-1} w_i = 1 : (v_{min_i} \leq w_i \leq v_{max_i}) \wedge (0 \leq u_{min_i} \leq v_{min_i} \leq v_{max_i} \leq u_{max_i} - 1)$$
$$(4)$$

Finally, the new optimal solution found will be used again to determine the positions and sizes of the search spaces for the following generation.

The absence of random features in this custom adaptation of the SA algorithm allows us to run SEFISA just one time in order to supply the optimal solution; this way, no statistical analysis (different number of runs, average error, etc) is required. In addition, we have limited SEFISA to four generations in general, taking into account the real-time constraint and because there is a soon stagnation of the optimal solution found in all the experiments done. Nevertheless, in few cases SEFISA could stop and finish the executions because of several reasons, that we name *stop criterion*. The stop criterion for SEFISA are:

- Predefined: When a determined value for the computation time or the search precision has been reached.
- Compulsory: There are not solutions found in a generation, so we cannot center in the optimal weights the next reduced search spaces. The absence of solutions can be often stated quickly, allowing to SEFISA to reinitialize with other settings that can offer a better performance.

4   Experimental Results

We have performed a wide set of experiments with SEFISA taking into account the mentioned datasets and profiles, reduction factor equal to 2, division factor equal to 3, 5, 10, 20, 50, 100 or 150, and $NQoS$ equal to 3, 4 or 5.

4.1   Computing Time

We state again, as we have proven with SEFI, that the computing time increases with the number of QoS parameters and the division factor (that defines the search precision) for the search spaces. This way, if we consider a high number of QoS parameters (greater or equal to five), we cannot use high division factors (as 50, 100 or 150); on the other hand, if $NQoS$ is low, we can perform experiments with any of the considered division factors.

4.2   Optimal Solution Improvement along Generations

We can see in Figure 3 how the optimal solution found by SEFI (lower fitness values that could imply different optimal networks) is improved in the successive SEFISA generations, by means of a representative case ($DS1$, $NQoS = 5$, $div = 3$). This evolvable feature of SEFISA is based on:

- The successive adaption of the search spaces, that are reduced in half and centered on the optimal weights found in the previous generation.
- The direct search of the solutions inside the search spaces. This is an exact technique driven by SEFI.

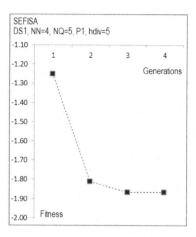

Fig. 3. Optimal solutions found by SEFISA up to four generations (stagnation is reached soon), for dataset $DS1$, five QoS parameters, and a division factor of five

The analysis of the experiments about this feature indicates us:

- In general, SEFISA evolves finding better optimal solutions, although sometimes the optimal solution holds up in the last generations. In order to prevent a holding up, it is better to use lower values for div (3 or 5).
- The stop criterion by computing time usually happens from six QoS parameters and for very high values of the division factor (more than 50). The stop criterion by excess precision or absence of solutions usually happens for high values of the division factor (from 20).

Summarizing, these observations moves us to consider lower values for the division factor ($div = 3$ or 5).

4.3   Performance of SEFISA against SEFI

We compare the performance of SEFISA against SEFI in order to validate the goodness of the adaptive heuristic in relation with the expectations pointed out in the first paragraph in Section 3.

This performance can be studied from Figure 4, understanding the performance as the best fitness found versus the same number of generated combinations (plots in the first column) or solutions (plots in the second column). Each row of plots matches with a different number of QoS parameters. The Figure 4 is representative of other cases, and it has been built from dataset DS1, profile P1, and three, four and five QoS parameters.

The marks on the SEFISA plot are the optimal fitness found in those runs of SEFISA with values for the division factor of 3, 5, 10, 20, 50 and 100 (counting from the first on the left). We have pointed out with a greater circle those values for div of 3 and 5, which represent a better behavior of SEFISA against SEFI (for the same number of combinations or solutions, the fitness found by SEFISA

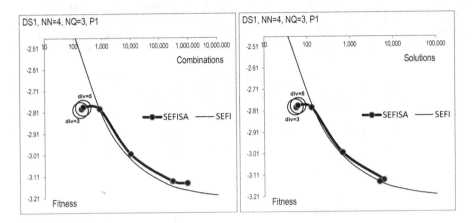

Fig. 4. Best fitness found by SEFI and SEFISA versus the same number of combinations or solutions, for dataset DS1, profile P1 and different number of QoS parameters

is lesser than the one found by SEFI). This way, we can state that SEFISA has equal or better performance than SEFI for low values of the division factor and, therefore, low values for the search precision.

This is the goal we were looking for. The fact of using low values for the division factor has two advantages:

- We can fit the real-time constraint (times lesser than one second) because the computing time increases too much when the values of div are high. Working with the lowest value of div has an additional advantage: we can consider a higher $NQoS$, because we know already that the more QoS parameters, the higher computing time, for medium or high search precisions.
- We can reach more generations in lesser time, allowing us to obtain better solutions, as we can see in Figure 3.

4.4   Performance of SEFISA against Other Algorithms

We have looked for other studies where the research area is similar and the datasets are explained in detail, in order to make a comparison of performance for SEFISA. Unfortunately, we have only found one article with these requirements [8], allowing us to build a common experimental framework that considers dataset DS2 and profiles P2 and P3.

In [8], the best combination of weights for a mobile terminal is calculated by means of the AHP algorithm [5], and its fitness value has been calculated by means of (2). We have considered the same experimental inputs for SEFISA, and we have pointed out the reported results in Table 2 for the second generation of the algorithm. As we can see, the solutions found by SEFISA have lower fitness than AHP in all the cases, even with low precision, thus proving the good performance of our heuristic.

Table 2. Solutions for DS2. The solutions given by Lassoued have been calculated by AHP for profiles P2 (conversational, $0.5 < D < 0.7$) and P3 (streaming, $0.5 < B < 0.7$ and $0.1 < D < 0.3$). The solutions given by SEFISA in the second generation for the same inputs have better fitness in all the cases, even when low precision.

	solution						Best Fitness
Weights:	w0	w1	w2	w3	w4	net	
QoS parameter:	B	E	D	S	C		
Lassoued:P2	0.065	0.065	0.614	0.128	0.128	4	1.25963
Lassoued:P3	0.545	0.035	0.178	0.121	0.121	4	-4.42821
SEFISA:P2,div=3	0.34	0.09	0.5	0.03	0.04	4	-1.41071
SEFISA:P2,div=10	0.44	0.02	0.5	0.02	0.02	4	-1.89888
SEFISA:P3,div=3	0.7	0.1	0.1	0.05	0.05	4	-6.17108
SEFISA:P3,div=10	0.7	0.16	0.1	0.02	0.02	4	-6.36615

5   Conclusions

We have designed an adaptive heuristic to obtain the optimal network in the Vertical Handoff decision phase, for a mobile terminal that moves in an scenery with heterogeneous wireless networks. This is an optimization problem because of the very high number of possible solutions.

Since this problem is formulated characterizing the networks by their QoS values, our heuristic proposal can be applied to any technology of wireless network. The heuristic is designed to return quickly a solution for the optimization problem, in order to allow a real-time behavior. It combines an exact technique from a former developed algorithm and an evolutionary feature inspired on the Simulated Annealing algorithm that must be performed just one time. We have proven the goodness of the heuristic for obtaining optimal solutions, improving the performance offered by the former exact technique and other algorithms.

Some tasks are scheduled to be addressed in the future. We will study the application of SEFISA to give a sort of networks, rather than an unique optimal network according to the fitness values. Also, we want to do a depth comparative study of our proposal against other techniques (AHP, SAW, TOPSIS) using a suitable and common experimental framework. Finally, we are planning a hardware implementation of this heuristic in the way of a custom coprocessor that can release the microprocessor from the load of running the associated computation.

Acknowledgments. This work was funded by the Spanish Ministry of Economy and Competitiveness under the contract TIN2012-30685 (BIO project), and by the Government of Extremadura, Spain, with the aid GR10025 to the group TIC015.

References

1. McNair, J., Zhu, F.: Vertical Handoffs in Fourth-Generation Multinetwork Environments. IEEE Wireless Communications 11(3), 8–15 (2004)
2. Chiasserini, C.F., Cuomo, F., Piacentini, L., Rossi, M., Tinirello, I., Vacirca, F.: Architectures and protocols for mobile computing applications: a reconfigurable approach. Computer Networks: The International Journal of Computer and Telecommunications Networking 44(4), 545–567 (2004)
3. Kassar, M., Kervella, B., Pujolle, G.: An overview of vertical handover decision strategies in heterogeneous wireless networks. Computer Communications 31, 2607–2620 (2008)
4. Song, Q., Jamalipour, A.: A quality of service negotiation-based vertical handoff decision scheme in heterogeneous wireless systems. European Journal of Operational Research 191(3), 1059–1074 (2008)
5. Song, Q., Jamalipour, A.: A Network Selection Mechanism for Next Generation Networks. In: IEEE International Conference on Communications (ICC 2005), pp. 1418–1422. IEEE Press, New York (2005)
6. Zhu, F., McNair, J.: Optimizations for vertical handoff decision algorithms. In: Wireless Communications and Networking Conference (WCNC 2004), pp. 867–872. IEEE (2004)
7. Stevens-Navarro, E., Lin, Y., Wong, V.: An MDP-Based Vertical Handoff Decision Algorithm for Heterogeneous Wireless Networks. IEEE Transactions on Vehicular Technology 57(2), 1243–1254 (2008)
8. Lassoued, I., Bonnin, J.M., Hamouda, Z.B., Belghith, A.: A Methodology for Evaluating Vertical Handoff Decision Mechanisms. In: Seventh International Conference on Networking (ICN 2008), pp. 377–384 (2008)
9. European Telecommunications Standards Institute: Quality of Service (QoS) concept and architecture. Technical report, 3rd Generation Partnership Project (3GPP), TS 23.107 V8.0.0 (2008)
10. Chen, J.C., Gilbert, J.M.: Measured Performance of 5-GHz 802.11a Wireless LAN Systems. Technical report Atheros Communications, Inc. (2001)
11. Betancur, L., Hincapie, R., Bustamante, R.: WiMAX Channel - PHY Model in Network Simulator 2. In: 2006 Workshop on ns-2: the IP Network Simulator (WNS2 2006) (2006)
12. Jaraz-Simon, M.D., Gomez-Pulido, J.A., Vega-Rodriguez, M.A., Sanchez-Perez, J.M.: Fast Decision Algorithms in Low-Power Embedded Processors for Quality-of-Service Based Connectivity of Mobile Sensors in Heterogeneous Wireless Sensor Networks. Sensors 12(2), 1612–1624 (2012)
13. Yick, J., Mukherjee, B., Ghosal, D.: Wireless sensor network survey. Computer Networks 52(12), 2292–2330 (2008)
14. Shah, G.A., Akan, O.B.: Timing-Based Mobile Sensor Localization inWireless Sensor and Actor Networks. Mobile Netw. Appl. 15, 664–679 (2010)
15. Hauck, S., DeHon, A.: Reconfigurable Computing, The Theory and Practice of FPGA-Based Computation. Morgan Kaufmann (2008)
16. Kirkpatrick, S., Gelatt, D., Vecchi, M.P.: Optimization by Simulated Annealing. Science 220, 671–680 (1983)
17. Cerny, V.: A Thermodynamical Approach to the Travelling Salesman Problem: an efficient Simulation Algorithm. Journal of Optimization Theory and Applications 45(1), 41–51 (1985)

A Fuzzy Tabu Search Approach to Solve a Vehicle Routing Problem

Kaj-Mikael Björk[1,2] and József Mezei[3]

[1] Åbo Akademi University, IAMSR
[2] Arcada University of Applied Sciences, Finland
kaj-mikael.bjork@abo.fi
[3] Åbo Akademi University, IAMSR, Finland
jmezei@abo.fi

Abstract. In this paper, we develop a framework to solve a multi-objective fuzzy vehicle routing problem. The decision variables in the problem are found in the routing decisions and the determination of the pickup order for a set of loads and available trucks. The objective to minimize is both the total time and distance traveled by all the vehicles. The uncertainty in the model is inspired from a timber transportation context, where times are, and sometimes even distances, uncertain. Because of lack of statistical data the uncertainties are sometimes best described as fuzzy numbers. The model developed is solved with a tabu search method, allowing for the above mentioned uncertainties. Finally, the framework is also illustrated with a numerical example.

Keywords: Tabu search, Vehicle routing, Fuzzy numbers, Optimization.

1 Introduction

Vehicle routing problems are well studied in the literature. Their importance is also explained by the vast amount of applications utilizing different vehicle routing schemes. Most transportation companies face problems that are related to the vehicle routing problem or some of its extensions. Garbage collection, postal logistics are only some examples. Tabu search have been implemented in forest industry applications; for instance, [1] presented case study in the newspaper business, where for an operational planning optimization problem was solved utilizing tabu search. A problem context also studied, but not in great detail is found in the timber collection transportation problem. However, recently [2] did an interesting computational study of neighborhood searches in timber transportation. But already [3] solved a timber vehicle routing problem successfully with a tabu search method.

Sometimes uncertainty is too prominent to be ignored. In timber transportation within the Nordic countries decisions are often to be done with imprecise information. The timber pickup transportation decisions are special in the sense that the timber needs to be transported from the forest along very small forest roads. Some of these roads may not be possible to use during certain periods of year, and it is not clear how long it is going to take due to the condition of the (dirt) road. Sometimes the driver

I. Rojas, G. Joya, and J. Cabestany (Eds.): IWANN 2013, Part I, LNCS 7902, pp. 210–217, 2013.

needs to take alternative routes (on the fly), so even the distances are allowed in this paper to be uncertain. For example, based on the experience of the truck driver, the delivery time for one shipment can be specified as "around 4hours"and based on the choice of the alternative routes, the distance can be "between 300 and 325 kms". In this type of systems, it is very difficult (if possible) to model the uncertain parameters as random variables since there is not enough data available. To handle uncertain information of this form, fuzzy set theory, introduced by [4] can be used. Fuzzy ve- hicle routing problems are studied in only a limited fashion [5]. However, there are a number of interesting papers that offer heuristic methods in order to find a near- optimal solution to the fuzzy vehicle routing problem;. (a) ant colony optimization [6] (b) genetic algorithm [7] (c) particle swarm optimization [8] (d) fuzzy simulation and genetic algorithm [9].

Previously [10] solved the crisp timber vehicle routing problem using an exact op- timization approach. In this paper, the combined truck assignment and route optimiza- tion was performed. The original problem was nonlinear, but it was possible to re- write the problem into a linear problem (MILP, Mixed Integer Linear Programming). Recently [11] solved the similar problem taking uncertainty of distances in both time and kilometers. This method was also a MILP-based method that guaranteed the global optimal solution. It was not possible to solve real-sized problems, even if the exact approaches can solve smaller instances and work as comparisons for other in- exact approaches. In addition, multi-objective optimization was desired in the sense that both time and distance should be optimized simultaneously. This paper extends the research track with a fuzzy tabu search method to solve a part of the vehicle routing problems for timber transportation. The method proposed is not very sensitive to the problem size as the case for [10] and [11]. This paper is outlined as follows: First the settings for the problem are described. Then some basics in fuzzy number theory are given along with the tabu search method. Finally an example is solved with some final remarks.

2   The Problem Formulation

The basic problem formulation for the problem consists of the following: There are a number of trucks located at possibly different geographical points. There are also a number of loads that need to be transported, each one to only one of the existing deli- very points (i.e. factories). The trucks are available for transporting each of the loads that needs to transported. However, it is possible to group together several loads if the total weight is not exceeded (creating a route). The pickup order should be deter- mined as well as which loads should be grouped together (of practical reasons only loads that have the same destination can be grouped together). Also the truck assign- ment should be determined (i.e. which truck should take care of a certain set of loads that are grouped together). The truck assignment and routing decisions are interde- pendent problems, however. After the truck has delivered all loads for a route, it will return to its home base. In these setting, it is of interest to find the optimal assignment of loads onto the trucks, and to plan the order of the pickups, route of each truck so that all the loads are shipped to its destinations and that the total distance travelled as well as the total time are minimized under the conditions that each truck may not exceed a certain weight and time limit (there is a time limit due to the working agreements for the truck drivers). The distances and times are allowed to be asymmetrical fuzzy numbers. The weight is assumed crisp, however.

The problem above is small but illustrative. Two trucks with two different home bases can be used to take care of six loads. The two problems, truck assignment and the routing problem can be tackled separately. We have decided to do so of two reasons. First the realistic problems are so large that only heuristic methods can solve them (without any guarantee of global convergence). Therefore solving an interdependent problem separately can be viewed as heuristic approach in itself. In addition, it is possible to create an "outer loop", where the truck assignment and the routing optimization problems are solved iteratively. Secondly, in some discussions with transportation companies, it has come forth that the planners are also interested in solutions only grouping together the loads and creating routes (without consideration of the trucks available). In this paper we therefore focus on only the routing decisions and grouping together the loads.

3   Preliminaries for Fuzzy Numbers and the Tabu Search Method

To model the uncertainty present in the delivery time and distance will be modeled by a specific type of fuzzy sets, namely triangular fuzzy numbers.

Definition 1. The fuzzy set $\tilde{A} = (a, b, c)$ defined on R is called a triangular fuzzy number, if the membership function of $\tilde{A}$ is 0 when $x \leq b - a$ or $b + c \leq x$, it is 1 when x=b, it increases linearly between b-a and b, and it decreases linearly between b and b+c.

Definition 2. Let $\tilde{A}$ be a fuzzy set on R and $0 \leq \alpha \leq 1$. The α-cut of $\tilde{A}$ is the set of all the points x such that $\mu_{\tilde{A}}(x) \geq \alpha$, i.e. $\tilde{A}(\alpha) = \{x | \mu_{\tilde{A}}(x) \geq \alpha\}$. The α-cut of a fuzzy number is a real interval and we will use the notation: $\tilde{A}(\alpha) = [\tilde{A}_{-}(\alpha), \tilde{A}_{+}(\alpha)]$ for $0 \leq \alpha \leq 1$.

The fuzzy number $\tilde{A} = (a, b, c)$ can be interpreted as "the quantity x is approximately b". To compare fuzzy numbers, a defuzzification method has to be employed: we use a function which assigns a real number to every fuzzy number and the ranking of the corresponding real numbers determines the ranking of the fuzzy numbers. In this paper, we use the method suggested by Yager [12]:

$$F(\tilde{A}) = \int_{0}^{1} \frac{\tilde{A}_{-}(\alpha) + \tilde{A}_{+}(\alpha)}{2} d\alpha$$

If $\tilde{A} = (a, b, c)$ is a triangular fuzzy number, the defuzzification function takes the value b+(c-a)/4.

The (non-symmetrical) triangular fuzzy numbers describing the distance and the travel time between loads i and j will be denoted by $\tilde{\lambda}_{(i,j)} = (\lambda_{l,(i,j)}, \lambda_{(i,j)}, \lambda_{u,(i,j)})$ and $\tilde{\eta}_{(i,j)} = (\eta_{l,(i,j)}, \eta_{(i,j)}, \eta_{u,(i,j)})$, respectively. The distance and delivery time from load I to the predefined destination will be denoted by $\tilde{\lambda}_{p(i,j)} = (\lambda_{pl,(i,j)}, \lambda_{p(i,j)}, \lambda_{pu,(i,j)})$ and $\tilde{\eta}_{p(i,j)} = (\eta_{pl,(i,j)}, \eta_{p(i,j)}, \eta_{pu,(i,j)})$, respectively. As the first step of the algorithm, an initial solution is identified. The weight of load i is denoted by π_i (crisp number). As the first step of the algorithm, an initial solution is identified. Since the parameters take the form of fuzzy quantities, the comparison of the actual solution with the predefined maximal time and distance performed using possibility theory: we require that the possibility of the new solution obtained by including a new load not exceeding the maximal time and distance should be greater than 0.80.

After calculating the fuzzy distance and delivery time of the initial solution (which will be stored as the best solution in the beginning), the tabu search algorithm proceeds by randomly choosing a destination point and two loads which are delivered to this destination but in different shipments by two different trucks: these two loads are swapped with each other to obtain a new potential solution (we also have to ensure that these new solution does not violate the requirements for maximal time, distance and weight). In other words, the shipments which belong to the same destination point and only differ in the position of two loads can be considered as neighbors in the tabu search algorithm. The random swapping of solutions takes place iteratively until a predefined stopping criterion is reached. If the new solution dominates the actual best solution in terms of time and distance, it will replace the best solution. If the new solution does not dominate the best solution but it is not worse than the best one by more than a predefined value (for example, 1%), it will not be set as the best solution but will be used as the initial solution of the next iteration in the tabu search. The comparison of different solutions is always performed by employing the defuzzification method described in the previous section. While performing the algorithm, a tabu list is created to store the previously visited solutions which were used as basis for random swapping of loads to ensure that we do not visit the same neighbors again.

When the stopping condition is reached (in our algorithm, we specify a number of iterations to be performed), the actual best solution is chosen. We obtain required delivery time and distance in the form of fuzzy numbers, and the defuzzification function can be used to associate crisp values to the identified solution.

4   The Example

In this chapter the framework is illustrated with a medium sized problem. This problem is fictional but will work as an example of timber transportation (timber pickups) from 25 different locations to three different destinations (factories). Each load has a designated destination, given *a priori*. This problem size is too big to be solved with the direct approaches (i.e. found in [13] and [14], for instance). The geographical data is given in the form of distances between each pair of loads. The distance is given in both kilometers (λ in the appendix) and minutes of driving (η in the appendix). These numbers are asymmetric triangular fuzzy numbers. The lower and upper part of the fuzzy distribution is given in the appendix as well (as λ_l, λ_u, η_l and η_u respectively). The distances for each load to the destination are given as triangular fuzzy numbers as well (λ_p and η_p). Finally the weight of each load is given as the parameter Π. The distances from the point of origin of the trucks (i.e. the home base) to the pickup points are not needed in this example since the truck assignment of the vehicle routing is left for further research at this stage.

In the tabu search method, the first solution is created simply by adding loads to a route until no load can be added anymore due to the total weight limit (of 42 tons in this example). This solution is evaluated saved as the current best one. Then two loads in different routes are swapped (in order to find a neighbor). If this is a dominant solution (in a fuzzy sense) in both total time elapsed and total distance travelled for all the routes, the current best node is updated. If the neighbor is (in a fuzzy sense) only 1 % worse (or better) than the current node (in both time and distance) then the neighbor is becoming the current node and previous node is put in the tabu list. In the tabu search algorithm, the potential solutions are evaluated by using a fuzzy weighted average. We calculate the weighted average of the time and distance values represented by fuzzy numbers and we employ the described defuzzification method to compare the alternatives. This iteration procedure is repeated 5000 times and the length of the tabu list is 10. The maximum length of a route is 500 km and effective time for a route is limited to 480 minutes. The gamma cut value used in the fuzzy comparisons is 0.8. After 5000 iterations, the best solution found had an objective value (fuzzy) of 1058 km (with 224.1 and 72.4 as the lower and upper part to the fuzzy distribution, respectively) and 1024 minutes (with 219.5 and 70.4 as the lower and upper parts, respectively). The solution consisted of a total amount of 10 routes. This solution can be compared to first iteration, where the objective value was of 1299 km (with 226.9 and 108.3 as the lower and upper part to the fuzzy distribution, respectively) and 1276 minutes (with 223.6 and 108.7 as the lower and upper parts, respectively). In other words, the tabu search method have improved the objective value approx. 20 %.

5   Summary and Further Research

In this paper, a fuzzy tabu search framework for the optimization of a vehicle routing problem has been presented. This method is inspired by the settings found in the forest industry application of timber pickup. The method is general purpose, however, and can be used in many application areas. Using fuzzy numbers are useful in order to capture the uncertainties found in the distances in time, and sometimes also distance. Therefore, the tabu search method allowed for asymmetric triangular fuzzy distances. Whereas the weight was limited by a crisp number, the tabu search method was implemented to optimize both total time and kilometers travelled. The framework was illustrated with a medium sized problem and shown to work well. However, the tabu search method is not very sensitive to the problem size, so really big problems can be expected to be solved as well. Genetic algorithms could possibly also be of interest, but this track is left for further research since we have not found a good chromosome structure yet.

Being the first results of the framework, some important parts are left for further research. Naturally, the truck assignment, left out in this model, need to be incorporate into the framework. In addition, other neighborhood approaches should be explored in the tabu search method as well as comparisons with both smaller instances of the problem (solved to global optimum in [11]) and bigger problems (solved with different crisp tabu search methods in the literature).

References

1. Russel, R., Chiang, W., Zepeda, D.: Integrating multi-product production and distribution in newspaper logistics. Computers and Operations Research 35(5) (2008)
2. Derigs, U., Pullmann, M., Vogel, U., Oberscheider, M., Gronalt, M., Hirsch, P.: Multilevel neighborhood search for solving full truckload routing problems arising in timber transportation. Electronic Notes in Discrete Mathematics 39, 281–288 (2012)
3. Gronalt, M., Hirsch, P.: Log-truck scheduling with a tabu search strategy. In: Doerner, K.F., Gendreau, M., Greistorfer, P., Gutjahr, W.J., Hartl, R.F., Reimann, M. (eds.) Metaheuristics - Progress in Complex Systems Optimization, pp. 65–88. Springer, New York (2007)
4. Zadeh, L.A.: Fuzzy Sets. Information and Control 8, 338–353 (1965)
5. Brito, J., Moreno, J.A., Verdegay, J.L.: Fuzzy Optimization in Vehicle Routing Problems, ISFA-EUSFLAT, pp. 1547–1552 (2009)
6. Kuo, R.J., Chiu, C.Y., Lin, Y.J.: Integration of fuzzy theory and ant algorithm for vehicle routing problem with time window. In: Processing NAFIPS 2004, IEEE Annual Meeting of the Fuzzy Information, vol. 2, pp. 925–930 (2004)
7. Jia, J., Liu, N., Wang, R.: Genetic algorithm for fuzzy logistics distribution vehicle routing problem. In: IEEE International Conference on Service Operations and Logistics, and Informatics, IEEE/SOLI 2008, pp. 1427–1432 (2008)
8. Xu, J., Yan, F., Li, S.: Vehicle routing optimization with soft time windows in a fuzzy random environment. Transportation Research Part E: Logistics and Transportation Review 47(6), 1075–1091 (2011)
9. Zheng, Y., Liu, B.: Fuzzy vehicle routing model with credibility measure and its hybrid intelligent algorithm. Applied Mathematics and Computation 176(2), 673–683 (2006)
10. Björk, K.-M.: A MILP - Model for the Optimization of Transports. In: Proceedings of the 8th International Conference of Numerical Analysis and Applied Mathematics, Rhodes, Greece (2010)
11. Björk, Mezei: A fuzzy MILP-model for the optimization of transports. Submitted to Journal of Intelligent and Fuzzy Systems (2012)
12. Yager, R.R.: Ranking fuzzy subsets over the unit interval. In: IEEE Conference on Decision and Control including the 17th Symposium on Adaptive Processes, Iona College, New Rochelle, New York, pp. 1435–1437 (1978)

Appendix

The data for the medium sized example problem in tables

DISTANCE (in km) between each load, λ

Load	1	2	3	4	5	6	7	8	9	10	11	12	13	14	15	16	17	18	19	20	21	22	23	24	25
1	0	35	59	86	91	115	97	54	105	37	42	110	19	83	25	41	67	41	21	130	9	92	51	119	102
2	35	0	28	77	103	102	109	78	26	103	103	94	22	44	20	60	121	24	44	126	60	18	54	112	72
3	59	28	0	51	85	68	119	68	13	28	112	22	24	75	121	69	66	32	67	10	101	11	56	86	71
4	86	77	51	0	35	57	46	70	20	28	119	107	79	71	123	40	31	24	111	62	46	68	24	117	62
5	91	103	85	35	0	89	104	68	12	21	130	41	100	12	13	38	73	82	96	90	131	39	63	88	123
6	115	102	68	57	89	0	20	110	128	108	15	88	71	18	65	21	27	95	41	44	89	25	92	107	131
7	97	109	119	46	104	20	0	61	19	3	79	22	19	16	15	36	91	110	104	39	59	42	61	9	11
8	54	78	68	70	68	110	61	0	49	97	121	66	58	22	110	45	121	100	114	111	81	120	77	57	44
9	105	26	13	20	12	128	19	49	0	116	94	102	17	54	22	1	7	10	96	73	39	91	78	60	72
10	37	103	28	28	21	108	3	97	116	0	19	50	9	60	94	51	22	52	92	6	111	83	110	4	90
11	42	103	112	119	130	15	79	121	94	19	0	107	22	1	77	70	72	79	76	71	95	33	59	100	63
12	110	94	22	107	41	88	22	66	102	50	107	0	60	108	67	99	72	4	83	98	100	66	28	110	117
13	19	22	24	79	100	71	19	58	17	9	22	60	0	16	85	42	94	61	57	97	109	110	58	105	78
14	83	44	75	71	12	18	16	22	54	60	1	108	16	0	86	41	82	12	102	64	12	24	8	92	111
15	25	20	121	123	13	65	15	110	22	94	77	67	85	86	0	9	58	67	117	93	113	35	81	65	117
16	41	60	69	40	38	21	36	45	1	51	70	99	42	41	9	0	16	39	64	39	37	68	62	31	80
17	67	121	66	31	73	27	91	121	7	22	72	72	94	82	58	16	0	23	80	97	22	119	16	1	22
18	41	24	32	24	82	95	110	100	10	52	79	4	61	12	67	39	23	0	65	75	107	74	59	24	16
19	21	44	67	111	96	41	104	114	96	92	76	83	57	102	117	64	80	65	0	63	72	22	108	18	7
20	130	126	10	62	90	44	39	111	73	6	71	98	97	64	93	39	97	75	63	0	45	110	103	56	100
21	9	60	101	46	131	89	59	81	39	111	95	100	109	12	113	37	22	107	72	45	0	45	34	54	45
22	92	18	11	68	39	25	42	120	91	83	33	66	110	24	35	68	119	74	22	110	45	0	56	45	58
23	51	54	56	24	63	92	61	77	78	110	59	28	58	8	81	62	16	59	108	103	34	56	0	32	47
24	119	112	86	117	88	107	9	57	60	4	100	110	105	92	65	31	1	24	18	56	54	45	32	0	116
25	102	72	71	62	123	131	11	44	72	90	63	117	78	111	117	80	22	16	7	100	45	58	47	116	0

Distance to delivery in km, λ_p

Load	middle	low	upper
1	97	15	16
2	108	8	31
3	91	15	16
4	68	0,5	5,9
5	95	2,3	11
6	60	1,5	22
7	108	18	19
8	122	23	43
9	37	0,1	9,8
10	94	5,6	8,1
11	140	13	46
12	70	5,3	27
13	79	13	26
14	58	6,3	4,2
15	103	15	11
16	42	2	5,1
17	30	1,9	3,2
18	84	9,1	12
19	94	1,5	24
20	85	11	6
21	67	7,5	25
22	119	19	21
23	67	5,5	14
24	88	6,3	5,1
25	85	2,4	0,7

The lower distribution of the DISTANCE (in km) between each load, λ_l

Load	1	2	3	4	5	6	7	8	9	10	11	12	13	14	15	16	17	18	19	20	21	22	23	24	25
1	0	5,7	4,8	11	12	8,3	8,9	9,3	20	1,1	3,5	13	2,5	2,2	1	3,7	8	8,2	4	13	1,4	6,6	1	10	18
2	0,4	0	4,7	0,5	11	1,2	2,2	5,3	0,8	0,4	8,2	12	0,7	6,8	2,5	6,1	24	0,4	5,6	18	0	0,3	0,2	13	11
3	3	3,4	0	5,1	2,9	8,2	7,6	5,7	0,5	0,5	14	1,8	3,1	14	8,5	4,4	5,1	3,5	8,2	1,5	9,7	1,7	1,9	15	7
4	6,7	0,6	2	0	5,2	8,2	7,7	12	1	3,2	11	20	14	11	12	7	1,6	3,1	21	11	6,8	9,5	2,7	22	5,2
5	15	8,4	4,7	0,3	0	12	4,2	13	1,8	0,3	10	5,3	19	1,4	1,5	5,4	0,3	6,1	3,3	6,7	24	5	3,3	1,2	8,6
6	14	2,9	14	6,8	15	0	0,4	21	11	18	2,1	1,9	1,8	3,1	9	4,1	3,7	7	3,4	1,8	2	3,1	7,4	14	21
7	6,2	21	22	7,6	14	3,2	0	8,2	1,9	0,1	9,3	0,1	2,2	0,2	1,8	3,5	1,3	10	20	5	10	7,2	0,1	0,3	0,8
8	0,9	3	0,4	4,9	14	19	3,8	0	7,4	11	4,6	6,1	4,6	1,9	4,8	4	5,6	1,4	22	18	7,6	18	14	8,1	3,4
9	19	5,2	1,5	2,8	1,3	16	1,1	7,5	0	3,2	17	12	2,8	1,8	0,1	0,2	1,3	0,9	18	4,2	3,7	6,7	4,4	3,2	7,3
10	7	13	4,5	0,6	2,9	13	0,1	13	7,7	0	2,1	4,5	0,5	7,1	1,3	1	0,7	5,8	16	1	6,4	8,1	18	0,5	12
11	5	18	11	14	13	1,1	13	15	13	2,8	0	18	3,8	0	9,7	2,2	11	7	0	13	8,9	4,4	2,4	9,4	4,3
12	15	12	0,5	6,6	6,9	10	2,6	9,5	2,2	4,5	12	0	12	1,9	6,6	5	1,9	0,5	0	15	13	4,2	4,6	9,5	1,9
13	0,8	4,4	1	8,4	1,2	2,3	1	10	0,3	1,8	1,3	12	0	2,2	1	4,3	11	9,2	8,1	15	7,2	7,3	5,7	17	0,9
14	9,1	3,5	6,5	11	2,1	2,1	1	3,9	3,9	9,2	0,2	5,4	0,9	0	4,1	6,6	14	0,6	4,1	10	0,4	3,9	1,1	11	22
15	2,8	3,7	14	22	1,6	11	0,2	20	1,7	15	13	7,2	16	4	0	0,3	7,3	5,8	17	12	15	6,4	2,1	12	22
16	2,5	12	1,9	7,6	4,2	0,5	2,7	6,7	0,1	7,1	9,2	5,7	0,2	7,2	0,2	0	0,8	5,3	9,9	6,1	1,1	1,5	3,3	2,2	15
17	13	21	3,8	3	11	4,4	8,6	1,7	1,2	3,3	4,3	6	3,2	0,3	7,3	1,1	0	1	0	2,9	4,4	19	1,4	0,2	1,5
18	3,4	2,9	1,1	2,6	0,8	4,6	6,8	17	2	0,3	11	0,1	1,5	2	7	7,6	2	0	0,7	5,1	16	6,5	11	0,3	1
19	3	4,6	6,6	22	1,2	4,8	14	6,7	8,6	0,4	2,7	0,7	6,3	9,6	15	4	1,1	9,4	0	7,4	11	2,9	1,9	3	0,9
20	6,2	20	1,6	6,8	12	6,3	1,5	9,3	1,2	0,8	0	0,6	3,1	9,9	16	2,4	16	4,8	5,9	0	6,1	19	14	8,5	17
21	1,6	0,2	16	9	25	8	11	2,8	5,3	16	15	20	12	0,4	0,7	2,1	3,3	9,4	0,1	5,9	0	1,6	5,4	3,8	8,1
22	3,9	1,2	1,8	7,8	3,5	2,1	5,6	22	8,6	15	1,7	0,7	3,5	4,7	2	0,1	7,4	5,3	3,4	21	1,9	0	5,5	4,2	1,6
23	1,3	1,2	7,8	1,6	5,2	12	3,9	0,3	14	12	3,2	5	3	0,7	13	3,3	0,2	3,1	11	20	0,4	10	0	0,9	2,2
24	22	12	1,4	1,6	2,1	1,1	0,1	9,5	2,4	0,7	20	5,7	10	5,2	9	1,7	0,2	1,9	1,9	0,6	4,1	1,7	4,2	0	21
25	7,1	0,7	3,4	2,1	15	9,2	1,2	0,4	2,6	5	6,3	20	7,5	0,2	6,8	12	2,9	1,1	1,2	1,4	5,3	8,6	7,6	23	0

Dist. to delivery in minutes, η_p

Load	middle	low	upper
1	90	13	15
2	130	9,6	37
3	66	11	12
4	70	0,6	6
5	114	2,7	14
6	65	1,7	24
7	117	19	21
8	127	24	44
9	38	0,2	10
10	96	5,8	8,2
11	132	12	43
12	81	6,1	31
13	55	9,2	18
14	45	4,8	3,2
15	104	15	11
16	49	2,3	5,9
17	30	1,9	3,3
18	78	8,4	11
19	82	1,3	21
20	75	9,3	5,2
21	64	7,2	24
22	138	22	24
23	59	4,8	12
24	87	6,3	5,1
25	60	1,7	0,5

The upper distribution of the DISTANCE (in km) between each load, λ_u

Load	1	2	3	4	5	6	7	8	9	10	11	12	13	14	15	16	17	18	19	20	21	22	23	24	25
1	0	0,8	19	2,1	32	12	25	17	33	6,3	4,1	34	5,1	6,1	0,5	12	4,6	13	4,6	35	0,5	17	8,5	9	23
2	4,9	0	5,9	0,9	6,6	39	20	28	0	31	22	27	2,1	13	7,9	21	2,4	5,3	13	25	0,1	6,4	18	20	21
3	10	9,1	0	14	9,5	18	33	3,5	2	6,5	40	2,6	0,1	25	9,4	8	12	0,8	14	1,3	27	1	20	29	0,3
4	9,3	19	2,1	0	11	7,4	4,8	27	3,9	7,3	27	18	2,7	27	10	7,1	8	0,1	37	13	1,5	10	9,2	47	20
5	13	2,1	8,8	6,7	0	4,1	8,9	18	1,9	1,5	40	3	4,4	2,8	0,7	5,2	22	27	25	7,9	1,3	12	17	32	47
6	25	34	14	12	8,2	0	6,5	36	26	40	4,9	5,5	1,4	3,4	8,7	5	3,6	17	1	8,4	3	4,5	14	32	23
7	28	10	1,9	12	6,4	0,5	0	3,7	0,8	0,9	16	4,2	1,3	0,3	1,4	4,4	12	8,4	6,9	7,8	21	0,5	6,5	3	4,2
8	17	7	22	11	12	33	10	0	2,4	37	25	25	7,3	2,4	19	6,8	11	15	25	44	4,2	41	26	7,8	7,9
9	3,2	8,6	3,2	6,1	2	27	5,7	11	0	8,1	7	33	2,7	0,6	8,3	0,3	0,3	0,8	12	28	6	15	27	21	26
10	2,1	1,6	1,3	6,2	3,7	13	0,7	18	14	0	2,1	14	2,1	11	3,2	8,4	8,2	0,6	19	0,9	28	0,8	1,8	0,3	34
11	7,9	30	16	14	9,1	1,1	8,8	1	23	1	0	26	3,7	0	27	2,1	23	8,5	21	17	0,6	8,6	2,7	21	13
12	27	10	5,5	6,4	4,6	9,9	4,4	26	4,7	2,2	39	0	15	8,2	10	10	0	0,7	2,5	11	21	4,8	11	33	14
13	2,4	5,6	7,7	24	4	28	6,8	10	5	1,5	3,6	6	0	0,7	29	7,9	18	5,2	11	12	12	27	13	26	12
14	4,2	2,7	25	21	4,8	1,3	4,9	4,3	15	24	0,2	28	4,5	0	3,1	2,1	13	3,3	15	2,8	0,4	6	1	26	13
15	9,2	8	38	43	3	6	2,7	28	6,5	22	18	15	11	28	0	0,6	3,2	16	5,6	27	2,3	0,8	20	24	24
16	1,3	2,3	27	15	2,4	5,2	4,8	13	0,3	0,2	5,6	18	11	16	0,2	0	3,5	6,7	1,9	2,6	3,8	26	10	7,8	29
17	23	30	20	3,1	19	1,2	16	0	0,8	8,3	0	15	23	29	10	4,4	0	4,6	14	28	8,6	13	0,3	0,3	2,3
18	12	4,2	4,9	1,1	11	6,7	37	34	2,2	7,8	16	0,2	21	1,5	26	6,3	9,2	0	3,1	17	15	8,9	6,3	5,1	6,3
19	0,9	11	19	17	37	9,6	3,7	43	19	5,5	0,5	20	23	29	4,4	23	22	24	0	21	21	8,4	11	0,4	1,1
20	8,6	50	3,8	14	33	9,4	2,7	43	17	1,7	27	12	21	21	9,7	5,9	38	28	14	0	4,6	39	10	21	26
21	1,2	20	7,1	18	3,1	5,9	15	5,8	1,2	23	22	0,2	25	1,1	39	6,4	2,3	34	8,5	4,6	0	1,2	8,9	14	4,4
22	6,3	4,6	4,2	3,3	13	3,7	17	32	14	2	2,5	3,6	20	7,7	14	7,2	22	1,3	8,8	14	7,5	0	17	2,1	20
23	8	7,1	18	3,3	11	32	7,7	1,8	2,2	40	6,3	2,2	10	0,3	29	19	1,1	16	41	0	2,9	17	0	5,7	2,2
24	27	31	24	14	31	5,8	2,6	9,1	16	0	8	40	6,7	17	22	12	0,2	6,2	5,1	14	5,3	18	10	0	0,5
25	2	7,8	22	21	32	4,7	4,2	11	21	1,4	21	1,2	15	35	37	3,2	1,1	1,7	0,3	1,4	17	15	0,7	44	0

Weight in tons, Π

Load	weight
1	8
2	11
3	5
4	16
5	21
6	7
7	23
8	6
9	14
10	18
11	20
12	16
13	14
14	17
15	9
16	19
17	16
18	11
19	10
20	10
21	17
22	18
23	5
24	24
25	8

DISTANCE (in minutes) between each load, η

Load	1	2	3	4	5	6	7	8	9	10	11	12	13	14	15	16	17	18	19	20	21	22	23	24	25
1	0	33	59	91	107	101	69	48	90	27	41	123	15	100	19	46	78	36	24	125	6,8	71	39	131	117
2	27	0	22	66	104	88	110	77	22	81	118	71	16	41	15	71	94	27	48	131	70	14	60	101	71
3	60	28	0	61	81	48	118	81	9,8	22	115	19	18	67	90	80	79	26	53	11	74	8,3	48	78	70
4	94	83	51	0	30	59	35	55	20	28	88	93	93	53	117	41	32	22	123	55	32	77	25	117	54
5	97	113	69	35	0	76	75	52	13	22	112	40	87	8,4	9,1	27	73	89	75	73	141	43	59	92	143
6	121	91	80	49	66	0	18	120	150	123	13	63	50	18	76	18	31	67	42	45	73	27	74	104	122
7	69	107	117	45	89	23	0	73	20	3	55	19	21	16	17	40	72	109	109	35	59	43	72	6,8	11
8	47	63	75	69	49	79	50	0	48	81	102	64	56	17	79	52	98	90	103	129	92	134	81	54	33
9	99	31	15	21	11	90	55	56	0	122	93	91	19	43	17	0,9	5,6	9,8	114	65	41	66	81	55	76
10	44	91	32	24	16	89	2,7	106	123	0	22	35	8,3	64	90	61	16	51	109	6,7	119	85	95	3	106
11	37	116	122	87	95	17	74	90	74	22	0	124	22	1,2	75	83	77	90	65	72	86	38	63	74	76
12	91	81	19	119	32	100	76	52	96	45	107	0	54	106	58	117	76	4,6	89	94	104	48	32	127	136
13	17	23	21	87	88	78	13	59	18	9,2	21	56	0	17	88	37	102	52	52	81	110	113	62	110	72
14	69	52	63	84	9,8	13	22	50	42	0,9	76	13	0	88	48	80	11	103	64	11	18	7,6	105	102
15	21	16	91	89	12	55	14	95	20	104	85	54	83	89	0	7,5	63	64	133	106	132	29	88	49	140
16	33	46	66	32	45	23	26	44	0,9	49	78	115	34	33	7,4	0	12	28	58	31	31	58	73	30	57
17	63	143	51	22	71	26	77	113	5,2	20	66	58	78	66	41	7	0	20	94	96	29	106	19	1,2	20
18	35	21	27	21	72	92	119	86	9,7	38	93	3,9	46	9,2	62	35	27	0	76	53	93	81	44	23	14
19	18	40	49	78	88	36	108	128	106	98	86	60	68	119	109	58	78	73	0	75	55	24	120	15	5
20	146	115	11	57	89	49	34	115	67	4,5	62	79	99	70	90	30	71	87	62	0	54	81	118	58	73
21	8,4	52	104	47	100	69	58	72	28	123	102	96	102	11	80	43	17	119	81	49	0	41	33	49	52
22	93	15	11	73	34	28	104	88	94	26	75	125	22	42	77	126	56	23	95	54	0	49	44	45
23	45	41	45	18	47	77	63	85	89	110	43	27	46	6,6	58	45	12	57	104	77	38	67	0	35	52
24	83	78	103	85	92	127	7,2	64	68	4,5	92	94	86	72	59	36	1	23	16	66	41	41	30	0	128
25	84	58	84	63	109	141	12	35	80	108	58	119	66	107	136	70	15	15	5,2	96	37	49	52	131	0

Delivery point

Load	1	2	3
1	1	0	0
2	1	0	0
3	1	0	0
4	1	0	0
5	0	1	0
6	0	1	0
7	1	0	0
8	1	0	0
9	1	0	0
10	1	0	0
11	0	1	0
12	0	1	0
13	0	0	1
14	0	0	1
15	0	0	1
16	0	0	1
17	1	0	0
18	1	0	0
19	1	0	0
20	1	0	0
21	0	1	0
22	0	1	0
23	0	0	1
24	1	0	0
25	1	0	0

The lower distribution of the DISTANCE (in minutes) between each load, η_l

Load	1	2	3	4	5	6	7	8	9	10	11	12	13	14	15	16	17	18	19	20	21	22	23	24	25
1	0	5,3	4,8	12	14	7,3	6,3	8,3	16	0,8	3,4	15	2	2,6	0,7	4,1	9,3	7,1	4,5	12	1	5,1	0,8	11	20
2	0,3	0	3,7	0,4	11	1,1	2,2	5,3	0,7	0,3	9,5	9,3	0,5	6,4	1,9	7,3	19	0,5	6,1	19	0	0,2	0,2	11	11
3	3,1	3,4	0	6,1	2,8	5,7	7,5	6,8	0,4	0,4	15	1,6	2,3	13	6,3	5,1	6,2	8,5	1,7	7,1	1,1	1,3	6,3	11	8,5
4	7,4	0,7	2	0	4,4	8,6	5,9	9,5	1,1	3,3	7,9	18	17	8	11	7,2	1,7	2,9	23	10	4,8	11	2,8	22	4,5
5	16	9,3	3,8	0,3	0	11	3	10	2	0,3	8,7	5,2	16	1	1	3,8	0,3	6,6	2,5	5,4	26	5,5	3,1	1,3	10
6	15	2,5	16	5,9	11	0	0,4	23	13	21	1,8	1,4	1,3	0	3,6	4,3	4,9	3,4	1,8	16	3,3	5,9	13	19
7	4,4	21	21	7,5	12	3,7	0	9,8	2	0,1	6,5	0,1	2,4	0,2	1,9	3,9	1	10	21	4,4	8,8	7,4	0,1	0,3	0,8
8	0,8	2,4	0,4	4,8	9,8	14	3,1	0	7,2	9,2	9,9	5,9	4,5	1,1	4,5	4,5	1,3	20	21	8,6	21	15	7,7	2,6
9	18	6,2	1,7	3	1,2	11	0,9	8,7	0	3,4	16	11	3	1,5	0,1	0,2	1	0,9	21	3,8	3,8	4,8	4,5	2,9	7,8
10	8,3	11	5,1	0,5	2,2	10	0,1	15	8,1	0	2,4	3,2	0,4	7,6	1,3	1,2	0,5	5,8	19	1,1	6,9	8,4	16	0,4	14
11	4,4	20	11	10	9,3	1,2	12	11	10	3,2	0	20	3,9	0	9,5	2,6	11	7,9	0	13	8	5,1	2,5	7	5,1
12	12	10	0,4	7,4	5,4	12	1,9	7,5	2,1	4	12	0	11	1,9	5,6	5,8	2	0,6	0	15	14	3,1	5,2	11	2,2
13	0,7	4,6	0,9	9,2	1,1	2,5	0,7	11	0,3	1,8	2,1	1	0	2,4	1,1	3,7	11	7,9	7,4	13	7,3	7,5	6,1	18	0,9
14	7,6	4,2	5,4	13	1,8	1,5	0,8	4,3	3,6	6,5	0,1	3,8	0,7	0	4,2	7,7	13	0,6	4,1	10	0,4	3	1	13	20
15	2,3	2,9	10	16	1,5	9,5	0,2	17	1,6	17	14	5,8	15	4,1	0	0,3	7,9	5,5	19	14	18	5,4	2,3	9,1	26
16	2	8,9	1,8	6,1	5	0,5	1,9	6,5	0,1	6,9	10	6,7	0,1	5,8	0,1	0	0,8	3,9	8,9	4,8	0,9	1,3	4	2,1	10
17	12	24	3	2,1	11	4,3	7,3	1,6	0,9	2,9	3,9	4,9	2,7	0,3	5,8	1,1	0	0,9	0	2,9	4,1	17	1,6	0,2	1,3
18	2,9	2,6	0,9	2,3	0,7	4,4	7,4	15	1,9	0,3	0,1	1,1	1,5	6,4	7	2,3	0	0,8	3,6	14	7,1	8,3	0,3	0,8
19	2,6	4,2	4,8	15	1,1	4,3	14	9,7	9,5	0,4	3,1	0,5	7,5	11	14	3,6	1,1	10	0	8,8	8,1	3,2	2,2	2,6	0,6
20	7	19	1,8	6,3	12	7	1,3	9,7	1,1	0,6	0	0,5	3,2	11	16	1,9	12	5,6	5,9	0	7,3	14	16	8,9	13
21	1,5	0,2	16	9,2	19	6,2	11	2,5	8	11	3,2	0	3,9	0,6	0,4	1	0,3	0,5	2,5	2,6	0	1,5	5,2	3,4	9,3
22	3,9	0,9	1,7	8,4	3,1	2,4	6,5	19	8,3	17	1,3	0,8	4	4,2	2,4	0,2	7,8	4	3,6	18	2,2	0	4,8	4,1	1,3
23	1,2	0,9	6,3	1,2	3,8	9,9	4,1	0,3	16	12	2,3	6	9	2,4	0,1	2,9	11	15	0,5	12	0	1	2,4
24	15	8,3	1,7	1,2	2,2	1,3	0,1	11	2,7	0,8	18	4,9	8,4	4	2,2	13	2,8	1,7	0,7	3,1	1,6	3,9	0	23
25	5,9	0,6	4	2,1	14	9,9	1,3	0,3	2,9	6	5,8	20	6,3	0,2	7,9	10	2	1	0,9	1,3	4,4	7,2	8,4	26	0

The upper distribution of the DISTANCE (in minutes) between each load, η_u

Load	1	2	3	4	5	6	7	8	9	10	11	12	13	14	15	16	17	18	19	20	21	22	23	24	25
1	0	0,8	19	2,2	37	10	18	15	29	4,6	4	38	4,1	7,4	0,3	14	5,3	12	5,2	33	0,4	13	6,5	9,9	27
2	3,7	0	4,7	0,6	6,7	34	20	28	0	25	26	6	14	26	0,1	4,8	20	18	21
3	11	9,3	0	16	9,1	13	33	4,2	1,5	5	42	2,2	0,1	23	7	9,3	15	0,7	11	1,5	20	0,7	17	26	0,3
4	10	21	2,1	0	9,7	7,7	3,6	21	4	7,4	20	16	3,2	20	9,6	7,3	8,3	0,1	41	11	1	11	9,6	47	18
5	14	2,3	7,1	6,7	0	3,5	6,4	14	2,1	1,6	35	2,9	3,8	1,9	0,5	3,7	22	29	19	8,4	1,4	14	18	34	55
6	26	31	17	10	6,1	0	5,9	39	30	46	4,3	3,9	1	3,4	10	4,3	4,1	12	1	8,7	2,5	4,8	11	31	21
7	20	10	1,9	12	5,5	0,8	0	4,4	0,9	0,9	11	3,6	5,5	8,3	7,2	6,9	18	0,5	7,6	2,2	4,2
8	15	5,7	24	10	8,4	24	8,4	0	2,3	31	21	24	7,1	1,8	14	7,8	9	14	22	51	4,8	46	27	7,4	6
9	3	10	3,7	6,5	1,8	19	4,3	0	8,5	6,9	30	2,9	0,5	6,5	0,3	0,2	0,8	15	25	6,3	11	28	19	27
10	2,5	1,5	1,5	5,4	2,8	10	0,6	20	15	0	2,4	9,5	1,9	12	3,1	10	6	0,8	23	0,9	30	0,9	1,5	0,2	40
11	6,9	34	17	10	6,6	1,3	8,3	0,7	18	1,2	0	30	3,7	0	26	2,5	24	9,7	18	17	0,5	9,9	2,9	16	16
12	22	8,7	4,7	7,1	3,6	11	3,3	20	4,4	2	14	8	9	12	0,8	2,7	11	21	3,5	13	38	16
13	2,1	5,9	6,8	26	3,5	30	4,7	10	5,1	1,5	3,5	5,6	0	0,7	30	6,9	19	4,5	9,8	9,7	12	27	14	28	11
14	3,4	3,2	21	25	3,9	0,9	3,7	4,8	14	17	0,2	20	7,3	0	3,2	2,4	12	2,9	15	2,8	0,4	4,5	0,9	29	12
15	7,5	6,2	28	31	2,8	5	2,6	24	6,1	25	20	12	11	29	0	0,5	3,5	15	6,4	30	2,8	0,7	22	18	29
16	1	1,7	26	12	2,9	5,7	3,4	13	0,2	0,2	6,2	21	9,4	13	0,2	0	2,7	4,9	1,7	2	3,2	23	12	7,4	21
17	21	35	16	2,2	19	1,2	13	0	0,6	7,4	0	12	39	7,9	4,6	0	4,1	16	28	11	8,1	11	0,4	0,3	21
18	10	3,7	4	1	9,4	6,5	40	29	2,1	5,8	19	0,2	16	1,2	24	5,7	11	0	3,7	12	13	9,7	4,6	4,8	5,5
19	0,8	9,7	14	12	34	8,4	3,9	48	21	5,9	0,5	15	27	34	4,1	21	21	27	0	25	16	9,2	12	0,3	0,8
20	9,6	46	4,4	13	33	10	2,3	45	15	1,3	24	9,8	21	23	9,4	4,5	28	33	13	0	5,5	29	12	22	19
21	1,1	17	7,3	18	2,4	4,6	14	5,2	0,9	25	24	0,2	23	0,9	27	7,5	1,8	38	9,6	5	0	1,1	8,6	13	5,1
22	6,3	3,7	4,1	3,5	11	4,1	19	28	13	2,3	2	4,1	23	7	17	0,1	24	1	9,2	12	8,9	0	15	2	15
23	7	5,4	15	2,5	8	27	8	2,1	2,5	40	4,6	2,1	8,1	0,3	21	14	0,8	16	40	0	3,2	20	0	6,2	2,4
24	19	22	29	9,9	33	6,9	2,1	10	18	0	7,4	34	5,5	13	20	14	0,2	6	4,5	17	4	16	9,6	0	0,5
25	1,7	6,2	26	21	28	5,1	4,4	8,8	24	1,7	20	1,2	13	34	42	2,8	0,8	1,6	0,2	1,3	14	13	0,7	50	0

Improved Particle Swarm Optimization Method in Inverse Design Problems

Y. Volkan Pehlivanoglu

Turkish Air Force Academy, Istanbul, Turkey
vpehlivan@hho.edu.tr

Abstract. An improved particle swarm optimization algorithm is proposed and tested for two different test cases: surface fitting of a wing shape and an inverse design of an airfoil in subsonic flow. The new algorithm emphasizes the use of an indirect design prediction based on a local surrogate modeling in particle swarm optimization algorithm structure. For all the demonstration problems considered herein, remarkable reductions in the computational times have been accomplished.

Keywords: PSO, shape optimization, inverse problems.

1   Introduction

An inverse design problem is a type of indirect problem and it is widely known in natural sciences. Any closed system contains three elements: these are a cause, a model, and an effect. We may call these factors as an input, a process, and an output, respectively [1]. Most of the formulations of inverse problems may proceed to the setting of an optimization problem. In general, an inverse design problem can be expressed as follows:

$$find \; \{x \in R^d\} \tag{1}$$

$$\min f(x, y) \tag{2}$$

Subject to

$$g(x, y) \leq 0 \tag{3}$$

$$x^L \leq x \leq x^U \tag{4}$$

where x is an input that is the design parameter vector whose values lie in the range given by upper and lower borders in equation (4). The objective function $f(x,y)$ in an inverse design problem is used to bring the computed response from the model as close as possible to the target output, y. In some problems, it may be necessary to satisfy certain inequality constraints given by $g(x,y)$. The objective function is usually a least-squares function given by

$$f(x, y) = \sum_{i=1}^{n} (y_i^c - y_i^t)^2 \tag{5}$$

I. Rojas, G. Joya, and J. Cabestany (Eds.): IWANN 2013, Part I, LNCS 7902, pp. 218–231, 2013.
© Springer-Verlag Berlin Heidelberg 2013

where y_i^t is i^{th} value of the target response and y_i^c is i^{th} value of the computed response obtained from the simulation model.

In most engineering problems, computational methods are gradually replacing empirical methods; and design engineers are spending more time in applying computational tools instead of conducting physical experiments to design and analyze engineering components. Computational optimization efforts may be divided into Gradient-Based (GB) and non-gradient methods [2]. GB methods give more accurate results; and they are usually efficient methods in terms of computational effort. However, they may have some drawbacks [3]. The demand for a method of operations research, which is capable of escaping local optima, has led to the development of non-traditional search algorithms. Non-gradient based methodologies, such as Genetic Algorithms (GAs) or Particle Swarm Optimization (PSO) algorithms, which are less susceptible to pitfalls of convergence to local optima, suggest a good alternative to conventional optimization techniques. These algorithms are population based, and they include a lot of design candidates waiting for the objective function computations in each generation. The major weakness of population based algorithms lies in their poor computational efficiency, because the evaluation of objective function is sometimes very expensive [4]. Despite the considerably improved computer power over the past few decades, computational simulation can still be prohibitive for a large number of executions in practical engineering design. Therefore, improving the efficiency of evolutionary search algorithms has become a key factor in their successful applications to real-world problems. Two categories of techniques have been proposed to tackle the efficiency issue of evolutionary search methods; the first type is focused on devising more efficient variants of the canonical algorithms, the second type involves using a surrogate model which is a kind of approximation in lieu of the exact and often expensive function evaluations [5].

In literature, there are a lot of surrogate model-based optimization algorithms. The details of these algorithms can be found in Pehlivanoglu and Yagiz [6]. The key idea in these methods is to parameterize the space of possible solutions via a simple, computationally inexpensive model, and to use this model to generate inputs in terms of predicted objective function values for the optimization algorithm. Therefore, the whole optimization process is managed by surrogate model outputs. Such a model is often referred to as the response surface of the system to be optimized, leading to the definition of a so-called surrogate-model based optimization methodology [7]. Major issues in surrogate model-based design optimization are the approximation efficiency and accuracy. In case of the problem which has a high number of design variables, the construction of surrogate model may cause extremely high computational cost, which means inefficient approximation. On the other hand, it is possible to miss the global optimum, because the approximation model includes uncertainty at the predicted point, and this uncertainty may mislead the optimization process in a wrong way.

The present paper introduces the application of an improved PSO to speed up the optimization algorithm and overcome problems such as inaccuracy and premature convergence during the optimization. To demonstrate the efficiency of the proposed PSO algorithm, it is applied to two different test cases, and the results were compared with four different PSOs, including constriction factor PSO (c-PSO), inertia weight

PSO (w-PSO), vibrational PSO (v-PSO), and comprehensive learning PSO (cl-PSO). The test bed selected herein includes surface fitting of a wing shape and an inverse design of an airfoil in subsonic flow.

2   Surrogate Modeling

The stages of surrogate-based modeling approach include a sampling plan for design points, numerical simulations at these design points, construction of a surrogate model based on simulations, and model validation [8]. There are both parametric and non-parametric alternatives to construct the surrogate model. The parametric approaches such as polynomial regression and Kriging presume the global functional form between the samples and corresponding responses. The non-parametric ones such as neural networks use simple local models in different regions of a sample plan to construct an overall model. After surrogate-based modeling is completed, the optimization problem is described as follows

$$\text{Minimize } \hat{f}(\pmb{x}) \tag{6}$$
$$\text{Subject to } \hat{g}_i(\pmb{x}) \leq 0, i = 1,2, \dots, I$$
$$\pmb{x}^L \leq \pmb{x} \leq \pmb{x}^U$$

This is where the functions are the approximation models. The main purpose of constructing approximate models in this framework is to predict the value of objective and constraints. The relationship between the true response and the approximation can be expressed as follows:

$$f(\pmb{x}) = \hat{f}(\pmb{x}) + \Delta(\pmb{x}) \tag{7}$$
$$\Delta(\pmb{x}) = \epsilon(\pmb{x}) + \delta(\pmb{x}) \tag{8}$$

The total error, $\Delta(\pmb{x})$, includes two types of errors: the first one is system error, $\epsilon(x)$, which exists because of the incompleteness of the surrogate model; and the second one is random error, $\delta(x)$, which exists because of uncontrollable factors such as discretization and round off errors in computational studies.

Many different surrogate-model based optimization algorithms were applied to decrease the level of $\Delta(\pmb{x})$ in engineering problems. Examples are commonly from GA applications such as: an iterative response surface based optimization scheme [3], a statistical improvement criteria with Kriging surrogate modeling [9], more accurate Kriging modeling by using a dynamic multi-resolution technique [7], the use of multiple surrogates [10, 11], a multistage meta-modeling approach [12], and an iteratively enhanced Kriging meta-model [13]. There are also a few applications from PSO studies. Praveen and Duvigneau [14] have constructed radial basis function approximations and used them in conjunction with particle swarm optimization in an inexact evaluation procedure for the objective function values of candidate aerodynamic designs. They showed that the new strategy based on the use of mixed evaluations by metamodels and real CFD solvers could significantly reduce the computational cost of PSO. Khurana et al. [15] developed an artificial neural network and validated with a relationship between the mapped PARSEC (a kind of geometry parameterization method) solution space and the aerodynamic coefficients of lift and drag. The validated surrogate model was used for airfoil shape optimization by replacing the flow solver from the direct numeric optimization loop. Similar to previous study, significant time savings were established with the aerodynamic performance of the output solution in line with the results of the direct PSO and real flow solver combination. Multi-fidelity simulation and surrogate models were employed by Singh and Grandhi [16] in mixed-variable optimization problem. In that research, a progressive mixed-variable optimization strategy is developed and low- and high-fidelity simulations and their respective surrogate models are combined to solve impulse-type problems such as laser peening of a structural component. On the other hand, using PSO type algorithms in multi-objective optimization problems for high-fidelity shape design is computationally more challenging than single-objective optimization problems. This is due to the excessive number of high-fidelity simulations required to identify a host of Pareto-optimal solutions. To get benefits from surrogate models in multi-objective optimization problems, Carrese *et al.* [17-18] presented the Kriging-assisted user-preference multi-objective particle swarm heuristic method. In that implementation, less accurate but inexpensive surrogate models were used cooperatively with the precise but expensive objective functions to ease the computational load. By doing this, the swarm is guided toward the preferred regions of the Pareto frontier.

In addition to the classical surrogate modeling approach, another methodology was also used in a few GA based studies. The main purpose of constructing approximate models in this framework is to predict the positions of new design points, rather than to make inexact computational evaluations as in the surrogate model. An example given by Ong *et al.* [19] presented an Evolutionary Algorithm (EA) that leverages surrogate models. The essential backbone of the framework is an EA coupled with a feasible Sequential Quadratic Programming (SQP) solver in the spirit of Lamarckian learning. Pehlivanoglu and Baysal [20] and Pehlivanoglu and Yagiz [6] have also suggested a novel usage of regression model and neural networks in GA architecture. They used a new technique to predict better solution candidates using local response surface approximation based on neural networks inside the population for the direct shape optimization of an airfoil in transonic flow conditions. Another novel example is given by Hacioglu [21]. A new hybridization technique has been proposed to employ NNs and EAs together to solve the inverse design of an airfoil problem. Similar to the previous technique, the essential backbone of the framework is GA coupled with NN.

3   Present Framework

As in other evolutionary algorithms, PSO method is a population-based stochastic optimization algorithm that originates from "nature". PSO algorithms search the optimum within a population called "swarm." It benefits from two types of learning, such as "cognitive learning" based on an individual's own history and "social learning" based on swarm's own history accumulated by sharing information among all particles in the swarm. Since its development in 1995 by Eberhart and Kennedy [22], it has attracted significant attention. Let s be the swarm size, d be the particle dimension space, and each particle of the swarm has a current position vector x_i, current velocity vector v_i, individual best position vector p_i found by particle itself. The swarm also has the global best position vector p_g found by any particle during all prior iterations in the search space. Assuming that the function f is to be minimized and describing the following notations in t^{th} iteration, then the definitions are as follows:

$$x_i(t) = \left(x_{i,1}(t), x_{i,2}(t), \dots, x_{i,d}(t)\right), \qquad x_{i,j}(t) \in R^d, i = 1,2,\dots,s \tag{9}$$

where each dimension of a particle is updated using the following equations:

$$
\begin{aligned}
v_{i,j}(t) = v_{i,j}(t-1) + c_1 r_1 \left(p_{i,j}(t-1) - x_{i,j}(t-1)\right) \\
+ c_2 r_2 \left(p_{g,j}(t-1) - x_{i,j}(t-1)\right)
\end{aligned}
\tag{10}
$$

$$x_{i,j}(t) = x_{i,j}(t-1) + v_{i,j}(t) \tag{11}$$

In Equation (10), c_1 and c_2 denote constant coefficients, r_1 and r_2 are elements from random sequences in the range of (0, 1). The personal best position vector of each particle is computed using the following expression:

$$p_i(t) = \begin{cases} p_i(t-1) & if \ f(x_i(t)) \geq f(p_i(t-1)) \\ x(t) & if \ f(x_i(t)) < f(p_i(t-1)) \end{cases} \tag{12}$$

Then, the global best position vector is found by

$$p_g(t) = arg \min_{P_i(t)} f(p(t)) \tag{13}$$

3.1   Comparative PSO Algorithms

Four well known PSO algorithms are selected as comparative optimization algorithms. These are c-PSO, w-PSO, v-PSO, and cl-PSO. In c-PSO algorithm the particle swarm with a constriction factor is introduced by Clerc and Kennedy [23], which investigated the use of a parameter called the constriction factor. With the constriction factor K, the particle velocity and position dimensions are updated via:

$$
\begin{aligned}
v_{i,j}(t) = K((v_{i,j}(t-1) + c_1 r_1(p_{i,j}(t-1) - x_{i,j}(t-1)) + c_2 r_2(p_{g,j}(t-1) - \\
x_{i,j}(t-1)))
\end{aligned}
$$

$$K = \frac{2}{\left|2 - \psi - \sqrt{\psi^2 - 4\psi}\right|}, \psi = c_1 + c_2, \psi > 4 \tag{14}$$

$$x_{i,j}(t) = x_{i,j}(t-1) + v_{i,j}(t) \tag{15}$$

A particularly important contribution of this factor is that if it is correctly chosen, it guarantees the stability of PSO without the need to bind the velocities. Typically, values of *2.05* are used for c_1 and c_2, making ψ is equal to *4.1* and K is equal to *0.729*. In the second algorithm called w-PSO Shi and Eberhart [24] introduced the idea of a time-varying inertia weight. The idea was based on the control of the diversification and intensification behavior of the algorithm. The velocity is updated in accordance with the following expressions:

$$v_{i,j}(t) = w(t)v_{i,j}(t-1) + c_1 r_1 \left(p_{I,j}(t-1) - x_{i,j}(t-1) \right)$$
$$+ c_2 r_2 (p_{g,j}(t-1) - x_{i,j}(t-1)) \tag{16}$$
$$x_{i,j}(t) = x_{i,j}(t-1) + v_{i,j}(t)$$

The inertia weight, w, is decreased linearly starting from initial point, w_{ini}, and ending to last point, w_{end}, related to maximum iteration number, T. Normally, the starting value of the inertia weight is set to 0.9 and the final to 0.4. However, we tuned them to $[0.6, 0.2]$ range for better performance. In v-PSO Pehlivanoglu [2] proposed periodic mutation activation based on the wavelet analysis of diversity in the swarm. A generalized mutation operation including mutation strategy can be described as follows:

$$x_{i,j}(t) = F(g(x_{i,j}(t)), fr) \tag{17}$$

where F is the generalized mutation function, $g(x_{i,j}(t))$ is the mutation operator providing the new vector, and fr is a user defined application frequency. Mutation strategy focuses on investigating how to apply mutation operators during the optimization process. Right after updating applications, in every fr^{-1} period of the generations applying the mutation operator to all particle dimensions of the whole swarm, particles in the swarm spread throughout the design space. This operator is called global mutation operator and given by

$$x_{i,j}(t) = x_{i,j}(t)[1 + A(0.5 - rand)\delta], i = 1,2,\ldots,s, j = 1,2,\ldots,d$$
$$\delta = \begin{cases} 1 & if\ t = nfr, n = 1,2,\ldots \\ 0 & if\ t \neq nfr \end{cases} \tag{18}$$

where A is an amplitude factor defined by the user; $rand$ is a random number specified by random number generator in accordance with $N[0, 1]$. In the applications, Gaussian probability density function is used. The velocity and the positions are updated via Equation (16) except the generations corresponding to the mutation period. The comprehensive learning particle swarm optimizer (cl-PSO) is proposed by Liang et al. [25], which uses all other particles' historical best information to update a particle's velocity. This approach keeps the diversity of the swarm in high level to be preserved to discourage the premature convergence. A particle's velocity and its position are updated by the following equations:

$$v_{i,j}(t) = w(t)v_{i,j}(t-1) + cr \left(p_{I(f_i(j)),j}(t-1) - x_{i,j}(t-1) \right)$$
$$x_{i,j}(t) = x_{i,j}(t-1) + v_{i,j}(t) \tag{19}$$

where f_i defines which particles' best position vector the particle i should follow, c is the constant value, and r is a random number drawn from a random sequence in the range of $(0,1)$. The decision about f_i depends on the learning probability value, Pc_i which is defined as the following:

$$Pc_i = 0.05 + 0.45 \frac{(\exp\left(\frac{10(i-1)}{s-1}\right)-1)}{(\exp(10)-1)} \tag{20}$$

For each dimension of particle i, a random number is generated and compared with the value of Pc_i. If a random number is larger than the learning probability value, the related dimension will learn from its own best position vector; otherwise, it will learn from another particle's p_i. A tournament selection procedure is taken into consideration to determine the particle i. The inertia weight w is decreased linearly starting from initial point w_{ini}, and ending to last point w_{end}, related to maximum iteration number T.

3.2   Proposed PSO Algorithm

The proposed algorithm is named vh-PSO. The backbone of the new algorithm is PSO coupled with single or multiple surrogate models and a periodic mutation. The basic steps of the proposed algorithm are outlined here:

```
┌ Initialization          ┐
S^I                       | Determination of initial swarm using random number operator
└ 1^st Swarm              ┘
┌ Design cycle            ┐
f                         | Computation by high-fidelity solver
ε                         | Convergence check
RSM                       | Response Surface Model fitting
p_i                       | Updating of particle best position
p_g                       | Updating of global best position
S^U                       | Updating of swarm by updating equations
S^M                       | Periodic mutation applications
S^P                       | New particle prediction by RSM
New swarm                 | S = S^U + S^M + S^P
└ t^th Swarm              ┘
```

At first, we generate the initial swarm of designs including the particles, S^I, computed by using random number operator. After initiation, all particles in the swarm are evaluated by using high-fidelity objective function solver. By the way, the convergence check is done whether the determined criteria such as the tolerance, $f(x, t)-f(x, t-1) < \varepsilon$, is satisfied or not. After that, all of the design points and the associated exact values of the objective function are archived in the database. In the next step, the input-output couples are used to construct Response Surface Model (RSM). For a local response surface, Radial Basis Neural Network (RBNN) approximates the response values as a weighted sum of radial basis functions. Matlab routine of *newrb* was used to construct RBNN [26]. Then, particle best position vectors and the global best position vector are determined. The updating equations are applied for the new particles. If necessary, mutation operations based on vibrational mutation operator given in equation (18) are applied to particles generated by updating equations. This application provides a random but global diversity within the population. The present indirect prediction strategy is applied right after this updating phase. In classical surrogate modeling approach x_i particle position vectors in each swarm are used as input values and f_i or y_i^c values computed by high-fidelity model are used as output values. These couples are sample points and used to train RBNN. During the optimization process, some particle's objective function ($\hat{f}_i$) or response values ($\hat{y}_i^c$) are predicted by trained neural net(s) to shorten the computation time. On the contrary, it is possible to use the computed response values (y_i^c) as input values and particle position vectors (x_i) as output values in neural network training process. Furthermore, we may predict a new design vector by using the target value(s) in inverse design problem as input for the trained neural network. A new particle predicted by indirect surrogate model can be randomly or by a certain way placed into the swarm. This application provides a local but controlled diversity within the population. At the next design cycle, all particles in the new swarm are evaluated by using high-fidelity objective function solver. All the design points and the associated exact values of the objective function are added to the database. This cycle is repeated until the convergence criterion is satisfied.

4   Numerical Studies

4.1   Surface Fitting of a Wing

One of the important issues in computer graphics is a surface reconstruction and it consists of obtaining a smooth surface that approximates a set of points given in three-dimensional (3D) space. It has a significant role in real engineering problems such as the design of ground, naval, or air vehicle surfaces. A typical application is a reverse engineering where free-form parametric surfaces are constructed from a set of points obtained from surface scanning process. This issue is not a trivial problem and several optimization algorithms including PSO were used to solve the surface reconstruction issue [28-29]. A set of surface points belong to wing in 3D can be modeled by using Bezier surface functions. Example wing surface is depicted in Fig. 1. This wing surface has different airfoil sections in each station through the x_2 axis. The root airfoil is selected as $RAE2822$ airfoil and the tip airfoil is chosen as $NACA0012$ symmetric airfoil. The wing is a rectangular wing and there is no any swept or dihedral angle. The length of chord is fixed to 1 unit. A general form of Bezier surface [27] is given below:

$$x_1(u,v) = \sum_{i=0}^{n}\sum_{j=0}^{m} B_i^n(u)B_j^m(v)\, x_{1,i,j} ,$$

$$x_2(u,v) = \sum_{i=0}^{n}\sum_{j=0}^{m} B_i^n(u)B_j^m(v)\, x_{2,i,j}$$

$$x_3(u,v) = \sum_{i=0}^{n}\sum_{j=0}^{m} B_i^n(u)B_j^m(v)\, x_{3,i,j}$$

$$B_i^n(u) = \frac{n!}{i!\,(n-i)!} u^i(1-u)^{n-i}, B_j^m(v) = \frac{m!}{j!\,(m-j)!} v^j(1-v)^{m-j},$$

$$0 \le u \le 1, 0 \le v \le 1$$

(21)

where $x_1(u,v)$, $x_2(u,v)$, and $x_3(u,v)$ are surface coordinates, u and v are parametric coordinates, n and m are the degrees of Bezier surface and they are fixed to 1 by 12, respectively. $x_{1,i,j}$, $x_{2,i,j}$, and $x_{3,i,j}$ are the control points of Bezier surface and only $x_{3,i,j}$ the third coordinates of control points are selected as the design parameters. The number of design parameters is fixed to 44. A half of them are used to parameterize the upper surface of the wing and the remaining 22 parameters are used to parameterize the lower surface of the wing. The control points are placed only on the root and tip sections of the wing. Phenotype of an example initial swarm and a particle from an example initial swarm are depicted in Fig. 2.

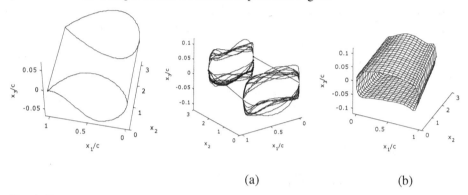

(a)   (b)

Fig. 1. Target wing surface in   **Fig. 2.** (a) Phenotype of an initial swarm, (b) an example
3D environment   surface of a particle from an initial swarm

The objective function value is based on the difference between the target surface points and the particle surface points. However, to facilitate the computation of the objective function only the surface points on the root and tip sections are considered. The objective function f is given below:

$$f = \sum_{j=1}^{k} (x_{3_j}^{c} - x_{3_j}^{t})^2 \tag{22}$$

where k is the number of target points and fixed to 256. We need to point out that this number contains both upper and lower surface pints of the root and tip sections. The swarm particles are optimized in accordance with given objective function by using five aforementioned PSO algorithms. The swarm size is selected as 10; the maximum generation number is selected as 500. Peculiar settings are the following: c_1 and c_2 are equal to 2.05 for c-PSO; c_1 and c_2 are equal to 2.05; w_{ini} and w_{end} are equal to 0.6, 0.2, respectively for w-PSO; c is equal to 1.49445; w_{ini} and w_{end} are equal to 0.9, 0.4, respectively for cl-PSO; c_1 and c_2 are equal to 2.05, w_{ini} and w_{end} are equal to 0.6, 0.2, respectively; fr is equal to 50, A is equal to 0.5 for v-PSO; c_1 and c_2 are equal to 2.05; fr is equal to 50, A is equal to 0.5, and N is equal to 20 which means the last 2 generations for vh-PSO. Additionally, the particles in vh-PSO algorithm are updated in accordance with equation (14). In a comparative study, all algorithms are run 40 times and the averaged global best particle values versus generations are taken into consideration for a fair comparison.

Optimization Results

The optimization results including convergence histories and an example surface optimized by vh-PSO are depicted in Fig. 3 and Fig. 4, respectively.

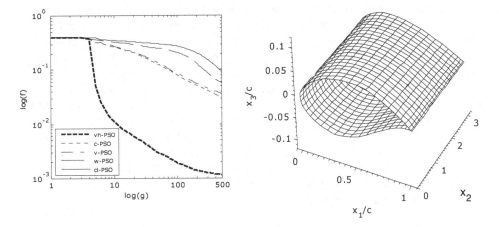

Fig. 3. Convergence histories for Bezier surface fitting problem

Fig. 4. Optimized wing surface model

Among the classical PSO algorithms the best performance belongs to c-PSO algorithm. It reaches the value of 0.0312 at 500th generation. On the other hand, vh-PSO does again outperform the regular algorithms. It reaches the value of 0.0011 at 500th generation. This result means an approximately 98% decrease in the required generations as compared with c-PSO.

4.2   Inverse Design Based on C_p Distribution

In inverse design problem from aerodynamics, the pressure distribution around the shape is known or predicted and the geometry of the shape is investigated. This approach recognizes that the designer usually has an idea of the kind of pressure distribution that will lead to the desired performance. Thus, it is useful to consider the inverse problem of calculating the shape that will lead to a given pressure distribution [30]. Within the second inverse design test case, *RAE2822* airfoil is selected as the test airfoil and the pressure coefficient (C_p) distribution of this airfoil under subsonic flow conditions is chosen as the target C_p distribution. An airfoil shape can be represented using the Bezier curves with a set of control points given in equation (21). The initial swarm is generated by using random number operator. The objective function value is based on the difference between the target C_p points and the computed particle C_p points. The angle of attack is assumed to be zero during the optimization process. The fitness function *f is* defined as,

$$f = \sum_{j=1}^{k} (C_{p_j}^{c} - C_{p_j}^{t})^2 \tag{23}$$

where k is the number of panels and it is fixed to 128. The pressure coefficient is computed by using panel solver [31]. The reference C_p distribution and the initial swarm are depicted in Fig. 5. The swarm particles are optimized in accordance with given objective function by using five PSO algorithms including c-PSO, w-PSO, cl-PSO, v-PSO, and vh-PSO. The swarm size is selected as 10; the maximum generation number is selected as 500. The problem dimension is fixed to 22 as the control points of Bezier curves. Peculiar settings are the following: c_1 and c_2 are equal to 2.05 for c-PSO; c_1 and c_2 are equal to 2.05; w_{ini} and w_{end} are equal to 0.6, 0.2, respectively for w-PSO; c is equal to 1.49445; w_{ini} and w_{end} are equal to 0.9, 0.4, respectively for cl-PSO; c_1 and c_2 are equal to 2.05, w_{ini} and w_{end} are equal to 0.6, 0.2, respectively; fr is equal to 20, A is equal to 0.5 for v-PSO; c_1 and c_2 are equal to 1.5, w_{ini} and w_{end} are equal to 0.6 and 0.2, fr is equal to 20, A is equal to 0.5, and N is equal to 50 which means the last five generations for vh-PSO. In a comparative study, all algorithms are run 40 times and the averaged global best particle values versus generations are taken into consideration for a fair comparison.

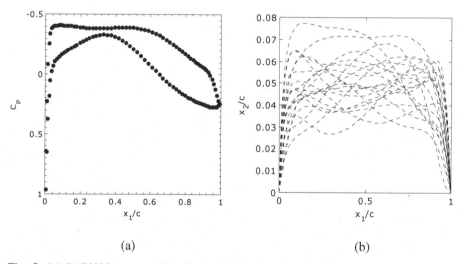

(a)   (b)

Fig. 5. (a) *RAE2822* target airfoil C_p distribution in subsonic flow, and (b) initial swarm particles

Optimization Results

The optimization results including convergence histories and an example airfoil curve optimized by vh-PSO are depicted in Fig. 6 and 7, respectively.

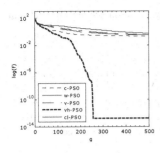
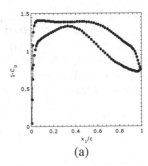
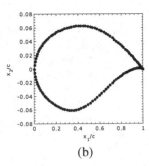

(a)   (b)

Fig. 6. Convergence histories for the inverse design of an airfoil problem

Fig. 7. (a) Target C_p points ($\bullet$ markers) and optimized particle's C_p points (solid line), (b) Target airfoil points ($\bullet$ markers) and optimized particle's curve points (solid line)

Among the classical PSO algorithms including c-PSO, w-PSO, cl-PSO, and v-PSO the best performance belongs to w-PSO algorithm. It reaches the value of 0.2122 at 500th generation. On the other hand, vh-PSO does outperform the regular algorithms. It reaches the value of 1.57×10^{-13} at 500th generation. This result means an approximately 78% decrease in the required generations as compared with w-PSO.

5   Conclusions

The present paper introduced a new use of surrogate modeling in PSO algorithm structure to speed up the optimization algorithm and overcome problems such as low efficiency and premature convergence. Then, depending on the nature of the problem at hand, the present approach employed a local response surface approximation constructed by using neural networks to provide a local but controlled diversity within the population. The average best-individual-fitness values of the algorithms were recorded for a fair comparison among them. To demonstrate their merits, a new approach and four comparative algorithms such as c-PSO, w-PSO, cl-PSO, and v-PSO were applied to two different test scenarios. The principal role of the use of a surrogate model was to answer the question of which individual(s) should be placed into the next swarm. Therefore, the main purpose of the surrogate model is to predict a new design point instead of determining an objective function value. Additionally, periodic mutation operations were applied to all variables of the whole swarm, and this application provided global but random diversity in the swarm. Due to still being a PSO based technique, this method was as robust as the plain PSO algorithms. Based on the results obtained, it was concluded that the proposed PSO algorithm approach is an efficient and fast algorithm in inverse design problems.

References

[1] Groetsch, C.W.: Inverse Problems: Activities for Undergraduates, p. 3. Cambridge University Press (1999)
[2] Pehlivanoglu, Y.V.: Hybrid Intelligent Optimization Methods for Engineering Problems Ph.D. Dissertation, Dept. of Aerospace Engineering, Old Dominion Univ., Norfolk, VA (2010)

[3] Vavalle, A., Qin, N.: Iterative response surface based optimization scheme for transonic airfoil design. Journal of Aircraft 44(2), 365–376 (2007)

[4] Peigin, S., Epstein, B.: Robust optimization of 2D airfoils driven by full Navier– Stokes computations. Computers & Fluids 33(9), 1175–1200 (2004)

[5] Song, W., Keane, A.J.: A new hybrid updating scheme for an evolutionary search strategy using genetic algorithms and Kriging. In: 46th AIAA/ASME/ASCE/AHS/ASC Structures, Structural Dynamics & Materials Conference AIAA paper 2005-1901 (2005)

[6] Pehlivanoglu, Y.V., Yagiz, B.: Aerodynamic design prediction using surrogatebased modeling in genetic algorithm architecture. Aerospace Science and Technology 23, 479–491 (2011)

[7] Jouhaud, J.C., Sagaut, P., Montagnac, M., Laurenceau, J.: A surrogate-model based multidisciplinary shape optimization method with application to a 2D subsonic airfoil. Computers & Fluids 36(3), 520–529 (2007)

[8] Qoeipo, N.V., Haftka, R.T., Shyy, W., Goel, T., Vaidyanathan, R., Tucker, P.K.: Surrogatebased analysis and optimization. Progress in Aerospace Sciences 41(1), 1–28 (2005)

[9] Keane, A.J.: Statistical improvement criteria for use in multi objective design optimization. AIAA Journal 44(4), 879–891 (2006)

[10] Glaz, B., Goel, T., Liu, L., Friedmann, P.P., Haftka, R.T.: Multiple-surrogate approach to helicopter rotor blade vibration reduction. AIAA Journal 47(1), 271–282 (2009)

[11] Papila, N., Shyy, W., Griffin, L., Dorney, D.J.: Shape optimization of supersonic turbines using global approximation methods. Journal of Propulsion and Power 18(3), 509–518 (2002)

[12] Xiong, C.Y., Chen, W.: Multi-response and multistage meta-modeling approach for design optimization. AIAA Journal 47(1), 206–218 (2009)

[13] Duchaine, F., Morel, T., Gicquel, L.Y.M.: Computational fluid dynamics based Kriging optimization tool for aeronautical combustion chambers. AIAA Journal 47(3), 631–645 (2009)

[14] Praveen, C., Duvigneau, R.: Low cost PSO using metamodels and inexact preevaluation: application to aerodynamic shape design. Comput. Methods Appl. Mech. Engrg. 198, 1087–1096 (2009)

[15] Khurana, M.S., Winarto, H., Sinha, A.K.: Airfoil optimization by swarm algorithm with mutation and artificial neural networks. In: 47th AIAA Aerospace Sciences Meeting Including the New Horizons Forum and Aerospace Exposition, AIAA 2009-1278, Orlando, Florida (2009)

[16] Singh, G., Grandhi, R.V.: Mixed-variable optimization strategy employing multifidelity simulation and surrogate models. AIAA Journal 48(1), 215–223 (2010)

[17] Carrese, R., Winarto, H., Li, X.: Integrating user-preference swarm algorithm and surrogate modeling for airfoil design. In: 49th AIAA Aerospace Sciences Meeting including the New Horizons Forum and Aerospace Exposition, AIAA 2011-1246, Orlando, Florida (2011)

[18] Carrese, R., Sobester, A., Winarto, H., Li, X.: Swarm heuristic for identifying preferred solutions in surrogate-based multi-objective engineering design. AIAA Journal 49(7), 1437–1449 (2011)

[19] Ong, Y.S., Nair, P.B., Keane, A.J.: Evolutionary optimization of computationally expensive problems via surrogate modeling. AIAA Journal 41(4), 687–696 (2003)

[20] Pehlivanoglu, Y.V., Baysal, O.: Vibrational genetic algorithm enhanced with fuzzy logic and neural networks. Aerospace Science and Technology 14(1), 56–64 (2010)

[21] Hacioglu, A.: Fast evolutionary algorithm for airfoil design via neural network. AIAA Journal 45(9), 2196–2203 (2007)

[22] Eberhart, R.C., Kennedy, J.: A new optimizer using particle swarm theory. In: Proc. 6th Int. Symp. Micromachine Human Sci., Nagoya, Japan, pp. 39–43 (1995)

[23] Clerc, M., Kennedy, J.: The particle swarm-explosion, stability, and convergence in a multidimensional complex space. IEEE Trans. Evol. Comput. 6(1), 58–73 (2002)

[24] Shi, Y., Eberhart, R.: A modified particle swarm optimizer. In: Proc. of the World Congr. Comput. Intell., pp. 69–73 (1998)

[25] Liang, J.J., Qin, A.K., Suganthan, P.N., Baskar, S.: Comprehensive learning particle swarm optimizer for global optimization of multimodal functions. IEEE Trans. Evol. Comput. 10(3), 281–295 (2006)

[26] Neural Network Toolbox, Matlab the language of technical computing Version R2007b The MathWorks, Inc. (2007)

[27] Farin, G.: Curves and surfaces for computer aided geometric design; a practical guide, pp. 41–42. Academic Press Inc. (1993)

[28] Gálvez, A., Cobo, A., Puig-Pey, J., Iglesias, A.: Particle Swarm Optimization for Bézier Surface Reconstruction. In: Bubak, M., van Albada, G.D., Dongarra, J., Sloot, P.M.A. (eds.) ICCS 2008, Part II. LNCS, vol. 5102, pp. 116–125. Springer, Heidelberg (2008)

[29] Gálvez, A., Iglesias, A., Cobo, A., Puig-Pey, J., Espinola, J.: Bézier curve and surface fitting of 3D point clouds through genetic algorithms, functional networks and leastsquares approximation. In: Gervasi, O., Gavrilova, M.L. (eds.) ICCSA 2007, Part II. LNCS, vol. 4706, pp. 680–693. Springer, Heidelberg (2007)

[30] Jameson, A.: Essential Elements of Computational Algorithms for Aerodynamic Analysis and Design NASA/CR-97-206268 ICASE Report No. 97-68, pp. 34–35 (1997)

[31] Anderson, J.D.: Fundamentals of Aerodynamics, pp. 217–222. Mc-Graw Hill, Inc. (1984)

Solving the Unknown Complexity Formula Problem with Genetic Programming

Rayco Batista, Eduardo Segredo, Carlos Segura,
Coromoto León, and Casiano Rodríguez

Dpto. Estadística, I. O. y Computación. Universidad de La Laguna
La Laguna, 38271, Santa Cruz de Tenerife, Spain
raycobd@gmail.com, {esegredo,csegura,cleon,casiano}@ull.es

Abstract. The Unknown Complexity Formula Problem (UCFP) is a particular case of the symbolic regression problem in which an analytical complexity formula that fits with data obtained by multiple executions of certain algorithm must be given. In this work, a set of modifications has been added to the standard Genetic Programming (GP) algorithm to deal with the UCFP. This algorithm has been applied to a set of well-known benchmark functions of the symbolic regression problem. Moreover, a real case of the UCFP has been tackled. Experimental evaluation has demonstrated the good behaviour of the proposed approach in obtaining high quality solutions, even for a real instance of the UCFP. Finally, it is worth pointing out that the best published results for the majority of benchmark functions have been improved.

Keywords: Genetic Programming, Symbolic Regression, Unknown Complexity Formula Problem.

1 Introduction

Symbolic regression is a process for analysing and modelling numeric multi-variate data sets by specifying mathematical models that fit such data sets. It is an optimisation problem in which the best combination of variables, symbols, and coefficients is looked for in order to develop an optimum model satisfying a set of fitness cases.

It is important to remark that the task of regression consist on identify the variables (inputs) in the data that are related to changes in the important control variables (outputs), to express these relationships in mathematical models, and to analyse the quality and generality of the constructed models.

Evolutionary Computing [1] (EC) draw inspiration from the process of natural evolution. In EC, a given environment (problem) is filled with a population of individuals (candidate solutions) which fight for surviving and reproducing. The fitness (quality) of such individuals indicates how well they are able to adapt to the environment. EC include a set of problem solving techniques such as evolution strategies, evolutionary programming, genetic algorithms, genetic programming, differential evolution, learning classifier systems, and also swarm based algorithms. These approaches have been successfully applied to different problems [2,3,4] related to optimisation, industrial design, data mining, symbolic regression, signal processing, and bioinformatics, among others.

I. Rojas, G. Joya, and J. Cabestany (Eds.): IWANN 2013, Part I, LNCS 7902, pp. 232–240, 2013.

Different EC techniques have been applied to the symbolic regression problem [5,6,7]. However, the most popular one is Genetic Programming (GP) [8,9].

Besides its representation, GP approaches differs from other Evolutionary Algorithms (EAs) in the fields it can be applied. Usually, GP is used to look for models with maximum fit. The individuals represent models and their fitness is the model quality to be maximised. In GP the individuals are encoded as parse trees (non-linear structures) that represent expressions belonging to certain language (arithmetic expressions, code written in some programming language, etc). In addition, the size of the individuals is variable, while in other EAs the size of the individuals is fixed. Such individuals are evolved by the use of selection and genetic operators that are responsible for guiding the search process. Finally, it is worth mentioning that the genetic operators must be able to work with parse trees.

The symbolic regression problem falls into the category of *data fitting* problems. During the last years, the number of publications of GP applied to this kind of problems, and particularly, to the symbolic regression problem, has increased [10,11,12]. Another important kind of GP applications falls into the category of *problems involving "physical" environments* [13]. This two categories group the majority of GP applications. A particular case of the symbolic regression problem has been addressed in this work. It is known as the *Unknown Complexity Formula Problem* (UCFP). Starting from data obtained during the execution of an algorithm, an analytical complexity formula which fits with the data must be given. This problem has not been directly tackled in the literature. However, some works based on obtaining analytical models of applications in parallel environments have been proposed [14,15,16].

The main contributions of this work are the following. Firstly, a set of modifications and optimisations has been incorporated to the standard GP algorithm to successfully deal with the UCFP. In order to validate the proposed approach several studies have been carried out with a set of well-known benchmark problems [17,18]. The best published results for the majority of these problems [18] have been improved. In addition, the formula of the standard matrix product algorithm has been inferred, starting from data obtained in multiple executions of such an algorithm. The rest of the paper is structured as follows: the mathematical formulation for the UCFP is given in Section 2. In Section 3, the applied optimisation method is detailed. Then, the experimental evaluation is described in Section 4. Finally, the conclusions and some lines of future work are given in Section 5.

2   UCFP: Formal Description

In this work, a particular case of the symbolic regression problem has been tackled. Starting from data obtained during the execution of certain algorithm (time, memory, etc.) , an analytical complexity formula which fits with such data must be found. This problem has awoken great interest and it is known as the Unknown Complexity Formula Problem (UCFP). The considered input data for this problem is the following:

- The source code of an algorithm *A* written in some programming language and the set of input parameters *I* of such an algorithm.

Algorithm 1. GP algorithm pseudocode

1: Generate an initial population with N individuals
2: Evaluate all individuals in the population
3: **while** (not stopping criterion) **do**
4:     Mating selection: select parents to generate the offsprings
5:     Variation: Apply genetic operators to the mating pool to create a child population
6:     Evaluate the offsprings
7:     Select individuals for the next generation
8: **end while**
9: Return the best individual

– A matrix M with the results of m executions (where m is large enough) of the algorithm A on a given set of machines taking the set of input parameters I. This matrix contains the values $(\vec{P}^i, T^i)_{i=1...m} \in (\Re^n \times \Re)^m$ of both an observable $T \in \Re$ and a vector of independent variables $\vec{P} = (P_1 \ldots P_n) \in \Re^n$. Usually, the quantitative attribute $T \in \Re$ represents the execution time. However, other interesting measures can be taken into account (consumed memory, heuristic optimal values, etc). Algorithm designers define the domain of the vector $\vec{P}$ which contains the parameters $(P_1, P_2, \ldots, P_n)$ they believe have some influence on the complexity behaviour of T.
– Algorithm designers also define the family of operators, functions, and constants $O = \{+, -, *, **, log, exp, \ldots\}$ which are allowed in the analytical formula.

Therefore, the optimisation problem consists in looking for the analytical formula

$$T = f(P_1, P_2, \ldots, P_n) \tag{1}$$

which minimises the error of the predictions of f for the values in $(P_1^i \ldots P_n^i, T^i)_{i=1...m}$. In this work, the error has been defined as follows:

$$Error = \sum_{i=1}^{m} |f(\vec{P}^i) - T^i| \tag{2}$$

3   Optimisation Scheme

This section describes the algorithm which has been used in order to solve the UCFP. It is an standard GP algorithm that incorporates a set of modifications and optimisations to improve its behaviour and performance when it is applied to the UCFP. Algorithm 1 shows the pseudocode of such an approach. Analysing such a pseudocode, different problems could appear when it is applied to the UCFP:

– Low quality individuals might be generated at the beginning of the execution.
– If the number of individuals in the population is too large, the algorithm may suffer from stagnation.
– If the population is filled with huge individuals, the time invested to evaluate it could increase, and consequently the optimisation process might be harder.

Algorithm 2. Generation of the initial population

1: $pop = $ random_population(N)
2: **for** (r times) **do**
3:     $random = $ generate ($k \cdot N$) new random individuals
4:     $best = $ select the $((1 - k) \cdot N)$ best individuals from pop
5:     $new\_pop = random + best$
6:     $pop = $ select_best_population(pop, new_pop)
7: **end for**
8: return pop

Taking into account the UCFP, the following methods have been implemented into the GP standard algorithm in order to solve the aforementioned problems:

- The initial population is randomly filled with N individuals. However, it is modified following the pseudocode shown in the Algorithm 2, in order to improve the quality of the new individuals.
- In order to avoid big populations, a ratio g which controls the population size is established. Thus, the population size of the current generation is multiplied by such a ratio, and the result is added up to the population size of the current generation in order to constitute the population size for the next generation, until a maximum population size M specified by the user is reached.
- Finally, a maximum depth d is allowed for the new generated individuals. This fact avoid the appearance of huge individuals in the population.

It is worth mentioning that the function *select_best_population* (line 6 in the Algorithm 2) calculates the mean objective value of the individuals for each population to select the best one. Moreover, the parameter k (lines 3-4) controls the number of individuals which are randomly generated during the r iterations (line 2) of such a process. When an individual is randomly generated, a parameter t allows to fix the probability of appearance of terminal nodes.

The genetic operators have been the ones proposed in [19,17]. The crossover operator has been the *Semantic Similarity based Crossover* (SSC), while the mutation operator has been the *Semantic Similarity based Mutation* (SSM). Both operators use the parameters α and β to establish the level of semantic similarity between two individuals. Particularly, such parameters set the minimum and maximum values allowed for the Sample Semantic Distance [19,17] (SSD).

4   Experimental Evaluation

In this section, the experiments conducted with the optimisation scheme presented in Section 3 are described. The optimisation scheme has been implemented using METCO (*Metaheuristic-based Extensible Tool for Cooperative Optimisation*) [20]. Tests have been run on a Debian GNU/Linux computer with four AMD ® Opteron ^{TM} (model number 6164 HE) at 1.7 GHz and 64 GB RAM. The compiler has been GCC 4.4.5.

Since experiments have involved the use of stochastic algorithms, each execution has been repeated 30 times. In order to provide the results with confidence, comparisons

Table 1. Constants, operators, and variables allowed in the formulas

Type	Value
Constants	1
Operators	$+, -, *, /, sin, cos, sqrt, log$
Variables	x, y (for multivariate benchmarks)

Table 2. Parameterisation of the optimisation scheme

Parameter	Values	Parameter	Values
Selection operator	Binary tournament	r	$1 \cdot 10^3$
Generational criterion	Elitism	N	30 individuals
Number of evaluations	$1 \cdot 10^5$	M	100, 200 individuals
Crossover probability	0.5	d	7, 15
Mutation probability	0.8	t	0.8, 0.2
g	3/4	α	1, 2, 3
k	2/3	β	5, 8

have been made by applying the following statistical analysis [21]. First, a *Shapiro-Wilk test* is performed in order to check whether the values of the results follow a normal (Gaussian) distribution. If so, the *Levene test* is used to check for the homogeneity of the variances. If samples have equal variance, an ANOVA *test* is done. Otherwise, a *Welch test* is performed. For non-Gaussian distributions, the non-parametric *Kruskal-Wallis* test is used to compare the medians of the algorithms. A confidence level of 95% has been fixed.

Each experiment has been carried out for the set of benchmark functions used in [17,18]. Due to space restrictions, results will be shown only for the benchmark function shown in Equation 3. However, the same conclusions can be extracted for the majority of such benchmarks functions. In order to obtain the instance of the benchmark function F4, 20 points (20 fitness cases) have been uniformly selected from the range $[-1, 1]$. On the other hand, the matrix product algorithm has been considered as a real case of the UCFP. To obtain this instance, the matrix product algorithm has been executed 30 times (30 fitness cases), varying the matrix size from 100 to 3000 considering increments equal to 100, and measuring the time invested to complete each execution. In this particular case, the main aim is to obtain an analytical formula which predicts the time invested by the matrix product algorithm if the matrix size is known.

$$F4 = x^6 + x^5 + x^4 + x^3 + x^2 + x \tag{3}$$

In order to study the robustness of the proposed optimisation scheme, 48 different configurations of such an approach have been applied to the benchmark problems and the matrix product algorithm instance. Table 1 shows the different constants, operators, and variables which have been allowed to appear in the formulas, while the parameterisation of the optimisation scheme is described in Table 2. Such values for the parameters have been selected since in a preliminary study they provided promising results. The 48 configurations of the optimisation scheme have been obtained by combining all the possible

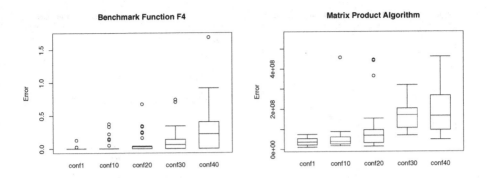

Fig. 1. Boxplots for the benchmark function F4 (left-hand side) and for the matrix product algorithm (right-hand side)

values of the the parameters M, d, t, α, and β. The configurations of the optimisation scheme have been sorted in ascending order considering the mean of the objective value (mean of the error) achieved at the end of the executions. Thus, the configuration which has obtained the lowest mean error has been named *conf1*, the second one *conf2*, and so on. In the case of the benchmark function F4, *conf1* has applied the parameterisation $M = 200$, $d = 15$, $t = 0.8$, $\alpha = 2$, and $\beta = 8$, while for the matrix product algorithm, the parameterisation of *conf1* has been $M = 100$, $d = 15$, $t = 0.2$, $\alpha = 1$, and $\beta = 8$.

The left-hand side of Fig. 1 shows the boxplots for the benchmark function F4, while the right-hand side shows the boxplots for the matrix product algorithm, considering different configurations of the optimisation scheme. In this case, a boxplot represents data about the objective value or error achieved at the end of each one of the 30 runs carried out by a particular configuration of the optimisation scheme. Circles represent outliers in the data, i.e. observations which are numerically distant from the rest of the data. It can be noted for both problems that, depending on the parameterisation, the performance of the optimisation scheme can vary significantly. The corresponding configuration *conf1* for both, the benchmark function F4 and the matrix product instance, has obtained better results in terms of the error than the remaining configurations. In fact, the statistical analysis explained at the beginning of the current section has shown for the benchmark function F4 that *conf1* has presented statistically significant differences with 34 configurations of the optimisation scheme, and there have not been statistically significant differences with other 13 configurations. In the case of the matrix product instance, there have been statistically significant differences with 36 configurations, while 11 configurations have not presented statistically significant differences with *conf1*. Consequently, it is very important to perform the right selection of the values for the different parameters of the proposed optimisation scheme in order to increase its performance and to improve the quality of the obtained solutions.

Taking into account the quality of the obtained solutions, the percentage of successful runs obtained by the configuration *conf1* for the benchmark function F4 has been equal to 96%. A successful run has been defined as in [17,18]. Since the input data of the matrix product instance have been obtained by experimental evaluation, it has no sense to calculate the percentage of successful runs for this real case. Finally, in

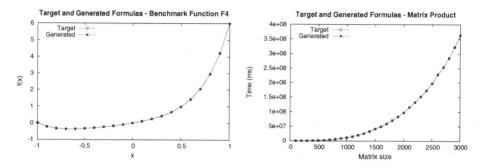

Fig. 2. Target and generated formulas for the benchmark function F4 (left-hand side) and for the matrix product algorithm (right-hand side)

order to compare the target formula with the one generated by the proposed optimisation scheme, the best individual in the population, i.e. the one which has achieved the lowest error in the 30 executions performed by the configuration *conf1*, has been selected. Fig. 2 shows the target formula and the one represented by such a best individual for the benchmark function F4 (left-hand side) and for the matrix product algorithm (right-hand side). It can be observed that the target and generated formulas are quite similar for both problems. In fact, in the case of the problem F4, the proposed approach has been able to generate exactly the target formula. Therefore, this fact demonstrates that high quality solutions can be obtained by the proposed optimisation scheme, even for a real case of the UCFP.

5   Conclusions and Future Work

In this work, the UCFP, a particular case of the symbolic regression problem, has been tackled. In order to deal with it, a set of modifications has been incorporated into the standard GP algorithm. The proposed approach has been applied to a set of well-known benchmark functions of the symbolic regression problem. In addition, such an algorithm has also been applied to a real instance of the UCFP, whose input data have been obtained by multiple executions of the standard matrix product algorithm. The best published results for the majority of the benchmarks have been outperformed. Moreover, the formula of the matrix product algorithm has been successfully inferred. The robustness and statistical analyses have revealed that the values for the different parameters of the proposed optimisation scheme must be properly selected. Otherwise, the performance of the approach could be seriously affected. It is worth pointing out that when the right parameterisation has been carried out, high quality solutions have been achieved. The formulas generated by the proposal and the target ones have been quite similar for the benchmark functions and for the matrix product algorithm. In fact, for the benchmark function F4, the target and generated formulas have been the same.

Lines of future work will include the application of the proposed optimisation scheme to other real instances of the UCFP. In addition, since the performance of the proposed approach highly depends on its parameterisation, it would be interesting to apply parameter setting strategies to such an algorithm.

Acknowledgements. This work was supported by the EC (FEDER) and the Spanish Ministry of Science and Innovation as part of the 'Plan Nacional de I+D+i', with contract number TIN2011-25448. The work of Eduardo Segredo was funded by grant FPU-AP2009-0457

References

1. Eiben, A.E., Smith, J.E.: Introduction to Evolutionary Computing (Natural Computing Series). Springer (2008)
2. Burke, E.K., Newall, J.P., Weare, R.F.: Initialization strategies and diversity in evolutionary timetabling. Evol. Comput. 6(1), 81–103 (1998)
3. Keane, A., Brown, S.: The design of a satellite boom with enhanced vibration performance using genetic algorithm techniques. In: Proceedings of Acedc 1996 PEDC, pp. 107–113 (1996)
4. Freitas, A.A.: Data Mining and Knowledge Discovery with Evolutionary Algorithms. Springer-Verlag New York, Inc., Secaucus (2002)
5. Cerny, B.M., Nelson, P.C., Zhou, C.: Using differential evolution for symbolic regression and numerical constant creation. In: Proceedings of the 10th Annual Conference on Genetic and Evolutionary Computation, GECCO 2008, pp. 1195–1202. ACM, New York (2008)
6. Johnson, C.: Artificial immune system programming for symbolic regression. In: Ryan, C., Soule, T., Keijzer, M., Tsang, E., Poli, R., Costa, E. (eds.) EuroGP 2003. LNCS, vol. 2610, pp. 345–353. Springer, Heidelberg (2003)
7. Poli, R., Langdon, W., Holland, O.: Extending particle swarm optimisation via genetic programming. In: Keijzer, M., Tettamanzi, A.G.B., Collet, P., van Hemert, J., Tomassini, M. (eds.) EuroGP 2005. LNCS, vol. 3447, pp. 291–300. Springer, Heidelberg (2005)
8. Cramer, N.L.: A representation for the adaptive generation of simple sequential programs. In: Proceedings of the 1st International Conference on Genetic Algorithms, pp. 183–187. L. Erlbaum Associates Inc., Hillsdale (1985)
9. Koza, J.R.: Genetic programming: on the programming of computers by means of natural selection. MIT Press, Cambridge (1992)
10. Korns, M.: Large-scale, time-constrained symbolic regression-classification. In: Riolo, R., Soule, T., Worzel, B. (eds.) Genetic Programming Theory and Practice V. Genetic and Evolutionary Computation Series, pp. 53–68. Springer, US (2008)
11. Korns, M., Nunez, L.: Profiling symbolic regression-classification. In: Genetic Programming Theory and Practice VI. Genetic and Evolutionary Computation, pp. 1–14. Springer, US (2009)
12. Korns, M.: Abstract expression grammar symbolic regression. In: Riolo, R., McConaghy, T., Vladislavleva, E. (eds.) Genetic Programming Theory and Practice VIII. Genetic and Evolutionary Computation, vol. 8, pp. 109–128. Springer, New York (2011)
13. Luke, S.: Genetic programming produced competitive soccer softbot teams for RoboCup97. In: Koza, J.R., Banzhaf, W., Chellapilla, K., Deb, K., Dorigo, M., Fogel, D.B., Garzon, M.H., Goldberg, D.E., Iba, H., Riolo, R. (eds.) Genetic Programming 1998: Proceedings of the Third Annual Conference, University of Wisconsin, Madison, Wisconsin, USA, pp. 214–222. Morgan Kaufmann (1998)
14. García, L., González, J.A., González, J.C., León, C., Rodríguez, C., Rodríguez, G.: Complexity driven performance analysis. In: Dongarra, J., Laforenza, D., Orlando, S. (eds.) EuroPVM/MPI 2003. LNCS, vol. 2840, pp. 55–62. Springer, Heidelberg (2003)
15. Martínez, D.R., Blanco, V., Boullón, M., Cabaleiro, J.C., Rodríguez, C., Rivera, F.F.: Software tools for performance modeling of parallel programs. In: IEEE International Parallel and Distributed Processing Symposium, IPDPS 2007, pp. 1–8 (2007)

16. Martínez, D.R., Blanco, V., Boullón, M., Cabaleiro, J.C., Pena, T.F.: Analytical performance models of parallel programs in clusters. In: Bischof, C., Bücker, M., Gibbon, P., Joubert, G.R., Lippert, T., Mohr, B., Peters, F. (eds.) Parallel Computing: Architectures, Algorithms, and Applications. Advances in Parallel Computing, vol. 15, pp. 99–106. IOS Press (2008)
17. Uy, N.Q., Hoai, N.X., O'Neill, M., Mckay, R.I., Galván-López, E.: Semantically-based crossover in genetic programming: application to real-valued symbolic regression. Genetic Programming and Evolvable Machines 12(2), 91–119 (2011)
18. Karaboga, D., Ozturk, C., Karaboga, N., Gorkemli, B.: Artificial bee colony programming for symbolic regression. Information Sciences 209, 1–15 (2012)
19. Uy, N.Q., Hoai, N.X., O'Neill, M.: Semantics Based Mutation in Genetic Programming: The case for real-valued symbolic regression. In: Mendel 2009, 15th International Conference on Soft Computing, Brno, Czech Republic (2009)
20. León, C., Miranda, G., Segura, C.: METCO: A Parallel Plugin-Based Framework for Multi-Objective Optimization. International Journal on Artificial Intelligence Tools 18(4), 569–588 (2009)
21. Demšar, J.: Statistical comparison of classifiers over multiple data sets. Journal of Machine Learning Research 7, 1–30 (2006)

Three Alternatives for Parallel GPU-Based Implementations of High Performance Particle Swarm Optimization

Rogério M. Calazan[1], Nadia Nedjah[2], and Luiza de Macedo Mourelle[3]

[1] Department of Telecommunications and Information Technology,
Brazilian Navy, Brazil
[2] Department of Electronics Engineering and Telecommunication
[3] Department of System Engineering and Computation,
Engineering Faculty, State University of Rio de Janeiro, Brazil
{rogerio,nadia,ldmm}@eng.uerj.br

Abstract. Particle Swarm Optimization (PSO) is heuristics-based method, in which the solution candidates of a problem go through a process that simulates a simplified model of social adaptation. In this paper, we propose three alternative algorithms to massively parallelize the PSO algorithm and implement them using a GPGPU-based architecture. We aim at improving the performance of computationally demanding optimizations of many-dimensional problems. The first algorithm parallelizes the particle's work. The second algorithm subdivides the search space into a grid of smaller domains and distributes the particles among them. The optimization subprocesses are performed in parallel. The third algorithm focuses on the work done with respect to each of the problem dimensions and does it in parallel. Note that in the second and third algorithms, all particles act in parallel too. We analyze and compare the speedups achieved by the GPU-based implementations of the proposed algorithms, showing the highlights and limitations imposed.

1 Introduction

Particle Swarm Optimization (PSO) was introduced by Kennedy and Eberhart [1] and is based on collective behavior, social influence and learning. Many successful applications of PSO have been reported, in which this algorithm has shown many advantages over other algorithms based on swarm intelligence, mainly due to its robustness, efficiency and simplicity. Moreover, it usually requires less computational effort when compared to other stochastic algorithms [2]. The PSO algorithm maintains a swarm of particles, where each of which represents a potential solution. In analogy with evolutionary computation, a *swarm* can be identified as the population, while a *particle* with an individual. In general terms, the particle flows through a multidimensional search space, where the corresponding position is adjusted according to its own experience and that of its neighbors [2].

I. Rojas, G. Joya, and J. Cabestany (Eds.): IWANN 2013, Part I, LNCS 7902, pp. 241–252, 2013.

Several works show that PSO implementation in GPGPU provide a better performance than CPU-based implementations [9] [10] [11]. In contrast, the purpose of this paper is to implement the Global Best version of PSO in GPGPUs. In order to take full advantage of the massively parallel nature of GPGPUs, we explore three different scenarios: *(i)* In the first proposed approach, the work done by the particles of the swarm is performed in parallel until a synchronization is required. Nonetheless, the work done by the particle itself is performed sequentially. Hence, here each thread is associated with a given particle of the swarm. *(ii)* In the second approach, the search space is divided into a grid of smaller subspaces. Then, swarms of particles are formed and assigned to search the subdomains. The swarms act simultaneously. Moreover, within each swarm, particles act in parallel until a synchronization point, during which they exchange knowledge acquired so far, individually. It is worth noting that there is no cooperative work among the swarms. So, there is no exchange of information about best position found by the groups. *(iii)* The third approach explores a fine-grained parallelism, which consists of doing the computational work with respect to each of the problem dimensions in parallel. As in the first approach, this one also handles a single swarm of particles. Nonetheless, here a thread corresponds to a given dimension of the problem and a block of threads to a given particle. This approach should favor optimization problems with high dimensionality.

An analysis is done in order to identify the number of swarms and particles per swarm as well as how to map the swarms into blocks and particles into threads, aiming at maximizing performance. Furthermore, we study the impact of the grid resolution on the convergence time. The grid resolution is defined by the number of cells used. It coincides with the number of swarms invested in the search. Finally, we study the change of the number of dimensions between the implementations.

This paper is organized as follows: First, in Section 2, we sketch briefly the PSO process and the algorithm; After that, in Section 3, we describe the first approach: PPSO; In the sequel, in Section 4, we describe the second approach: SGPSO; Then, in Section 5, we describe the third approach: PDPSO; Subsequently, in Section 6, we present and analyze the obtained results; Finally, in Section 7, we draw some concluding remarks and point out directions for future work.

2   Particle Swarm Optimization

The main steps of the PSO algorithm are described in Algorithm 1. Note that, in this specification, the computations are executed sequentially. In this algorithm, each particle has a *velocity* and an *adaptive direction* [1] that determine its next movement within the search space. The particle is also endowed with a memory that makes it able to remember the best previous position it passed by.

In this variation of the PSO algorithm, the neighborhood of each particle is formed by all the swarm's particles. Using this strategy, the social component of the particle's velocity is influenced by all other particles [2] [3]. The velocity

Algorithm 1. PSO

---

for $i = 1$ *to* n **do**
   randomly initialize position and velocity of particle i
repeat
   for $i = 1$ *to* n **do**
      compute the $Fitness_i$ of particle i
      if $Fitness_i \leq Pbest$ **then**
         update $Pbest$ using the position of particle i
      if $Fitness_i \leq Gbest$ **then**
         update $Gbest$ using the position of particle i
      update the velocity of particle i; **update** the position of particle i
   until stopping criterion
   return $Gbest$ and corresponding position

---

is the element that promotes the capacity of particle locomotion and can be computed as described in (1) [1] [2], wherein w is called *inertia weight*, r_1 and r_2 are random numbers in [0,1], c_1 and c_2 are positive constants, y_{ij} is the best position $Pbest$ found by the particle i so far, w.r.t. dimension j, and y_j is the best position $Gbest$, w.r.t. dimension j, found so far, considering all the population's particles. The position of each particle is also updated as described in (1). Note that $x_{i,j}^{(t+1)}$ is the current position and $x_{i,j}^{(t)}$ is the previous position.

$$v_{i,j}^{(t+1)} = wv_{i,j}^{(t)} + c_1 r_1 \left(y_{i,j} - x_{i,j}^{(t)} \right) + c_2 r_2 \left(y_j - x_{i,j}^{(t)} \right), \quad x_{i,j}^{(t+1)} = v_{i,j}^{(t+1)} + x_{i,j}^{(t)} \quad (1)$$

The velocity component drives the optimization process, reflecting both the experience of the particle and the exchange of information between the particles. The particle's experimental knowledge is referred to as the cognitive behavior, which is proportional to the distance between the particle and its best position found, with respect to its first iteration [3]. The maximum velocity $v_{k,max}$ is defined for each dimension k of the search space. It can be expressed as a percentage of this space by $v_{k,max} = \delta(x_{k,max} - x_{k,min})$, wherein $x_{k,max}$ and $x_{k,min}$ are the maximum and minimum limits of the search space explored, with respect to dimension k, respectively and $\delta \in [0, 1]$.

3 First Algorithm: PPSO

The first proposed algorithm, called PPSO, follows from the idea that the work performed by a given particle is independent of that done by the other particles of the swarm, except in terms of $Gbest$, and thus the computation done by the particles could be executed simultaneously. This algorithm has a synchronization point at the election of $Gbest$, wherein $p_1, \ldots p_n$ denote the n particles of the swarm, and $v^{(p_1)}, \ldots v^{(p_n)}$ and $x^{(p_1)}, \ldots x^{(p_n)}$ the respective velocities and positions. Each particle computes the corresponding fitness, velocity and position, independently and in parallel with the other particles, until the election of $Gbest$. In order to synchronize the process and prevent using incorrect values of $Gbest$,

---

Algorithm 2. CUDA Pseudo-code for PPSO

---

let b = number of blocks; **let** t = number of threads
kernel$\langle b, t \rangle$ position and velocity random generators
repeat
   kernel$\langle b, t \rangle$ fitness calculator; **kernel**$\langle b, t \rangle$ velocity and position calculator
until stopping condition
transfer result back to CPU
return *Gbest* and corresponding position;

Algorithm 3. CUDA Pseudo-code of kernel fitness calculator

---

let $tid = threadIdx + blockIdx \times blockDim$
compute *fitness* of particle *tid*; **update** *Pbest* of particle *tid*
if $(tid = 0)$ **then**
   compute *Gbest* of swarm

---

the velocity and position computations can only commence once *Gbest* has been chosen among the *Pbest* values of all particles of the swarm [4] [5]. Note that the verification of the stopping criterion achievement is also done synchronously by the parallel processes, but it does not hinder the performance of the algorithm.

The CUDA pseudo-code of algorithm PPSO is shown in Algorithm 2. Algorithm 3 shows the code executed by thread *tid* associated with a given particle of the swarm. Note that the processes corresponding to the n threads launched within a kernel are executed in parallel. Recall that, in this first approach, each particle is mapped onto a single thread. The algorithm uses b blocks and t threads per block. Thus, the total number of particles is $b \times t$. In Algorithm 3, a particle thread *tid* identification is done relatively to the associated thread, identified by *threadIdx*, and block, identified by *blockIdx*, and number of of threads per block, identified by *blockDim*.

4   Second Algorithm: SGPSO

The main idea behind the second approach consists of subdividing the search space into a grid of cells, where each cell is searched by an independent swarm of particles. This approach should favor optimization problems with large search space. In [6], we studied the impact of the number and size of the swarms on the optimization process, in terms of the execution time, convergence and quality of the solution found.

The dimension and size of blocks per grid and the dimension and size of threads per block are both important factors. The number of blocks in a grid should be at least the same or larger than the number of streaming multiprocessors (SMs), so that all available SM have at least one block to execute. Furthermore, there should be multiple active blocks per SM, so that blocks that are not waiting, due to a synchronization point, can keep the hardware busy. This recommendation is subject to resource availability. Therefore, it should be determined in the context of the second execution parameter, which is the number of threads per block, or block size, as well as shared memory usage.

In the proposed parallel implementation, the maximum velocity $v_{i,max}$ with respect to dimension i is formulated as a percentage of the considered search subspace of size D_i for that dimension, as defined in (2), wherein x_{max} and x_{min} are the maximum and minimum values of the whole search space, N_s represents the number of swarms, that work in parallel, and $0 \leq \delta \leq 1$. Moreover, the search space for a given swarm i is delimited by $x_{i,min}$ and $x_{i,max}$. In order to increase the efficiency of the algorithm in high dimensions, we use dynamic update of the inertia weight w.

$$
\begin{aligned}
D_i &= (x_{max} - x_{min})/N_s, \; v_{i,max} = \delta * D_i \\
x_{i,min} &= i \times D_i + x_{min}, \qquad x_{i,max} = x_{i,min} + D_i
\end{aligned}
\tag{2}
$$

In order to implement the SGPSO approach using CUDA, we opted to exploit two kernels. The first kernel generates random numbers and stores them in the GPU global memory. The second kernel runs all the steps of the parallel PSO. This way, the host CPU, after triggering the PSO process, becomes totally free. Using a single kernel for the whole PSO, excluding the random number generation, allows us to optimize the implementation, as there is no need for host/device communications. Recall that kernel *particle swarm optimizer* updates the inertia weight dynamically.

As introduced earlier, the problem search domain is organized into a grid of swarms, wherein each swarm is implemented as a block and each particle as a thread. The grid size is the number of swarms and block size is the number of particles. So, population size can be defined as the product of the grid size and block size, and this coincides with the total number of threads run by the GPU infrastructure. In this implementation, the position, velocity and *Pbest* of all the particles are kept in the global memory on the GPU chip.

Nonetheless, the *Gbest* obtained for all the grid's swarms are stored in the shared memory of the respective SM. The CUDA pseudo-code for the approach behind SGPSO is shown in Algorithm 4, wherein s denotes the number of segments into which each of the d dimensions of the problem is divided. Note that this subdivision generates s^d voxels which are the search subspaces. The code launches t threads per block, which means that it starts t particles per subspace. The total number of particles is thus $t \times s^d$. Kernel *particle swarm optimizer* proceeds as described in Algorithm 5. Note, that in this approach, the number of blocks coincides with that of swarms and the number of threads coincides with that of particles in each swarm.

The initialization of positions and velocity as well as the maximal velocity allowed for a particle within a swarm is done as described in Algorithm 6.

5   Third Algorithm: PDPSO

The third approach considers the fact that in some computationally demanding optimization problems the objective function is based on a large number of

Algorithm 4. CUDA Pseudo-code for SGPSO

---

let s = the number of segments; let d = the number of dimensions
let $b = s^d$ be the number of blocks ; let t = the number of threads
generate the swarm grid according the s and d
transfer data of the grid from CPU to GPU
kernel$\langle b, t \rangle$ random number generator; **kernel**$\langle b, t \rangle$ particle swarm optimizer
transfer result back to CPU
return *Gbest* and corresponding position;

Algorithm 5. CUDA Pseudo-code for particle swarm optimizer

---

initialize randomly position and velocity of particles according to subspace of respective swarm *blockIdx*
repeat
    compute *fitness* of particle *threadIdx*; **update** *Pbest* of particle *threadIdx*
    if *(threadIdx = 0)* **then**
        update *Gbest* of respective swarm *blockIdx*
    synchronize all threads of swarm *blockIdx*
    update velocity and position of particle *threadIdx*
until stopping condition

---

dimensions. Here, we are talking about more than thirty different dimensions and can even reach 100. Therefore, in this approach, the parallelism is more fine-grained as it is associated with the problem dimensions. The algorithm is called PDPSO (*Parallel Dimension PSO*). In contrast with SGPSO, this algorithm handles only one swarm and its main characteristic is the parallelism at the dimension level. Thus, the particle is now implemented as a block wherein each dimension is a thread of the block. This should favor optimization problems that exhibit a very high dimensionality. The GPU grid size is the number of particles and block size is the number of dimensions. For example, if the number of dimensions of the problem is 100, the SGPSO needs 100 iterations to compute the fitness values. The PDPSO will do the job using a single iteration to obtain the fitness values with respect to each of the problem dimensions plus an extra 10 iterations to summarize these intermediary results in order to get a single value which is the particle fitness. We call this process the *fitness reduction*. Thus, after 11 steps the result will be ready. Thus, it is possible to distribute the computational load at a lower degree of granularity, which can be up to one thread per problem dimension.

The PDPSO algorithm written in a CUDA-based pseudo-code is given in Algorithm 7. It uses four kernels: The first one launches the random number generators, i.e. one for each particle dimension and initialize the positions and velocities of the particles; The second kernel generates the threads that compute the fitness according to the corresponding dimension, perform the reduction process to get the fitness value of the particle that is represented by the block, and when this is completed, checks whether *Pbest* needs to be updated.

Algorithm 6. CUDA Pseudo-code position and maximum velocity initialization in subspace $blockIdx$

---

let $k = blockIdx$
for $i = 1$ to d do
$\quad x_i := (k \times d + i)(rand(max_k - min_k) + min_k); v_i := 0.0f$
$\quad vmax_i := \delta(k \times d + i)(max_k - min_k)$

Algorithm 7. CUDA Pseudo-code for PDPSO

---

let t = number of threads (dimensions); let b = number of blocks (particles)
kernel$\langle b, t \rangle$ position and velocity random generators (one for each dimension)
repeat
$\quad$**kernel**$\langle b, t \rangle$ fitness and *Pbest* calculator (one for each dimension)
$\quad$**kernel**$\langle b, t \rangle$ *Gbest* elector
$\quad$**kernel**$\langle b, t \rangle$ velocity and position calculator (one for each dimension)
until stopping condition
transfer result back to CPU
return *Gbest* and corresponding position

Algorithm 8. CUDA Pseudo-code for fitness, *Pbest* calculator

---

let $j = blockIdx$, $k = threadIdx$ and $b = blockDim$; $tid = k + j * b$
let cache be the shared memory of the GPU where $cache[k] = x[tid]$
compute fitness with respect to dimension k; $i := t/2$
while $(i \neq 0)$ **do**
$\quad$**if** $(k < i)$ **then**
$\quad\quad$**reduce** $fitness[k]$ and $fitness[k + i]$ according to objective function
$\quad i := i/2$
synchronize all threads
if $(fitness[j] < Pbest[j])$ **then**
$\quad Pbestx[(j \times d) + k] := cache[(j \times d) + k]$
$\quad$**if** $k = 0$ **then**
$\quad\quad Pbest[j] := fitness[j]$

Algorithm 9. CUDA Pseudo-code for kernel *Gbest* elector

---

let $tid = threadIdx + blockIdx * blockDim$; $i := b/2$
while $(i \neq 0)$ **do**
$\quad$**if** $(tid < i)$ **and** $Gbest[tid + i] < Gbest[tid]$ **then**
$\quad\quad Gbest[tid] := Gbest[tid + i]$
$\quad i := i/2$

If this is the case, the threads update the coordinates associated with this new *Pbest*. Note that there is a synchronization point of all threads so as to use the fitness value only when all the fitness reduction process has been completed.

6 Performance Results

The three proposed approaches were implemented on a NVDIA GeForce GTX 460 GPU [7]. This GPU contains 7 SMs with 48 CUDA cores each, hence a total of 336 cores. Three classical benchmark functions, as listed in Table 1, were used to evaluate the implementations performance. Function f_1 defines a Sphere, f_2 is Griewank function and f_3 is the Rastrigin function.

In the following, we report on the experiments performed to analyze the impact of each one of the proposed approaches. In all experiments, we always run the PSO algorithms for 2000 iterations.

Table 1. Fitness Functions

Function	Domain	f_{min}
$f_1(x) = \sum\limits_{i=1}^{n} (x_i^2)$	$(-100, 100)^n$	0
$f_2(x) = 1 + \frac{1}{4000} \sum\limits_{i=1}^{n} x_i^2 - \prod\limits_{i=1}^{n} cos\left(\frac{x_i}{\sqrt{i}}\right)$	$(-600, 600)^n$	0
$f_3(x) = \sum\limits_{i=1}^{n} (x_i^2 - 10cos(2\pi x_i) + 10)$	$(-10, 10)^n$	0

6.1 Impact of the Swarm Number

Using the CUDA Occupancy calculator [7], the GPU occupancy, which depends on the number of threads per block and that of register as well as the size of the kernel shared memory, amounts to 67%. Note that in all verified cases of different pairs of number and size of blocks per SM, the total number of 7168 threads was kept constant. In the case of SGPSO, this means a total number of particles of 7168 was used, as it is also the case of PPSO. However, in the case of PDPSO, as threads correspond to dimensions in the particles, which is 32 in this experiment, hence the number of particles sums up to 224 only. This explains the poor performance presented by this algorithm. Nonetheless, the disposition of block and thread numbers had significant impact on the performance in the case of SGPSO. Fig. 1(a) shows that despite the fact that the total number of particles is the same in all checked dispositions of number of swarms and particles per swarm, the combination 56×128 leads to the lowest execution time for SGPSO.

The increase in execution time can be explained by the work granularity level that each block of threads is operating at. Parallel computation of position coordinates and subsequently the velocity are performed by all threads within a block, but conditional branches, used to elect *Pbest* and *Gbest*, as well as loops that allow the iteration of the work for each one of the problem dimensions, dominate most part of the thread computation. It is well-known that conditional constructions are not well suited for the Stream Processing model.

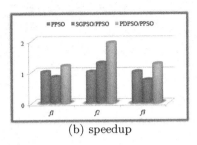

(a) Execution time   (b) speedup

Fig. 1. Execution times for different configurations of swarms for SGPSO and Impact of the total number of particles

Also, the performance degenerates because more blocks of threads are competing for the resources available to the SMs. A GPU offers a limited amount of shared memory, which limits the number of threads that can be simultaneously executed in the SM for a given application. In general, the more memory each thread requires, the fewer the number of threads that can reside in the processor [8]. Therefore, the choice of pair (block number and block size) has the kind of effect illustrated in Fig. 1(a) on the execution time. This experiment was repeated for different problem dimensions. The observed behavior is confirmed independently of this parameter. The case reported here is for dimension 32. Figure 1(b) shows the speedup achieved. Note that due to the stochastic nature of PSO, we run the same optimization 50 times.

6.2 Impact of the Swarm Size

It is expected that the number of particles influences positively the convergence speed of the optimization process, yet it has a negative impact on the

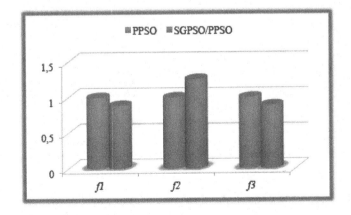

Fig. 2. Impact of the total number of particle

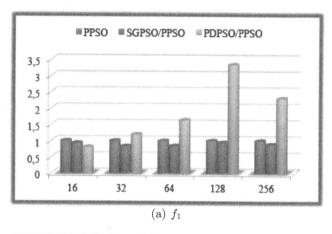

(a) f_1

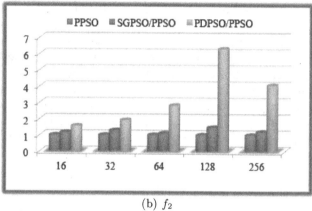

(b) f_2

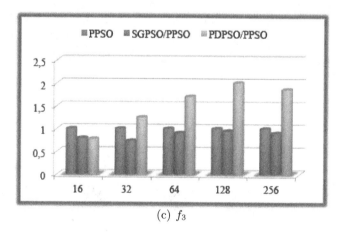

(c) f_3

Fig. 3. Impact of the number of dimensions for functions used as benchmarks

corresponding processing time. In SGPSO, increasing the number of particles can be achieved by either increasing the number of swarms and/or the number of particles per swarm. In order to study the impact of this parameter on the performance of SGPSO, we opted to keep the number of particles in a given swarm constant, i.e. 128, and increase the number of swarms. The latter was set as a multiple of the available streaming multiprocessors. Recall that the GPU used here includes 7 SMs.

Considering the optimization time comparison for the different studied configurations, with respect to the three used functions, we could easily observe that, in the case of SGPSO, for at most 56 swarms, which entails $56 \times 128 = 7168$ particles, the increase in terms of processing time is justified as the quality of the best solution is improved proportionally. PPSO presented a similar behavior as SGPSO when varying upwards the number of particles. Nonetheless, one can notice that for function f_2, SGPSO performs better, which in our opinion is due to the large search space and thus SGPSO takes advantage of the topology of distributed swarms. In the case of PDPSO, because of the explosion in terms of number of required threads, even in the first case, wherein a total of 28672 threads are required, the computational work surely ends up being sequentialized. Therefore, we do not show all the results for this approach. Figure 2 shows the speedup achieved.

6.3   Impact of the Number of Dimensions

Surely, the increase in terms of problem dimensions has an impact on the execution time. Recall that, in PPSO and SGPSO approaches, the computation with respect to the many dimensions of the objective function are performed sequentially, while in PDPSO, this is done concurrently. Figures 3(a) – 3(c) show a positive speedup for at most 256 dimensions. Nonetheless, for 512 dimensions the rate of increase of the performance deteriorates for all three implementations. The implementation of PDPSO performed much better than PPSO and SGPSO, even though the latter (PPSO and SGPSO) are handling 7168 threads while PDPSO 14336. This is twice the whole capacity of the GPU.

7   Conclusion

This paper presents three implementations of parallel PSO using GPGPU: PPSO, SGPSO and PDPSO. The first approach explores the parallelism between particles. In the second approach, the algorithm divides the search space into a grid of subspaces and assigns a swarm to each and every one of them. The implementation exploits the parallelism of the particle computation of the corresponding position and velocity as well as the fitness value of the solution associated. This is performed independently of the others particles of the swarm. A swarm of particles was implemented as a block of threads, wherein each thread simulates a single particle. This has a positive impact on the performance of the optimization of problems with large search space. In the third implementation, the particle is implemented as a block of threads and each dimension as one thread. This allows the distribution of the computational load at a finer degree of granularity, which is up to one thread per problem dimension. This has a positive impact on the performance of the optimization of large dimension problems.

A three-fold analysis was carried out to evaluate the performance of the proposed parallel implementation: first, the impact of the number of invested swarms; second the impact of their size; then the impact of the number of dimensions.

References

1. Kennedy, J., Eberhart, R.: Particle Swarm Optimization. In: Proc. of IEEE International Conference on Neural Network, Australia, pp. 1942–1948. IEEE Press (1995)
2. Engelbrecht, A.P.: Fundamentals of Computational Swarm Intelligence. John Wiley & Sons Ltd., New Jersey (2005)
3. Nedjah, N., Coelho, L.S., Mourelle, L.M.: Multi-Objective Swarm Intelligent Systems – Theory & Experiences. Springer, Berlin (2010)
4. Calazan, R.M., Nedjah, N., Mourelle, L.M.: Parallel co-processor for PSO. Int. J. High Performance Systems Architecture 3(4), 233–240 (2011)
5. Calazan, R.M., Nedjah, N., Mourelle, L.M.: A Massively Parallel Reconfigurable Co-processor for Computationally Demanding Particle Swarm Optimization. In: 3rd International Symposium of IEEE Circuits and Systems in Latin America, LASCAS 2012. IEEE Computer Press, Los Alamitos (2012)
6. Calazan, R.M., Nedjah, N., de Macedo Mourelle, L.: Swarm Grid: A Proposal for High Performance of Parallel Particle Swarm Optimization Using GPGPU. In: Murgante, B., Gervasi, O., Misra, S., Nedjah, N., Rocha, A.M.A.C., Taniar, D., Apduhan, B.O. (eds.) ICCSA 2012, Part I. LNCS, vol. 7333, pp. 148–160. Springer, Heidelberg (2012)
7. NVIDIA: NVIDIA CUDA C Programming Guide, Version 4.0 NVIDA Corporation (2011)
8. Kirk, D.B., Hwu, W.-M.W.: Programming Massively Parallel Processors. Morgan Kaufmann, San Francisco (2010)
9. Veronese, L., Krohling, R.A.: Swarm's flight: accelerating the particles using C-CUDA. In: 11th IEEE Congress on Evolutionary Computation, pp. 3264–3270. IEEE Press, Trondheim (2009)
10. Zhou, Y., Tan, Y.: GPU-based parallel particle swarm optimization. In: 11th IEEE Congress on Evolutionary Computation (CEC 2009), pp. 1493–1500. IEEE Press, Trondheim (2009)
11. Cádenas-Montes, M., Vega-Rodríguez, M.A., Rodríguez-Vázquez, J.J., Gómez-Iglesias, A.: Accelerating Particle Swarm Algorithm with GPGPU. In: 19th Euromicro International Conference on Parallel, Distributed and Network-Based Processing (PDP), pp. 560–564. IEEE Press, Cyprus (2011)

A Particle-Swarm-Optimized Fuzzy Classifier Used for Investment Decision Support

Lars Krueger and Matthias Walter

Abstract. We propose a Particle Swarm-based optimization technique to enhance the quality and performance of our Investment decision system (IDSS). It allows the classification of future performances of high-technology venture investments on the basis of very limited information. Our system thus helps investors to decide whether to invest in a young High-Technology Venture (HTV) or not. In order to cope with uncertain data we apply a Fuzzy Rule based Classifier. As we want to attain an objective and clear decision making process we implement a learning algorithm that learns rules from given real-world examples. The availability of data on early-stage investments is typically limited. For this reason we equipped our system with a bootstrapping mechanism which multiplies the number of examples without changing the quality. We show the efficacy of this approach as by comparing the classification power and other metrics of the PSO-optimized system with the respective characteristics of the conventionally built IDSS.

Keywords: Particle Swam Optimization, Fuzzy Classifier, Membership Tuning, Rule Base learning, Adaptive Fuzzy System.

1 Introduction

We use a self-learning rule–based Investment decision Support System (IDSS) that help make investment decisions venture capital investments into young high technology ventures [1]. By implementing a deterministic learning algorithm [2] the system can handle successively available data. This is of particular importance as data on venture capital investments in young High-Technology Ventures (HTV) in the very early stage of their life cycle is typically limited. Such a system must be functional already when there is only a limited number of examples available. For this reason we developed and tested a bootstrapping procedure [1]. Although being robust and functional, we aspire to improve the IDSS's accuracy, i.e., the number of correctly classified patterns. Furthermore, the classification power of the rule base needs to be augmented. This intended optimization is a multi-dimensional challenge since not only formal restrictions but also boundaries with regards to content and plausibility need to be taken into account. Meta-heuristic optimization procedures such as particle swarm optimization [3] offer a resort to this dilemma. Against that background, we propose a PSO algorithm that enhances the performance of rule base of a fuzzy classifier which can handle limited availability of data and is used as an investment decision support system. We build on our previous works on the use of bio-optimized Fuzzy Approaches to evaluate investments high-technology ventures [4].

I. Rojas, G. Joya, and J. Cabestany (Eds.): IWANN 2013, Part I, LNCS 7902, pp. 253–261, 2013.

2   Investment Decision Support System

2.1   Investment Decision Making

There is no practical, unambiguous, and objective conventional method that helps make investment decisions under uncertainty, incomplete and limited data as well as vagueness. Since these circumstances apply particularly for investments in young High-technology ventures, the need for alternative decision support system in this domain is obvious. We assume that the success of a HTV is dominantly influenced by five distinct features. This perception is backed by literature [5]. As there is no exact math or causal model known which fully describes the connection between these five parameters and a success or failure metric. The input parameters comprise [5]:

X1: Industry experience founder
X2: Product Status (Maturity, Innovation Level)
X3: IP Status (License, IP Type, Exclusivity,…)
X4: Business Development Capabilities
X5: Infrastructure Support (R&D)

A common metric to measure the success of a value is the Return-on-Investment (RoI) r. Because the internal rate of return is a rate quantity, it is an indicator of the efficiency, quality, or yield of an investment [5]. According to the observed RoI averaged over three years, the HTV is classified into one of the three classes (1=loss, 2= mediocre, 3= promising case).

2.2   General Structure as a Fuzzy Classifier

Rule Structure
The IDDS uses the algorithm described in [2]. The fuzzy IF-THEN rules used are of the type-2:

R_q: IF x_1 is A_{q1} and … and x_n is A_{qn} THEN Class C_q with CF_q, where R_q is the label of the q-th IF-THEN rule, $\vec{A}_q = (A_{q1},...,A_{qn})$ represents a set of antecedent fuzzy set, Cq the consequent class, CFq with $q = 1,..., N$ is the confidence of the rule Rq and N is the total number of generated fuzzy IF-THEN rules. The confidence CFq of a Rule q expresses with which certainty the consequent class Cq was assigned to the rule through the rule learning process. It is completely based on the rule compatibility $\mu_{A_{qi}}(x_{pi})$ via the auxiliary parameter β_h^q. It is determined as follows:

$$CF_q = \frac{\beta_{C_q} - \overline{\beta}}{\sum_{h=1}^{m} \beta_h^q} \text{ (1) where } \overline{\beta} = \frac{1}{M-1} \sum_{h \neq C_q} \beta_h^q \text{ .(2)}$$

Classification Process

A rule Rq has a summed compatibility β_h^q for each class C. The class C which has the highest summed compatibility β_h^q over all classes is assigned to the Rule R_q. The consequent parts of the Fuzzy IF-Then C_q and C_{Fq} are obtained as follows:

$$C_q = \arg \max_{h=1,...,M} \beta_h^q, \text{ (3) where } \beta_h^q = \sum_{x_p \in Classh} \mu_{A_q}(x_p) \text{ (4)}$$

The rule base generated from the training examples is subsequently used to assign a class label C to a new pattern by the following equation:

$$C = \arg \max_{h=1,...,M} \{\alpha_h\} \text{ (5) where } \alpha_h = \max_{q=1,...,N} \left\{\mu_{A_q(x)} \cdot CF_q\right\} \text{ (6)}$$

3   Particle Swarm Optimized-Based Fuzzy Classifier

3.1   PSO Preliminaries

General Structure. Similar to other bio-inspired meta-heuristic optimization approaches such as Genetic Algorithm (GA) or Evolutionary Algorithms (EA), the PSO places a number of candidate solutions (particles) randomly in a D-dimensional search space. For each individual particle, the fitness function or problem solving capacity (position in the search space) is evaluated. Subsequently, each particle determines its way through the search space on the basis of its own current and (previous) best position (fitness) *pbest*. This movement is intentionally influenced by the best positions (fitness) of its neighbors and amended by some random factors. The next iteration takes place after all particles moved. The population of the particles, the swarm, moves eventually towards an optimum of the fitness function. This is usually reached after a predetermined number of iterations or when a certain minimum threshold of the fitness function is exceeded [6].

Specific Parameters. The PSO has only few parameters that need to be controlled.

— Size of Population n. It is often set empirically depending on the complexity and dimensionality of the problem. A number between 20 and 50 is considered as common in the current literature.
— The often called *acceleration coefficients* $\phi1$ and $\phi2$ determine the impact of the random factors on the velocity, i.e., step size by which the particle's position pi in the search space is changed. By changing these parameters, one can control the responsiveness of the PSO and even provoke unstable behavior. Current research suggests $\phi = \phi1 + \phi2 > 4$, thus ϕ is usually set to 4.1, with $\phi1 = \phi2$.
— *Constriction coefficients* are needed to damp the PSO system. These coefficients prevent it from exploding and showing unstable behavior and ensuring convergence of the swarm towards an optimum. Recent literature present an analytical model yielding approx. $\chi = 0.7298$.

The Position and the Velocity of a Particle Pi. At each iteration, the current position of a particle p_i is evaluated as a problem solution. New positions are obtained by adding component-wise the velocity vector v_i to the particle or position vector p_i. On each iteration, the velocity vector v_i is also updated (see equation (7) and (8)):

$$\vec{v}_i^{t+1} \leftarrow \chi(\vec{v}_i + \vec{U}(0,\phi_1) \otimes (\vec{p}_i - \vec{x}_i) + \vec{U}(0,\phi_2) \otimes (\vec{p}_g - \vec{x}_i)) \tag{7}$$

$$\vec{x}_i^{t+1} \leftarrow \vec{x}_i^{t} + \vec{v}_i^{t+1} \tag{8}$$

Where $\vec{U}(0,\phi_i)$ is a vector of random numbers which is uniformly distributed in $[0, \phi_i]$ and randomly generated at each iteration for each particle pi.

Topology. A static topology, the so-called *gbest* topology (global best) is used here. That means, that each particle is affected only by the global best position *gbest* of the entire swarm and this type of affection is not changed over time. The *gbest* topology is considered to converge relatively quickly towards an optimum. In addition, it is simpler to implement as only one single parameter needs to stores and evaluated.

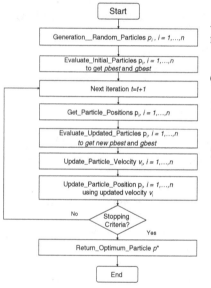

Algorithm. The flowchart of the PSO algorithm is depicted in Fig. 1.

Our Approach

1. Training of the rule base. During this procedure, the membership functions remain unchanged.
2. Tuning of the Membership functions. The optimum rule base is further enhanced through an tuning of the membership functions. Expected outcome: membership example to rule grows thus confidence grows

Fig. 1. Flowchart of the generic PSO Algorithm

4 PSO-Based Knowledge Acquisition / Training of the Rule Base

4.1 Coding of the Rule Base

Each particle p_i represents a rule base. The rule base is coded as a matrix in which every row represents a fuzzy rule. The structure of a particle pi is:

$$p_i = \begin{bmatrix} A_{1,1}^i & A_{1,2}^i & \cdots & A_{1,n}^i & C_1^i & CF_1^i \\ A_{2,1}^i & A_{2,2}^i & \cdots & A_{2,n}^i & C_2^i & CF_2^i \\ \cdots & \cdots & \cdots & \cdots & \cdots & \cdots \\ A_{q,1}^i & \cdots & \cdots & A_{q,n}^i & C_q^i & CF_q^i \end{bmatrix} \tag{9}$$

where q is the number of rules of the RB, n is the number of input variables, A represents the fuzzy sets of the input variables, C the corresponding consequent of the rule q and CFq the confidence of the rule R_q (see Eq.1). It is assumed that the connector is always AND. According to the model described in 2.1, certain constrain need to be imposed in order to ensure compatibility with the real world problem:

- $A_{q,n}^i \in \mathbb{N} \wedge \{1, 2, ..., 5\}$ as the input variables are each segmented into five equal fuzzy sets A,
- $C_q^i \in \mathbb{N} \wedge \{1, 2, 3\}$ as the consequence C_q of a rule R_q can be only one of out of three classifications.

These boundaries need already to be accounted for when the initial rule bases are randomly generated.

4.2 Coding of the Update Process

The velocity vi at every iteration is also represented as a matrix with the same number of rows like the particle matrix pi. Note that the update process does not alter the confidence CF_q of a rule. The measure CF_q is calculated from the compatibility of the rule q to the example e and is furthermore used as an independent performance measure. Again, to ensure the algorithm's compatibility with the real-world problem, certain boundaries need to be set:

- $v_{q,n}^i \in \mathbb{N} \wedge \{1, 2, ..., 5\}$ as the input variables are each segmented into five equal fuzzy sets described by discrete numbers,
- $v_{q,C}^i \in \mathbb{N} \wedge \{1, 2, 3\}$ as the consequence Cq can be only one of out of three classifications.

As shown in Eq. (7) and (8), the entire update process is driven by the particles best position so far, denoted as $p_i\#$, and the global best position of the swarm, p^*.

4.3   Calculation of the Fitness

Aim of the PS-driven optimization is to maximize CF_g. The fitness of a rule base is determined by its capability to correctly classify a given set of training examples. This measure, denoted as global confidence CFg, is obtained by:

$$CF_g = \sum_{\forall Rq:\, Cq=Cp} \mu_{\vec{A}_q}(\vec{x}_p) \ (13) \quad \text{where } \mu_{\vec{A}_q}(\vec{x}_p) = \prod_{i=1}^{n} \mu_{A_{qi}}(x_{pi}), q = 1,2,..,N \ (14)$$

where C_q is the consequent of the rule R_q and Cp and x_p are the consequent and the input vector of the example x_p, respectively.

5   PSO-Based Tuning of the Fuzzy Classifier

After first optimum fitness value has been reached (Eq. 13), the optimized rule base RB* is further improved by a PSO-based tuning algorithm that aims to design the shape of the given membership functions. We prefer an automatic tuning over an expert humane tuning process as we need as much objectivity as possible to foster the validity and thus the acceptance of the system. The assumptions for the 2nd PSO algorithm are:

- Triangular Membership functions are used (input variables),
- a triangular fuzzy set is determined by three parameters: left (l), center (c), right (r)
- First and last membership functions of each input variable are represented with left- and right-skewed triangles thus $l_1=c_1=0$ and $c_5=r_5=1$.

In order to ensure compatibility with the underlying real-world problem, certain boundaries need to be defined again:

- the initial particle is not generated randomly, but it is given by beforehand,
- the universe of discourse must be covered completely by fuzzy sets, thus $\forall x_i : \mu A_i(x_i) \geq 0$,
- the order and the number (no 'shared' centers) of membership functions must remain unchanged, thus $0 \leq c_1 < c_2 < c_3 < c_4 < c_5 \leq 1$.

5.1   Coding of Membership Functions

Each particle pi contains the parameters l,c,r of all used triangular membership functions. Each row represents an input parameter, each column of the parameters of a input fuzzy set. The structure of a particle p_{membi} is:

$$p_{membi} = \begin{bmatrix} l_{11} & c_{11} & r_{11} & l_{12} & c_{12} & r_{12} & \cdots & \cdots & l_{1n} & c_{1n} & r_{1n} \\ l_{21} & c_{21} & r_{21} & l_{22} & c_{22} & r_{22} & \cdots & \cdots & l_{2n} & c_{2n} & r_{2n} \\ \cdots & \cdots & \cdots & \cdots & \cdots & \cdots & \cdots & \cdots & \cdots & \cdots & \cdots \\ \cdots & \cdots & \cdots & \cdots & \cdots & \cdots & \cdots & \cdots & \cdots & \cdots & \cdots \\ l_{m1} & c_{m1} & r_{m1} & l_{m2} & c_{m2} & r_{m2} & \cdots & \cdots & l_{mn} & c_{mn} & r_{mn} \end{bmatrix} \quad (15$$

with $(l,c,r)_{mn}$ being the parameter of the n-th fuzzy set of the m-th input.

5.2   Fitness Function

The update process is realized similar to sub section 4.3. The fitness function is calculated using eq. (8) and (7) (subsection 4.3) using the changed membership functions. The membership degree of a given input xi to a triangular fuzzy set Ai is calculated as follows:

$$\mu A_i(x_i) = \begin{cases} \dfrac{1}{c-l} \cdot x_i + \dfrac{l}{l-c} & if\ l \le x_i < c \\ 1 & if\ x_i = c \\ \dfrac{1}{c-r} \cdot x_i + \dfrac{r}{r+c} & if\ c < x_i \le r \\ 0 & else \end{cases} \tag{16}$$

6   Experiments

We use two distinct set of HTV patterns. The training set consists of 50 real-world examples provided by an early-stage investor. We test the optimized rule-based IDSS against a set of real-world examples consisting of seven HTVs. Afterwards, we compare the accuracy of both, the PSO optimized IDSS and the conventionally trained IDSS. We conducted n=20 experiments.

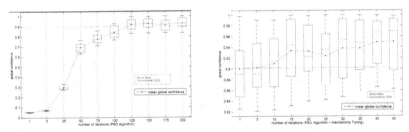

Fig. 2. Global Confidence after rule base training (right) and membership functions tuning (left) compared to conventionally built IDSS

The Fig 2 proves the growth of the quality metric CFg after certain PSO iterations achieved by both, the PSO-learning of the rule base (left) and the tuning of the membership functions (right). The detailed alterations of the membership functions are depicted in Fig. 3.

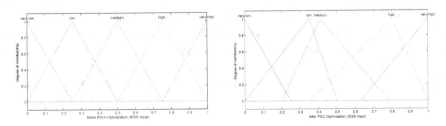

Fig. 3. Membership function – before and after PSO optimization

As mention, we tried to improve the systems capabilities to cope with limited availability by sampling the limited data. We used a Gaussian distributed sampling procedure. We found a significant change of the overall quality of the system (Fig 4). However, there is no clear linearity. A more of sampling does not necessarily lead to a 'more' of performance growth.

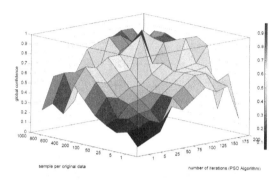

Fig. 4. Confidence landscape of the PSO algorithm

In tab 1, the results of the classification test are shown. We fond clear evidence that the PSO –optimized IDSS performs better in classifying unseen patterns compared to the IDSS trained with the limited data set or the sampled data set.

Table 1.

	Classified correctly [#]	Classified correctly[%]
RB original data	1	5,89
IDSS conv.	10	58,8
IDSS PSO	15	88,3

7   Conclusion

We have shown that we can improve our IDSS that we built up by learning from very limited real world data. As suspected, the PSO algorithm proved to be efficient and powerful to raise the IDSS quality metric, the Confidence of the rule base. This was achieved by a two-step approach: to prevent unnecessary computing we split the procedure into improving the rule base (first step) and fine tuning the membership function (second step). This turned out to be appropriate as we needed significantly less iteration steps in the second part. Furthermore, we have shown that the PSO help to improve the classification performance of the IDSS significantly. All in all, by applying PSO we made a significant leap ahead to a practical Investment Decision Support System that is suited to work even under the condition of very limited data. We also hope that we could enlarge the *flock* of applications of the PSO algorithm.

References

1. Krueger, L., Walther, M., Heydebreck, P.: Future Performance Classification of High-Technology Venture Investments with limited Data - The Application of Self-Learning and Incremental Update Algorithms for a Fuzzy Classifier in an Investment Decision Support System (IDSS) (unpublished)
2. Nakashima, T., Sumitani, T., Bargiela, A.: Incremental Learning of fuzzy rule-based Classifiers for large Data Sets. In: World Automation Congress (WAC), June 24-28, pp. 1–5 (2012)
3. Kennedy, J., Eberhart, R.: Particle Swarm Optimization. In: Proceedings of IEEE Int. Conf. Neural Networks, vol. 4, pp. 1942–1948 (1995)
4. Krüger, L., Heydebreck, P., Salomon, R.: Bio-Inspired Optimization of Fuzzy Cognitive Maps for their Use as a Means in the Pricing of Complex Assets. In: Proceedings of the IEEE-CIFER, Computational Intelligence for Financial Engineering & Economics 2012, New York, USA, March 29-30 (2012)
5. Song, M., Podoynitsyna, K., van der Bij, H., Halman, J.: Success Factors in New Ventures. A Meta Analysis, Journal of Product Innovation Management 25, 7–27 (2008)
6. Poli, R., Kennedy, J., Blackwell, T.: Particle Swarm Optimization – An overview. Swarm Intelligence 1, 33–57 (2007)

Ant Colony Optimization Inspired Algorithm for 3D Object Segmentation

Rafael Arnay and Leopoldo Acosta

La Laguna University, Department of System Engineering and Control and Computer Architecture, Avda. Francisco Sánchez S/N, 38204, La Laguna, Spain

Abstract. In this work, an ACO-based approach to the problem of 3D object segmentation is presented. Ant Colony Optimization (ACO) meta-heuristic uses a set of agents to explore a search space, gathering local information and utilizing their common memory to obtain global solutions. In our approach to the 3D segmentation problem, the artificial ants start their exploratory movements in the outer contour of the object. They explore the surface of the object influenced by its curvature and by the trails followed by other agents. After a number of generations, particular solutions of the agents converge to the best global paths, which are used as borders to segment the object's parts. This convergence mechanism avoids over-segmentation, detecting regions based on the global structure of the object and not on local information only.

Keywords: Ant Colony Optimization, 3D object segmentation.

1 Introduction

In this paper, a segmentation algorithm for 3D data as a prior step to a part-based object classification method is presented. The proposed method is based on an Ant Colony Optimization meta-heuristic (ACO) [1,2,3]. Traditionally, ACO-based algorithms try to find the best path from a known position to another following certain criteria. However, it is our goal to obtain a set of paths that start and end on positions that depend on the nature of the object. The artificial ants, in their exploratory movement will travel those paths, reinforcing the best ones and defining the borders that will separate the segmented regions. ACO agents tend to converge to a limited set of paths, delimiting a reduced number of regions and avoiding over-segmentation. Moreover, artificial agents make use of their common memory that are the pheromone trails, to construct paths even when there is no heuristic information, segmenting regions where other algorithms, especially those based on region growing, will encounter more difficulties. This paper is organized as follows: Section 2 gives an overview on approaches in the field of 3D data segmentation. Section 3 provides details on the proposed algorithm. Section 5 shows some results of our algorithm tested on some objects of the RGB-D dataset [4]. Finally, Section 6 presents the conclusions.

I. Rojas, G. Joya, and J. Cabestany (Eds.): IWANN 2013, Part I, LNCS 7902, pp. 262–269, 2013.

2   Related Work

Image segmentation is a topic studied for decades. However, in the last years some effort has been done to extend traditional computer vision segmentation techniques to work with 3D data. A region growing approach is used in [5,6,7]. Such methods based on curvature often lead to over segmentation. Marton et al.[5] construct their classification system assuming over-segmentation, because, as they say, segmenting objects accurately is not always robust but over-segment is easily realizable. The use of a region growing approach that starts at random points is not a completely reproducible process. They rely on a large amount of train data to cover all segmentation cases. Over-segmentation is avoided in [6] using smoothness constraints: local connectivity and surface smoothness, which find smoothly connected areas. Ellipsoidal region growing is performed in [7], where two distance criterions for merging the ellipses are presented, a shape distance and a density distance. Graph clustering approaches are used in [8,9,10]. In both [8] and [9], a 3D graph is constructed using k-nearest neighbors. In [8] Golovinskiy and Funkhouser assign weights according to an exponential decay in lengths, and use a min-cut algorithm to separate objects in foreground from the background. In [9] Zhu et al. assume the ground to be flat to segment objects in outdoors environments. In [10] the segmentation process is performed on a registered camera-laser pair to work with 3D colored data. A set of segmentation algorithms are presented in [11] for different types of point clouds. These methods rely on a prior non-flat ground detection to use it as a separator, as they are designed for outdoor environments. Probabilistic reasoning and conditional random fields are employed in [12] to detect objects of similar type that occur more than once in the scene.While our approach is still based on the object's surface curvature to detect the segmentation borders, the main contribution of this work is the use of the ACO agents common memory to construct global paths that not only rely on local curvature information, but on the global structure of the object also.

3   ACO-Based Segmentation Algorithm

In order to utilize the ACO metaheuristic, the problem of segmentation has to be mapped to an optimization one. Artificial ants will be exploring in a 2D graph where the nodes are the pixels of the depth image. They will start on the outer contour of the object and will explore its surface until they return to the outer contour again. Surface attractiveness is influenced by its curvature (heuristic information) and by other agents' pheromone trails.

Working with 3D data simplifies the process of segmenting the objects from the background as it is not influenced by lightning conditions or homogeneous colors or texture. We are working with the RGB-D dataset [4] in which every object has it associated pre-segmented boundary. Agents will start their exploratory movements in any pixel of the external boundary of the object, oriented in a perpendicular direction to the tangent line of the boundary in that pixel and facing to the interior of the contour. When an agent that is not in an external boundary pixel moves to one of them, it reaches the stop condition.

In order to obtain the heuristic information for the problem, Radial Surface Descriptors (RSD) [13] are extracted from the object's point cloud. These descriptors give the minimum and maximum surface curvature radii for each point. Depending on the values of the radii, the surface can be classified as one of several primitive types like cylindrical, spherical, planar or edge for example. A 2D heuristic function is obtained where the value in each position is directly proportional to the curvature value in the corresponding pixel of the depth image.

Given a state, agents will only be able to move to one of their feasible neighbors. Feasible neighbors are the pixels that intersect with the arc of a circumference centered in the agent's position with a fixed angle, radius and oriented in the heading direction of the agent. The motion rule is the random-proportional rule for artificial ants [2] and determines which element among the feasible neighbors will be the next state of the agent, see Eq. 1.

$$p_{ij}^k(t) = \begin{cases} \frac{[\tau_{ij}]^\alpha [\eta_{ij}]^{1-\alpha}}{\sum_{l \in N_i^k}([\tau_{ij}]^\alpha [\eta_{ij}]^{1-\alpha})}, j \in N_i^k \\ 0, \text{ otherwise} \end{cases} \tag{1}$$

The probability that a given state is the next one depends on a parameter $\alpha \in [0, 1]$ with which it is possible to tune the balance between heuristic information exploitation and pheromone exploitation in the agents behavior. i is the current state, τ_{ij} is the amount of pheromone in state $< \sigma_i, j >$, η_{ij} is the heuristic value of state $< \sigma_i, j >$, p_{ij}^k is the probability of state $< \sigma_i, j >$ of being the next state for the agent k and N_i^k are the neighbors of the agent k in state i. When heuristics values and pheromone trails among the feasible neighbors of a given state are zero, a pure random movement to one of the feasible neighbors is performed.

Initially, the cost of moving towards a pixel is inversely proportional to the curvature in the corresponding position of the surface. However, as can be seen in Eq. 2 the cost function is also influenced by the α parameter with which it is possible to tune the importance of heuristics and pheromone information in the calculation of the cost.

$$cost_{path} = \frac{\sum_{ij \in path} (\alpha(\eta_{max} - \eta_{ij}) + (1 - \alpha)(\tau_{max} - \tau_{ij}))}{l} \tag{2}$$

Where l is the length of the path in pixels, η_{max} is the ceil value of the heuristic function and τ_{max} is the ceil value of the pheromone deposit.

The colony has N agents divided into a set of generations $\{N_1, N_2, \ldots, N_m\}$ which will be exploring iteratively. The motion rule and cost function parameter α is modified to make successive generations of agents more sensitive to pheromone trails and less to heuristics. The motivation of this mechanism is to make the agents explore only based on heuristics at first as there is no other kind of information and then to converge to other agent's pheromone trails as they represents the accumulated experience of the colony and they are more reliable.

Seed points are the pixels where the agents can start their exploration. Initially, any external boundary pixel is a good candidate to be a seed point. However, those pixels have an associated *life value* ($L \in [0, L_{max}]$) to them. Initially, all seed points have an associated life value of L_{max}. When the cost of a path that starts in a seed point S_{ij} is above a certain threshold L_t, L_{ij} is reduced. In the same way, when the cost of that paths is under L_t, L_{ij} is augmented. If L_{ij} reaches zero, S_{ij} will be considered invalid and no more agents will start in this pixel. This is done in order to optimize the resources of the colony, making the agents to explore in the areas where the most promising paths are being discovered. Each seed point has an associated *bridge*. For a given seed point S_{ij}, its associated bridge (B_{ij}) is the best path found so far that starts in this seed point.

Initially, the pheromone is set to zero in every component of the exploration graph. When the agents reach a final state, they deposit an amount of pheromone in the components of the path that they have followed, see Eq. 3. In every generation, all bridges associated to valid seed points receive a pheromone contribution also. This is done in order to maintain the best paths found so far.

$$\tau_{ij}(t+1) = (1-\rho).\tau_{ij}(t) + \rho.\sum_{k=1}^{N_x} \Delta_{ij}^{k}(t) \tag{3}$$

Where t is a time measure, $\rho \in (0, 1]$ represents the pheromone evaporation ratio and $\Delta_{ij}^{k}(t)$ is the k-ant contribution constant. When all generations of agents have obtained their solutions, the bridges that start in seed points that have an amount of pheromone above a certain threshold are selected. In order to make the segmentation borders narrower, a local average across the points of those bridges is calculated. In Fig. 4,(row c) pheromone trails are shown in red and local average of best bridges is shown in violet. The extracted solution is used as borders to obtain contours to segment the object, see Fig. 4,(row d).

In Algorithm 1 pseudocode for the main ant managment algorithm is shown.

4   Experimental Results

To carry out the experiments with the proposed algorithm, the RGB-D dataset [4] has been used. This dataset contains RGB, depth images and their associated point clouds of everyday objects. Firstly, a quantitative comparison with a region growing approach like the one used in [5] is presented. The region growing algorithm implementation is the one from the Point Cloud Library (PCL) [14]. The metric used to perform this comparison is the number of detected regions for small variations of the point of view from which the object is being observed. In Fig. 1, Fig. 2 and Fig. 3, the number of segmented parts for each view of different sample objects comparing the proposed ACO-based algorithm to a region growing approach is shown. As can be seen, heterogeneous objects like the cap (Fig. 1) or the coffe mug (Fig. 2) are segmented in a different number of parts depending on the point of view. This variations are normal and depend on the

Data: heuristic information
Result: solution paths
initialize pheromone();
initialize seed points();
while *(not all generations of ants have explored)* **do**
⎢   explore(ants in current generation);
⎢   calculate path cost(ants in current generation);
⎢   bridges update();
⎢   seed points update();
⎢   pheromone update();
end
extract solution();

Algorithm 1. Pseudocode for the main ant management algorithm where agent's exploration, path cost calculation and pheromone, seed points and bridges update are performed.

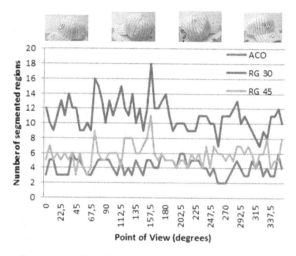

Fig. 1. Number of segmented regions for an object of class *cap*. Data instances are obtained from a 360° around the object point of view. RG 30 and RG 45 goes for region growing with an angular threshold of 30° and 45° respectively.

structure of the object. However, the rapid variation in the number of detected regions (for small variations on the point of view) is an undesirable effect that makes more difficult to classify the object correctly. As can be seen in Fig. 1, Fig. 2 and Fig. 3, the proposed algorithm segments the objects in a fewer number of parts than the region growing approach. The pheromone-convergence philosophy of the ACO metaheuristic tends to avoid over-segmentation. Stability is also higher with the proposed approach.

In Fig. 4 a sample of object segmentations for a qualitative comparison is shown. As can be seen, the segmented parts using the proposed algorithm are intuitively more meaningful than the ones detected by the region growing approach. For example, in Fig. 4 *2b*, part of a panel and part of the bill of the

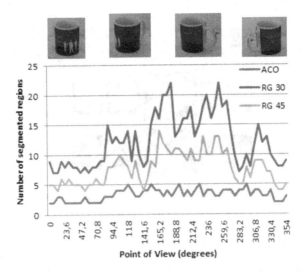

Fig. 2. Number of segmented regions for an object of class *coffee mug*. Data instances are obtained from a 360° around the object point of view. RG 30 and RG 45 goes for region growing with an angular threshold of 30° and 45° respectively.

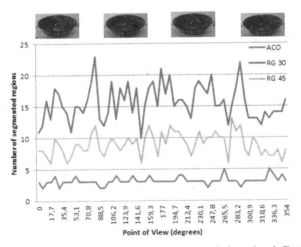

Fig. 3. Number of segmented regions for an object of class *bowl*. Data instances are obtained from a 360° around the object point of view. RG 30 and RG 45 goes for region growing with an angular threshold of 30° and 45° respectively.

cap are segmented under the same region. This is produced because the region growing algorithm works locally and is agnostic of the global structure of the object. The paths explored by the ACO approach, on the other hand, store global information about the structure of the object. This allows the artificial agents to fill the lack of curvature variation with pheromone information, segmenting

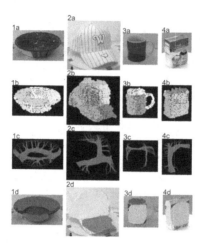

Fig. 4. Segmentation results for some sample objects. 50000 agents divided into 500 generations are employed for the detection. In row *a* an RGB image of the object is shown. In row *b* segmented point clouds using a region growing approach can be seen. In row *c* the pheromone trails are shown in red and the final solution in violet. In row *d* final segmentations using our approach are shown. Solutions obtained with the ACO-based algorithm are used as borders to segment the object.

regions where other algorithms have difficulties. Other problem of region growing approaches is the over-segmentation of cylindrical surfaces, as can be seen in Fig. 4 *1b and 3b*. The convergence of the artificial ants to the most appealing global paths avoids this type of local, smooth connected over-segmentation.

In Table 1 computation times of the proposed algorithm for different colony sizes are shown.

Table 1. Computation times of the ACO-based segmentation approach running on a Intel(R) Core(TM) i5-2450M CPU @ 2.50GHz. Implementation is done in C. Average input depth image size: 170 x 170. Average input point cloud size: 2000 points.

Agents No.(x1000)	500	100	50	20	10	5
Time(sec.)	7.120	1.670	0.970	0.450	0.370	0.150

5   Conclusions

In this paper, a novel ACO-based approach for 3D data segmentation is presented. The main contribution of our work is the incorporation of global information to complement the local search in the segmentation process. This is achieved through the pheromone trails of the artificial ants which encode global surface information and directs the exploration of agents even when there is no heuristic information to guide the search. Our approach gives a good trade-off between over- and under- segmentation and a good stability level in terms of number of detected regions for slightly different point of views of the same object.

Acknowledgments. The authors gratefully acknowledge the contribution of the Spanish Ministry of Science and Technology under Project SAGENIA DPI2010-18349 and the funds from the Agencia Canaria de Investigación, Innovación y Sociedad de la Información (ACIISI).

References

1. Dorigo, M., Stützle, T.: The Ant Colony Optimization Metaheuristic: Algorithms, Applications and Advances. In: Glover, F., Kochenberger, G.A. (eds.) Handbook of Metaheuristics. Springer, New York (2002)
2. Dorigo, M., Stützle, T.: Ant Colony Optimization. MIT Press, Cambridge (2004)
3. Dorigo, M., Maniezo, V., Colorni, A.: The Ant System: Optimization by a Colony of Cooperating Agents. IEEE Trans. on Systems, Man and Cybernetics Part B 26, 29–41 (1996)
4. Lai, K., Bo, L., Ren, X., Fox, D.: A large-scale hierarchical multi-view rgb-d object dataset. In: Proc. of International Conference on Robotics and Automation, ICRA (2011)
5. Marton, Z.-C., Balint-Benczedi, F., Seidel, F., Goron, L.C., Beetz, M.: Object Categorization in Clutter using Additive Features and Hashing of Part-graph Descriptors. In: Stachniss, C., Schill, K., Uttal, D. (eds.) Spatial Cognition 2012. LNCS (LNAI), vol. 7463, pp. 17–33. Springer, Heidelberg (2012)
6. Rabbani, T., Vosselman, G.: Segmentation of point clouds using smoothness constraint. In: ISPRS Commission V Symposium 'Image Engineering and Vision Metrology' (2006)
7. Pauling, F., Bosse, M., Zlot, R.: Automatic Segmentation of 3D Laser Point Clouds by Ellipsoidal Region Growing. In: Proc. of the Australasian Conference on Robotics & Automation (ACRA) (2009)
8. Golovinskiy, A., Funkhouser, T.: Min-cut based segmentation of point clouds. Princeton University
9. Moosmann, F., Pink, O., Stiller, C.: Segmentation of 3D Lidar Data in non-flat Urban Environments using a Local Convexity Criterion. In: IEEE Intelligent Vehicles Symposium, pp. 215–220 (2009)
10. Strom, J., Richardson, A., Olson, E.: Graph-based segmentation of colored 3d laser point clouds. In: Proc. of the IEEE/RSJ International Conference on Intelligent Robots and Systems (IROS) (2010)
11. Douillard, B., Underwood, J., Kuntz, N., Vlaskine, V., Quadros, A., Morton, P., Frenkel, A.: On the Segmentation of 3D LIDAR Point Clouds. In: International Conference on Robotics and Automation (ICRA) (2011)
12. Triebel, R., Shin, J.: Siegwart. R.: Segmentation and unsupervised part-based discovery of repetitive objects. In: Proceedings of Robotics: Science and Systems (2010)
13. Marton, Z.C., Pangercic, D., Blodow, N., Kleinehellefort, J., Beetz, J.: General 3D Modelling of Novel Objects from a Single View. In: Proceedings of the 2010 IEEE/RSJ International Conference on Intelligent Robots and Systems (IROS) (2010)
14. Rusu, R.B., Cousin, S.: 3D is here: Point Cloud Library (PCL). In: IEEE International Conference on Robotics and Automation (ICRA) (2011)

Kernelizing the Proportional Odds Model through the Empirical Kernel Mapping*

María Pérez-Ortiz, Pedro Antonio Gutiérrez, Manuel Cruz-Ramírez,
Javier Sánchez-Monedero, and César Hervás-Martínez

University of Córdoba, Dept. of Computer Science and Numerical Analysis
Rabanales Campus, Albert Einstein building, 14071 - Córdoba, Spain

Abstract. The classification of patterns into naturally ordered labels is referred to as ordinal regression. This paper explores the notion of kernel trick and empirical feature space in order to reformulate the most widely used linear ordinal classification algorithm (the Proportional Odds Model or POM) to perform nonlinear decision regions. The proposed method seems to be competitive with other state-of-the-art algorithms and significantly improves the original POM algorithm when using 8 ordinal datasets. Specifically, the capability of the methodology to handle non-linear decision regions has been proven by the use of a non-linearly separable toy dataset.

Keywords: Proportional Odds Model, Ordinal Regression, Kernel Trick.

1 Introduction

Here, we consider the specific problem of ordinal regression, which shares properties of classification and regression. Formally, $\mathcal{Y}$ (the labelling space) is a finite set, but there exists some ordering among its elements. In contrast to regression, $\mathcal{Y}$ is a non-metric space, thus distances among categories are unknown. Besides, the standard zero-one loss function does not reflect the ordering of $\mathcal{Y}$. Ordinal regression (or classification) problems arise in fields as information retrieval, preference learning, economy, and statistics and nowadays it is considered as an emerging field in the areas of machine learning and pattern recognition research.

A great number of statistical methods for categorical data treat all response variables as nominal, in such a way that the results are invariant to order permutations on those variables. However, there are many advantages in treating an ordered categorical variable as ordinal rather than nominal [1], a statement applicable to classification as well. In this vein, several approaches to tackle ordinal regression have been proposed in the domain of machine learning over the years, since the first methodology (the Proportional Odds Model or POM) dating back to 1980 [2]. Indeed, the most popular approach in this paradigm is the

---

* This work has been partially subsidized by the TIN2011-22794 project of the Spanish Ministerial Commission of Science and Technology (MICYT), FEDER funds and the P2011-TIC-7508 project of the "Junta de Andalucía" (Spain).

I. Rojas, G. Joya, and J. Cabestany (Eds.): IWANN 2013, Part I, LNCS 7902, pp. 270–279, 2013.

use of threshold models, which are based on the assumption that an underlying real-valued outcomes exist (also known as latent variables), although they are unobservable. Thus, these methods try to determine the nature of those underlying real-valued outcomes through a function $f(\cdot)$ and a set of bias to represent intervals in the range of this function. Although very sophisticated and successful learning techniques have been developed recently for ordinal regression, the use of the POM method is widespread, despite the fact that it is linear. To deal with this issue, this paper makes use of the notion of the so-called kernel trick, which implicitly maps inputs into a high-dimensional feature space via a function $\Phi(\cdot)$ in order to compute non-linear decision regions; and the idea of empirical feature space [3,4], which preserves the geometrical structure of the original feature space, given that distances and angles in the feature space are uniquely determined by dot products and that dot products of the corresponding images are the original kernel values. This empirical feature space is Euclidean, so it is useful for the kernelization of all kinds of linear machines [5,6], with the advantage that the algorithm does not need to be formulated to deal with dot products between data points. Indeed, by the use of this methodology, the dimensionality of the space can be controlled, in such a way that only r-dominant dimensions can be chosen, an advantage that will also be a key factor in the POM algorithm, whose computational cost is closely related to the data dimensionality. Because of that, this paper explores the kernelization of the POM method through the use of the empirical feature space in order to provide the opportunity of computing nonlinear decision regions at a limited computational cost and while leading very naturally to probabilistic outputs.

A similar work to the proposal of this paper can be found in [7], where the POM method is naturally extended for non-crisp ordinal regression tasks, since the underlying latent variable is not necessarily restricted to the class of linear models. This work can be said to be similar in the sense that it also proposes to kernelize the POM model through the use of the well-known kernel trick, nevertheless, it makes reference to a conceptually different setup, as partial class memberships are given for the patterns. However, in our case, we are provided with crisp ordinal targets.

The paper is organized as follows: Section II shows a description of the method; Section III describes the experimental study and analyses the results; and finally, Section IV outlines some conclusions and future work.

2   Methodology

The goal in classification is to assign an input vector $\mathbf{x}$ to one of K discrete classes $\mathcal{C}_k, k \in \{1, \ldots, K\}$. A formal framework for the ordinal regression problem can be introduced considering an input space $\mathcal{X} \in \mathbb{R}^d$, where d is the data dimensionality and N is the number of patterns. To do so, an outcome space $\mathcal{Y} = \{\mathcal{C}_1, \mathcal{C}_2, \ldots, \mathcal{C}_K\}$ is defined, where the labels are ordered (i.e. $\mathcal{C}_1 \prec \mathcal{C}_2 \prec \cdots \prec \mathcal{C}_K$, where $\prec$ denotes this order information). The objective then is to find a prediction rule $f : \mathcal{X} \to \mathcal{Y}$ by using an i.i.d. sample $D = \{\mathbf{x}_i, y_i\}_{i=1}^N \in \mathcal{X} \times \mathcal{Y}$.

2.1 Empirical Kernel Mapping

In this section, the empirical feature space spanned by the training data is defined. Let $\mathcal{H}$ denote a high-dimensional or infinite-dimensional Hilbert space. Then, for any mapping of patterns $\Phi : \mathcal{X} \to \mathcal{H}$, the inner product $\mathcal{K}(\mathbf{x}, \mathbf{x}') = \langle \Phi(\mathbf{x}), \Phi(\mathbf{x}') \rangle_{\mathcal{H}}$ of the mapped inputs is known as a kernel function, giving rise to a symmetric and positive semidefinite matrix (known as Gram or kernel matrix $\mathbf{K}$) from a given input set $\mathcal{X}$. By definition, these matrices can be diagonalised as follows:

$$\mathbf{K}_{(m \times m)} = \mathbf{P}_{(m \times r)} \cdot \mathbf{M}_{(r \times r)} \cdot \mathbf{P}_{(r \times m)}^{\top}, \tag{1}$$

where r is the rank of $\mathbf{K}$, $\mathbf{M}$ is a diagonal matrix containing the r positive eigenvalues of $\mathbf{K}$ in decreasing order and $\mathbf{P}$ consists of the eigenvectors associated with those r eigenvalues. Note that this mapping corresponds to the principal component analysis *whitening* step [8], but applied to the kernel matrix, instead of the covariance one. Then, the empirical feature space can be defined as a Euclidean space preserving the dot product information about $\mathcal{H}$ contained in $\mathbf{K}$ (i.e. this space is isomorphic to the embedded feature space $\mathcal{H}$, but being Euclidean). That is, since distances and angles of the vectors in the feature space are uniquely determined by dot products, the training data has the same geometrical structure in both the empirical feature space and the feature space. The map from the input space to this r-dimensional empirical feature space is defined as $\Phi_r^e : \mathcal{X} \to \mathbb{R}^r$. More specifically:

$$\Phi_r^e : \mathbf{x}_i \to \mathbf{M}^{-1/2} \cdot \mathbf{P}^{\top} \cdot (\mathcal{K}(\mathbf{x}_i, \mathbf{x}_1), \dots, \mathcal{K}(\mathbf{x}_i, \mathbf{x}_N))^{\top}. \tag{2}$$

It can be checked that the kernel matrix of training images obtained by this transformation corresponds to $\mathbf{K}$, when considering the standard dot product [3,4]. Therefore, this methodology provides us with the opportunity to limit the dimensionality of the space by choosing the r dominant eigenvalues (and their associated eigenvectors) to project the data while maintaining the structure of $\mathcal{H}$. However, the correct choice of r is still a major issue to be resolved.

Far beyond the definition of this empirical feature space, it is well-known that the kernel trick turns a linear decision region in $\mathcal{H}$ into a nonlinear decision in $\mathcal{X}$, allowing the formulation of nonlinear variants of any algorithm which can be cast in terms of the inner products between patterns. Furthermore, if the empirical feature space is used, any standard linear decision algorithm can be used without any loss of generality. Fig. 1 shows the case of a synthethic dataset concerning a non-linearly separable classification task and its transformation to the two-dimensional empirical feature space, which is linearly separable.

2.2 Proportional Odds Model

This is one of the first models specifically designed for ordinal regression, and it was arisen from a statistical background [2]. Let h denote an arbitrary monotonic link function. The model

$$h\left(P(y \leq \mathcal{C}_j | \mathbf{x})\right) = b_j - \mathbf{w}^{\top} \mathbf{x}, \quad j = 1, \dots, K - 1, \tag{3}$$

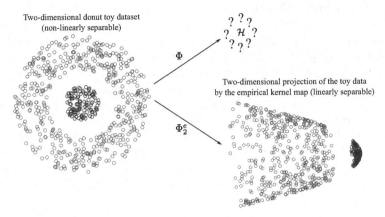

Fig. 1. Synthethic two-dimensional dataset representing a non-linearly separable classification problem and its transformation to the 2 dominant dimensions of the empirical feature space induced by the Gaussian kernel function (linearly separable problem).

links the cumulative probabilities to a linear predictor and imposes an stochastic ordering of the space $\mathcal{X}$, where b_j is the threshold separating $\mathcal{C}_j$ and $\mathcal{C}_{j+1}$ and $\mathbf{w}$ is a linear projection. This model is naturally derived from the latent variable motivation; then instead of fitting a decision rule $f : \mathcal{X} \to \mathcal{Y}$ directly, this model defines a probabilty density function over the class labels for a given feature vector $\mathbf{x}$. Let us assume that the ordinal response is a coarsely measured *latent* continuous variable $f(\mathbf{x})$. Thus, label $\mathcal{C}_i$ in the training set is observed if and only if $f(\mathbf{x}) \in [b_{i-1}, b_i]$, where the function f (latent utility) and $b = (b_0, b_1, ..., b_{K-1}, b_K)$ are determined from data. By definition, $b_0 = -\infty$ and $b_K = +\infty$ and the real line $f(\mathbf{x})$ is divided into K consecutive intervals, where each interval corresponds to a category $\mathcal{C}_i$.

Now, let us define a model of the latent variable, $f(\mathbf{x}) = \mathbf{w}^\top \mathbf{x} + \epsilon$, where ϵ is the random component with zero expectation, $\mathbf{E}[\epsilon] = 0$, and distributed according to the distribution function F_ϵ. Then, it follows that:

$$P(y \leq \mathcal{C}_j | \mathbf{x}) = \sum_{k=1}^{j} P(y = \mathcal{C}_k | \mathbf{x}) = \sum_{k=1}^{j} P(f(\mathbf{x}) \in [b_{k-1}, b_k]) =$$

$$= P(f(\mathbf{x}) \in [-\infty, b_j]) = P(\mathbf{w}^\top \mathbf{x} + \epsilon \leq b_j) = P(\epsilon \leq b_j - \mathbf{w}^\top \mathbf{x}) = F_\epsilon(b_j - \mathbf{w}^\top \mathbf{x}).$$

If a distribution assumption F_ϵ is made for ϵ, the cumulative model is obtained by choosing, as the inverse link function h^{-1}, the inverse distribution F_ϵ^{-1} (quantile function). Note that $F_\epsilon^{-1} : [0, 1] \to (-\infty, +\infty)$ is a monotonic function.

Now, consider the use of the transformed $\Phi_r^e(\mathbf{x})$ (instead of $\mathbf{x}$) in (3). In this case, the model of the latent variable will submit the formulation $f(\Phi_r^e(\mathbf{x})) = \mathbf{w}^\top \cdot \Phi_r^e(\mathbf{x}) + \epsilon$, where $\mathbf{w}$ will be a linear projection but performing as a non-linear decision function in $\mathcal{X}$, since the kernel trick is being used.

3   Experimental Results

Several benchmark datasets have been tested in order to validate the methodology proposed; some publicly available real ordinal classification datasets (squash-unstored, bondrate and eucalyptus) were extracted from UCI repository [9] and some of the ordinal regression benchmark datasets (pyrim, machine, housing and abalone) provided by Chu et. al [10] were considered due to their widespread use in ordinal regression [11,12], although they do not originally represent ordinal classification tasks but regression ones instead. To turn regression into ordinal classification, the target variable is discretized into K different bins (representing classes), with equal frequency, as proposed in previous works [10,11,12]. Table 1 presents the main characteristics of the datasets used for the experimentation.

Table 1. Characteristics of the benchmark datasets

Dataset	#Pat.	#Attr.	#Classes	Class distribution
squash-unstored	52	52	3	$(24, 24, 4)$
bondrate	57	37	5	$(6, 33, 12, 5, 1)$
pyrim	74	27	5	$(15, 15, 15, 15, 14)$
machine	209	7	5	$(42, 42, 42, 42, 41)$
toy	300	2	5	$(35, 87, 79, 68, 31)$
eucalyptus	736	91	5	$(180, 107, 130, 214, 105)$
housing	506	14	5	$(101, 101, 101, 101, 101)$
abalone	4177	11	5	$(836, 836, 835, 835, 835)$

All nominal variables are transformed into binary ones.

Concerning evaluating measures, several metrics can be considered for the evaluation of ordinal classifiers, the most common ones in machine learning being the Mean Absolute Error (MAE) and the Mean Zero-one Error (MZE) [11,12,13], being $MZE = 1 - Acc$, where Acc is the accuracy or correct classification rate and the Mean Absolute Error (MAE) is the average deviation in absolute value of the predicted class from the true class [14]: $MAE = (1/N) \sum_{i=1}^{N} e(\mathbf{x}_i)$, where $e(x_i) = |r(y_i) - r(y_i^*)|$ is the distance between the true and the predicted ranks ($r(y)$ being the rank for a given target y), and, then, MAE values range from 0 to $K - 1$ (maximum deviation in number of ranks between two labels).

Regarding the experimental setup, a 30-holdout stratified technique was applied to divide the real datasets, using 75% of the patterns for training the model, and the remaining 25% for testing it. For the regression datasets provided by Chu et. al [10] (pyrim, machine, housing and abalone), the number of random splits was 20 and the number of training and test patterns are the same as those presented in the corresponding works [10,12]. Since the methods tested are all deterministic, one model was obtained and evaluated, for each split. The results are taken as the mean and standard deviation over each one of the test sets. For model selection, a stratified nested 5-fold cross-validation was used on the training sets, with kernel width and C parameter for SVM selected within the values $\{10^{-3}, 10^{-2}, \ldots, 10^3\}$. The cross-validation criterium is the MAE measure since it can be considered the most common one in ordinal regression. The kernel selected for all the algorithms is the Gaussian one, $K(\mathbf{x}, \mathbf{y}) = \exp\left(-\frac{\|\mathbf{x}-\mathbf{y}\|^2}{\sigma^2}\right)$ where σ is the width of the kernel. Due to the choice of the Gaussian kernel, the probit function is used with our proposal and the logit one when comparing to the original POM algorithm. The number of dimensions for the empirical feature space (r) has been crossvalidated within the values $\{10,20,30\}$.

Two well-known kernel methods for ordinal regression have been chosen for comparison purposes (Kernel Discriminant Learning for Ordinal Regression or KDLOR [11] and Support Vector for Ordinal Regression with Implicit Constraints or SVORIM [12]).

3.1 Results

The results of the experiments can be seen in Table 2, where the proposal (Kernelized Proportional Odds Model or KPOM), the original linear methodology (Proportional Odds Model or POM) and two ordinal state-of-the-art algorithms (KDLOR and SVORIM) are tested. Reported metrics are MZE, MAE and the time needed to build the model (train, cross-validation and test).

Table 2. Results obtained for each method reported in terms of MZE, MAE and time

Dataset	KPOM	POM	KDLOR	SVORIM
		MZE		
squash-unstored	**0.248 ± 0.114**	0.651 ± 0.142	_0.249 ± 0.127_	0.264 ± 0.121
bondrate	**0.431 ± 0.045**	0.656 ± 0.161	0.469 ± 0.083	_0.464 ± 0.077_
pyrim	_0.504 ± 0.070_	**0.485 ± 0.118**	0.527 ± 0.096	0.508 ± 0.096
machine	_0.397 ± 0.059_	**0.394 ± 0.065**	0.412 ± 0.066	0.415 ± 0.056
toy	_0.040 ± 0.020_	0.711 ± 0.026	0.114 ± 0.030	**0.023 ± 0.014**
eucalyptus	_0.363 ± 0.024_	0.851 ± 0.016	0.367 ± 0.029	**0.360 ± 0.030**
housing	**0.325 ± 0.033**	0.355 ± 0.018	0.363 ± 0.037	_0.328 ± 0.028_
abalone	**0.524 ± 0.009**	0.539 ± 0.005	0.548 ± 0.010	_0.525 ± 0.008_
Ranking	**1.50**	3.00	3.25	_2.25_
		MAE		
squash-unstored	**0.250 ± 0.118**	0.826 ± 0.230	_0.251 ± 0.132_	0.264 ± 0.121
bondrate	**0.604 ± 0.084**	0.947 ± 0.321	0.629 ± 0.082	_0.613 ± 0.081_
pyrim	**0.606 ± 0.109**	0.700 ± 0.198	0.669 ± 0.189	_0.638 ± 0.141_
machine	_0.445 ± 0.084_	**0.425 ± 0.079**	0.486 ± 0.101	0.459 ± 0.082
toy	_0.040 ± 0.020_	0.981 ± 0.039	0.114 ± 0.030	**0.023 ± 0.014**
eucalyptus	_0.400 ± 0.031_	1.939 ± 0.254	0.401 ± 0.032	**0.394 ± 0.036**
housing	_0.359 ± 0.037_	0.400 ± 0.024	0.392 ± 0.045	**0.358 ± 0.035**
abalone	_0.657 ± 0.010_	0.690 ± 0.007	0.758 ± 0.017	**0.654 ± 0.006**
Ranking	**1.62**	3.50	3.13	_1.75_
		Time		
squash-unstored	5.8 ± 5.4	**0.6 ± 0.0**	6.1 ± 3.0	10.6 ± 0.3
bondrate	8.0 ± 0.7	**0.5 ± 0.1**	9.2 ± 3.2	10.6 ± 0.4
pyrim	9.8 ± 3.5	**0.4 ± 0.1**	16.0 ± 4.2	10.9 ± 0.3
machine	63.5 ± 8.6	**0.3 ± 0.0**	74.2 ± 4.8	_23.6 ± 1.6_
toy	101.9 ± 51.7	**0.3 ± 0.0**	109.5 ± 42.5	_23.3 ± 1.0_
eucalyptus	379.8 ± 77.4	**9.0 ± 9.7**	912.5 ± 104.5	_139.8 ± 5.2_
housing	148.1 ± 11.8	**0.4 ± 0.0**	199.1 ± 7.5	_49.2 ± 4.8_
abalone	_967.4 ± 29.0_	**4.1 ± 0.1**	1530.1 ± 38.9	1331.4 ± 61.7
Ranking	2.50	**1.00**	3.75	2.75

The best method is in **bold** face and the second one in _italics_.

The results show that the proposal is competitive with the selected ordinal state-of-the-art methods and is able to outperform the standard linear POM algorithm in most cases. Indeed, in those datasets where the POM has achieved better results, the proposed method also obtained a comparable performance. Specifically, the good performance of KPOM in the toy dataset, which is a synthetically generated non-linearly separable set of data (representation in Fig. 2), has demonstrated that the proposal is able to capture the nonlinearity present in the data. Concerning time, the proposed method achieves better results than the KDLOR and comparable results to the SVORIM.

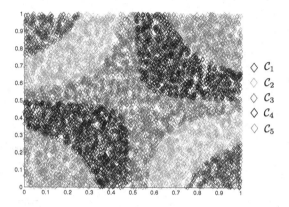

Fig. 2. Representation of the structure of the non-linearly separable toy dataset

In order to graphically clarify the concept of the empirical kernel map, Fig. 3 has been included, where it can be seen that, despite the fact that the three most representative dimensions are not enough to linearly separate the data, they actually include more useful information about the order of the classes and the separation between them.

In order to determine the statistical significance of the differences observed in the methodology constructed, statistical tests have been performed for MZE and MAE and the KPOM and POM algorithms. First of all, there has been an analysis to determine whether each of the different performance metrics followed a normal distribution. In none of these cases can a normal distribution be assumed by using a Kolmogorov-Smirnov's test (KS-test) at a significance level $\alpha = 0.05$. As a consequence, the algorithms are compared by means of the Wilcoxon test at a level of significance of $\alpha = 0.05$. Using this test, the KPOM and POM were compared for each dataset and the number of statistically significant wins or losses were recorded, together with the number of draws. The results obtained show that the KPOM outperforms the results of the POM methodology in 6 datasets and obtains similar performance in 2 of them. Furthermore, in order to compare the 4 methodologies, the non-parametric Friedman's test [15] (with $\alpha = 0.05$) has been applied to the mean MZE, MAE and time rankings, rejecting the null-hypothesis that all algorithms perform similarly for all the

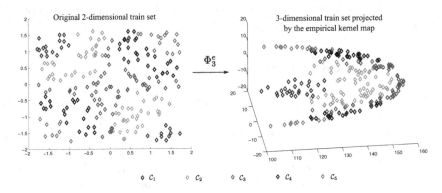

Fig. 3. Graphic showing the 3-dimensional approximation of the empirical feature space induced by a gaussian kernel and the non-linearly separable synthetic toy dataset

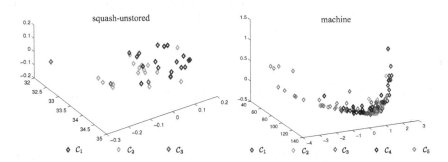

Fig. 4. Graphic showing the 3-dimensional approximation of the empirical feature space induced by a gaussian kernel and the squash-unstored and machine datasets

metrics. The confidence interval was $C_0 = (0, F_{(\alpha=0.05)} = 3.07)$ and the corresponding F-values were $9.00 \notin C_0$ for MZE, $8.34 \notin C_0$ for MAE and $24.11 \notin C_0$ for time. The Holm test has also been applied and the test concluded that there were statistically significant differences for $\alpha = 0.1$ when selecting the KPOM as the control method. These differences were found for MZE and MAE (and only for POM and KDLOR). Thereby, as a conclusion, it can be stated that the use of the empirical feature space in conjunction with the POM method helps to improve its efficiency, performing well in non-linearly separable cases and showing competitiveness when compared to other ordinal state-of-the-art methodologies.

Finally, a graphic experiment has been included to visualize the behaviour of the empirical feature space method in real datasets. To do so, the experiments in Table 2 have been repeated but now considering the first 3 dominant eigenvalues for the computation of the empirical feature space, thus considering only the mapping Φ_3^e. The datasets performing similarly to the results in Table 2 for this set of eigenvalues were graphically represented (Fig. 4), where it can be seen that the classes followed an ordered structure and maintained an acceptable separation, which could be useful to the classifier.

4   Conclusions and Future Work

This paper explores the notion of empirical feature space (an isomorphic space to the original feature space induced by the kernel trick) to reformulate a well-known ordinal regression method (the Proportional Odds Model) to handle non-linearly separable classification tasks. The proposed method seems to significantly outperform the original algorithm and is competitive with other ordinal state-of-the-art algorithms. As future work, several promising lines can be introduced. Firstly, a different kernelized version of the POM algorithm could be constructed by the use of the Nyström approximation for low-rank decomposition [16] to solve the dimensionality problem with the POM method. Furthermore, in the same vein as this paper, an analytical methodology [17] could be used to compute the optimal number of relevant dimensions for the empirical feature space (note that in our case this value was obtained by cross-validation).

References

1. Agresti, A.: Categorical Data Analysis, 2nd edn. Wiley Series in Probability and Statistics. Wiley-Interscience (2002)
2. McCullagh, P.: Regression models for ordinal data. Journal of the Royal Statistical Society 42(2), 109–142 (1980)
3. Schölkopf, B., Mika, S., Burges, C.J.C., Knirsch, P., Müller, K.R., Rätsch, G., Smola, A.J.: Input space versus feature space in kernel-based methods. IEEE Transactions on Neural Networks 10, 1000–1017 (1999)
4. Xiong, H., Swamy, M.N.S., Ahmad, M.O.: Optimizing the kernel in the empirical feature space. IEEE Transactions on Neural Networks 16(2), 460–474 (2005)
5. Abe, S., Onishi, K.: Sparse least squares support vector regressors trained in the reduced empirical feature space. In: de Sá, J.M., Alexandre, L.A., Duch, W., Mandic, D.P. (eds.) ICANN 2007. LNCS, vol. 4669, pp. 527–536. Springer, Heidelberg (2007)
6. Xiong, H.: A unified framework for kernelization: The empirical kernel feature space. In: Chinese Conference on Pattern Recognition (CCPR), pp. 1–5 (November 2009)
7. Verwaeren, J., Waegeman, W., De Baets, B.: Learning partial ordinal class memberships with kernel-based proportional odds models. Comput. Stat. Data Anal. 56(4), 928–942 (2012)
8. Schölkopf, B., Smola, A., Müller, K.R.: Nonlinear component analysis as a kernel eigenvalue problem. Neural Computation 10(5), 460–474 (1998)
9. Asuncion, A., Newman, D.: UCI machine learning repository (2007)
10. Chu, W., Ghahramani, Z.: Gaussian processes for ordinal regression. Journal of Machine Learning Research 6, 1019–1041 (2005)
11. Sun, B.Y., Li, J., Wu, D.D., Zhang, X.M., Li, W.B.: Kernel discriminant learning for ordinal regression. IEEE Transactions on Knowledge and Data Engineering 22, 906–910 (2010)

12. Chu, W., Keerthi, S.S.: Support vector ordinal regression. Neural Computation 19(3), 792–815 (2007)
13. Gutiérrez, P.A., Pérez-Ortiz, M., Fernández-Navarro, F., Sánchez-Monedero, J., Hervás-Martínez, C.: An Experimental Study of Different Ordinal Regression Methods and Measures. In: Corchado, E., Snášel, V., Abraham, A., Woźniak, M., Graña, M., Cho, S.-B. (eds.) HAIS 2012, Part II. LNCS, vol. 7209, pp. 296–307. Springer, Heidelberg (2012)
14. Baccianella, S., Esuli, A., Sebastiani, F.: Evaluation measures for ordinal regression. In: Proceedings of the Ninth International Conference on Intelligent Systems Design and Applications (ISDA 2009), Pisa, Italy (2009)
15. Demsar, J.: Statistical comparisons of classifiers over multiple data sets. Journal of Machine Learning Research 7, 1–30 (2006)
16. Drineas, P., Mahoney, M.W.: On the nyström method for approximating a gram matrix for improved kernel-based learning. J. Mach. Learn. Res. 6 (2005)
17. Braun, M.L., Buhmann, J.M., Müller, K.R.: On relevant dimensions in kernel feature spaces. J. Mach. Learn. Res. 9, 1875–(1908)

Parallel Support Vector Data Description

Phuoc Nguyen, Dat Tran, Xu Huang, and Wanli Ma

Faculty of Education Science Technology and Mathematics
University of Canberra, ACT 2601, Australia
dat.tran@canberra.edu.au

Abstract. This paper proposes an extension of Support Vector Data Description (SVDD) to provide a better data description. The extension is called Distant SVDD (DSVDD) that determines a smallest hypersphere enclosing all normal (positive) samples as seen in SVDD. In addition, DSVDD maximises the distance from centre of that hypersphere to the origin. When some abnormal (negative) samples are introduced, the DSVDD is extended to Parallel SVDD that also determines a smallest hypersphere for normal samples and at the same time determines a smallest hyperphere for abnormal samples and maximises the distance between centres of these two hyperspheres. Experimental results for classification show that the proposed extensions provide higher accuracy than the original SVDD.

Keywords: Support vector data description, spherically shaped boundary, one-class classification, novelty detection.

1 Introduction

Unlike two-class classification problem which aims at determining the discrimination function that best separates the two classes, the target of one-class classification or data description problem is to make a description of a target (positive) data set and to test whether a new data sample belongs to this target data set or outlier (negative data) [1]. There are several approaches to the data description problem. The density-based approach aims to estimate a probability density of the target data set, then the likelihood of a test data sample given the target density is computed. The low likelihood indicates a possible outlier. As pointed out in [1], the Bayesian approach for data description problem has several drawbacks, for example many samples are required in higher dimensional space and only high density areas are modelled although low density areas may define legal target data. In [2], the author estimated distribution of the target data set by first mapping the data into the feature space by a kernel function then a hyperplane will separate them from the origin with maximum margin resulting a complex boundary of the target data in the input space. A test sample is determined by evaluating whether it falls on side of the target data. This is a form of Support Vector Machine (SVM) known as one-class SVM (OCSVM) [3].

In [4] [1], the author used a small hyperphere to describe the target data set instead of a hyperplane in the feature space. This approach was called Support

I. Rojas, G. Joya, and J. Cabestany (Eds.): IWANN 2013, Part I, LNCS 7902, pp. 280–290, 2013.
© Springer-Verlag Berlin Heidelberg 2013

vector Data Description (SVDD). Minimising the hypersphere volume will also minimise the chance of accepting outliers. The author argued that the data description method should use mainly the target data and do not require outlier data. This is true when one of the classes is sampled very well while the other class is severely under sampled. However we will show that if we know some information about distribution of outliers, we can utilise it to find a better data description for the target data.

There are several extensions to SVDD for one-class classification or data description. For example the authors in [5] introduced a new distance measure based on the relative density degree for each data sample which reflects the target data distribution. Another extension is in [6] where the author used a hypersphere with the maximum separation ratio between the sum of its radius and a margin and the subtraction of its radius and that margin. Additionally, the ratio of the radius of the sphere to the separation margin can be adjusted to provide a series of solutions ranging from spherical to linear decision boundaries.

In this paper, inspired by the OCSVM and the SVDD, we propose an extension to SVDD to provide a better data description. The extension is called Distant SVDD (DSVDD) that determines a smallest hypersphere enclosing all normal (positive) samples as SVDD does, and in addition, DSVDD maximises the distance from centre of that hypersphere to the origin. When some abnormal (negative) samples are introduced, the DSVDD is extended to Parallel SVDD that also determines a smallest hypersphere for normal samples and at the same time determines a smallest hyperphere for abnormal samples and maximises the distance between centres of these two hyperspheres. In PSVDD, we propose a method to translate the origin to one of the two centres in the feature space and the PSVDD problem will become the DSVDD problem. In addition, if information about distribution of the abnormal data set is known, we can utilise it to find a better data description for the normal data.

Experimental results for classification on 8 UCI data sets showed that the proposed extensions provide higher accuracy than the original SVDD and OCSVM. We also compare our proposed extensions with Gaussian mixture model which is a statistical method.

The remaining of this paper is as follows. In Section 2 we summarise the SVDD method. In Section 3 we present the theory of DSVDD for the case of positive data only and that for both positive and negative data. Then in Section 4 we develop the PSVDD method and the translation in feature space. In Section 5 we present visual experiments on artificial data set and experiments on UCI datasets. Finally we conclude in Section 6.

2   Support Vector Data Description (SVDD)

Let $\mathbf{X} = \{\mathbf{x}_1, \mathbf{x}_2, \ldots, \mathbf{x}_n\}$ be the normal data set. SVDD [1] aims at determining an optimal hypersphere that encloses all normal data samples in this data set $\mathbf{X}$ while abnormal data samples are not included. The optimisation problem is formulated as follows

$$\min_{R,c,\xi} \left(R^2 + C \sum_{i=1}^{n} \xi_i \right) \tag{1}$$

subject to

$$||\phi(\mathbf{x}_i) - \mathbf{c}||^2 \leq R^2 + \xi_i \quad i = 1, \ldots, n$$
$$\xi_i \geq 0, \quad i = 1, \ldots, n \tag{2}$$

where R is radius of the hypersphere, C is a constant, $\xi = [\xi_i]_{i=1,\ldots,n}$ is vector of slack variables, $\phi(.)$ is the nonlinear function related to the symmetric, positive definite kernel function $K(\mathbf{x}_1, \mathbf{x}_2) = \phi(\mathbf{x}_1) \cdot \phi(\mathbf{x}_2)$, and $\mathbf{c}$ is centre of the hypersphere.

For classifying an unknown data sample $\mathbf{x}$, the following decision function is used: $f(\mathbf{x}) = sign(R^2 - ||\phi(\mathbf{x}) - \mathbf{c}||^2)$. The unknown data sample $\mathbf{x}$ is normal if $f(\mathbf{x}) = +1$ or abnormal if $f(\mathbf{x}) = -1$.

3   Distant SVDD (DSVDD)

3.1   Problem Formulation

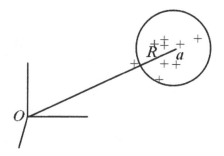

Fig. 1. DSVDD determines a smallest hypersphere enclosing all normal (positive) data samples and maximises the distance from its centre to the origin

Let $\mathbf{X} = \{\mathbf{x}_1, \mathbf{x}_2, \ldots, \mathbf{x}_n\}$ be the normal data set. The task of DSVDD is to determine an optimal hypersphere that encloses all normal data samples in this data set $\mathbf{X}$ and maximise the distance from centre of the hypersphere to the origin as shown in Figure 1. The optimisation problem is formulated as follows:

$$\min_{R,a,\xi} \left(R^2 - k \, ||\mathbf{a}||^2 + \frac{1}{\nu n} \sum_i \xi_i \right) \tag{3}$$

subject to

$$||\mathbf{x}_i - \mathbf{a}||^2 \leq R^2 + \xi_i \quad i = 1, \ldots, n$$

$$\xi_i \geq 0 \qquad i = 1, \ldots, n \tag{4}$$

where R is radius of the sphere, ν and k are parameters, $\xi = [\xi_i]_{i=1,\ldots,n}$ is vector of slack variables and $\mathbf{a}$ is centre of the hypersphere. We can construct the following Lagrange function using Lagrange multipliers α_i, β_i:

$$L(R, \mathbf{a}, \xi, \alpha, \beta) = R^2 - k\|\mathbf{a}\|^2 + \frac{1}{\nu n}\sum_i \xi_i - \sum_i \alpha_i[R^2 + \xi_i - \|\mathbf{x}_i - \mathbf{a}\|^2] - \sum_i \beta_i \xi_i \tag{5}$$

Using KKT conditions, we have:

$$\frac{\partial L}{\partial R} = 0 \Rightarrow 2R(1 - \sum_i \alpha_i) = 0 \Rightarrow \sum_i \alpha_i = 1 \tag{6}$$

$$\frac{\partial L}{\partial \mathbf{a}} = 0 \Rightarrow -2k\mathbf{a} - 2\sum_i \alpha_i(\mathbf{x}_i - \mathbf{a}) = 0 \Rightarrow \mathbf{a} = \frac{\sum_i \alpha_i \mathbf{x}_i}{1 - k} \tag{7}$$

$$\frac{\partial L}{\partial \xi_i} = 0 \Rightarrow \frac{1}{\nu n} - \alpha_i - \beta_i = 0 \Rightarrow \alpha_i + \beta_i = \frac{1}{\nu n} \quad i = 1, \ldots, n \tag{8}$$

By connecting KKT conditions and Lagrangian function we can achieve:

$$
\begin{aligned}
L &= -k\|\mathbf{a}\|^2 + \sum_i \alpha_i \|\mathbf{x}_i - \mathbf{a}\|^2 \\
&= -k\|\mathbf{a}\|^2 + \sum_i \alpha_i(\|\mathbf{x}_i\|^2 - 2\mathbf{x}_i \cdot \mathbf{a} + \|\mathbf{a}\|^2) \\
&= (1 - k)\|\mathbf{a}\|^2 + \sum_i \alpha_i \|\mathbf{x}_i\|^2 - 2\sum_i \alpha_i \mathbf{x}_i \cdot \mathbf{a} \\
&= \frac{1}{1 - k}\left\|\sum_i \alpha_i \mathbf{x}_i\right\|^2 + \sum_i \alpha_i \|\mathbf{x}_i\|^2 - \frac{2}{1 - k}\sum_i \alpha_i \mathbf{x}_i \cdot (\sum_i \alpha_i \mathbf{x}_i) \\
&= \frac{-1}{(1 - k)}\sum_{i,j} \alpha_i \alpha_j \mathbf{x}_i \cdot \mathbf{x}_j + \sum_i \alpha_i \mathbf{x}_i \cdot \mathbf{x}_i
\end{aligned}
\tag{9}
$$

The Lagrange function L should be maximised with respect to α_i or negative of L should be minimised, we have the new optimisation problem:

$$\min \frac{1}{(1 - k)}\sum_{i,j} \alpha_i \alpha_j \mathbf{x}_i \cdot \mathbf{x}_j - \sum_i \alpha_i \mathbf{x}_i \cdot \mathbf{x}_i \tag{10}$$

subject to:

$$\sum_i \alpha_i = 1 \quad i = 1, \ldots, n$$

$$0 \leq \alpha_i \leq \frac{1}{\nu n} \quad i = 1, \ldots, n \tag{11}$$

where the Lagrange multipliers $\beta_i \geq 0$ have been removed from Equation (8) In the test phase, a sample $\mathbf{x}$ can be determined whether it belongs to the hypersphere, i.e. normal class $+1$, by the following decision function:

$$sign(R^2 - d^2(\mathbf{x})) \tag{12}$$

where $d^2(\mathbf{x})$ is its distance to the hypersphere centre and can be determined by:

$$d^2(\mathbf{x}) = \|\mathbf{x} - \mathbf{a}\|^2 = \mathbf{x} \cdot \mathbf{x} - \frac{2}{1-k} \sum_i \alpha_i \mathbf{x} \cdot \mathbf{x}_i + \frac{1}{(1-k)^2} \sum_{i,j} \alpha_i \alpha_j \mathbf{x}_i \cdot \mathbf{x}_j \tag{13}$$

We select the support vector that lies on hypersphere $\mathbf{x}_t$ and corresponds to the smallest α_t, $0 < \alpha_t < \frac{1}{\nu n}$, to determine the radius R:

$$R^2 = \mathbf{x}_t \cdot \mathbf{x}_t - \frac{2}{1-k} \sum_i \alpha_i \mathbf{x}_t \cdot \mathbf{x}_i + \frac{1}{(1-k)^2} \sum_{i,j} \alpha_i \alpha_j \mathbf{x}_i \cdot \mathbf{x}_j \tag{14}$$

These vectors $\mathbf{x}_i$ only appear in inner product form in the above problem, therefore a more complex decision boundary than the hypersphere can be achieved by replacing the inner product with a kernel function to transform $\mathbf{x}_i$ to high dimension space as follows [7]. Let $\mathbf{x}_i, \mathbf{x}_j \in \mathbf{R}^d$, ϕ is some function that maps the data to some other Euclidean space H:

$$\phi : \mathbf{R}^d \mapsto H \tag{15}$$

then the training algorithm use data through inner products $\phi(\mathbf{x}_i) \cdot \phi(\mathbf{x}_j)$ in H. If we use a kernel function K such that $K(\mathbf{x}_i, \mathbf{x}_j) = \phi(\mathbf{x}_i) \cdot \phi(\mathbf{x}_j)$ then we can ignore the form of ϕ. Gaussian radial basis function (RBF) kernel is an example:

$$K(\mathbf{x}_i, \mathbf{x}_j) = e^{-\gamma \|\mathbf{x}_i - \mathbf{x}_j\|^2} \tag{16}$$

3.2   Distant SVDD with Negative Data Samples

When there are negative data samples, the above problem can be reformulated as follows. Let $\{(\mathbf{x}_i, y_i)\}, y_i \in \{+1, -1\}, i = 1, \ldots, n$ be the data set including positive data $y_i = +1$ and negative data $y_i = -1$. The DSVDD with negative samples aims at determining a smallest hypersphere that encloses all positive data samples and at the same time maximising the distance from centre of the hypersphere to the origin, as seen in Figure 3. The optimisation problem is formulated as follows:

$$\min_{R,a,\xi} \left(R^2 - k\|\mathbf{a}\|^2 + \frac{1}{\nu n} \sum_i \xi_i \right) \tag{17}$$

subject to

$$\|\mathbf{x}_i - \mathbf{a}\|^2 \leq R^2 + \xi_i \quad y_i = +1$$

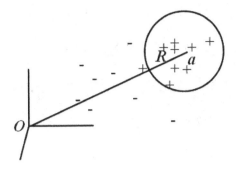

Fig. 2. DSVDD with negative data samples determines a smallest hypersphere that encloses all positive samples and discards all negative samples while at the same time maximises the distance from its centre to the origin

$$\|\mathbf{x}_i - \mathbf{a}\|^2 > R^2 - \xi_i \quad y_i = -1$$
$$\xi_i \geq 0 \qquad i = 1, \dots, n \tag{18}$$

or

$$y_i \|\mathbf{x}_i - \mathbf{a}\|^2 \leq y_i R^2 + \xi_i \quad i = 1, \dots, n$$
$$\xi_i \geq 0 \qquad i = 1, \dots, n \tag{19}$$

where R is radius of the hypersphere, ν and k are parameters, $\xi = [\xi_i]_{i=1,\dots,n}$ is vector of slack variables and $\mathbf{a}$ is centre of the hypersphere. The problem of DSVDD with negative samples is quite similar to the DSVDD without negative samples except for the extra factor y_i in the first constraint in Equation (19). Using Lagrange multipliers, we reconstruct the Lagrange function:

$$L(R, \mathbf{a}, \xi, \alpha, \beta) = R^2 - k\|\mathbf{a}\|^2 + \frac{1}{\nu n}\sum_i \xi_i - \sum_i \alpha_i [y_i R^2 + \xi_i - y_i \|\mathbf{x}_i - \mathbf{a}\|^2] - \sum_i \beta_i \xi_i \tag{20}$$

Using KKT conditions, we have:

$$\frac{\partial L}{\partial R} = 0 \;\Rightarrow\; 2R(1 - \sum_i \alpha_i y_i) = 0 \Rightarrow \sum_i \alpha_i y_i = 1 \tag{21}$$

$$\frac{\partial L}{\partial \mathbf{a}} = 0 \;\Rightarrow\; -2k\mathbf{a} - 2\sum_i \alpha_i y_i(\mathbf{x}_i - \mathbf{a}) = 0 \Rightarrow \mathbf{a} = \frac{\sum_i \alpha_i y_i \mathbf{x}_i}{1 - k} \tag{22}$$

$$\frac{\partial L}{\partial \xi_i} = 0 \Rightarrow \frac{1}{\nu n} - \alpha_i - \beta_i = 0 \;\Rightarrow\; \alpha_i + \beta_i = \frac{1}{\nu n} \quad i = 1, \dots, n \tag{23}$$

By connecting KKT conditions and Lagrangian function we can achieve:

$$L = -k\|\mathbf{a}\|^2 + \sum_i \alpha_i y_i \|\mathbf{x}_i - \mathbf{a}\|^2$$

$$= -k\|\mathbf{a}\|^2 + \sum_i \alpha_i y_i (\|\mathbf{x}_i\|^2 - 2\mathbf{x}_i \cdot \mathbf{a} + \|\mathbf{a}\|^2)$$

$$= (1-k)\|\mathbf{a}\|^2 + \sum_i \alpha_i y_i \|\mathbf{x}_i\|^2 - 2\sum_i \alpha_i y_i \mathbf{x}_i \cdot \mathbf{a}$$

$$= \frac{1}{1-k}\left\|\sum_i \alpha_i y_i \mathbf{x}_i\right\|^2 + \sum_i \alpha_i y_i \|\mathbf{x}_i\|^2 - \frac{2}{1-k}\sum_i \alpha_i y_i \mathbf{x}_i \cdot \left(\sum_i \alpha_i y_i \mathbf{x}_i\right)$$

$$= \frac{-1}{(1-k)}\sum_{i,j} \alpha_i \alpha_j y_i y_j \mathbf{x}_i \cdot \mathbf{x}_j + \sum_i \alpha_i y_i \mathbf{x}_i \cdot \mathbf{x}_i \tag{24}$$

The Lagrange function L should be maximised with respect to α_i or negative of L should be minimised, we have the new optimisation problem:

$$\min \frac{1}{(1-k)}\sum_{i,j} \alpha_i \alpha_j y_i y_j \mathbf{x}_i \cdot \mathbf{x}_j - \sum_i \alpha_i y_i \mathbf{x}_i \cdot \mathbf{x}_i \tag{25}$$

subject to:

$$\sum_i \alpha_i y_i = 1 \quad i = 1, \ldots, n$$

$$0 \le \alpha_i \le \frac{1}{\nu n} \quad i = 1, \ldots, n \tag{26}$$

where the Lagrange multipliers $\beta_i \ge 0$ have been removed from Equation (23). The distance from a test sample $\mathbf{x}$ to the centre becomes:

$$d^2(\mathbf{x}) = \|\mathbf{x} - \mathbf{a}\|^2 = \mathbf{x} \cdot \mathbf{x} - \frac{2}{1-k}\sum_i \alpha_i y_i \mathbf{x} \cdot \mathbf{x}_i + \frac{1}{(1-k)^2}\sum_{i,j} \alpha_i \alpha_j y_i y_j \mathbf{x}_i \cdot \mathbf{x}_j \tag{27}$$

The radius R can be determined by:

$$R^2 = \mathbf{x}_t \cdot \mathbf{x}_t - \frac{2}{1-k}\sum_i \alpha_i y_i \mathbf{x}_t \cdot \mathbf{x}_i + \frac{1}{(1-k)^2}\sum_{i,j} y_i y_j \alpha_i \alpha_j \mathbf{x}_i \cdot \mathbf{x}_j \tag{28}$$

where $\mathbf{x}_t$ is support vector with $0 < \alpha_t < \frac{1}{\nu n}$

4 Parallel SVDD

Based on the idea of DSVDD, we develop Parallel SVDD (PSVDD) as follows. Instead of maximising the distance from centre of the hypersphere to the origin, the PSVDD maximises the distance to the center $\mathbf{b}$ of the abnormal (negative) class. We can solve the new problem directly in the input space but we may face difficulties in the feature space due to the unknown mapping ϕ in Equation (15). However, if we apply a translation that maps the origin to the centre $\mathbf{b}$ to form the new axes, the PSVDD problem becomes the DSVDD problem under the new axes. The translation mapping is straightforward in the input space

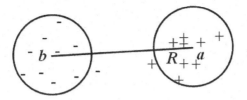

Fig. 3. PSVDD determines a smallest hypersphere enclosing all positive samples and another smallest hypersphere for all negative samples and maximises the distance between two centres

but it needs the following derivations in feature space: Let $K(.,.)$ be the kernel that implements the mapping $\phi(.)$ implicitly, $K'(.,.)$ and $\phi'(.)$ be respectively the kernel and the mapping after translation. If $\mathbf{x}$, $\mathbf{y}$ are two data samples in the input space we have the following formula for the new kernel:

$$
\begin{aligned}
K'(\mathbf{x}, \mathbf{y}) &= \phi'(\mathbf{x}) \cdot \phi'(\mathbf{y}) \\
&= (\phi(\mathbf{x}) - \mathbf{b}) \cdot (\phi(\mathbf{y}) - \mathbf{b}) \\
&= K(\mathbf{x}, \mathbf{y}) - \phi(\mathbf{x}) \cdot \mathbf{b} - \phi(\mathbf{y}) \cdot \mathbf{b} + \|\mathbf{b}\|^2
\end{aligned}
\tag{29}
$$

The centre $\mathbf{b}$ can be determined by training a SVDD hypersphere enclosing the negative samples using the same mapping ϕ or kernel K, we have its kernel expansion $\mathbf{b} = \sum_i \gamma_i y_i \phi(\mathbf{x}_i)$. The above kernel formula becomes:

$$
K'(\mathbf{x}, \mathbf{y}) = K(\mathbf{x}, \mathbf{y}) - \sum_i \gamma_i y_i K(\mathbf{x}_i, \mathbf{x}) - \sum_i \gamma_i y_i K(\mathbf{x}_i, \mathbf{y}) + \|\mathbf{b}\|^2
\tag{30}
$$

where $x_i, i = 1, \ldots, p$ are p support vectors of the hypersphere with centre $\mathbf{b}$.

5   Experimental Results

Figures 4 and 5 show visual results for experiments performed on artificial datasets using DSVDD and PSVDD, respectively. When parameter $k = 0$ the optimisation function in Equation (17) for DSVDD becomes the optimisation function for SVDD. Figure 4 shows that when k increases, the centre of the hypersphere moves away from the origin while all the negative samples are still outside the hypersphere. The first row in Figure 5 shows that when parameter k increases, the hypersphere enclosing positive samples is moving away from negative samples while keeping all the positive samples inside it. The second row in Figure 5 shows that when ν increases, more positive samples are outside the hypersphere.

One class classification experiments were conducted on 8 UCI datasets having two classes. Details of these datasets are listed in Table 1. The datasets were divided in to 2 subsets, the subset contained 60% of the data is for training and the other 40% for testing. The training was done using 3-fold cross validation. The best parameter values searched in the training phase are $\gamma = 2^{-13}, 2^{-11}, \ldots, 2^1$ $\nu = 2^{-8}, 2^{-7}, \ldots, 2^{-2}$, and $k = 0, 0.1, \ldots, 0.9$. Experiments were repeated 10

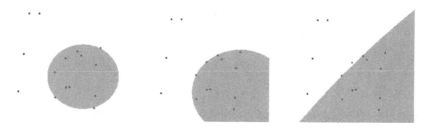

Fig. 4. Screenshots for DSVDD with parameter $\nu = 0.2$ and $k = 0, 0.3$ and 0.9, linear kernel was used. Red points are positive samples and blue points are negative samples.

Fig. 5. The first row contains screenshots for PSVDD when $k = 0, 0.3$ and 0.6, and $\nu = 0.2$. The second row contains screenshots for PSVDD when $\nu = 0.1, 0.2$ and 0.5, and $k = 0.9$. Gaussian RBF kernel was used with $\gamma = 5$. Red points are positive samples and blue points are negative samples.

times and the results were averaged with standard deviations given. The GMM was trained using 64 Gaussian mixtures.

Table 2 shows the prediction rates in cross validation training. Table 3 shows the prediction rates in test phase with best parameters selected. Overall the one class techniques perform worse than classification techniques that include the negative data information. The DSVDD with negative data samples show a slightly better performance than the SVDD. The PSVDD with negative data samples shows best performance due to its ability to push the hypersphere further from the negative samples while maintaining the smallest hypersphere enclosing the positive data samples.

Table 1. Number of data samples in 8 data sets. #normal: number of normal data samples, #abnormal: number of abnormal data samples and d: dimension.

Data set	#normal	#abnormal	d
Australian	383	307	14
Breast Cancer	444	239	10
Diabetes	500	268	8
Fourclass	307	555	2
German numer	700	300	24
Heart	303	164	13
Ionosphere	255	126	34
Letter	594	567	16
Liver disorders	200	145	6

Table 2. Prediction rates in cross validation training of one class classification on 8 datasets

Dataset	GMM	OCSVM	SVDD	SVDD_NEG	DSVDD_NEG	PSVDD_NEG
australian	77.44±5.14	71.98±0.74	73.13±1.4	82.11±0.93	83.86±0.78	88.22±0.25
breast-cancer	50.23±9.52	93.63±0.34	93.88±0.49	94.19±0.48	94.94±0.11	97.29±0.11
diabetes	59.4±2.2	62.05±0.11	62.1±0.22	68.45±0.89	69.6±0.45	75.2±0.84
fourclass	58.56±0.89	80.08±0.36	79.68±1.04	94.83±0.59	95.59±0.4	98.92±0.11
german.numer	66.24±2.17	69.42±0.14	69.38±0.33	68.65±1.47	70.48±1.02	75.29±0.72
heart	70.6±5.74	71.14±3.22	69.53±1.02	76.51±1.57	77.72±1.67	85.64±0.37
ionosphere	76.84±2.92	87.94±0.62	87.66±0.62	89.19±0.73	90.62±0.64	94.16±0.62
liver-disorders	51.1±4.08	55.69±1.29	57.03±2.12	65.74±0.93	67.94±0.76	72.44±0.8

Table 3. Prediction rates in test phase of one-class classification on 8 datasets

Dataset	GMM	OCSVM	SVDD	SVDD_NEG	DSVDD_NEG	PSVDD_NEG
australian	76.1±5.5	67.6±0.8	59.9±6.4	80.4±1.9	82.0±0.3	83.2±1.7
breast-cancer	53.8±33.4	95.1±1.0	94.3±0.8	92.4±1.3	90.9±5.8	97.4±1.7
diabetes	46.3±12.3	69.1±1.5	68.2±0.7	71.2±3.4	65.0±13.3	77.5±2.9
fourclass	57.6±0.9	76.1±0.9	76.2±0.9	93.7±0.3	93.6±0.4	97.7±0.3
german.numer	67.7±1.4	69.3±0.3	68.9±0.7	58.3±9.1	52.1±10.8	71.7±2.7
heart	66.8±7.8	58.4±2.3	63.6±9.6	72.9±1.8	73.7±3.2	83.3±1.8
ionosphere	61.3±10.7	84.5±0	83.9±0.8	94.9±0.9	95.2±0.9	95.5±2.4
liver-disorders	54.3±5.9	43.7±11.4	41.3±2.6	59.0±2.8	59.9±1.9	65.6±3.5

6 Conclusion

We have presented the Distant SVDD as an extension of the well-known SVDD model. When negative samples are introduced, the Distant SVDD is extended to Parallel SVDD to provide a data description for both positive and negative samples. The Parallel SVDD provides a very good data description since it can maximise the distance between centres of two hyperspheres that best separate the positive and negative classes. The results show that our proposed models provide better classification rates than the standard SVDD model.

References

1. Tax, D.M.J., Duin, R.P.W.: Support vector data description. Machine Learning 54(1), 45–66 (2004)
2. Schlkopf, B., Platt, J.C., Shawe-Taylor, J., Smola, A.J., Williamson, R.C.: Estimating the support of a high-dimensional distribution. Neural Computation 13(7), 1443–1471 (2001)
3. Chang, C.C., Lin, C.J.: LIBSVM: a library for support vector machines. ACM Transactions on Intelligent Systems and Technology (TIST) 2(3), 27 (2011)
4. Tax, D.M.J., Duin, R.P.W.: Support vector domain description. Pattern Recognition Letters 20(11), 1191–1199 (1999)
5. Lee, K.Y., Kim, D.W., Lee, K.H., Lee, D.: Density-induced support vector data description. IEEE Transactions on Neural Networks 18(1), 284–289 (2007)
6. Wang, J., Neskovic, P., Cooper, L.N.: Pattern classification via single spheres. In: Hoffmann, A., Motoda, H., Scheffer, T. (eds.) DS 2005. LNCS (LNAI), vol. 3735, pp. 241–252. Springer, Heidelberg (2005)
7. Burges, C.J.C.: A tutorial on support vector machines for pattern recognition. Data Mining and Knowledge Discovery 2(2), 121–167 (1998)

Antinoise Texture Retrieval Based on PCNN and One-Class SVM

Le Tian *, Yi-De Ma, Li Liu, and Kun Zhan

School of Information Science and Engineering,
Lanzhou University, Lanzhou 730000, China
`tianl10@lzu.cn`

Abstract. By training and predicting the features that are extracted by pulse coupled neural network (PCNN), a noise immunity texture retrieval system combined with PCNN and one-class support vector machine (OCSVM) is proposed in this paper, which effectively improve the anti-noise performance of image retrieval system. The experiment results in different noise environment show that our proposed algorithm is able to obtain higher retrieval accuracy and better robustness to noise than traditional Euclidean distance based system.

Keywords: Pulse-coupled neural network, Texture retrieval, One-class support vector machine, Feature extraction.

1 Introduction

The content-based image retrieval technology carry on analysis of image content, such as color, texture, shape and so on, which is widely applied in the image retrieval fields. How to extract the effective texture characteristics has been the hotspot in the field of texture retrieval and pattern recognition. Early classical approaches for texture feature retrieval is Gabor filter[1]. A rotation-invariant and scale-invariant Gabor representation was proposed in [2,3]. As image number in the base increasing, retrieval rate drops and the computation increases dramatically. In addition, the method like Gabor filter is sensitive to noise and it is lack of robustness. PCNN has been widely applied in image processing from 1990s called as the main mark of the third generation of artificial neural network. Output pulses sequence of PCNN contains the unique character of the original stimulus, it has translation, rotation, scale and twist invariance, especially the well robust to noise. Thus PCNN is suitable for feature extraction and the retrieval rate is obviously improved[3-5]. As a simplified Model of the PCNN, Intersecting Cortical Model (ICM) is first proposed by Kinser[6]. Support vector machine (SVM) solved the problem of limited number of samples and dimension disasters, and can get the global optimal solutions, especially one-class SVM

---

* Tian Le, master student of the Information Science and Engineering School, Lanzhou University, China. Her research interest covers artificial neural network, image retrieval and image processing.

I. Rojas, G. Joya, and J. Cabestany (Eds.): IWANN 2013, Part I, LNCS 7902, pp. 291–298, 2013.

(OCSVM) is suitable to solve the one class problem and widely used in image retrieval[7-9].

In this paper, OCSVM is utilized to train and predict the features which are output of PCNN or ICM. Computational anti-noise experiments show that the local-connected neural networks, such as PCNN and ICM, have better robustness to noise than the previous feature extraction methods, such as the Gabor filter[2], and the proposed texture retrieval system based on OCSVM is superior to ED based system.

2   PCNN AND ICM

Pulse coupled neural network (PCNN) is a single layer neural network model, which is given priority to iterative algorithm, and has the property of self-supervision and self-learning. It is widely applied in image segmentation, edge detection, image denoising, image enhancement, feature extraction and so on. The complete mathematical description of PCNN for image processing as follows[5]:

$$F_{ij}[n] = e^{-\alpha_F \Delta_t} F_{ij}[n-1] + V_F \sum_{kl} M_{ijkl} Y_{kl}[n-1] + S_{ij} \tag{1}$$

$$L_{ij}[n] = e^{-\alpha_L \Delta_t} L_{ij}[n-1] + V_L \sum_{kl} W_{ijkl} Y_{kl}[n-1] \tag{2}$$

$$U_{ij}[n] = F_{ij}[n]\{1 + \beta L_{ij}[n]\} \tag{3}$$

$$E_{ij}[n] = e^{-\alpha_E \Delta_t} E_{ij}[n-1] + V_E Y_{ij}[n] \tag{4}$$

$$Y_{ij}(n) = \begin{cases} 1, & if\ U_{ij}(n) > E_{ij}(n) \\ 0, & otherwise \end{cases} \tag{5}$$

The subscript ij is the label of the neurons, S_{ij}, F_{ij}, L_{ij}, U_{ij} and E_{ij} is the signal of the external stimulus, the input of feedback, the input of connecting, internal activity and a dynamic threshold respectively. M and W represent the constant synaptic weights which are computed by inverse square rule (generally $M=W$). V_F, V_L, V_E is amplitude constant; α_F, α_L, α_E is attenuation coefficient; Δt is time constant; β is the connection coefficient; n is iteration times. Y_{ij} is output of PCNN which can be only 0 or 1. Each iteration, when the internal activity value U is greater than dynamic threshold E, PCNN produces the output pulse.

ICM is one of the successful simplified model from PCNN and other visual cortex, which inherits good characteristics from PCNN and is simpler than PCNN. The mathematical model of the ICM is described as follows[6]:

$$F_{ij}(n) = f F_{ij}(n-1) + \sum_{kl} M_{ijkl} Y_{kl}(n-1) + S_{ij} \tag{6}$$

$$E_{ij}(n) = gE_{ij}(n-1) + hY_{ij}(n) \tag{7}$$

$$Y_{ij}(n) = \begin{cases} 1, \; if \; F_{ij}(n) > E_{ij}(n) \\ \quad 0, \; otherwise \end{cases} \tag{8}$$

where f and g is the attenuation coefficient of the units of the threshold functions and the units of the threshold functions respectively, the size of them determine the decay rate; Generally speaking, $g < f < 1$. The value of h is large since h can increase the dynamical threshold of the neurons quickly to make sure each neuron fires only one time.

3   OCSVM

One-class SVM (OCSVM) algorithm was first proposed by Schölkopf et al. [7] to solve the one-class classification problem. The OCSVM algorithm maps training data into a high-dimensional feature space corresponding to a kernel and finds the optimal hyper plane to separates the training data from the origin with maximum margin. The OCSVM can be viewed as a regular two-class SVM when considering the origin as the only member of the second class. OCSVM has been widely used in the identification of the character classification and image retrieval[8,9].

Suppose that a given data set $X = \{x_1, x_2, ..., x_d\}$, $X \in R^d$. Φ is characteristics of feature mapping from the original space to the feature space: $X \rightarrow F$. For the purpose of leaving origin of the optimal hyper plane away from the feature space with the maximum interval, then the problem can be attributed to the following quadratic programming problem:

$$\begin{cases} \min_{w,b,\xi_i} \frac{1}{2} \| w \|^2 + C \sum_{i=1}^{N} \xi_i + b \\ \langle w \cdot \Phi(x_i) \rangle + b \geq 0 - \xi_i, \xi_i \geq 0, i = 1, \ldots, N \end{cases} \tag{9}$$

Where w is the optimal weights, C is the penalty parameter defined, b is the classification threshold, ξ_i is the introduction of slack variables, which allow a certain degree of violation the interval constraints.

Solve the above equation can get the optimal classification:

$$f(x) = sign(\langle w \cdot \Phi(x) \rangle + b) \tag{10}$$

Equation(10) will be positive for most example x_i in the training set. Using the Lagrangian theorem, we can formulate the dual problem as:

$$\begin{cases} \min_{\alpha_i} \frac{1}{2} \sum_{i=1}^{N} \sum_{j=1}^{N} \alpha_i \alpha_j k(x_i, x_j) \\ 0 \leq \alpha_i \leq C, i = 1, \ldots, N, \sum_i \alpha_i = 1, \end{cases} \tag{11}$$

Where α_i is the nonnegative Lagrange multipliers.

Introduce the kernel function $k(x, y) = (\Phi(x) \cdot \Phi(y))$, and the optimal classification function change to:

$$f(x) = \text{sgn}\{\sum_i \alpha_i{}^* K(x_i, x) + b^*\} \qquad (12)$$

Where α_i^* is the optimal solution, b^* is the classification threshold. x_i is the corresponding support vector of the training samples, x is the unknown sample vector. By calculating the value of the optimal classification function (12), we can judge whether the unknown samples belong to the database, if $f(x)$ with a value of +1, then x belongs to the database; if $f(x)$ is -1, then x is not in the database.

4   Experiment

The proposed anti-noise texture retrieval algorithm in this paper, which is based on the framework of the PCNN and OCSVM as shown in Fig.1:

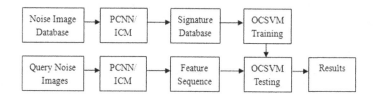

Fig. 1. Diagram for anti-noise texture retrieval system

First, 112 pictures of the Brodatz texture image database[10] are standardized for size of 128×128 pixels and added different noise as the image database, and in accordance with the entropy feature extraction method to get the signature database. Second, the query images are standardized for size of 128×128 pixels and added different noise too. Here in Fig.1, OCSVM trains the features in signature database to achieve a training model. Then OCSVM uses the trained model to predict whether the features of the query image belong to the database, which returns the decision function that takes the value +1 in the database and -1 elsewhere. Euclidean distance (ED) similarity measure method is used to do comparison with OCSVM.

4.1   Parameter Selection in the Experiment

Parameters in the experiment mainly include the model of PCNN or ICM, OCSVM and each noise. The setting of parameters of PCNN is shown in Table 1.

Internal connection weight matrix M and W is a 3 by 3 matrix, and the numerical value of every element is in eq.(13):

Table 1. Parameters of PCNN

Parameter	α_L	α_E	α_F	V_F	V_L	V_E	β	n
Value	1.0	1.0	0.1	0.5	0.2	20	0.1	37

$$M = W = \begin{bmatrix} 0.5 & 1 & 0.5 \\ 1 & 0 & 1 \\ 0.5 & 1 & 0.5 \end{bmatrix} \tag{13}$$

According to the results of several experiments, the parameters of ICM model are set to: $f=0.9$, $g=0.8$, $h=20$, $n=37$.

The part of classification retrieval after feature extraction experimental adopts the OCSVM model in LIBSVM[11]. Because RBF kernel function has its advantages in comparison with other kernel function, such as nonlinear, less parameters, etc[12], the paper uses RBF kernel function. The key parameters of the experiment are the C in OCSVM and the γ in RBF kernel function. The values of parameters C is the penalty factor of the false sample, which can control the balance between the sample bias and the generalization ability of the machine, we take $C=100$ in this paper. Generally, parameter γ is the reciprocal of dimension of the input data [11,12], here $\gamma=1/37$.

The settings of different noise's parameters are as follows:

(1)Parameters of the salt & pepper noise

Salt & pepper noise performance for black and white pixels, we uses the function imnoise(I, 'salt & pepper', d) in MATLAB to add salt & pepper noise in image I. where d is noise density parameter, the default value of which is 0.05. Salt & pepper noise in the experiments joins according to parameter d from 0 to 0.1 by 0.01.

(2)Parameters of the Gaussian noise

The mean and variance of Gaussian noise are constant, we uses the function imnoise(I, 'gaussian', M, V) in MATLAB to add Gaussian noise in image I. where M is the mean, V is the variance, usually the default value are 0 and 0.01. Gaussian noise in the experiment joins according to parameter M $=0$, and V from 0 to 0.01 by 0.001.

(3)Parameters of the speckle noise

Usually the speckle noise performances is multiplicative noise, we uses the function imnoise(I, 'speckle', V) in MATLAB to add speckle noise in image I, which using type (I+n*I) to add the noise in the image. Where n is evenly distributed random noise (mean value is 0, the variance is V). Speckle noise in the experiment joins according to parameter V from 0 to 0.1 by 0.01.

4.2   Experimental Results and Analysis

Different degree of salt & pepper noise, Gaussian noise and speckle noise are added to image respectively in the experiments. These noise images are input to the proposed retrieval system as the test sample. In order to facilitate the observation, the output is shown in the form of graphs, which can be seen in Fig.2, Fig.3 and Fig.4. Fig.2, Fig.3 and Fig.4 show the comparison retrieval results based on different retrieval models. It can be seen from the figures that the difference is very obvious. The analysis and summarize are as follows:

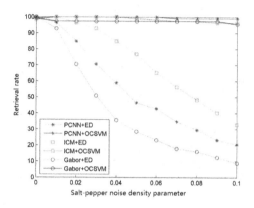

Fig. 2. Retrieval results of images influenced by salt-pepper noise

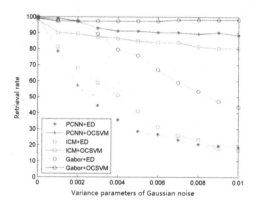

Fig. 3. Retrieval results of images influenced by Gaussian noise

(1) It can be seen overall by the three figures that OCSVM feature classification method is superior to the ED similarity measure method. Of which, PCNN+OCSVM, ICM+OCSVM, and Gabor+OCSVM method have good anti-noise performance to salt & pepper noise, while Gabor+OCSVM has a better anti-noise performance to Gaussian noise. As to speckle noise, the anti-noise performance of the ICM+OCSVM declined apparently.

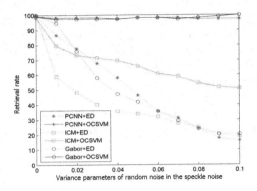

Fig. 4. Retrieval results of images influenced by speckle noise

(2) It is clearly from Fig.2 that the anti-noise performance to salt & pepper noise of PCNN+OCSVM and ICM+OCSVM are better than Gabor +OCSVM. And the anti-noise performance of the ICM+OCSVM declined slightly with the increase of the noise density, while the anti-noise performance of PCNN+OCSVM has been very stable. At the same time, the anti-noise performance of the three models based on the ED similarity measure by descending order: ICM+OCSVM, PCNN+OCSVM and Gabor+OCSVM.

(3) In Fig.3, as for the Gaussian noise, whether it is based on OCSVM or ED model, the Gabor filters has an outstanding performance, which indicated that the Gabor features has a good anti-noise performance of the Gaussian noise.

(4) In Fig.4, as to the speckle noise, PCNN+OCSVM and Gabor+OCSVM have better anti-noise performance than the ICM+OCSVM, and the anti-noise performance of the PCNN+ED and Gabor+ED is also superior to the ICM+ED, which indicated that the ICM has a poor anti-noise performance to the speckle noise.

(5) Based on the above points, we can draw a conclusion that the feature extraction techniques and feature matching techniques are the two key technologies for image retrieval. If there is only a good feature extraction technique without the right feature matching techniques, the final retrieval results will be seriously affected. Similarly, feature matching technology with a good generalization performance also based on an outstanding feature extraction technique. Therefore, for a good image retrieval system, it is necessary to take effective extraction technology, and also to have a superior performance feature matching techniques, so as to achieve the best anti-noise performance.

5 Conclusion

In this paper, we use the entropy sequences of the binary image which are output by the pulse-coupled neural network as the features to reflect the information of the original images. With global optimal solution and good generalization ability, One-class support vector machine can solve one class problems. Using one-class support vector machine to train and predict the features has higher feature matching accuracy than traditional distance matching method. The experimental results in Bro-datz texture image library show that the proposed anti-noise texture retrieval system based on pulse-coupled neural network and one-class support vector machine has good robustness and stability to noise. As the main symbol of the third generation of the artificial neural network, PCNN has good development potential and application prospect. It is good direction of de-velopment to make OCSVM combine with fuzzy theory and uncertainty theory, and there are many issues we need to study and explore in the future.

References

1. Zhang, D.S., Wong, A., Indrawan, M., et al.: Content-based image retrieval using Gabor texture features. In: IEEE Pacific-Rim Conference on Multimedia (PCM 2000), pp. 392–395 (2000)
2. Han, J., Ma, K.K.: Rotation-invariant and scale-invariant gabor features for texture image retrieval. Image and Vision Computing 25, 1474–1481 (2007)
3. Johnson, J.L.: Pulse-coupled neural nets: translation, rotation, scale, distortion and intensity signal invariance for images. Applied Optics 33(26), 6239–6253 (1994)
4. Zhang, J.W., Zhan, K., Ma, Y.D.: Rotation and scale invariant antinoise PCNN features for content-based image retrieval. Neural Network World 2, 121–132 (2007)
5. Ma, Y.D., Li, L., Zhan, K., et al.: Pulse coupled neural network and digital image processing. Science Press, Beijing (2008)
6. Kinser, J.M.: Simplified pulse-coupled neural network. In: Proceedings of SPIE, vol. 2760(3), pp. 563–569. SPIE Press, Orlando (1996)
7. Schölkopf, B., Platt, J.C., Shawe-Taylor, J., et al.: Estimating the support of a high-dimensional distribution. Neural Computation 13(7), 1443–1471 (2001)
8. Chen, Y.Q., Zhou, X.S., Huang, T.S.: One-class SVM for learning in image retrieval. IEEE Transactions on Image Processing 1, 34–37 (2001)
9. Wu, R.S., Chung, W.H.: Ensemble one-class support vector machines for content-based image retrieval. Expert Systems with Applications 36, 4451–4459 (2009)
10. Brodatz, P.: Textures: A photographic album for artists and designers. Dover Publications, New York (1996)
11. Chang, C.C., Lin, C.J.: Libsvm: A library for support vector machines. Department of Computer Science and Information Engineering, National TaiWan University (2006), http://www.csie.ntu.edu.tw/cjlin/libsvm
12. Hsu, C.W., Chang, C.C., Lin, C.-J.: A practical guide to support vector classification. Department of Computer Science and Information Engineering, National TaiWan University (2008), http://www.csie.ntu.edu.tw/cjlin/papers/guide/guide.pdf

A FPGA Spike-Based Robot Controlled with Neuro-inspired VITE[*]

Fernando Perez-Peña[1], Arturo Morgado-Estevez[1], Alejandro Linares-Barranco[2],
Angel Jiménez-Fernández[2], Juan Lopez-Coronado[3], and Jose Luis Muñoz-Lozano[3]

[1] Applied Robotics Research Lab, University of Cadiz, Spain
[2] Robotic and Technology of Computers Lab, University of Seville, Spain
[3] Automation and System Engineering Department,
University Polytechnics of Cartagena, Spain
fernandoperez.pena@uca.es

Abstract. This paper presents a spike-based control system applied to a fixed robotic platform. Our aim is to take a step forward to a future complete spikes processing architecture, from vision to direct motor actuation. This paper covers the processing and actuation layer over an anthropomorphic robot. In this way, the processing layer uses the neuro-inspired VITE algorithm, for reaching a target, based on PFM taking advantage of spike system information: its frequency. Thus, all the blocks of the system are based on spikes. Each layer is implemented within a FPGA board and spikes communication is codified under the AER protocol. The results show an accurate behavior of the robotic platform with 6-bit resolution for a 130° range per joint, and an automatic speed control of the algorithm. Up to 96 motor controllers could be integrated in the same FPGA, allowing the positioning and object grasping by more complex anthropomorphic robots.

Keywords: Spike systems, Motor control, VITE, Address Event Representation, Neuro-inspired, Neuromorphic engineering, Anthropomorphic robots.

1 Introduction

Movement generation is one of the most studied topics at science and engineering. The community known as neuro-engineers has a look into biological movement, which is supposed to have nearly the perfect behavior, with the aim to mimic the process [1]. The nervous system is the driver for movement generation in humans.

From the very beginning it is known that the nervous system uses spikes or action-potentials to carry the information across the organism [2]. The excellent behavior of those systems leads us to mimic them into electronic devices based on interconnected neuron systems; they are called neuromorphic systems. Therefore, the challenge of

---

[*] This work was supported by the Spanish grant (with support from the European Regional Development Fund) VULCANO (TEC2009-10639-C04-02) and BIOSENSE (TEC2012-37868-C04-02).

I. Rojas, G. Joya, and J. Cabestany (Eds.): IWANN 2013, Part I, LNCS 7902, pp. 299–308, 2013.

the neuroengineering community is to create architectures of neuromorphic chips with the same properties of human neural system: low power consumption, compact size and scalability.

In our aim of generating intended movements towards a target in a biological, neural way with electronic devices, we have to deal with several problems when implementing the spike processing blocks:

- How to consider the information: In these systems each neuron fires a spike when it reaches a specific threshold in a completely asynchronous way. There are several ways to encode these spikes; for example, the rate coded [3]: when the excitation is low, the spike rate is low and thus the time between spikes is high; however, when the signal excitation increases, the inter-spikes interval time (ISI) decreases, while the spike rate increases. Consequently, the information is codified as the firing rate or frequency.
- The way to implement this architecture: We have implemented it into a FPGA and apparently it is not an asynchronous system but the clock frequency of these digital systems is high enough to allow us to consider an asynchronous behavior for the neurons.
- The other problem is related to the manner of holding communication between different neuromorphic devices. Since neurons communicate in a point-to-point manner and it is possible to integrate several thousands of artificial neurons into the same electronic device (VLSI chip or FPGA), new communication strategies have been adopted, such as the Address-Event-Representation (AER) protocol [4]. AER maps each neuron with a fixed address which is transmitted through the interconnected neuron system.

In this way, with these three considerations, a neuromorphic chip is continuously sending information about its excitation level to the system [5]. Thus, connecting several of them with a parallel AER bus, all the information is available for real time processing. Just by adding chips to the bus, it is possible to enlarge the system. That is one of the most important reasons for using AER, i.e. the scalability allowed by parallel connections. Since each chip has an internal arbiter to access the AER bus [5], real time is limited by the digital clock.

Previous works show that the spikes paradigm in conjunction with AER technology is a suitable join. There are VLSI chips for sensors [5-6], extended systems like the spike-based PID motor controller [7], neuro-inspired robotics [8] and bio-inspired systems for processing, filtering or learning [9-10].

At previous works we can find an approach to a spike processing architecture but not entirely [11] and in [12] a complete one for real-time objects tracking with an AER retina.

Our motivation for the entire research in progress is to succeed in integrating the visual information from an AER retina to a bio-inspired robot by using just spikes for the whole process. That is, to set up a complete neuro-inspired architecture to generate intended movements.

In this paper we have developed the processing layer which generates the trajectory and the actuation layer that applies the commands to reach a target by the motors that mimic the biological muscles (Fig. 1). Both layers use the spikes processing blocks presented in [13]. The processing stage is implemented in a Spartan-6 board with a micro controller plug-in to send configuration parameters to the spike blocks. This layer uses the neuro-inpired VITE (Vector Integration To Endpoint) algorithm developed by Daniel Bullock and Stephen Grossberg [14], although reformulated into the spikes paradigm. The target position, at this moment, is fed manually and the speed of the movement produced can be adjusted by a signal called GO also implemented as spike streams [15].

The second layer is the actuation and it has been implemented in a Spartan-3 board [7]. It has two different parts, the control and the power stage. At the first one we adequate the signal (expanding the spikes) to feed the motors and the power stage transforms the signal to the motors. The motors are controlled with PFM taking advantage of the way we have chosen for codifying the information: the frequency. It operates in an open loop until we integrate the proprioceptive sensors to close the loop. This integration will be carried out with other algorithm also purposed by Daniel Bullock and Stephen Grossberg [14], the Factorization of Length and Tension (FLETE) algorithm.

The robotic platform used is a fixed stereo vision head with two arms with two degrees of freedom for vision sensors holding (Fig. 2).

In section two the first layer is presented: details of VITE algorithm transformed into spikes processing blocks and hardware details in the FPGA. Then in section three we describe the second layer and the advantages of using PFM modulation. In section four a block diagram shows and explains the real hardware used. Then, the characterization of the robotic platform and their limits will show the range of configurable parameters for the first layer. Finally the results of different movements are presented with the main conclusions.

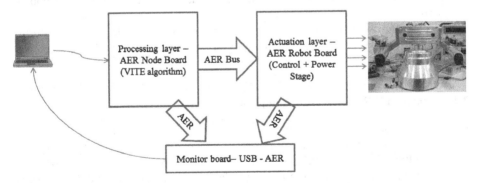

Fig. 1. Layer diagram for the system. The PC sends several configuration parameters to the processing layer across a microcontroller who communicates with the FPGA. Also, the PC sets the target, shoots the movement and receives the spikes to monitor the complete architecture.

2   Processing Layer

This layer is responsible for planning the movement. It receives the target position and generates a spike stream for the actuation layer. The VITE algorithm implemented with spikes ensures a synchronized movement of several joints in order to reach the target position. AER has been used for the communication with previous and next layers.

The hardware used consists of a Xilinx Spartan-6 FXT 1500 FPGA platform developed by RTC lab under the VULCANO project (called AER-node board), that allows high speed serial AER communications over RocketIO transceivers, and adaptation to particular scenarios through plug-in hardware PCBs connected on the top. It includes a plug-in with a USB microcontroller that communicates with the FPGA using SPI (Serial Peripheral Interface). This USB interface has been used for configuring the spike-based blocks of the VITE algorithm.

2.1   Spike-Based VITE Algorithm

The VITE algorithm [14] calculates a non-planned trajectory by computing the difference between the target and the present position. It also introduces the problem to deal with different frames of reference, one for the visual sensor, another one for the central processing (typically the head) and the last one for the actuator. It solves the problem by using the motor frame for all the system.

This algorithm introduced a non-specific control signal called GO. This signal allows separating the spatial pattern characteristics, such as distance and direction, from the energy of the movement. Thus, this manages the movement rate. This signal is introduced as a gate for the movement.

References [16] and [17] justify this algorithm. They show by means of electromyogram how the activity is present at any area of the motor cortex before the muscles initiate the movement. More specifically, the activity is present at the premotor cortex area.

The algorithm will be replicated for each motor present at the robotic platform. Consequently it is very important to analyze the consumption of hardware resources for the algorithm.

2.2   Hardware Resources Consumption

In general, in order to measure the hardware consumption in a FPGA, two points should be considered: the dedicated resources included to build up complex devices such as multipliers and the configurable logic blocks (CLBs) for general purposes.

The presented block does not use any multiplier or memory available. It just needs counters and simple arithmetic operations. Therefore the measurements are focused into the available slices at the FPGA.

The Spartan-6 FPGA present in the AER node board is the XC6SLX150T. It has two slices per CLB, reaching a total of 23,038 slices. We have implemented the system also in a Xilinx Virtex-5 prototyping board (XC5VFX30T FPGA) which has 5120 slices because this board was the first option for the whole architecture.

The VITE algorithm requires around 240 slices (VITE and AER bus interface) and 533 slices with a spikes-monitor block. Therefore, the AER node board is able to implement up to 95 and 43 spike-based VITE in parallel respectively in front of 21 and 9 for the Virtex-5 prototyping board.

The results obtained let us control complex robotic structures with up to 240 degrees of freedom just with one board.

3   Actuation Layer

This layer will adapt the spike-based input signal in order to feed the motors of the robot. It receives the AER output of the processing layer and adapts these addresses to produce the output frequency signal (PFM) for the corresponding motor.

We propose to use PFM to drive the motors becasuse it is intrinsically a spike-based solution almost identical to the solution that animals and humans use in their nervous systems for controlling the muscles.

If we make a comparison between the common used modulation PWM (Pulse Wide Modulation) and the PFM being proposed, we can find some advantages: a typical use of microcontrollers with PWM output generators limits the performance by the hardware timers and its bit resolution. But if PFM is used instead, the system frequency is only limited by the input signal frequency, and the duty-cycle would be limited by the motor driver (optical isolator and H bridges) which implies a low pass filter. Thereby, the use of microcontrollers implies resource sharing, which is not desirable for multi-motor controllers.

Moreover, the use of PFM instead of PWM considerably improves the power consumption when driving the motors because PFM, in average, will produce a lower commutation rate on the power stages. This is because PWM has a constant commutation rate while with PFM the commutation depends on the input of the system, thus it can be adjusted for low power.

Besides, there are more advantages of using PFM instead of PWM for motor control. Resource consumption is half for PFM than PWM when using spike-based controllers, and the power consumption is also much lower for PFM, as expressed in [7].

For the control, right now, it operates in an open loop until we integrate the proprioceptive robot sensor information to close the loop.

The hardware used to implement the actuation layer is a Spartan-3 family FPGA by Xilinx. The board also includes a power stage that consists of optical isolators and H bridges to feed up to four DC motors.

The board is called AER Robot [7].

4   Experimental Section

In this section we present the hardware scenario to develop the tests of the architecture designed, the characteristics that fix the functionality of the system and the techniques for the test carried out.

The boards, robotic platform and power supply are shown in Fig. 2.

To carry out the test, first of all it is necessary to characterize the architecture: the power stage limits the actuation layer, the dc motors used and the relation between the targets fixed and the movement of the robotic platform, that is, the resolution that can reach our system:

- The power stage uses an optical isolator that limits the frequency up to 48.8 KHz and the H bridge can reach this level. Thus this data give us the operation region.
- The DC motors need at least 15.4 μs of pulse width to start-up. They include an encoder with a resolution of 25K pulses/rev.
- Experimental findings with the saturation value (48.8 KHz) fixed as the input for the system, the global resolution can be calculated as:

$$\text{Resolution} = 65 \times 2^{10} / 2^{\text{NBITS} - 1} \text{ (degrees / generator step)} \tag{1}$$

Where the parameter NBITS is the number of bits selected to implement the spike generator that supply the target. For example, if we consider 16 bits the resolution will be 2.031 degrees / step.

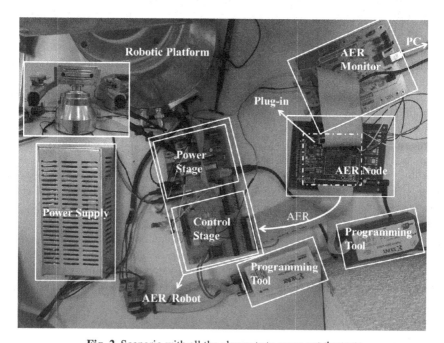

Fig. 2. Scenario with all the elements to carry out the tests

We have carried out several tests changing the speed profile and filling the complete movement range of the robotic platform. As it has been commented, the achieved results are with open loop control. Consequently, the signal sent to the motors is the position instead of the speed. The next section shows the results.

5   Results

The tests are restricted by the monitor board [18] and its maximum spikes firing rate, set to 5Mevps (Mega events per second).

Fig. 3, 4 and 5 show the results achieved. Both VITE behaviors are shown: spike-based on real application (solid lines) and non-spike-based simulation with MATLAB® (dotted lines). The speed of the movement is controlled by a slope profile signal called GO which multiplies the error inside the VITE algorithm [15]. This multiplication provides the speed at which it could provoke instability for the system.

We have plotted the results for one motor. The input signal is shared by them as the target to reach. In the graphs, the red line represents the input generated with the synthetic spikes generator, the purple line shows the speed profile and the green line is the output delivered to the motor. The higher the slope, the faster the fixed target is reached.

The figures show three different slope profiles that confirm the bell shape profiles predicted by [17].

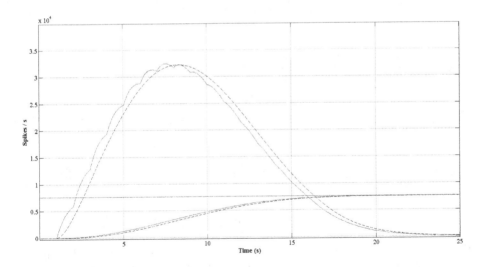

Fig. 3. Performance achieved corresponding to 1 % GO signal slope. The bell shape profile signals represent the speed. The ripple in the spike-base behavior is due to the function that transforms the spikes into a continuous signal. It takes a total of 17 seconds to reach the target if we look through the position.

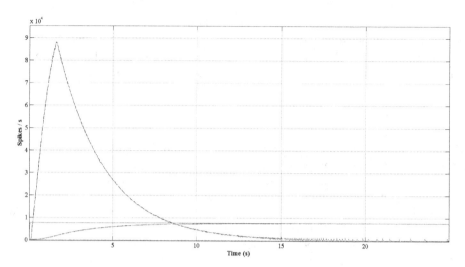

Fig. 4. Performance achieved corresponding to 10 % GO signal slope. The bell shape profile signals represent the speed. The ripple in the spike-base behavior is due to the function that transforms the spikes into a continuous signal. It takes a total of 11 seconds to reach the target if we look through the position.

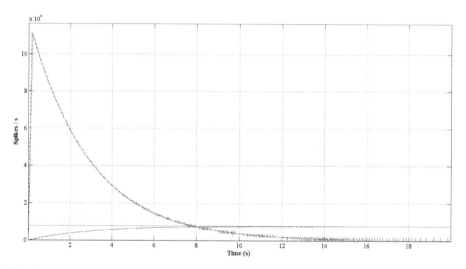

Fig. 5. Performance achieved corresponding to 100 % GO signal slope. The bell shape profile signals represent the speed. With this high slope, the ripple in the spike-base behavior is more significant than in the others. It takes a total of 9 seconds to reach the target if we look through the position.

6   Discussion and Conclusions

A complete spike processing architecture is proposed with excellent results in a fixed robotic platform. The bell shape profiles achieved with the spikes VITE algorithm implemented confirm the studies in [17] about the asymmetric speed profiles for higher speeds. The internal signal "GO" of the VITE algorithm is responsible for putting the movement on speed. Moreover, the signal has a temporal effect. Consequently, if its effect is not enough, the target could not be reached. But the temporal effect avoids instability.

The latency present in all results at the beginning can be interpreted as the previous activity detected at the premotor cortex in humans. In the hardware architecture, this latency is due to the counter that generates the slope profile signal. It can also be understood as the motor priming.

It is possible to reach a high resolution (0.25 degrees by step of the generator with 19 bits) but it always depends on the robot used.

We have fed the motors with the position according to an open loop control and the results fit with the expected behavior. If we change the robotic platform or include some control, it is possible to use the speed profile (available due to an integrator at the final processing blocks chain of the algorithm) for the motors. Nevertheless, the next step is to close the loop, although from a biological point of view, which means using the proprioceptive sensors of length and tension at the joints of the robot. The algorithm FLETE considers both sensors called neurotendinous spindle and muscles spindles and also the gamma neurons. The last step will be to include the feedback of the retina for fine tuning and passive movements updates.

References

1. Bullock, D., Grossberg, S.: Neural dynamics of planned arm movements: Emergent invariants and speed-accuracy properties during trajectory formation. Psychological Review 95, 49–90 (1988)
2. Sherrington, C.E.: Notes and Records of the Royal Society of London 30(1), 45–63 (1975)
3. Linares-Barranco, A., Jimenez-Moreno, G., Linares-Barranco, B., Civit-Balcells, A.: On algorithmic rate-coded AER generation. IEEE Transactions on Neural Networks 17(3), 771–788 (2006)
4. Sivilotti, M.: Wiring Considerations in Analog VLSI Systems with Application to Field-Programmable Networks, Ph.D. Thesis, California Institute of Technology, Pasadena CA (1991)
5. Lichtsteiner, P., Posch, C., Delbruck, T.: A 128×128 120 dB 15 μs latency asynchronous temporal contrast vision sensor. IEEE J. Solid-State Circuits 43, 566–576 (2008)
6. Chan, V., Liu, S.C., van Schaik, A.: AER EAR: A matched silicon cochlea pair with address event representation interface. IEEE Trans. Circuits Syst. 54, 48–59 (2007)
7. Jimenez-Fernandez, A., Jimenez-Moreno, G., Linares-Barranco, A., Dominguez-Morales, M., Paz-Vicente, R., Civit-Balcells, A.: A Neuro-Inspired Spike-Based PID Motor Controller for Multi-Motor Robots with Low Cost FPGAs. Sensors 12(4), 3831–3856 (2012)

8. Linares-Barranco, A., Gomez-Rodriguez, F., Jimenez-Fernandez, A., Delbruck, T., Lichtensteiner, P.: Using FPGA for visuo-motor control with a silicon retina and a humanoid robot. In: Proceedings of ISCAS 2007 IEEE International Symposium on Circuits and Systems, New Orleans, LA, USA, pp. 1192–1195 (2007)

9. Barranco, F., Diaz, J., Ros, E., del Pino, B.: Visual system based on artificial retina for motion detection. IEEE Trans. Syst. Man Cybern. Part B: Cybern. 39, 752–762 (2009)

10. Indiveri, G., Chicca, E., Douglas, R.: A VLSI array of low-power spiking neurons and bistable synapses with spike-timing dependent plasticity. IEEE Trans. Neural Netw. 17, 211–221 (2006)

11. Linares-Barranco, A., Paz-Vicente, R., Jimenez, G., Pedreno-Molina, J.L., Molina-Vilaplana, J., Lopez-Coronado, J.: AER neuro-inspired interface to anthropomorphic robotic hand. In: Proceedings of International Joint Conference on Neural Networks, Vancouver, Canada, pp. 1497–1504 (2006)

12. Gómez-Rodríguez, F., Miró-Amarante, L., Rivas, M., Jimenez, G., Diaz-del-Rio, F.: Neuromorphic Real-Time Objects Tracking using Address Event Representation and Silicon Retina. In: Cabestany, J., Rojas, I., Joya, G., et al. (eds.) IWANN 2011, Part I. LNCS, vol. 6691, pp. 133–140. Springer, Heidelberg (2011)

13. Jimenez-Fernandez, A., Domínguez-Morales, M., Cerezuela-Escudero, E., Paz-Vicente, R., Linares-Barranco, A., Jimenez, G.: Simulating building blocks for spikes signals processing. In: Cabestany, J., Rojas, I., Joya, G. (eds.) IWANN 2011, Part II. LNCS, vol. 6692, pp. 548–556. Springer, Heidelberg (2011)

14. Bullock, D., Grossberg, S.: The VITE model: A neural command circuit for generating arm and articulator trajectories. In: Kelso, J.A.S., Mandell, A.J., Shlesinger, M.F. (eds.) Dynamic Patterns in Complex Systems, pp. 305–326. World Scientific Publishers, Singapore (1988)

15. Perez-Peña, F., Morgado-Estevez, A., Linares-Barranco, A., Jimenez-Fernandez, A., Lopez-Coronado, J., Muñoz-Lozano, J.L.: Towards AER VITE: building spike gate signal. In: 19th IEEE International Conference on Electronics, Circuits, and Systems, Seville, Spain, pp. 881–884 (2012)

16. Georgopoulos, A.P.: Neural integration of movement: role of motor cortex in reaching. The FASEB Journal 2(13), 2849–2857 (1988)

17. Nagasaki, H.: Asymmetric velocity and acceleration profiles of human arm movements. Experimental Brain Research 74(2), 319–326 (1989)

18. Berner, R., Delbruck, T., Civit-Balcells, A., et al.: A 5 Meps $100 USB2.0 Address-Event Monitor-Sequencer Interface. In: IEEE International Symposium on Circuits and Systems, ISCAS, New Orleans, LA, pp. 2451–2454 (2007)

A Cognitive Approach for Robots' Autonomous Learning

Dominik M. Ramík, Kurosh Madani, and Christophe Sabourin

Signals, Images, and Intelligent Systems Laboratory (LISSI / EA 3956),
University Paris-Est Creteil, Senart-FB Institute of Technology, 36-37 rue Charpak,
77127 Lieusaint, France
{dominik.ramik,madani,sabourin}@u-pec.fr

Abstract. In this work we contribute to development of a real-time intelligent system allowing to discover and to learn autonomously new knowledge about the surrounding world by semantically interacting with human. The learning is accomplished by observation and by interaction with a human tutor. We provide experimental results as well using simulated environment as implementing the approach on a humanoid robot in a real-world environment including every-day objects. We show, that our approach allows a humanoid robot to learn without negative input and from small number of samples.

Keywords: intelligent system, visual saliency, autonomous learning, learning by interaction.

1   Introduction

In recent years, there has been a substantial progress in robotic systems able to robustly recognize objects in real world using a large database of pre-collected knowledge (see [1] for a notable example). There has been, however, comparatively less advance in autonomous acquisition of such knowledge. In fact, if a robot is required to share the living space with its human counterparts, to learn and to reason about it in "human terms", it has to face at least two important challenges. One is the vast number of objects and situations, the robot may encounter in the real world. The other one comes from humans' richness concerning various ways they use to address those objects or situations using natural language. Moreover, the way we perceive the world and speak about it is strongly culturally dependent ([2] and [3]). A robot supposed to defeat those challenges, cannot rely solely on a priori knowledge given to it by a human expert. It should be able to learn on-line and by interaction with the people it encounters in its environment ([4] for a survey on human-robot interaction and learning and [5] for an overview of the problem of anchoring). This will inherently require that the robot has the ability of learning without an explicit negative evidence or "negative training set" and from a relatively small number of samples. This important capacity is observed in children learning the language [6] and has been addressed on different degrees in various works. For example, in [7] a computational model of word-meaning by interaction is presented. [8] presents a computational model for describing simple objects. In [9] and [10], a humanoid robot is taught to associate simple shapes to human lexicon or grasp different objects. More advanced works on robots' autonomous learning and dialog are given by [11] and [12].

I. Rojas, G. Joya, and J. Cabestany (Eds.): IWANN 2013, Part I, LNCS 7902, pp. 309–320, 2013.

In this paper, we describe an intelligent system, allowing robots (as for example humanoid robots) to learn and to interpret the world, in which it evolves, using appropriate terms from human language, while not making use of a priori knowledge. This is done by word-meaning anchoring based on learning by observation and by interaction with its human tutor. Our model is closely inspired by learning behaviour of human infants ([13] or [14]). The goal of this system is to anchor the heard terms to its sensory-motor experience and to flexibly acquire knowledge about the world.

In this Section 2, we detail our approach by outlining its architecture and principles, we explain how beliefs about the world are generated and evaluated by the system and we describe the role of human-robot interaction in the learning process. Validation of the presented system on colors' learning and interpretation, using simulation facilities, is reported in Section 3. Section 4 focuses the validation of the proposed approach on a real robot in real world. Finally Section 5 discusses the achieved results and outlines the future work.

2   Interpretation and Knowledge Acquisition from Observation

The problem of learning brings an inherent problem of distinguishing the pertinent sensory information and the impertinent one. The solution to this task is not obvious. This is illustrated on Fig. 1. If a tutor points to one object (e.g. a toy-frog) among many others, and describes it as "green", the robot still has to distinguish, which of the detected colors and shades of the object the human is referring to. To achieve correct anchoring, we adopt the following strategy. The robot extracts features from important objects found in the scene along with the words the tutor used to describe the objects. Then, the robot generates its beliefs about which word could describe which feature. The beliefs are used as organisms in a genetic algorithm. To calculate the fitness, a classifier is trained and used to interpret the objects the robot has already seen. The utterances pronounced by the human tutor for of each object are compared with those the robot would use to describe it based on its current belief. The closer the robot's description is to that given by the human, the higher the fitness is. Once the evolution has been finished, the belief with the highest fitness is adopted by the robot and is used to interpret occurrences of new (unseen) objects. Fig. 2 depicts important parts of the proposed system.

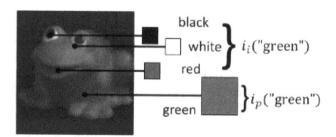

Fig. 1. A human would describe this toy-frog as green in spite of the fact, that this is not the only visible color

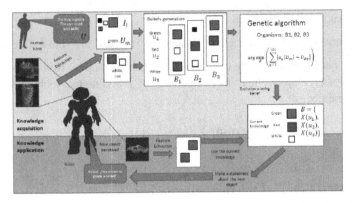

Fig. 2. Example of the system's operation in the case of autonomous learning of colors

2.1   From Observation to Interpretation

Let us suppose a robot equipped by a sensor observing the surrounding world. The world is represented as a set of features $I = \{i_1, i_2, \cdots, i_k\}$, which can be acquired by this sensor. Each time the robot makes an observation o, a human tutor gives it a set of utterances U_H describing the found important (e.g. salient) objects. Let us denote the set of all utterances ever given about the world as U. The observation o is defined as an ordered pair $o = \{I_l, U_H\}$, where $I_l \subseteq I$, expressed by (1), stands for the set of features obtained from observation and $U_H \subseteq U$ is a set of utterances (describing o) given in the context of that observation. i_p denotes the pertinent information for a given u (i.e. features that can be described semantically as u in the language used for communication between the human and the robot), i_i the impertinent information (i.e. features that are not described by the given u, but might be described by another $u_i \in U$) and sensor noise $\mathcal{E}$. The goal is to distinguish the pertinent information from the impertinent one and to correctly map the utterances to appropriate perceived stimuli (features). Let us define an interpretation $X(u) = \{u, I_j\}$ of an utterance u as an ordered pair where $I_j \subseteq I$ is a set of features from I. So, the belief B is defined accordingly to (2) as an ordered set of $X(u)$ interpreting utterances u from U.

$$I_l = \bigcup_{U_H} i_p(u) + \bigcup_{U_H} i_i(u) + \mathcal{E} \qquad (1)$$

$$B = \{X(u_1), \cdots, X(u_n)\} \text{ with } n = |U| \qquad (2)$$

Accordingly to the criterion expressed by (3), one can calculate the belief B, which interprets in the most coherent way the observations made so far: in other words, by looking for such a belief, which minimizes across all the observations $o_q \in O$ the difference between the utterances U_{Hq} made by human, and those utterances U_{Bq}, made by the system by using the belief B. Thus, B is a mapping from the set U to I: all members of U map to one or more members of I and no two members of U map to the same member of I.

$$\arg\min_{B} \left(\sum_{q=1}^{|O|} |U_{Hq} - U_{Bq}| \right) \tag{3}$$

2.2   The Most Coherent Interpretation Search

The system has to look for a belief B, which would make the robot describing a particular scene with utterances as close and as coherent as possible to those made by a human on the same scene. For this purpose, instead performing the exhaustive search over all possible beliefs, we propose to search for a suboptimal belief by means of a genetic algorithm. For doing that, we assume that each organism within it has its genome constituted by a belief, which, results into genomes of equal size $|U|$ containing interpretations $X(u)$ of all utterances from U. The task of coherent belief generation is to generate beliefs, which are coherent with the observed reality.

In our genetic algorithm, the genomes' generation is a belief generation process generating genomes (e.g. beliefs) as follows. For each interpretation $X(u)$ the process explores whole the set O. For each observation $o_q \in O$, if $u \in U_{Hq}$ then features $i_q \in I_q$ (with $I_q \subseteq I$) are extracted. As described in (1), the extracted set of features contains as well pertinent as impertinent features. The coherent belief generation is done by deciding, which features $i_q \in I_q$ may possibly be the pertinent ones. The decision is driven by two principles. The first one is the principle of "proximity", stating that any feature i is more likely to be selected as pertinent in the context of u, if its distance to other already selected features is comparatively small. The second principle is the "coherence" with all the observations in O. This means, that any observation $o_q \in O$, corresponding to $u \in U_{Hq}$, has to have at least one feature assigned into I_q of the current $X(u) = \{u, I_q\}$.

To evaluate a given organism, a classifier is trained, whose classes are the utterances from U and the training data for each class $u \in U$ are those corresponding to $X(u) = \{u, I_q\}$, i.e. the features associated with the given u in the genome. This classifier is used through whole set O of observations, classifying utterances $u \in U$ describing each $o_q \in O$ accordingly to its extracted features. Such a classification results in the set of utterances U_{Bq} (meaning that a belief B is tested regarding the qth observation).

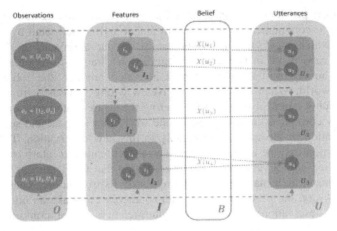

Fig. 3. Relations between observations, features, beliefs and the most coherent utterances

On Fig. 3 gives an alternative view on the previously defined notions and their relationships. It depicts an example where three observations O_1, O_2, and O_3 are made. Features i_1 and i_2 were extracted along with utterances u_1 and u_2 likewise for the second and the third observation. Accordingly to the above-defined notions, the entire set of features is $I = \{i_1, i_2, i_3, i_4, i_5, i_6\}$, while sub-sets I_1, I_2 and I_3 refer to features corresponding to each particular observation. Similarly the set of utterances $U_H = \{u_1, u_2, u_3, u_4, u_5, u_6\}$ is the set of all utterances made by human and the sub-sets U_1, U_2, and U_3 refer to U_{Hq} with $q \in \{1,2,3\}$. In this view an interpretation $X(u_1)$ is a relation of u_1 with the subset of features I_1 from I. Then a belief B is a mapping (relation) from the set of U to I. All members of U map to one or more members of I and no two members of U map to the same member of I.

The fitness function evaluating the fitness of each above-mentioned organism is defined as "disparity" between U_{Bq} and U_{Hq} (defined in previous subsection) which is computed accordingly to the equation (4), where v is the number of utterances that are not present in both sets U_{Bq} and U_{Hq} (e.g. either missed or are superfluous utterances interpreting the given features). The globally best fitting organism is chosen as the belief that best explains observations O made (by robot).

$$D(v) = \frac{1}{1+v} \quad \text{with} \quad v = \left| U_{Hq} \bigcup U_{Bq} \right| - \left| U_{Hq} \bigcap U_{Bq} \right| \tag{4}$$

2.3 Role of Human-Robot Interaction

Human beings learn both by observation and by interaction with the world and with other human beings. The former is captured in our system in the "best interpretation search" outlined previous subsections. The latter type of learning requires that the robot be able to communicate with its environment and is facilitated by learning by observation, which may serve as its bootstrap. In our approach, the learning by interaction is carried out in two kinds of interactions: human-to-robot and robot-to-human. The human-to-robot interaction is activated anytime the robot interprets wrongly the world. When the human receives a wrong response (from robot), he provides the robot a new observation by uttering the desired interpretation. The robot takes this new corrective knowledge about the world into account and searches for a new interpretation of the world conformably to this new observation. The robot-to-human interaction may be activated when the robot attempts to interpret a particular feature classified with a very low confidence: a sign that this feature is a borderline example. In this case, it may be beneficial to clarify its true nature. Thus, led by the epistemic curiosity, the robot asks its human counterpart to make an utterance about the uncertain observation. If the robot's interpretation is not conforming to the utterance given by the human (robot's interpretation was wrong), this observation is recorded as a new knowledge and a search for the new interpretation is started.

3   Simulation Based Validation

The simulation based validation finds its pertinence in assessment of the investigated cognitive-system's performances. In fact, due to difficulties inherent to organization of strictly same experimental protocols on different real robots and within various realistic contexts, the simulated validation becomes an appealing way to ensure that the protocol remains the same. For simulation based evaluation of the behaviour of the above-described system, we have considered color names learning problem. In everyday dialogs, people tend to describe objects, which they see, with only a few color terms (usually only one or two), although the objects in itself contains many more colors. Also different people can have slightly different preferences on what names to use for which color. Due to this, learning color names is a difficult task and it is a relevant sample problem to test our system.

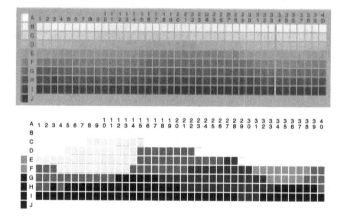

Fig. 4. Original WCS table (upper image), its system's made interpretation (lower image)

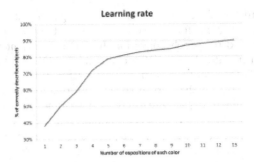

Fig. 5. Evolution of number of correctly described objects with increasing number of exposures of each color to the simulated robot

In the simulated environment, images of real-world objects were presented to the system alongside with textual tags describing colors present on each object. The images were taken from the Columbia Object Image Library (COIL: contains 1000 color images of different views of 100 objects) database. Five fluent English speakers were asked to describe each object in terms of colors. We restricted the choice of colors to "Black", "Gray", "White", "Red", "Green", "Blue" and "Yellow", based on the color opponent process theory [15]. The tagging of the entire set of images was highly coherent across the subjects. In each run of the experiment, we have randomly chosen a tagged set. The utterances were given in the form of text extracted from the descriptions. The object was accepted as correctly interpreted if the system's and the human's interpretations were equal. The rate of correctly described objects from the test set was approximately 91% with the robot fully learned. Fig. 4 gives the result of interpretation by the system of the colors of the WCS table. Fig. 5 shows the learning rate versus the increasing number of exposures of each color.

4 Implementation and Validation on Real Robot

Although the usage of the presented system is not specifically bound to humanoid robots, it is pertinent to state two main reasons why a humanoid robot is used for the system's validation. The first reason for this is that from the definition of the term "humanoid", a humanoid robot is aspired to make its perception close to the human's one, entailing a more human-like experience of the world. This is an important aspect to be considered in context of sharing knowledge between a human and a robot. Some aspects of this problem are discussed in [16]. The second reason is that humanoid robots are specifically designed to interact with humans in a "natural" way (fr example by using a loudspeaker and microphone set) in order to allow for bi-directional communication with human by speech synthesis and speech analysis and recognition. This is of importance when speaking about "natural human-robot interaction" during learning.

The designed system has been implemented on NAO robot (from Aldebaran Robotics). It is a small humanoid robot which provides a number of facilities such as onboard camera (vision), communication devices and onboard speech generator. The fact that the above-mentioned facilities been already available offers a huge save of time, even if those faculties remain quite basic in that kind of robots. If NAO robot integrates an onboard speech-recognition algorithm (e.g. some kind of speech-to-text converter) which is sufficient for "hearing" the tutor, however its onboard speech generator is a basic text-to-speech converter. It is not sufficient to allow the tutor addressing the robot in natural speech. To overcome NAO's limitations relating this purpose, the TreeTagger tool[1] was used in combination with robot's speech-recognition system to obtain the part-of-speech information from situated dialogs. Standard English grammar rules were used to determine whether the sentence is demonstrative (e.g. for example: "This is an apple."), descriptive (e.g. for example: "The apple is red.") or an order (e.g. for example: "Describe this thing!"). To communicate with the tutor, the robot used its text-to-speech engine.

The total of 25 every-day objects was collected for purposes of the experiment (Fig.6-a). They have been randomly divided into two sets for training and for testing. The learning set objects were placed around the robot and then a human tutor pointed to each of them calling it by its name. Using its 640x480 monocular color camera, the robot discovered and learned the objects around it by the salient object detection approach we have described earlier (please see [17] and [18]). Here, this approach has been extended by detecting the movement of the tutor's hand to achieve joint attention. In this way, the robot was able to determine what object the tutor is referring to and to learn its name. Fig. 6 shows the experimental setup as well as two photographs of the above-reported experimental validation.

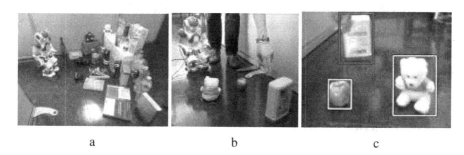

a   b   c

Fig. 6. Experimental setup showing collected set of objects (a), tutor pointing a yellow chocolate box which has been seen, interpreted and learned (by the robot) in term of colors then asking the robot to describe the chosen object (b) and the ground truth detected objects as the robot perceives them (c)

4.1   First Experiment involving the Humanoid Robot

In this first experiment, the robot has been asked to learn a subset among the 25 considered objects: in term of associating the name of each detected object to that object. At the same time, a second learning has been performed involving the interaction with the tutor who has successively pointed the above-learned objects describing (e.g. telling) to the robot the color of each object. Here-bellow an example of the Human-Robot interactive learning is reported:

---

[1] Developed by the ICL at University of Stuttgart, available online at:
http://www.ims.uni-stuttgart.de/projekte/corplex/TreeTagger

- **Human** [*pointing a red aid-kit*]: "This is a first-aid-kit!"
- **Robot:** "I will remember that this is a first-aid-kit."
- **Human:** "It is red and white".
- **Robot:** "OK, the first-aid-kit is red and the white."

After learning the names and colors of the discovered objects, the robot is asked to describe a number of objects including as well some of already learned objects but in different posture (for example the yellow box presented in reverse posture) as a number of still unseen objects (as for example a red apple or a white teddy-bear). The robot has successfully described, in a coherent linguistics, the presented seen and unseen objects. Here-bellow is the Human-Robot interaction during the experiment:

- **Human** [*pointing the unseen white teddy-bear*]: "Describe this!"
- **Robot:** "It is white!"
- **Human** [*pointing the already seen, but reversed, yellow box*]: "Describe this!"
- **Robot:** "It is yellow!"
- **Human** [*pointing the unseen apple*]: "Describe this!"
- **Robot:** "It is red!"

4.2   Second Experiment Involving the Humanoid Robot

In another experiment, after having learned objects (among which a black book: the only available book in the learning set), the robot has been asked to search for the "book" in a new (unseen) environment where a number of additional objects, among which two new books and other objects looking closely to a book, have been placed. The robot's behavior and Human-Robot interaction during this experiment is given here-bellow:

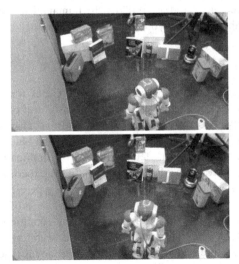

Fig. 7. Images from a video sequence showing the robot searching for the book (1st row) and localizing several of them (2nd row). The right column shows robot's camera view and visualization of color interpretation of the searched object.

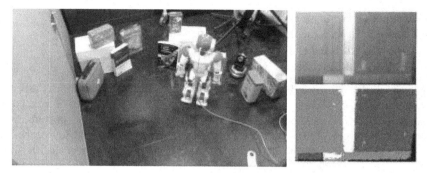

Fig. 8. Images from a video sequence showing the robot finding the required "red" book (left-side picture) and how the robot perceives that required "red" book (right-side picture)

- **Human**: "Go find the book!"
- **Robot**: "OK, I am searching for the book!"
- **Robot** [*after visually exploring its new surroundings*]: "I can see three books!"
- **Human**: "Get the red one!"
- **Robot** [*after turning toward the "red and grey" book and after walking and reaching the aforementioned book's location*] "I am near the red book!"

Fig. 7 and Fig. 8 show photographs of the robot during the above-described experiment. Beside the correct recognition of the desired object (asked by the tutor), what is pertinent to note is that robot found also two other unseen books. What is also greatly interesting is that even if there is no "red" book, the robot has correctly interpreted the fact that the red book required by the human was the "red and grey" book: the only book may coherently be considered as "red" by the human. The video of the experimental validation is available on: http://youtu.be/W5FD6zXihOo.

4.3   Robustness Regarding Illumination Conditions

Robustness regarding ambient conditions' changes is one of the key issues for any system supposed to be deployed in real environment. We have been insightful on the subject of this foremost issue, especially concerning illumination conditions' changes. (e.g. luminance scarcity). Fig. 9 shows the robot perceiving an object (here a blue box of milk) shown to it by the human tutor with two different settings of illumination. The upper pictures of this figure show the robot inside the room where the experiment has been performed and the lower pictures of the same figure show robot's perception of the beheld object. The illumination in the left-side pictures is an artificial illumination from ceiling. It is clearly visible that this illumination condition is causing reflections on floor and glossy objects like the one the robot is searching for (a blue box of milk). The right-side picture shows the robot in same environment illuminated by natural ambient light, while during a cloudy day.

Apart of the above-mentioned scrounging reflections, the robot's camera was obviously having difficulties with white balance for this particular color brightness.

This is alternating the color balance, rendering whole the image "yellowish". Both effects combined make this illumination particularly challenging. On the other hand, concerning right-side pictures, due to cloudy weather the amount of light coming to the room was insufficient and the robot's camera was producing images with significantly more noise and with an accentuated "bluish" tint. Although the system has been tested several times in such greatly varying conditions of illumination, no visible impact on the behavior of the system itself has been observed and the robot has been fully able to pursue its normal cycle of operation.

Fig. 9. Robot in the same room searching for the same object in two different illumination conditions: direct artificial illumination (left-side) and. natural ambient light (right-side)

5   Conclusion and Further Work

In this paper, we have detailed and validated an intelligent system for high-level knowledge acquisition from visual patterns. The presented system allow to learn in an autonomous manner new knowledge about the surrounding world and to complete (enrich or correct) it by interacting with a human. Experimental results, performed as well on a simulation platform as using the NAO robot show the pertinence of the investigated concepts.

Several appealing perspectives are pursuing to push further the presented work. Regarding the versatility of the presented concepts, the short-term perspective will focus on integration of the investigated concepts on other kinds of robots, such as mobile robots. Also, current implemented version allows the robot to work with a single category or property at a time (e.g. for example the color in utterances like "it is red"). We are working on extending its ability to allow the learning of multiple categories at the same time and to distinguish which of the used words are related to which category. While, concerning the middle-term perspectives of this work, they will focus aspects reinforcing the autonomy of such cognitive robots. The ambition here is integration of the designed system to a system of larger capabilities realizing multi-sensor artificial machine-intelligence. There, it will play the role of an underlying part for machine cognition and knowledge acquisition.

References

1. Meger, D., Forssén, P.E., Lai, K., Helmer, S., McCann, S., Southey, T., Baumann, M., Little, J.J., Lowe, D.G.: Curious George: An attentive semantic robot. Robot. Auton. Syst. 56(6), 503–511 (2008)
2. Kay, P., Berlin, B., Merrifield, W.: Biocultural Implications of Systems of Color Naming. Journal of Linguistic Anthropology 1(1), 12–25 (1991)
3. Bowerman, M.: How Do Children Avoid Constructing an Overly General Grammar in the Absence of Feedback about What is Not a Sentence? Papers and Reports on Child Language Development (1983)
4. Goodrich, M.A., Schultz, A.C.: Human-robot interaction: a survey. Found. Trends Hum.-Comput. Interact. 1(3), 203–275 (2007)
5. Coradeschi, S., Saffiotti, A.: An introduction to the anchoring problem. Robotics & Autonomous Sys. 43, 85–96 (2003)
6. Regier, T.: A Model of the Human Capacity for Categorizing Spatial Relations. Cognitive Linguistics 6(1), 63–88 (1995)
7. Wellens, P., Loetzsch, M., Steels, L.: Flexible word meaning in embodied agents. Connection Science 20(2-3), 173–191 (2008)
8. de Greeff, J., Delaunay, F., Belpaeme, T.: Human-robot interaction in concept acquisition: a computational model. In: Proc. of Int. Conf. on Develop. & Learning, pp. 1–6 (2009)
9. Saunders, J., Nehaniv, C.L., Lyon, C.: Robot learning of lexical semantics from sensorimotor interaction and the unrestricted speech of human tutors. pp. 1–8
10. Lütkebohle, I., Peltason, J., Schillingmann, L., Wrede, B., Wachsmuth, S., Elbrechter, C., Haschke, R.: The curious robot - structuring interactive robot learning, pp. 2154–2160 (2009)
11. Araki, T., Nakamura, T., Nagai, T., Funakoshi, K., Nakano, M., Iwahashi, N.: Autonomous acquisition of multimodal information for online object concept formation by robots. In: Proc. of IEEE/ IROS, pp. 1540–1547 (2011)
12. Skocaj, D., Kristan, M., Vrecko, A., Mahnic, M., Janicek, G.-J., Kruijff, M., Hanheide, M., Hawes, N., Keller, T., Zillich, M., Zhou, K.: A system for interactive learning in dialogue with a tutor. In: Proc. IEEE/ IROS, pp. 3387–3394 (2011)
13. Yu, C.: The emergence of links between lexical acquisition and object categorization: a computational study. Connection Science 17(3-4), 381–397 (2005)
14. Waxman, S.R., Gelman, S.A.: Early word-learning entails reference, not merely associations. Trends in Cognitive Science (2009)
15. Schindler, M., von Goethe, J.W.: Goethe's theory of colour applied by Maria Schindler. New Knowledge Books, East Grinstead, Eng. (1964)
16. Klingspor, V., Demiris, J., Kaiser, M.: Human-Robot-Communication and Machine Learning. In: Applied Artificial Intelligence, pp. 719–746 (1997)
17. Ramík, D.M., Sabourin, C., Madani, K.: A Real-time Robot Vision Approach Combining Visual Saliency and Unsupervised Learning. In: Proc. of 14th Int. Conf. on Climbing & Walking Robots & the Support Technologies for Mobile Machines, Paris, pp. 241–248 (2011)
18. Ramík, D.M., Sabourin, C., Madani, K.: Hybrid Salient Object Extraction Approach with Automatic Estimation of Visual Attention Scale. In: Proc. of Seventh Int. Conf. on Signal Image Technology & Internet-Based Systems, Dijon, pp. 438–445 (2011)

Self-Organizing Incremental Neural Network (SOINN) as a Mechanism for Motor Babbling and Sensory-Motor Learning in Developmental Robotics

Tarek Najjar and Osamu Hasegawa

Tokyo Institute of Technology,
Dept. of Computational Intelligence and Systems Science,
4259 Nagatsuta-cho, Midori-ku, Yokohama, 226-8503, Japan
http://www.titech.ac.jp/english/

Abstract. Learning how to control arm joints for goal-directed reaching tasks is one of the earliest skills that need to be acquired by Developmental Robotics in order to scaffold into tasks of higher Intelligence. Motor Babbling seems as a promising approach toward the generation of internal models and control policies for robotic arms. In this paper we propose a mechanism for learning sensory-motor associations using layered arrangement of Self-Organizing Neural Network (SOINN) and joint-egocentric representations. The robot starts off by random exploratory motion, then it gradually shift into more coordinated, goal-directed actions based on the measure of error-change. The main contribution of this research is in the proposition of a novel architecture for online sensory-motor learning using SOINN networks without the need to provide the system with a kinematic model or a preprogrammed joint control scheme. The viability of the proposed mechanism is demonstrated using a simulated planar robotic arm.

Keywords: Developmental Robotics, SOINN, Self-organizing Neural Network, Motor Babbling, Sensory-Motor Learning, Incremental Learning.

1 Introduction

Inspired by Both Developmental Psychology and Cognitive neuroscience, developmental robotics has gained considerable interest among roboticists recently, [1]. The basic concern in this discipline is to formulate embodied Artificial Agents that are capable of autonomous mental development[2],which is the ability of the agent to adapt and grow mentally in the way it perceive, represent and process its experiences and the way it acts in the world around . This development must take place through interaction with the environment, using the agent's sensors and actuators, in a continuous life-long and open-ended manner[3].

Evidence from developmental psychology literature[4][5] suggests the presence of exploratory learning processes in the behavior of infants during the first

I. Rojas, G. Joya, and J. Cabestany (Eds.): IWANN 2013, Part I, LNCS 7902, pp. 321–330, 2013.

months of motor-ability development. During the repetitive random motion of the arm, that is considered as a characteristic pattern of infant motor behavior, babies are believed to keep their hand constantly in visual field, which is supposed to serve the goal of building internal associations between actions and consequences in one's own body [6]. So Motor Babbling is described as the exploratory learning process of generating sensory-motor associations through continues random motions with ballistic trajectories. These motions serve the purpose of sampling representative data points that bootstrap the learning system into incremental generation of internal model and implicit control policy for the system at hand.

Many roboticists have attempted to mimic this developmental process using robotic platforms. An example is found in the work of the group[7], here a gradient descent method is used in order to enable the system to learn some of the unprovided elements of the system's kinematic model where the rest of the elements were already provided and preprogrammed. A more efficient approach than gradient descent was taken by group in[8] where the system starts off by a population of candidate possible models then, and through interaction with environment, the system evolve in approximating a more accurate model that represents the system in hand. Beside the explicit dependency, in this system, on artificial visual tags that are attached to segments of the robotic arm , this approach make use of Bayesian learning and Gaussian regression, the mechanism actually is very expensive on the computational side. A rather different approach was taken by the group[9]. Here a camera calibration based method were adopted together with open loop mechanism for generation of an implicit body schema model, this system made use of look-up table learning mechanism which naturally requires longer time for learning. The research group in[10]used a more biologically inspired approach by incorporating concepts like population code and equilibrium-point hypothesis in order to enable the system to achieve reaching tasks. In a different approach[11]the research group used both Bayesian belief functions and social learning mechanisms to facilitate learning-by-imitation competence. This approach actually made use of hard-wired motor primitives that were encoded manually into the robot. A Reinforcement learning approach together with imitation methods using locally weighted regression was facilitated by[12]where a robot was taught specific motor primitives, that are specific to given task sittings, then the robot generated policies that enable it to learn those primitives in an episodic manner . Although the robot managed to perform the given tasks but it seemed like the system was kind of a task-specific oriented in the way it learned each motor primitive.

2   Methodologies

The mechanism we are proposing is based on the idea of autonomous, incremental generation of implicit system model and control policy using layers of self-organizing maps and joint-egocentric representation of reaching experiences. The robot is not provided with any control models or methods for calculating Inverse and Forward kinematics.

2.1 SOINN

The core associative learning mechanism that is adopted in this research is based on Self-Organizing Incremental Neural Network (SOINN)[13]. SOINN is a Self-organizing map that does not require any presumption to be made about the topology or the distribution, of data.

Basically SOINN works by propagating network topology in a way that would self-organize as to resemble the "hot zones" of perception. For example, if a new data point is presented to SOINN then the algorithm would find the closest two network nodes to this newly presented data point ,Fig. 1.a. once these, most closest, nodes are found , SOINN determines whether the newly presented data point is within the coverage zone of these nodes. If yes then these nodes would be now connected by an edge to make up a single cluster of nodes and then they would be altered as to reflect the current blobs of persistent activity,Fig. 1.b.

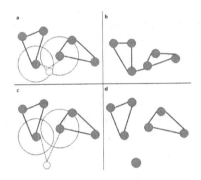

Fig. 1. SOINN dynamics

In the other case where the newly presented data point is out of the coverage zone of the closet nodes, Fig. 1.c, then this data point itself would be stored by SOINN as a node that represents a possible independent zone of activity, Fig.1.d. For a detailed explanation of the algorithm see[13].

SOINN has the feature of eliminating noisy and non-stable representations by checking the level of activity of each stored cluster of nodes and then discarding those stored clusters that doesn't represent regions of input space with high activity. So if a cluster of nodes has not been referenced frequently as being a coverage zone for input data points, then this cluster would eventually fade away and removed from the network.

2.2 The Architecture

First of all it is important to mention that each sensory-motor experience is represented and learned as a pairing between joint angle and the resultant gripper location in space. This pairing is joint-related, or joint-egocentric, i.e. for a given joint this sensory-motor learning experience would be $[\theta_i, L_i]$ where θ_i is the angle of joint i, and L_i is the resultant location of the gripper represented in relevance to the joint i ,hence, in the Peripersonal space of Joint i. Representing the location in the Peripersonal space of a given joint could be achieved by using a receptive field or mathematical transformation method. The purpose of this joint-egocentric representation is to make sure that learning is achieved on the joint level, where each joint would learn, the required associations, in manner that is independent from the other joints.

Each joint has its own associative learner, implemented as self-organizing map (SOINN)Fig. 2. This learner is responsible of learning sensory-motor pairs of the form $[\theta_i, L_i]$ that are related to the joint to which the self-organizing map belong to ,as mentioned above .

When a new target is presented to the system, the location of this target is represented in relevance to the first joint. Then the system would ask the self-organizing map, of the first joint, for the best angle that would achieve as close gripper location as possible to the given target . Depending on the joint's previous experience, the self-organizing map would respond by retrieving the joint angle that is associated with the closest gripper location to the target.

Now this angle, would be used to actuate the first joint of the manipulator even before passing the control to the next joint. This means that after the system has found out the suitable joint angle for the first joint, the target perception would be altered for the rest of joints on the manipulator, so in order for the next joint "$joint_{i+1}$" to ask its associative learner for suitable joint angle, θ_{i+1}, the robot must check the new altered location of target, L_{i+1}, in relevance to the next joint i.e. in the Peripersonal space of the next joint.

Fig. 2 reveals the iterative nature of the solution proposed here, where the problem of finding the best set of joint angles for multi-joint manipulator is solved by breaking down the reaching task into smaller sub problems, each handled by an independent subsystem that consist of single joint with its own perceptual space and its own associative learner.

2.3   From Exploration into Coordinated Reaching

In the approach we are proposing, training and learning take place in a real-time manner. The system itself decides when an exploration action is needed and when actual goal-reaching can be performed while the system is being trained continuously in both cases. So initially when we run the robot for the first time, the robot actions would be random ballistic trajectories similar to the ones performed by infants at early stages of motor development[5]. During this random motor babbling behavior the robot starts to generate an internal model for the control policy of its joints, through action-consequences coupling, which result in an increased ability to control these joints in coordinated manner, hence a less resultant error in reaching a target. To control the balance between motor babbling and target-reaching behaviors the following equation is used:

$$P(rnd) = 0.5 + \xi(m_{cp} - m_{fp}) \tag{1}$$

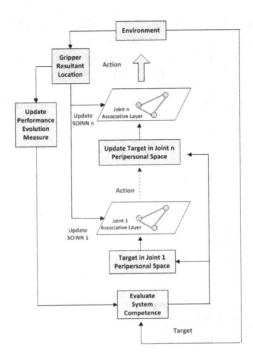

Fig. 2. The system Architecture

Where $P(rnd)$ is the probability of performing a random action, and $\xi(x)$ is the normalized value of x. The quantity m_{cp} is the mean error in the close past and m_{fp} is mean error in the far past.

The concept of close past and far past is generated by making the system maintains, at each time step, a list of measured error, described as the distance between the target location and the resultant gripper location, during the last n steps. This list then is divided in two halves. The most recent half, which consist of set of errors between $j = t$ and $j = t - (n/2)$, is considered as a set of errors in the close past. The other half, that consists of set of errors between $j = t - (n/2)$ and $j = t - n$, is considered as a set of errors in the far past.

Dividing the most recent n time steps into close past and far past serves the goal of altering the frequency and the necessity for random actions. So when error is reducing, and the robot performance is getting better, a negative value of $\xi(x)$ would result, which would decrease the random action probability, $P(rnd)$. on the other hand, when the error is increasing, a positive value of $\xi(x)$ would be generated resulting in higher motor-babbling probability, equation 1.

3   The Experiments

In this experimental setup, a simulated 2DoF planar robotic arm is used to demonstrate the developmental sensory-motor learning process, starting by random motor-babbling actions and then shifting gradually toward performing more coordinated target-reaching trajectories.

It is crucial to mention that the robot was not provided with any knowledge about how to control its joints, besides no action-consequence model was preprogrammed by the designer beforehand of learning.

A red ball is used as the target that the robot is required to reach at any given time. The ball location is generated randomly and then the robot is asked to reach it with its end effector, then, after the robot trail to reach the target, a new location is generated whether the robot has managed actually to reach the target or not, Fig. 3.

As mentioned above, and illustrated in Fig. 2, the trajectories that are performed by the robot, whether target-directed or random, are always used as a training signal for the learning system, which implies a continues adaptation and learning of the generated implicit model of control.

In Fig. 4, a gradual decrease in error is noticed with more practicing of the learned model that was initiated by the babbling actions.

Fig. 3. The Experimental setup

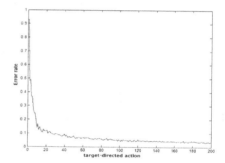

Fig. 4. resultant error during learning

The robot performance starts with high error rate. But with more training experiences the multilayer architecture of self-organizing map, SOINN, starts gradually to capture the contingencies behind joints angles and resultant end effector location. This incremental self-organizing process results in the observed decrease of the anticipated error of generated actions.

3.1   A Sudden Change

In this second experiment we demonstrate the system's reaction to a sudden unexpected change in the physical structure of the robot. This sudden change could account for a breakage in a joint, increased length of a link or a displacement of the end effector location in relevance to the arm links.

In this experimental setup we still have the same task of reaching a red ball, but now , after the system has learned its own implicit model, we suddenly

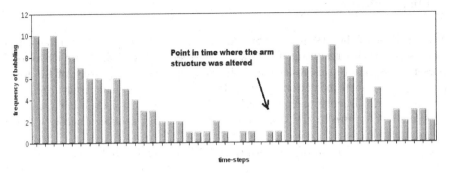

Fig. 5. A real-time reaction to an unexpected change in the physical structure

increased the length of the arm's second link by 10% of its original one. Altering the physical structure of the system means that now the learned implicit model does not accurately reflect the actual system nature. So if the system, before this unexpected change, had already reached a level of stability in term of the frequency of babbling actions, where a lower rate of random motion could be noticed, then now this stability won't last, and the robot would need to re-explore the contingencies of its action-consequence relation.

In Fig. 5, the horizontal axis shows a sequence of groups of time-steps ,each consist of 10 actions, that depicts the transition of the robots performance between motor babbling and target-directed actions. The vertical axis shows the number of babbling motions that was performed in each group of 10 time-steps.

As expected, most of the robot's actions, when it starts learning, are babbling ones and that is because the robot is not aware of its kinematic model. But then gradually this rate of babbling actions would decrease as the system proceed in building an implicit model of its control. Eventually we notice that almost no babbling actions are being performed but rather almost all of the taken actions are goal-directed.

During the robot's performance we altered the second link length, as mentioned above. This change would increase the resultant error in the robot's reaching accuracy because the learned sensory-motor associations does not accurately reflect the actual current status of the system. This increased error would generate a positive difference between m_{cp} and m_{fp} from equation 1, what eventually results in a higher $P(rnd)$, which is the probability of performing a babbling action.

This change in the behavior of the system can be observed in Fig. 5 where a peak in the frequency of babbling actions is clearly noticed around the point in time where the physical structure of the robot was altered. What can be noticed also is that the domination of babbling actions won't last forever, but rather it would be there as long as the system hasn't fully recaptured the Contingent action-consequence relation of its recently altered physical structure.This observation emphasizes the impact of the concept of learning through babbling on the ability of the system to adapt and react to unanticipated changes and conditions.

4 Discussion

A visualization of the first layer of SOINN is depicted in, Fig. 6. A 3-dimensional visualization of this resultant network can be seen in Fig. 6.a, where each node represents a single representative associative sensory-motor pairing of the form $[\theta_1, L_1]$, as described in section2.2. if we look at the topological structure of this network from 2-dimensional perspective, Fig. 6.b, we notice that it captures a very similar structure to the Cartesian work space but spawned across a third dimension of the associated angles of $joint_1$.

Next is a visualization of the learned SOINN network but for $joint_2$, Fig. 7. Again the network to the right, Fig. 7.b, is the 2-dimensional perspective of the 3-dimensional SOINN network, Fig. 7.a, that represents the sensory-motor associative model for $joint_2$. Notice, from Fig. 7.b, the egocentric characteristic

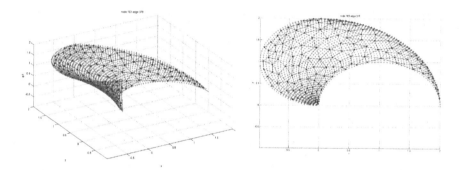

Fig. 6. A 3-dimensional representation of the approximated sensory-motor associations that correspond to $joint_1$ (left). the same network but in 2-dimensional perspective (right)

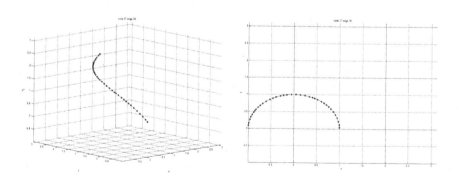

Fig. 7. A 3-dimensional representation of the approximated sensory-motor associations that correspond to $joint_2$ (left). the same network but in 2-dimensional perspective (right)

of the learned model since the Cartesian part of the associative data points does not reflect the whole work space but rather it captures only locations that are taken from the perspective of $joint_2$.

In both learned networks, Fig. 6 and Fig. 7, we notice that SOINN has the ability to cover the whole input training space with consistent distribution of nodes that enables the system to generalize even for unseen data points that was not provided during the process of network generation. This was demonstrated in Fig. 4 where the ball location was generated in continues input space rather than discrete one, but yet the system managed to generate trajectories of decreasing error even without the need for separated training and testing phases.

5   Conclusion

In this paper we have presented an architecture for learning sensory-motor associations for coordinated reaching tasks, using Self Organizing Neural Network (SOINN). The approach that was taking is inspired by developmental psychology where motor learning starts by babbling-like ballistic trajectories, similar to the ones observed during early stages of motor development in human infants, then the robot shifts toward coordinated actions with continuously decreasing error. This Developmental approach toward robot learning was demonstrated by the fact that no preprogrammed control policy was provided beforehand of learning. But rather the robot explored, on its own, the action-consequences contingencies of its joints and then, autonomously, generated an implicit control model through Motor babbling actions.

Acknowledgments. This work was sponsored by the Japan Science and Technology Agency's CREST project.

References

1. Lungarella, M., Metta, G., Pfeifer, R., Sandini, G.: Developmental robotics: a survey. Connection Science 15(4), 151–190 (2003)
2. Weng, J., Hwang, W.: From neural networks to the brain: Autonomous mental development. IEEE Computational Intelligence Magazine 1(3), 15–31 (2006)
3. Stoytchev, A.: Five basic principles of developmental robotics. In: NIPS 2006 Workshop on Grounding Perception, Knowledge and Cognition in Sensori-Motor Experience (2006)
4. Piaget, J.: The origins of intelligence in
5. Van der Meer, A., Van der Weel, F., Lee, D., et al.: The functional significance of arm movements in neonates. Science-New York Then Washington, 693–693 (1995)
6. von Hofsten, C.: Eye–hand coordination in the newborn. Developmental Psychology 18(3), 450 (1982)
7. Hersch, M., Sauser, E., Billard, A.: Online learning of the body schema. International Journal of Humanoid Robotics 5(02), 161–181 (2008)

8. Sturm, J., Plagemann, C., Burgard, W.: Body schema learning for robotic manipulators from visual self-perception. Journal of Physiology-Paris 103(3), 220–231 (2009)
9. Metta, G., Sandini, G., Konczak, J.: A developmental approach to visually-guided reaching in artificial systems. Neural Networks 12(10), 1413–1427 (1999)
10. Caligiore, D., Parisi, D., Baldassarre, G.: Toward an integrated biomimetic model of reaching. In: IEEE 6th International Conference on Development and Learning, ICDL 2007, pp. 241–246. IEEE (2007)
11. Demiris, Y., Dearden, A.: From motor babbling to hierarchical learning by imitation: a robot developmental pathway (2005)
12. Kober, J., Peters, J.: Learning motor primitives for robotics. In: IEEE International Conference on Robotics and Automation, ICRA 2009, pp. 2112–2118. IEEE (2009)
13. Furao, S., Ogura, T., Hasegawa, O.: An enhanced self-organizing incremental neural network for online unsupervised learning. Neural Networks 20(8), 893–903 (2007)

Alternative OVA Proposals for Cooperative Competitive RBFN Design in Classification Tasks

Francisco Charte Ojeda[1], Antonio Jesús Rivera Rivas[2],
María Dolores Pérez-Godoy[2], and María Jose del Jesus[2]

[1] Dept. of Computer Science and Artificial Inteligence
University of Granada, Spain
francisco@fcharte.com
[2] Dept. of Computer Science
University of Jaén, Spain
{arivera,lperez,mjjesus}@ujaen.es

Abstract. In the Machine Learning field when the multi-class classification problem is addressed, one possibility is to transform the data set in binary data sets using techniques such as One-Versus-All. One classifier must be trained for each binary data set and their outputs combined in order to obtain the final predicted class. The determination of the strategy used to combine the output of the binary classifiers is an interesting research area.

In this paper different OVA strategies are developed and tested using as base classifier a cooperative-competitive RBFN design algorithm, CO^2RBFN. One advantage of the obtained models is that they obtain as output for a given class a continuous value proportional to its level of confidence. Concretely three OVA strategies have been tested: the classical one, one based on the difference among outputs and another one based in a voting scheme, that has obtained the best results.

Keywords: OVA, RBFNs, Multi-class classification.

1 Introduction

A general approach to tackle several kind of classification problems is data transformation. For example, in multi-class classification, One-Versus-All (OVA) [1] is one of the most well-known.

The OVA strategy obtains a data set for each class included in the original data set. Thereby, each obtained data set contains two classes: the positive class or the class to predict, and the negative class that comprise the rest of classes. A classifier is trained for each binary data set and finally the outputs of these classifiers are combined in order to obtain the resulting class. In most of cases, this class correspond to the classifier with higher output for the positive class.

Radial Basis Function Networks (RBFNs) are one of the most important Artificial Neural Network (ANN) paradigms in the machine learning field. An

I. Rojas, G. Joya, and J. Cabestany (Eds.): IWANN 2013, Part I, LNCS 7902, pp. 331–338, 2013.

RBFN is a feed-forward ANN with a single layer of hidden units, called radial basis functions (RBFs) [2]. The overall efficiency of RBFNs has been proved in many areas [3] such as pattern classification, function approximation and time series prediction.

An important paradigm for RBFN design is Evolutionary Computation [4]. There are different proposals in this area with different scheme representations: Pittsburgh [5], where each individual is a whole RBFN, and cooperative-competitive [6], where an individual represents a single RBF.

Authors have developed an algorithm for the cooperative-competitive design of Radial Basis Functions Networks, CO^2RBFN [7], that has been successfully used in multi-class classification.

As demonstrated in [1] the use of OVA strategies can improve the results of a multi-class classifier. Thus, the aim of this paper is testing different OVA techniques with the RBFN design algorithm, CO^2RBFN. Concretely three OVA approaches have been implemented: the classical one, one based on the difference among outputs and another one based in a voting scheme.

The text is organized as follows. In Section 2, OVA methodology to multi-class classification is described as well as the concrete three methods to obtain the output class. The cooperative-competitive evolutionary model for the design of RBFNs applied to classification problems, CO^2RBFN, is described in Section 3. The analysis of the experiments and the conclusions are shown in Sections 4 and 5, respectively.

2   The OVA Approach to Multi-class Classification

There are many situations in which the class associated to a set of input attributes is not binary, but one of a set of outputs with more than two options. When it is necessary to work with a data set of this kind, a multi-class data set, there are two main methods to follow: design a classifier able to work with several classes, or split the original problem, applying the divide-and-conquer technique, by transforming the data set so that it can be processed with binary classifiers.

The decomposition of a multi-class data set in binary ones can be done using different approaches, being One-vs-All (OVA) one of the best known. The basic idea is to produce as many data sets as classes exist in the original multi-class data set, taking in each one of them a certain class as *positive* (P) and the rest as *negative* (N). Each of these data sets will be used to train a binary-independent classifier, therefore obtaining several predictions as output: one for each class.

The final predicted class could change depending on how the binary outputs obtained are combined. The kind of output generated by the binary classifiers will also influence this result; a rule based system will only indicate if the output is P or N without any additional information, on the other hand a neural network will give a weight associated to each of the two possible outputs, not a simple P or N. In the following subsections the traditional OVA approach will be exposed, along with the specific variations used within the experimentation of this proposal.

2.1 How Is Predicted the Output in Traditional OVA

Assuming that the underlying binary classifier B produces as output a value expressing a weight or likelihood associated to the positive class, and being X an instance and C the total number of classes, equation 1 will give as result the index of the class to predict following the traditional OVA method. This method is denoted as Classic OVA in the experimentation section.

$$I(X) = argmax(B_i(X))   i = 1 \ldots C; \tag{1}$$

It is as simple as taking the class associated to the binary classifier which has generated the maximum output. For this method to work it is necessary that the values given by the binary classifiers are comparable, applying previously a normalization process if is it required. Usually it is accepted a range between 0 and 1.

2.2 Global and Local Normalization of Outputs from Binary Classifiers

In order to normalize the values obtained from CO^2RBFN, as they are not normalized internally, two different methods has been used. The influence of the normalization method in the final results is important enough to warrant special attention.

The first method explores the outputs obtained for all the instances, gets the maximum and the minimum values, and uses this information to adjust these outputs before entering the final OVA prediction process. Therefore, it is a global normalization. In contrast, the second method does a local normalization using only the values associated to each sample. In both cases the final values will be in the range 0 to 1, as has been said above.

In the experimentation the traditional OVA approach explained before has been used in two variations, global and local, which only differ in the normalization method used.

2.3 Alternative Methods to OVA Prediction

Aiming to improve the prediction made by the Classic OVA approach, always working with the same set of output values obtained from the binary classification, we have defined and tested two alternative interpretations of these values once they have been normalized.

In the first alternative, Difference OVA, each classified instance has two values incoming from each binary classifier: the weight associated to the positive class and the one which belongs to the negative class. Instead of looking for the maximum positive value, as it is done in the traditional OVA, it is possible to calculate the difference between these two weights in order to obtain a unique value. The class predicted will be that which has the maximum difference, discarding those cases in which the positive and negative weights are very near, even though the positive could be the absolute maximum.

The second alternative method proposed to do the OVA prediction, Voting OVA, is based in the idea of a majority-voting system. Given that there are several individual predictions for each instance, coming from the use of two normalization techniques and the repetitions made in the execution over the partitioned data sets, we have taken each of those predictions as a vote for a class. The votes are summarized and the class with the higher count is the final prediction.

In the experimentation, Difference OVA are used in combination with the two normalization methods described above, giving as result two final predictions. Voting OVA approach incorporates one more prediction in the set of results to analyze.

3   CO²RBFN: An Evolutionary Cooperative-Competitive Hybrid Algorithm for RBFN Design

CO²RBFN [7] is an evolutionary cooperative-competitive hybrid algorithm for the design of RBFNs. In this algorithm each individual of the population represents, with a real representation, an RBF and the entire population is responsible for the final solution.

The individuals cooperate towards a definitive solution, but they must also compete for survival. In this environment, in which the solution depends on the behavior of many components, the fitness of each individual is known as credit assignment. In order to measure the credit assignment of an individual, three factors have been proposed: the RBF contribution to the network output, the error in the basis function radius, and the degree of overlapping among RBFs.

The application of the operators is determined by a Fuzzy Rule-Based System. The inputs of this system are the three parameters used for credit assignment and the outputs are the operators' application probability.

The main steps of CO²RBFN, explained in the following subsections, are shown in the pseudocode, in Algorithm 1. For a wider explanation of the algorithm see reference [7].

Algorithm 1. Main steps of CO²RBFN

```
1. Initialize RBFN
2. Train RBFN
3. Evaluate RBFs
4. Apply operators to RBFs
5. Substitute the eliminated RBFs
6. Select the best RBFs
7. If the stop condition is not verified go to step 2
```

RBFN Initialization. To define the initial network a specified number m of neurons (i.e. the size of population) is considered. The center of each RBF is randomly allocated to a different pattern of the training set. The RBF widths,

d_i, will be set to half the average distance between the centres. Finally, the RBF weights, w_{ij}, are set to zero.

RBFN Training. The Least Mean Square algorithm [8] is used to calculate the RBF weights.

RBF Evaluation. A credit assignment mechanism is required in order to evaluate the role of each RBF ϕ_i in the cooperative-competitive environment. For an RBF, three parameters, a_i ,e_i ,o_i are defined:

- The contribution, a_i, of the RBF ϕ_i, is determined by considering the weight, w_i, and the number of patterns of the training set inside its width, pi_i:

$$a_i = \begin{cases} |w_i| & if \quad pi_i > q \\ |w_i| * (pi_i/q) & otherwise \end{cases} \tag{2}$$

  where q is the average of the pi_i values minus the standard deviation of the pi_i values.
- The error measure, e_i, for each RBF ϕ_i, is obtained by counting the wrongly classified patterns inside its radius:

$$e_i = \frac{pibc_i}{pi_i} \tag{3}$$

  where $pibc_i$ and pi_i are the number of wrongly classified patterns and the number of all patterns inside the RBF width respectively.
- The overlapping of the RBF ϕ_i and the other RBFs is quantified by using the parameter o_i. This parameter is computed by taking into account the fitness sharing methodology [4], whose aim is to maintain the diversity in the population.

Applying Operators to RBFs. In CO^{2}RBFN four operators have been defined in order to be applied to the RBFs:

- Operator Remove: eliminates an RBF.
- Operator Random Mutation: modifies the centre and width of an RBF in a random quantity.
- Operator Biased Mutation: modifies, using local information, the RBF trying to locate it in the centre of the cluster of the represented class.
- Operator Null: in this case all the parameters of the RBF are maintained.

The operators are applied to the whole population of RBFs. The probability for choosing an operator is determined by means of a Mandani-type fuzzy rule based system [9]. The inputs of this system are parameters a_i, e_i and o_i used for defining the credit assignment of the RBF ϕ_i. These inputs are considered as linguistic variables va_i, ve_i and vo_i. The outputs, p_{remove}, p_{rm}, p_{bm} and p_{null}, represent the probability of applying Remove, Random Mutation, Biased

Table 1. Fuzzy rule base representing expert knowledge in the design of RBFNs

	Antecedents			Consequents					Antecedents			Consequents			
	v_a	v_e	v_o	p_{remove}	p_{rm}	p_{bm}	p_{null}		v_a	v_e	v_o	p_{remove}	p_{rm}	p_{bm}	p_{null}
R1	L			M-H	M-H	L	L	R6	H			M-H	M-H	L	L
R2	M			M-L	M-H	M-L	M-L	R7		L		L	M-H	M-H	M-H
R3	H			L	M-H	M-H	M-H	R8		M		M-L	M-H	M-L	M-L
R4		L		L	M-H	M-H	M-H	R9		H		M-H	M-H	L	L
R5		M		M-L	M-H	M-L	M-L								

Mutation and Null operators, respectively. Table 1 shows the rule base used to relate the antecedents and consequents described.

Introduction of New RBFs. In this step, the eliminated RBFs are substituted by new RBFs. The new RBF is located in the centre of the area with maximum error or in a randomly chosen pattern with a probability of 0.5 respectively.

Replacement Strategy. The role of the mutated RBF in the network is compared with the original one to determine the RBF with the best behavior in order to include it in the population.

4 Experimentation

In order to test in a multi-class classification scenario the different OVA approaches developed and using as classifier our cooperative-competitive algorithm for RBFN design, CO^2RBFN, ten different data sets have been chosen from KEEL data set repository [10]. The properties of these data sets are shown in table 2. With these data sets, a typical experimental framework has been established with ten-fold cross validation (90% for training data set, 10% for test data set) and three repetitions for obtaining the results.

Table 2. Data set properties

Data-set	Instances	Attributes	Classes
Balance	625	4	3
Cleveland	467	13	5
Dermatology	358	33	6
Ecoli	336	7	8
Glass	214	9	6
Hayes-Roth	160	4	3
New-thyroid	215	5	3
Lymphography	148	18	4
Wine	178	13	3
Yeast	1484	8	10

The same configuration parameters are set up for all the CO^2RBFN versions: 200 iterations are established for the main loop and the number of individuals or RBFs are set to the twice of the number of classes existing in the processed data set.

In table 3 the average correct classification rate for test data sets of the different proposals are shown. Specifically the Base column shows the results obtained for the multi-class version of CO^2RBFN, without preprocessing the data set. In the following columns the results of different OVA strategies (Classic, Difference and Voting) are shown. For the Classic and Difference techniques two normalization alternatives are exhibited. All the OVA strategies are described in the section 2. For a given data set the best result is in bold.

Table 3. Average correct classification rate of different OVA strategies against the base version

Datasets	Base	Classic OVA		Difference OVA		Voting OVA
		Global	Local	Global	Local	
Balance	0.8907	0.6525	0.8810	0.9018	0.8864	**0.9071**
Cleveland	**0.5766**	0.5701	0.4940	0.5095	0.5547	0.5546
Dermatology	**0.9524**	0.6428	0.6401	0.9265	0.9247	0.9443
Ecoli	0.8167	0.5724	0.7930	0.7703	0.7781	**0.8200**
Glass	**0.6669**	0.4549	0.5703	0.5594	0.6244	0.6399
Hayes-Roth	0.6688	0.5396	0.6625	0.6938	0.7375	**0.7750**
New-thyroid	0.9511	0.8206	0.9584	0.9509	0.9556	**0.9677**
Lymphography	0.7298	0.3235	0.3374	0.6910	0.7173	**0.8165**
Wine	**0.9616**	0.6671	0.9328	0.9366	0.9385	0.9556
Yeast	**0.5780**	0.1787	0.4230	0.4569	0.5095	0.5377

From the results obtained we can conclude that OVA strategies as Classic OVA or Difference OVA do not achieve any best result with respect to the base version of CO^2RBFN (without OVA preprocessing). This fact underpins the good behavior of the base CO^2RBFN algorithm, correctly designing RBFNs for multi-class data sets.

Nevertheless, this trend changes when the more innovative OVA strategy, Voting, is applied. In fact, Voting outperforms the base version of the CO^2RBFN in five of the ten data sets. It must be also highlighted that for certain data sets, such as Hayes-Roth or Lymphography, Voting OVA has obtained significantly better results than CO^2RBFN with differences around ten points. Besides this, Voting OVA can outperforms in data sets with interesting properties, such as a moderate number of instances (Balance), attributes (Lymphography) or classes (Ecoli).

Thus, although there is tie between base CO^2RBFN and Voting OVA, the results obtained leads to carry out a more deep research about the OVA Voting strategy.

5 Conclusions

With the aim of improving the performance obtained in the classification of multi-class data sets OVA transformations can be used. The resulting binary data sets are processed by binary classifiers and the output of these ones must be combined in order to obtain the final predicted class.

In this paper different combination OVA strategies are tested using CO^2RBFN, a cooperative-competitive evolutionary algorithm for the design of RBFNs, as base classifier.

The results show that while most classic OVA strategies do not improve the performance of the base version of CO^2RBFN, the developed voting strategy outperforms this base version in certain data sets. These results encourage us to carry out a more in-deep research over the last strategy.

Acknowledgments. F. Charte is supported by the Spanish Ministry of Education under the F.P.U. National Program (Ref. AP2010-0068). This paper is partially supported by the Spanish Ministry of Science and Technology under the Project TIN 2012-33856, FEDER founds, and the Andalusian Research Plan TIC-3928.

References

1. Rifkin, R., Klautau, A.: In defense of one-vs-all classification. Journal of Machine Learning Research 5, 101–141 (2004)
2. Broomhead, D., Lowe, D.: Multivariable functional interpolation and adaptive networks. Complex Systems 2, 321–355 (1988)
3. Buchtala, O., Klimek, M., Sick, B.: Evolutionary optimization of radial basis function classifiers for data mining applications. IEEE Transactions on System, Man, and Cybernetics, B 35(5), 928–947 (2005)
4. Goldberg, D.: Genetic Algorithms in Search, Optimization, and Machine Learning. Addison-Wesley, Reading (1989)
5. Harpham, C., Dawson, C., Brown, M.: A review of genetic algorithms applied to training radial basis function networks. Neural Computing and Applications 13, 193–201 (2004)
6. Whitehead, B., Choate, T.: Cooperative-competitive genetic evolution of radial basis function centers and widths for time series prediction. IEEE Transactions on Neural Networks 7(4), 869–880 (1996)
7. Pérez-Godoy, M., Rivera, A., del Jesus, M., Berlanga, F.: co^2rbfn: An evolutionary cooperative-competitive RBFN design algorithm for classification problems. Soft Computing 14(9), 953–971 (2010)
8. Widrow, B., Lehr, M.: 30 years of adaptive neural networks: perceptron, madaline and backpropagation. Proceedings of the IEEE 78(9), 1415–1442 (1990)
9. Mandani, E., Assilian, S.: An experiment in linguistic synthesis with a fuzzy logic controller. International Journal of Man-Machine Studies 7(1), 1–13 (1975)
10. Alcalá-Fdez, J., Luengo, J., Derrac, J., García, S., Sánchez, L., Herrera, F.: Keel data-mining software tool: Data set repository, integration of algorithms and experimental analysis framework. Journal of Multiple-Valued Logic and Soft Computing 17(2-3), 255–287 (2011)

Committee C-Mantec: A Probabilistic Constructive Neural Network

Jose Luis Subirats, Rafael Marcos Luque-Baena, Daniel Urda, Francisco Ortega-Zamorano, Jose Manuel Jerez, and Leonardo Franco

Department of Computer Science, University of Málaga, Málaga, Spain
{jlsubirats,rmluque,durda,fortega,jja,lfranco}@lcc.uma.es

Abstract. C-Mantec is a recently introduced constructive algorithm that generates compact neural architectures with good generalization abilities. Nevertheless, it produces a discrete output value and this might be a drawback in certain situations. We propose in this work two approaches in order to obtain a continuous output network such as the output can be interpreted as the probability of a given pattern to belong to one of the output classes. The CC-Mantec approach utilizes a committee strategy and the results obtained both with the XOR Boolean function and with a set of benchmark functions shows the suitability of the approach, as an improvement over the standard C-Mantec algorithm is obtained in almost all cases.

Keywords: Committee networks, Supervised classification, Constructive neural networks.

1 Introduction

Neural computing techniques offer attractive alternatives over other classical techniques, particularly when data is noisy or in cases when no explicit knowledge is known. In practical applications, the most important criterion to evaluate the performance of trained Artificial Neural Networks (ANNs) is its ability to generalize knowledge. Although properly trained ANNs may offer very good results, they will inevitably overfit. Therefore, other techniques should be developed in order to improve ANNs generalization capabilities. A widespread approach involves training several networks (varying topologies, initialization of synaptic weights, etc.) and then choosing the one that offers the greatest generalization capacity. Under this approach, the acquired knowledge by non-optimal networks is lost, whereas, in principle, this information should not be discarded. One way to avoid this, is the use of a committee of ANNs, and it has been shown that by combining several ANNs the generalization can be improved [6]. The different approaches that exists for applying Committee machines can be classified into two broad categories [4]: Static and dynamic structures.

The generalization error generated by several ANNs that form a committee are not necessarily related. In this sense, when a committee based on different

I. Rojas, G. Joya, and J. Cabestany (Eds.): IWANN 2013, Part I, LNCS 7902, pp. 339–346, 2013.

networks is created, the generalization error of a single ANN can be corrected by the remaining networks.

A very important issue related to the application of ANNs is the selection of a proper neural architecture for each network in the committee [1,3]. Despite the existence of several proposals to solve or alleviate this problem [4], there is no general agreement on the strategy to follow in order to select an optimal neural network architecture. Constructive algorithms have been proposed in recent years [5,8] with the aim of dynamically estimating the neural network topology. In general, constructive methods start with a small network, adding new units as needed until a stopping criteria is met. In [7], the Competitive MAjority Network Trained by Error Correction (C-Mantec) algorithm was introduced, with the novelty in comparison to existing approaches that C-Mantec incorporates competition between neurons and thus all neurons can learn at any stage of the procedure. Based on this previous algorithm, a new Committee C-Mantec (CC-Mantec) method is proposed in this work, in order to obtain a probabilistic version of the algorithm.

The remainder of this paper is organized as follows: Section 2 provides a description of C-Mantec algorithm, Section 3 shows two novel approaches, the HC-Mantec based in a Hyperbolic tangent sigmoid transfer function, and the CC-Mantec based in a Committee of networks. Section 4 shows the experimental results using several prediction, and finally, Section 5 concludes the article.

2   The C-Mantec Algorithm

C-Mantec is a constructive neural network algorithm that creates architectures containing a single layer of hidden nodes with sign activation functions. For binary classification tasks, the constructed networks have a single output neuron computing the majority function of the responses of the N hidden nodes:

$$CMantec\left(\boldsymbol{\psi}\right) = sign\left(\sum_{n=1}^{N} sign\left(h_n\left(\boldsymbol{\psi}\right)\right)\right) \tag{1}$$

$$h_n\left(\boldsymbol{\psi}\right) = \sum_{i=1}^{M} w_{n,i}\psi_i + \theta_n \tag{2}$$

$$sign\left(x\right) = \begin{cases} 1 & x \geq 0 \\ -1 & in\,other\,case \end{cases} \tag{3}$$

where M is the number of inputs of the target function, w_i are the synaptic weights, θ is the neuron threshold, and ψ_i indicates the set of inputs. The learning procedure starts with an architecture comprising a single neuron in the hidden layer and continues by adding a neuron every time the present ones are not able to learn the whole set of training examples. The hidden layer neurons learn according to the thermal perceptron learning rule proposed by Frean [2]. The thermal perceptron can be seen as a modification of the standard perceptron rule that incorporates a modulation factor, which forces the neurons to learn only target examples close to the already learned ones.

The network generated by the algorithm has an output neuron that computes the majority function of the activation of the neurons belonging to the single hidden layer. If the target of a given example is not matched by the network output, this implies that more than half of the neurons in the hidden layer classify incorrectly the current input. In these cases, the algorithm, in the training phase selects one of the 'wrong' neurons in a competitive process in order to retrain it. For a deeper analysis of the C-Mantec algorithm, see the original paper[7].

3 Probabilistic C-Mantec Approaches

In this section, two different versions of a C-Mantec algorithm with continuous output are proposed, namely: Hyperbolic tangent sigmoid C-Mantec (HC-Mantec) and Committee C-Mantec (CC-Mantec).

3.1 HC-Mantec

HC-Mantec is a simple continuous version of C-Mantec where a Hyperbolic tangent sigmoid transfer function is used in all the neurons of the hidden layer:

$$HCMantec\left(\boldsymbol{\psi}\right) = \frac{\sum_{n=1}^{N} tansig\left(h_n\left(\boldsymbol{\psi}\right)\right)}{N} \tag{4}$$

$$tansig\left(x\right) = \frac{2}{\left(1 + exp\left(-2x\right)\right)} - 1 \tag{5}$$

The output of the network is normalized by a factor equal to the the number of neurons of the hidden layer, N, so it belongs to the interval $[-1, 1]$.

3.2 CC-Mantec

CC-Mantec is a constructive neural network algorithm that uses the power of committee methods with the advantage of dynamically estimating each network architecture. C-Mantec can be seen as a neurons committee that finds one approximated solution to the problem, and in this sense, different executions of the algorithm would give different committee for solving the problem. Let K be the number of total neurons generated by all the single C-Mantec networks which compose our CC-Mantec approach. CC-Mantec use all K generated neurons to create a new single committee of neurons with no need to retrain this new model, such that, no relevant information is missed and, on average, the generated hyperplanes are quite close to the optimal solution (see Equation 6). Figure 1 shows the CC-Mantec network topology as a result of combining two single C-Mantec networks.

$$CCMantec\left(\boldsymbol{\psi}\right) = \frac{\sum_{n=1}^{K} sign\left(h_n\left(\boldsymbol{\psi}\right)\right)}{K} \tag{6}$$

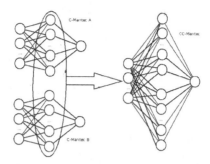

Fig. 1. The CC-Mantec topology obtained from combining two independent C-Mantec networks

3.3 Multiclass Classification

A $K - Class$ pattern recognition task can be implemented by a system formed by M CC-Mantec and an additional decision module. The M CC-Mantec are trained separately, and a decision module is used to select the final classification result based on the outputs of the M neural networks. The value of M and the training methodology depends on the modeling scheme used. In this work, three different approaches (One-Against-All, One-Against-One, and P-Against-Q) are applied.

One-Against-All scheme utilizes $M = K$ CC-Mantec, where K is the number of output classes of the original problem. Each CC-Mantec is trained with the same training dataset but with different objective values. On each of the K networks, one of the $K - classes$ is assigned the target value 1 while the rest of classes is assigned the value 0. The decision module computes the belonging probability for class i:

$$CCMantec_{OAA}(\mathbf{x}, i) = \frac{CCMantec_i(\mathbf{x})}{\sum_{j=1}^{M} CCMantec_j(\mathbf{x})} \tag{7}$$

One-Against-One scheme transforms a $K - Class$ pattern classification problem into $M = K(K-1)/2$ two-classes sub-problems. Each CC-Mantec solves a classification problem of an individual class against another and is trained only with a subset of the dataset where these two classes are active. A simple voting scheme can be used for the decision module, that computes the belonging probability for each class based on the continuous outputs from the M CC-Mantec networks.

$$CCMantec_{OAO}(\mathbf{x}, i) = \frac{\sum_{j=1, j \neq i}^{M} CCMantec_{i,j}(\mathbf{x})}{\sum_{i=1}^{M-1} \sum_{j=i+1}^{M} CCMantec_{i,j}(\mathbf{x})} \tag{8}$$

In the P-Against-Q classification scheme, the original classes are grouped in M different two class problems, in a way that from the output of these M groups is possible to infer the output class. The implementation can be considered as

M binary codes of length K, where each code has P bits equal to one and $Q = M - P$ bits equal to zero. One type of P-against-Q encoding consists in using the shorter code that specify all classes, $M = \log_2 K$ bits. This dense encoding is efficient in terms of the resulting size of the architecture but not in terms of the generalization obtained, as some redundancy on the encoding is usually beneficial. An Euclidean distance scheme is implemented by the decision module. The M CC-Mantec generate an output vector $\mathbf{v}$, and the class with code nearest to $\mathbf{v}$ will be the chosen output.

$$CCMantec_{PAQ}(\mathbf{x}, i) = \frac{Distance(\mathbf{v}, Code_i)}{\sum_{i=1}^{M} Distance(\mathbf{v}, Code_i)} \qquad (9)$$

$$v_i = CCMantec_i(\mathbf{x}) \qquad (10)$$

4 Experimental Results

4.1 Detailed Analysis on the XOR Function

We have carried out a detailed analysis about the functioning of C-Mantec, HC-Mantec and CC-Mantec on the clasical XOR problem. Figure 2 represents some solutions obtained with the different approaches. The top two inset figures (Figures 2a and 2b), show two possible C-Mantec solutions, as different solutions are proposed depending on the order in which the training patterns are presented. These figures show that the 'zero' and 'one' classes are not balanced, since the position of the separating cutting planes are not optimal. In addition, the binary nature of the C-Mantec method gives no information about how close the points are to these planes. HC-Mantec (cf. 5) provides some improvement on the classification probabilities. Figures 2c and 2d show the belonging probability of each class in two possible solutions of the XOR problem. In this case, the method provides some information about how close is each point to the plane, but classes are still unbalanced. Figure 2e shows the result of the CC-Mantec approach. The figure was obtained using 2000 C-Mantec networks, and each represented point indicates the belonging probability to each class. For a better visualization purpose, the results were discretized in 10 regions and shown in Figure 2f, where it can be shown that the CC-Mantec output is very close to the optimal solution for the XOR problem.

4.2 Tests on Benchmark Datasets

Seven benchmark data sets were used to analyze the performance of the introduced CC-Mantec algorithm, in comparison to the standard C-Mantec version.

Table 1 and 2 shows the generalization results for the CC-Mantec and C-Mantec algorithms. For every data set, three multiclass approaches have been launched (OAA, OAO and PAQ), and the prediction accuracy was calculated as the average from 50 independent runs using a 80/20 training/generalization splitting sets with the standard C-Mantec parameter setting ($I_{max} = 10000, g_{fac} = 0.05$ & $\phi = 2$).

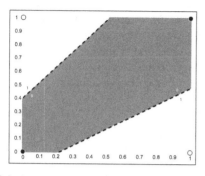
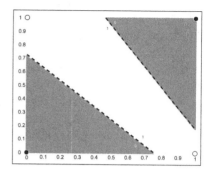

(a) A possible C-Mantec solution to the XOR problem.

(b) Another C-Mantec possible XOR solution.

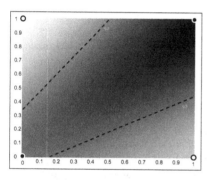
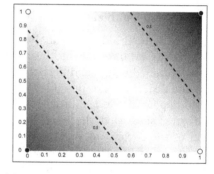

(c) HC-Mantec possible solution.

(d) Another HC-Mantec possible solution.

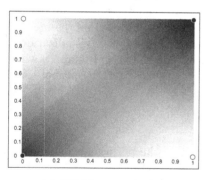
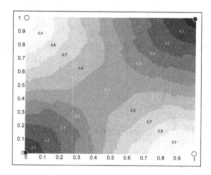

(e) CC-Mantec solution.

(f) CC-Mantec result discretized in 10 levels for better visualization.

Fig. 2. Solutions to the XOR problem obtained by the C-Mantec, HC-Mantec and CC-Mantec algorithms

Table 1. Results obtained by our CC-Mantec approach over several benchmark datasets. The generalization rate is shown using the mean and standard deviation.

	CC-Mantec		
	OAA	OAO	PAQ
balance-scale	91.392 ± 1.70	90.896 ± 2.57	91.488 ± 1.94
glass	**68.698 ± 6.17**	58.512 ± 8.39	59.163 ± 7.60
iris	**96.067 ± 3.51**	96.800 ± 2.98	93.733 ± 4.14
soybean	**92.486 ± 2.20**	90.397 ± 2.72	88.307 ± 3.25
vowel	**88.283 ± 2.58**	88.141 ± 2.58	56.626 ± 3.52
heart-statlog	**83.556 ± 4.10**	80.444 ± 5.30	83.407 ± 3.95
ionosphere	**88.732 ± 3.40**	83.408 ± 3.25	85.606 ± 3.31

Table 2. Results obtained by our C-Mantec approach over several benchmark datasets

	C-Mantec		
	OAA	OAO	PAQ
balance-scale	90.576 ± 2.20	89.872 ± 3.00	**92.432 ± 2.21**
glass	65.023 ± 7.51	66.930 ± 7.95	60.605 ± 6.87
iris	95.533 ± 3.10	95.933 ± 3.35	94.333 ± 3.61
soybean	91.037 ± 2.47	90.600 ± 3.20	81.861 ± 2.35
vowel	79.141 ± 4.19	87.182 ± 2.68	71.980 ± 3.63
heart-statlog	79.593 ± 4.62	79.778 ± 4.81	75.444 ± 5.50
ionosphere	87.465 ± 3.76	88.028 ± 3.51	86.620 ± 4.74

5   Conclusion

We propose in this work two possible extensions to the C-Mantec algorithm in order to obtain a continuous output value that approximates the probability of a given pattern to belong to one of many possible classes. The first proposal named HC-Mantec uses a sigmoidal activation function in the output layer to obtain a continuous response, but does not really work differently to the standard C-Mantec algorithm. The second proposal, named CC-Mantec, tries to take advantange of the potential of committee networks, and creates a network that combines several independent trained C-Mantec neurons, whose outputs are combined to obtain a continuous value in the range [-1, 1]. A detailed analysis of the performance of the new algorithms on the classic XOR problem is presented, showing that the behaviour of the CC-Mantec is very close to the optimal solution expected, as the output value can be interpreted as the probability of a given input pattern to belong to one of the output classes.

Using a set of 7 multiclass benchmark functions from the UCI repository an evaluation of the generalization ability of the CC-Mantec is carried out and compared to the standard C-Mantec implementation (results from the

HC-Mantec were not shown as they were almost indistinguishable with those from C-Mantec). The results show that in 6 out of the 7 data sets the CC-Mantec lead to a clear improvement in predictive accuracy (3.8 % of average improvement), suggesting the suitability of the developed approach. Some preliminary tests (not shown) done with other alternative classification algorithms (J48, SVM, Naive Bayes & MLP) confirms also that CC-Mantec performs better or at the level of bagging versions of these alternative algorithms, but these results will be the subject of further studies.

Acknowledgements.   The authors acknowledge support through grants TIN2010-16556 from MICINN-SPAIN and, P08-TIC-04026 and P10-TIC-5770 (Junta de Andalucía), all of which include FEDER funds.

References

1. Baum, E.B., Haussler, D.: What size net gives valid generalization? Neural Comput. 1(1), 151–160 (1989)
2. Frean, M.: A thermal perceptron learning rule. Neural Comput. 4(6), 946–957 (1992)
3. Gómez, I., Franco, L., Jerez, J.M.: Neural network architecture selection: can function complexity help? Neural Process. Lett. 30(2), 71–87 (2009)
4. Haykin, S.: Neural Networks: A Comprehensive Foundation. Macmillan, New York (1994)
5. do Carmo Nicoletti, M., Bertini Jr., J.R.: An empirical evaluation of constructive neural network algorithms in classification tasks. Int. J. Innov. Comput. Appl. 1(1), 2–13 (2007)
6. Perrone, M.P., Cooper, L.N.: When networks disagree: Ensemble methods for hybrid neural networks, pp. 126–142. Chapman and Hall (1993)
7. Subirats, J.L., Franco, L., Jerez, J.M.: C-mantec: A novel constructive neural network algorithm incorporating competition between neurons. Neural Netw. 26, 130–140 (2012)
8. Subirats, J.L., Jerez, J.M., Franco, L.: A new decomposition algorithm for threshold synthesis and generalization of boolean functions. IEEE Trans. on Circuits and Systems 55-I(10), 3188–3196 (2008)

Secure Semi-supervised Vector Quantization for Dissimilarity Data

Xibin Zhu, Frank-Michael Schleif, and Barbara Hammer

CITEC - Centre of Excellence,
Bielefeld University, 33615 Bielefeld, Germany
{xzhu,fschleif,bhammer}@techfak.uni-bielefeld.de

Abstract. The amount and complexity of data increase rapidly, how-ever, due to time and cost constrains, only few of them are fully la-beled. In this context non-vectorial relational data given by pairwise (dis-)similarities without explicit vectorial representation, like score-values in sequences alignments, are particularly challenging. Existing semi-supervised learning (SSL) algorithms focus on vectorial data given in Euclidean space. In this paper we extend a prototype-based classifier for dissimilarity data to non i.i.d. semi-supervised tasks. Using confor-mal prediction the 'secure region' of unlabeled data can be used to im-prove the trained model based on labeled data while adapting the model complexity to cover the 'insecure region' of labeled data. The proposed method is evaluated on some benchmarks from the SSL domain.

Keywords: Semi-Supervised Learning, Proximity Data, Dissimilarity Data, Conformal Prediction, Learning Vector Quantization.

1 Introduction

Big data are getting more and more challenging by means of storage and analysis requirements. Besides the amount of data, only few of these data are totally labeled, and labeling of all these data is indeed very costly and time consuming. Techniques of data mining, visualization, and machine learning are necessary to help people to analyse such data. Especially semi-supervised learning techniques, which integrate the structural and statistical knowledge of unlabeled data into the training, are widely used for this setting. A variety of SSL methods has been published [1]. They all focus on vectorial data given in Euclidean space or representations by means of positive semi-definite (psd) kernel matrices.

Many real world data are non-vectorial, often non-euclidean and given in the form of pairwise proximities between objects. Such data are also referred to as *proximity* or *relational data*, which are based on pairwise comparisons of objects providing some score-value of the (dis-)similarity of the objects. For such data, a vector space is not necessarily available and there is no guarantee of metric conditions. Examples of such proximity or (dis-)similarity measures are edit distance based measures for strings or images [5] or popular similarity measures in bioinformatics such as scores obtained by the Smith-Waterman,

I. Rojas, G. Joya, and J. Cabestany (Eds.): IWANN 2013, Part I, LNCS 7902, pp. 347–356, 2013.

FASTA, or blast algorithm [4]. Such partially labeled relational data are not widely addressed in the literature of SSL, yet. Only few methods consider SSL for classification of proximity data without an explicit underlying vector space and without requesting a metric space [9,13], this is the topic of this paper.

In this paper we extend a prototype-based classifier proposed in [3] for semi-supervised tasks of non i.i.d. data employing conformal prediction [14] technique. For SSL tasks, conformal prediction is used to determine the *secure region* of unlabeled data, which can potentially enhance the performance of the training, and at the same time estimates a so-called *insecure region* of labeled data which helps to adapt the model complexity. The proposed method can directly deal with non-psd proximity multi-class data.

First we will review relational supervised prototype-based learning as recently introduced by the authors in a specific model, employing conformal prediction concepts as discussed in [11]. Thereafter we introduce an extension to semi-supervised learning. We show the effectiveness of our technique on simulated data, well-known vectorial data sets and biomedical dissimilarity data which are not psd. Finally we summarize our results and discuss potential extensions.

2   Semi-supervised Prototype-Based Relational Learning

Prototype-based relational learning for unsupervised and supervised cases has been investigated by [3]. For semi-supervised problems, first we will briefly review the idea of prototype-based learning for relational data, then we will give a short introduction about conformal prediction for prototype-based learning and finally show how to extend it for semi-supervised problems.

2.1   Prototype-Based Relational Learning

As mentioned before, in the relational setting, data is not given as vectors, but as pairwise relation(s) between data points, e.g. distances between two points or some scores that describe some relations between the data. Let $\mathbf{v}_j \in \mathbb{V}$ be a set of objects defined in some data space, with $|\mathbb{V}| = N$. We assume, there exists a dissimilarity measure such that $D \in \mathbb{R}^{N \times N}$ is a dissimilarity matrix measuring the pairwise dissimilarities $D_{ij} = d(\mathbf{v}_i, \mathbf{v}_j)$ between all pairs $(\mathbf{v}_i, \mathbf{v}_j) \in \mathbb{V} \times \mathbb{V}$. Any reasonable (possibly non-metric) distance measure is sufficient. We assume zero diagonal $d(\mathbf{v}_i, \mathbf{v}_i) = 0$ for all i and symmetry $d(\mathbf{v}_i, \mathbf{v}_j) = d(\mathbf{v}_j, \mathbf{v}_i)$ for all $\{i, j\}$.

We assume a training set is given where data point $\mathbf{v}_j$ is labeled $\mathbf{l}_j \in \mathbb{L}, |\mathbb{L}| = L$. The objective is to learn a classifier f such that $f(\mathbf{v}_k) = \mathbf{l}_k$ for any given data point. We use a recently published prototype classifier for dissimilarity data [3] as basic method in the following. As detailed in [3], these data can always be embedded in pseudo-euclidean space in such a way that $d(\mathbf{v}_i, \mathbf{v}_j)$ is induced by a synthetic (but possibly not psd) bilinear form.

Classification takes place by means of k prototypes $\mathbf{w}_j \in W$ in the pseudo-Euclidean space, which are priorly labeled. Typically, a winner takes all rule is assumed, i.e. a data point is mapped to the label assigned to the prototype which is closest to the data in pseudo-Euclidean space, taking the bilinear form in pseudo-Euclidean space to compute the distance. For relational data classification, the key assumption is to restrict prototype positions to linear combinations of data points of the form $\mathbf{w}_j = \sum_i \alpha_{ji} \mathbf{v}_i$ with $\sum_i \alpha_{ji} = 1$. Then dissimilarities between data points and prototypes can be computed implicitly by means of

$$d(\mathbf{v}_i, \mathbf{w}_j) = [D \cdot \alpha_j]_i - \frac{1}{2} \cdot \alpha_j^t D \alpha_j \tag{1}$$

where $\alpha_j = (\alpha_{j1}, \ldots, \alpha_{jn})$ refers to the vector of coefficients describing the prototype $\mathbf{w}_j$, as shown in [3].

Using this observation, prototype classifier schemes which are based on cost functions can be transferred to the relational setting. We use the cost function defined in [10]. The corresponding cost function of the *relational prototype-based classifier* (RPC) becomes:

$$E_{\text{RPC}} = \sum_i \Phi \left(\frac{[D\alpha^+]_i - \frac{1}{2} \cdot (\alpha^+)^t D\alpha^+ - [D\alpha^-]_i + \frac{1}{2} \cdot (\alpha^-)^t D\alpha^-}{[D\alpha^+]_i - \frac{1}{2} \cdot (\alpha^+)^t D\alpha^+ + [D\alpha^-]_i - \frac{1}{2} \cdot (\alpha^-)^t D\alpha^-} \right) ,$$

where the closest correct and wrong prototypes are referred to, $\mathbf{w}^+$ and $\mathbf{w}^-$, respectively, corresponding to the coefficients α^+ and α^-, respectively and $\Phi(x) = (1 + \exp(-x))^{-1}$. A simple stochastic gradient descent leads to adaptation rules for the coefficients α^+ and α^- in RPC: component k of these vectors is adapted as

$$\Delta\alpha_k^+ \sim -\Phi'(\mu(\mathbf{v}_i)) \cdot \mu^+(\mathbf{v}_i) \cdot \frac{\partial \left([D\alpha^+]_i - \frac{1}{2} \cdot (\alpha^+)^t D\alpha^+\right)}{\partial \alpha_k^+}$$

$$\Delta\alpha_k^- \sim \Phi'(\mu(\mathbf{v}_i)) \cdot \mu^-(\mathbf{v}_i) \cdot \frac{\partial \left([D\alpha^-]_i - \frac{1}{2} \cdot (\alpha^-)^t D\alpha^-\right)}{\partial \alpha_k^-}$$

with

$$\mu(\mathbf{v}_i) = \frac{d(\mathbf{v}_i, \mathbf{w}^+) - d(\mathbf{v}_i, \mathbf{w}^-)}{d(\mathbf{v}_i, \mathbf{w}^+) + d(\mathbf{v}_i, \mathbf{w}^-)}$$

$$\mu^+(\mathbf{v}_i) = \frac{2 \cdot d(\mathbf{v}_i, \mathbf{w}^-)}{(d(\mathbf{v}_i, \mathbf{w}^+) + d(\mathbf{v}_i, \mathbf{w}^-))^2}$$

$$\mu^-(\mathbf{v}_i) = \frac{2 \cdot d(\mathbf{v}_i, \mathbf{w}^+)}{(d(\mathbf{v}_i, \mathbf{w}^+) + d(\mathbf{v}_i, \mathbf{w}^-))^2}$$

The partial derivative yields

$$\frac{\partial \left([D\alpha_j]_i - \frac{1}{2} \cdot \alpha_j^t D\alpha_j\right)}{\partial \alpha_{jk}} = d_{ik} - \sum_l d_{lk}\alpha_{jl}$$

After every adaptation step, normalization takes place to guarantee $\sum_i \alpha_{ji} = 1$. This way, a learning algorithm which adapts prototypes in a supervised manner is given for general dissimilarity data, whereby prototypes are implicitly embedded in pseudo-Euclidean space.

The prototypes are initialized as random vectors corresponding to random values α_{ij} which sum to one. It is possible to take class information into account by setting all α_{ij} to zero which do not correspond to the class of the prototype. Out-of-sample extension of the classification to new data is possible based on the following observation [3]: For a novel data point $\mathbf{v}$ characterized by its pairwise dissimilarities $D(\mathbf{v})$ to the data used for training, the dissimilarity of $\mathbf{v}$ to a prototype α_j is $d(\mathbf{v}, \mathbf{w}_j) = D(\mathbf{v})^t \cdot \alpha_j - \frac{1}{2} \cdot \alpha_j^t D \alpha_j$.

2.2   Conformal Prediction for RPC

RPC can be effectively transferred to a conformal predictor which will be useful to extend it in a non-trivial way to semi-supervised learning. Conformal predictor introduced in [14] aims at the determination of confidence and credibility of classifier decisions. Thereby, the technique can be accompanied by a formal stability analysis. In the context of vectorial data, sparse conformal predictors have been recently discussed in [6], which we review now briefly.

Conformal Prediction. Denote the labeled training data $\mathbf{z}_i = (\mathbf{v}_i, l_i) \in \mathbb{Z} = \mathbb{V} \times \mathbb{L}$. Furthermore let $\mathbf{v}_{N+1}$ be a new data point with unknown label l_{N+1}, i.e. $\mathbf{z}_{N+1} := (\mathbf{v}_{N+1}, l_{N+1})$. For given training data $(\mathbf{z}_i)_{i=1,\ldots,N}$, an observed data point $\mathbf{v}_{N+1}$, and a chosen error rate ϵ, the *conformal prediction* computes an $(1 - \epsilon)$-*prediction region* $\Gamma^\epsilon(\mathbf{z}_1, \ldots, \mathbf{z}_l, \mathbf{v}_{N+1}) \subseteq \mathbb{L}$ consisting of a number of possible label assignments. The applied method ensures that if the data $\mathbf{z}_i$ are *exchangeable*[1] then

$$P(l_{N+1} \notin \Gamma^\epsilon(\mathbf{z}_1, \ldots, \mathbf{z}_l, \mathbf{v}_{N+1})) \le \epsilon$$

holds asymptotically for $N \to \infty$ for each distribution of $\mathbb{Z}$ [14].

To compute the conformal prediction region Γ^ϵ, a *non-conformity measure* is fixed $A(\mathcal{D}, \mathbf{z})$. It is used to calculate a non-conformity value μ that estimates how an observation $\mathbf{z}$ fits to given representative data $\mathcal{D} = \{\mathbf{z}_1, \ldots, \mathbf{z}_N\}$. The conformal algorithm for classification is as follows: given a non-conformity measure A, significance level ϵ, examples $\mathbf{z}_1, \ldots, \mathbf{z}_N$, object $\mathbf{v}_{N+1}$ and a possible label l, it is decided whether l is contained in $\Gamma^\epsilon(\mathbf{z}_1, \ldots, \mathbf{z}_N, \mathbf{v}_{N+1})$, see algorithm 1.

For given $\mathbf{z} = (\mathbf{x}, l)$ and a trained relational prototype-based model, we choose as non-conformity measure

$$\mu := \frac{d^+(\mathbf{x})}{d^-(\mathbf{x})} \tag{2}$$

with $d^+(\mathbf{x})$ being the distance between $\mathbf{x}$ and the closest prototype labeled l, and $d^-(\mathbf{x})$ being the distance between $\mathbf{x}$ and the closest prototype labeled differently than l where distances are computed according to Eq. (1).

---

[1] *Exchangeability* is a weaker condition than data being i.i.d. which is readily applicable to the online setting as well, for example [14].

Algorithm 1. Conformal Prediction (CP)

```
1: function CP(D, v_{N+1}, ε)
2:     for all l ∈ L do
3:         z_{N+1} := (v_{N+1}, l)
4:         for i = 1, ..., N + 1 do
5:             D_i := {z_1, ..., z_{N+1}}\{z_i}
6:             μ_i := A(D_i, z_i)                                    ▷ eq. 2
7:         end for
8:         r_l := |{i=1,...,N+1 | μ_i ≥ μ_{N+1}}| / (N+1)
9:     end for
10:    return Γ^ε := {l : r_l > ε}
11: end function
```

Confidence and Credibility. The prediction region $\Gamma^\epsilon(z_1, \ldots, z_N, v_{N+1})$ stands in the center of conformal prediction. For a given error rate ϵ it contains the possible labels of $\mathbb{L}$. But how can we use it for prediction?

Suppose we use a meaningful non-conformity measure A. If the value ϵ is approaching 0, a conformal prediction with almost no errors is required, which can only be satisfied if the prediction region contains all possible labels. If we raise ϵ we allow errors to occur and as a benefit the conformal prediction algorithm excludes unlikely labels from our prediction region, increasing its information content. In detail those l are discarded for which the r-value is less or equal ϵ. Hence only a few z_i are as non conformal as $z_{N+1} = (v_{N+1}, l)$. This is a strong indicator that z_{N+1} does not belong to the distribution $\mathbb{Z}$ and so l seems not to be the right label. If one further raises ϵ only those l remain in the conformal region that can produce a high r-value meaning that the corresponding z_{N+1} is rated as very typical by A.

So one can trade error rate against information content. The most useful prediction is those containing exactly one label. Therefore, given an input v_i two error rates are of particular interest, ϵ_1^i being the smallest ϵ and ϵ_2^i being the greatest ϵ so that $|\Gamma^\epsilon(D, v_i)| = 1$. ϵ_2^i is the r-value of the best and ϵ_1^i is the r-value of the second best label. Thus, typically, a conformal predictor outputs the label l which describes the prediction region for such choices ϵ, i.e. $\Gamma^\epsilon = \{l\}$, and the classification is accompanied by the two measures

$$\text{confidence} : cf_i := 1 - \epsilon_1^i = 1 - r_{l_{2nd}} \tag{3}$$

$$\text{credibility} : cr_i := \epsilon_2^i = r_{l_{1st}} \tag{4}$$

Confidence says something about being sure that the second best label and all worse ones are wrong. *Credibility* says something about to be sure that the best label is right respectively that the data point is (un)typical and not an outlier.

2.3 Semi-supervised Conformal RPC

In semi-supervised learning unlabeled data are used to enhance the learned model based on only labeled data (denoted as T_1). A very naive approach is so-called *self-training*, which takes iteratively a part of the unlabeled data (denoted as T_2) as new training data into the retraining process until all labeled data are considered [15]. The problem of self-training is how to determine the labels of the unlabeled data which will be taken into the retraining, a simple idea is using *k-NN*, i.e. label the k nearest unlabeled data by the trained model and the predicted labels serve as 'true' labels of the unlabeled data in the retraining. For safety normally small k is used to avoid the degeneration of the learning performance, which can also cause very high computational effort for large data.

In order to get over this problem we combine the self-training approach with conformal prediction. First of all, to identify the unlabeled data with high confidence and credibility values defined by cc_i. For a given data $\mathbf{v}_i \in T_2$,

$$cc_i := cf_i \times cr_i \tag{5}$$

High cc-values of unlabeled data indicate that with high probability their predicted labels are the true underlying labels. That means only the unlabeled data with predicted labels of high probability will be taken into the next retraining. The region which consists of these unlabeled data with high cc_i is referred as '*secure region*' (denoted as $\mathcal{SR}$). Therefrom to identify $\mathcal{SR}$ we take a fraction (prc) of the top cc-values of the unlabeled data[2].

On the other hand in the retraining the '*insecure region*' ($\mathcal{ISR}$) of the training data can be found by

$$\mathcal{ISR} := \left\{ \mathbf{v}_i \in T_1 : cf_i \leq \left(1 - \frac{1}{L}\right) \vee cr_i \leq \frac{1}{L} \right\}. \tag{6}$$

and represented by a new prototype as the median of $\mathcal{ISR}$. This step automatically adapts the complexity of the model, i.e. the number of prototypes. For the next retraining this new prototype will be also trained with the new training data. The proposed method is referred to as *secure semi-supervised conformal relational prototype-based classifier* (SSC-RPC). See algorithm 2.

During the self-training process the training set T_1 is expanded by adding the secure region $\mathcal{SR}$ of unlabeled data to itself while the unlabeled data T_2 is shrunk by discarding its secure region $\mathcal{SR}$. The performance of the retaining is evaluated based on only labeled data. The method terminates if the improvement of the performance is not significant (less than 1%) after a given number of iterations ($win_{\mathrm{max\_itr}}$) or the maximal iterations are reached (max_{itr}) or the insecure region ($\mathcal{ISR}$) is too small or the unlabeled set T_2 is empty. Since the size of $\mathcal{ISR}$ controls the complexity of the model, we found by some independent experiments, that $|\mathcal{ISR}| \leq 5$ is a good compromise between too dense or too sparse models.

---

[2] prc is customizable and in our experiments we set $prc = 5\%$ which is a good compromise between learning performance and efficiency.

Algorithm 2. Secure semi-supervised conformal RPC

```
 1: init: W := ∅, W_new := ∅, W_best := ∅, ISR := ∅; SR := ∅
 2: T_1 := labeled data;    T_2 := unlabeled data
 3: improve = 1%                                    ▷ threshold of improvement: default 1%
 4: EvalSet = T_1                                          ▷ Evaluation set, i.e. labeled data
 5: itr = 0                                                        ▷ iteration counter
 6: ctn_best = 0                                             ▷ counter for best result
 7: max_itr = 100                                         ▷ maximal total iterations
 8: win_max_itr = 10                         ▷ maximal iterations for a result as winner
 9: acc_best = 0
10: repeat                                                   ▷ self-training process
11:     W := W ∪ W_new
12:     T_1 := T_1 ∪ SR, T_2 := T_2 \ SR
13:     W := train T_1 by RPC given W             ▷ training with given prototypes
14:     acc := evaluation of W on EvalSet;
15:     if acc − acc_best ≥ improve then
16:         W_best = W, acc_best = acc, ctn_best = 0
17:     else
18:         ctn_best = ctn_best + 1
19:     end if
20:     A_{T_1} := {μ_i, ∀i ∈ T1}                              ▷ μ-values of T_1: eq. (2)
21:     A_{T_2} := {μ_i, ∀i ∈ T2}
22:     CF_{T_2} := {cf_i, ∀i ∈ T_2}; CR_{T_2} := {cr_i, ∀i ∈ T_2};          ▷ eq. (3),(4)
23:     CF_{T_1} := {cf_i, ∀i ∈ T_1}; CR_{T_1} := {cr_i, ∀i ∈ T_1};
24:     generate ISR of T_1 based on CF_{T_1} and CR_{T_1}                    ▷ eq. (6)
25:     generate SR of T_2 based on CF_{T_2} and CR_{T_2}         ▷ eq. (5) and prc = 5%
26:     generate W_new from SR
27:     itr = itr + 1
28: until |ISR| ≤ 5 or itr = max_itr or ctn_best = win_max_itr or T_2 = ∅
29: return W_best;
```

3   Experiments

We compare SSC-RPC for SSL and RPC (trainded only on labeled data) on a large range of tasks including, five well-known UCI binary data sets[3], four SSL binary benchmark data sets[4], and two real life non-vectorial multi-class data sets from bioinformatic domain. Except for i.i.d. labeled data, we also demonstrate an artificial data set to show the ability of dealing with non i.i.d. labeled data of SSC-RPC. For vectorial data dissimilarity matrices D have been generated by using the squared-Euclidean distance. SSC-RPC has been initialized with one prototype per class, selected randomly from the labeled data set. In order to keep the comparisons fair we set the number of prototypes for each class for RPC to the number of prototypes for each class from SSC-RPC's final result.

Benchmarks and Real Life Data Sets

First we evaluate the methods on different UCI data sets, i.e. Diabetes(D1), German(D2), Haberman(D3), Voting(D4), WDBC(D5), and typical SSL benchmarks, i.e. Digit1(D6), USPS(D7), G241c(D8), COIL(D9) [1] [7]. For Digit1, USPS, G241c, COIL, the archive includes twelve data splits with 100 i.i.d. labeled data points. In oder to keep the same experimental setting, as for UCI

---

[3] http://archive.ics.uci.edu/ml/datasets.html
[4] http://www.kyb.tuebingen.mpg.de/ssl-book

Table 1. Classification results for different vectorial and non-vectorial data

Data	D1	D2	D3	D4	D5	D6	D7	D8	D9	D10	D11
SSC-RPC	70.17	71.61	**73.30**	89.20	92.34	83.57	**79.47**	**73.64**	**59.24**	**81.06**	78.88
	(2.32)	(1.14)	(5.02)	(0.89)	(1.19)	(8.49)	(1.44)	(3.53)	(5.50)	(5.53)	(3.28)
RPC	70.00	71.44	70.27	89.20	92.29	83.55	78.25	72.31	57.00	79.37	78.78
	(2.20)	(1.30)	(7.29)	(0.90)	(1.64)	(8.62)	(2.43)	(5.13)	(2.89)	(4.78)	(3.70)

data sets (as well as for the real life data sets later on), we randomly select 100 examples of the data to be used as labeled examples, and use the remaining data as unlabeled data. The experiments are repeated for 12 times and the average test-set accuracy (on the unlabeled data) and standard deviation are reported.

Further we evaluate the methods on two real life relational data sets, where no direct vector embedding exists and the data are given as (dis-)similarities. The *SwissProt* data set (D10) consists of 5,791 samples of protein sequences in 10 classes taken as a subset from the popular SwissProt database of protein sequences [2] (release 37). The 10 most common classes such as Globin, Cytochrome b, etc. provided by the Prosite labeling. These sequences are compared using Smith-Waterman[4]. The *Copenhagen Chromosomes* data (D11) constitute a benchmark from cytogenetics [8]. 4,200 human chromosomes from 21 classes are represented by grey-valued images. These are transferred to strings measuring the thickness of their silhouettes. These strings can directly be compared using the edit distance based on the differences of the numbers and insertion/deletion costs 4.5 [8]. The classification problem is to label the data according to the chromosome type. The results are shown in Table 1. In half of all cases, semi-supervised learning improves the result, and in the remaining cases it never degenerates the learning performance, which is also an very important issue in SSL [12,15].

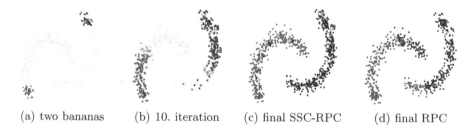

(a) two bananas   (b) 10. iteration   (c) final SSC-RPC   (d) final RPC

Fig. 1. (a) The data consist of green/blue labeled data and gray unlabeled data. Two prototypes are trained by only labeled data and marked with squares. (b) The secure region $\mathcal{SR}$ consists of the unlabeled data marked by stars and the insecure region $\mathcal{ISR}$ contains labeled data rounded by red circles. The new prototype taken from $\mathcal{ISR}$ is marked with a big red cross. During the self-training process additional prototypes are created. (c) the final result of SSC-RPC (d) the final result of RPC based only on labeled data.

Artificial Data Set: Two Banana-Shaped Data Clouds

This data set contains two banana-shaped data clouds indicating two classes. Each banana consists of 300 2-D data points, Fig. 1(a). We select randomly non i.i.d. a small fraction (ca. 5%) of each banana as labeled data. RPC is trained only on labeled data with the same number of prototype for each class which SSC-RPC finally outcomes and can not learn the whole data space very well (see e.g. 1(d)). However, by means of $\mathcal{SR}$ of SSC-RPC the unlabeled data are considered iteratively by the self-training procedure. Figure 1(b), 1(c) shows some intermediate results up to convergence. The average accuracy (on unlabeled data) of 10 times randomly non i.i.d. selected labeled data is reported: SSC-RPC: **94.55%**(8.38), RPC: 77.29%(13.13).

4 Conclusions

We proposed an extension of conformal RPC for SSL by means of 'secure region' of unlabeled data to improve the classifier and 'insecure region' of labeled data to adapt the model complexity. It is a natural multi-class semi-supervised learner for vectorial and non-vectorial data sets. As a wrapper method it can also be integrated with other prototype-based methods. Our experiments show that the approach demonstrates in general superior results compared to standard RPC based on the labeled data alone, especially for non i.i.d. labeled data. Due to the lack of classical SSL benchmarks for non i.i.d. data, we will provide more detailed experiments for these relevant data in later work. Also additional parameter studies for SSC-RPC focusing on the *prc* parameter and sparsity aspects to address large scale problem will be addressed in the future.

Acknowledgments. Financial support from the Cluster of Excellence 277 Cognitive Interaction Technology funded by the German Excellence Initiative is gratefully acknowledged. F.-M. Schleif was supported by the "German Sc. Found. (DFG)" (HA-2719/4-1).

References

1. Chapelle, O., Schölkopf, B., Zien, A. (eds.): Semi-Supervised Learning. MIT Press, Cambridge (2006)
2. Boeckmann, B., et al.: The swiss-prot protein knowledgebase and its supplement trembl in 2003. Nucleic Acids Research 31, 365–370 (2003)
3. Gisbrecht, A., Mokbel, B., Schleif, F.-M., Zhu, X., Hammer, B.: Linear time relational prototype based learning. J. of Neural Sys. 22(5), 72–84 (2012)
4. Gusfield, D.: Algorithms on Strings, Trees, and Sequences: Computer Science and Computational Biology. Cambridge University Press (1997)
5. Haasdonk, B., Bahlmann, C.: Learning with distance substitution kernels. In: Rasmussen, C.E., Bülthoff, H.H., Schölkopf, B., Giese, M.A. (eds.) DAGM 2004. LNCS, vol. 3175, pp. 220–227. Springer, Heidelberg (2004)

6. Hebiri, M.: Sparse conformal predictors. Statistics and Computing 20(2), 253–266 (2010)
7. Li, Y.-F., Zhou, Z.-H.: Towards making unlabeled data never hurt. In: Getoor, L., Scheffer, T. (eds.) ICML, pp. 1081–1088. Omnipress (2011)
8. Neuhaus, M., Bunke, H.: Edit distance based kernel functions for structural pattern classification. Pattern Recognition 39(10), 1852–1863 (2006)
9. Rajadell, O., Garcia-Sevilla, P., Dinh, V.C., Duin, R.P.W.: Semi-supervised hyperspectral pixel classification using interactive labeling. In: 2011 3rd Workshop on WHISPERS, pp. 1–4 (June 2011)
10. Sato, A., Yamada, K.: Generalized learning vector quantization. In: Touretzky, D.S., Mozer, M., Hasselmo, M.E. (eds.) NIPS, pp. 423–429. MIT Press (1995)
11. Schleif, F.-M., Zhu, X., Hammer, B.: A conformal classifier for dissimilarity data. In: Iliadis, L., Maglogiannis, I., Papadopoulos, H. (eds.) AIAI 2012. IFIP AICT, vol. 381, pp. 234–243. Springer, Heidelberg (2012)
12. Singh, A., Nowak, R.D., Zhu, X.: Unlabeled data: Now it helps, now it doesn't. In: Koller, D., Schuurmans, D., Bengio, Y., Bottou, L. (eds.) NIPS, pp. 1513–1520. Curran Associates, Inc. (2008)
13. Trosset, M.W., Priebe, C.E., Park, Y., Miller, M.I.: Semisupervised learning from dissimilarity data. Computational Statistics and Data Analysis 52(10), 4643–4657 (2008)
14. Vovk, V., Gammerman, A., Shafer, G.: Algorithmic Learning in a Random World. Springer, New York (2005)
15. Zhu, X., Goldberg, A.B.: Introduction to semi-supervised learning. Synthesis Lectures on Artif. Intell. and Machine Learning 3(1), 1–130 (2009)

Border-Sensitive Learning in Kernelized Learning Vector Quantization

Marika Kästner[1], Martin Riedel[1], Marc Strickert[2], Wieland Hermann[3], and Thomas Villmann[1],[*]

[1] Computational Intelligence Group,
University of Applied Sciences Mittweida, 09648 Mittweida, Germany
thomas.villmann@hs-mittweida.de
[2] Computational Intelligence Group,
Philipps-University Marburg, 35032 Marburg, Germany
[3] Department of Neurology,
Paracelsus Hospital Zwickau, Zwickau, Germany

Abstract. Prototype based classification approaches are powerful classifiers for class discrimination of vectorial data. Famous examples are learning vector quantization models (LVQ) and support vector machines (SVMs). In this paper we propose the application of kernel distances in LVQ such that the LVQ-algorithm can handle the data in a topologically equivalent data space compared to the feature mapping space in SVMs. Further, we provide strategies to force the LVQ-prototypes to be class border sensitive. In this way an alternative to SVMs based on Hebbian learning is established. After presenting the theoretical background, we demonstrate the abilities of the model for an illustrative toy example and for the more challenging task of classification of Wilson's disease patients according to their neurophysiological impairments.

1 Introduction

Classification of vectorial data is still a challenging topic. If the class distributions are non-linear manifolds or distributions, traditional statistical methods like linear discriminant analysis (LDA) frequently fail. Adaptive models from machine leaning like Learning Vector Quantizers (LVQ,[16]), support vector machines (SVM, [21]) or multilayer perceptrons (MLP,[7]) promise alternatives. LVQs as well as SVMs belong to prototype based classifiers. LVQ algorithms frequently (under certain conditions) generate class typical prototypes whereas in SVMs the resulting prototypes determine the class borders and are here called support vectors. These support vectors are data points identified by convex optimization. Yet, LVQs as introduced by KOHONEN realize a Hebbian learning. Further, LVQs handle the prototypes in the data space such that they are easy to interpret. In contrast, SVMs implicitly map the data into the feature mapping space (FMS). This FMS is high-dimensional, maybe infinite, and the mapping is generally non-linear. These properties frequently lead to a superior performance of SVMs compared to other models. However, the number of support

---

[*] Corresponding author.

I. Rojas, G. Joya, and J. Cabestany (Eds.): IWANN 2013, Part I, LNCS 7902, pp. 357–366, 2013.

vectors, which can be taken as model complexity, may become large and cannot be explicitly controlled.

Recently, a kernelized variant of LVQ was proposed [31], which is based on the generalized LVQ (GLVQ,[19]). The GLVQ provides a cost function to the heuristically motivated LVQ, such that a stochastic gradient descent Hebbian learning can be applied. For this purpose, the classification error is approximated by a differentiable parametrized function based on distance evaluations between prototypes and data. The kernelized GLVQ (KGLVQ) replaces the usually applied Euclidean metric by a kernel metric. However, the prototypes remain class-typical and, in consequence, class border insensitive.

We propose two different methods to establish class border sensitivity in GLVQ. The first one uses an additional penalty term for the cost function such that prototypes move closer to the class borders such that a better sesibility is implicitly achieved. The second one controls the sensitivity by the parameter of the classifier function. This approch leads to an adaptation of prototypes only for those datapoints, which are close to the class borders. Hence, the prototypes learn only these data and, therfore, are sensitized for the class decision boundaries. Both methods are demonstrated for an artificial, illustrating data set and a real world medical classification problem.

2   Generalized Learning Vector Quantization (GLVQ)

2.1   The Basic GLVQ

Basic GLVQ was published by SATO & YAMADA in [19]. The aim was to keep the basic principle of attraction and repulsion in prototype based classification learning in LVQ2.1 but vanquishing the problem of the adaptation heuristic in standard LVQ as suggested by KOHONEN [16]. To be precisely, given a set $V \subseteq \mathbb{R}^D$ of data vectors $\mathbf{v}$ with class labels $x_\mathbf{v} \in \mathcal{C} = \{1, 2, \ldots C\}$ and N prototypes $\mathbf{w}_j \in W \subset \mathbb{R}^D$ with class labels $y_j \in \mathcal{C}$ $(j = 1, \ldots, N)$, the GLVQ introduces a cost function

$$E_{GLVQ}(W) = \sum_{\mathbf{v} \in V} f(\mu(\mathbf{v})) \qquad (1)$$

where the dependence on W is implicitly given by the classifier function

$$\mu(\mathbf{v}) = \frac{d^+(\mathbf{v}) - d^-(\mathbf{v})}{d^+(\mathbf{v}) + d^-(\mathbf{v})} \qquad (2)$$

with $d^+(\mathbf{v}) = d(\mathbf{v}, \mathbf{w}^+)$ denoting the distance between the data vector $\mathbf{v}$ and the closest prototype $\mathbf{w}^+$ with the same class label $y^+ = x_\mathbf{v}$, and $d^-(\mathbf{v}) = d(\mathbf{v}, \mathbf{w}^-)$ is the distance to the best matching prototype $\mathbf{w}^-$ of a different class $(y^- \neq x_\mathbf{v})$. Frequently, the squared Euclidean distance is used. We remark that $\mu(\mathbf{v}) \in [-1, 1]$ holds. The *transfer function* f is the monotonically increasing and commonly a sigmoid function

$$f_\theta(\mu) = \frac{1}{1 + \exp\left(\frac{-\mu}{\theta}\right)} \qquad (3)$$

is taken with $0 < f_\theta(\mu) < 1$. For $\theta \to 0$ the logistic function $f_\theta(\mu)$ converges to the Heaviside function

$$H(x) = \begin{cases} 0 & if \ x \le 0 \\ 1 & else \end{cases}. \tag{4}$$

In this limit, the cost functions $E_{GLVQ}(W)$ counts the misclassifications.

Learning takes place as stochastic gradient descent on $E_{GLVQ}(W)$. In particular, we have

$$\triangle \mathbf{w}^+ \sim \xi^+(\mathbf{v}) \cdot \frac{\partial d^+(\mathbf{v})}{\partial \mathbf{w}^+} \tag{5}$$

and

$$\triangle \mathbf{w}^- \sim \xi^-(\mathbf{v}) \cdot \frac{\partial d^-(\mathbf{v})}{\partial \mathbf{w}^-} \tag{6}$$

with the scaling factors

$$\xi^+(\mathbf{v}) = f'(\mu(\mathbf{v})) \cdot \frac{2 \cdot d^-(\mathbf{v})}{(d^+(\mathbf{v}) + d^-(\mathbf{v}))^2} \tag{7}$$

and

$$\xi^-(\mathbf{v}) = -f'(\mu(\mathbf{v})) \cdot \frac{2 \cdot d^+(\mathbf{v})}{(d^+(\mathbf{v}) + d^-(\mathbf{v}))^2}. \tag{8}$$

For the quadratic Euclidean metric we obtain a vector shift $\frac{\partial d^\pm(\mathbf{v})}{\partial \mathbf{w}^\pm} = -2(\mathbf{v} - \mathbf{w}^\pm)$ for the prototypes.

2.2 GLVQ and Non-euclidean Distances

Depending on the classification problem other (differentiable) dissimilarity measures than the Euclidean may be more appropriate [6,30,28,27,3,15]. Quadratic forms $d_\Lambda(\mathbf{v}, \mathbf{w}) = (\mathbf{v}, \mathbf{w})^\top \Lambda (\mathbf{v}, \mathbf{w})$ are discussed in [4,22,23,24]. Here, the positive semi-definite matrix Λ is decomposed into $\Lambda = \Omega^\top \Omega$ with arbitrary matrices $\Omega \in \mathbb{R}^{m \times D}$ which can be adapted during the training. For classification visualization the parameter m has to be two or three, the full problem is obtained for $m = D$.

Recent considerations deal with kernel distances $d_{\kappa_\Phi}(\mathbf{v}, \mathbf{w})$ [5,20,31] defined by the kernel κ_Φ determining the so-called kernel map Φ. More precisely we have

$$d_{\kappa_\Phi}(\mathbf{v}, \mathbf{w}) = \sqrt{\kappa_\Phi(\mathbf{v}, \mathbf{v}) - 2\kappa_\Phi(\mathbf{v}, \mathbf{w}) + \kappa_\Phi(\mathbf{w}, \mathbf{w})}, \tag{9}$$

where the kernel $\kappa_\Phi(\mathbf{v}, \mathbf{w})$ is assumed to be universal and differentiable [31]. The kernel $\kappa_\Phi(\mathbf{v}, \mathbf{w})$ implicitly defines a generally non-linear mapping

$$\Phi : V \to \mathcal{I}_{\kappa_\Phi} \subseteq \mathcal{H} \tag{10}$$

of the data and prototypes into a high- maybe infinite-dimensional function Hilbert space $\mathcal{H}$ with the metric $d_\mathcal{H}(\Phi(\mathbf{v}), \Phi(\mathbf{w})) = d_{\kappa_\Phi}(\mathbf{v}, \mathbf{w})$ [1,18]. The image $\mathcal{I}_{\kappa_\Phi} = span(\Phi(V))$ forms a subspace of $\mathcal{H}$ for universal kernels [25]. For differentiable universal kernels we can define an accompanying transformation $\Psi : V \longrightarrow \mathcal{V}$, where in $\mathcal{V}$ the data are equipped with the kernel metric d_{κ_Φ}. The map Ψ is bijective and non-linear iff Φ does [25]. It turns out that $\mathcal{V}$ is an isometric isomorphism to $\mathcal{I}_{\kappa_\Phi}$, and the differentiability of the kernel ensures the applicability of the stochastic gradient learning of GLVQ in $\mathcal{V}$ for the kernel distance [31]. Hence, the resulting kernel GLVQ (KGLVQ) is running in the new data space $\mathcal{V}$, which offers the same topological structure and richness as the image $\mathcal{I}_{\kappa_\Phi}$ as known from SVMs. We denote this new data space $\mathcal{V}$ as *kernelized data space*.

3   Class Border Sensitive Learning in GLVQ

As we have seen in the previous section, GLVQ can be extended using kernel distances in the new data space $\mathcal{V}$. However, in general, the prototypes of GLVQ are not particularly sensitized to detect the class borders. This might be a disadvantage for KGLVQ compared to support vector machines, if precise decisions are favored. In this section, we provide two possibilities to integrate class border sensitivity in GLVQ or KGLVQ. The first one applies an additive attraction force for prototypes with different class responsibilities, such that the prototypes move closer to each other, which *implicitly* leads to an improved classborder sensitivity. The second approach uses a parametrized sigmoid transfer functions $f_\theta(\mu)$ in (1), where the θ-parameter controls the class border sensitivity via so-called active set. These active sets are subsets of the whole dataset containing only datpoints close to the class decision borders.

3.1   Border Sensitive Learning in GLVQ by a Penalty Function

Class border sensitivity learning by an additive penalty term was proposed for classification problems using unsupervised fuzzy-c-means models in [29,33]. Here we adopt these ideas for class border sensitive learning in GLVQ (BS-GLVQ). In particular, we extend the cost function of GLVQ (1) by a convex sum

$$E_{BS-GLVQ}(W,\gamma) = (1-\gamma) \cdot E_{GLVQ}(W) + \gamma \cdot F_{neigh}(W,V) \qquad (11)$$

with the new *neighborhood-attentive attraction force* (NAAF)

$$F_{neigh}(W,V) = \sum_{\mathbf{v} \in V} \sum_{k:\mathbf{w}_k \in W^-(\mathbf{v})} h_\sigma^{NG}\left(k, \mathbf{w}^+, W^-(\mathbf{v})\right) d\left(\mathbf{w}^+, \mathbf{w}_k\right) \qquad (12)$$

and the sensitivity control parameter $\gamma \in (0,1)$. The set $W^-(\mathbf{v}) \subset W$ is the set of all prototypes with incorrect class labels for a given data vector $\mathbf{v}$. The neighborhood function

$$h_{\sigma_-}^{NG}\left(k, \mathbf{w}^+, W^-(\mathbf{v})\right) = c_{\sigma_-}^{NG} \cdot \exp\left(-\frac{\left(rk_k\left(\mathbf{w}^+, W^-(\mathbf{v})\right) - 1\right)^2}{2\sigma_-^2}\right) \qquad (13)$$

defines a neighborhood of the prototypes in $W^-(\mathbf{v})$ with respect to the best matching correct prototype $\mathbf{w}^+$ known from *Neural Gas* (NG,[17]). Here, $rk_k(\mathbf{w}^+, W^-(\mathbf{v}))$ is the dissimilarity rank function of the prototypes $\mathbf{w}_k \in W^-(\mathbf{v})$ with respect to $\mathbf{w}^+$ defined as

$$rk_k\left(\mathbf{w}^+, W^-(\mathbf{v})\right) = \sum_{\mathbf{w}_l \in W^-(\mathbf{v})} H\left(d\left(\mathbf{w}^+, \mathbf{w}_k\right) - d\left(\mathbf{w}^+, \mathbf{w}_l\right)\right) \qquad (14)$$

with H being the Heaviside function (4). The NAAF causes an additional gradient term

$$\frac{\partial F_{neigh}(W, V)}{\partial \mathbf{w}_j} = h_{\sigma_-}^{NG}\left(j, \mathbf{w}^+, W^-(\mathbf{v})\right) \cdot \frac{\partial d\left(\mathbf{w}^+, \mathbf{w}_j\right)}{\partial \mathbf{w}_j} \qquad (15)$$

for a given input vector $\mathbf{v}$ and $\mathbf{w}_j \in W^-(\mathbf{v})$, i.e. all incorrect prototypes are gradually moved towards the correct best matching prototype $\mathbf{w}^+$ according to their dissimilarity rank with respect to $\mathbf{w}^+$. Thus, σ_- adjusts the neighborhood cooperativeness while the weighting coefficient γ controls the influence of border sensitive learning in this model.

Obviously, this method enhances prototypes positioned close to the decision borders to move closer together. Hence, an implicit better classborder sensitivity is obtained.

3.2 Class Border Sensitive Learning by Parametrized Transfer Functions in GLVQ

Following the explanations in [26,32], we investigate in this subsection the influence of an appropriately chosen parametrized transfer function f to be applied in the cost function (1) of GLVQ. For the considerations here, the logistic function f_θ from (3) is used. It is well-known that the derivative $f'_\theta(\mu(\mathbf{v}))$ of the logistic function can be expressed as

$$f'_\theta(\mu(\mathbf{v})) = \frac{f_\theta(\mu(\mathbf{v}))}{2\theta^2} \cdot (1 - f_\theta(\mu(\mathbf{v}))), \qquad (16)$$

which appears in the scaling factors for $\xi^\pm$ in (7) and (8) for the winning prototypes $\mathbf{w}^\pm$. Looking at these derivatives (see Fig. 1) we observe that a significant prototype update only takes place for a small range of the classifier values μ in (2) depending on the parameter θ. Hence, we consider the *active set* of the data contributing significantly to a prototype update as

$$\hat{\Xi} = \left\{ \mathbf{v} \in V | \mu(\mathbf{v}) \in \left[-\frac{1 - \mu_\theta}{1 + \mu_\theta}, \frac{1 - \mu_\theta}{1 + \mu_\theta} \right] \right\} \qquad (17)$$

with μ_θ chosen such that $f'_\theta(\mu) \approx 0$ is valid for $\mu \in \Xi = V \backslash \hat{\Xi}$, see Fig. 1.

Obviously, the active set is distributed along the class decision boundaries, because $f'_\theta(\mu) \gg 0$ is valid only there. This corresponds to $\mu(\mathbf{v}) \approx 0$. Hence, this active set $\hat{\Xi}$ can be understood as another formulation of KOHONEN's window rule in LVQ2.1

$$\min\left(\frac{d^+(\mathbf{v})}{d^-(\mathbf{v})}, \frac{d^-(\mathbf{v})}{d^+(\mathbf{v})}\right) \geq \frac{1 - w}{1 + w} \qquad (18)$$

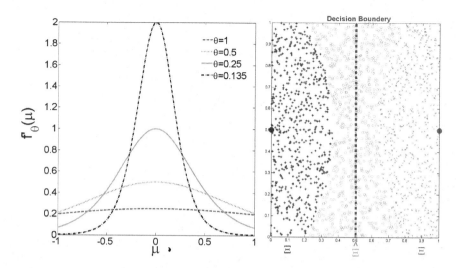

Fig. 1. left): derivatives $f'_\theta(\mu)$ for different θ-values; **right)** Visualization of the active set $\hat{\Xi}$ (green points) for a simple example. The prototypes are the big dots.

taking there $w = \mu_\theta$ [16]. The learning of the parameter θ in GLVQ was explicitly addressed in [32]. Optimization for accuracy improvement was discussed in [26].

Here we emphasize the aspect that the parameter θ allows a *control of the width of the active set* surrounding the class borders. Small θ-values define small stripes as active sets. In consequence, only these data contribute to the prototype updates. In other words, according to (17), the active set is crisp but the possibilities for control are smooth such that we could speak about *thresholded active sets*. Therefore, border sensitivity leads to prototypes sensibilized to those datapoints close to the class borders in dependence on the control parameter θ. In this sense, the active set learning can be seen as a kind of *attention based learning* [14].

4 Illustrative Example and Application

In the following we give an illustrative example for the above introduced concepts of class border sensitivity for two-dimensional data for better visualization. Thereafter we present results from a medical application.

4.1 Illustrative Toy Example

For illustration we consider a two-dimensional three-class problem, Fig.2.

We compare both border sensitive approaches with a standard GLVQ network with the identity transfer function $f(\mu) = \mu$. In case of the parametrized transfer function, we used the logistic function (3) with initial parameter $\theta_{init} = 1.0$ decreased to $\theta_{fin} = 0.1$ during learning (*sigmoid GLVQ*). We observe that both border sensitive models place the prototypes closer to the class borders than standard GLVQ, see Fig. 2. Moreover, the classification accuracy is improved: For the BS-GLVQ we achieved 91.1% and the sigmoid variant results 97.2%

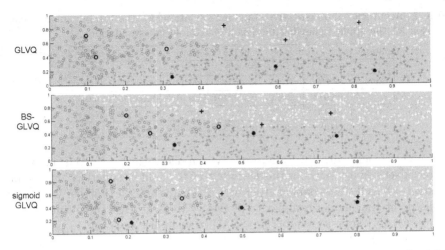

Fig. 2. Obtained prototype positions for standard GLVQ (top), BS-GLVQ (middle) and sigmoid GLVQ (down). Further explanations, see text.

whereas standard GLVQ gets only 89.7%. Thus class border sensitive models detect the noisy class borders more accurately.

4.2 Classification of Wilson's Disease Based on Neurological Data

Wilson's disease is an autosomal-recessive disorder of copper metabolism, which leads to disturbances in liver function and basal ganglia showing hepatic and extrapyramidal motor symptoms. This disorder causes neurological as well as fine-motoric impairments [8,9,11]. The fine-motoric symptoms cna be used for classification of the patiens with respect to their extrapyramidal motor symptom level [13].

According to a clinical scheme based on the neurological disturbances suggested by KONOVALOV, patients can be partitioned into two main groups: patients with neurological and without neurological manifestation denoted as neurological and non-neurological group, respectively [2,8]. In addition to hepatolenticular degeneration in Wilson's disease, sensory and extrapyramidal motoric systems are also disturbed. The impairments of these nervous pathways can be detected investigation of latencies of evoked potentials collected in a data vector denoted as electrophysiological impairment profil (EIP). The database here contains $M = 122$ five-dimensional EIPs described in [12]. Yet, it is not clear so far, whether a precise classification of the EIPs according to their underlying neurological type is possible [10].

We applied with and without border sensitive learning KGLVQ via the θ-parameter of the sigmoid transfer function as well as BS-KGLVQ. For all these models we used 6 prototypes per class. For comparison, a SVM with radial basis function kernel (rbf) was trained. The data were preprocessed by a z-score-transformation and classification results are obtained as 10-fold cross validation. The results are depicted in Tab. 1.

Table 1. Accuracies and respective standard deviations for the Wilson's disease classification for the applied classifier models after 10-fold crossvalidation

Dataset	KGLVQ					SVM
	$\theta = 1$	$\theta = 2.5$	$\theta = 3.5$	$\theta = 5$	BS	rbf
training	87.8%	91.9%	90.0%	90.4%	90.1 %	87.5%
	(±0.013)	(±0.015)	(±0.015)	(±0.014)	(±0.011)	(±0.015)
test	81.9%	82.6%	89.2%	87.4%	91.0 %	87.4%
	(±0.086)	(±0.086)	(±0.083)	(±0.090)	(±0.090)	(±0.137)

KGLVQ achieves drastically improved accuracies, which is further improved by border sensitivity. Without this feature, SVMs would be superior. Yet, adding this strategy, KGLVQ variants outperform SVMs in this case. Further we remark at this point that model complexity of the SVMs is at least three times larger (in average 45.5 support vectors for SVM) in comparison to the 12 prototypes used for the GLVQ models.

Althoug we obtained a quite high performance, the obtained classification accuracies are not sufficiently high for a secure clinical discrimination. For this purpose, further investigations including an improved database and/or other dissimilarity measures are mandatory.

5   Conclusion and Outlook

In this paper we investigate two strategies for class border sensitive learning in GLVQ. The first one adds a penalty term to the cost function to force class border sensitivity of the prototypes, the second uses a parameter control of the sigmoid transfer function defining active sets of data as a kind of attention based learning. These strategies together with a kernelized variant of GLVQ offer a powerful alternative to SVMs. An advantage of the introduced approaches compared to SVM is the explicit control of the model complexity in GLVQ/KGLVQ, because the number of prototypes has to be chosen in advance for these models whereas in SVMs the number of support vector may become quite large in case of difficult classification tasks. We applied and compared the approaches for a medical data set of neurophysiological data in case of Wilson's disease. Border sensitive KGLVQ variants achieve better results than SVMs with significant lower model complexity. Further, the classification results indicate that a discrimination between neurological and non-neurological type of Wilson's disease can be performed on the basis of electrophysiological impairment profiles. However, this needs further investigations.

References

1. Aronszajn, N.: Theory of reproducing kernels. Transactions of the American Mathematical Society 68, 337–404 (1950)
2. Barthel, H., Villmann, T., Hermann, W., Hesse, S., Kühn, H.-J., Wagner, A., Kluge, R.: Different patterns of brain glucose consumption in Wilsons disease. Zeitschrift für Gastroenterologie 39, 241 (2001)

3. Biehl, M., Hammer, B., Schneider, P., Villmann, T.: Metric learning for prototype-based classification. In: Bianchini, M., Maggini, M., Scarselli, F., Jain, L. (eds.) Innovations in Neural Information Paradigms and Applications. SCI, vol. 247, pp. 183–199. Springer, Berlin (2009)

4. Bunte, K., Schneider, P., Hammer, B., Schleif, F.-M., Villmann, T., Biehl, M.: Limited rank matrix learning, discriminative dimension reduction and visualization. Neural Networks 26(1), 159–173 (2012)

5. Hammer, B., Strickert, M., Villmann, T.: Supervised neural gas with general similarity measure. Neural Processing Letters 21(1), 21–44 (2005)

6. Hammer, B., Villmann, T.: Generalized relevance learning vector quantization. Neural Networks 15(8-9), 1059–1068 (2002)

7. Haykin, S.: Neural Networks. A Comprehensive Foundation. Macmillan, New York (1994)

8. Hermann, W., Barthel, H., Hesse, S., Grahmann, F., Kühn, H.-J., Wagner, A., Villmann, T.: Comparison of clinical types of Wilson's disease and glucose metabolism in extrapyramidal motor brain regions. Journal of Neurology 249(7), 896–901 (2002)

9. Hermann, W., Eggers, B., Barthel, H., Clark, D., Villmann, T., Hesse, S., Grahmann, F., Kühn, H.-J., Sabri, O., Wagner, A.: Correlation between automated writing movements and striatal dopaminergic innervation in patients with Wilson's disease. Journal of Neurology 249(8), 1082–1087 (2002)

10. Hermann, W., Günther, P., Wagner, A., Villmann, T.: Klassifikation des Morbus Wilson auf der Basis neurophysiologischer Parameter. Der Nervenarzt 76, 733–739 (2005)

11. Hermann, W., Villmann, T., Grahmann, F., Kühn, H., Wagner, A.: Investigation of fine motoric disturbances in Wilson's disease. Neurological Sciences 23(6), 279–285 (2003)

12. Hermann, W., Villmann, T., Wagner, A.: Elektrophysiologisches Schädigungsprofil von Patienten mit einem Morbus Wilson'. Der Nervenarzt 74(10), 881–887 (2003)

13. Hermann, W., Wagner, A., Kühn, H.-J., Grahmann, F., Villmann, T.: Classification of fine-motoric disturbances in Wilson's disease using artificial neural networks. Acta Neurologica Scandinavia 111(6), 400–406 (2005)

14. Herrmann, M., Bauer, H.-U., Der, R.: The 'perceptual magnet' effect: A model based on self-organizing feature maps. In: Smith, L.S., Hancock, P.J.B. (eds.) Neural Computation and Psychology, Stirling, pp. 107–116. Springer (1994)

15. Kästner, M., Hammer, B., Biehl, M., Villmann, T.: Functional relevance learning in generalized learning vector quantization. Neurocomputing 90(9), 85–95 (2012)

16. Kohonen, T.: Self-Organizing Maps (Second Extended Edition). Springer Series in Information Sciences, vol. 30. Springer, Heidelberg (1997)

17. Martinetz, T.M., Berkovich, S.G., Schulten, K.J.: 'Neural-gas' network for vector quantization and its application to time-series prediction. IEEE Trans. on Neural Networks 4(4), 558–569 (1993)

18. Mercer, J.: Functions of positive and negative type and their connection with the theory of integral equations. Philosophical Transactions of the Royal Society, London, A 209, 415–446 (1909)

19. Sato, A., Yamada, K.: Generalized learning vector quantization. In: Touretzky, D.S., Mozer, M.C., Hasselmo, M.E. (eds.) Proceedings of the 1995 Conference on Advances in Neural Information Processing Systems 8, pp. 423–429. MIT Press, Cambridge (1996)

20. Schleif, F.-M., Villmann, T., Hammer, B., Schneider, P.: Efficient kernelized proto-type based classification. International Journal of Neural Systems 21(6), 443–457 (2011)

21. Schölkopf, B., Smola, A.: Learning with Kernels. MIT Press (2002)

22. Schneider, P., Bunte, K., Stiekema, H., Hammer, B., Villmann, T., Biehl, M.: Regularization in matrix relevance learning. IEEE Transactions on Neural Networks 21(5), 831–840 (2010)

23. Schneider, P., Hammer, B., Biehl, M.: Adaptive relevance matrices in learning vector quantization. Neural Computation 21, 3532–3561 (2009)

24. Schneider, P., Hammer, B., Biehl, M.: Distance learning in discriminative vector quantization. Neural Computation 21, 2942–2969 (2009)

25. Steinwart, I.: On the influence of the kernel on the consistency of support vector machines. Journal of Machine Learning Research 2, 67–93 (2001)

26. Strickert, M.: Enhancing M|G|RLVQ by quasi step discriminatory functions using 2nd order training. Machine Learning Reports 5(MLR-06-2011), 5–15 (2011), http://www.techfak.uni-bielefeld.de/~fschleif/mlr/mlr$_$06$_$2011.pdf, ISSN:1865-3960

27. Strickert, M., Schleif, F.-M., Seiffert, U., Villmann, T.: Derivatives of Pearson correlation for gradient-based analysis of biomedical data. Inteligencia Artificial, Revista Iberoamericana de Inteligencia Artificial (37), 37–44 (2008)

28. Villmann, T.: Sobolev metrics for learning of functional data - mathematical and theoretical aspects. Machine Learning Reports, 1(MLR-03-2007), 1–15 (2007), http://www.uni-leipzig.de/~compint/mlr/mlr_01_2007.pdf, ISSN:1865-3960

29. Villmann, T., Geweniger, T., Kästner, M.: Border sensitive fuzzy classification learning in fuzzy vector quantization. Machine Learning Reports, 6(MLR-06-2012), 23–39 (2012), http://www.techfak.uni-bielefeld.de/~fschleif/mlr/mlr$_$06$_$2012.pdf, ISSN:1865-3960

30. Villmann, T., Haase, S.: Divergence based vector quantization. Neural Computation 23(5), 1343–1392 (2011)

31. Villmann, T., Haase, S., Kästner, M.: Gradient based learning in vector quantization using differentiable kernels. In: Estevez, P.A., Principe, J.C., Zegers, P. (eds.) Advances in Self-Organizing Maps. AISC, vol. 198, pp. 193–204. Springer, Heidelberg (2013)

32. Witoelar, A., Gosh, A., de Vries, J., Hammer, B., Biehl, M.: Window-based example selection in learning vector quantization. Neural Computation 22(11), 2924–2961 (2010)

33. Yin, C., Mu, S., Tian, S.: Using cooperative clustering to solve multiclass problems. In: Wang, Y., Li, T. (eds.) Foundations of Intelligent Systems. AISC, vol. 122, pp. 327–334. Springer, Heidelberg (2011)

Smoothed Emphasis for Boosting Ensembles[*]

Anas Ahachad, Adil Omari, and Aníbal R. Figueiras-Vidal

Department of Signal Theory and Communications,
Universidad Carlos III de Madrid, 28911 Leganés-Madrid, Spain
{anas,aomari,arfv}@tsc.uc3m.es
http://www.tsc.uc3m.es

Abstract. Real AdaBoost ensembles have exceptional capabilities for success-fully solving classification problems. This characteristic comes from progres-sively constructing learners paying more attention to samples that are difficult to be classified.

However, the corresponding emphasis can be excessive. In particular, when the problem to solve is asymmetric or includes imbalanced outliers, even the previously proposed modifications of the basic algorithm are not as effective as desired.

In this paper, we introduce a simple modification which uses the neighbor-hood concept to reduce the above drawbacks. Experimental results confirm the potential of the proposed scheme.

The main conclusions of our work and some suggestions for further research along this line close the paper.

Keywords: Boosting, classification, emphasized samples, machine ensembles, nearest neighbors.

1   Introduction

Decision making (or classification) is a very frequent and remarkably important hu-man activity. Learning machines [1] [2] provide a useful support for it. In particular, machine ensembles offer high performance with a moderate design effort, thanks to their appropriate use of diverse learners and adequate aggregation mechanisms [3] [4]. Among them, boosting ensembles, that appeared under a filtering form [5] to arrive to the basic Adaboost [6] for hard output learners and Real Adaboost [7] for soft output units, merit much attention because their surprising resistance to overfit-ting, that allows impressive results in solving decision and classification problems. Nevertheless, these ensembles are sensitive to very noisy samples and outliers, just because they progressively emphasize the samples that are more difficult to classify for the previously constructed (partial) ensembles. Detailed research served to reach the conclusion of that this resistance is decreased when the samples are very noisy or outliers appear.

---

[*] This work has been partly supported by grant TIN 2011-24533 (Spanish MEC).

I. Rojas, G. Joya, and J. Cabestany (Eds.): IWANN 2013, Part I, LNCS 7902, pp. 367–375, 2013.

Many techniques have been proposed to limit the performance reduction that intensive noise and outliers produce, from simply deleting clearly incorrect samples [8] to using hybrid weighting methods according to the error size and the proximity of each sample to the decision border [9] [10], including regularization [11], soft-marging methods [12] [13], data skewness penalization [14], margin optimization [15], by applying subsampling [16] and other procedures.

As the own boosting algorithms, the above mentioned modifications to fight against noise and outliers do not pay attention to the local characteristics of the problem under analysis. In fact, the use of local learners for Real Adaboost designs is not very frequent, because its systematic versions demand the introduction of kernels, and this and the subsequent Support Vector Machine (SVM) or Maximal Margin (MM) formulations create difficulties, because the learners become strong, and this is not adequate for the boosting principles. Most of the proposed solutions [17] [18] are not successful, and using Linear Programming and subsampling [19] is a must. There is also the possibility of applying a kernel gate for aggregation in order to provide local sensitivity to Real Adaboost [20]. In any case, we repeat, boosting algorithms are essentially not sensitive to the local characteristics of the databases.

Consequently, the above mentioned modifications to keep the good properties of boosting are not useful against imbalanced situations, as the appearance of outliers for just a class or in asymmetric forms. Since outliers are exceptional samples, a possibility to deal with these situations (that are not easy to perceive) is to emphasize the samples according to their errors and those of their neighbors.

In this paper, we propose an elementary form of applying this approach, and check its usefulness by considering a number of benchmark binary classification problems.

The rest of the paper is organized as follows. Section 2 presents the modification of Real Adaboost we propose, and discuss its main characteristics. Section 3 checks the usefulness of that modification in a toy problem which allows to perceive the difficulties and to visualize the different results coming from applying Real Adaboost (RAB) and our modified algorithm. Section 4 shows the comparative performance of the proposed modification versus Real Adaboost in a number of benchmarking problems, discussing the corresponding results. A brief summary of the conclusions from our work plus some possibilities for further research along this line close the contribution.

2   The Proposed Modified RAB

For a binary classification problem and a dataset $\left\{\mathbf{x}^{(l)}, d^{(l)}\right\}$, $l=1,...,L$, RAB sequentially trains learners $t=1,...,T$ with outputs $o_t(x)$ by minimizing the weighted quadratic cost

$$E_t = \sum_{l=1}^{L} D_t(\mathbf{x}^{(l)})[d^{(l)} - o_t(\mathbf{x}^{(l)})]^2 \qquad (1)$$

where

$$D_t(\mathbf{x}^{(l)}) = \frac{D_{t-1}(\mathbf{x}^{(l)})\exp[-\alpha_{t-1}o_{t-1}(\mathbf{x}^{(l)})d^{(l-1)}]}{Z_t} \tag{2}$$

starting with $D_1(x^{(l)}) = 1/L$, Z_t being a normalization constant, and

$$\alpha_t = \frac{1}{2}\ln\frac{1+\gamma_t}{1-\gamma_t} \tag{3}$$

where

$$\gamma_t = \sum_{l=1}^{L} D_t(\mathbf{x}^{(l)})o_t(\mathbf{x}^{(l)})d^{(l)} \tag{4}$$

is the edge parameter. The final decision is given by

$$\hat{d}(\mathbf{x}) = sgn\left[\sum_{t=1}^{T}\alpha_t o_t(\mathbf{x})\right] \tag{5}$$

We propose to replace $D_t(x^{(l)})$ in (1) for the convex combination

$$Q_t(\mathbf{x}^{(l)}) = \beta D_t(\mathbf{x}^{(l)}) + (1-\beta)\frac{\sum_{i \in V^{(l)}} D_t(\mathbf{x}^{(i)})}{K} \tag{6}$$

where $\beta \in [0,1]$ is a convex combination parameter, and $V^{(l)}$ indicates the K-NN neighborhood of the sample $\mathbf{x}^{(l)}$.

It is evident that (6) creates an emphasis for each sample that is a combination of the value that corresponds to that sample and those of the K nearest neighbors. Obviously, if $\mathbf{x}^{(l)}$ is an outlier, most the neighbors will be correctly classified, and (6) will reduce the value of the emphasis to be applied to $\mathbf{x}^{(l)}$. When $\mathbf{x}^{(l)}$ is near the classification border, its neighbor samples will also be near that border, and the net effect will be minor. The same will occur when $\mathbf{x}^{(l)}$ is a typical sample lying on the correct side of the classification frontier.

It is true that other methods could offer similar effects, such as mixed emphasis techniques [9] [10] according to the error value and the proximity to the border, but they will be less effective when there are outliers or asymmetries in the samples that are impossible for a correct classification. On the other hand, according to the previous discussion, (6) has also a "mixed emphasis" effect.

We will denote our design as KRAB (K-nearest neighbor RAB). In Sections 3 and 4, we will check its usefulness to reduce the difficulties we have addressed in our discussion.

3   A Toy Problem

To visualize and understand the advantage of KRAB with respect to RAB, we present the classification results of both algorithms using MultiLayer Perceptrons (MLPs) as learners, with M=3 hidden units, and β =0.3, K=3, as values for the parameters of (6), in solving an easy classification problem: C_1 and C_{-1} are samples of two circular Gaussian distributions with the same variance v = 0.04 and means [0.3,0] and [-0.3,0], respectively. We take 100 + 100 samples, plus 5 outliers appearing around the center of the first Gaussian.

Fig. 1 shows how KRAB creates a border more similar to the theoretical frontier (vertical line X_2 = 0), because it avoids the "attraction" of the border by the outliers.

Similar effects appear, for example, when the Gaussian distributions are not circular and one of them presents a more acute form towards the center of the other. All these imbalances and asymmetries can appear in practical problems, but are difficult to visualize. We will check experimentally if KRAB gives good results when applied to a number of benchmark datasets.

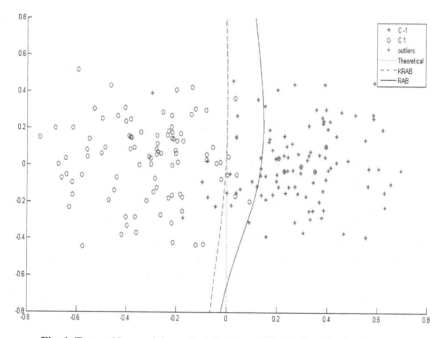

Fig. 1. Toy problem and theoretical, RAB and KRAB classification borders

4   Experiments

4.1   Databases

We will compare our smoothed emphasis RAB with the standard RAB for twelve binary benchmark problems: Nine from [17]: Abalone, Breast, Credit, Diabetes,

German, Hepatitis, Image, Ionosphere, and Waveform; plus Crabs and Ripley [18] and Kwok [19].

Table 1 presents their main characteristics and their three letter names to be used from now here.

Table 1. Characteristics of the benchmark problems

Dataset	Notation	#Train C_1 / C_{-1}	#Test C_1 / C_{-1}	Dimension (D)
Abalone	aba	2507 1238/1269	1670 843/827	8
Breast	bre	420 145/275	279 96/183	9
Crabs	cra	120 59/61	80 41/39	7
Credit	cre	414 167/247	276 140/136	15
Diabetes	dia	468 172/296	300 96/204	8
German	ger	700 214/486	300 86/214	20
Hepatitis	hep	93 70/23	62 53/9	19
Image	ima	1300 736/564	1010 584/426	18
Ionosphere	ion	201 101/100	150 124/26	34
Kwok	kwo	500 300/200	10200 6120/4080	2
Ripley	rip	250 125/125	1000 500/500	2
Waveform	wav	400 124/276	4600 1523/3077	21

4.2   Classifier Design

As learners, we use MultiLayer Perceptrons (MLPs) with one hidden layer with M activation units.

We need to select M and, for our design, β and K. We use a 20 runs (with different initial weights for the MLPs) 5-fold cross validation (CV) to select these parameters, exploring them in the following manner:

– M: from 2 to M_{max} (the value which ensures that there are at least four training samples per each trainable parameter) in steps of 1

– K: 1, 2, 3, …, 11

– β : 0, 0.1, 0.2, …, 1

(hep and rip require to explore M until 20 and 70, respectively).

MLPs are trained with the backpropagation algorithm using a decreasing learning step from 0.01 to 0 along 100 training cycles (that are enough for convergence), and T is selected, following the ideas presented in [9] [10], as the first value for which

$$\frac{\sum_{t'=T-9}^{T} \alpha_{t'}}{\sum_{t'=1}^{T} \alpha_{t'}} < C_{stop}$$

where the threshold C_{stop} has been empirically established to 0.1 for all the algorithms and all the problems. Let us mention that the selection of this stopping criterion is due to its robustness. Nor in [9] [10], nor in our experiments there have been significant overfitting problems due to this simple stopping procedure.

4.3   Results and Their Discussion

Table 2 presents the average classification error rates (CE (%)) and their standard deviations, the average number of learners and their standard deviation, the CV selected parameters for each design, and the results of T-tests to evaluate the statistical significance of performance differences (when T-test<0.05), all obtained from 50 runs on the test sets. Note that there is a tie for cra (a curious dataset, that many classifiers lead to CE=2.5%), and that the proposed modification KRAB is never worse that the standard RAB, KRAB being better for 7 out of the 12 datasets without relevantly increasing the operation computational cost but in aba and hep. In some cases, such as bre and hep, the CE differences are clearly important.

With respect to the reliability of the CV process for designing the ensembles (three parameters are involved for KRAB), Table 2 presents the "omniscient" references (invalid designs that select non-trainable parameters according to the test sets). It is easy to check that there are not unexpected performance differences with respect to the CV designs (parameter differences only indicates a moderate sensitivity with respect to their selection for the corresponding problems). Naturally, omniscient results are (slightly) better than those of the cross-validated designs, but note that the difference is relatively small. So, we can conclude that CV has successfully found good non-trainable parameter values in all the problems and designs – even for our proposed KRAB, which needs to cross-validate 3 non-trainable parameters.

Table 2. Classification error rates (CE) and ensemble sizes (T) for RAB and KRAB with CV, and classification error rates (CE) achieved by the omniscient designs

		RAB M	KRAB M / K / β	Ttest	Omni RAB M	Omni KRAB M / K / β
aba	CE (%)	19.40 ± 0.02	**18.98 ± 0.18**		19.32 ± 0.26	18.88 ± 0.15
	T	31.20 ± 0.40	26.64 ± 2.31	**7.7597e-025**		
	Param.	4	27/9/0.6		7	21/5/0.4
bre	CE (%)	2.60 ± 0.40	**2.19 ± 0.34**		2.29 ± 0.38	1.97 ± 0.22
	T	21.30 ± 4.20	30.32 ± 7.12	**7.0811e-008**		
	Param.	6	2/7/0.9		7	2/10/0.5
cra	CE (%)	2.50 ± 0	2.50 ± 0		2.50 ± 0	2.50 ± 0
	T	11.10 ± 0.80	20.42 ± 1.58	--		
	Param.	2	2/1/0		2	2/1/0
cre	CE (%)	10.14 ± 0.74	**9.86 ± 0.81**		9.40 ± 0.76	9.18 ± 0.52
	T	29.36 ± 6.49	21.60 ± 3.43	0.0726		
	Param.	2	4/1/0.5		3	5/1/0.8
dia	CE (%)	20.61 ± 0.78	**20.39 ± 0.66**		20.05 ± 0.44	20.00 ± 0.81
	T	33.20 ± 4.98	33.68 ± 4.41	0.2279		
	Param.	2	2/3/0.8		8	11/3/0
ger	CE (%)	22.27 ± 0.71	**22.11 ± 0.80**		21.67 ± 0.63	21.57 ± 0.55
	T	33.60 ± 5.92	18.18 ± 1.16	0.0731		
	Param.	2	2/2/0		2	2/5/0.8
hep	CE (%)	8.90 ± 1.80	**6.55 ± 0.89**		7.26 ± 1.85	6.05 ± 1.27
	T	22.20 ± 3.90	60.06 ± 17.52	**7.6872e-013**		
	Param.	17	18/11/0.3		11	2/6/0.2
ima	CE (%)	2.99 ± 0.43	**2.86 ± 0.46**		2.82 ± 0.40	2.63 ± 0.36
	T	21.16 ± 3.08	21.62 ± 2.81	0.0807		
	Param.	11	15/11/0.6		15	8/8/0.5
ion	CE (%)	4.90 ± 0.90	**4.20 ± 0.90**		4.50 ± 1.06	4.07 ± 0.75
	T	13.40 ± 4.50	22.98 ± 4.08	**1.4028e-004**		
	Param.	5	2/9/0.1		2	2/9/0.2
kwo	CE (%)	11.70 ± 0.01	**11.59 ± 0.06**		11.70 ± 0.12	11.52 ± 0.07
	T	29.30 ± 0.1	29.88 ± 2.50	**5.1686e-009**		
	Param.	15	13/1/0.8		9	25/5/0.9
Rip	CE (%)	9.70 ± 0.01	**9.16 ± 0.21**		9.00 ± 0.20	9.00 ± 0.20
	T	28.90 ± 0.90	33.12 ± 4.52	**2.6148e-018**		
	Param.	48	53/2/0.9		61	61/1/1
Wav	CE (%)	11.65 ± 0.36	**11.31 ± 0.42**		11.57 ± 0.26	11.31 ± 0.25
	T	30.08 ± 6.05	15.98 ± 1.13	**3.8557e-005**		
	Param.	2	3/10/0 .		2	2/10/0

5   Conclusions and Further Work

In this contribution, we have checked the usefulness of smoothing the Real AdaBoost emphasis function by considering the neighborhood of each sample. This idea comes from the purpose of limiting the influence of isolated imbalanced outliers, and a simple implementation consisting of applying an emphasis which is a convex combination of the RAB values for each sample and the average of the RAB values for its neighborhood, improves the performance of the direct Real AdaBoost in a number of benchmark problems: KRAB is never worse than RAB, and it is significantly better in the majority of the considered benchmark datasets.

Further work will explore more general formulations of the smoothing procedure and analyze the characteristics of the problems that are better solved by them. It is particularly attractive the possibility of combining these smoothing techniques with other kinds of modifications and extensions of boosting algorithms. How to extend the smoothing ideas to other machine and machine ensemble designs is also a relevant research subject.

References

1. Bishop, C.M.: Neural Networks for Pattern Recognition. Oxford University Press, Oxford (1995)
2. Bishop, C.M.: Pattern Recognition and Machine Learning. Springer, New York (2006)
3. Kuncheva, L.I.: Combining Pattern Classifiers: Methods and Algorithms. Wiley, Hoboken (2004)
4. Rokach, L.: Pattern Classification Using Ensemble Methods. World Scientific, Singapore (2010)
5. Schapire, R.E.: The strength of weak learnability. Machine Learning 5, 197–227 (1990)
6. Freund, Y., Schapire, R.E.: A decision-theoretic generalization of on-line learning and an application to boosting. J. Computer and System Sciences 55(1), 119–139 (1997)
7. Schapire, R.E., Singer, Y.: Improved boosting algorithms using confidence-rated predictions. Machine Learning 37(3), 297–336 (1999)
8. Freund, Y.: An adaptive version of the boost by majority algorithm. Machine Learning 43(3), 293–318 (2001)
9. Gómez-Verdejo, V., Ortega-Moral, M., Arenas-García, J., Figueiras-Vidal, A.R.: Boosting by weighting critical and erroneous samples. Neurocomputing 69(7-9), 679–685 (2006)
10. Gómez-Verdejo, V., Arenas-García, J., Figueiras-Vidal, A.R.: A dynamically adjusted mixed emphasis method for building boosting ensembles. IEEE Trans. Neural Networks 19(1), 3–17 (2008)
11. Rätsch, G., Onoda, T., Müller, K.R.: Regularizing Adaboost. In: Kears, M., Solla, S., Cohn, D. (eds.) Advances in Neural Information Processing Systems, vol. 11, pp. 564–570. Cambridge University Press, Cambridge (1999)
12. Rätsch, G., Onoda, T., Müller, K.R.: Soft margins for AdaBoost. Machine Learning 42(3), 287–320 (2001)
13. Rätsch, G., Warmuth, M.K.: Efficient margin maximizing with boosting. Journal of Machine Learning Research 6, 2131–2152 (2005)

14. Sun, Y., Todorovic, S., Li, J.: Reducing the overfitting of AdaBoost by controlling its data distribution skewness. International Journal of Pattern Recognition and Artificial Intelligence 20(7), 1093–1116 (2006)
15. Shen, C., Li, H.: Boosting through optimization of margin distributions. IEEE Trans. Neural Networks 21(4), 659–666 (2010)
16. Zhang, C.-X., Zhang, J.-S., Zhang, G.-Y.: An efficient modified boosting method for solving classification problems. J. Computational and Applied Mathematics 214(2), 381–392 (2008)
17. Kim, H.C., Pang, S., Je, H.M., Kim, D., Bang, S.Y.: Constructing support vector machine ensemble. Pattern Recognition 26, 2757–2767 (2003)
18. Li, X., Wang, L., Sung, E.: Adaboost with SVM-based component classifiers. Engineering Applications of Artificial Intelligence 21, 785–795 (2008)
19. Mayhua-López, E., Gómez-Verdejo, V., Figueiras-Vidal, A.R.: Boosting ensembles with subsampling LPSVM learners. Submitted to IEEE Trans. Neural Networks and Machine Learning
20. Mayhua-López, E., Gómez-Verdejo, V., Figueiras-Vidal, A.R.: Real Adaboost with gate controlled fusion. IEEE Trans. Neural Networks and Learning Systems 23(12), 2003–2009 (2012)
21. UCI Machine Learning Repository. School Information & Computer Sciences, Univ. California, Irvine, http://archive.ics.uci.edu/ml
22. Ripley, B.D.: Pattern Recognition and Neural Networks. Cambridge University Press, Cambridge (1996)
23. Kwok, J.T.Y.: Moderating the outputs of support vector machine classifiers. IEEE Trans. Neural Networks 10(5), 1018–1031 (1999)

F-Measure as the Error Function to Train Neural Networks

Joan Pastor-Pellicer[1], Francisco Zamora-Martínez[2],
Salvador España-Boquera[1], and María José Castro-Bleda[1]

[1] Departament de Sistemes Informàtics i Computació,
Universitat Politècnica de València, Valencia, Spain
[2] Departamento de Ciencias Físicas, Matemáticas y de la Computación,
Universidad CEU Cadenal Herrera, Alfara del Patriarca (Valencia), Spain

Abstract. Imbalance datasets impose serious problems in machine learning. For many tasks characterized by imbalanced data, the F-Measure seems more appropiate than the Mean Square Error or other error measures. This paper studies the use of F-Measure as the training criterion for Neural Networks by integrating it in the Error-Backpropagation algorithm. This novel training criterion has been validated empirically on a real task for which F-Measure is typically applied to evaluate the quality. The task consists in cleaning and enhancing ancient document images which is performed, in this work, by means of neural filters.

Keywords: Neural Networks, Error-Backpropagation algorithm, F-Measure, Imbalanced datasets.

1 Introduction

It is not uncommon in many real tasks that the number of patterns of one class is significantly lower than other classes. Examples of tasks with very imbalanced data are information retrieval (a lot of information and very few useful data) or medical diagnosis (less ill than healthy patients). Imbalance datasets impose serious problems in machine learning and, particularly, in Artificial Neural Networks (ANN) training. Some authors have addressed this problem by resampling the data in order to balance the occurrences, others have modified the training algorithm [2,15].

We have followed the second approach, designing a new training algorithm which uses the F-Measure [13] as an objective error function for the Backpropagation (BP) algorithm. The F-Measure (FM) may be more suitable than the Mean Square Error (MSE) or other error measures, for problems with imbalanced data, because it is a quality measure of the performed task as a combination between Precision and Recall. Though there are different approaches for the optimization of the F-Measure using supervised techniques like SVMs [9], logistic regression [8] and other approaches [1], no such algorithm exists for ANNs to the best of our knowledge.

I. Rojas, G. Joya, and J. Cabestany (Eds.): IWANN 2013, Part I, LNCS 7902, pp. 376–384, 2013.

In order to illustrate the interest of this proposal in a real task, we have studied a problem for which the F-Measure has been typically used to assess the quality: image cleaning and enhancement of ancient document images. This task consists in estimating the probability of ink in each pixel of the cleaned image given the noisy counterpart, and it can be considered to be an imbalanced problem since only a few percentage of pixels in an image corresponds to ink. This task is required not only to improve the readability of these documents by humans but is also the first stage in most preprocessing pipelines applied in text recognition systems. This stage is quite critical because any mistake could be propagated to the following ones. The relevance of this stage depends on the quality of the documents and it is particularly important in historical documents which suffer many types of degradation.

The rest of the paper is structured as follows. First, Section 2 defines the F-Measure for continuous values and explains how this measure can be used as the objective error function in the Backpropagation algorithm. In order to illustrate the performance of the new training criteria, a task of image cleaning and enhancement of ancient printed and handwritten documents is proposed (see Section 3). The proposed method is successfully applied to different competition datasets and experimental results are presented in Section 4. The conclusions are finally drawn at the end.

2   Error-Backpropagation with F-Measure

The Backpropagation (BP) algorithm updates the ANN weights following the derivative of a given error function. The MSE function is widely used. For a given ANN output layer with n neurons $o_1, o_2, \ldots, o_n$ and its corresponding target output $t_1, t_2, \ldots, t_n$, the MSE for one pattern is computed as $MSE = 1/2 \cdot \sum_{i=1}^{n} (o_i - t_i)^2$, and its derivative is $\partial MSE / \partial o_i = (o_i - t_i)$.

This equation for training ANNs is well known, and has been successfully applied in several pattern recognition tasks (classification, regression, forecasting, ...). Different weight updating modes exists [4]:

- the *batch* training mode, which computes and sums the derivatives of all training patterns and updates weights once every epoch;
- the *on-line* training mode, which computes the derivative of one training pattern and updates weights once for each pattern every epoch; and finally
- the *mini-batch* or *bunch* training mode [3], which computes and sums the derivatives of a few training patterns, updating weights once for the mini-batch size, but several times for one epoch.

Mini-batch and on-line training modes have some advantages compared with the batch mode: convergence is faster and the result is equal or even more accurate.

As it was previously stated, for tasks with imbalanced data, the F-Measure (FM) function is more appropriate than the MSE or other error measures. F-Measure is a quality measure computed as a combination between Precision (PR) and Recall (RC). It is possible to compute a version of the F-Measure interpreting the output of the model as a binary value (for 2-class problems: 1 for relevant and 0 for non-relevant), being $o^{(i)}$ the output of the model for pattern i and $t^{(i)}$ the real-class value (0 or 1) for pattern i. The computation of FM is a harmonic mean of PR and RC, and leads to the final formulation of FM in terms of *true positives* (TPs), *false positives* (FPs) and *false negatives* (FNs). TPs, FPs, and FNs are computed over a dataset of m patterns:

$$TP = \sum_{i=1}^{m} o^{(i)} \cdot t^{(i)}$$

$$FP = \sum_{i=1}^{m} o^{(i)} \cdot (1 - t^{(i)})$$

$$FN = \sum_{i=1}^{m} (1 - o^{(i)}) \cdot t^{(i)} \tag{1}$$

FM is formalized for positive real β as usual, although the formula can be simplified by substituting TPs, FPs and FNs with previous definitions:

$$FM_\beta = \frac{(1 + \beta^2) \cdot PR \cdot RC}{\beta^2 \cdot PR + RC} = \frac{(1 + \beta^2) \cdot TP}{(1 + \beta^2) \cdot TP + \beta^2 \cdot FN + FP} = \tag{2}$$

$$= \frac{(1 + \beta^2) \cdot \sum_{i=1}^{m} o^{(i)} \cdot t^{(i)}}{\sum_{i=1}^{m} \left(o^{(i)} + \beta^2 \cdot t^{(i)} \right)} \tag{3}$$

In order to use the F-Measure as the objective error function in BP algorithm it is required to derive it by $o^{(i)}$:

$$\frac{\partial FM_\beta}{\partial o^{(i)}} = \frac{(1 + \beta^2) t^{(i)}}{\sum_{j=1}^{m} \left(o^{(j)} + \beta^2 \cdot t^{(j)} \right)} - \frac{(1 + \beta^2) \cdot \sum_{j=1}^{m} o^{(j)} \cdot t^{(j)}}{\left[\sum_{j=1}^{m} \left(o^{(j)} + \beta^2 \cdot t^{(j)} \right) \right]^2} \tag{4}$$

Since BP is defined for minimization, the sign of the F-Measure function has to be inverted. Note that the F-Measure derivative of pattern i depends on the others $m - 1$ patterns, so it is not separable as the MSE. Therefore, the exact computation of this derivative forces to use batch training mode. However, batch training is slow and inaccurate when the number of patterns m is large (in the reported experiments, millions of patterns). Because of these issues, we decided to use a mini-batch training mode, which leads to an approximation highly correlated with the true F-Measure computed on the entire dataset. Also, for large training partitions, it is better to train each epoch with a shorter random replacement sampled from training data. In this way, the error of one epoch will be the mean of each mini-batch FM. Let b be the size of the mini-batch, and R the replacement sample size, then weights are updated $\lceil R/b \rceil$ times every epoch.

The derivative of F-Measure has some issues that need to be discussed: the use of mini-batch mode combined with random replacement makes it possible to sample a bunch of patterns where every target is *false* (class 0). In this case, the FM and its derivative are both zero, meaning that these mini-batch presentations does not update the weights. This problem becomes more likely the lower the mini-batch size and the more imbalanced the data are. Since each sample selection is independent of others, the probability of occurrence of this situation can be easily computed from the mini-batch size b and the proportion of 0's in the entire training dataset (of size m) as $(F/m)^b$, where $F = \sum_{j=1}^{m}(1-t^{(j)})$. This issue reduces convergence speed because mini-batch presentations suffering this problem do not update weights even if the output of the model is not correct.

3 Cleaning and Enhancement as a Probability Pixel Estimation Problem

Image cleaning and enhancement, specially for ancient documents, are common and crucial steps for any document recognition system. Traditionally, the output of image cleaning is a binarized image where black pixels mean the presence of ink in this region. Nevertheless, since many preprocessing techniques can also deal with gray level images, it is possible to consider the gray level of cleaned image pixels as the probability of ink. Thus, cleaned images are not considered as arbitrary gray level images but, rather, as a soft estimation of a black and white image which tries to represent, in a limited resolution, the set of ideal strokes. Gray values are a way to represent the probability of picking a black sub-pixel in this pixel, so intermediate gray values are expected to be found in the borders of strokes. This idea has a correspondence with the desirable anti-aliasing property of geometrical transformations applied in most common preprocessing stages such as the correction of the skew of the page, the slope of the words in the lines and the slant of the strokes.

This problem can be considered as the joint estimation of the probability of finding ink in pixel areas, as the classification of pixels into two classes or as the retrieval of ink areas in the whole document.

It is not easy to classify image cleaning and binarization techniques, many are based on geometrical heuristics, but we propose the use of supervised machine learning techniques. Although other machine learning techniques exists (e.g. based on Markov Random Fields [14]) we have used neural network filters [7] which estimate the probability of ink for each pixel given a window of the original image centered at the pixel to be cleaned (see Figure 1). This has two main limitations: each pixel is estimated independently, and only local information is taken into account. The first limitation is alleviated by the high correlation of window inputs of neighboring pixels.

There exists many assessment measures to evaluate the quality of image cleaning and binarization which can be classified into three categories [10]: by means of human supervision, indirectly by evaluating the overall performance of a recognition system and, finally, by comparing the cleaned image with a reference or *ground truth*. Several measures have been proposed in the literature for the last option which is the only considered in this work. Note that, since this task is imbalanced (the percentage of ink pixels ranges between 5 and 15 percent), the F-Measure seems quite suitable in this case. As a matter of fact, many prestigious image binarization contests [6] employ the F-Measure to rank contenders.

MSE is a common error metric in BP training algorithms which has been applied in probability estimation tasks and which can also be used to measure the quality of image cleaning, but the use of F-Measure seems more appropriate. Indeed, a cleaned image with lower MSE may appear subjectively of lower quality than another one which may assign little mistakes on white pixels which are the majority. That is why we have opted for the use of this measure as the training criterion in BP. Since the output of the neural network is represented as a real-value, it is straightforward to compute a "soft" F-Measure and error derivatives interpreting the output of the model as a binary probability where the value for a pattern i is set as $o^{(i)} = P(\text{relevant}|\text{sample})$ and $1 - o^{(i)} = P(\text{non-relevant}|\text{sample})$. The soft F-Measure is computed as stated in Section 2.

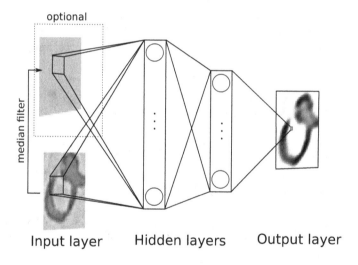

Fig. 1. Scheme of a neural network filter: a feedforward neural network estimates the probability of ink of a pixel on the cleaned image from a window centered at the same pixel in the original image. Another optional window receives an estimation of background computed by means of a median filter.

4 Experimental Setup and Results

The ANNs for enhancement and cleaning have been applied to the four Document Image Binarization COntest (DIBCO) datasets (DIBCO 2009 [5], DIBCO 2010 [12], DIBCO 2011 [6], and DIBCO 2012 [11]), partitioned as follows:

- Training set: includes DIBCO 2009 and DIBCO 2010 datasets. A total of 24 images (19.8 Mpx, 6.6% classified as ink).
- Validation set: includes DIBCO 2011 dataset. A total of 12 images (10.0 Mpx patterns, 9.0% classified as ink).
- Test: includes DIBCO 2012 dataset. A total of 14 images. (19.2 Mpx, 6.7% classified as ink).

In order to compare the proposed technique, different configurations have been tried which differ in the error criteria:

- Logistic output unit ANN trained using the MSE error criteria.
- Logistic output unit ANN trained using the FM error criteria.

The training and validation datasets have been used to find a common topology which works fine with both error criteria and to adjust parameters afterwards. Finally, the trained networks have been used to compute the performance on the test set. The quality of ANN filters has been measured by comparing the cleaned images with the ground-truth.

Each type of error criterion has been tested on networks which share the same input, hidden and output topology. The input layer is composed by 90 input neurons: 81 pixels corresponding to a window of size 9×9 centered at the pixel to be cleaned and 9 additional context pixels associated to a 3×3 window with an estimation of background using a median filter. Relating the hidden layers, the best configuration was two hidden layers of sizes 64 and 16, respectively. Nine different random initialized networks have been trained in order to reduce the effect of local minima. Table 1 shows the average of the FM and MSE measures, along with the standard deviation on validation and test sets for a training with a mini-batch of 32 samples. Also, an example of a test set image cleaned with both ANNs is depicted in Figure 2. In general, both training techniques perform quite well when measured either on MSE or FM, since a well cleaned image gives good results on both metrics. Comparing these results with [11], which is measured on the same test, the results are not competitive related to the best models, but they are better than method 1 which is also based on neural networks (they obtain a FM 0.82 on binarized data, and we report a FM 0.836 measured with a soft version of FM, which is a lower bound of the former binarized FM). We can also observe, from Table 1, that ANNs trained with the FM error function obtain better F-Measure than ANNs trained with the MSE error function. Conversely, the second model outperforms the first one in MSE in both validation and test sets. As expected, each training criteria prioritizes a different goal.

In order to study the influence of the size of the mini-batch, different trainings have been carried out varying this parameter and the reported F-Measure is illustrated in Figure 3. Two different factors may influence the results in opposite ways: on the one side, the larger the mini-batch size, the more accurate the approximation to the true F-Measure should be. On the other side, a smaller mini-batch size corresponds to a training scheme closer to the online version of BP which may have faster convergence. As it can be observed, as the mini-batch

Table 1. Average and Standard Deviation of the MSE and FM

	Validation Data		Test Data	
	$\mu \pm \sigma$ MSE	$\mu \pm \sigma$ FM	$\mu \pm \sigma$ MSE	$\mu \pm \sigma$ FM
MSE training	0.0254 ± 0.0010	0.708 ± 0.013	0.0165 ± 0.0004	0.754 ± 0.007
FM training	0.0376 ± 0.0036	0.774 ± 0.012	0.0181 ± 0.0006	0.836 ± 0.009

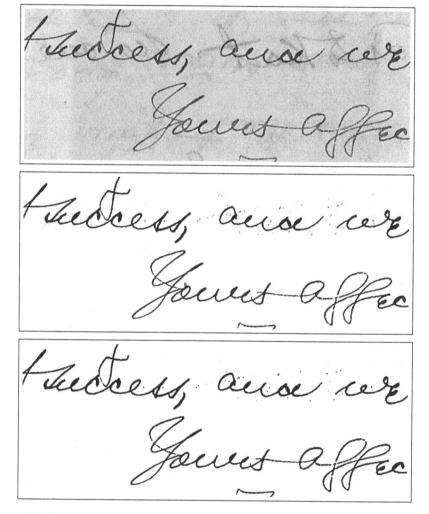

Fig. 2. (Top) Example of a noisy test image. (Middle) The same image cleaned with the ANN trained with the MSE error criteria. (Bottom) The same image cleaned with the F-Measure error criteria.

size is increased, the F-Measure performs worse, which means that even smaller mini-batch sizes may be highly correlated with the global F-Measure.

In order to study the correlation between mini-batch size and FM, a statistic experiment has been carried out (see Figure 4) obtaining a Pearson

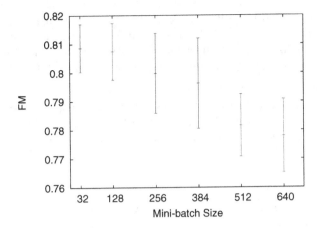

Fig. 3. Influence of mini-batch size

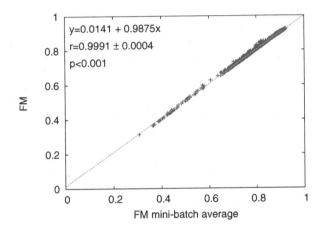

Fig. 4. Correlation between the average FM of 100 000 mini-batch presentations of size 32 taken randomly and the FM value computed on the concatenation of all validation images

product-moment correlation coefficient $r = 0.9991 \pm 0.0004$ with a confidence interval 99.9% ($p < 0.001$).

5   Conclusions

In this work, a novel objective error function for the Error-Backpropagation algorithm is proposed based on the F-Measure. Additionally, it has been explained how it can be adapted to mini-batch training mode of BP. In order to empirically validate this training mode, a real task using an imbalanced dataset of several millions of patterns has been carried out. The task consisted in the estimation of a cleaned image from a noisy one by means of neural network filters. Experimental results show that, although ANNs trained with MSE or with FM performs quite similar, each training mode prioritizes its corresponding assessment measure. This error criteria can be used in tasks where F-Measure makes sense, as is the case of information retrieval or document classification. As a future work, we plan to extend this work to other symmetrical measures such as the Matthews correlation coefficient.

Acknowledgments. This work has been partially supported by MICINN project HITITA (TIN2010-18958) and by the FPI-MICINN (BES-2011-046167) scholarship from Ministerio de Ciencia e Innovación, Gobierno de España.

References

1. Dembczyński, K., Waegeman, W., Cheng, W., Hüllermeier, E.: An exact algorithm for f-measure maximization. Advances in Neural Information Processing Systems 24, 223–230 (2011)
2. Al-Haddad, L., Morris, C.W., Boddy, L.: Training radial basis function neural networks: effects of training set size and imbalanced training sets. J. of Microbiological Methods 43(1), 33–44 (2000)
3. Bilmes, J., Asanovic, K., Chin, C.W., Demmel, J.: Using PHiPAC to speed error back-propagation learning. In: Proc. of ICASSP, vol. 5, pp. 4153–4156 (1997)
4. Duda, R.O., Hart, P.E., Stork, D.G.: Pattern Classification, 2nd edn. Wiley (2001)
5. Gatos, B., Ntirogiannis, K., Pratikakis, I.: ICDAR 2009 document image binarization contest (DIBCO 2009). In: Proc. of ICDAR, pp. 1375–1382 (2009)
6. Gatos, B., Ntirogiannis, K., Pratikakis, I.: DIBCO 2009: document image binarization contest. Int. J. on Document Analysis and Recognition 14(1), 35–44 (2011)
7. Hidalgo, J.L., España, S., Castro, M.J., Pérez, J.A.: Enhancement and cleaning of handwritten data by using neural networks. In: Marques, J.S., Pérez de la Blanca, N., Pina, P. (eds.) IbPRIA 2005. LNCS, vol. 3522, pp. 376–383. Springer, Heidelberg (2005)
8. Jansche, M.: Maximum expected f-measure training of logistic regression models. In: Proc. of HLT & EMNLP, pp. 692–699 (2005)
9. Musicant, D.R., Kumar, V., Ozgur, A.: Optimizing f-measure with support vector machines. In: Proc. of Int. Florida AI Research Society Conference, pp. 356–360 (2003)
10. Ntirogiannis, K., Gatos, B., Pratikakis, I.: A Performance Evaluation Methodology for Historical Document Image Binarization (2012)
11. Pratikakis, I., Gatos, B., Ntirogiannis, K.: ICFHR 2012 Competition on Handwritten Document Image Binarization (H-DIBCO 2012) (2012)
12. Pratikakis, I., Gatos, B., Ntirogiannis, K.: H-DIBCO 2010-handwritten document image binarization competition. In: Proc. of ICFHR, pp. 727–732 (2010)
13. van Rijsbergen, C.J.: A theoretical basis for the use of co-occurrence data in information retrieval. J. of Documentation 33(2), 106–119 (1977)
14. Wolf, C.: Document Ink Bleed-Through Removal with Two Hidden Markov Random Fields and a Single Observation Field. IEEE PAMI 32(3), 431–447 (2010)
15. Zhou, Z.H., Liu, X.Y.: Training cost-sensitive neural networks with methods addressing the class imbalance problem. IEEE Trans. on Knowledge and Data Engineering 18(1), 63–77 (2006)

Isomorphisms of Fuzzy Sets and Cut Systems[*]

Jiří Močkoř

Centre of Excellence IT4Innovations
Division of the University of Ostrava
Institute for Research and Applications of Fuzzy Modeling
30. dubna 22, 701 03 Ostrava 1, Czech Republic
Jiri.Mockor@osu.cz

Abstract. Any fuzzy set X in a classical set A with values in a complete (residuated) lattice Ω can be identified with a system of α-cuts X_α, $\alpha \in \Omega$. Analogical results were proved for sets with similarity relations with values in Ω (e.g. Ω-sets) which are objects of two special categories $\mathbf{K} = \mathbf{Set}(\Omega)$ or $\mathbf{SetR}(\Omega)$ of Ω-sets and for fuzzy sets defined as morphisms from Ω-set into a special Ω-set $(\Omega, \leftrightarrow)$. These fuzzy sets can be defined equivalently as special cut systems $(C_\alpha)_\alpha$, called f-cuts. That equivalence then represents a natural isomorphism between covariant functor of fuzzy sets $\mathcal{F}_\mathbf{K}$ and covariant functor of f-cuts $\mathcal{C}_\mathbf{K}$. In the paper we are interested in relationships between sets of fuzzy sets and sets of f-cuts in an Ω-set (A, δ) in corresponding categories $\mathbf{Set}(\Omega)$ and $\mathbf{SetR}(\Omega)$, which are endowed with binary operations extended either from binary operations in the lattice Ω, or from binary operations defined in a set A by the generalized Zadeh's extension principle. We prove that the final binary structures are (under some conditions) isomorphic.

1 Introduction and Notations

It is well known that any fuzzy set (with values in a residuated lattice Ω) can be alternatively expressed as a system of α-cuts. In fact, recall that a nested system of α-cuts in A is a system $(C_\alpha)_\alpha$ of subsets of A such that $C_\alpha \subseteq C_\beta$ if $\alpha \geq \beta$ and the set $\{\alpha \in \Omega : a \in C_\alpha\}$ has the greatest element for any $a \in A$. Then for any nested system of α-cuts $\mathbf{C} = (C_\alpha)_\alpha$, a classical fuzzy set $\mu_\mathbf{C} : A \to \Omega$ can be constructed such that $\mu_\mathbf{C}(x) = \bigvee_{\{\beta : x \in C_\beta\}} \beta$ and, conversely, for any fuzzy set μ in A, a nested system of α-cuts is defined by $C_\alpha = \{x \in A : \mu(x) \geq \alpha\}$. Between nested systems of α-cuts in A and fuzzy sets in A there are some interesting relationships, and from some point of view an investigation of fuzzy sets can be substituted by an investigation of nested systems of α-cuts (see e.g. [1,2]).

In our previous papers [10,11,7], we introduced a notion of a fuzzy set in sets with similarity relation (A, δ) (the so called Ω-sets), where values of a similarity relation $\delta : A \times A \to \Omega$ are from the residuated lattice Ω. Ω-sets then represent objects in various categories $\mathcal{K}$ with differently defined morphisms. Among these

---

[*] This work was supported by the European Regional Development Fund in the IT4Innovations Centre of Excellence project (CZ.1.05/1.1.00/02.0070).

I. Rojas, G. Joya, and J. Cabestany (Eds.): IWANN 2013, Part I, LNCS 7902, pp. 385–392, 2013.

categories we are interested in two special categories: the category $\mathbf{Set}(\Omega)$ with morphisms $(A,\delta) \to (B,\gamma)$ defined as special maps $A \to B$ and the category $\mathbf{SetR}(\Omega)$ with morphisms defined as special relations $A \times B \to \Omega$. A notion of a fuzzy set in (A,δ) then depends on a category $\mathcal{K}$, i.e. f is a fuzzy set in an Ω-set (A,δ) in a category $\mathcal{K}$ (shortly, $f \subseteq_{\mathcal{K}} (A,\delta)$), if $f : (A,\delta) \to (\Omega,\leftrightarrow)$ is a morphism in $\mathcal{K}$, where $\leftrightarrow$ is the biresiduation operation in Ω (= special similarity relation in Ω). This formal extension of classical fuzzy sets enables us to develop the fuzzy set theory in any category of Ω-sets, with a lot of properties similar to those of classical fuzzy sets. In papers [8,9], we proved that fuzzy sets in the categories $\mathbf{Set}(\Omega)$ and $\mathbf{SetR}(\Omega)$ can be represented by some cut systems. Namely, any fuzzy set $f \subseteq_{\sim\mathbf{Set}(\Omega)} (A,\delta)$ can be represented by the so called f-cut $\mathbf{C} = (C_\alpha)_{\alpha\in\Omega}$, where C_α are subsets of A with some special properties, and, any fuzzy set $g \subseteq_{\sim\mathbf{SetR}(\Omega)} (A,\delta)$ can be analogously represented by an f-cut $\mathbf{D} = (D_\alpha)_{\alpha\in\Omega}$, where D_α are subsets of $A \times \Omega$ also with some special properties. We also proved that these representation theorems can be expressed as natural isomorphisms between two special functors $\mathcal{F}_{\mathbf{K}}$ and $\mathcal{C}_{\mathbf{K}}$ from the corresponding category $\mathbf{K} = \mathbf{Set}(\Omega)$ or $\mathbf{SetR}(\Omega)$ into the category $\mathbf{Set}$ of classical sets.

In this paper we are also interested in relationships between sets of fuzzy sets and sets of f-cuts in an Ω-set (A,δ) in corresponding categories $\mathbf{Set}(\Omega)$ and $\mathbf{SetR}(\Omega)$, respectively, which are endowed with binary operations extended either from binary operations in the lattice Ω, or from binary operations defined in a set A by the generalized Zadeh's extension principle. For any binary operation $\oplus_\Omega$ defined in Ω, for any Ω-set (A,δ) and for any category $\mathcal{K} = \mathbf{Set}(\Omega), \mathbf{SetR}(\Omega)$, the operation $\oplus_\Omega$ is extended onto a binary operation $\oplus_{\mathcal{K},f}$ in the corresponding set $\mathcal{F}_{\mathcal{K}}(A,\delta)$ of fuzzy sets in a category $\mathcal{K}$ and onto a binary operation $\oplus_{\mathcal{K},c}$ in the corresponding set $\mathcal{C}_{\mathcal{K}}(A,\delta)$ of f-cuts in a category K. In that way we will receive the following semigroup structures:

$$(\mathcal{F}_{\mathcal{K}}(A,\delta), \oplus_{\mathcal{K},f}), \quad (\mathcal{C}_{\mathcal{K}}(A,\delta), \oplus_{\mathcal{K},c}).$$

Moreover, any binary operation $+$ defined in an Ω-set (A,δ) can be extended onto a binary operation $+_{\mathcal{K},f}$ defined in $\mathcal{F}_{\mathcal{K}}(A,\delta)$, and a binary operation $+_{\mathcal{K},c}$ defined in $\mathcal{C}_{\mathcal{K}}(A,\delta)$, for any category $\mathcal{K} = \mathbf{Set}(\Omega), \mathbf{SetR}(\Omega)$. The principal results of the paper then show that, under some conditions, these binary structures are isomorphic. It means that relationships between fuzzy sets and corresponding f-cuts is really deep and enables us to use equivalently either of these structures.

It should be mentioned that another more categorical approach to the f-cut theory is also possible and used by other authors (see e.g. [4,5,6,3]). That approach is based on a generalization of sheaves of sets on Ω and generalized subobject classifier constructions which is, in fact, an analogy of an f-cut construction in the category $\mathbf{Set}(\Omega)$. Nevertheless we will not used that approach since it will require a deeper knowledge from the category theory and for most of specialists in fuzzy set theory it would be probably difficult to read.

To be more self-contained we will introduce several notations which will be used in the paper, and we also recall several results from previous papers which can be useful for full understanding and notation of our results. Recall that a set with similarity relation (or Ω-set) is a couple (A, δ), where $\delta : A \times A \to \Omega$ is a map such that

(a) $(\forall x \in A)$ $\delta(x, x) = 1$,
(b) $(\forall x, y \in A)$ $\delta(x, y) = \delta(y, x)$,
(c) $(\forall x, y, z \in A)$ $\delta(x, y) \otimes \delta(y, z) \leq \delta(x, z)$ (generalized transitivity).

In the paper, we will use two principal categories with Ω-sets as objects and with differently defined morphisms. A morphism $f : (A, \delta) \to (B, \gamma)$ in the first category $\mathbf{Set}(\Omega)$ is a map $f : A \to B$ such that $\gamma(f(x), f(y)) \geq \delta(x, y)$ for all $x, y \in A$. The other category $\mathbf{SetR}(\Omega)$ is an analogy of the category of sets with relations between sets as morphisms. Objects of the category $\mathbf{SetR}(\Omega)$ are the same as in the category $\mathbf{Set}(\Omega)$ and morphisms $f : (A, \delta) \to (B, \gamma)$ are maps $f : A \times B \to \Omega$ (i.e. Ω-valued relations) such that

(a) $(\forall x, z \in A)(\forall y \in B)$ $\delta(z, x) \otimes f(x, y) \leq f(z, y)$,
(b) $(\forall x \in A)(\forall y, z \in B)$ $f(x, y) \otimes \gamma(y, z) \leq f(x, z)$.

As we mentioned, a fuzzy set f in an Ω-set (A, δ) in a category $\mathcal{K}$ (shortly, $f \subseteq_{\mathcal{K}} (A, \delta)$) is a morphism $f : (A, \delta) \to (\Omega, \leftrightarrow)$, where $\leftrightarrow$ is the biresiduation operation in Ω defined by $\alpha \leftrightarrow \beta = (\alpha \to \beta) \wedge (\beta \to \alpha)$. Any classically defined fuzzy set X in a set A with values in Ω can be defined equivalently by a system of level sets $X_\alpha, \alpha \in \Omega$, where $X_\alpha = \{a \in A : X(a) \geq \alpha\}$. Conversely, any (nested) system $(Y_\alpha)_\alpha$ of subsets of A such that for any $a \in A$ the set $\{\alpha \in \Omega : a \in Y_\alpha\}$ has the greatest element, defines a fuzzy set Y such that $Y(a) = \bigvee_{\{\beta : a \in Y_\beta\}} \beta$. In our previous papers [9,8], we proved that analogously any fuzzy set in (A, δ) in the category $\mathbf{Set}(\Omega)$ or in the category $\mathbf{SetR}(\Omega)$ can be defined equivalently by a system of some special subsets of A or $A \times \Omega$, respectively, called **f-cut**, which is defined as follows:

Definition 1. *Let (A, δ) be an Ω-set. Then a system $\mathbf{C} = (C_\alpha)_\alpha$ of subsets of A is called an **f-cut** in (A, δ) in the category $\mathbf{Set}(\Omega)$ if*

(a) $\forall a, b \in A,$ $a \in C_\alpha \Rightarrow b \in C_{\alpha \otimes \delta(a,b)}$,
(b) $\forall a \in A, \forall \alpha \in \Omega,$ $\bigvee_{\{\beta : a \in C_\beta\}} \beta \geq \alpha \Rightarrow a \in C_\alpha$.

In the category $\mathbf{Set}(\Omega)$ a tensor product $\odot$ exists for any two objects (A, δ), (B, γ) such that

$$(A, \delta) \odot (B, \gamma) = (A \times B, \delta \odot \gamma),$$
$$\delta \odot \gamma((a_1, b_1), (a_2, b_2)) = \delta(a_1, a_2) \otimes \gamma(b_1, b_2).$$

By using the tensor product in the category $\mathbf{Set}(\Omega)$ we can now define an f-cut in the category $\mathbf{SetR}(\Omega)$ as follows

Definition 2. *Let (A, δ) be an Ω-set. Then a system $(C_\alpha)_{\alpha \in \Omega}$ is an **f-cut** in (A, δ) in the category $\mathbf{SetR}(\Omega)$ if it is an f-cut in $(A, \delta) \odot (\Omega, \leftrightarrow)$ in the category $\mathbf{Set}(\Omega)$.*

It is then clear that a system $(C_\alpha)_{\alpha \in \Omega}$ is an f-cut in (A, δ) in the category $\mathbf{SetR}(\Omega)$, if

(a) $C_\alpha \subseteq A \times \Omega$, for any $\alpha \in \Omega$,
(b) $\forall a, b \in A, \quad (a, \beta) \in C_\alpha \Rightarrow (b, \beta) \in C_{\alpha \otimes \delta(a,b)}$,
(c) $\forall a \in A, \forall \gamma \in \Omega, \quad \bigvee_{\{\beta : (a,\gamma) \in C_\beta\}} \beta \geq \alpha \Rightarrow (a, \gamma) \in C_\alpha$,
(d) $\forall a \in A, \forall \alpha, \gamma \in \Omega, \quad (a, \alpha) \in C_\beta \Rightarrow (a, \gamma) \in C_{\beta \otimes (\alpha \leftrightarrow \gamma)}$.

In the next parts, for any category $\mathcal{K} = \mathbf{Set}(\Omega), \mathbf{SetR}(\Omega)$, we will use the following notations.

$$\mathcal{C}_\mathcal{K}(A, \delta) = \{\mathbf{C} : \mathbf{C} \text{ is an f-cut in } (A, \delta) \text{ in } \mathcal{K}\},$$
$$\mathcal{F}_\mathcal{K}(A, \delta) = \{s : s \underset{\sim \mathcal{K}}{\subseteq} (A, \delta)\}.$$

It should be mentioned that analogously as for f-cut systems, fuzzy sets in the category $\mathbf{SetR}(\Omega)$ can be defined by fuzzy sets in the category $\mathbf{Set}(\Omega)$. Namely, the following are equivalent statements:

(a) $s \underset{\sim \mathbf{SetR}(\Omega)}{\subseteq} (A, \delta)$,
(b) $s \underset{\sim \mathbf{Set}(\Omega)}{\subseteq} (A, \delta) \odot (\Omega, \leftrightarrow)$.

By using the above notations we can write

$$\mathcal{F}_{\mathbf{SetR}(\Omega)}(A, \delta) := \mathcal{F}_{\mathbf{Set}(\Omega)}((A, \delta) \odot (\Omega, \leftrightarrow)),$$
$$\mathcal{C}_{\mathbf{SetR}(\Omega)}(A, \delta) := \mathcal{C}_{\mathbf{Set}(\Omega)}((A, \delta) \odot (\Omega, \leftrightarrow)).$$

In the paper [10], for any Ω-set (A, δ) we introduced the following extensional maps

$$\widehat{} : \{s : s : A \to \Omega \text{ is a map}\} \to \mathcal{F}_{\mathbf{Set}(\Omega)}(A, \delta),$$
$$\widetilde{} : \{s : s : A \times \Omega \to \Omega \text{ is a map}\} \to \mathcal{F}_{\mathbf{SetR}(\Omega)}(A\delta).$$

The first one is defined by the following lemma:

Lemma 1. *Let (A, δ) be an Ω-set and let $s : A \to \Omega$ be a map. Then we define a map $\widehat{s} : A \to \Omega$ such that $\widehat{s}(a) = \bigvee_{x \in A} \delta(a, x) \otimes s(x)$ for all $a \in A$. Then*

(a) $\widehat{s} \underset{\sim \mathbf{Set}(\Omega)}{\subseteq} (A, \delta)$,
(b) If $s \underset{\sim \mathbf{Set}(\Omega)}{\subseteq} (A, \delta)$, then $\widehat{s} = s$,
(c) $\widehat{s} = \bigwedge \{t : t \underset{\sim \mathbf{Set}(\Omega)}{\subseteq} (A, \delta), t \geq s \text{ in } \mathbf{Set}(\Omega)\}$.

Now, let $g : A \times \Omega \to \Omega$ be a map. Then by $\widetilde{g}$ we denote a fuzzy set $\widehat{\widetilde{g}} \underset{\sim \mathbf{Set}(\Omega)}{\subseteq} (A, \delta) \odot (\Omega, \leftrightarrow)$. Hence, $\widetilde{g} \underset{\sim \mathbf{SetR}(\Omega)}{\subseteq} (A, \delta)$.

Recall (see [9,8]) that also any system of subsets indexed by elements from Ω can be extended onto f-cut system. In fact, for any Ω-set (A, δ) there exist maps

$$\overline{(-)} : \{\mathbf{C} : \mathbf{C} = (C_\alpha)_{\alpha \in \Omega}, C_\alpha \subseteq A\} \to \mathcal{C}_{\mathbf{Set}(\Omega)}(A, \delta),$$
$$\overline{(-)} : \{\mathbf{D} : \mathbf{D} = (D_\alpha)_{\alpha \in \Omega}, D_\alpha \subseteq A \times \Omega\} \to \mathcal{C}_{\mathbf{SetR}(\Omega)}(A, \delta),$$

such that $\overline{\mathbf{C}} = (\overline{C_\alpha})_\alpha$ and $\overline{\mathbf{D}} = (\overline{D_\alpha})_\alpha$, where

$$\overline{C_\alpha} = \{a \in A : \bigvee_{\{(x,\beta):x \in C_\beta\}} \beta \otimes \delta(a,x) \geq \alpha\}, \text{ for any } (C_\alpha)_\alpha, C_\alpha \subseteq A,$$

$$\overline{D_\alpha} = \{(a,\beta) \in A \times \Omega : \bigvee_{\{(x,\tau,\rho):(x,\tau) \in D_\rho\}} \rho \otimes \delta(a,x) \otimes (\tau \leftrightarrow \beta) \geq \alpha\}$$

for any $(D_\alpha)_\alpha, D_\alpha \subseteq A \times \Omega$. It is clear that the second extension procedure can be derived from the first one by using the equality $\mathcal{C}_{\mathbf{SetR}(\Omega)}(A, \delta) = \mathcal{C}_{\mathbf{Set}(\Omega)}((A, \delta) \odot (\Omega, \leftrightarrow))$.

For basic information about the category theory see [13].

2   Binary Operations on Fuzzy Sets and F-cuts

2.1   Operations Derived from Operations on Ω

Let $\oplus_\Omega$ be a binary operation in Ω. For any Ω-set (A, δ) and for any category $\mathcal{K} = \mathbf{Set}(\Omega), \mathbf{SetR}(\Omega)$ we can extend the operation $\oplus_\Omega$ onto a binary operation $\oplus_{\mathcal{K},f}$ in a corresponding set of fuzzy sets in a category $\mathcal{K}$ and onto a binary operation $\oplus_{\mathcal{K},c}$ in a corresponding set of f-cuts in a category K, respectively. In that way we will receive the following binary structures:

$$(\mathcal{F}_{\mathbf{Set}(\Omega)}(A, \delta), \oplus_{\mathbf{Set}(\Omega),f}), \quad (\mathcal{C}_{\mathbf{Set}(\Omega)}(A, \delta), \oplus_{\mathbf{Set}(\Omega),c}),$$
$$(\mathcal{F}_{\mathbf{SetR}(\Omega)}(A, \delta), \oplus_{\mathbf{SetR}(\Omega),f}), \quad (\mathcal{C}_{\mathbf{SetR}(\Omega)}(A, \delta), \oplus_{\mathbf{SetR}(\Omega),c}).$$

We will define these extended operations in the category $\mathbf{Set}(\Omega)$ firstly. Let $s, v : A \to \Omega$ be maps and let $\mathbf{C} = (\mathbf{C}_\alpha)_\alpha, \mathbf{D} = (\mathbf{D}_\alpha)_\alpha$ be such that $C_\alpha, D_\alpha \subseteq A$. For any $x \in A$, we firstly set

$$(s \oplus v)(x) := s(x) \oplus_\Omega v(x),$$
$$C_\alpha \oplus D_\alpha := \{a \in A : (\bigvee_{\{\beta:a \in C_\beta\}} \beta) \oplus_\Omega (\bigvee_{\{\gamma:a \in D_\gamma\}} \gamma) \geq \alpha\}.$$

Then the operations $\oplus_{\mathbf{Set}(\Omega),f}$ and $\oplus_{\mathbf{Set}(\Omega),c}$ are defined as follows:

$$s \oplus_{\mathbf{Set}(\Omega),f} v := \widehat{s \oplus v},$$
$$\mathbf{C} \oplus_{\mathbf{Set}(\Omega),c} \mathbf{D} := (\overline{C_\alpha \oplus D_\alpha})_\alpha.$$

Now, by using the identities

$$\mathcal{C}_{\mathbf{SetR}(\Omega)}(A, \delta) = \mathcal{C}_{\mathbf{Set}(\Omega)}((A, \delta) \odot (\Omega, \leftrightarrow)),$$
$$\mathcal{F}_{\mathbf{SetR}(\Omega)}(A, \delta) = \mathcal{F}_{\mathbf{Set}(\Omega)}((A, \delta) \odot (\Omega, \leftrightarrow)),$$

for any $r, w \in \mathcal{F}_{\mathbf{SetR}(\Omega)}(A, \delta)$ and $\mathbf{T} = (T_\alpha)_\alpha, \mathbf{U} = (U_\alpha)_\alpha \in \mathcal{C}_{\mathbf{SetR}(\Omega)}(A, \delta)$ we define

$$(r \oplus w)(x, \alpha) := r(x, \alpha) \oplus_\Omega w(x, \alpha),$$

$$T_\alpha \oplus U_\alpha := \{(a, \beta) \in A \times \Omega : (\bigvee_{\{\gamma : (a,\beta) \in T_\gamma\}} \gamma) \oplus_\Omega (\bigvee_{\{\omega : (a,\beta) \in U_\omega\}} \gamma) \geq \alpha\},$$

$$r \oplus_{\mathbf{SetR}(\Omega), f} w := \widetilde{r \oplus w},$$

$$\mathbf{T} \oplus_{\mathbf{SetR}(\Omega), c} \mathbf{U} := (\overline{T_\alpha \oplus U_\alpha})_\alpha.$$

It means that

$$(\mathcal{F}_{\mathbf{SetR}(\Omega)}(A, \delta), \oplus_{\mathbf{SetR}(\Omega), f}) := (\mathcal{F}_{\mathbf{Set}(\Omega)}((A, \delta) \odot (\Omega, \leftrightarrow)), \oplus_{\mathbf{Set}(\Omega), f})$$

$$(\mathcal{C}_{\mathbf{SetR}(\Omega)}(A, \delta), \oplus_{\mathbf{SetR}(\Omega), c}) := (\mathcal{C}_{\mathbf{Set}(\Omega)}((A, \delta) \odot (\Omega, \leftrightarrow)), \oplus_{\mathbf{Set}(\Omega), c}).$$

Proposition 1. *Let $\oplus_\Omega$ be a binary operation in Ω such that $(\Omega, \oplus_\Omega)$ is a semigroup. Then for any category $\mathcal{K} = \mathbf{Set}(\Omega), \mathbf{SetR}(\Omega)$ and any Ω-set (A, δ), the binary structures $(\mathcal{F}_\mathcal{K}(A, \delta), \oplus_{\mathcal{K}, f})$ and $(\mathcal{C}_\mathcal{K}(A, \delta), \oplus_{\mathcal{K}, c})$ are semigroups.*

The proof is straightforward.

Theorem 1. *Let $\oplus_\Omega$ be a binary operation in Ω such that it is completely distributive with respect to $\bigvee$ in Ω. Then for any category $\mathcal{K} = \mathbf{Set}(\Omega), \mathbf{SetR}(\Omega)$ and any Ω-set (A, δ), the following semigroup structures are isomorphic.*

$$(\mathcal{F}_\mathcal{K}(A, \delta), \oplus_{\mathcal{K}, f}) \cong (\mathcal{C}_\mathcal{K}(A, \delta), \oplus_{\mathcal{K}, c}).$$

Hence, the theorem holds for $\oplus_\Omega = \wedge, \vee$ or $\otimes$ operations in Ω, for example.

2.2   Operations Derived from Operations on Underlying Sets

It is well known that any binary operation $+$ defined in a set A can be simply extended by the so called *Zadeh's extension principle* onto a binary operation defined in the set of all classical fuzzy sets in A with values in Ω. In fact, if s, v are classical fuzzy sets in A, i.e. maps from A into Ω, then a fuzzy set $s + v$ can be defined such that for any $a \in A$, $(s + v)(a) = \bigvee_{\{(x,y) \in A \times A : x + y = a\}} s(x) \wedge v(y)$. It is then a natural process to extend the Zadeh's extension principle onto fuzzy sets in Ω-sets or even onto f-cuts in categories $\mathcal{K} = \mathbf{Set}(\Omega), \mathbf{SetR}(\Omega)$ and for any binary operation $+$ in A to define a binary operation $+_{\mathcal{K}, f}$ onto the set $\mathcal{F}_\mathcal{K}(A, \delta)$ and a binary operation $+_{\mathcal{K}, c}$ onto the set $\mathcal{C}_\mathcal{K}(A, \delta)$. We show firstly how these extensions can be defined. Let us start with the category $\mathbf{Set}(\Omega)$.

Definition 3. *Let $+$ be a binary operation in a set A, let $s, v \subseteq_{\sim \mathbf{Set}(\Omega)} (A, \delta)$ be fuzzy sets in the category $\mathbf{Set}(\Omega)$ and let $\mathbf{C}$ and $\mathbf{D}$ be f-cuts in (A, δ) in the category $\mathbf{Set}(\Omega)$. Then*

(i) a fuzzy set $s +_{\mathbf{Set}(\Omega), f} v \subseteq_{\sim \mathbf{Set}(\Omega)} (A, \delta)$ is defined such that for any $a \in A$,

$$(s +_{\mathbf{Set}(\Omega), f} v)(a) = \bigvee_{x, y \in A} (s(x) \wedge v(y)) \otimes \delta(a, x + y),$$

(ii) an f-cut $\mathbf{C} +_{\mathbf{Set}(\Omega),c} \mathbf{D}$ *is defined such that for any* $\alpha \in \Omega$,

$$(\mathbf{C} +_{\mathbf{Set}(\Omega),c} \mathbf{D})_\alpha = \{a \in A : \bigvee_{\{(x,y,\beta):x\in C_\beta,y\in D_\beta\}} \beta \otimes \delta(a,x+y) \geq \alpha\}.$$

To show that the definition is correct, let us mention that we have $s +_{\mathbf{Set}(\Omega),f}$ $v = \widehat{s+v}$, where a map $s+v : A \to \Omega$ is defined such that $(s+v)(a) = \bigvee_{\{(x,y):x+y=a\}} s(x) \wedge v(y)$ and, analogously, $(\mathbf{C} +_{\mathbf{Set}(\Omega),c} \mathbf{D})_\alpha = \overline{C_\alpha + D_\alpha}$ in the category $\mathbf{Set}(\Omega)$, where $C_\alpha + D_\alpha = \{x+y : x \in C_\alpha, y \in D_\alpha\}$.

An analogical extension of a binary operation can be also defined in the category $\mathbf{SetR}(\Omega)$.

Definition 4. *Let* $+$ *be a binary operation in a set* A, *let* $r, w \subseteq_{\sim \mathbf{SetR}(\Omega)} (A, \delta)$ *be fuzzy sets in the category* $\mathbf{SetR}(\Omega)$ *and let* $\mathbf{E}$ *and* $\mathbf{F}$ *be f-cuts in* (A, δ) *in the category* $\mathbf{SetR}(\Omega)$. *Then*

(i) a fuzzy set $r +_{\mathbf{SetR}(\Omega),f} w \subseteq_{\sim \mathbf{SetR}(\Omega)} (A, \delta)$ *is defined such that for any* $a \in A, \alpha \in \Omega$,

$$(r +_{\mathbf{SetR}(\Omega),f} w)(a,\alpha) = \bigvee_{x,y\in A, \beta\in\Omega} (r(x,\beta) \wedge w(y,\beta)) \otimes \delta(a,x+y) \otimes (\beta \leftrightarrow \alpha),$$

(ii) an f-cut $\mathbf{E} +_{\mathbf{SetR}(\Omega),c} \mathbf{F}$ *is defined such that for any* $\alpha \in \Omega$,

$$(\mathbf{E} +_{\mathbf{SetR}(\Omega),c} \mathbf{F})_\alpha =$$
$$\{(a,\beta) \in A \times \Omega : \bigvee_{\{(x,y,\tau,\rho):(x,\tau)\in E_\rho,(y,\tau)\in F_\rho\}} \rho \otimes \delta(a,x+y) \otimes (\tau \leftrightarrow \beta) \geq \alpha\}.$$

To show that the definition is correct, let us mention that we have $r +_{\mathbf{SetR}(\Omega),f}$ $w = \widetilde{r+w}$, where a map $r+w : A \times \Omega \to \Omega$ is defined such that $(r+w)(a,\alpha) = \bigvee_{\{(x,y):x+y=a\}} r(x,\alpha) \wedge w(y,\alpha)$ and, analogously, $(\mathbf{E}+_{\mathbf{SetR}(\Omega),c}\mathbf{F})_\alpha = \overline{E_\alpha + F_\alpha}$ in the category $\mathbf{SetR}(\Omega)$, where $E_\alpha + F_\alpha = \{(x+y,\alpha) : (x,\alpha) \in E_\alpha, (y,\alpha) \in F_\alpha\}$.

Theorem 2. *Let* $+$ *be a binary operation in a set* A. *Then for any category* $\mathcal{K} = \mathbf{Set}(\Omega), \mathbf{SetR}(\Omega)$ *and any* Ω-set (A, δ), *the following binary structures are isomorphic:*

$$(\mathcal{F}_\mathcal{K}(A,\delta), +_{\mathcal{K},f}) \cong (\mathcal{C}_\mathcal{K}(A,\delta), +_{\mathcal{K},c}).$$

It should be mentioned that there is another possibility how to define binary operations in $\mathcal{F}_{\mathbf{SetR}(\Omega)}(A,\delta)$ and $\mathcal{C}_{\mathbf{SetR}(\Omega)}(A,\delta)$, respectively. In fact, let $\oplus_\Omega$ be a binary operation in Ω and let $+$ be a binary operation in A. For any such pair $(\oplus_\Omega, +)$ we can define a binary operation $\oplus = \oplus(\oplus_\Omega, +)$ on $A \times \Omega$ such that $(a,\alpha) \oplus (b,\beta) = (a+b, \alpha \oplus_\Omega \beta)$. The operation $\oplus$ can be extended (by the above procedures) on binary operations $\oplus_{\mathbf{Set}(\Omega),f}$ and $\oplus_{\mathbf{Set}(\Omega),c}$ defined onto

$\mathcal{F}_{\mathbf{Set}(\Omega)}((A,\delta)\odot(\Omega,\leftrightarrow))$ and $\mathcal{C}_{\mathbf{Set}(\Omega)}((A,\delta)\odot(\Omega,\leftrightarrow))$, respectively. We can then define the operations $\oplus_{\mathbf{SetR}(\Omega),f}$ and $\oplus_{\mathbf{SetR}(\Omega),c}$ as follows:

$$(\mathcal{F}_{\mathbf{SetR}(\Omega)}(A,\delta),\oplus_{\mathbf{SetR}(\Omega),f}) := (\mathcal{F}_{\mathbf{Set}(\Omega)}((A,\delta)\odot(\Omega,\leftrightarrow)),\oplus_{\mathbf{Set}(\Omega),f})$$
$$(\mathcal{C}_{\mathbf{SetR}(\Omega)}(A,\delta),\oplus_{\mathbf{SetR}(\Omega),c}) := (\mathcal{C}_{\mathbf{Set}(\Omega)}((A,\delta)\odot(\Omega,\leftrightarrow)),\oplus_{\mathbf{Set}(\Omega),c}).$$

Theorem 3. *Let (A,δ) be an Ω-set and let $\oplus = \oplus(\oplus_\Omega,+)$ be the binary operation defined by a binary operation $+$ in A and a binary operation $\oplus_\Omega$ in Ω which is completely distributive with respect to $\bigvee$ in Ω. Then*

$$(\mathcal{F}_{\mathbf{SetR}(\Omega)}(A,\delta),\oplus_{\mathbf{SetR}(\Omega),f}) \cong (\mathcal{C}_{\mathbf{SetR}(\Omega)}(A,\delta),\oplus_{\mathbf{SetR}(\Omega),c}).$$

References

1. Bělohlávek, R.: Fuzzy relational systems, Foundations and Principles. Kluwer Academic Publ., Dordrecht (2002)
2. Bělohlávek, R., Vychodil, V.: Fuzzy equational logic. Springer, Heidelberg (2005)
3. Fourman, M.P., Scott, D.S.: Sheaves and logic. Lecture Notes in Mathematics, vol. 753, pp. 302–401. Springer, Heidelberg (1979)
4. Höhle, U.: M-Valued sets and sheaves over integral, commutative cl-monoids. In: Applications of Category Theory to Fuzzy Subsets, pp. 33–72. Kluwer Academic Publ., Dordrecht (1992)
5. Höhle, U.: Fuzzy sets and sheaves. Part I, Basic concepts. Fuzzy Sets and System 158, 1143–1174 (2007)
6. Höhle, U.: Fuzzy sets and sheaves Part II: Sheaf-theoretic foundations of fuzzy set theory with applications to algebra and topology. Fuzzy Sets and System 158, 1175–1212 (2007)
7. Močkoř, J.: Fuzzy Sets in Categories of Sets with Similarity Relations. In: Computational Intelligence, Theory and Applications, pp. 677–682. Springer, Heidelberg (2006)
8. Močkoř, J.: Cut systems in sets with similarity relations. Fuzzy Sets and Systems 161(24), 3127–3140 (2010)
9. Močkoř, J.: Fuzzy sets and cut systems in a category of sets with similarity relations. Soft Computing 16, 101–107 (2012)
10. Močkoř, J.: Morphisms in categories of sets with similarity relations. In: Proceedings of IFSA Congress/EUSFLAT Conference, Lisabon, pp. 560–568 (2009)
11. Močkoř, J.: Fuzzy objects in categories of sets with similarity relations. In: Computational Intelligence, Theory and Applications, pp. 677–682. Springer, Heidelberg (2006)
12. Novák, V., Perfilijeva, I., Močkoř, J.: Mathematical principles of fuzzy logic. Kluwer Academic Publishers, Boston (1999)
13. Mac Lane, S.: Categories for the Working Mathematician. Springer, Heidelberg (1971)

Hierarchical Modified Regularized Least Squares Fuzzy Support Vector Regression through Multiscale Approach

Arindam Chaudhuri

Faculty of Post Graduate Studies and Research, Computer Engineering and Technology,
Marwadi Education Foundation's Group of Institutions, Rajkot, India
arindam_chau@yahoo.co.in

Abstract. Support vector regression (SVR) is a promising regression tool based on support vector machine (SVM). It is a paradigm for identifying estimated models that are based on minimizing Vapnik's loss function of residuals. It is based on linear combination of displaced replicas of kernel function. Single kernel is ineffective when function approximated is non stationary. This problem is taken care of by hierarchical modified regularized least squares fuzzy support vector regression (HMRLFSVR). It is developed from modified regularized least squares fuzzy support vector regression (MRLFSVR) and regularized least squares fuzzy support vector regression (RLFSVR). HMRLFSVR consists of a set of hierarchical layers each containing MRLFSVR with Gaussian kernel at given scale. On increasing scale layer by layer details are incorporated inside regression function. It adapts local scale to data keeping number of support vectors and configuration time comparable with classical SVR. It considers disadvantages when approximating non stationary function using single kernel approach where it is not able to follow variations in frequency content in different regions of input space. The approach is based on interleaving regression estimate with pruning activity. It denoises original data obtaining an effective multiscale reconstruction. The tuning of SVR configuration parameters becomes simplified in HMRLFSVR. Favourable results over noisy synthetic and real datasets are obtained when compared with multikernel approaches.

Keywords: SVM, SVR, HMRLFSVR, MRLFSVR, multiple kernels, multiscale regression.

1 Introduction

Support vector machine (SVM) introduced by Vapnik et al [1], [2] is an excellent machine learning tool for pattern classification and regression. SVMs have emerged from research in statistical learning theory. They are based on formalizing classification boundary which segregates points having different labels such that distance of boundary from closest data point is maximized. The boundary between classes is defined by hyperplane leading to different support vectors. To recognize support vectors the problem is restructured as quadratic optimization problem that is

I. Rojas, G. Joya, and J. Cabestany (Eds.): IWANN 2013, Part I, LNCS 7902, pp. 393–407, 2013.

convex, guarantees uniqueness and optimality of solution. It was identified that SVM classifies data that exhibits linear separability [3] which is often not suitable for many applications. For this reason mapping machinery transforms classification problem into higher dimensional space called feature space which achieves linear separability in higher dimensional space. This is not a hindrance in obtaining solution because determining coefficients does not require computing the mapped value of any single data points.

SVM has been extended to several regression problems [4] such as support vector regression (SVR). SVR fits linear regressor through given set of data points where points may be in higher dimensional space. It requires solving quadratic minimization problem subject to linear inequality constraints which is convex programming task [3]. The output of SVR is computed as:

$$Z_{SVR}(x) = \sum_{i=1}^{n} w_i k(x, x_i) + bi \tag{1}$$

In equation (1) w_i and x_i are weight and position of each support vector, n is number of support vectors, bi is bias and $k(x, x_i)$ is kernel function. The quality of regression depends on the choice of kernel function and its parameters. The selection of kernel selection [5] is difficult task. It is done by trial and error, genetic optimization or prior expertise. When data is characterized by space varying frequency, usage of single kernel cannot produce accurate solutions. Thus, approaches based on multiple kernels have been recently studied [6–8]. On incorporating multiple kernels, output given by equation (1) becomes:

$$Z_{SVR}(x) = \sum_{i=1}^{n} \left(w_i \sum_{j=1}^{m} \eta_j k_j(x, x_i) \right) + bi \tag{2}$$

In equation (2) type and number m of kernels $k_j(x, x_i)$ used are chosen *apriori* and mixing coefficients η_j are determined in optimization phase. Multiple kernels play a vital role in determining the quality of solution. One possible solution is to adapt multiple kernels to local frequency as suggested in [7] and [8] where parameters η_j are function of both support vector and data position. A second solution is based on using same kernel with different scales as given in [9] and [10]. In [9] solution is computed using a set of groups of kernels where kernels in each group have same scale which increases computational complexity. In [10] multiple scales originate from wavelet decomposition which requires that coefficients of all levels are estimated together, thus increasing the number of parameters. A more affordable solution is obtained when regression is built through hierarchical iterative approximation as explored in neural networks [11–13] and machine learning [14] domains where uniform residual error is guaranteed using limited number of units. In [15] an approximation at coarse level is first produced with SVR featuring kernel with large scale and then refined using kernels with smaller scales. Another approach is based on boosting used in classification domain [16] and then extended to regression [17], [18] which produces good results with long configuration time.

Another important challenge with SVR when using multiscale kernels is the number of support vectors. This problem has been addressed in [19–23]. These approaches are classified into two major categories. The first redefines optimization problem in order to control number of support vectors inserted [20], [21]. The second category realizes reduction of support vectors in two steps: (a) standard training of SVM is performed first and (b) pruning procedure is applied which involve a tradeoff between reduction of support vectors and accuracy loss. In [24], hyperplane is formed by SVM from mechanical point of view. Each support vector exerts force on hyperplane and minimum of optimization problem is determined keeping the system in equilibrium state. However, new support vectors do not exert an exact equivalent force of those deleted. The stopping criterion is based on monitoring accuracy loss when it goes over given threshold. This approach controls loss of accuracy in each pruning step.

In this work, we first illustrate Modified Regularized Least Squares Fuzzy Support Vector Regression (MRLFSVR) [25] which is an extension of Regularized Least Squares Fuzzy Support Vector Regression (RLFSVR) [26]. MRLFSVR is then remodeled as hierarchical modified regularized least squares fuzzy support vector regression (HMRLFSVR) which consists of a set of hierarchical layers each containing MRLFSVR with Gaussian kernel at given scale. On increasing scale layer by layer details are incorporated inside regression function. It adapts local scale to data keeping number of support vectors and comparable configuration time. It considers disadvantages encountered when approximating non stationary function using single kernel approach where it does not follow variations in frequency content in different regions of input space. The approach is based on interleaving regression estimate with pruning activity. HMRLFSVR has been applied to noisy synthetic and real datasets. It denoises original data obtaining an effective multiscale reconstruction of better quality. Results also compare favorably with multikernel approaches. Further, tuning SVR configuration parameters is strongly simplified in HMRLFSVR. This paper is presented as follows. In section 2, we introduce MRLFSVR. This is followed by HMRLFSVR in section 3. In next section experimental results and corresponding discussions are highlighted. Finally, in section 5 conclusions are illustrated.

2   Modified Regularized Least Squares Fuzzy Support Vector Regression

In this section, we present MRLFSVR which is an extension of RLFSVR approach proposed by Khemchandani et al [26]. Taking a set of M data samples and assuming i^{th} sample represented by the tuple (P_i, y_i) where $P_i = (P_{i1}, \ldots \ldots, P_{iN})$ is point in $\mathbf{R}^N$ and y_i is the value at P_i. Let P denote the matrix comprising of row vectors P_i; $i = 1, \ldots \ldots, M$. Here nonlinear transformation ϕ is used to transform data points from input space of dimension N into a feature space having a higher dimension V. The nonlinear mapping is given by $\phi: \mathbf{R}^N \rightarrow \mathbf{R}^V$ such that $\phi(P_i) \in \mathbf{R}^V$. MRLFSVR achieves a regressor by solving the following quadratic programming problem:

$$MRLFSVR\ Minimize_{q,w,b}\ \frac{c}{2} \parallel S^n q \parallel^2 + \frac{1}{2}(w^T w + b^2) \qquad (3)$$

$$subject\ to\ (w^T \phi(P_i) + b) - y_i + q_i = 0 \qquad i = 1, \ldots \ldots, M; n \geq 2$$

In equation (3) $S = diag(s_1, \ldots \ldots, s_M)$ and $C > 0$. The diagonal elements of S correspond to fuzzy membership values of data samples. The significance of S^n lies in the fact that it addresses higher dimensionality issues of data samples for nonlinear regression to be accomplished. The main idea of using Fuzzy Sets in MRLFSVR is that if the input is detected as an outlier, one input's membership value is reduced such that its contribution to total error term is decreased. MRLFSVR also treats each input as an input of opposite class with higher membership. In such way we expect the new fuzzy machine to make full use of data and achieve better generalization ability. The approach is adapted from [33] where one instance contributes its errors to total error term in objective function. Based on this idea, we treat every data point in training dataset as both positive and negative class but with different memberships. Memberships are assigned to both classes for every data point. The membership values are bounded from above and below as:

$$s_j \leq \sigma \leq s_i; i, j = 1, \ldots \ldots, M \tag{4}$$

The representation given by equations (3) and (4) is significant because error variables q cannot be non-negative. By selecting an appropriate value of q the above constraints can be met while minimizing objective function. From equations (3) and (4) it is evident that $\left(\frac{50}{n}\right)\%$ less error variables are required as compared to classical SVR. The choice of parameter n usually exceeds 2 but its maximum value depends on the type of dataset used. Here we consider $n = 2$. The Lagrangian corresponding to MRLFSVR given by equations (3) and (4) is as follows:

$$L(w, b, q, \gamma) = \frac{1}{2}(w^T w + b^2) + \frac{c}{2}(S^n q)^T (S^n q) - \gamma^T (\phi(P)w + eb - Y + q) \tag{5}$$

In equation (5) $\gamma = (\gamma_1, \ldots \ldots, \gamma_M)^T$. The Karush Kuhn Tucker (KKT) conditions are given by:

$$w = \phi(P)^T \gamma \tag{6}$$

$$b = e^T \gamma \tag{7}$$

$$q = \frac{(S^{nT} S^n)^{-1} \gamma}{C} \tag{8}$$

$$(\phi(P)w + eb) - Y + q = 0 \tag{9}$$

Substituting w, b and q from equations (6), (7) and (8) in equation (9) we have,

$$(\phi(P)\phi(P)^T \gamma + ee^T \gamma) - Y + \frac{(S^{nT} S^n)^{-1} \gamma}{C} = 0 \tag{10}$$

$$Y = ((\phi(P)\phi(P)^T \gamma + ee^T \gamma) + \frac{(S^{nT} S^n)^{-1} \gamma}{C} \tag{11}$$

The matrix $K = \phi(P)\phi(P)^T$ is always positive semi-definite. However, it can be made positive definite by introducing regularization term $\epsilon I, \epsilon > 0$ where I is an identity matrix of appropriate dimension. This modification affected is similar to ridge regression approaches [27] and it allows decompose the matrix $(K + \epsilon I)$ as ZZ^T

where Z is a positive definite matrix. Let $H = [Z\ e]$, then equation (11) can be written as:

$$\left(HH^T + \frac{(S^{nT}S^n)^{-1}}{C}\right)\gamma = Y \tag{12}$$

$$\gamma = \left(HH^T + \frac{(S^{nT}S^n)^{-1}}{C}\right)^{-1}Y \tag{13}$$

We observe that in equation (13) the matrix being inverted is always positive definite for any $C > 0$. Determining w requires a single matrix inversion as compared to the solution of quadratic programming problem in conventional SVR model [28]. In LSSVM, an iterative technique such as conjugate gradient or successive over relaxation method is to be used. In case of linear kernels added benefits can be obtained. From [29] and applying Sherman Morrison Woodbury (SMW) formula [30] we have following expression for γ:

$$\gamma = [C(S^{nT}S^n) - \{C(S^{nT}S^n)H(I + H^TC(S^{nT}S^n)H)^{-1}\} \times CH^T(S^{nT}S^n)Y] \tag{14}$$

$$\gamma := \left[I - (S^{nT}S^n)H\left(\frac{I}{C} + H^T(S^{nT}S^n)H\right)^{-1}H^T\right]C(S^{nT}S^n)Y \tag{15}$$

The equation (15) requires inversion of matrix of dimension $(N + 1)$ instead of $(M + 1)$. In case of nonlinear kernel the order of matrix H is $M \times (M + 1)$. In this case, the use of SMW formula to solve equation (15) provides no computational benefit. Considering the following optimization problem in similar direction of proximal support vector machine classifier [29] we have,

$$MRLFSVR_{New}\ Minimize_{q,u,b}\ \frac{C}{2}\parallel S^n q \parallel^2 + \frac{1}{2}(u^T u + b^2) \tag{16}$$

$$subject\ to\ Y - (ku + eb) + q = 0\ \ n \geq 2$$

Here, $(K)_{ij} = \phi(P_i)^T\phi(P_j), C > 0$. The solution is obtained by solving KKT conditions yields:

$$\left(KK^T + ee^T + \frac{(S^{nT}S^n)^{-1}}{C}\right)\gamma = Y \tag{17}$$

$$\gamma = \left(JJ^T + \frac{(S^{nT}S^n)^{-1}}{C}\right)^{-1}Y \tag{18}$$

In equation (18) $J = [K\ e]$. Using rectangular kernels [31], $K = K(P, P^T)$ can be replaced by rectangular kernel viz. $K = [K(P, Q^T)]_{M \times T}$; Q is a $T \times N$ random sub matrix of P with $T \ll M$ (T may be $0.01M$). Rectangular kernels provide significant computational savings and reduce over fitting. In effect, Q uses subset of vector set P similar to pruning approach [32]. Next we address membership values s_i that constitute diagonal matrix S. Chaudhuri et al [33] suggested an exponential function in computing fuzzy membership value of P_i which is:

$$s_i = \frac{1}{1+e^{\left(a-2a\left(\frac{t_i-t_1}{t_l-t_1}\right)\right)}} \tag{19}$$

The characteristics of equation (19) are: (i) As $a \to 0$ then $lim_{a \to 0} s_i = \frac{1}{2}$ such that $s_i \to \frac{1}{2}$. Here, all training samples are assigned same fuzzy membership values.

(ii) As $a \to \infty$ then $lim_{a \to \infty} s_i = \begin{cases} 0, & t_i < \frac{t_i + t_1}{2} \\ 1, & t_i > \frac{t_i + t_1}{2} \end{cases}$. Here, fuzzy membership values for data points arriving in first and second half are zero and one respectively. (iii) When $a \in (0, \infty)$ then as a increases, fuzzy membership values for data points arriving in first and second half are smaller and relatively larger respectively.

Some other membership functions we use here are extensions of functions suggested by [33] are:

(a) The membership function considered as n^{th} degree polynomial function of time given by:

$$s_i = g(t_i) = t_i \left(\frac{1-\sigma}{t_M - t_1}\right)^n + \left(\frac{t_M \sigma - t_1}{t_M - t_1}\right)^n \tag{20}$$

Here, generally $n > 2$ and we assume that last point x_M is most important and consider $s_M = g(t_M) = 1$ while first sample x_1 is considered to be least important i.e. $s_1 = g(t_1) = \sigma$.

(b) The membership function considered as $2n^{th}$ degree polynomial function of time given by:

$$s_i = g(t_i) = (1 - \sigma)\left(\frac{t_i - t_1}{t_M - t_1}\right)^{2n} + \sigma \tag{21}$$

3   Hierarchical Modified Regularized Least Squares Fuzzy Support Vector Regression

Based on MRLFSVR, HMRLFSVR is formulated in this section. HMRLFSVR is constituted into a pool of L layers, each comprising of a single kernel MRLFSVR $\{v_l(\circ)\}$ characterized by suitable scale. The different layers are placed in a hierarchy having a scale determined by parameter σ_l which increases when layer number decreases such that $\sigma_l \leq \sigma_{l+1}$ holds. The output of HMRLFSVR model is obtained as sum of output of layers:

$$g(x) = \sum_{l=1}^{L} v_l(x; \sigma_l) \tag{22}$$

HMRLFSVR configuration proceeds by adding and configuring one layer at a time. It initiates from layer featuring smallest scale to that featuring largest one. The first layer is trained such that distance between the regression curve produced by first layer itself and data is minimized [4]. It plays a significant role in the success of HMRLFSVR. It is trained heuristically such that number of used layers is reduced. All other layers are trained to approximate the residual. The residual measure for each layer is given as:

$$rm_l(x_i) = rm_{l-1}(x_i) - v_l(x_i) \tag{23}$$

The lth layer is configured with training set $TS_l = \{(x_1, rm_{l-1}(x_1)), \ldots\ldots$ $(x_n, rm_{l-1}(x_n))\}$. The value of scale parameter of first layer σ_1 is chosen to be proportional to the size of input domain. The parameter σ can be decreased arbitrarily. Generally the most preferred value of σ for each layer is $\sigma_{l+1} = \sigma_l/n$; $n \geq$ 2 which produces satisfactory results. On decreasing σ slowly the accuracy of solution improves but number of layers and number of support vectors increases. New layers are added during training until stopping criterion is satisfied. The two other parameters are defined for each layer viz. (a) C_l the tradeoff between regression error and smoothness of solution and (b) ε which controls amplitude of ε-insensitivity tube around solution itself. The value of C is usually set experimentally by trial and error [34], [35]. Here, C_l is chosen for each layer as J times variance of residuals used to configure the lth layer as:

$$C_l = Jvar(r_{l-1}(x_i)) \tag{24}$$

In equation (24) C_l assumes the value taken by Lagrange multipliers associated to support vectors of the lth layer which represents the maximum weight that can be associated to each kernel. For input space regions where Gaussians associated to support vectors have no significant overlap. This depends both on Gaussian scale parameter and on data density. The value of C_l is approximately the maximum value that can be assumed by regression function in those regions as Gaussian kernel is equal to 1. For this reason C_l should be large enough to allow regression curve reaching maximum or minimum value of data points inside whole input domain. However, a large value of C_l favors over fitting. Experimental results on different datasets suggest the value of J lying in interval $(0, 5]$ which represents a tradeoff between these two requirements. Similar to MRLFSVR, the parameter ε cannot be determined from dataset. Rather ε could be set proportional to the accuracy required for regression as its value is related to noise amplitude [36].

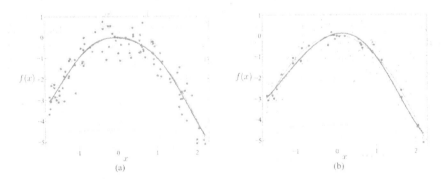

Fig. 1. Data points reduction. A set of 200 randomly sampled points through Gaussian quantity are displayed as dots. Thick dots represent all points used by optimization engine to determine regression. Circled dots represent support vectors. (a) SVR curve is obtained through standard SVR (b) Regression curve is obtained considering only the points in TS_l'. Both regression curves are contained inside ε-tube around real function.

Experiments show that in HMRLFSVR, the layers with larger value of σ have number of support vectors similar to layers with smaller σ. There appears some contradiction as fewer units are required to realize a reconstruction at larger scale, but it is due to the fact that all data points distant from regression curve by more than ε are selected as support vectors as shown in Fig 1. Hence, in first layers where HMRLFSVR output has low frequency content many data points lie far from curve and are still selected as support vectors. This leads to high number of support vectors. To avoid this after each layer has been configured, a pruning step is carried out to reduce number of support vectors. The cost function [4] is then minimized a second time considering only reduced training set to obtain final approximation for each current layer. To reduce the number of support vectors we notice that distance of training point from regression curve measures suitability of current curve to describe information conveyed by that data point. In this sense, points too distant from regression curve cannot be explained by the curve and their utility can be at stake. They can be regarded as outliers. For these reasons, acceptable approximation of regression curve should be obtained using only those points that lie close to curve. This has been confirmed experimentally. We have observed that the quality of regression at a given scale does not degrade significantly if we compute regression considering only points close to ε-tube. The closeness of a point to ε-tube ψ can be assessed only after computation of regression itself which is done by considering all training points. In second pass, regression is computed again considering only points close to ε-tube. Consider the lth layer and regression computed for the layer $v_l(x)$ using complete training set TS_l. Let us define TS_l' the set constituting only of those support vectors that lie on the border of ε-tube and those whose distance from $v_l(x)$ is less than ε/n as:

$$TS_l' = \left\{ (x_i, r_{l-1}(x_i)) \middle| \left| |r_l(x_i)| - \varepsilon \right| < \tau \vee |r_l(x_i)| < \frac{\varepsilon}{n} \right\} \tag{25}$$

In equation (25) τ is tolerance parameter that determines thickness of ε-tube margin. We have structured the configuration phase of each layer in two sequential steps: (a) the first provides regression curve v_l considering all training points and (b) the second v_l' realizes an efficient regression curve by considering only a selected subset of points. To cope with diminished point density in TS_l', value of parameter C_l is increased proportionally in second optimization step as:

$$C_l' = C_l \frac{|TS_l|}{|TS_l'|} = Jvar(r_{l-1}(x_i)) \frac{|TS_l|}{|TS_l'|} \tag{26}$$

4   Experimental Results

We present here the results obtained using HMRLFSVR on both simulated and real data and compare them with those obtained with MRLFSVR and SVR [4] in terms of number of support vectors, computational time and accuracy. The accuracy is assessed through root mean square error *RMSE*, mean absolute error *MAE* and variance *VAR*. These are computed over a test set with respect to the optimal value of parameters which constitutes (a) ε, σ and C for MRLFSVR and (b) ε and C only for HMRLFSVR. The parameter C is set according to equations (24) and (26). The value of σ for first layer σ_0 is equal to the size of input for HMRLFSVR in order to cover the entire domain. The optimization problem in equation (16) for both HMRLFSVR and MRLFSVR is solved through LibCVM Toolkit Version 2.2 [37] on PC having Intel P4 processor with 2.80 GHz, 512 MB DDR RAM @ 400 MHz with 512 KB cache.

4.1   Regression on Synthetic Data

We consider space varying function $h\colon \mathbb{R} \longrightarrow \mathbb{R}$ defined by equation (27) and shown in Fig. 2.

$$h(x) = \sin(2\pi x^4) + x \qquad (27)$$

The training dataset is obtained from equation (27) into 275 points such that sampled data is proportional to local frequency and random uniform quantity $[-0.1, 0.1]$ is added to simulate error. The regression is evaluated using a test set comprising 500 points sampled from $h(\cdot)$. MRLFSVR's accuracy is evaluated for all combinations of ε, σ and $C\colon \varepsilon \in \{0, 0.005, 0.01, 0.025,\ 0.05, 0.075, 0.1, 0.2\};\ \sigma \in \{0.015, 0.022,$ $0.0313, 0.0625, 0.125, 0.25\}$ and $C \in \{0.5, 1, 1.5, 2, 5, 10, 20\}$.

The best value of MRLFSVR is considered after comparison which is obtained with $\varepsilon = 0.075$, $\sigma = 0.0313$ and $C = 20$. This is duly supported by data presented in Table 1 based on the accuracy of different models. The computational time for HMRLFSVR is referred to the entire process of configuring nine layers required before the growth stops while for MRLFSVR it does not consider the process of searching for optimum values of ε, σ and C, but the solution is computed with best parameters as $\varepsilon = 0.075$, $\sigma = 0.0313$ and $C = 20$. Setting search space $C = 100000$ the accuracy of SVR improves. In fact, test error decreases from 0.0816 to 0.0517 although it remains higher than HMRLFSVR. The time required to compute this solution increases to 3340 seconds.

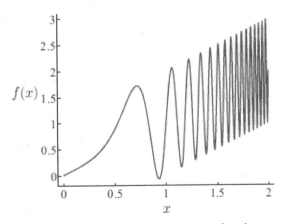

Fig. 2. Nonstationary frequency content function

As evident from Table 1 using a single kernel function provides regression of good quality when structure of data changes in input domain. In particular, spurious oscillations are produced in region of low frequency when high frequency kernel is adopted. In the multiscale approach investigated here this low frequency region is reconstructed by first layers while higher layers improve reconstruction in high frequency region.

Table 1. Accuracy of the models on synthetic dataset

	MAE	*VAR*	*RMSE*	No. of support vectors	Time (seconds)
HMRLFSVR	0.0140	0.0033	0.0189	1696	0.287
MRLFSVR	0.0269	0.0075	0.0272	275	0.316
SVR	0.0816	0.0278	0.186	149	0.451

On investigating the role of ε, it is observed that test error produced by HMRLFSVR is below ε for $\varepsilon > 0.05$. This is smaller than optimal value 0.075. This implies that data points on average lie inside ε-tube. For SVR optimal value of ε is 0.05 which is smaller than test error achieved (0.0816). This means that a relatively large number of data points are not contained inside ε-tube. As expected the number of support vectors decreases with increase of ε. The results are summarized in Table 2. The reduction of support vectors that effectively contribute to output of each layer does not degrade reconstruction and produces an identical output with fewer support vectors. The same is true for output of intermediate layers as given in Table 2. From this we conclude that most of support vectors employed in first layers of multiscale reconstruction are wasted in an attempt of approximating details with a kernel that operates at a large scale. These support vectors can be pruned without degrading output quality.

4.2   Regression on Real Data

Now we consider real artifact data viz. a panda mask obtained through a 3D scanner [13]. The points are sampled from a single point of view. They belong to 2.5D surface that can be described as a $\mathbb{R}^2 \longrightarrow \mathbb{R}$ function making these points suitable to SVM regression. The dataset is composed of 22000 points, 18000 of which are randomly chosen and used for training and remaining 4000 for testing. The experimental data is normalized to fit inside the range $[-1, 1]$ before optimization as input features can be measured in different domains. In order to limit border effects test error have been computed in the inner region of dataset considering only points distant from closest boundary by more than 0.1. MRLFSVR is computed with all combinations of following values of ε, σ and J:$\varepsilon \in \{0, 0.0025, 0.005, 0.01, 0.02\}$; $\sigma \in \{0.188, 0.0938, 0.0469, 0.0234\}$ and $J \in \{0.5, 1, 2, 5\}$. The parameter C identical to equation (24) is set proportionally to variance of height h of points through proportionality factor J:

$$C = Jvar(h) \tag{28}$$

Table 2. Performance of HMRLFSVR and MRLFSVR ($\varepsilon = 0.075$) on synthetic dataset

No. of Layers	HMRLFSVR					MRLFSVR				
	MAE	VAR	RMSE	No. of support vectors	Time (seconds)	MAE	VAR	RMSE	No. of support vectors	Time (seconds)
1	0.436	0.108	0.556	240	0.013	0.476	0.107	0.576	15	0.013
2	0.415	0.109	0.540	246	0.018	0.416	0.110	0.542	22	0.017
3	0.369	0.125	0.524	235	0.016	0.393	0.124	0.540	27	0.018
4	0.316	0.124	0.474	222	0.019	0.330	0.130	0.495	30	0.020
5	0.251	0.123	0.433	205	0.042	0.281	0.127	0.469	42	0.048
6	0.179	0.107	0.354	180	0.072	0.205	0.113	0.383	70	0.096
7	0.072	0.047	0.205	136	0.056	0.109	0.052	0.270	96	0.069
8	0.022	0.002	0.012	90	0.025	0.026	0.007	0.020	124	0.056
9	0.011	0.001	0.005	52	0.005	0.015	0.002	0.009	45	0.010

HMRLFSVR is computed for all combinations of above values of J and ε. Table 3 presents best test error results in all three models in the range $[0.0107, 0.0111]$. This consideration can be extended to all best models obtained for a given value of ε. On the contrary, average and worst case test error computed over all other parameter combinations are much higher for MRLFSVR than for HMRLFSVR. Although test error for all three optimal models is identical, the configuration time is different. It is 609, 904 and 382 seconds for HMRLFSVR, MRLFSVR and SVR respectively. This large difference becomes smaller when suboptimal configuration parameters are considered. Time required exploring the parameters space is added to this time. The dimensionality of this space is smaller for HMRLFSVR as σ is not to be considered. Here, 80 and 20 parameter combinations have been used for SVR and HMRLFSVR respectively. For SVR total configuration time including parameters space search is 87410 seconds while for MRLFSVR and HMRLFSVR total configuration time is 15696 and 20405 seconds respectively.

Table 3. Results on panda mask dataset

	MAE	VAR	RMSE	No. of support vectors	Time (seconds)
HMRLFSVR	0.0107	0.00009	0.0156	114999	975
MRLFSVR	0.0109	0.00012	0.0157	100889	783
SVR	0.0111	0.00013	0.0161	12442	382

As obvious from Table 3 the accuracy of HMRLFSVR and number of support vectors used are similar to SVR. The analysis of configuration time is more critical. When configuration parameters ε, σ and C are fixed optimally, configuration time for

SVR is less than HMRLFSVR. Efforts are devoted towards selection of optimum parameters because quality of regression depends critically on them [3]. This makes SVR regression extremely time consuming as 3D parameter space has to be searched optimally. This search space is reduced to 2D space in HMRLFSVR as we need not optimize σ. This allows reduction in configuration time which largely over compensates computing SVR twice for each level. Moreover, HMRLFSVR is much less sensitive to initial value of these two parameters. This is due to the robustness of hierarchical structure in which data not explained in one layer can be recovered by next layers. The parameter C is set proportionally to variance of data points from equation (24) through factor J by computing experimentally on different datasets. The values in the range $[0.1, 21]$ for parameter C and values of J revolving 1 have always yielded very good results where best, average and worst case error with different combinations of J and ε are almost similar. As kernels with different σ are chosen in different layers, the criticality in choosing single value of σ which is adequate for data often disappears. In case of equally spaced support vectors [4], [37], [38] represent approximation in terms of Riesz basis. Moreover, angle between the spaces spanned in two consecutive layers is quite small which allows slowly incorporating details on one side but stop adding new layers, when regression starts becoming noisy [12].

Table 4. Results on motorcycle dataset

	σ	ε	C	No. of support vectors	*RMSE*
SVR	1	0.1436	10	51.8	0.2272
MRLFSVR	10	0.1436	1	33.2	0.2209
SVR [9]	1	0.1436	10	49.3	0.2334
MSSVR (E) [9]	0.7 – 2.8	0.1436	5	8	0.2322
MSSVR (Q) [9]	0.7 – 2.8	–	5	7.5	0.2329
MSSVR (H) [9]	0.7 – 2.8	0.1436	1	9	0.2329

We investigated HMRLFSVR also on other available datasets. Table 4 summarizes results on motorcycle dataset [39]. Results of HMRLFSVR are averaged over 10 different randomizations of dataset to create different training and test sets. They are also compared with multiscale approach of [9]. We observe that both HMRLFSVR and multiscale SVR (MSSVR) produce a regression error smaller than SVR. HMRLFSVR uses fewer support vectors than SVR demonstrating its ability to catch local frequency although spending more time for configuration (0.003 seconds for SVR versus 0.009 seconds for HMRLFSVR). It is noted that no search for hyper parameters is carried out since we have used same values given in [9]. The approach in [9] uses fewer support vectors but it requires solving global optimization problem in one pass for all scales. This requires large configuration time and makes it feasible for small datasets. Moreover, number of scales has to be defined *apriori* which can lead to over or under fitting the data.

HMSLFSVR thus cannot be considered direct result of minimization of cost function [9], [40]. However, cost function constitutes regularization function that produces smooth regression penalizing both norms of error and kernel coefficients.

We consider two norms at each layer with the aim of producing smooth regression incrementally. We favor smoothness also by decreasing maximum amplitude of coefficients in each layer from equation (24). This allows good regression with limited computational time. The number of layers may be critical to avoid over fitting. Different strategies can be used to stop configuration procedure. If no information is available *apriori*, use of test error guarantees that a good generalization is obtained. The introduction of new layer can be stopped when test error does not decrease with new layer. Alternatively, growth can be stopped when residual drops below a threshold which is associated to measurement noise [13]. Any other · criteria depending on application context can be adopted.

Although in principle a linear combination of kernels featuring single large scale can be used to realize an accurate estimate of function with high frequency content, computational effort required would be too high. The use of large scale kernels involves large coefficients viz. higher C which may cause numerical instability. Also configuration time tends to increase with C which makes use of high values almost infeasible for real cases. HMRLFSVR configuration scheme can also work with other kernels. The Gaussian kernel allows parameter σ to shape the kernel such that support vectors are sensitive to different frequency ranges and its non orthogonality recovers reconstruction errors left by previous layers.

5   Conclusion

In this work, we have developed a novel multiscale kernel regression approach i.e. HMRLFSVR. The method gradually evolves from MRLFSVR and RLFSVR. It is based on two key elements viz. (a) multiscale incremental approximation and (b) reduction in number of data points passed to optimization engine. HMRLFSVR consists of a set of hierarchical layers each containing a MRLFSVR with Gaussian kernel at given scale. The technique adds layers that employ kernels at increasing scales until an adequate output is obtained. The number of support vectors used is identical to SVR. It approximates non stationary function using single kernel approach. This is based on interleaving regression estimate with pruning activity. The configuration time is comparatively less as full search in parameters space is not required. The technique incorporates data reduction step where data considered meaningful for output of each layer are passed to optimization procedure with substantial reduction in number of support vectors making it suitable for very large datasets. It denoises original data obtaining a multiscale reconstruction of better quality. The results show improvement when compared with multikernel approaches.

References

1. Vapnik, V.N.: Statistical Learning Theory. Wiley, New York (1998)
2. Vapnik, V.N.: The Natural of Statistical Learning Theory. Springer, New York (1995)
3. Burges, C.: A Tutorial on Support Vector Machines for Pattern Recognition. Data Mining and Knowledge Discovery 2(2), 121–167 (1998)

4. Smola, A.J., Schölkopf, B.: A Tutorial on Support Vector Regression. Statistics and Computing 14(3), 199–222 (2004)
5. Lanckriet, G., Cristianini, N., Bartlett, P., Ghaoui, L.E., Jordan, M.I.: Learning the Kernel Matrix with Semi Definite Programming. Journal of Machine Learning Research 5, 27–72 (2004)
6. Wang, Z., Chen, S., Sun, T.: MultiK-MHKS: A Novel Multiple Kernel Learning Algorithm. IEEE Transactions on Pattern Analysis and Machine Intelligence 30(2), 348–353 (2008)
7. Gönen, M., Alpaydin, E.: Localized Multiple Kernel Learning. In: 25th International Conference on Machine Learning, pp. 352–359 (2008)
8. Gönen, M., Alpaydin, E.: Localized Multiple Kernel Regression. In: 20th IAPR International Conference Pattern Recognition, pp. 1425–1428 (2010)
9. Zheng, D., Wang, J., Zhao, Y.: Non-flat Function Estimation with a Multi-scale Support Vector Regression. Neurocomputing 70(1-3), 420–429 (2006)
10. Peng, H., Wang, J.: Nonlinear System Identification based on Multiresolution Support Vector Regression. In: International Conference on Neural Networks and Brain, vol. 1, pp. 240–243 (2005)
11. Moody, J.E.: Fast Learning in Multi-resolution Hierarchies. In: Neural Information Processing Systems, pp. 29–39. Morgan Kaufmann, San Francisco (1988)
12. Ferrari, S., Maggioni, M., Borghese, N.A.: Multi-scale Approximation with Hierarchical Radial Basis Functions Networks. IEEE Transactions on Neural Networks 15(1), 178–188 (2004)
13. Ferrari, S., Bellocchio, F., Piuri, V., Borghese, N.A.: A Hierarchical RBF Online Learning Algorithm for Real Time 3D Scanner. IEEE Transactions on Neural Networks 21(2), 275–285 (2010)
14. Reddy, C.K., Park, J.-H.: Multi-resolution Boosting for Classification and Regression Problems. In: Theeramunkong, T., Kijsirikul, B., Cercone, N., Ho, T.-B. (eds.) PAKDD 2009. LNCS, vol. 5476, pp. 196–207. Springer, Heidelberg (2009)
15. Steinke, F., Schölkopf, B., Blanz, V.: Support Vector Machines for 3D Shape Processing. Computer Graphics Forum 24(3), 285–294 (2005)
16. Schapire, R.E.: A Brief Introduction to Boosting. In: International Joint Conference on Artificial Intelligence, pp. 1401–1406 (1999)
17. Freund, Y., Schapire, R.E.: A Decision-theoretic Generalization of On-line Learning and an Application to Boosting. Journal of Computer and System Sciences 55, 119–139 (1997)
18. Duffy, N., Helmbold, D.: Boosting Methods for Regression. Machine Learning 47, 153–200 (2002)
19. Liang, X.: An Effective Method of Pruning Support Vector Machine Classifiers. IEEE Transactions on Neural Networks 21(1), 26–38 (2010)
20. Fung, G.M., Mangasarian, O.L., Smola, A.J.: Minimal Kernel Classifiers. Journal of Machine Learning Research 3, 2303–2321 (2002)
21. Keerthi, S.S., Chapelle, O., Coste, D.D.: Building Support Vector Machines with Reduced Classifier Complexity. Journal of Machine Learning Research 7, 1493–1515 (2006)
22. Guo, J., Takahashi, N., Nishi, T.: An Efficient Method for Simplifying Decision Functions of Support Vector Machines. IEICE Transactions on Fundamentals of Electronics, Communications and Computer Science E89-A(10), 2795–2802 (2006)
23. Zeng, X., Chen, X.: SMO-based Pruning Methods for Sparse Least Squares Support Vector Machines. IEEE Transactions on Neural Networks 16(6), 1541–1546 (2005)
24. Nguyen, D., Ho, T.: An Efficient Method for Simplifying Support Vector Machines. In: 22nd International Conference on Machine Learning, pp. 617–624 (2005)

25. Chaudhuri, A.: Forecasting Rice Production in West Bengal State in India: Statistical vs. Computational Intelligence Techniques. International Journal of Agricultural and Environmental Information Systems 4(2) (in press, 2013)
26. Khemchandani, R., Jayadeva, Chandra, S.: Regularized Least Squares Fuzzy Support Vector Regression for Financial Time Series Forecasting. Expert Systems with Applications 36(1), 132–138 (2009)
27. Saunders, C., Gammerman, A., Vovk, V.: Ridge Regression Learning Algorithm in Dual Variables. In: 15th International Conference on Machine Learning, pp. 515–521. Madison, Wisconsin (1998)
28. Gunn, S.R.: Support Vector Machines for Classification and Regression. School of Electronics and Computer Science, University of Southampton, Southampton, Technical Report (1998)
29. Fung, G., Mangasarian, O.L.: Proximal Support Vector Machine Classifiers. In: International Conference of Knowledge Discovery and Data Mining, pp. 77–86. Association for Computing Machinery, New York (2001)
30. Golub, G.H., Van Loan, C.F.: Matrix Computations, 3rd edn. The Johns Hopkins University Press (1996)
31. Lee, Y.J., Mangasarian, O.L.: RSVM: Reduced Support Vector Machines. Data Mining Institute, Computer Sciences Department, University of Wisconsin, Madison, Wisconsin, Technical Report 00-07 (2000)
32. Brabanter, B., Lukas, L., Vandewalle, J.: Weighted Least Squares Support Vector Machines: Robustness and Sparse Approximation. Neurocomputing 48(1-4), 85–105 (2002)
33. Chaudhuri, A., De, K.: Fuzzy Support Vector Machine for Bankruptcy Prediction. Applied Soft Computing 11(2), 2472–2486 (2011)
34. Wang, D., Wu, X.B., Lin, D.M.: Two Heuristic Strategies for Searching Optimal Hyper parameters of C-SVM. In: 8th International Conference on Machine Learning and Cybernetics, pp. 3690–3695 (2009)
35. Tang, Y., Guo, W., Gao, J.: Efficient Model Selection for Support Vector Machine with Gaussian Kernel Function. In: IEEE Symposium on Computational Intelligence and Data Mining, pp. 40–45 (2009)
36. Smola, A.J., Murata, N., Schölkopf, B., Müller, K.R.: Asymptotically optimal choice of ε-loss for Support Vector Machines. In: 8th International Conference on Artificial Neural Networks, Perspectives on Neural Computing, pp. 105–110. Springer, Berlin (1998)
37. Tsang, I.W., Kwok, J.T., Cheung, P.M.: Core Vector Machines: Fast SVM Training on very large Data Sets. Journal of Machine Learning Research 6, 363–392 (2005)
38. Joachims, T.: Making Large Scale SVM Learning Practical. In: Schölkopf, B., Burges, C., Smola, A. (eds.) Advances in Kernel Methods - Support Vector Learning, ch. 11, pp. 169–184. MIT Press, Cambridge (1999)
39. Heteroscedastic Kernel Ridge Regression Demo,
    http://theoval.cmp.uea.ac.uk/matlab/hkrr_demo/hkrr_demo.m
40. Qiu, S., Lane, T.: Multiple Kernel Learning for Support Vector Regression. Department of Computer Science, University of New Mexico, Albuquerque, Technical Report, TR-CS-2005-42 (2005)

Minimal Learning Machine: A New Distance-Based Method for Supervised Learning

Amauri Holanda de Souza Junior[1,*], Francesco Corona[3], Yoan Miche[3], Amaury Lendasse[3], Guilherme A. Barreto[2], and Olli Simula[3]

[1] Federal Institute of Ceará, Department of Computer Science
Av. Contorno Norte, 10 - Maracanaú, Ceará, Brazil
[2] Federal University of Ceará, Department of Teleinformatics Engineering
Av. Mister Hull, S/N - Campus of Pici, Center of Technology, Fortaleza, Ceará, Brazil
[3] Aalto University, Department of Information and Computer Science
Konemiehentie 2, Espoo, Finland

Abstract. In this work, a novel supervised learning method, the Minimal Learning Machine (MLM), is proposed. Learning a MLM consists in reconstructing the mapping existing between input and output distance matrices and then estimating the response from the geometrical configuration of the output points. Given its general formulation, the Minimal Learning Machine is inherently capable to operate on nonlinear regression problems as well as on multidimensional response spaces. In addition, an intuitive extension of the MLM is proposed to deal with classification problems. On the basis of our experiments, the Minimal Learning Machine is able to achieve accuracies that are comparable to many *de facto* standard methods for regression and it offers a computationally valid alternative to such approaches.

1 Introduction

In this paper, we present a new supervised method, the Minimal Learning Machine (MLM). The basic idea behind the Minimal Learning Machine is the existence of a mapping between the geometric configurations of points in the input and output space. On the basis of our experiments, such a mapping can be accurately reconstructed by learning a multi-response linear regression model between distance matrices. Under these conditions, for an input point with known configuration in the input space, its corresponding configuration in the output space can be easily estimated after learning a simple linear model between input and output distance matrices. The resulting estimate is then used to locate the output point and thus provide an estimate for the response. In its basic formulation, the MLM closely resembles a classical unsupervised dimensionality reduction method, Multidimensional Scaling (MDS, [1]), and more specifically its variant known as Landmark MDS [2], the main difference being that the output configuration in MLM is known beforehand.

---

* The author would like to thank the financial support received from the Brazilian Agency of Post-Graduate Studies (CAPES) under the grant number 9147-12-8.

I. Rojas, G. Joya, and J. Cabestany (Eds.): IWANN 2013, Part I, LNCS 7902, pp. 408–416, 2013.

The remainder of the paper is organized as follows. In Section 2, the Minimal Learning Machine is presented; the MLM is formulated (Section 2.1), its properties discussed (Section 2.2) and two illustrative examples presented (Section 2.3) along with a simple extension of MLM that renders it suitable also for classification tasks. In Section 3, a thorough experimental assessment of the Minimal Learning Machine is conducted to evaluate its performance and to compare it with state-of-the-art approaches in regression.

2   Minimal Learning Machine

We are given a set of N input points $X = \{\mathbf{x}_i\}_{i=1}^N$, with $\mathbf{x}_i \in \mathbb{R}^D$, and the set of their corresponding outputs $Y = \{\mathbf{y}_i\}_{i=1}^N$, with $\mathbf{y}_i \in \mathbb{R}^S$. Assuming the existence of a continuous mapping $f : \mathcal{X} \to \mathcal{Y}$ between the input and the output space, we want to estimate it from data using a multi-response model

$$\mathbf{Y} = f(\mathbf{X}) + \mathbf{R},$$

where the columns of the matrix $\mathbf{X} = [\mathbf{x}_1, \dots, \mathbf{x}_D]$ correspond to the input variables and the rows to the observations, analogously the columns of the matrix $\mathbf{Y} = [\mathbf{y}_1, \dots, \mathbf{y}_S]$ correspond to the output variables and the rows to the observations. The $N \times S$ matrix $\mathbf{R}$ denotes the output residual vectors.

2.1   Formulation

Provided that the input space $\mathcal{X}$ is well sampled and f is smooth, we expect that for each pair of input points $(\mathbf{x}_i, \mathbf{x}_j)$ and for every $\varepsilon_y > 0$, there exists a $\varepsilon_x > 0$ such that for $d(\mathbf{x}_i, \mathbf{x}_j) < \varepsilon_x$ we have that $\delta(f(\mathbf{x}_i)), \delta(f(\mathbf{x}_j)) < \varepsilon_y$, where $d(\cdot, \cdot)$ and $\delta(\cdot, \cdot)$ are distance functions in $\mathcal{X}$ and $\mathcal{Y}$, respectively. Under this condition, we are interested in reconstructing the mapping $g : \mathcal{D_X} \to \mathcal{D_Y}$ between input distance matrices $\mathbf{D}_x$ and the corresponding output distance matrices $\mathbf{\Delta}_y$. The availability of the geometrical configurations of the points in the input and in the output space is then used to estimate the response $\mathbf{y}$ of a query input $\mathbf{x}$.

Distance Regression. For a selection of reference input points $R = \{\mathbf{m}_k\}_{k=1}^K$ with $R \subseteq X$ and corresponding outputs $T = \{\mathbf{t}_k\}_{k=1}^K$ with $T \subseteq Y$, we define $\mathbf{D}_x \in \mathbb{R}^{N \times K}$ in such a way that its kth column contains the distances $d(\mathbf{x}_i, \mathbf{m}_k)$ between the $i = 1, \dots, N$ input points $\mathbf{x}_i$ and the kth reference point $\mathbf{m}_k$. Analogously, we define $\mathbf{\Delta}_y \in \mathbb{R}^{N \times K}$ in a way that its kth column contains the distances $\delta(\mathbf{y}_i, \mathbf{t}_k)$ between the output points $\mathbf{y}_i$ and the output $\mathbf{t}_k$ of the kth reference point. The associated multi-response regression model for estimating g is thus

$$\mathbf{\Delta}_y = g(\mathbf{D}_x) + \mathbf{E}, \tag{1}$$

where the columns of the matrix $\mathbf{D}_x$ correspond to the K input vectors and the columns of the matrix $\mathbf{\Delta}_y$ correspond to the K response vectors. As usual, the K columns of the $N \times K$ matrix $\mathbf{E}$ correspond to the residuals.

We assume that the mapping g between input and output distances is linear, thus the multi-response regression model between distance matrices becomes

$$\mathbf{\Delta}_y = \mathbf{D}_x \mathbf{B} + \mathbf{E}, \tag{2}$$

where the regression matrix $\mathbf{B} \in \mathbb{R}^{K \times K}$ has to be solved from data. Under the normal conditions where the number of equations in (2) is larger to the number of unknowns, the problem is overdetermined and, usually, with no solution. This corresponds to the case where the number of selected reference points is smaller than the number of points available for solving the model (i.e., $K < N$) and we have to rely on the approximate solution provided by the least squares estimate:

$$\hat{\mathbf{B}} = (\mathbf{D}'_x \mathbf{D}_x)^{-1} \mathbf{D}'_x \mathbf{\Delta}_y. \tag{3}$$

On the other hand, if in (2) the number of equations equals the number of unknowns (i.e., all the learning points are also selected as reference points and $K = N$), the problem is uniquely determined and, usually, with a single solution $\hat{\mathbf{B}} = (\mathbf{D}_x)^{-1} \mathbf{\Delta}_y$. Clearly less interesting is the case where in (2) the number of equations is smaller than then number of unknowns (i.e., for $K > N$, corresponding to the situation where, after selecting the reference points, only a smaller number of learning points is used), for it leads to an underdetermined problem with, usually, infinitely many solutions.

Given the possibility for $\mathbf{B}$ to be either uniquely solvable or estimated (Equation 3), for a test point $\mathbf{x} \in \mathbb{R}^D$ whose distances from the K reference input points $\{\mathbf{m}_k\}_{k=1}^K$ are collected in the vector $\mathbf{d}(\mathbf{x}, R) = [d(\mathbf{x}, \mathbf{m}_1), \ldots, d(\mathbf{x}, \mathbf{m}_K)]$, the corresponding distances between its unknown output $\mathbf{y}$ and the known outputs of the reference points, the vector $\boldsymbol{\delta}(\mathbf{y}, T) = [\delta(\mathbf{y}, \mathbf{t}_1), \ldots, \delta(\mathbf{y}, \mathbf{t}_K)]$, are

$$\hat{\boldsymbol{\delta}}(\mathbf{y}, T) = \mathbf{d}(\mathbf{x}, R)\hat{\mathbf{B}}. \tag{4}$$

The vector $\hat{\boldsymbol{\delta}}(\mathbf{y}, T)$ provides an estimate of the geometrical configuration in $\mathcal{D}_y$ of $\mathbf{y}$ with respect to all the reference points $\{\mathbf{t}_k\}_{k=1}^K$ and thus can be used to estimate its location in $\mathcal{Y}$.

Output Estimation. Estimating $\mathbf{y}$ is equivalent to solve the overdetermined set of nonlinear equations corresponding to the K $(S+1)$-dimensional hyperspheres centered in $\mathbf{t}_k$ and all passing through $\mathbf{y}$, that is with a radius equal to $\hat{\delta}(\mathbf{y}, \mathbf{t}_k)$:

$$(\mathbf{y} - \mathbf{t}_k)'(\mathbf{y} - \mathbf{t}_k) = \hat{\delta}^2(\mathbf{y}, \mathbf{t}_k), \quad \forall k = 1, \ldots, K. \tag{5}$$

The problem in Equation 5 can be formulated as an optimization problem where an estimate $\hat{\mathbf{y}}$ can be obtained by the following minimization:

$$\hat{\mathbf{y}} = \underset{\mathbf{y}}{\operatorname{argmin}} \sum_{k=1}^K \left((\mathbf{y} - \mathbf{t}_k)'(\mathbf{y} - \mathbf{t}_k) - \hat{\delta}^2(\mathbf{y}, \mathbf{t}_k) \right)^2. \tag{6}$$

The objective has a minimum equal to 0 that can be achieved if and only if $\mathbf{y}$ is the solution of Equation 5. If it exists, such a solution is global and unique.

Due to the uncertainty introduced by the estimates $\hat{\delta}(\mathbf{y}, \mathbf{t}_k)$, an optimal solution to Equation 6 can still be achieved using gradient descent methods or the Levenberg-Marquardt algorithm. This method is used in our experiments.

2.2   Parameters and Computational Complexity

On the basis of the aforementioned overview, the number of reference points K is virtually the only hyper-parameter that the user needs to select in order to optimize a Minimal Learning Machine. As always, a selection based on standard resampling methods for cross-validation could be adopted for the task and thus optimize the MLM against over-fitting. Two figures of merit can be used for selecting K; one for the distance regression step and another one for the output estimation. In this work, we use the Average Mean Squared Error for the output distances $(AMSE(\delta) = 1/K \sum_{k=1}^{K}(1/N \sum_{i=1}^{N} (\delta(y_i, t_k) - \hat{\delta}(y_i, t_k))^2))$ and the Mean Squared Error for the responses $(MSE(y) = 1/N \sum_{i=1}^{N} (y_i - \hat{y}_i)^2)$.

The computation for learning a MLM can be decomposed into two steps: i) calculations of the pairwise distance matrices in the output and input space and ii) calculation of the least-square solution for the multi-response linear regression problem on distance matrices (Equation 3). The first procedure takes $\Theta(KN)$ time. The computational cost of the second step is driven by the calculation of the Moore-Penrose pseudo-inverse matrix. One of the most used method for the task is the Singular Value Decomposition, which runs in $\Theta(K^2N)$ time. The time complexity of the overall learning phase is thus driven by the computation of the Moore-Penrose matrix and then it is given by $\Theta(K^2N)$. However, because the optimal number of reference points might not grow at the same rate of the number of learning points, then such complexity can be reduced to $\mathcal{O}(N)$ if one considers $K = \mathcal{O}(1)$, or $\mathcal{O}(N^2)$ if $K = \mathcal{O}(N^{0.5})$. In addition, large pairwise distance matrices could be approximated using Nyström methods and matrix multiplication operations could be parallelized using multicore architectures.

2.3   Two Illustrative Examples

In this section, we illustrate the effectiveness of the Minimal Learning Machine using two synthetic problems. The first one is related to nonlinear regression (the smoothed parity function) and the second one (the Tai Chi) is used to introduce an intuitive extension that allows the MLM to deal with classification problems.

The Smoothed Parity. To illustrate the behavior of the Minimal Learning Machine for regression, we generated 2^{13} bidimensional input points uniformly distributed in the unit-square, $\mathbf{x} \in [0, 1]^2$, and built the response using the model $y = f(x) + \varepsilon$ with $f = \sin(2\pi x_1)\sin(2\pi x_2)$ and $\varepsilon \sim \mathcal{N}(0, 0.1^2)$, Figure 1(a).

We analyzed the performance of the MLM for N learning points ranging from 2^1 to 2^{12} and K reference points such that always $K \leq N$. A common set of $N_v = 2^{12}$ independent points is used for validating the MLM in terms of its

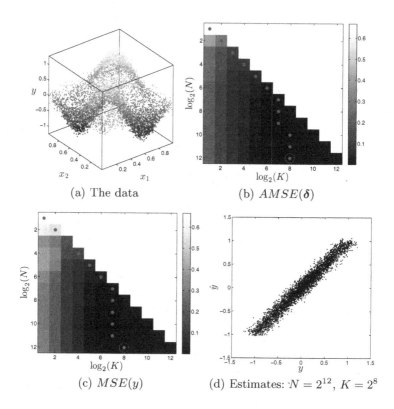

(a) The data

(b) $AMSE(\boldsymbol{\delta})$

(c) $MSE(y)$

(d) Estimates: $N = 2^{12}$, $K = 2^8$

Fig. 1. A regression example: The *smoothed* parity function

hyper-parameter K. Two figures of merit are considered for selecting K, the $AMSE(\boldsymbol{\delta})$ for the output distances and the $MSE(y)$ for the response.

As expected, for each size of the learning set it is possible to select an optimal number of reference points that minimizes the validation error. Figure 1(b) and 1(c) depicts such optimal models with red dots. In these two figures, the circle is used to depict the best model overall ($N = 2^{12}$ and $K = 2^8$, for both $AMSE(\boldsymbol{\delta})$ and $MSE(y)$). Figure 1(d) illustrates the validation results when estimating the response with such a model. Interestingly, the MSE achieved by this MLM is 0.011, which tends to the variance of the noise (0.010) and thus also to the smallest MSE that any regression model can achieve without over-fitting.

The Tai Chi. Since the Minimal Learning Machine is able to deal with multidimensional response spaces, it can be easily extended to multi-class classification problems by representing the classes through binary output encoding schemes.

For compactness, here we illustrate the behavior of the MLM for binary classification. We generated 2^{13} bidimensional input points uniformly distributed in the Tai Chi symbol and, after assigning the class labels to the Yin an Yang areas, we purposely mislabeled 10% of the observations, Figure 2(a). A binary output

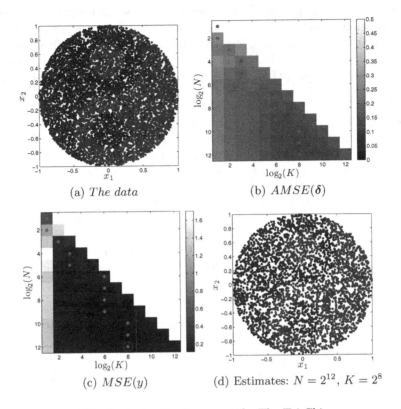

(a) *The data*   (b) $AMSE(\delta)$

(c) $MSE(y)$ (d) Estimates: $N = 2^{12}$, $K = 2^8$

Fig. 2. A classification example: The Tai Chi

encoding that assigns $y_i = 0$ to the points in the class 'Yin' and $y_i = 1$ to those in the class 'Yang' is used. The output distances are calculated accordingly.

The performance of the MLM for classification with N and K ranging from 2^1 to 2^{12} and $K \leq N$ is presented. For validation purposes, a set of 2^{12} observations is used. The results in terms of $AMSE(\delta)$ and $MSE(y)$ are reported in Figure 2(b) and 2(c), respectively. The red dots denote the best model per number of learning points and the circle depicts the best model overall. Figure 2(d) shows the estimated classes in validation using the best model; the accuracy is 88%.

3 Experiments

In this section, we present results obtained with six real-world regression datasets used for benchmarking purposes (UCI Repository: www.ics.uci.edu/~mlearn/). The datasets have been chosen to object heterogeneity in the number of samples and inputs: 1) Breast Cancer (32 inputs, 194 samples); 2) Boston Housing (13 inputs, 506 samples); 3) Servo (4 inputs, 167 samples); 4) Abalone (8 inputs, 4177 samples); 5) Stocks (9 inputs, 950 samples); and 6) Auto Price (15 inputs, 159 samples). For each problem, ten different random permutations of the whole dataset are taken, two thirds are used for learning and the rest for testing. The learning sets are normalized to have zero mean and unit variance, and the test sets are normalized using the corresponding mean and variance from the learning set.

The Minimal Learning Machine is compared to five other methods: The Extreme Learning Machine (ELM, [3]), the Optimally Pruned ELM (OP-ELM, [4]), the Support Vector Machine for Regression (SVM, [5]), Gaussian Processes (GP, [6]) and the MultiLayer Perceptron (MLP, [7]).

The hyper-parameters for the SVM and the MLP are selected using 10-fold cross-validation. The SVM is learned using the SVM toolbox [8] with default settings for the hyper-parameters and grid search, with a radial basis kernel. The MLP is optimized using Levenberg-Marquardt and validated on a range of hidden units from 1 to 20. The learning of GP is based on the default settings in the Matlab Toolbox [6]. The ELM and OP-ELM have been validated using sigmoid, gaussian and linear kernels, and a maximum number of 100 hidden units. The only hyper-parameter of the Minimal Learning Machine (the number of reference points) has also been selected through 10-fold cross-validation, for a K ranging from 5% to 100% (with a step size of 5%) of the learning samples.

Table 1. Test results: MSE, standard deviations (below the MSE) and t-test results ($\checkmark$ for accept, $\times$ for reject and p-values). The best performing models are in bold.

Datasets	Models					
	MLM	ELM	OP-ELM	SVM	GP	MLP
Breast Cancer	**1.1e+3**	7.7e+3	1.4e+3	1.2e+3	1.3e+3	1.5e+3
	2.1e+2	2.0e+3	3.6e+2	7.2e+1	1.9e+2	4.4e+2
		$\times$ (e-9)	$\checkmark$ (.955)	$\checkmark$ (.913)	$\checkmark$ (.109)	$\times$ (.027)
Boston	2.3e+1	1.2e+2	1.9e+1	3.4e+1	**1.1e+1**	2.2e+1
	1.2e+1	2.1e+1	2.9	3.1e+1	3.5	8.8
		$\times$ (e-10)	$\checkmark$ (.374)	$\checkmark$ (.314)	$\times$ (.008)	$\checkmark$ (.851)
Servo	4.9e−1	7.1	8.0e−1	6.9e−1	**4.8e−1**	6.0e−1
	2.9e−1	5.5	3.3e−1	3.2e−1	3.5e−1	3.2e−1
		$\times$ (.001)	$\times$ (.037)	$\checkmark$ (.164)	$\checkmark$ (.961)	$\checkmark$ (.427)
Abalone	4.6	8.3	4.9	**4.5**	**4.5**	4.6
	2.9e−1	7.5e−1	6.6e−1	2.7e−1	2.4e−1	5.0e−1
		$\times$ (e-11)	$\checkmark$ (.353)	$\checkmark$ (.378)	$\checkmark$ (.206)	$\checkmark$ (.844)
Stocks	**4.1e−1**	3.4e+1	9.8e−1	5.1e−1	4.4e−1	8.8e−1
	5.8e−2	9.35	1.1e−1	9.8e−2	5.0e−2	2.1e−1
		$\times$ (e-9)	$\times$ (e-11)	$\times$ (.016)	$\checkmark$ (.329)	$\times$ (e-6)
Auto Price	5.1e+7	7.9e+9	9.5e+7	9.8e+7	2.0e+7	**1.0e+7**
	7.4e+7	7.2e+9	4.0e+6	8.4e+6	1.0e+7	3.9e+6
		$\times$ (.003)	$\checkmark$ (.096)	$\checkmark$ (.346)	$\checkmark$ (.205)	$\checkmark$ (.103)

All the models are evaluated using the mean and standard deviation of the resulting MSE over 10 independently drawn test sets. We also carried out a statistical evaluation of the MLM performance against those achieved by the other models using the two-sample t-test [9] with a significance level equal to

5%. The null hypothesis is that the MSE distributions are independent random samples from normal distributions with equal means and equal but unknown variances, against the alternative that the means are not equal. On the basis of the experimental results (Table 1), we can observe that the state-of-the-art models seem to be able to achieve similar accuracies. In this regard, also the MLM achieves performances that are comparable to such methods. The table also shows that most of the models are equivalent for the Auto Price and Abalone datasets, except of the ELM. The MLM is not among the best models only for the Boston dataset where the GP is the most reliable option. In addition, the most similar performances to MLM are those achieved by the SVM and GP models, whose null hypotheses were accepted for five different datasets.

4 Conclusions

This work presents a novel method for supervised learning, the Minimal Learning Machine, MLM. Learning a MLM consists in reconstructing the mapping existing between input and output distance matrices and then exploiting the geometrical arrangement of the output points for estimating the response. Based on our experiments, a multiresponse linear regression model is capable to reconstruct the mapping existing between the aforementioned distance matrices. The MLM has only one hyper-parameter to be optimized. Given its general formulation, the Minimal Learning Machine is also inherently capable to operate on multidimensional responses and it can be extended to classification problems.

On a large number of real-world problems, the Minimal Learning Machine has achieved accuracies that are comparable to what is obtained using state-of-the art nonlinear regression methods. For compactness, we have reported the performances on a selection of six datasets from the UCI Repository and comparisons with five reference regression approaches. The results highlight the potentiality of the MLM and we are currently further investigating its properties and the ties with classical dimensionality reduction methods based on distances.

References

1. Cox, T., Cox, M.: Multidimensional Scaling. Chapman & Hall, London (1994)
2. de Silva, V., Tenenbaum, J.B.: Global versus local methods in nonlinear dimensionality reduction. In: Advances in Neural Information Processing Systems, vol. 15, pp. 705–712 (2002)
3. Huang, G.B., Zhu, Q.Y., Ziew, C.K.: Extreme Learning Machine: Theory and applications. Neurocomputing 70(1-3), 489–501 (2006)
4. Miche, Y., Sorjamaa, A., Bas, P., Simula, O., Jutten, C., Lendasse, A.: OP-ELM: Optimally Pruned Extreme Learning Machine. IEEE Transactions on Neural Networks 21(1), 158–162 (2010)
5. Smola, A.J., Schölkopf, B.: A tutorial on support vector regression. Statistics and Computing 14(3), 199–222 (2004)

6. Rasmussen, C.E., Williams, C.K.I.: Gaussian Processes for Machine Learning. The MIT Press, Cambridge (2006)
7. Bishop, C.M.: Neural Networks for Pattern Recognition. Oxford University Press, New York (1995)
8. Chang, C.-C., Lin, C.-J.: LIBSVM: A Library for Support Vector Machines. ACM Transactions on Intelligent Systems and Technology 2(27), 1–27 (2011)
9. Lehmann, E.L., Romano, J.P.: Testing statistical hypotheses. Springer, New York (2005)

Extending Extreme Learning Machine with Combination Layer

Dušan Sovilj[1], Amaury Lendasse[1,2,3], and Olli Simula[1]

[1] Aalto University School of Science, Espoo, Finland
[2] IKERBASQUE, Basque Foundation for Science, Bilbao, Spain
[3] University of The Basque Country, Donostia-San Sebastián, Spain
{dusan.sovilj,amaury.lendasse,olli.simula}@aalto.fi

Abstract. We consider the Extreme Learning Machine model for accurate regression estimation and the related problem of selecting the appropriate number of neurons for the model. Selection strategies that choose "the best" model from a set of candidate network structures neglect the issues of model selection uncertainty. To alleviate the problem, we propose to remove this selection phase with a combination layer that takes into account all considered models. The proposed method in this paper is the Extreme Learning Machine(Jackknife Model Averaging), where Jackknife Model Averaging is a combination method based on leave-one-out residuals of linear models. The combination approach is shown to have better predictive performance on several real-world data sets.

1 Introduction

Accurate predictions of future instances are becoming a recurring problem in scientific research. The problem is addressed by forming a model and then making all subsequent inferences on that constructed model. Prediction of continuous values, such as daily temperature or stock market prices, is considered a *regression* problem or estimation of a regression function.

In the paper, we are concerned with regression problems of the form

$$y_i = f(\mathbf{x}_i) + \epsilon_i \qquad (1)$$

where $\{(\mathbf{x}_i, y_i) \,|\, 1 \leq i \leq N\}$ are data samples with $\mathbf{x}_i$ consisting of several explanatory features or variables and y_i the target variable, while ϵ_i is the noise term. Usually, the noise is assumed to be homoskedastic with a Gaussian distribution $\mathcal{N}(0, \sigma^2)$ with known variance. The problem is finding a model $\hat{f}$ that can best approximate the target function f. This is a general setting, and many such models exist, including, but not limited to, linear regression, neural networks, support vector regression, kernel regression, nearest neighbour estimators and fuzzy regression. Each model class is characterized by the approximation capabilities and the training algorithms with different computational complexities.

This paper focuses on a specific type of neural network that is gaining popularity in recent years, namely the Extreme Learning Machine (ELM). It is

I. Rojas, G. Joya, and J. Cabestany (Eds.): IWANN 2013, Part I, LNCS 7902, pp. 417–426, 2013.

shown in [1] that a single feedforward hidden layer with input weights randomly assigned has universal approximation capability for any target function, that is, the estimation can be made as small as possible considering standard squared error loss function. The advantage of ELM is its very fast training time, since the output weights are found in a simple linear setting between the hidden feature space and the target variable.

One drawback concerning the ELM model is the selection of the starting number of neurons to adequately capture the overall variations in data. To solve this issue, two approaches have been proposed: 1) *selection* methods [2–6] focus on choosing a single model from a candidate set by optimizing some criterion; 2) *ensemble of many ELMs* [7], i.e., the candidate models *all* contribute to the weighted average as the final model. The second strategy tries to avoid the problem by considering much larger candidate set, and focusing on finding appropriate model weights, aiming to assign zero weights for poor models and non-zero weights for better models.

We propose a mixture between the two mentioned strategies. That is, when constructing a *single* ELM that considers several competing models, instead of picking only one model, use all models and find appropriate weights to separate the models based on their *generalization ability*. Two reasons for such an approach are: 1) empirical success of combination methods over the selection ones when it comes to prediction accuracy [8]; 2) philosophical issue of ignoring model selection uncertainty by making inference solely on a single model once it is identified [9]. Combination methods have been proposed both in Bayesian statistics, where Bayesian Model Averaging [10] is considered a natural approach to model selection uncertainty by considering models as another nuisance parameter, and in a frequentist spirit with weights computed based on bootstrapping or perturbation of data [11]. The proposed idea is to *remove* the procedure "selection of the best model" and consider the ELM as a set of models altogether.

The paper is organized as follows. Section 2 describes the proposed method alongside its main parts: Extreme Learning Machine, Jackknife Model Averaging and Leave-one-out Cross-validation. Section 3 shows one variant of the ELM which can be extended to include the proposed strategy. In Section 4, we show results on several UCI Machine Learning Repository data sets. Conclusions are summarized in Section 5.

2   Combining Extreme Learning Machine(s)

Two approaches have been adopted for selection strategy: *pruning* − starting from a large number of neurons and then removing unnecessary ones [5, 6] and *constructive* − building up from smaller pool of neurons until some condition (usually error) does not improve with the additional complexity [2–4]. Both approaches consider several alternative network structures and finally output "the best" model. The idea is to consider them all together and form a weighted average.

The proposed method Extreme Learning Machine-*combination*, denoted by ELM(c), is an ELM model where the selection phase is replaced with a *combination layer*, i.e., additional layer of hierarchy. The parentheses denote that combination is taking place, while c denotes the method used to produce model weights. The following subsections present the building blocks which constitute the ELM(Jackknife Model Averaging), or ELM(JMA) for short.

2.1   Extreme Learning Machine Overview

Extreme Learning Machine (ELM) network presents a new way of building a neural structure. The idea is in random initialization of input weights and biases for a single hidden layer, which leads to removal of any kind of iterative training algorithm. As shown in [1], with this randomization and under certain constraints on transfer functions, the output weights of a hidden layer can be computed with simple linear regression and the model has universal approximation capabilities.

Consider a data set $\{(\mathbf{x}_i, y_i) \mid 1 \leq i \leq N\}$ with $\mathbf{x}_i \in \mathbb{R}^d$ and $y_i \in \mathbb{R}$. The ELM network with M neurons is constructed by first computing the hidden matrix $\mathbf{H}$

$$\mathbf{H} = \begin{bmatrix} g_1(\mathbf{w}_1^i \mathbf{x}_1 + b_1) & \cdots & g_M(\mathbf{w}_M^i \mathbf{x}_1 + b_M) \\ \vdots & \ddots & \vdots \\ g_1(\mathbf{w}_1^i \mathbf{x}_N + b_1) & \cdots & g_M(\mathbf{w}_M^i \mathbf{x}_N + b_M) \end{bmatrix}$$

with j-th neuron having activation function g_j, input weights $\mathbf{w}_j^i$, bias b_j and both $\mathbf{w}_j^i$ and b_j are randomly generated. Hidden layer output weights β are found by solving the linear system $\mathbf{H}\beta = \mathbf{y}$, with the Moore-Penrose generalized inverse of the matrix $\mathbf{H}$ and the target values $- \beta = (\mathbf{H}^T\mathbf{H})^{-1}\mathbf{H}^T\mathbf{y}$. The matrix $\mathbf{H}$ is sometimes called feature mapping or feature space of the ELM. Any function that is a bounded non-constant piecewise continuous function can be taken as activation function in the ELM.

2.2   Leave-One-Out Cross-Validation

Cross-validation (CV) is one of the most used strategies for evaluating regression models, and provides immediate comparison between a wide range of different model classes. k-fold CV splits the data set into k parts, and each part plays the validation role once the model is trained on the remaining $k - 1$ parts. The average error of all k parts is taken as a measure of generalization ability of the model. For selection strategy, the model with the smallest average validation error is assumed to be the most suitable for the given data set.

The extreme case is $k = N$ or leave-one-out (LOO) CV, where each data sample is taken to be a sole sample in the validation set. In the case of least squares linear regression, the LOO error can be computed with a single fit of the model using the PRESS statistic [12], which removes the computational burden of training N separate models. If we denote with $\mathbf{P} = \mathbf{H}(\mathbf{H}^T\mathbf{H})^{-1}\mathbf{H}^T$, then the leave-one-out residuals for all samples are computed with the PRESS formula

$$\tilde{\mathbf{e}}^{\mathrm{LOO}} = \frac{\mathbf{y} - \mathbf{P}\mathbf{y}}{1 - \mathrm{diag}(\mathbf{P})}. \tag{2}$$

where $\mathrm{diag}(\mathbf{P})$ denotes the main diagonal of $\mathbf{P}$, $\mathbf{1}$ a vector of ones and division is performed element by element.

2.3   Jackknife Model Averaging

Jackknife model averaging (JMA) is a model combining method minimizing the LOO-CV error which has recently been proposed in [13]. The authors show that considering the LOO residuals of all models, the combination asymptotically achieves the lowest possible expected squared error out of any model considered. That is, the combination is conditioned on the candidate set, and can only be better than those in the set itself. The theory in [13] is restricted to linear models and random samples, but it allows heteroskedastic noise term ϵ_i in Eq. (1) and *unbounded* number of models to be present in the candidate set.

JMA is a linear combination of leave-one-out residuals off all models, and for the purpose of presentation we use similar notation as in [13]. Denote with $\tilde{\mathbf{e}}^m = [\tilde{e}_1^m, \dots, \tilde{e}_N^m]^{\mathrm{T}}$ the leave-one-out residual vector of the m-th model in the candidate set. Then, the jackknife averaging residual vector is

$$\tilde{\mathbf{e}}(\mathbf{w}) = \sum_{m=1}^{M} w^m \tilde{\mathbf{e}}^m = \tilde{\mathbf{e}}\mathbf{w}$$

where $\tilde{\mathbf{e}} = [\tilde{\mathbf{e}}^1, \dots, \tilde{\mathbf{e}}^M]$ and jackknife estimate of the generalization error is given by

$$\mathrm{CV}(\mathbf{w}) = \frac{1}{N}\tilde{\mathbf{e}}(\mathbf{w})^{\mathrm{T}}\tilde{\mathbf{e}}(\mathbf{w}) = \mathbf{w}^{\mathrm{T}}\mathbf{S}\mathbf{w} \tag{3}$$

where $\mathbf{S} = \tilde{\mathbf{e}}^{\mathrm{T}}\tilde{\mathbf{e}}/N$ is an $M \times M$ matrix. The choice of $\mathbf{w}$ is the one that minimizes the cross-validation criterion defined in Eq. (3) with the weights constrained on a unit simplex $\mathcal{H} = \{\mathbf{w} \in \mathbb{R}^M | w^m \geq 0, \sum_{m=1}^{M} w^m = 1\}$. This is a quadratic programming problem with respect to $\mathbf{w}$ and can easily be solved with publicly available software packages. All that is needed to solve the minimization problem are the LOO residuals of every model in the set.

2.4   ELM(JMA)

The ELM(JMA) model is constructed as follows. Start with a fixed number of neurons, say M, and compute the matrix $\mathbf{H}$. Then, train M models, where the output weights for each model are computed with $\beta^m = (\mathbf{H}_m^{\mathrm{T}}\mathbf{H}_m)^{-1}\mathbf{H}_m^{\mathrm{T}}\mathbf{y}$, $1 \leq m \leq M$, and $\mathbf{H}_m$ consists of the first m columns of $\mathbf{H}$. That is, keep increasing the number of neurons/columns by 1 (starting from 1 until M) and compute the new model. During this training phase, gather LOO residual vectors $\tilde{\mathbf{e}}^m$ for each model using Eq. (2). The final step is JMA combining of all M

models using $\tilde{\mathbf{e}}$ in the minimization problem defined by Eq. (3) which gives model weights $\mathbf{w} = [w^1, \ldots, w^M]^{\mathrm{T}}$. The function estimate of the target variable $\mathbf{y}$ is then $\sum_{m=1}^{M} w^m \mathbf{H}_m \boldsymbol{\beta}^m$. Prediction for a fresh sample $\mathbf{x}_n$ is straightforward: compute the output $\hat{f}^{i_l}(\mathbf{x}_n)$ for those models whose weights are greater than zero $\mathcal{M} = \{w^m > 0 | 1 \leq m \leq M\} = \{i_1, \ldots, i_L\}$ and output the weighted average of those models $\hat{f}(\mathbf{x}_n) = \sum_{l=1}^{L} w^{i_l} \hat{f}^{i_l}(\mathbf{x}_n)$.

Reason for considering only M models is that there is an exponential number of possible models, i.e., all subsets selection problem with M variables. The other issue is the ordering of neurons with different activation functions. In our approach, we randomly permute the order to prevent only one type of function from dominating the network structure and to allow more variability.

It should be noted that a linear combination of linear models can be seen as *one* linear model by summation of appropriate weight vectors. The proposed method then returns a single model, but the model where selection uncertainty has been accounted for. In this view, ELM(c) introduces another form of regularization on the weights $\boldsymbol{\beta}$.

3   TROP-ELM

Tikhonov Regularized Optimally Pruned Extreme Learning Machine (TROP-ELM) [6] brings two adjustments to the original ELM. First phase is ranking of neurons by LARS and selecting the appropriate number of neurons by minimizing the LOO error (OP part). The other improvement is L_2 regularization on the weights $\boldsymbol{\beta}$, by introducing a slight bias which is reflected on the LOO residuals (TR part). The adjusted LOO residuals are computed with a new formula where matrix $\mathbf{P}$ is replaced by $\mathbf{P}(\lambda) = \mathbf{H}(\mathbf{H}^{\mathrm{T}}\mathbf{H} + \lambda\mathbf{I})^{-1}\mathbf{H}^{\mathrm{T}}$, where $\mathbf{I}$ denotes the identity matrix. The parameter λ is locally optimized, providing the solution in a few iterations, with slight increase in computational time.

Inclusion of the TROP-ELM in experiments is two-fold. First, to demonstrate that the model combination can easily be applied to other variants of the ELM model. Second, the TROP-ELM uses a ranking algorithm to produce a different ordering than our suggested random permutation, and as such can give insight whether a heuristic approach to ordering is better than a random one.

4   Experiments

This section shows the comparison between the two groups of ELM models: one group employs the selection strategy (ELM* and TROP-ELM) while the other consists of our proposed method with the added combination layer (ELM(JMA) and TROP-ELM(JMA)). The comparison is done via squared error risk which is estimated as an average of 10 Monte-Carlo runs on a test set. The test set consists of one third of the data set in consideration, while the other two thirds constitute the training set. The training set is standardized to zero mean and unit variance, and the same parameters (mean and variance) are used to scale

Table 1. Data sets used in the experiments. N indicates the total number of samples in the data set without division into the training and test parts and d denotes the number of features.

Name	N	d	Name	N	d
Abalone	4177	8	Computer activity	8192	12
Bank_8FM	4500	8	Delta ailerons	7129	5
Boston housing	506	13	Servo	167	4
Breast cancer	194	32	Stocks	950	9

the test set. For each run, a total of 1000 models are trained to account for randomness of the ELM algorithm and the average of 1000 test errors is taken to be the estimated error for that run.

Experiments are performed on eight data sets from two repositories: the UCI machine learning repository [14] and the LIACC regression repository [15]. The data sets are listed in Table 1.

Three types of activation functions g are used: sigmoid, Gaussian and linear. For Gaussian functions, the centres of kernels are taken randomly from data points $\mathbf{x}_i$, while linear activations functions are simply identity functions for each feature, i.e., $g_{j_k}(\mathbf{x}_i) = x_i^k$, $k \in \{1, \ldots, d\}$, the k-th feature of sample i. A total of 50 sigmoid and 50 Gaussian kernels are used, which gives $M = 100 + d$ neurons, and the same number of models to train.

4.1   Performance Comparison

Tables 2 and 3 show the estimated squared error risk for the ELM and TROP-ELM variants respectively, and the improvement obtained when the combination is taken into account. The ELM* method selects the model that has the smallest LOO error in the candidate set (the best model) and is used instead of the full model with M neurons in order to provide a fair comparison between two strategies. The proposed combination approach always produces an improvement compared to a single model for both ELM variants. The reduced risk can range from small values of 1–2% to much larger improvement of around 25% (Abalone). Smaller achievements are expected in data sets with a high number of samples as a single ELM is able to capture all complexity present in the data. Larger jumps are seen for data with moderate sample sizes, where more variability in the training phase is expected, and the combination alleviates that increased variability. The only exception is ELM*/ELM(JMA) for Breast cancer data. In this case, the increased risk estimate comes from LOO computation of larger models where the number of samples for training ($2/3 \cdot 194 \approx 129$) and models considered $M = 132$ pose problems for accurate estimation, and may lead to overfitting. Such extreme cases are potential pitfalls for JMA, since the models with overconfident LOO estimates are selected during the combination phase. This is where TROP-ELM plays an important role and provides the JMA with more stable results.

Table 2. Estimated risk (average test mean-squared error) for ELM* and ELM(JMA) models and the improvement (in percent) of the combination approach over the selection strategy

Data set	ELM*	ELM(JMA)	(%)
Abalone	12.1	9.14	24.77
Bank_8FM	1.085e−3	1.044e−3	3.79
Boston housing	18.0	15.2	15.92
Breast cancer	1.19e+3	1.43e+3	−20.59
Computer activity	35.8	31.1	12.97
Delta ailerons	2.81e−8	2.74e−8	2.57
Servo	0.729	0.614	15.76
Stocks	0.831	0.716	13.85

Table 3. Estimated risk (average test mean-squared error) for TROP-ELM and TROP-ELM(JMA) models and the improvement (in percent) of the combination approach over the selection strategy

Data set	TROP-ELM	TROP-ELM(JMA)	(%)
Abalone	6.39	5.95	6.97
Bank_8FM	1.081e−3	1.057e−3	2.23
Boston housing	18.7	15.9	15.20
Breast cancer	1.31e+3	1.19e+3	8.82
Computer activity	33.6	30.5	9.07
Delta ailerons	2.75e−8	2.71e−8	1.46
Servo	0.748	0.652	12.90
Stocks	0.926	0.781	15.68

The issue of ordering of neurons is less obvious. The ELM*/ELM(JMA) pair outperforms the TROP versions in some data sets (Boston housing, partially Breast cancer, Servo and Stocks), while it is inferior for the other data sets. The only noticeable difference is for Abalone, where ordering improves performance dramatically, while for the other data sets the increase (Bank and Delta ailerons) is quite small. This suggest that the ordering of neurons might not be so critical for accurate prediction.

4.2   Run Times

A great advantage of ELM is its low computational cost. The question is whether proposed combination procedure takes too much time compared to the original ELM. Table 4 summarizes the execution time for the ELM* and ELM(JMA) models. The execution time for the ELM* is computed as the amount of time required to train all M linear systems. The computational increase mostly

Table 4. Run times in *seconds* for ELM* and ELM(JMA) models and the increase in computation time (in percent) for the combination strategy with respect to the selection strategy

Data set	ELM*	ELM(JMA)	(%)
Abalone	0.85	0.89	4.48
Bank_8FM	0.98	1.00	1.82
Boston housing	0.13	0.15	22.42
Breast cancer	0.08	0.12	42.17
Computer activity	1.98	2.02	2.05
Delta ailerons	1.39	1.41	1.25
Servo	0.05	0.07	37.82
Stocks	0.21	0.23	13.34

depends on the sample size of the data set, and for larger ones the increase is quite small, while for data sets with a couple of hundred samples there is substantial extra cost (around 40%), but such cost is still affordable and on a scale of less than one second.

Run times are computed in the Matlab environment and carried out on an Intel Xeon processor (E3-1200 family; 3.20GHz) using a single core. Each model is trained independently of the other models, and quadratic programming for JMA is solved with the *quadprog* function from the Optimization Toolbox.

It should be stressed that quadratic programming is only dependent on the number of models considered M, i.e., independent of the sample size N. This means that the combination phase takes almost the same amount of time for all data sets in our case. The extra cost is even more negligible for the TROP-ELM since there is additional optimization for the λ parameter. Solving quadratic problems for even larger cases when $M \approx 1000$ is still fast, but the actual bottleneck becomes the training phase for all 1000 models.

5   Conclusions

This paper addresses the issue of selection strategy for the Extreme Learning Machine and its variants. As explicated, this approach neglects the issue of model selection uncertainty which leads to degraded prediction accuracy. Instead, a combination procedure taking into account all available models must be considered to combat the problem. The proposed approach ELM(JMA) with Jackknife Model Averaging as the combination method of the LOO residuals shows better results than a single "best" model based on the same LOO errors. Extension to TROP-ELM(JMA) shows that the method can be easily adapted to the other ELM variants that are based on the selection strategy. The extra computational cost is quite low and only depends on the number of models considered.

Notation ELM(c) signifies that other criteria and combination methods can be paired instead of the leave-one-out residuals and Jackknife Model Averaging, such as the Bayesian Information Criterion (BIC) where the model weights can easily be derived from the BIC scores.

In the experiments, we have used fixed starting number of neurons, but the question remains whether that number can automatically be selected based on the combination weights. One approach would be to start with some small number, say $M_b = 10$, train models and perform the combination, and check if the full model (or the most complex from this set) is included in the combination, i.e., $w^{M_b} > 0$. If it is, then add new batch of neurons and repeat the procedure until the most complex model is not present in the combination or if all newly added models are excluded.

Another question worth examining is the *pruning/screening step* or removal of poor models from the candidate set prior to combination. From a theoretical perspective this issue is addressed via weight assignment, but in practice due to finite sample size of the data there are difficulties in stable estimation of the parameters of larger models which is a potential pitfall for model combining.

Nevertheless, the success of the proposed combination strategy suggests that the practice of selecting one model from a candidate set leads to less accurate inference, and that some form of weighted average is required even in the case when building a single ELM network.

References

1. Huang, G.B., Zhu, Q.Y., Siew, C.K.: Extreme learning machine: Theory and applications. Neurocomputing 70(1-3), 489–501 (2006)
2. Huang, G.B., Chen, L., Siew, C.K.: Universal approximation using incremental constructive feedforward networks with random hidden nodes. IEEE Transactions on Neural Networks 17(4), 879–892 (2006)
3. Feng, G., Huang, G.B., Lin, Q., Gay, R.: Error minimized extreme learning machine with growth of hidden nodes and incremental learning. IEEE Transactions on Neural Networks 20(8), 1352–1357 (2009)
4. Lan, Y., Soh, Y.C., Huang, G.B.: Constructive hidden nodes selection of extreme learning machine for regression. Neurocomputing 73(16-18), 3191–3199 (2010)
5. Rong, H.J., Ong, Y.S., Tan, A.H., Zhu, Z.: A fast pruned-extreme learning machine for classification problem. Neurocomputing 72(1-3), 359–366 (2008)
6. Miche, Y., van Heeswijk, M., Bas, P., Simula, O., Lendasse, A.: TROP-ELM: A double-regularized ELM using LARS and Tikhonov regularization. Neurocomputing 74(16), 2413–2421 (2011)
7. van Heeswijk, M., Miche, Y., Lindh-Knuutila, T., Hilbers, P., Honkela, T., Oja, E., Lendasse, A.: Adaptive ensemble models of extreme learning machines for time series prediction. In: Alippi, C., Polycarpou, M., Panayiotou, C., Ellinas, G. (eds.) ICANN 2009, Part II. LNCS, vol. 5769, pp. 305–314. Springer, Heidelberg (2009)
8. Breiman, L.: Stacked regressions. Machine Learning 24(1), 49–64 (1996)
9. Draper, D.: Assessment and propagation of model uncertainty (with discussion). Journal of the Royal Statistical Society: Series B 57(1), 45–97 (1995)
10. Hoeting, J.A., Madigan, D., Raftery, A.E., Volinsky, C.T.: Bayesian model averaging: A tutorial. Statistical Science 14(4), 382–417 (1999)
11. Breiman, L.: Bagging predictors. Machine Learning 24(2), 123–140 (1996)

12. Allen, D.M.: The relationship between variable selection and data augmentation and a method for prediction. Techometrics 16(1), 125–127 (1974)
13. Hansen, B.E., Racine, J.S.: Jackknife model averaging. Journal of Econometrics 167(1), 38–46 (2012)
14. Frank, A., Asuncion, A.: UCI machine learning repository (2010),
    http://archive.ics.uci.edu/ml
15. Torgo, L.: LIACC regression data sets,
    http://www.dcc.fc.up.pt/~ltorgo/Regression/DataSets.html

Texture Classification Using Kernel-Based Techniques

Carlos Fernandez-Lozano[1], Jose A. Seoane[2], Marcos Gestal[1],
Tom R. Gaunt[2], and Colin Campbell[3]

[1] Information and Communications Technologies Department, Faculty of Computer Science,
University of A Coruña, Campus Elviña s/n, 15071, A Coruña, Spain
[2] MRC Centre for Causal Analyses in Translational Epidemiology,
School of Social and Community Medicine, University of Bristol,
Oakfield House, Oakfield Grove, Bristol BS82BN, UK
[3] Department of Engineering Mathematics,
University of Bristol, Merchant Venturer's Building, Bristol BS81UB, UK
carlos.fernandez@udc.es, j.seoane@bristol.ac.uk,
mgestal@udc.es, {tom.gaunt,c.campbell}@bristol.ac.uk

Abstract. In this paper, a high-dimensional textural heterogenous dataset is evaluated. This problem should be studied with specific techniques or a solution for decreasing dimensionality should be applied in order to improve the classification results. Thus, this problem is tackled by means of three differente techniques: an specific technique such as Multiple Kernel Learning, and two different feature selection techniques such as Support Vector Machines-Recursive Feature Elimination and a Genetic Algorithm-based approaches. We found that the best technique is Support Vector Machines-Recursive Feature Elimination, with a AUROC score of 92,45%.

Keywords: Multiple Kernel Learning, Support Vector Machines, Recursive Feature Elimination, Genetic Algorithms.

1 Introduction

One of the most important characteristics used for identifying objects or regions of interest in an image is texture, related with the spatial (statistical) distribution of the grey levels within an image [1]. Texture is a surface's property and can be regarded as the almost regular spatial organization of complex patterns, always present even if they could exist as a non-dominant feature. A good review for different Texture approaches can be found in [2].Thus texture analysis is one of the central concepts regarding computer vision. We recognize texture when we see it but there is no generally accepted definition of texture in the literature, in general different researchers use different definitions depending upon the particular area of application [3]. There are four major issues one can identify as the main problems to solve using a texture analysis: texture synthesis, classification, segmentation and shape from texture [3, 4]. This work is focused in texture classification.

There exist several groups of textural features: Statistical, Fractal, Geometrical, Model Based or Signal Processing textural features [3, 5] among others. In this paper

I. Rojas, G. Joya, and J. Cabestany (Eds.): IWANN 2013, Part I, LNCS 7902, pp. 427–434, 2013.

we evaluate first and second-order statistics, spatial frequencies, co-occurrence matrices, autoregressive models, wavelet based analysis, etc. These features are based on image histogram, co-ocurrence matrix (information about the grey level value distribution of pairs of pixels with a preset angle and distance between each other), runlength matrix (information about sequences of pixels with the same grey level values in a given direction), image gradients (spatial variation of grey level values), autoregressive models (description of texture based on statistical correlation between neighboring pixels) and wavelet analysis (information about image frequency content at different scales). Features are reported in Table 1. Histogram-related measures conform the first-order statistics proposed by Haralick [1] but second-order statistics are those derived from the Spatial Distribution Grey-Level Matrices (SDGM). First-order statistics depend only on individual pixel values and can be computed from the histogram of pixel intensities in the image. Second-order statistics depend on pairs of grey values and on their spatial resolution.

With a specialized software called Mazda [6], textural features are computed. Various approaches have demonstrated the effectiveness of this software extracting textural features in different types of medical images [7-9].

Table 1. Textural Features extracted and used in this work

Group	Features	Num. of Feat.
Histogram	Mean, variance, skewness, kurtosis, percentiles 1%, 10%, 50%, 90% and 99%	9
Absolute Gradient	Mean, variance, skewness, kurtosis and percentage of pixels with nonzero gradient	5
Run-length Matrix	Run-Length non-uniformity, grey-level non-uniformity, long-run emphasis, short-run emphasis and fraction of image in runs	20
Co-ocurrence Matrix	Angular second moment, contrast, correlation, sum of squares, inverse difference moment, sum average, sum variance, sum entropy, entropy, difference variance and difference entropy	221
Autoregressive Model	Theta: model parameter vector, four parameters; Sigma: standard deviation of the driving noise	5
Wavelet	Energy of wavelet coefficients in subbands at successive scales; max four scales, each with four parameters	14

Such a high-dimensional and heterogeneous set of textural features should be studied with specific techniques or a solution for decreasing dimensionality should be applied in order to improve the classification results. In this paper we study three kernel-based possibilities: a Multiple Kernel Learning and two different kernel-based feature selection techniques.

The paper is organized as follows. In Section 2 provides a brief theoretical background on Support Vector Machines (SVM), Multiple Kernel Learning (MKL),

Support Vector Machine-Recursive Feature Elimination (SVM-RFE) and Genetic Algorithms (GA). In Section 3 describes the materials and methods used in this work. Section 4 is dedicated to analyse the results and in Section 5 conclusions are presented.

2   Theoretical Background

Vapnik introduces SVMs in the late 1970s on the foundation of statistical learning theory [10]. The basic implementation deals with two-class problems in which data are separated by a hyperplane defined by a number of support vectors. In classification task, the objective of the SVM is to find the hyperplane that separates the positives and the negative samples. This hyperplane separates the positive from the negative examples, to orient it such that the distance between the boundary and the nearest data point in each class is maximal; the nearest data points are used to define the margins, known as support vectors [11]. In machine learning, kernel methods and SVMs specifically have proven to be exceptionally efficient in classification problems of higher dimensionality [12, 13] because of their ability to generalize in high-dimensional spaces, such as the ones spanned by texture patterns. SVM involves the optimization of a concave function so, in contrast with other classification algorithms, there are only one solution. As most of the complex datasets are not linearly separable, the SVM introduces the concept of kernel. A kernel is a function that maps the input into a higher dimensionality, where the data are linearly separable. However, the inclusion of these kernels functions requires a new level of parameterization, where the kernel functions and its parameters must be carefully selected. In order to automatically select these kernels and these parameters, a new kernel-based method, called Multiple Kernel Learning, has been developed in last decade, for binary classification, has been introduced by Lanckriet et al. [14].

MKL basically learn an optimal linear or non-linear combination of kernels. This combination of kernel consists in several kinds of kernel and different parameterization for each kernel. MKL provides a general framework for learning from multiple heterogeneous data sources [15]. MKL works by first constructing a kernel from each of the data sources and then combining these kernels based on a certain criterion for improved classification performance. The sparse nature of some optimization algorithms can reduce the number of features by turning to zero the kernel coefficient. This optimized combination of coefficients sigma can be used to understand which features are important for the classification purposes. Although the kernel and its parameters can be selected through a single kernel approach and cross validation, multiple kernels methods have been proposed to solve these problems directly through an optimization process. Also, MKL has been proved to be very useful providing an efficient way to integrate feature representation coming from different data sources [14]. A comprehensive tutorial on SVM [11] and MKL [16].

Guyon et al. [17] proposed the SVM-RFE for selecting relevant genes in a cancer classification problem. The SVM-RFE ranks all the genes according to some score function and eliminate one or more genes with the lowest scores by training a SVM with a linear kernel. This removing criterion is the w value of the decision hyperplane given by the SVM. Thus, this algorithm analyses the relevance of input variables by estimating changes in the cost function. This process is repeated until the highest classification accuracy. This technique was modified for its application in nonlinear classification problems subsequently by [18].

2.1   Feature Selection Techniques

Feature selection techniques emerge to deal with high dimensional input spaces. Performance of supervised algorithms can be affected by the number and relevance of input variables. The aim of these techniques is to find a subset of input variables that best describes the structure of the data as well or better than the original data. These techniques can be divided into three major categories [19, 20]: filter methods, wrapper methods and embedded methods. In this work we test a wrapper method [21] such as a Genetic Algorithm in combination with SVM and a particular case of the embedded methods [22], a nested method SVM-RFE.

GA are search techniques inspired by Darwinian Evolution and developed by Holland in the 1970s [23]. GAs for feature selection were first proposed by Siedlecki and Skalansky [24]. Many studies have been done on GA for feature selection since then [25-27], concluding that GA is suitable for finding optimal solutions to large problems with more than 40 features to select from. GA for feature selection could be used in combination with a classifier such SVM, optimizing it.

3   Materials and Methods

In order to generate the dataset, ten two dimensional gel electrophoresis images of different types of tissues and different experimental conditions were used. These images are similar to the ones used by G.-Z. Yang (Imperial College of Science, Technology and Medicine, London). For each image 100 regions of interest were selected to build a training set with 1000 samples and 274 variables. Hunt et al. [28] determined that 7-8 is the minimum acceptable number of samples for a proteomic study.

This dataset is tackled by means of three differente techniques: an specific technique such as Multiple Kernel Learning suitable for high-dimensional problems, and two different feature selection techniques such as Support Vector Machines-Recursive Feature Elimination and a Genetic Algorithm-based approaches.

Simple MKL [29] is a gradient descent based implementation of MKL with an SVM as solver of a single kernel, which is actually a linear combination of multiple kernels. The algorithm performs a minimization of both the parameters of the SVM (weight and bias) and the parameters of the kernel (sigmas), using an iterative gradient descent algorithm. This approach is very efficient in problems with high dimensionality, because the memory consumption remains stable during the minimization, compared with other implementations based in quadratically constrained quadratical programming method by [14] or the semi-infinite linear programming method by [30]. This particular MKL implementation uses a 2-norm regularization with and additional constrain to the weights, that leads to a sparse solution for which the algorithm performs kernel selection (leading the sigma values to 0).

The Caret package [31] and pROC [32] of the R Statistical Package [33] was used to conduct SVM-RFE with linear regression functions. This package implements a backwards selection of predictors based on predictor importance ranking. The predictors are ranked and the less important ones are sequentially eliminated prior to modelling. To goal is to find a subset of predictors that can be used to produce an accurate model.

There is no fixed number of variables in the GA-based approach [27], and a pruned search is implemented. The fitness function considers not only the classification results but also the number of variables used for such a classification, so it is defined by the sum of two factors, one related to the classification results (F-measure) and another to the number of variables selected.

4   Results

To evaluate the performance, there are several number of well-known accuracy measures for a two-class classifier in the literature such as: classification rate (accuracy), Area Under an ROC Curve (AUC), etc. An experimental comparison of performance measures for classification could be found in [34]. In [35], the authors proposed that AUC is a better measure in general than accuracy when comparing classifiers and in general. Despite this, the most common measures used for their simplicity and successful application are the classification rate and Cohen's kappa measures.. This work requires the division of the pattern set into two halves: training and validation. The pattern set has been divided extracting a total of 20% of the data for validation purposes. To test the performance of these techniques, 10 trails of ten-fold cross validation was conducted for each one with the training set. Results in validation for the MKL, SVM-RFE and GA-SVM are shown in Table 2, 3 and 4, separated by iterations and in columns the accuracy, ROC, sensibility and specificity measures. The last column in Tables 3 and 4 is the number of features used for in this iteration for each To test the performance of these techniques, 10 trails of ten-fold cross validation was conducted for each one.

Table 2. MKL Results					**Table 3.** SVM-RFE Results				
Iter.	Accu	ROC	Sensi	Speci	Iter.	Accu	ROC	Sensi	Speci	Feat.
1	81.50	89.56	69.20	93.80	1	93.00	93.43	98.00	88.00	103
2	80.80	89.73	68.60	93.00	2	93.00	93.43	98.00	88.00	138
3	81.10	86.58	69.00	93.20	3	89.00	89.78	96.00	82.00	113
4	81.30	89.68	69.20	93.40	4	93.00	93.02	94.00	92.00	145
5	80.90	89.56	68.40	93.40	5	92.00	92.61	98.00	86.00	164
6	81.10	**89.82**	68.60	93.60	6	90.00	91.05	98.00	82.00	114
7	81.40	89.79	69.20	93.60	7	93.00	93.43	98.00	88.00	109
8	80.80	89.20	68.40	93.20	8	93.00	**93.83**	1	86.00	175
9	81.10	89.63	69.00	93.20	9	93.00	93.02	94.00	92.00	107
10	81.30	89.11	68.80	93.80	10	89.00	90.98	1	78.00	108

Table 4. GA-SVM Results

Iter.	Accu	ROC	Sensi	Speci	Feat
1	74.75	74.86	65.21	84.83	5
2	84.50	84.68	92.35	76.96	5
3	89.00	88.94	94.39	83.33	9
4	90.50	**90.50**	95.30	85.61	8
5	84.00	83.76	89.85	78.03	5
6	85.75	85.96	89.67	81.89	8
7	84.13	84.31	87.79	80.59	5
8	85.38	85.43	86.26	84.52	4
9	88.75	88.94	92.77	84.71	7
10	88.75	88.71	94.26	83.21	7

Global results for the 10 trials of 10-fold CV for each technique are in mean of error and standard deviation in brackets shown in Table 5.

Table 5. Performance of ten 10-folds runs for MKL, SVM-RFE and GA-SVM algorithms. Standard deviation is in brackets.

Method	Accuracy	ROC	Sensitivity	Specificity
MKL	81.10 (0.24)	89.56 (0.22)	68.84 (0.32)	93.42 (0.27)
SVM-RFE	91.80 (1.75)	**92.45 (1.36)**	97.40 (2.11)	86.20 (4.46)
GA-SVM	85.55 (4.46)	85.60 (4.43)	88.78 (8.80)	82.36 (2.97)

5   Conclusions

The feature-based texture image classification produces a high number of variables, so some feature selection technique or high dimensional efficient classification techniques must be applied in order to improve the classification accuracy. In this work several kernel based methods has been applied to a texture-based features extracted from two dimensional gel electrophoresis images. Kernel based methods are suitable to work with high-dimensionality and non-linear separable dataset. The results show that SVM-RFE have better results, improving the AUROC score of the other two methods. The results show that SVM-RFE needs around hundred features whilst the GA-SVM approach needs around ten. MKL shows also good results, but is highly biased to specificity, with a poor sensibility performance. However, MKL shows a high stability in the different executions of the algorithm, presenting a very low deviation, in contrast to the GAFS approach, with high difference between iterations. In future work in order to exploit the high stability observed, a feature selection mechanism can be added to the MKL technique.

Acknowledgements. This work is supported by the General Directorate of Culture, Education and University Management of the Xunta de Galicia (Ref. 10SIN105004PR). Jose A. Seoane, Tom Gaunt and Colin Campbell acknowledge Medical Research Council Project Grant G1000427.

References

1. Haralick, R.M., Shanmugam, K., Dinstein, I.: Textural features for image classification. IEEE Transactions on Systems, Man and Cybernetics smc 3, 610–621 (1973)
2. Materka, A., Strzelecki, M.: Texture analysis methods-A review. Technical University of Lodz, Institute of Electronics. COST B11 report (1998)
3. Tuceryan, M., Jain, A.: Texture analysis. In: Handbook of Pattern Recognition and Computer Vision, vol. 2, World Scientific Publishing Company, Incorporated (1999)
4. Levina, E.: Statistical Issues in Texture Analysis. University of California, Berkeley (2002)
5. Henry, W.: Texture Analysis Methods for Medical Image Characterisation (2010)
6. Szczypinski, P.M., Strzelecki, M., Materka, A.: MaZda - A software for texture analysis, pp. 245–249 (Year)
7. Bonilha, L., Kobayashi, E., Castellano, G., Coelho, G., Tinois, E., Cendes, F., Li, L.M.: Texture Analysis of Hippocampal Sclerosis. Epilepsia 44, 1546–1550 (2003)
8. Mayerhoefer, M.E., Breitenseher, M.J., Kramer, J., Aigner, N., Hofmann, S., Materka, A.: Texture analysis for tissue discrimination on T1-weighted MR images of the knee joint in a multicenter study: Transferability of texture features and comparison of feature selection methods and classifiers. Journal of Magnetic Resonance Imaging 22, 674–680 (2005)
9. Szymanski, J.J., Jamison, J.T., DeGracia, D.J.: Texture analysis of poly-adenylated mRNA staining following global brain ischemia and reperfusion. Computer Methods and Programs in Biomedicine 105, 81–94 (2012)
10. Vapnik, V.N.: Estimation of dependences based on empirical data. Nauka (1979) (in Russian) (English Translation Springer Verlang, 1982)
11. Burges, C.J.C.: A tutorial on support vector machines for pattern recognition. Data Mining and Knowledge Discovery 2, 121–167 (1998)
12. Chapelle, O., Haffner, P., Vapnik, V.N.: Support vector machines for histogram-based image classification. IEEE Transactions on Neural Networks 10, 1055–1064 (1999)
13. Moulin, L.S., Alves Da Silva, A.P., El-Sharkawi, M.A., Marks Ii, R.J.: Support vector machines for transient stability analysis of large-scale power systems. IEEE Transactions on Power Systems 19, 818–825 (2004)
14. Lanckriet, G.R.G., Cristianini, N., Bartlett, P., Ghaoui, L.E., Jordan, M.I.: Learning the Kernel Matrix with Semidefinite Programming. J. Mach. Learn. Res. 5, 27–72 (2004)
15. Lanckriet, G.R.G., De Bie, T., Cristianini, N., Jordan, M.I., Noble, W.S.: A statistical framework for genomic data fusion. Bioinformatics 20, 2626–2635 (2004)
16. Campbell, C., Ying, Y.: Learning with Support Vector Machines. Synthesis Lectures on Artificial Intelligence and Machine Learning 5, 1–95 (2011)
17. Guyon, I., Weston, J., Barnhill, S., Vapnik, V.: Gene Selection for Cancer Classification using Support Vector Machines. Mach. Learn. 46, 389–422 (2002)
18. Rakotomamonjy, A.: Variable selection using svm based criteria. J. Mach. Learn. Res. 3, 1357–1370 (2003)
19. Saeys, Y., Inza, I., Larrañaga, P.: A review of feature selection techniques in bioinformatics. Bioinformatics 23, 2507–2517 (2007)
20. Alonso-Atienza, F., Rojo-Álvarez, J.L., Rosado-Muñoz, A., Vinagre, J.J., García-Alberola, A., Camps-Valls, G.: Feature selection using support vector machines and bootstrap methods for ventricular fibrillation detection. Expert Systems with Applications 39, 1956–1967 (2012)
21. Kohavi, R., John, G.H.: Wrappers for feature subset selection. Artif. Intell. 97, 273–324 (1997)

22. Guyon, I., Elisseeff, A.: An introduction to variable and feature selection. J. Mach. Learn. Res. 3, 1157–1182 (2003)
23. Holland, J.H.: Adaptation in natural and artificial systems: an introductory analysis with applications to biology, control, and artificial intelligence. University of Michigan Press (1975)
24. Siedlecki, W., Sklansky, J.: A note on genetic algorithms for large-scale feature selection. Pattern Recognition Letters 10, 335–347 (1989)
25. Kudo, M., Sklansky, J.: A comparative evaluation of medium- and large-scale feature selectors for pattern classifiers. Kybernetika 34, 429–434 (1998)
26. Aguiar, V., Seoane, J.A., Freire, A., Munteanu, C.R.: Data Mining in Complex Diseases Using Evolutionary Computation. In: Cabestany, J., Sandoval, F., Prieto, A., Corchado, J.M. (eds.) IWANN 2009, Part I. LNCS, vol. 5517, pp. 917–924. Springer, Heidelberg (2009)
27. Fernandez-Lozano, C., Seoane, J.A., Mesejo, P., Nashed, Y.S.G., Cagnoni, S., Dorado, J.: 2D-PAGE Texture classification using support vector machines and genetic algorithms. In: Proceedings of the 4th International Conference on Bioinformatics Models, Methods and Algorithms (in press, 2013)
28. Hunt, S.M.N., Thomas, M.R., Sebastian, L.T., Pedersen, S.K., Harcourt, R.L., Sloane, A.J., Wilkins, M.R.: Optimal Replication and the Importance of Experimental Design for Gel-Based Quantitative Proteomics. Journal of Proteome Research 4, 809–819 (2005)
29. Rakotomamonjy, A., Bach, F., Canu, S., Grandvalet, Y.: SimpleMKL. Journal of Machine Learning Research 9, 2491–2521 (2008)
30. Sonnenburg, S., Rätsch, G., Schäfer, C., Schölkopf, B.: Large Scale Multiple Kernel Learning. J. Mach. Learn. Res. 7, 1531–1565 (2006)
31. Kuhn, M.: Building Predictive Models in R Using the caret Package. Journal of Statistical Software 28, 1–26 (2008)
32. Robin, X., Turck, N., Hainard, A., Tiberti, N., Lisacek, F., Sanchez, J.-C., Müller, M.: pROC: an open-source package for R and S+ to analyze and compare ROC curves. BMC Bioinformatics 12, 77 (2011)
33. Development Core, T.: R: A language and environment for statistical computing. R Foundation for Statistical Computing, Vienna, Austria (2005)
34. Ferri, C., Hernandez-Orallo, J., Modroiu, R.: An experimental comparison of performance measures for classification. Pattern Recogn. Lett. 30, 27–38 (2009)
35. Huang, J., Ling, C.X.: Using AUC and accuracy in evaluating learning algorithms. IEEE Transactions on Knowledge and Data Engineering 17, 299–310 (2005)

A Genetic Algorithms-Based Approach for Optimizing Similarity Aggregation in Ontology Matching

Marcos Martínez-Romero[1], José Manuel Vázquez-Naya[2], Francisco Javier Nóvoa[2], Guillermo Vázquez[3], and Javier Pereira[1]

[1] IMEDIR Center, University of A Coruña, Campus de Elviña s/n, 15071, A Coruña, Spain
[2] Department of Information and Communication Technologies,
Computer Science Faculty, University of A Coruña, 15071, A Coruña, Spain
[3] Institute of Biomedical Research of A Coruña (INIBIC), Xubias de Arriba 84,
Hospital Materno Infantil (1ª planta), 15006 , A Coruña, Spain
{marcosmartinez,jmvazquez,fjnovoa,javierp}@udc.es,
Guillermo.Vazquez.Gonzalez@sergas.es

Abstract. Ontology matching consists of finding the semantic relations between different ontologies and is widely recognized as an essential process to achieve an adequate interoperability between people, systems or organizations that use different, overlapping ontologies to represent the same knowledge. There are several techniques to measure the semantic similarity of elements from separate ontologies, which must be adequately combined in order to obtain precise and complete results. Nevertheless, combining multiple similarity measures into a single metric is a complex problem, which has been traditionally solved using weights determined manually by an expert, or through general methods that do not provide optimal results. In this paper, a genetic algorithms-based approach to aggregate different similarity metrics into a single function is presented. Starting from an initial population of individuals, each one representing a combination of similarity measures, our approach allows to find the combination that provides the optimal matching quality.

Keywords: genetic algorithms, ontology matching, ontologies, Semantic Web.

1 Introduction

At present, the role of ontologies as the essential artifact for allowing a more effective data and knowledge sharing and reusing in the Semantic Web [1] is widely recognized [2] and a variety of public ontologies exist for different areas. This innovative knowledge representation method is considered to be an appropriate solution to the problem of heterogeneity in data, since ontological methods make it possible to reach a common understanding of concepts in a particular domain, supporting the exchange of information between people (or systems) that utilize different representations for the same or similar knowledge [3, 4].

Nevertheless, given that different tasks or different points of view usually require different conceptualizations, utilizing a single ontology is neither always possible nor

I. Rojas, G. Joya, and J. Cabestany (Eds.): IWANN 2013, Part I, LNCS 7902, pp. 435–444, 2013.

advisable. This can lead to the usage of different ontologies, although in some cases they might contain information that could be overlapping. This, in turn, represents another type of heterogeneity that can result in inefficient processing or misinterpretation of data, information, and knowledge. Addressing this problem requires to find the correspondences, or mappings, that exist between the elements of the different ontologies being used. This process is commonly known as *ontology matching, mapping or alignment* [5]. The resulting set of inter-ontology relations can be used to adequately exchange information between people, systems and organizations.

During the last years, multiple *ontology alignment techniques* have been conceived to identify these correspondences [6]. These methods are based on computing a similarity (or distance) value between elements of different ontologies. When computing the ontology alignment between two ontologies, it is frequently to use several ontology alignment techniques, based on different similarity approaches (e.g. lexical similarity, structural similarity, etc.) and then aggregating them into a unique similarity value. However, calculating the optimal similarity aggregation is a computationally expensive task that requires new, more efficient methods to get precise and complete alignments [5, 7, 8].

In this work, we propose an approach based on genetic algorithms (GAs) to ascertain how to combine multiple similarity measures into a single aggregated metric, in order to provide the optimal matching result. Our work can be useful to automatically tune an ontology matching system in environments where a reference matching is provided.

Qazvinian et al. [9] are among the small number of authors who have tried, up to the moment, to apply GAs to the ontology matching task. They considered ontology matching as an optimization problem in which the objective is maximizing the overall similarity value between the input ontologies, and they used a GA to find the optimal mapping. In a similar way, another interesting approach to ontology matching by using GAs is GAOM [10]. In this work, ontology features are defined from two aspects: intensional and extensional, and the ontology matching problem is modeled as a global optimization of a mapping between two ontologies. Then GAs are used to achieve an approximate optimal solution.

2   A New Approach to Optimize Similarity Aggregation

In this section, a genetic algorithm to find the optimal aggregation of multiple similarity measures is presented. The GA starts from a randomly generated aggregation of similarity measures (set of weights), and tries to find the weights that optimize the global matching quality. In order to reliably describe the proposed strategy, it is necessary to define the following elements (see Fig. 1):

- A and B are two ontologies with n and m elements (entities) respectively. A is composed by the entities $a_1,...,a_n$ while B has the entities $b_1,...,b_m$.

- S is an existing set of semantic mappings or correspondences s_{ij} between A and B, being s_{ij} a semantic mapping between the entity a_i from A and the entity b_j from B, with $0 < i \leq n$ and $0 < j \leq m$.

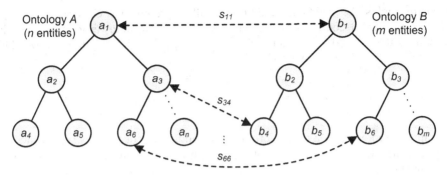

Fig. 1. Graphical representation of the ontology matching problem. The figure shows the taxonomy of two ontologies (*A* and *B*), and a set of semantic mappings (s_{ij}) between them.

- $F = \{F_1(a_i,b_j),..., F_p(a_i,b_j)\} = \{F_1(s_{ij}),..., F_p(s_{ij})\}$ is a set composed by p functions, or ontology matching metrics, to compute a value of semantic similarity (in the [0, 1] interval) between pairs of entities from separate ontologies.

- t is a similarity threshold belonging to the interval [0, 1], which indicates the minimum similarity value required to consider that exists a semantic correspondence between two different entities.

- $F_{agg}(a_i,b_j) = \sum_{k=1}^{p} w_k \cdot F_k(a_i,b_j)$, with $\sum_{k=1}^{p} w_k = 1$, is a function to compute an aggregated similarity value between two entities. This function combines the similarity values provided by p different similarity functions into a single value belonging to the interval [0, 1]. The aggregation is based on the values of a set of p weights w_k, which quantify the contribution of each separate similarity measure to the aggregated value.

- $Q(S) \rightarrow [0,1]$ is a function that measures the quality of a set of semantic correspondences between two ontologies. A good example of quality measure is the f-measure metric, which considers both the precision and the recall to compute the score.

The approach is addressed to find the values of the weights w_k that maximize the quality of the matching between the input ontologies *A* and *B*, that is, the function *Q(S)*. The obtained set of weights could be subsequently used to compute the matching of ontologies with similar characteristics, or belonging to the same domain as the ontologies whose matching was selected as a reference.

2.1   Encoding Mechanism and Initialization

Each individual in the population represents a potential solution to the problem, that is, a set of weights w_k that indicate the contribution of each similarity metric to the aggregated similarity function. We propose an encoding mechanism based on that each position in the chromosome contains a value in the interval [0, 1], which represents a *cut*, or *separation point* that limits the value of a weight (remember that the summation of all weights is equal to 1). Considering that p is the number of required weights, the set of cuts could be formally represented as $C' = \{c_{1'}, ..., c_{p-1'}\}$. The chromosome decoding is carried out by ordering C' from lower to higher, which constitutes the ordered set of values $C = \{c_{1}, ..., c_{p-1}\}$, and calculating the weights according to the following expression:

$$w_k = \begin{cases} c_1, & k = 1 \\ c_i - c_{i-1}, & 1 < k < p \\ 1 - c_{p-1}, & k = p \end{cases}$$

A graphical representation of the chromosome and the decoded values is presented in Fig. 2, while Fig. 3 shows an example that can be useful to understand the decoding mechanism.

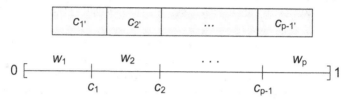

Fig. 2. Graphical representation of a chromosome and the set of weights obtained after decoding it. Each gene in the chromosome contains a value belonging to the interval [0, 1] that represents a cut, or separation point between weights. $C' = \{c_{1'}, ..., c_{p-1'}\}$ is an unordered set of cuts, while $C = \{c_{1}, ..., c_{p-1}\}$ is the result obtained after ordering C' from lower to higher. $W = \{w_1, ..., w_p\}$ is the set of weights that constitute the solution to the problem.

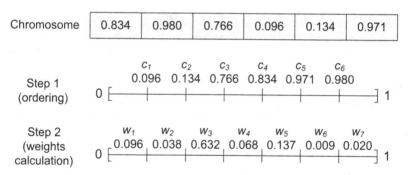

Fig. 3. Example of a specific individual and the weights obtained after decoding it. In this example, 7 different weights were considered.

2.2 Reproduction Methods

To go from one generation to the next one, we suggest using the following operators:

- **Selection.** We propose to use a roulette wheel selection method, which consists in that individuals are given a probability of being selected that is directly proportionate to their fitness, so the best individuals will have more opportunities of reproduction. Two individuals are then chosen randomly based on these probabilities and produce offspring.

- **Crossover.** Crossover will use a non-destructive strategy, in such a way that the descendants will pass to the following generation only if they exceed the fitness of their parents. A single-point crossover will be used, which consists in randomly selecting a crossover point on both parent chromosomes and then interchanging the two parent chromosomes to produce two new offspring.
- **Copy.** The best individual from one generation will be also copied to the following generation (elitist strategy). This decision has been taken to keep the best set of weights (best solution) that has been obtained up to the moment.
- **Mutation.** When the crossover has been achieved, genes will be mutated with a low probability. This mutation will consist in replacing the selected gene by a randomly generated one.

2.3   Fitness Function

For the smooth running of a GA, it is necessary to have a method that allows to show if the individuals of the population are or are not good solutions to the problem. That is the aim of the fitness or objective function. As our fitness function, we propose to use the *f-measure* [11], which is the uniformly weighted harmonic mean of precision and recall. F-measure will be used as the reference quality metric, in such a way that we will consider that the best alignment is the alignment with highest f-measure.

$$fitness = f - measure = 2 \cdot \frac{precision \cdot recall}{precision + recall}$$

2.4   Stop Criterion

We propose to use a hybrid stop criterion: the GA will stop when one of the following conditions is true: (1) A fixed number of iterations have been reached; (2) The value for the fitness function is higher than a particular threshold.

3   Execution Example

In this section we provide a "toy" example with two small ontologies, which can be useful to understand how the proposed GA works. We will assume that:

- A and B are two ontologies from a specific domain. Both ontology A and ontology B have 3 entities ($n = 3$, $m = 3$).
- S is the reference matching between A and B. In this example $S = \{s_{12}, s_{33}\}$, that is, it will be supposed that there is a semantic mapping between the pairs of entities (a_1, b_2) and (a_3, b_3), as shown in Fig. 4. We will also suppose that there are some similarity between the entities (a_1, b_1) and (a_2, b_2), but not enough to be considered semantic mappings.
- $F = \{F_1(s_{ij}), F_1(s_{ij}), F_2(s_{ij}), F_3(s_{ij}), F_4(s_{ij}), F_5(s_{ij})\}$ is a set composed by five different similarity functions. We need to aggregate the similarity values provided by these functions into a single measure. We will also suppose that the functions $F_3(s_{ij})$ and $F_5(s_{ij})$, due to the particular characteristics of A and B, are not adequate to align them, so they will not provide reliable similarity values.

- The similarity threshold t is set to 0.7, which means that the algorithm will consider that exists a mapping between a pair of entities (a_i, b_j) if the similarity function for such entities provides a value higher than 0.7. We will also suppose that the algorithm will finish if the value of the fitness function is higher than 0.8 (stop criterion).

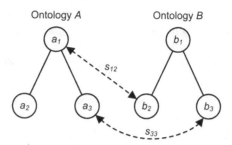

Fig. 4. Graphical view of ontologies A and B, and the reference matching S (gold standard)

Considering the previous information, the aggregated similarity function would be:
$F_{agg}(a_i, b_j) = w_1 \cdot F_1(s_{ij}) + w_2 \cdot F_2(s_{ij}) + w_3 \cdot F_3(s_{ij}) + w_4 \cdot F_4(s_{ij}) + w_5 \cdot F_5(s_{ij})$,   with
$w_1 + w_2 + w_3 + w_4 + w_5 = 1$. The GA will be used to find the values of weights w_1, w_2, w_3, w_4 and w_5 that provide the optimal matching.

Firstly, it is necessary to compute the values of similarity for the $n \times m$ possible correspondences between A and B, according to the five different similarity functions. In this example, we will suppose that the results of this computation are the ones in Table 1. Remember that it has been supposed that the correct mappings are s_{12} and s_{33}, and that $F_3(s_{ij})$ and $F_5(s_{ij})$ are not adequate to align the given ontologies, so they are not able to identify s_{12} and s_{33} as the valid mappings.

Table 1. Results of initial similarity computation

	s_{11}	s_{12}	s_{13}	s_{21}	s_{22}	s_{23}	s_{31}	s_{32}	s_{33}
$F_1(s_{ij})$	0.56	0.93	0.12	0.05	0.66	0.31	0.08	0.18	0.97
$F_2(s_{ij})$	0.65	0.99	0.20	0.03	0.68	0.49	0.03	0.23	0.81
$F_3(s_{ij})$	0.11	0.17	0.23	0.41	0.56	0.11	0.65	0.09	0.21
$F_4(s_{ij})$	0.72	0.72	0.44	0.50	0.45	0.11	0.01	0.13	0.98
$F_5(s_{ij})$	0.77	0.28	0.81	0.74	0.98	0.79	0.87	0.17	0.09

The following step would be to generate the initial population. In this case, it is composed by 10 randomly generated individuals, which are shown in Table 2.

Table 2. Initial population (first generation)

Individual	Values				Individual	Values			
1	0.37	0.62	0.23	0.43	6	0.30	0.27	0.92	0.71
2	0.32	0.08	0.07	0.56	7	0.69	0.22	0.17	0.94
3	0.53	0.91	0.11	0.73	8	0.22	0.66	0.45	0.21
4	0.65	0.63	0.01	0.70	9	0.20	0.14	0.25	0.12
5	0.86	0.19	0.59	0.21	10	0.85	0.53	0.41	0.19

The next step would be to calculate the fitness value for each individual. Each chromosome is decoded in order to obtain the values for the 5 weights (see Table 3).

Table 3. Weights for the 1st generation, obtained after decoding the chromosomes in Table 2

Individual	w_1	w_2	w_3	w_4	w_5
1	0.23	0.14	0.06	0.19	0.38
2	0.07	0.01	0.24	0.24	0.44
3	0.11	0.42	0.20	0.18	0.09
4	0.01	0.62	0.02	0.05	0.30
5	0.19	0.02	0.38	0.27	0.14
6	0.27	0.03	0.41	0.21	0.08
7	0.17	0.05	0.47	0.25	0.06
8	0.21	0.01	0.23	0.21	0.34
9	0.12	0.02	0.06	0.05	0.75
10	0.19	0.22	0.12	0.32	0.15

The obtained weights are then used to compute the aggregated similarity value for each possible correspondence. These values are shown in Table 4. As an example, the aggregated value for the correspondence (a_1, b_1) and the weights obtained after decoding the individual 1, would be calculated as:

$$F_{agg}(a_1, b_1) = w_1 \cdot F_1(s_{11}) + w_2 \cdot F_2(s_{11}) + w_3 \cdot F_3(s_{11}) + w_4 \cdot F_4(s_{11}) + w_5 \cdot F_5(s_{11}) =$$

$$= 0.23 \cdot 0.56 + 0.14 \cdot 0.65 + 0.06 \cdot 0.11 + 0.19 \cdot 0.72 + 0.38 \cdot 0.77 = 0.66$$

Table 4. Aggregated similarity values for the initial population. The table also shows the mappings that exceed the similarity threshold (0.70), which are used to calculate the fitness value for each individual.

Ind.	F_{agg11}	F_{agg12}	F_{agg13}	F_{agg21}	F_{agg22}	F_{agg23}	F_{agg31}	F_{agg32}	F_{agg33}	Mappings	Fitness
1	0.66	0.61	0.46	0.42	**0.74**	0.47	0.39	0.17	0.57	s_{22}	-
2	0.58	0.41	0.53	0.55	**0.73**	0.43	0.55	0.14	0.40	s_{22}	-
3	0.56	**0.71**	0.30	0.26	0.64	0.35	0.23	0.17	0.67	s_{12}	0.67
4	0.68	**0.75**	0.39	0.27	**0.76**	0.55	0.29	0.20	0.59	s_{12}, s_{22}	0.50
5	0.46	0.49	0.35	0.40	0.61	0.25	0.39	0.13	0.56	-	-
6	0.43	0.52	0.29	0.35	0.60	0.23	0.36	0.13	0.59	-	-
7	0.41	0.48	0.30	0.37	0.58	0.20	0.38	0.13	0.55	-	-
8	0.56	0.49	0.45	0.46	0.70	0.39	0.46	0.15	0.50	-	-
9	0.70	0.39	0.66	0.61	**0.88**	0.65	0.70	0.17	0.26	s_{22}	-
10	0.61	0.69	0.36	0.34	0.63	0.33	0.23	0.16	**0.71**	s_{33}	0.67

The correspondences with a similarity value higher than the given threshold (0.7) are considered valid semantic mappings. Using these mappings and the reference matching (gold standard), the fitness value (f-measure) is calculated. There are two individuals (3 and 10) that provide a fitness value of 0.67, but this value is not enough to stop the algorithm according to the fitness threshold that has been set (0.8). As a consequence, the next step is to select the individuals that will reproduce themselves to create the next generation.

The individuals that form the second generation are shown in Table 5. According to an elitist strategy, the individuals 3 and 10 are copied to the second generation (they are named 11 and 12). We suppose that the roulette selection method selects the individuals 3 and 10 to reproduce themselves and that a single-point crossover is applied between genes 1 and 2, giving as a result individuals 13 and 14; in the middle point (individuals 15 and 16); and between genes 3 and 4 (individuals 17 and 18). Individuals 19 and 20 are obtained by mutating one gene from the individuals 3 (gene 1) and 10 (gene 2), respectively. The corresponding weights are shown in Table 6.

Table 5. Second generation

Individual	Values				Individual	Values			
11	0.53	0.91	0.11	0.73	16	0.85	0.53	0.11	0.73
12	0.85	0.53	0.41	0.19	17	0.53	0.91	0.11	0.19
13	0.53	0.53	0.41	0.19	18	0.85	0.53	0.41	0.73
14	0.85	0.91	0.11	0.73	19	0.25	0.91	0.11	0.73
15	0.53	0.91	0.41	0.19	20	0.85	0.87	0.41	0.19

Table 6. Weights for the 2nd generation, obtained after decoding the chromosomes in Table 5

Individual	w_1	w_2	w_3	w_4	w_5
11	0.11	0.42	0.20	0.18	0.09
12	0.19	0.22	0.12	0.32	0.15
13	0.19	0.22	0.12	0.00	0.47
14	0.11	0.62	0.12	0.06	0.09
15	0.19	0.22	0.12	0.38	0.09
16	0.11	0.42	0.20	0.12	0.15
17	0.11	0.08	0.34	0.38	0.09
18	0.41	0.12	0.20	0.12	0.15
19	0.11	0.14	0.48	0.18	0.09
20	0.19	0.22	0.44	0.02	0.13

Table 7. Aggregated similarity values, mappings and fitness for the second generation

Ind.	F_{agg11}	F_{agg12}	F_{agg13}	F_{agg21}	F_{agg22}	F_{agg23}	F_{agg31}	F_{agg32}	F_{agg33}	Mappings	Fitness
11	0.56	**0.71**	0.30	0.26	0.64	0.35	0.23	0.17	0.67	s_{12}	0.67
12	0.61	0.69	0.36	0.34	0.63	0.33	0.23	0.16	**0.71**	s_{33}	0.67
13	0.62	0.55	0.48	0.41	**0.80**	0.55	0.51	0.18	0.43	s_{22}	-
14	0.59	**0.80**	0.26	0.17	0.68	0.43	0.18	0.20	0.70	s_{12}	0.67
15	0.61	**0.71**	0.33	0.32	0.60	0.29	0.18	0.16	**0.77**	s_{12}, s_{33}	1
16	0.56	0.68	0.32	0.27	0.67	0.39	0.28	0.18	0.62	-	-
17	0.49	0.54	0.35	0.40	0.58	0.22	0.31	0.13	0.62	-	-
18	0.53	0.66	0.29	0.28	0.67	0.34	0.30	0.16	0.67	-	-
19	0.40	0.48	0.30	0.36	0.61	0.25	0.41	0.13	0.51	-	-
20	0.41	0.52	0.28	0.30	0.66	0.32	0.42	0.15	0.49	-	-

The aggregated similarity values for the second generation are shown in Table 7. It is possible to see that the individual 15 has a fitness value of 1, which is the maximum value for the fitness function. Having reached this value, the GA stops (according to the stop criterion). The GA has provided the following solution to the problem, obtained after decoding the individual 15:

$$w_1 = 0.19;\ w_2 = 0.22;\ w_3 = 0.12;\ w_4 = 0.38;\ w_5 = 0.09$$

Given these weights, the aggregated similarity function for this example would be calculated according the following expression:

$$F_{agg}(s_{ij}) = 0.19 \cdot F_1(s_{ij}) + 0.22 \cdot F_2(s_{ij}) + 0.12 \cdot F_3(s_{ij}) + 0.38 \cdot F_4(s_{ij}) + 0.09 \cdot F_5(s_{ij})$$

As it can be observed, this function gives a low weight to the functions $F_3(s_{ij})$ and $F_5(s_{ij})$. We had supposed that F_3 and F_5 were not reliable, so the result provided by the approach makes sense. Using this function, we could align any pair of ontologies with similar characteristics to A and B.

4   Conclusion and Future Research

Although a lot has been done towards tackling ontology matching, the research community still reports open issues that impose new challenges for researchers and underline new directions for the future. One of these issues, which represents an emerging research area, is the aggregation of different similarity measures into a single one. In this work, we have proposed a GA-based approach to combine different measures into a single metric, optimizing the quality of the matching results. The presented GA can be useful to automatically configure the similarity aggregation process in ontology matching systems addressed to provide precise and complete results in domains that require rapid processing. Through a simple example, we have showed how the GA can find the similarity combination that provides an optimal matching result between two ontologies.

The most immediate future work is to embed our GA into a real existing ontology matching system that achieves similarity aggregation in a traditional manner (i.e., either through manual, user-based aggregation or by means of general methods), in order to measure the improvement of matching quality. We are also interested in extending our theory and mechanisms for providing an ontology matching system with full self-configuration capabilities, in order to obtain good results in dynamic environments that require immediate response, without requiring user interaction.

Acknowledgements. Work supported by the Carlos III Health Institute (grant FIS-PI10/02180), the Ibero-NBIC Network (ref. 209RT0366) funded by CYTED, and grants CN2012/217 (REGICC), CN2011/034 ("Programa de consolidación y estructuración de unidades de investigación competitivas") and CN2012/211 ("Agrupación estratégica") from the Xunta de Galicia. Work also co-funded by FEDER (European Union).

References

1. Bemers-Lee, T., Hendler, J., Lassila, O.: The semantic web. Scientific American 284, 34–43 (2001)
2. Staab, S., Studer, R.: Handbook on ontologies. Springer (2009)
3. Gruber, T.R.: A translation approach to portable ontology specifications. Knowledge Acquisition 5, 199–220 (1993)
4. Gomez-Perez, A., Fernández-López, M., Corcho, O.: Ontological Engineering: with examples from the areas of Knowledge Management, e-Commerce and the Semantic Web. Springer (2004)
5. Shvaiko, P., Euzenat, J.: Ontology matching: state of the art and future challenges (2012)
6. Martínez-Romero, M., Vázquez-Naya, J.M., Pereira, J., Ezquerra, N.: Ontology alignment techniques. Encyclopedia of Artificial Intelligence 3, 1290–1295 (2008)
7. Kalfoglou, Y., Schorlemmer, M.: Ontology mapping: the state of the art. The Knowledge Engineering Review 18, 1–31 (2003)
8. Shvaiko, P., Euzenat, J.: A survey of schema-based matching approaches. Journal on Data Semantics IV, 146–171 (2005)
9. Qazvinian, V., Abolhassani, H., Haeri, S.H., Hariri, B.B.: Evolutionary coincidence-based ontology mapping extraction. Expert Systems 25, 221–236 (2008)
10. Wang, J., Ding, Z., Jiang, C.: GAOM: genetic algorithm based ontology matching. In: IEEE Asia-Pacific Conference on Services Computing, APSCC 2006, pp. 617–620. IEEE (2006)
11. Rijsbergen, C.J.: Information Retrieval. Butterworth (1979, 1997)

Automatic Fish Segmentation on Vertical Slot Fishways Using SOM Neural Networks

Álvaro Rodriguez[1], Juan R. Rabuñal[2], María Bermúdez[3], and Alejandro Pazos[1]

[1] Dept. of Information and Communications Technologies, Faculty of Informatics
[2] Center of Technological Innovation in Construction and Civil Engineering (CITEEC)
[3] Dept. of Hydraulic Engineering, ETSECCP,
University of A Coruña, Campus Elviña s/n 15071 A Coruña
{Arodriguezta,juanra,mbermudez,apazos}@udc.es

Abstract. Vertical slot fishways are hydraulic structures which allow the upstream migration of fish through obstructions in rivers. The appropriate design of these should consider the behavior and biological variables of the target fish species and currently existing mechanisms to measure the behavior of the fish in these assays, such as direct observation or placement of sensors on the specimens, are impractical or unduly affect the animal behavior.

This paper studies the application of Artificial Neural Networks to the problem of automatic fish segmentation in vertical slot fishways. In particular, SOM Neural Networks have been used to detect fishes using visual information sampled by an underwater camera system. A ground true dataset was designed with experts and different approaches were tested providing promising results.

Keywords: ANN, Fish-Detection, Segmentation, SOM.

1 Introduction

The construction of engineering works in rivers, such as dams or weirs, alters the ecosystem of rivers, causing changes in the fauna and flora. One of the most important effects is the obstruction of fish migration.

Out of the various solutions employed to restore fish passage, some of the most versatile are known as vertical slot fishways. This type of fishway is basically a channel divided into several pools separated by slots.

An effective vertical slot fishway must allow fish to enter, pass through, and exit safely with minimum cost to the fish in time and energy. Thus, biological requirements should drive design and construction criteria for this type of structures. However, while some authors have characterized the flow in vertical slot fishways [1-4] and others have studied fish swimming performance [5, 6], the analysis of the real fish behavior in full-scale physical fishway models are scarce in the literature, so the actual behavior of the fish within a fishway is practically unknown.

The detection of fishes in images, is a problem widely studied in the literature.

I. Rojas, G. Joya, and J. Cabestany (Eds.): IWANN 2013, Part I, LNCS 7902, pp. 445–452, 2013.
© Springer-Verlag Berlin Heidelberg 2013

Some examples are techniques based on color segmentation such as [7] where fluorescent marks are used for the identification of fishes in a tank or in [8] where color properties and background subtraction are used to recognize live fish in clear water.

Other used techniques are stereo vision [9], background models [10], shape priors [11], local thresholding [12], moving average algorithms [13] or pattern classifiers applied to the changes measured in the background [14].

The works previously described, use color features, background or fish models, and stereo vision systems. These techniques are carried out in calm and low turbulent water where information about textures and color may be enough to discriminate the fish. However, these conditions are not fulfilled in the turbulent conditions of a vertical slot fishway.

In this work, the use of SOM Neural Networks will be studied to detect living fishes in vertical slot fishways from images obtained through a network of video cenital underwater video cameras (Fig. 1).

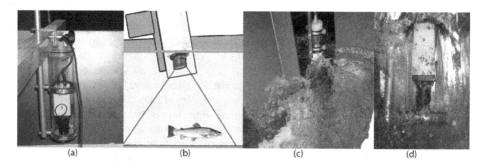

(a)   (b)   (c)   (d)

Fig. 1. Camera system. (a) Camera position in the fishway. (b) Recording scheme. (c), (d) Water conditions during assays.

This work is an extension of the carried out in [15] where an image system is designed to record assays with real fishes and a methodology to study fish behavior is carried out, consisting in analysing fish trajectories with computer vision techniques and computing fish statistics mixing observed trajectories with hydraulic properties of the fishway.

As a part of the system proposed in [15] a SOM based segmentation technique was defined. However, the segmentation question was not in depth analysed, and this technique was not compared with other techniques or tested with ground true data.

Furthermore, SOM networks have not been analysed or tested before in the problem of fish detection.

This paper is focused in studying the fish segmentation problem in a vertical slot fishway as a part of a tracking system with the objective of generate the trajectory of the fish over time by locating its position in every frame of the video.

The behavior of SOM Neural Networks is studied, analysing different approaches with a dataset recorded in experimental conditions and a true ground data created by experts. Additionally a representation and filtering shape-based technique has been defined and tested.

The efficiency and accuracy of SOM Neural Networks in the task of fish segmentation is properly analysed in this paper for the first time.

2   Fish Detection

In order to study the behavior of the fish from the sequence of images acquired, the area occupied by the fish should be separated from the rest of the image or background. This process is known as image segmentation.

The aim of image segmentation algorithms is to partition the image into perceptually similar regions. Every segmentation algorithm addresses two problems, the criteria for a good partition and the method for achieving efficient partitioning [16].

Therefore it is necessary to find a variable which allow a robust separation of the fish from the background. Then it is necessary to choose a technique to classify the image in different groups according to a criteria.

As discussed before, most common criteria to detect fishes in the image are the color features and a priori knowledge of the background.

Even analysing calm and low turbulent water in high quality images, simple image segmentation techniques such as thresholding and edge detection do not perform well due to the low levels of contrast typical of underwater images [14]. Additionally, in the images recorded in the high turbulent water of a vertical slot fishway, background subtraction techniques are usually ineffective alone. Furthermore, the images will be characterized by extreme luminosity changes, huge noise levels and poor contrast, being texture and color information useless (Fig. 2).

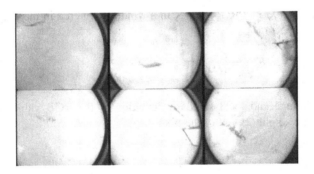

Fig. 2. Example of images recorded in the fishway during the assays

In this paper, a type of Artificial Neural Network (ANN) called Self-organizing map (SOM) [10] has been used for image classification.

The SOM network model is aimed at establishing a correlation between the number patterns supplied as an input and a two-dimensional output space (Topological map); thus, the input data with common features activate areas close to the map. This characteristic can be applied to image segmentation, also giving the following advantages: Adaptive Learning based on a training phase using input examples, generalization ability, error tolerance, highly operable in parallel and integrable into the existing technology [10].

Thus, works like those carried out by [17, 18] have successfully applied the SOM networks to image segmentation, and their importance in this field was pointed out by in [19, 20].

Finally, the SOM networks have the advantage of not needing to define a set of patterns to supervise the learning process. They can be straightforward created by defining a representative image or set of images.

However, their behavior in fish detecting problems was not properly studied before.

To test the proposed approach, different tests were realized with different SOM architectures, and using different combinations of features for classification, such as color and grey values, local average, local standard deviation or local entropy.

After preliminary assays, a three-layer topology with 3 processing elements in each layer was fixed for each network.

In all cases, input images were normalized and preprocessed to enhance the contrast of the image by using a contrast-limited adaptive histogram equalization (CLAHE).

Different SOM networks were considered, being $E_{i,j}$ the input data for the pixel (i, j) formed by one or more NxN vectors parameters defined in the neighborhood of the pixel defined. The following input values were considered:

- The RGB intensity values of the image I from the neighborhood of the pixel (i, j).

$$E_{i,j} = \left\{ I(x, y) \right\}_{(x,y)=\left[(i-N/2,\, j-N/2),\, (i+N/2,\, j+N/2) \right]} \tag{1}$$

- The local average of the RGB values in a window centered in the neighborhood of the pixel (i, j).

$$E_{i,j} = \left\{ \mu_{x,y} \right\}_{(x,y)=\left[(i-N/2,\, j-N/2),\, (i+N/2,\, j+N/2) \right]} \tag{2}$$

- The local average and local standard deviation of the RGB values in a window centered in the neighborhood of the pixel (i, j).

$$E_{i,j} = \left\{ \mu_{x,y} \sigma_{x,y} \right\}_{(x,y)=\left[(i-N/2,\, j-N/2),\, (i+N/2,\, j+N/2) \right]} \tag{3}$$

Two versions of each network were implemented; one, using information of the processed image and another including background modeling information, where a selected frame of the background was used to normalize the current image values.

Additionally, the SOM network proposed in [15] using the global and local average features from the current image I and a background reference image I' was considered. This technique will be tested for the first time with ground true data.

$$E_{i,j} = \left\{ \frac{\mu_{x,y}}{\mu_I} \quad \frac{\mu_{x,y}}{\mu_I} - \frac{\mu_{x,y}{}'}{\mu_I{}'} \right\}_{(x,y)=[(i-N/2,\,j-N/2),(i+N/2,\,j+N/2)]} \tag{4}$$

3   Filtering and Representation

Once the image segmentation is obtained with the SOM network, the objective of the segmentation system is to determine the position of the detected fish in the image.

Due to the characteristics of the image, where the fish is often partially hidden and where is expected the presence of shadows, bubbles and reflections, the algorithm should respond well to partial or abnormal detections.

A simple algorithm has been built, in order not to increase the computational burden of the process. The algorithm is based on obtaining the connected-body vector from the segmented image. Each body will be characterized by the vector of pixels that make it up and by a set of descriptive parameters: its area, its centroid and the minimum ellipse containing the body.

Then the detected object is classified as a potential fish or noise, according to the size of the object and the minimum and major axis of the ellipse. An iterative fusion operation has been defined to replace two close unconnected bodies for a new one, which will be formed by the points from the two previous ones while shape criteria are satisfied.

Thus, the algorithm determines the mass centers of the fishes and puts together those connected bodies whose characteristics can be matched to a fish or a part of it, and discards those bodies that, due to their size or shape, are regarded as noise. Subsequently, the mass center of each detected fish is obtained.

4   Experimental Results

To measure the performance of the SOM networks, a set of experiments were performed with living fishes of the salmo trutta specie in a 1:1 vertical slot fishway model located at Center for Studies and Experimentation of Public Works CEDEX, in Madrid.

A data set of 1000 images from 10 different cameras selected from different pool and fishway regions was defined and the corresponding ground true data was manually created by experts.

To evaluate the dependence of the results with the selected training patterns, all the networks were training using a single image, and results were analysed using 3 different trainings (Fig. 3).

The average obtained results are shown in Table 1 and can be observed in Fig. 3.

Analyzing the results it may be observed that the best precision has been obtained with the local average classifier with a background modeling. Therefore, the inclusion of temporal information appears to be a critical factor in the results. Additionally, none of the networks without background information obtained good results.

Table 1. Average Results

Avg.	Pixel	Pixel BackGnd.	Avg.	**Avg. BackGnd.**	Avg.- St.Dev.	Avg.- St.Dev. BackGnd.	Features
Detections	690	1094	542	**876**	949	767	894
True Pos.	328	754	554	**809**	446	694	808
False Pos.	362	340	624	**67**	265	73	85
True Neg.	35	42	30	**50**	36	49	49
False Neg.	753	331	579	**187**	564	300	189
Precision	0.48	0.69	0.47	**0.92**	0.63	0.90	0.90
Recall	0.30	0.69	0.49	**0.81**	0.44	0.70	0.81
F.P. Rate	0.52	0.31	0.53	**0.08**	0.37	0.10	0.10
F.N. Rate	0.70	0.31	0.51	**0.19**	0.56	0.30	0.19

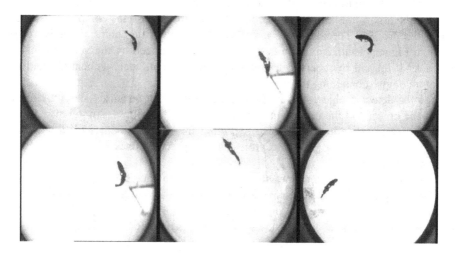

Fig. 3. Segmentation results

The SOM network proposed in [15] has obtained the second best result.

It can be seen that a very low false positive rate is achieved, so obtained results are very reliable, and they will represent true positions of the fish with a high probability. Furthermore, precision is the most important factor in a detection system because a high rate of false positives would make the results useless.

However, the false negative rate is much higher. Showing that, in almost a 20% of the measurements, the system will not provide information about fish position. This is due to the fact that the fish will be frequently occluded by the turbulence of water. Nevertheless, it must be taken in account that the system can execute up to 25 measurements per second, so it will provide a correct position more than 15 times per second in average, which is far beyond any detection system used at present in this field.

5   Conclusions

In this work, a solution to detect fishes in vertical slot fishways based on SOM networks and an algorithm to filter anomalous detections is analysed.

The accuracy and performance of different SOM networks have been tested, analysing and comparing for the first time, the behavior of these technique in the fish segmentation problem.

The results obtained with this system have been very promising, as they allowed us to obtain the fish position in the image with a low error rate.

In future stages of this work, a tracking algorithm will be defined to manage the detected positions of the fish or fishes obtaining the trajectories of the specimens.

Although further research is needed, the results obtained will be used to elaborate a new methodology to study fish behavior inside vertical slot fishways so it can contribute to develop robust guidelines for future fishway designs and to establish more realistic criteria for the evaluation of biological performance of the current designs.

Acknowledgments. This work was supported by the Dirección Xeral de Investigación, Desenvolvemento e Innovación (General Directorate of Research, Development and Innovation) de la Xunta de Galicia (Galicia regional government) (Ref. 10MDS014CT) and from the Ministerio de Economía y Competitividad (Spanish Ministry of Economy and Competitiveness) Ref. CGL2012-34688. The authors would also like to thank the Center for Studies and Experimentation on Public Works (CEDEX).

References

[1] Puertas, J., Pena, L., Teijeiro, T.: An Experimental Approach to the Hydraulics of Vertical Slot Fishways. Journal of Hydraulics Engineering 130 (2004)

[2] Rajaratnam, N., Van der Vinne, G., Katopodis, C.: Hydraulics of Vertical Slot Fishways. Journal of Hydraulic Engineering 112, 909–927 (1986)

[3] Tarrade, L., Texier, A., David, L.: Topologies and measurements of turbulent flow in vertical slot fishways. Hydrobiologia 609, 177–188 (2008)

[4] Wu, S., Rajaratma, N., Katopodis, C.: Structure of flow in vertical slot fishways. Journal of Hydraulic Engineering 125, 351–360 (1999)

[5] Dewar, H., Graham, J.: Studies of tropical tuna swimming performance in a large water tunnel– Energetics. Journal of Experimental Biology 192, 13–31 (1994)

[6] Blake, R.W.: Fish functional design and swimming performance. Journal of Fish Biology 65, 1193–1222 (2004)

[7] Duarte, S., Reig, L., Oca, J., Flos, R.: Computerized imaging techniques for fish tracking in behavioral studies. presented at the European Aquaculture Society (2004)

[8] Chambah, M., Semani, D., Renouf, A., Courtellemont, P., Rizzi, A.: Underwater color constancy enhancement of automatic live fish recognition. presented at the IS&T Electronic Imaging (SPIE), California, USA (2004)

[9] Petrell, R.J., Shi, X., Ward, R.K., Naiberg, A., Savage, C.R.: Determining fish size and swimming speed in cages and tanks using simple video techniques. Aquacultural Engineering 16, 63–84 (1997)

[10] Morais, E.F., Campos, M.F.M., Padua, F.L.C., Carceroni, R.L.: Particle filter-based predictive tracking for robust fish count. presented at the Brazilian Symposium on Computer Graphics and Image Processing (SIBGRAPI) (2005)

[11] Clausen, S., Greiner, K., Andersen, O., Lie, K.-A., Schulerud, H., Kavli, T.: Automatic segmentation of overlapping fish using shape priors. presented at the Scandinavian Conference on Image Analysis (2007)

[12] Chuang, M.-C., Hwang, J.-N., Williams, K., Towler, R.: Automatic fish segmentation via double local thresholding for trawl-based underwater camera systems. presented at the IEEE International Conference on Image Processing) (ICIP) (2011)

[13] Spampinato, C., Chen-Burger, Y.-H., Nadarajan, G., Fisher, R.: Detecting, Tracking and Counting Fish in Low Quality Unconstrained Underwater Videos. presented at the Int. Conf. on Computer Vision Theory and Applications (VISAPP) (2008)

[14] Lines, J.A., Tillett, R.D., Ross, L.G., Chan, D., Hockaday, S., McFarlane, N.J.B.: An automatic image-based system for estimating the mass of free-swimming fish. Computers and Electronics in Agriculture 31, 151–168 (2001)

[15] Rodriguez, A., Bermudez, M., Rabuñal, J.R., Puertas, J., Dorado, J., Balairon, L.: Optical Fish Trajectory Measurement in Fishways through Computer Vision and Artificial Neural Networks. Journal of Computing in Civil Engineering 25, 291–301 (2011)

[16] Yilmaz, A., Javed, O., Shah, M.: Object Tracking: A Survey. ACM Computing Surveys 38 (2006)

[17] Verikas, A., Malmqvist, K., Bergman, L.: Color image segmentation by modular neural networks. Pattern Recognition Letters 18, 173–185 (1997)

[18] Dong, G., Xie, M.: Color clustering and learning for image segmentation based on neural networks. IEEE Transactions on Neural Networks 16, 925–936 (2005)

[19] Egmont-Petersen, M., Ridder, D., Handels, H.: Image processing with neural networks-a review. Pattern Recognition 35, 2279–2301 (2002)

[20] Cristea, P.: Application of Neural Networks In Image Processing and Visualization. In: Amicis, R., Stojanovic, R., Conti, G. (eds.) GeoSpatial Visual Analytics, pp. 59–71. Springer, Netherlands (2009)

Clustering of Gene Expression Profiles Applied to Marine Research

Vanessa Aguiar-Pulido[1,*], Victoria Suárez-Ulloa[2,*], Daniel Rivero[1],
José M. Eirín-López[2], and Julián Dorado[1]

[1] Artificial Neural Networks and Adaptive Systems Laboratory (RNASA-IMEDIR),
Information and Communication Technologies Department, Faculty of Informatics,
University of A Coruña, Campus de Elviña, 15071 A Coruña, Spain
[2] Chromatin Structure and Evolution (CHROMEVOL) Group,
Department of Biological Sciences, Florida International University,
33181 North Miami, FL, USA
{vaguiar,v.ulloa,drivero,jeirin,julian}@udc.es

Abstract. This work presents the results of applying two clustering techniques to gene expression data from the mussel *Mytilus galloprovincialis*. The objective of the study presented in this paper was to cluster the different genes involved in the experiment, in order to find those most closely related based on their expression patterns. A self-organising map (SOM) and the k-means algorithm were used, partitioning the input data into nine clusters. The resulting clusters were then analysed using Gene Ontology (GO) data, obtaining results that suggest that SOM clusters could be more homogeneous than those obtained by the k-means technique.

Keywords: clustering, microarray, neural networks, data mining, bioinformatics, gene ontology.

1 Introduction

Gene expression can be defined as the process by which information from a gene is used in the synthesis of a functional gene product, which is often a protein. Measuring this activity or expression for thousands of genes at once enables creating a global picture of cellular function, which is known as gene expression profiling.

Microarrays [1, 2] are tools widely used to analyse gene expression profiles of a large number of genes simultaneously. This method is an approach to the quantitative analysis of the proteins being produced under given environmental and physiological circumstances, assuming that each gene would produce one single type of protein, which is not absolutely accurate but is widely accepted as an approximation [3].

The application of clustering techniques to this kind of data allows identifying non-obvious relationships between genes such as co-expression phenomena [4, 5]. This approach represents a valuable contribution to marine research since molecular data

---

* The first two authors contributed equally to this paper.

I. Rojas, G. Joya, and J. Cabestany (Eds.): IWANN 2013, Part I, LNCS 7902, pp. 453–462, 2013.

remains scarce for this kind of organisms, despite their environmental relevance, especially regarding seawater pollution monitoring.

The mussel *Mytilus galloprovincialis* is considered an excellent sentinel organism in coastal environmental control given its sessile condition, ubiquity and extremely high seawater filtering rate [6-8]. Although much effort is being placed on the sequencing of the genomes of mussels and other molluscs, there is still an important gap in the knowledge of these marine invertebrates [9].

The present work contributes to this goal by covering technical aspects of molecular data management which are relevant for marine biology research. Several helpful and widely known techniques are used here to get an insight into non obvious biological patterns, constituting the basis to many other more sophisticated methods of gene expression analysis that very often rely not just in the experimental quantitative data, but also in qualitative metadata by using Gene Ontology (GO) annotation statistics.

2   Materials and Methods

2.1   Mytilus Galloprovincialis Data

For this study, a dataset previously published by Banni et al. [10] was used. This data was obtained as a result of an expression profiling experiment by array and it represents temporal expression analysis of female digestive gland tissue from the mussel *M. galloprovincialis*. It contains 11 gene expression records from 295 genes. The data was retrieved from the Gene Expression Omnibus (GEO) database and it can be accessed at the following link: http://www.ncbi.nlm.nih.gov/projects/geo /query/acc.cgi?acc = GSE23052.

2.2   Clustering Techniques

In this study, the performance of two clustering techniques was compared: a *Self-Organising Map (SOM)* [11, 12] and the *k-means* [13] algorithm. These techniques have been applied over the time to solve a variety of problems in many different environments, obtaining good results [14-27]. Both techniques were implemented using Matlab and several configurations were tested to achieve the results shown in this paper.

2.2.1 Self-Organising Map (SOM)
A SOM is a type of Artificial Neural Network (ANN) which uses unsupervised learning to group instances taken as input, projecting these onto a regular, usually two-dimensional grid called map. In this technique, an instance will be mapped into the node which is nearer to it using some metric. Unlike other ANNs, a neighbourhood function is used in order to preserve the topological properties of the input space.

2.2.2 K-Means

K-means is a method designed for cluster analysis which partitions the instances taken as input into k clusters in such a way that each instance will belong to the cluster with the nearest mean.

2.3   Ontologies

An ontology [28] is a formal representation of knowledge, involving a set of concepts within a domain, and the relationships between pairs of concepts. It can be used to model a domain and support reasoning about entities. Ontologies can be graphically represented as graphs (nodes = concepts; edges = relationships) or trees (nodes and leaves = concepts; branches = relationships, including hierarchical relationships). Ontologies have been widely used, especially in fields related to biomedicine, gaining a lot of attention in the past few years [29-31].

2.3.1 Gene Ontology (GO)

Reported knowledge about genomic data and their products is wide and heterogeneous in nature, however, big efforts have been carried out by specialized consortiums in order to standardize this knowledge. This has been done by defining specific terms and relationships among them, so the gene attributes are then described in a more machine-like manner. This approach allows for the application of Knowledge Discovery in Databases techniques, very useful in functional analysis of massive genomic data. Furthermore, Evidence Codes are used to account for the reliability of the annotations, and weights are established for analysis automatization.

Gene Ontology covers three key aspects in gene description: the Biological Process it takes part in, Molecular Function of its corresponding gene product and Cellular compartment referring to the specific cellular location where it mainly displays its action. The GO terms belonging to any of these three categories have ancestor (lower level) - descendent (higher level) relationships between them, becoming more specific and informative the higher the term's level is. This ontology structure complies with the general build up of ontologies, with GO terms being represented as nodes and relationships as branches. A thumb rule in ontology, that also applies to GO, is that if a gene is annotated with a specific GO term, the correspondence with all its ancestor terms is automatically inferred. This has critical implications for the functional analysis of datasets that analysis tools, such as those embedded in the Blast2GO suite, have taken into account.

2.3.2 Blast2GO

Blast2GO [32-35] is a software suite designed for functional annotation of genomic sequences using GO terminology and for the analysis of such annotation data. In this paper, genes were annotated specifically with terms belonging to the Biological Process type, and the resulting clusters were analysed using the statistical tools provided by the Blast2GO software in order to obtain those terms that are more representative of each clustered gene set.

Blast2GO ranks the GO terms related to the sequences in each set based on scores. Scores are calculated out of the number of sequences annotated with a given term and the distance (number of intermediate nodes) from the GO term directly assigned to the sequence to that one that is being scored. This way, the fact that more general GO terms are more likely to get high scores as they add up all sequences annotated by descendent nodes, is compensated by the fact of getting penalized by the distance to the actual term reported as gene annotation. Therefore, the score is calculated according to the following formula:

$$score = \sum_{GOs} seq \times \alpha^{dist}$$

where *seq* is the number of different sequences annotated at a child GO term, *dist* the distance to the node of the child and α is a constant parameter set to default value 0.6.

3   Results and Discussion

Many tests were run until obtaining the best configuration parameters of the two techniques used in this paper. The same distance metric was used for both methods so that results could be fairly compared. The metric that was finally used was the Euclidean distance. In the case of the SOM technique, different architectures were tested in order to choose the one that obtained the best clusters, that is, an architecture involving 9 neurons. As for the k-means technique, several cluster numbers were tested and, finally, k=9 seemed to obtain the best partitions.

Banni et al. [10] also used the k-means algorithm for the computation of different gene expression trends, obtaining similar results to those presented here. The authors of this paper obtained ten clusters but concluded that two of those clusters could be merged into one.

Fig. 1 and Fig. 2 show the differences between the performances of the proposed models. The horizontal axis represents time (in months) and the vertical axis represents the expression level. Although there are resemblances in the results obtained by both techniques, some genes are clustered into different partitions. Observing these figures, we can study the behaviour of the different genes in terms of gene expression over the time for each cluster and for each technique.

Analysing these graphics and the genes contained in each cluster, we found the following:

- the k-means technique divided cluster 1 obtained by the SOM technique into two different clusters (clusters 8 and 9)
- most of the elements contained in cluster 2 obtained by the SOM, were part of k-means' cluster 4
- clusters 3 and 4 of SOM corresponded to cluster 6 of k-means
- clusters 4 and 7 of SOM corresponded to cluster 2 of k-means
- cluster 6 of SOM and cluster 5 of k-means were very similar, the same happens for cluster 8 of SOM and cluster 3 of k-means.

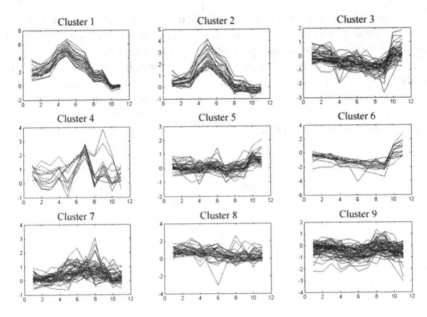

Fig. 1. SOM clustering

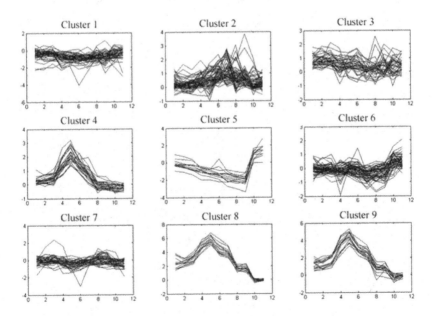

Fig. 2. K-means clustering

Results are mainly consistent between both techniques regarding the biological sense of the obtained clusters. Table 1 and Fig. 3 show the most relevant results of the GO term analysis for each cluster, presenting the levels they belong to as a way of measuring how specific the terms are and how informative they get. It is worth highlighting that in the case of k-means clusters, there are two gene sets that have not obtained any representative GO term by failing to achieve the minimum score threshold set by default in Blast2GO analysis tool, while this happens for only one of the clusters obtained by the SOM technique.

Table 1. TOP 3 Highest scored GO-terms level

Cluster #	K-means	SOM
1	10, 9, 8	N/A
2	6,5,2	6,5,1
3	5,4,3	10,8,8
4	6,5,1	7,6,5
5	6,5,4	8,3,2
6	2,3,8	6,5,4
7	6,5,4	6,5,2
8	N/A	4,5,3
9	N/A	6,3,1

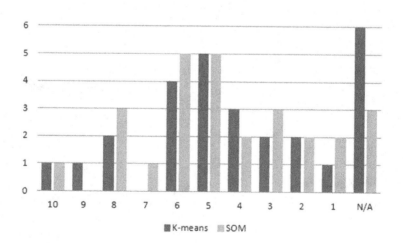

Fig. 3. GO-term level occurrence

Cases such as the SOM cluster 2 and k-means cluster 4, that are similar in terms of included genes and profile shape, obtain very similar GO terms statistics, having the same GO term "translation" (GO:0006412) as the most representative of the group. As an example of clusters displaying different ontological representation, we find the case of SOM clusters 6 and 3, and k-means clusters 5 and 6, all alike in the profile shapes. While SOM cluster 6 throws the same ontological results as k-means cluster number 5, there exist significant differences between SOM cluster 3 and k-means cluster 6. For the first one (SOM cluster 3), the most representative GO terms are terms of high specificity, belonging to the GO levels 10-8 (being the higher the GO level, the more specific the term), while for the latter (k-means cluster 6), the GO terms that obtained the highest representation belong to the levels 2-3, meaning that these are rather general terms not so informative of the biological meaning of this group of genes. This can be understood as being SOM clusters more homogeneous than those obtained by the k-means technique, since more specific GO terms are obtained meaning that more sequences are directly annotated by terms with a closer relationship. However, ontological analysis has the drawback of low statistical significance due to the general lack of functional information for these sequences.

4   Conclusions and Future Work

This work presents a study of gene expression analysis using data from a mussel species obtained from a year-long experiment. Two techniques, a self-organising map (SOM) and the k-means algorithm, were used in order to partition 295 genes with 11 gene expression records over the time into nine clusters. These clusters were then annotated and analysed, obtaining results that suggest that SOM clusters could be more homogeneous than those obtained by the k-means technique.

As future work, we plan to apply more techniques to this type of data, such as biclustering.

Acknowledgements. Vanessa Aguiar-Pulido acknowledges the funding support for a research position by the "Plan I2C" Program from Xunta de Galicia (Spain), partially funded by the European Social Fund (ESF). José M. Eirín-López was awarded with grants by the Spanish Ministry of Economy and Competitivity (CGL2011-24812 & Ramon y Cajal Subprogramme) and by the Xunta de Galicia (10-PXIB-103-077-PR). Finally, the following projects also supported the work presented: "Galician Network for Colorectal Cancer Research" (REGICC, Ref. 2009/58) from the General Directorate of Research, Development and Innovation of Xunta de Galicia, "Ibero-American Network of the Nano-Bio-Info-Cogno Convergent Technologies", Ibero-NBIC Network (209RT-0366) funded by CYTED (Spain), grant Ref. PIO52048, RD07/0067/0005 funded by the Carlos III Health Institute.

References

1. Schena, M., Shalon, D., Davis, R.W., Brown, P.O.: Quantitative monitoring of gene expression patterns with a complementary DNA microarray. Science 270, 467–470 (1995)
2. Lashkari, D.A., DeRisi, J.L., McCusker, J.H., Namath, A.F., Gentile, C., Hwang, S.Y., Brown, P.O., Davis, R.W.: Yeast microarrays for genome wide parallel genetic and gene expression analysis. Proc. Natl. Acad. Sci. U. S. A. 94, 13057–13062 (1997)
3. Gibson, G.: Microarray Analysis. PLoS Biol. 1, e15 (2003)
4. Lee, H.K., Hsu, A.K., Sajdak, J., Qin, J., Pavlidis, P.: Coexpression analysis of human genes across many microarray data sets. Genome Res. 14, 1085–1094 (2004)
5. Chou, J.W., Zhou, T., Kaufmann, W.K., Paules, R.S., Bushel, P.R.: Extracting gene expression patterns and identifying co-expressed genes from microarray data reveals biologically responsive processes. BMC Bioinformatics 8, 427 (2007)
6. Kock, W.C.: Monitoring bio-available marine contaminants with mussels (Mytilus edulis L) in the Netherlands. Environ. Monit. Assess. 7, 209–220 (1986)
7. Laffon, B., Rabade, T., Pasaro, E., Mendez, J.: Monitoring of the impact of Prestige oil spill on Mytilus galloprovincialis from Galician coast. Environ. Int. 32, 342–348 (2006)
8. Scarpato, A., Romanelli, G., Galgani, F., Andral, B., Amici, M., Giordano, P., Caixach, J., Calvo, M., Campillo, J.A., Albadalejo, J.B., Cento, A., BenBrahim, S., Sammari, C., Deudero, S., Boulahdid, M., Giovanardi, F.: Western Mediterranean coastal waters–monitoring PCBs and pesticides accumulation in Mytilus galloprovincialis by active mussel watching: the Mytilos project. J. Environ. Monit. 12, 924–935 (2010)
9. Zhang, G., Fang, X., Guo, X., Li, L., Luo, R., Xu, F., Yang, P., Zhang, L., Wang, X., Qi, H., Xiong, Z., Que, H., Xie, Y., Holland, P.W., Paps, J., Zhu, Y., Wu, F., Chen, Y., Wang, J., Peng, C., Meng, J., Yang, L., Liu, J., Wen, B., Zhang, N., Huang, Z., Zhu, Q., Feng, Y., Mount, A., Hedgecock, D., Xu, Z., Liu, Y., Domazet-Loso, T., Du, Y., Sun, X., Zhang, S., Liu, B., Cheng, P., Jiang, X., Li, J., Fan, D., Wang, W., Fu, W., Wang, T., Wang, B., Zhang, J., Peng, Z., Li, Y., Li, N., Chen, M., He, Y., Tan, F., Song, X., Zheng, Q., Huang, R., Yang, H., Du, X., Chen, L., Yang, M., Gaffney, P.M., Wang, S., Luo, L., She, Z., Ming, Y., Huang, W., Huang, B., Zhang, Y., Qu, T., Ni, P., Miao, G., Wang, Q., Steinberg, C.E., Wang, H., Qian, L., Liu, X., Yin, Y.: The oyster genome reveals stress adaptation and complexity of shell formation. Nature 490, 49–54 (2012)
10. Banni, M., Negri, A., Mignone, F., Boussetta, H., Viarengo, A., Dondero, F.: Gene expression rhythms in the mussel Mytilus galloprovincialis (Lam.) across an annual cycle. PloS One 6, e18904 (2011)
11. Kohonen, T.: Self-organized formation of topologically correct feature maps. Biol. Cybern. 43, 59–69 (1982)
12. Kohonen, T.: Essentials of the self-organizing map. Neural Netw. 37, 52–65 (2013)
13. MacQueen, J.: Some methods for classification and analysis of multivariate observations. In: Proc. Fifth Berkeley Symp. on Math. Statist. and Prob., pp. 281–297. Univ. of Calif. Press (1965)
14. Yan, A., Hu, X., Wang, K., Sun, J.: Discriminating of ATP competitive Src kinase inhibitors and decoys using self-organizing map and support vector machine. Mol Divers 17, 75–83 (2013)
15. Wang, L., Wang, M., Yan, A., Dai, B.: Using self-organizing map (SOM) and support vector machine (SVM) for classification of selectivity of ACAT inhibitors. Mol Divers 17, 85–96 (2013)

16. Zhu, D., Huang, H., Yang, S.X.: Dynamic Task Assignment and Path Planning of Multi-AUV System Based on an Improved Self-Organizing Map and Velocity Synthesis Method in Three-Dimensional Underwater Workspace. IEEE Trans. Syst. Man Cybern. B Cybern. (2012)

17. Piastra, M.: Self-organizing adaptive map: Autonomous learning of curves and surfaces from point samples. Neural Netw. (2012)

18. Marique, T., Allard, O., Spanoghe, M.: Use of Self-Organizing Map to Analyze Images of Fungi Colonies Grown from Triticum aestivum Seeds Disinfected by Ozone Treatment. Int. J. Microbiol., 865175 (2012)

19. Bae, M.J., Kim, J.S., Park, Y.S.: Evaluation of changes in effluent quality from industrial complexes on the Korean nationwide scale using a self-organizing map. Int. J. Environ. Res. Public Health 9, 1182–1200 (2012)

20. Wiggins, J.L., Peltier, S.J., Ashinoff, S., Weng, S.J., Carrasco, M., Welsh, R.C., Lord, C., Monk, C.S.: Using a self-organizing map algorithm to detect age-related changes in functional connectivity during rest in autism spectrum disorders. Brain Res. 1380, 187–197 (2011)

21. Yang, Z., Wu, Z., Yin, Z., Quan, T., Sun, H.: Hybrid Radar Emitter Recognition Based on Rough k-Means Classifier and Relevance Vector Machine. Sensors (Basel) 13, 848–864 (2013)

22. Sun, G., Hakozaki, Y., Abe, S., Vinh, N.Q., Matsui, T.: A novel infection screening method using a neural network and k-means clustering algorithm which can be applied for screening of unknown or unexpected infectious diseases. J. Infect. 65, 591–592 (2012)

23. Armstrong, J.J., Zhu, M., Hirdes, J.P., Stolee, P.: K-means cluster analysis of rehabilitation service users in the Home Health Care System of Ontario: examining the heterogeneity of a complex geriatric population. Arch. Phys. Med. Rehabil. 93, 2198–2205 (2012)

24. Stricker, M.D., Onland-Moret, N.C., Boer, J.M., van der Schouw, Y.T., Verschuren, W.M., May, A.M., Peeters, P.H., Beulens, J.W.: Dietary patterns derived from principal component- and k-means cluster analysis: Long-term association with coronary heart disease and stroke. Nutr. Metab. Cardiovasc. Dis. (2012)

25. Konicek, A.R., Lefman, J., Szakal, C.: Automated correlation and classification of secondary ion mass spectrometry images using a k-means cluster method. Analyst 137, 3479–3487 (2012)

26. Chang, N.B., Wimberly, B., Xuan, Z.: Identification of spatiotemporal nutrient patterns in a coastal bay via an integrated k-means clustering and gravity model. J. Environ. Monit. 14, 992–1005 (2012)

27. Zhang, S., Jin, W., Huang, Y., Su, W., Yang, J., Feng, Z.: Profiling a Caenorhabditis elegans behavioral parametric dataset with a supervised K-means clustering algorithm identifies genetic networks regulating locomotion. J. Neurosci. Methods 197, 315–323 (2011)

28. Gruber, T.R.: A translation approach to portable ontology specifications. Knowl. Acquis. 5, 199–220 (1993)

29. Warita, K., Mitsuhashi, T., Tabuchi, Y., Ohta, K., Suzuki, S., Hoshi, N., Miki, T., Takeuchi, Y.: Microarray and gene ontology analyses reveal downregulation of DNA repair and apoptotic pathways in diethylstilbestrol-exposed testicular Leydig cells. J. Toxicol. Sci. 37, 287–295 (2012)

30. Schaid, D.J., Sinnwell, J.P., Jenkins, G.D., McDonnell, S.K., Ingle, J.N., Kubo, M., Goss, P.E., Costantino, J.P., Wickerham, D.L., Weinshilboum, R.M.: Using the gene ontology to scan multilevel gene sets for associations in genome wide association studies. Genet. Epidemiol. 36, 3–16 (2012)

31. Ma, N., Zhang, Z.G.: Evaluation of clustering algorithms for gene expression data using gene ontology annotations. Chin. Med. J (Engl.) 125, 3048–3052 (2012)

32. Conesa, A., Gotz, S., Garcia-Gomez, J.M., Terol, J., Talon, M., Robles, M.: Blast2GO: a universal tool for annotation, visualization and analysis in functional genomics research. Bioinformatics 21, 3674–3676 (2005)

33. Conesa, A., Gotz, S.: Blast2GO: A comprehensive suite for functional analysis in plant genomics. Int. J. Plant Genomics, 619832 (2008)

34. Gotz, S., Garcia-Gomez, J.M., Terol, J., Williams, T.D., Nagaraj, S.H., Nueda, M.J., Robles, M., Talon, M., Dopazo, J., Conesa, A.: High-throughput functional annotation and data mining with the Blast2GO suite. Nucleic Acids Res. 36, 3420–3435 (2008)

35. Gotz, S., Arnold, R., Sebastian-Leon, P., Martin-Rodriguez, S., Tischler, P., Jehl, M.A., Dopazo, J., Rattei, T., Conesa, A.: B2G-FAR, a species-centered GO annotation repository. Bioinformatics 27, 919–924 (2011)

Genetic Programming to Improvement FIB Model
Bond and Anchorage of Reinforcing Steel in Structural Concrete

Juan Luis Pérez[1], Ismael Vieito[2], Juan Rabuñal[3], and Fernando Martínez-Abella[2]

[1] School of Building Engineering and Technical Architecture, University of A Coruña, Spain
[2] Department of Construction Technology, University of A Coruña, Spain
[3] Department of Information and Communication Technologies, University of A Coruña, Spain
{jlperez,camivr01,juanra,fmartinez}@udc.es

Abstract. Starting from the FIB database, this work is aimed to analyze the current equations which predict the main datum that can be provided by bond tests: the ultimate bar stress when the failure is reached. Furthermore, Genetic Programming (GP) techniques are also applied in order to enhance the expression of the FIB, which achieves the best adjustment so far, giving rise to the new Model Code 2010. The final result shown is a highly predictive equation. The results are compared with those included in the Model Code and it is showed the influence of the main variables on the phenomenon (concrete strength, yield strength of steel, concrete cover, transverse reinforcement and diameter of the bar).

1   Introduction

Since the dawn of the 20th century, when Abrams's tests were performed, the bond between concrete and steel led to numerous scientific papers, complex laboratory tests and many approaches of structural codes. There are very few expressions that are so different when comparing the various rules, such as those aimed at predicting the anchorage length of reinforcing bars in structural concrete. Two main lines were created starting from the works carried out by Orangun, Jirsa & Breen [1], precursors of the ACI code equation, and the studies performed by Tepfers [2], which inspired the guidelines of the Model Code, leading to the Eurocode. The tests carried out in Spain also were of great importance, as they gave rise to a specific formulation, extremely conservative for large-diameter reinforcement.

Despite the varied approaches, the three lines have a common nexus: the proposals are developed from the experimental evidence. From a basic expression of bond stress, dependent on a main variable are incorporated as a multiplicative factor the effect of other variables

When a pull-out test is performed, a state of radial tension is generated around the bar that can cause damage to the surrounding concrete. The damage can be mitigated by the placement of transverse reinforcement, and having adequate cover bar. Figure 1 shows graphically the phenomenon of a bar anchorage [3].

I. Rojas, G. Joya, and J. Cabestany (Eds.): IWANN 2013, Part I, LNCS 7902, pp. 463–470, 2013.
© Springer-Verlag Berlin Heidelberg 2013

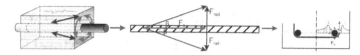

Fig. 1. Phenomenon of a bar anchorage

Various factors affect the bond capacity. They are generally associated with different origins: the materials used, the geometric conditions and, finally, the applied loads.

The design equations for the anchorage length determination are based in the basic straight anchorage length necessary to reach the break of the bar. On this equation, of experimental base, are added the effects of other variables, also obtained experimentally [1],[4].

2   FIB Model

The workgroup TG4.5 of the *Fédération Internationale du Beton* (FIB) [5] has been working for a long time in the analysis of the anchorage and the bond characteristics of reinforcement steel bars. Based in the works of Canbay and Frosch [6], the group has proposed two equations: the version 2006 (1) and the current version employed in the development of the Model Code 2010 (2), that provides the work stress that can be reached on an anchored bar. This equation depends on the parameters seen until now and is protected by a contrast with a strong experimental database.

The variables described in both formulas correspond to the bond stresses (σ_{su}), compressive strength in cylindrical specimen (f_c), diameter of the bar (d_b), length of anchored bar (l_s), maximum and minimum coatings of the bar (c_{min}, c_{max}) and the contribution of the transverse reinforcement (K_{tr}).

$$\sigma_{su} = 25 * f_c^{0.25} * \left(\frac{20}{d_b}\right)^{0.2} * \left(\frac{l_s}{d_b}\right)^{0.55} * \left(\frac{c_{min}}{d_b}\right)^{0.33} * \left(\frac{c_{max}}{c_{min}}\right)^{0.1} * (1 + 10 * K_{tr}) \tag{1}$$

$$\sigma_{su} = 54 * \left(\frac{f_c}{20}\right)^{0.25} * \left(\frac{20}{d_b}\right)^{0.2} * \left(\frac{l_s}{d_b}\right)^{0.55} * \left[\left(\frac{c_{min}}{d_b}\right)^{0.33} * \left(\frac{c_{max}}{c_{min}}\right)^{0.1} + 8K_{tr}\right] \tag{2}$$

2.1   Dataset

The database currently contains data (variables and results measured or calculated) corresponding to 813 trials. As will be applied GP techniques for analysis, so that the database range is consistent and frequencies of each of the data must be analyzed. Considering the frequency histogram data, several filters are applied and are accepted those recommended by the FIB [5]. One of the most important is related to the output data σ_{su}, the stress reached by the bar during the test. Since any bar limits its maximum stress f_y, the maximum value that can take the relationship σ_{su}/f_y is 1.05. In other variables their relative values are also limited, for example those related to the concrete cover. Thus, filters over c_{min}/c_{max} and c_{min}/d_b are applied. Table 1 shows the parameters used for filtering and the filter finally applied.

Table 1. Filters applied to the dataset

Variables	Filter
f_c	>15 and <115 MPa
d_b	< 37 mm
l_s	< 2100 mm
C_b	<136 mm
c_{min}/c_{max}	≥ 1.0 and ≤ 5.0
c_{min}/d_b	≥ 0.5 and ≤ 3.0
σ_{su}/f_y	≤ 1.05

After applying the filter, the BD is reduced to a total of 628 trials, of which 77.5% (487 trials), by random selection, are used for training, dedicating the remaining 22.5% (141 trials) to check. Table 2 shows the final range of the data in each of the subsets set (training and verification).

On the BD filtered FIB expressions produce results whose accuracy is presented in the following sections.

Table 2. Distribution of data in subsets defined over the BD filtered

	Training (#487)		Test (#141)	
	Mín	Max	Mín	Max
l_s (mm)	50	2095	120	2032
d_b (mm)	8	35.81	10	35.81
c_{min} (mm)	8	76	8	76
c_{max} (mm)	20	140	20	140
f_c (MPa)	15	114	20	110
K_{tr}	0	0.114	0	0.106
σ_{su}	126	788	182	814

3 Method

The method followed is oriented to improving the FIB equations developed for predicting the stress of bar anchored. The method used follows the same guidelines referred to in the paper developed by Pérez et al [7]. In summary, the method is based on GP techniques, imposing some restrictions based on knowledge of the problem provided by an expert. Symbolic regression data is one of the capabilities provided by the GP.

Having a data set (input-output), the GP is able to relate these data algebraically by an equation. Its complexity may vary, and dimensional integrity is not guaranteed.

This technique, applied in many cases in civil engineering, is one of those followed by Ashour al [8], for example, to predict shear strength in concrete beams. Naturally, the form of the equations obtained is very different from the ones in the common codes. The presented method improves the terms accepted by the scientific community, getting a better fit when the results are applied to a database.

It starts from the expression FIB-2006, because it shows better results over the database. The search expression will determine the bar stress predicted (spred) to be compared with the real stress test (σ_{test}). Firstly, it is necessary to define how individuals will be evaluated in the fitness function (equation 4). In this equation, σ_{test} is the bar stress at failure, α is the parsimony coefficient, s_i is the number of nodes in the expression and n is the number of cases of the database. It should set the parameters p_i and l_{bias} defined in equation 3.

After performing several tests, was adopted l_{bias} = 1.0, and equation 4 shows the value of p_i (DP). This equation is based on the use of the technique of "demerit points".

$$fitness(i) = \frac{\sum_{i=0}^{n} p_i * \left| l_{bias} - \frac{\sigma_{test}}{\sigma_{pred}} \right|}{n} + \alpha * s_i \tag{3}$$

$$DP = \sum_{i=1}^{n} p_i, \quad p_i = \begin{cases} 16, & \frac{\sigma_{test}}{\sigma_{pred}} < 0.5 \\ 8, & 0.5 \leq \frac{\sigma_{test}}{\sigma_{pred}} < 0.67 \\ 4, & 0.67 \leq \frac{\sigma_{test}}{\sigma_{pred}} < 0.85 \\ 2, & 0.85 \leq \frac{\sigma_{test}}{\sigma_{pred}} < 1.0 \\ 0, & 1.0 \leq \frac{\sigma_{test}}{\sigma_{pred}} < 1.3 \\ 3, & 1.3 \leq \frac{\sigma_{test}}{\sigma_{pred}} < 2 \\ 6, & \frac{\sigma_{test}}{\sigma_{pred}} \geq 2 \end{cases} \tag{4}$$

The technique was adapted for "oriented" searches were possible, with different purposes.

The orientation was introduced through impositions or restrictions, which include:

— restriction on the type of functions that link the variables
— preferred selection of individuals with the highest ratios $\sigma_{test}/\sigma_{pred}$. From the structural point of view, is much more appropriate this option for safety reasons

The method used starts with the establishment of a "framework" from wich genetic programming will make the evolutive process, taking into account the restrictions and impositions.

The framework is defined directly from the equation FIB-2006, which is divided into subexpressions. Also, each subexpression is written indicating which factor (branch) may change in the search process. The working lines can find:

— The optimization of the numerical coefficients of the equation. The branches will be Real values
— The introduction of a new subexpression. This can be a Real number or a function (new branch) linked to a variable

As mentioned, in this type of model is very important that the predicted stress is equal to or greater than the actual value.

In general, if an individual differs from the real value is penalized during training. From a mathematical point of view, S values equal to 0.5 or 1.5 should be penalized equally. To take into account the structural safety, the individual 0.5 should be penalized more than the individual 1.5, as it causes structural insecurity (collapse).

This is achieved through the technique of demerit points, whereby the error of the expression is weighted according to the ranges defined by Pérez [9]. The fitness function (3) shows how the p_i factor weights the prediction error, according to the intervals and values of the equation (4).

The method used starts with the establishment of a "framework" about whom genetic programming will make the evolutive process, taking into account the restrictions and impositions. Such "framework" is based on the 2006 FIB formulation, about which it will be introduced new variables or its coefficients will be modified.

In the searching process it has been proposed three basic equations (5)(6)(7). Each *branch* is designated as B_i. Table 3 shows the default settings implemented, based on the initial tests. The input data have not been standardized, so expressions can be used directly.

Table 3. Parameters used

Parameter	Default value	Other values
Population size	1000	
Crossover rate	80%	
No-terminal selection rate:	90%	
Mutation rate:	20%	
Algorithms:	Selection: Tournament Initialization: Ramped Half & Half Mutation& Crossover: Subtree	
Elitist strategy	Yes	
Parsimony	0	$0.0001, 1*10^{-6}$ ó $1*10^{-9}$
Initial tree depth	4	5
Maximum tree depth	6	7
Maximum mutation depth	4	5

$$\sigma_{su} = B_1^{B_2} * fc^{B_3} * \left(\frac{l_s}{d_b}\right)^{B_4} * \left(\frac{c_{min}}{d_b}\right)^{B_5} * \left(\frac{c_{max}}{c_{min}}\right)^{B_6} * (1 + B_7 * K_{tr})^{B_8} \tag{5}$$

$$\sigma_{su} = l s^{B_1} * B_2 * (1 + B_3 * K_{tr})^{B_4} \tag{6}$$

$$\sigma_{su} = B_1 * fc^{B_2} * \left(\frac{B_3}{d_b}\right)^{B_4} * \left(\frac{l_s}{d_b}\right)^{B_5} * \left(\frac{c_{min}}{d_b}\right)^{B_6} * \left(\frac{c_{max}}{c_{min}}\right)^{B_7} * (1 + B_8 * Ktr)^{B_9} \tag{7}$$

By default, addition, subtraction, multiplication and protected division were chosen as operators or non-terminal nodes. Variables from the data set (l_s, d_b, c_{min}, c_{max} and f_c), and integers in the range [-10, 10] were adopted as terminal nodes.

Constraints over the equations are showed in table 4. Equation (5) have three types of constraints ("A", "B" and "C"), the constraint "D" is imposed to equation (6) and finally the constraint "E" is imposed to equation (7).

Table 4. Constraints

Eq.	Const.	B_1	B_2	B_3	B_4	B_5	B_6	B_7	B_8	B_9
(5)	A	$l_s\, d_b\, c_{min}\, c_{max}\, f_c$	Const. 2 dec.	Const. 2 dec.	Const. 2 dec.	Const. 2 dec.	Const. 2 dec.	Const. 2 dec.	Const. 2 dec.	-
(5)	B	$d_b\, c_{min}\, c_{max}\, f_c$	Const. 2 dec.	Const. 2 dec.	Const. 2 dec.	Const. 2 dec.	Const. 2 dec.	Const. 2 dec.	Const. 2 dec.	-
(5)	C	$d_b\, c_{min}\, c_{max}\, f_c$	Const. 2 dec.	Const. 2 dec	0.5	Const. 2 dec	Const. Ent.	Const. 2 dec	Const. 2 dec	-
(6)	D	Const. 2 dec.	$d_b\, c_{min}\, c_{max}\, f_c$	Const. 2 dec	Const. 2 dec	-	-	-	-	-
(7)	E	Const. 2 dec.	Const. 2 dec.	Const. 2 dec.	Const. 2 dec.	Const. 2 dec.	Const. 2 dec.	Const. 2 dec.	Const. 2 dec.	Const. 2 dec.

4   Results

In total, more than 4,500 executions were carried out. The results are analyzed essentially through the following indicators: COV (variation coefficient), $\sigma_{test}/\sigma_{pred}$, R^2 (square root of Pearson product-moment correlation coefficient), MSE (mean square root error), ME (mean error), and finally demerit points calculated according to equation (4).

According to the best results, a select group of equations was chosen. If the denominator could be negative, expressions containing function "protected division" were rejected. Also too complex equations were also discarded.

PG_9RSC4 (8), PG_8v2R5 (9), PG_7v3F2 (10), PGcc6 (11) were more accurate equations. Since not provide substantial improvements, these equations do not contain the derivatives of the classic GP. The results are shown in Table 5. The significant improvement achieved is evident by comparing the results of the equations FIB.

Table 5. Results

	FIB (2006)	FIB CM2010	PG_7v3F2	PG_8v2R5	PG_9RSC4	PGcc6
COV	15.683	16.010	14,404	14,900	15,239	15,442
$\sigma_{test}/\sigma_{pred}$	0.9712	0.9748	1,0254	1,0082	0.9994	1,0079
Max ($\sigma_{test}/\sigma_{pred}$)	1.5091	1.5367	1,5228	1,5119	1,4815	1,5684
Min ($\sigma_{test}/\sigma_{pred}$)	0.4990	0.4885	0,5702	0,5137	0,5016	0,4873
R^2	0.7095	0.6938	0,7545	0,7400	0,7271	0,7193
MSE	4215	4343	3551	3608	3740	3847
ME	51.26	52.19	46.08	47.24	48.10	48.40
DP	2642	2646	2398	2492	2508	2486

Some of the expressions stand out by different appearances. PG_9RSC4 is a simple improvement of the FIB equation, achieved with better adjusts of the exponents and constants. To clear the value of the length, it is necessary to impose conditions to the search, proposing a first free function, not dependent on the length, and a adjust coefficient for the rest of parameters: the equation PG_8v2R5 arises this way. The marked tendency that exhibits the exponent (l_s/d_b) to the value 0.5, induces a new group of executions in which this constant is fixed. With this procedure, the PG_7v3F2 equation is obtained, achieving a very noticeable distribution.

$$PG9RSC4: \sigma_{su} = 27 * fc^{0.27} * \left(\frac{29}{d_b}\right)^{0.23} * \left(\frac{ls}{d_b}\right)^{0.47} * \left(\frac{c_{min}}{d_b}\right)^{0.16} * \left(\frac{c_{max}}{c_{min}}\right)^{0.13}$$
$$* (1 + 286 * Ktr)^{0.11}$$
(8)

$$PG8v2R5: \sigma_{su} = \left(\frac{c_{max}}{36} - \frac{fc}{36} + \frac{150}{c_{min}} + 19.294\right) * fc^{0.29} * \left(\frac{ls}{d_b}\right)^{0.49} * \left(\frac{c_{min}}{d_b}\right)^{0.29}$$
$$* \left(\frac{c_{max}}{c_{min}}\right)^{0.04} * (1 + 239.29 * Ktr)^{0.12}$$
(9)

$$PG7v3F2: \sigma_{su} = \left(\frac{48}{c_{min}} - \frac{fc - 44}{\frac{c_{min}}{48} + 20} + 9\right) * fc^{0.5} * \left(\frac{ls}{d_b}\right)^{0.5} * \left(\frac{c_{min}}{d_b}\right)^{0.34} * \left(\frac{c_{max}}{c_{min}}\right)^{0.01}$$
$$* (1 + 173 * Ktr)^{0.14}$$
(10)

$$PGcc6: \sigma_{su} = ls^{0.5} * \left(\left(\frac{75}{d_b} + \frac{19}{7}\right) * \left(\frac{fc}{5 * d_b} + \frac{c_{max} * fc}{360 * d_b} + 2\right)\right) * (1 + 165 * Ktr)^{0.15}$$
(11)

In the last remarkable groups, it is allowed the apparition of a free function (without l_s) that multiplies l_s with constant exponent and the classical term of transversal reinforcement contribution, improved with constants. This is the PGcc6 expression, which exhibit a strong concentration around the unit.

Next, the stresses that can be developed for some specific variables are compared in two of the equations found against the FIB deduced expressions. It can be observed the similarity of the approach, even for equations that are not born from the structure of the FIB.

As a result of the previously exposed, it can be recommended to adopt the expression PG_7v3F2 as a good equation to get the bond behavior of the passive reinforcement in a concrete element.

5   Conclusions

FIB equation to determine rebar tension stress was improved with the application of heuristic techniques.

In the applied method, structural safety was taken into account, through the weighting provided by demerit points

As a final conclusion and summary it should be noted that it has managed to implement a novel method based on genetic programming to extract knowledge from experimental data based on the experience. This experience is implemented through constraints that are induced in the algorithm.

Acknowledgements. This work was partially supported by the Spanish Ministry of Science and Innovation (Ref. BIA2010-21551) and grants from the Ministry of Economy and Industry (Consellería de Economía e Industria) of the Xunta de Galicia (Ref. 10MDS014CT, Ref. 10TMT042E, Ref. 10TMT118004PR and Ref. 10TMT034E).

References

1. Orangun, C.O., Jirsa, J.O., Breen, J.E.: A Reevaluation of Test Data on Development Length and Splices. ACI Journal 74(11), 114–122 (1977)
2. Tepfers, R.A.: Theory of bond applied to overlapped tensile reinforcement splices for deformed bars Ph.D. Thesis, Division of concrete structures, Chalmers University of Technology. Gothenburg, Sweden (1973)
3. Cairns, J.: Model for strength of lapped joints and anchorages. Paper presented at the meeting of the TG group 4.5 of the International Federation for Structural Concrete, Stuttgart, Germany (2006)
4. Abrams, D.A.: Tests of Bond Between Concrete and Steel. University of Illinois Bulletin N° 71. University of Illinois, Urbana (1913)
5. FIB Task Group 4.5 "Bond models" (s.f) (2007), http://fibtg45.dii.unile.it/ (January 25, 2013)
6. Canbay, E., Frosch, R.: Bond strength of lap-spliced bars. ACI Structural Journal 102(4), 605–614 (2005)
7. Pérez, J.L., Cladera, A., Rabuñal, J.R., Martínez-Abella, F.: Optimization of existing equations using a new genetic programming algorithm: Application to the shear strength of reinforced concrete beams. Advances in Engineering Software 50(1), 82–96 (2012)
8. Ashour, A.F., Alvarez, L.F.: Toropov VV.: Empirical modeling of shear strength of RC deep beams by genetic programming. Computers & Structures 81(5), 331–338 (2003)
9. Pérez, J.L.: Metodología para orientar procesos de extracción de conocimiento basados en Computación Evolutiva. Aplicación al desarrollo de modelos y formulaciones en el ámbito del hormigón estructural, Ph.D. Thesis, Department of Information and Communication Technologies, University of A Coruña (2010)

Rainfall Forecasting Based on Ensemble Empirical Mode Decomposition and Neural Networks

Juan Beltrán-Castro[1], Juliana Valencia-Aguirre[1], Mauricio Orozco-Alzate[1], Germán Castellanos-Domínguez[1], and Carlos M. Travieso-González[2]

[1] Universidad Nacional de Colombia - Sede Manizales, Colombia - Signal Processing and Recognition Group - Km. 7, Vía al Magdalena, Campus La Nubia - Manizales, Colombia {jdbeltranc,jvalenciaag,morozcoa,cgcastellanosd}@unal.edu.co
[2] Universidad de Las Palmas de Gran Canaria, Technological Centre for Innovation in Communications (CeTIC), Campus Universitario de Tafira, s/n, Las Palmas de Gran Canaria, Spain ctravieso@dsc.ulpgc.es

Abstract. In this paper a methodology for rainfall forecasting is presented, using the principle of decomposition and ensemble. In the proposed framework, the employed decomposition technique is the Ensemble Empirical Mode Decomposition (EEMD), which divides the original data into a set of simple components. Each component is modeled with a Feed Forward Neural Network (FNN) as a forecasting tool. Finally, the individual forecasting results for all components are combined to obtain the prediction result of the input signal. Experiments were performed on a real-observed rainfall data, and the attained results were compared against a single FNN model for the raw data, showing an improvement on the system performance.

Keywords: Forecasting, Neural Networks, Ensemble Empirical Mode Decomposition, Rainfall.

1 Introduction

Rainfall forecasting is an interesting field of research with important applications, such as management of water resources, and the evaluation of drought and flooding events. In particular, the analysis of this variable is relevant in tropical countries like Colombia where the agriculture is an essential part of the economy. Moreover, unanticipated flash floods are very destructive and threaten human lives and properties. Then, among all weather happenings, rainfall plays an imperative part in human life [1]. However, obtaining accurate forecast results is a challenging task, since rainfall is difficult to understand and model, due to the complexity of the atmospheric processes that generate it [3].

In the last decades, it has been shown that Artificial Neural Networks (ANNs) are suitable to predict weather variables with accurate results. This technique is

I. Rojas, G. Joya, and J. Cabestany (Eds.): IWANN 2013, Part I, LNCS 7902, pp. 471–480, 2013.

able to capture the complex nonlinear relation between input and output without the physics being explicitly provided. Another advantage of ANNs is that they avoid the use of differential equations [6, 11]. Regarding practical applications, ANNs have also been used in rainfall forecasting [1, 9, 13].

In spite of the generalization ability of ANNs and due to the nonlinear and non-stationary nature of the rainfall time series, it is necessary the search for analysis alternatives that improve the accuracy of predictions. For instance, it has been used the decomposition and ensemble principle introduced by Huang et al. in [7], which aims to simplify the forecasting task by dividing it into forecasting subtasks [2, 12]. The goal of the ensemble is to formulate a consensus forecasting on the input data. In general, this technique is suitable for time series analysis, presenting even better results than those obtained by other techniques such as Wavelet and Fourier decomposition [8, 10].

This paper proposes a methodology for rainfall forecasting, adopting the above-mentioned decomposition and ensemble principle. In the proposed framework, the employed decomposition technique is the Ensemble Empirical Mode Decomposition (EEMD), and as a forecasting tool, the Feed Forward Neural Network (FNN). The experiments are performed on a real-observed rainfall signal. The obtained results are compared against a single FNN model for the raw data.

The remaining part of the paper is organized as follows. The proposed methodology is detailed in Sect. 2, where a brief description of Empirical Mode Decomposition (EMD), EEMD and FFNs are also presented. Section 3 presents the experiments and the obtained results. Finally, the paper is concluded in Sect. 4.

2   Methodology

2.1   Empirical Mode Decomposition (EMD)

The EMD method was firstly introduced in [7]. The essence of the method is to empirically identify the intrinsic oscillatory modes by their characteristic time scales in the data in order to decompose them accordingly. This guided the authors to the definition of a class of functions, based on their local properties, designated as intrinsic mode function (IMF) for which the instantaneous frequency can be defined everywhere. According to [7], an IMF is a function that satisfies two conditions:

1. In the whole data set, the number of extrema and the number of zero crossings must either equal or differ at most by one.
2. At any point, the mean value of the envelope defined by the local maxima and the envelope defined by the local minima is zero.

Using the definition, any data series $x(t)(t = 1, 2, \ldots, n)$, can be decomposed according to the following sifting procedure.

1. Identify all the local extrema of $x(t)$.
2. Connect all local extrema by a cubic spline line to generate its upper and lower envelopes $x_{up}(t)$ and $x_{low}(t)$, respectively.
3. Compute the mean $m(t)$ as $m(t) = (x_{up}(t) + x_{low}(t))/2$.
4. Extract $m(t)$ from the data series and define $c(t) = x(t) - m(t)$.
5. Check the properties of $c(t)$: (i) if $c(t)$ meets the above two requirements, an IMF is derived and then replace $x(t)$ with the residual $r(t) = x(t) - c(t)$; (ii) if $c(t)$ is not an IMF, replace $x(t)$ with $c(t)$.
6. Repeat Steps 1 to 5 until some stopping criterion is satisfied.

Finally, after the above procedure is complete, the original data series $x(t)$ can be expressed by

$$x(t) = \sum_{j=1}^{n} c_j(t) + r_n(t), \tag{1}$$

where n is the number of IMFs, $r_n(t)$ is the residue, which is the main trend of $x(t)$, and $c_j(t)(j = 1, 2, \ldots, n)$ are the IMFs. All IMFs are nearly orthogonal to each other, and all have nearly zero means. Thus, the data series can be decomposed into n IMFs and one residue. The IMF components contained in each frequency band are different and change with variation of the data series $x(t)$, while $r_n(t)$ represents the central tendency of the data series $x(t)$.

2.2 Ensemble Empirical Mode Decomposition (EEMD)

One of the major drawbacks of the original EMD is the mode mixing. It is defined as any IMF consisting of oscillations of dramatically disparate scales, often caused by intermittency of the driving mechanisms. When mode mixing occurs, an IMF can cease to have physical meaning by itself, suggesting falsely that there may be different physical processes represented in a mode. Ensemble empirical mode decomposition (EEMD) represents a major improvement of the EMD method, eliminating largely the mode mixing problem and preserving physical uniqueness of decomposition [15]. One of the basic principles is that adding white noise to the data series will provide a relatively uniform reference scale distribution to facilitate EMD, making EEMD a truly noise-assisted data analysis method. The EEMD is developed as follows:

1. Add a white noise series to the targeted data.
2. Decompose the data with added white noise into IMFs usign EMD.
3. Repeat step 1 and step 2 again and again, but with different white noise series each time.
4. Obtain the (ensemble) means of corresponding IMFs of the decompositions as the final result.

One of the effects of the decomposition using the EEMD are that the added white noise series cancel each other in the final mean of the corresponding IMFs. This effect should decrease following the well-established statistical rule:

$$\varepsilon_n = \frac{\varepsilon}{\sqrt{N}}, \tag{2}$$

where N is the number of ensemble members, ε is the amplitude of the added noise, and ε_n is the final standard deviation of the error, which is defined as the difference between the input signal and the corresponding IMFs. In most cases, as indicated in [15], it is suggested to add noise of an amplitude that is about 20% of the standard deviation of the data. However, when the data is dominated by high-frequency signals, the noise amplitude may be smaller; similarly, when the data is dominated by low-frequency signals, the noise amplitude may be increased.

2.3   Feed Forward Neural Network (FNN)

The main reason for selecting a FNN as a predictor is that it is often viewed as a *"universal approximator"* [5]. In [5] and [14] it was found that a three-layer FNN with an identity transfer function in the output unit and logistic functions in the middle-layer units can approximate any continuous function arbitrarily well, given a sufficient amount of middle-layer units. That is, neural networks have the ability to provide a flexible mapping between inputs and outputs [16].

2.4   Proposed EEMD-FNN Scheme

The central idea of the proposed methodology is to address the task of predicting a complex signal, first decomposing it into simpler parts that can be modeled more accurately, and then aggregating these results into the final prediction of the original signal. The method follows these main steps:

1. Decompose the rainfall signal (time series) using EEMD to obtain a set of IMF components and a residue.
2. Use the FNN model as a forecasting tool for each extracted IMF and the residue component, and make the individual one step ahead prediction for each one.
3. Combine (aggregation) the individual forecasting results of all IMF components to obtain the final prediction result of the input signal.

The parameter determination of the neural networks are based on the methodology proposed in [4], in which the Partial Autocorrelation Function (PACF) is used to determine the input variables of each FNN model, looking for those lags at which the PACF is outside of the confidence interval, and the number of hidden nodes is equal to $2p + 1$, where p is the number of inputs.

3   Experimental Analysis

3.1   Data Description and Performance Measures

The rainfall data were registered in a meteorological station at Manizales city, Colombia. The observations were taken with a time step of 5 minutes, and then a daily time series is derived summing the whole register per day. The time series covers the period from 01-Jan-2005 to 31-Dec-2008, for a total of 1461 days, see Fig. 1. The data are randomly divided into three parts for training, validation and test, respectively assigning 70%, 15%, 15%. These divisions are used to train each FNN model with its respective IMF, referred hereafter as component, in a more general way.

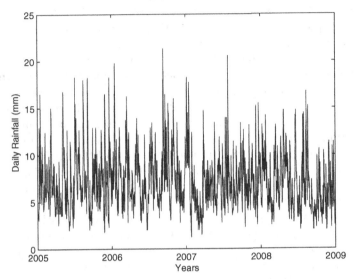

Fig. 1. Daily rainfall computed for a meteorological station at Manizales city

Three criteria are used for evaluating the forecasting performance: the Mean Absolute Percentage Error (MAPE), the Mean Square Error (MSE) and the Mean Absolute Error (MAE), calculated as

$$MAPE = \frac{1}{M} \sum_{t=1}^{M} \left| \frac{x_t - y_t}{x_t} \right| \cdot 100 \tag{3a}$$

$$MSE = \frac{1}{M} \sum_{t=1}^{M} (x_t - y_t)^2 \tag{3b}$$

$$MAE = \frac{1}{M} \sum_{t=1}^{M} |x_t - y_t| \tag{3c}$$

where M is the number of elements in the set, x_t is the actual value and y_t the predicted one. In these performance measures, the smaller, the better.

3.2   Experimental Results

The decomposition using EEMD is presented in Fig. 2, for a total of 9 components and the residue. The parameters used were an amplitude of added noise equals to 0.2 and an ensemble of 100 trials. Regarding to this technique, the EEMD code package proposed by [15] was used, which is available at http://rcada.ncu.edu.tw/ in MATLAB language. Minor changes were made to the code relative to enhance computation time, but in essence the original procedure was preserved. For the FNNs, the Neural Network Toolbox-Version 7.0.1 (R2011a) for MATLAB was used for the modeling task.

The entire process of training all models is repeated 10 times with a different division of the data. The box plot for each FNN model for the training set is shown in Fig. 3; on each box, the central mark is the median, the edges of the box are the 25th and 75th percentiles, the whiskers extend to the most extreme data points and outliers are plotted individually. The horizontal axis represents each predicted component. As shown, the high frequency components exhibit worse performance than the lower ones, probably because of their complexity. It can also be noted that, in general the lowest components have a higher dispersion than the highest ones, and that the outliers represent cases in which the tests were significantly worse than the others.

The final step in the proposed EEMD-FNN is the combination of the individual forecasting results of all components to obtain the final prediction result of the input signal. This can be done summing it all, according to (1). For all the 10 experiments, the mean for each performance measure and its Standard Error (SE) are computed and thus a reliable estimate of the performance is obtained. Table 1 summarizes these estimates for the proposed methodology, which is compared against a single FNN model for the original signal, as a basic benchmark. This FNN model was estimated in the same way for all components, i.e. using the PAFC to determine the inputs and setting the hidden nodes to $2p+1$, where p is the number of inputs. The proposed EEMD-FNN scheme outperforms the single FFN model for original signal, showing a considerable difference. The MAPE criterion is almost halved even when the SEs are quite similar, the MSE and MAE criteria and their SEs are reduced too. This proves that decomposing the original signal enhances the forecasting results, specifically, in the case of the studied rainfall data.

Table 1. Mean performance measures and their standard errors

	EEMD-FNN		Single FNN	
	Mean	SE	Mean	SE
MAPE(%)	**14.0908**	0.4365	28.1798	0.4378
MSE	**1.5126**	0.0637	5.4430	0.2131
MAE	**0.9177**	0.0191	1.7906	0.0314

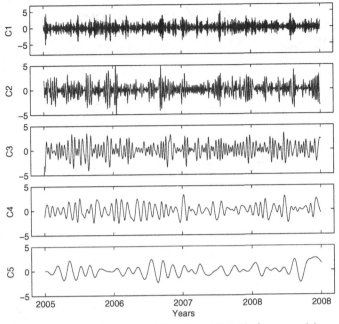

(a) Components 1 to 5 resulting from EEMD decomposition.

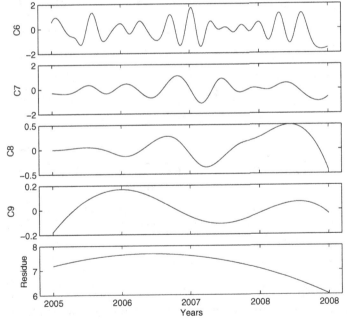

(b) Components 6 to 9 resulting from EEMD decomposition and the residue.

Fig. 2. Components resulting from the EEMD decomposition

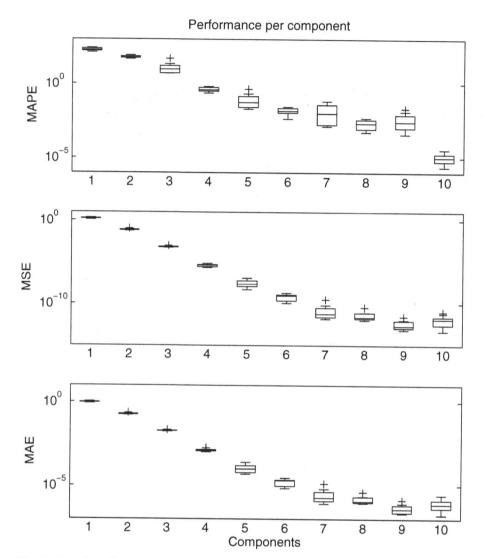

Fig. 3. Box plot of performance measures for the test data set per each FNN model or component predicted, the last one (10) is the residue. The vertical axis are in *log* scale.

4   Conclusions

In this paper a rainfall forecasting scheme has been proposed, which integrated EEMD and FNN. The proposed methodology decomposes the rainfall signal into more stationary and regular components (IMFs) using the EEMD technique. Moreover, the single model applied to each component is simple. According to the obtained results, the EEMD-FNN scheme improves the forecasting results and offers a simple approach for the stable prediction of non-stationary data.

The results were compared with a single FNN model for the original signal, using MAPE, MSE and MAE as their criteria. As a future work, it would be interesting to explore the possibility of employing different aggregation methods as well as performing an additional and more significant test that exposes more reliable results, maybe considering other data sets. In addition, it would be useful to perform a comparison with traditional forecasting techniques and testing the scheme with different long term predictions.

Acknowledgments. This study is supported by the "Programa Jóvenes Investigadores e Innovadores 2011, convenio especial de cooperación No. 0043 de 2012 suscrito entre la Fiduciaria Bogotá S.A. como vocera del patrimonio autónomo denominado Fondo Nacional de Financiamiento para la Ciencia, la Tecnología y la Innovación Francisco José de Caldas y la Universidad Nacional de Colombia", the research program "Fortalecimiento de capacidades conjuntas para el procesamiento y análisis de información ambiental" (code Hermes-12677) and "Grupo de Control y Procesamiento Digital de Señales Código 20501007205" funded by Universidad Nacional de Colombia. Instituto de Estudios Ambientales (IDEA) from Universidad Nacional de Colombia - Sede Manizales is also acknowledged for kindly supplied the data used in the experiments.

References

1. Abhishek, K., Kumar, A., Ranjan, R., Kumar, S.: A rainfall prediction model using artificial neural network. In: 2012 IEEE Control and System Graduate Research Colloquium (ICSGRC), pp. 82–87 (July 2012)
2. Chen, C.F., Lai, M.C., Yeh, C.C.: Forecasting tourism demand based on empirical mode decomposition and neural network. Knowledge-Based Systems 26, 281–287 (2012)
3. French, M.N., Krajewski, W.F., Cuykendall, R.R.: Rainfall forecasting in space and time using a neural network. Journal of Hydrology 137(1), 1–31 (1992)
4. Guo, Z., Zhao, W., Lu, H., Wang, J.: Multi-step forecasting for wind speed using a modified EMD-based artificial neural network model. Renewable Energy 37(1), 241–249 (2012)
5. Hornik, K., Stinchcombe, M., White, H.: Multilayer feedforward networks are universal approximators. Neural Networks 2(5), 359–366 (1989)
6. Hsu, K.L., Gupta, H.V., Sorooshian, S.: Artificial neural network modeling of the rainfall-runoff process. Water Resources Research 31(10), 2517–2530 (1995)
7. Huang, N.E., Shen, Z., Long, S.R., Wu, M.C., Shih, H.H., Zheng, Q., Yen, N.C., Tung, C.C., Liu, H.H.: The empirical mode decomposition and the Hilbert spectrum for nonlinear and non-stationary time series analysis. Proceedings of the Royal Society A Mathematical Physical and Engineering Sciences 454(1971), 903–995 (1998)
8. Huang, N., Shen, Z., Long, S.: A new view of nonlinear water waves: The Hilbert spectrum. Annual Review of Fluid Mechanics 31(1), 417–457 (1999)
9. Hung, N.Q., Babel, M.S., Weesakul, S., Tripathi, N.K.: An artificial neural network model for rainfall forecasting in Bangkok, Thailand. Hydrology and Earth System Sciences 13(8), 1413–1425 (2008)

10. Li, X.: Temporal structure of neuronal population oscillations with empirical model decomposition. Physics Letters A 356(3), 237–241 (2006)
11. Luk, K., Ball, J., Sharma, A.: A study of optimal model lag and spatial inputs to artificial neural network for rainfall forecasting. Journal of Hydrology 227(1), 56–65 (2000)
12. Tang, L., Yu, L., Wang, S., Li, J., Wang, S.: A novel hybrid ensemble learning paradigm for nuclear energy consumption forecasting. Applied Energy 93, 432–443 (2012)
13. Vovoras, D., Tsokos, C.P.: Statistical analysis and modeling of precipitation data. Nonlinear Analysis: Theory, Methods & Applications 71(12), e1169–e1177 (2009)
14. White, H.: Connectionist nonparametric regression: Multilayer feedforward networks can learn arbitrary mappings. Neural Networks 3(5), 535–549 (1990)
15. Wu, Z., Huang, N.E.: Ensemble empirical mode decomposition: A noise-assisted data analysis method. Advances in Adaptive Data Analysis 1(1), 1–41 (2009)
16. Yu, L., Wang, S., Lai, K.K.: Forecasting crude oil price with an EMD-based neural network ensemble learning paradigm. Energy Economics 30(5), 2623–2635 (2008)

Self-regulating Neurons in the Sensorimotor Loop

Frank Pasemann

Institute of Cognitive Science, University of Osnabrück, Germany
frank.pasemann@uni-osnabrueck.de

Abstract. Synaptic plasticity for recurrent neural networks is derived by introducing neurons as self-regulating units. These neurons have homeostatic properties for certain parameter domains. Depending on its underlying connectivity a neurocontroller endowed with the derived synaptic plasticity rule can generate a variety of different behaviors. The structure of these networks can be developed by evolutionary techniques. For demonstration, examples are given generating a walking behavior for a 3-joint single leg of a walking machine.

1 Introduction

Experience-dependent plasticity of neural control is one of the essential mechanisms to meet the survival conditions for living systems as well as the successful performance of autonomous robots acting in dynamically changing environments. This may be achieved by mechanisms for adjusting the intrinsic excitability of neurons, for developing the structure of networks and for regulating the synaptic strengths. In the following we concentrate on the regulation of synaptic efficacies.

The transmission of signals between biological neurons is mediated via the liberation of neurotransmitters from the pre-synaptic terminal and subsequent activation of receptors on the post-synaptic cell. It is by now obvious that intracellular mechanisms do modify pre-synaptic release and/or the abundance of functional post-synaptic receptors. This is often referred to as *homeostatic plasticity* and the various candidates for homeostatic mechanisms of biological neural systems have been reported [1,7]. On the other hand, theoretical approaches, relying on comparable mechanisms, have been taken up early in the context of Evolutionary Robotics [2,10,11].

In this paper we introduce artificial neurons as self-regulating units which try to maintain one of two desired activity states, referring to low and high activity. Such a neuron regulates its inputs by varying its *receptor strength*, and its output by varying its *transmitter strength*. Synaptic weights between two neurons are then defined as the product of pre-synaptic transmitter strengths and post-synaptic receptor strengths. Such a self-regulating neuron, called a *SR-neuron*, is thus described by 3-dimensional parametrized family of dynamical systems.

Depending on parameter domains, these neurons can display for instance a kind of homeostatic regulation which is assumed to be a basic mechanism for

I. Rojas, G. Joya, and J. Cabestany (Eds.): IWANN 2013, Part I, LNCS 7902, pp. 481–491, 2013.

the self-organization of behavior. Homeostatic regulation is essential for a system which undergoes certain kinds of disturbances. Such "perturbations" are for example given by sensory inputs during interaction of a system with its environment, and homeostasis then will stabilize a relevant behavior.

On this background, networks of SR-neurons are good candidates for controlling the behavior of simulated and physical robots. Working in the embodiment context and using Evolutionary Robotics and Artificial Life techniques, these controllers will operate continuously in the sensorimotor loop. For such systems the activity of neurons together with the synaptic efficacies will change over time, usually fluctuating around some average values.

For demonstration we choose a walking behavior of a single leg with three degrees of freedom, which is a leg model of the modular physical walking machine Octavio [8,9]. Already in [3], it was shown that bio-inspired homeostatic mechanisms can be effectively used to control legged robots.

In the following section the dynamics of SR-neurons is defined. Then examples of controllers are presented which lead to a walking behavior of the simulated 3-joint leg. Finally, there is a short discussion of the results.

2   Self-regulating Neurons

Given a neural network with n neurons. In what follows a single *self-regulating* neuron i will be described as a parametrized 3-dimensional dynamical system with state variables $(a_i, \xi_i, \eta_i) \in \mathbb{R}^3$, $i = 1, \ldots, n$, where a_i denotes its activation, ξ_i and η_i its *receptor* and *transmitter strength*, respectively. The output $o_i = \tau(a_i)$ of neuron i is given by the sigmoid transfer function $\tau := \tanh$.

In addition, θ_i denotes a fixed bias term, and c denotes the structure matrix of the network defined by $c_{ij} = \pm 1$ for an excitatory, respectively inhibitory connection from neuron j to neuron i; and $c_{ij} = 0$ if no such connection exists. The synaptic weight w_{ij} of the connection from neuron j to neuron i is then defined by

$$w_{ij} := c_{ij}\, \xi_i\, \eta_j \ . \tag{1}$$

Furthermore it is assumed that there exists a desirable state a^* for the activation of a neuron, and that the 3-dimensional dynamics to be defined is able to regulate towards the desirable state for a certain range of input signals. In principle there are several possible choices for such a desirable state. Here it is assumed that the neurons should preferably operate in the non-linear domain of their transfer function, because this will allow non-trivial dynamical properties to be adopted for behavior control. Thus, an activation a^* for which the non-linearity of the hyperbolic tangent τ is "maximal" is given by the condition that its third derivative satisfies $\tau'''(a^*) = 0$. Because τ''' is a symmetric function there are two such activations satisfying this condition, and they take values $a^* = a^*_\pm = \pm 0.658$, and correspondingly $\tau(a^*_\pm) = \pm 0.577$.

The dynamics of a SR-neuron is then defined by the following equations. The standard additive discrete-time dynamics for the activation of a neuron is given by:

$$a_i(t+1) = \theta_i + \xi_i(t) \sum_{j=1}^{n} c_{ij}\, \eta_j(t)\, \tau(a_j(t))\,. \tag{2}$$

Assuming that the receptor strength ξ_i and the transmitter strength η_i both are strictly positive for all times, the dynamics of ξ_i is defined by

$$\xi_i(t+1) = \xi_i(t)\left(1 + \beta\left(\tau^2(a^*) - \tau^2(a_i)\right)\right), \quad 0 < \beta < 1\,. \tag{3}$$

The transmitter strength η_i is assumed to have a decay rate $(1-\gamma)$ and it should increase with the activation a_i of the neuron. The dynamics of the transmitter strengths η_i is then defined by

$$\eta_i(t+1) = (1-\gamma)\,\eta_i(t) + \delta\,(1 + \tau(a_i)), \quad 0 < \gamma < \delta < 1\,. \tag{4}$$

The discrete-time dynamics $f : \mathbb{R}^3 \to \mathbb{R}^3$ given by equations (2) - (4) is called the SRN-dynamics for short, and we write

$$g(x) := \beta\left(\tau^2(a^*) - \tau^2(x)\right), \quad h(x) := \delta\,(1 + \tau(x))\,, \tag{5}$$

to determine the weight change per time step Δw_{ij} is then given in the form

$$\Delta w_{ij} = w_{ij} \cdot [F_i(a_i) + G_j(a_j) + H_{ij}(a_i, a_j)]\,, \tag{6}$$

with

$$F_i(x) := -\gamma + (1-\gamma)\,g(x), \quad x \in \mathbb{R},$$
$$G_j(x) := \frac{1}{\eta_j}\,h(x), \quad x \in \mathbb{R},$$
$$H_{ij}(x,y) := \frac{1}{\eta_j}\,g(x)\,h(y), \quad x, y \in \mathbb{R}.$$

This demonstrates that the synaptic SRN-dynamics depends on the activation a_i of the post-synaptic neuron as well as on the activation a_j of the pre-synaptic neuron; thus relating in part, by the function H_{ij}, to a Hebb-like synaptic dynamics.

Because $g(x)$ vanishes for $x = a_\pm^*$, and $G_j(x) > 0$ for all $x \in \mathbb{R}$ the condition for $\Delta w_{ij} = 0$ is given by $a_i = a_\pm^*$ for all i, and the resulting asymptotic transmitter strength is given by

$$\eta_\pm^* := \frac{\delta}{\gamma}(1 + \tau(a_\pm^*))\,.$$

With respect to the following experiments, the homeostatic regulatory property of SR-neurons is demonstrated in the following section for a single SR-neuron with self-connection.

2.1   SR-Neuron with an Excitatory Self-connection

As an example consider the dynamics $f : \mathbb{R}^3 \to \mathbb{R}^3$ of a single SR-neuron with excitatory self-connection $w := c\,\xi\,\eta$, $c = +1$, and bias term $\theta = 0$, where indices are dropped for convenience. Suppose this neuron gets an input I from other neurons of the network. Then its activation dynamics reads

$$a(t+1) = \xi(t)(I(t) + \eta(t)\,\tau(a(t)))\,. \tag{7}$$

Assuming that the input I is varying so slowly that a desired state a^* can be reached; then there will always exist a finite $\xi^* > 0$ such that the asymptotic equation

$$a^* = \xi^*(I + \eta^*\tau(a^*))$$

can be satisfied for all I; either for a_+^* or for a_-^*.

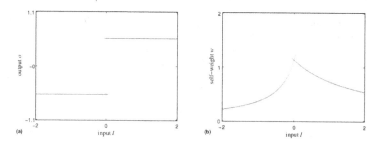

Fig. 1. Varying the input I of a neuron with excitatory self-connection w: a) the neuron output o, b) the self-connection w

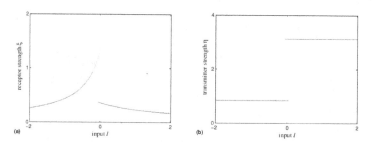

Fig. 2. Varying the input I of a neuron with excitatory self-connection w: a) the receptor strength ξ, b) the transmitter strength η

This is demonstrated by the bifurcation diagrams in figure 1 and 2 for SR-constants $\beta = 0.1$, $\gamma = 0.01$, $\delta = 0.02$, and corresponding $\eta_+^* = 3.16$, $\eta_-^* = 0.84$. These asymptotic values are seen also in figure 2b. From figure 1b one can read already that for a certain interval around $I = 0$ the self-weight satisfies $w > 1$, and therefore a hysteresis effect can be expected for varying inputs I crossing this interval [4].

3   Synaptic Dynamics in the Sensorimotor Loop

For the following experiments the sensor neurons are chosen to be linear buffers, here with a range $(-1, 1)$ corresponding to the output of neurons with *tanh* transfer function. All hidden and motor neurons will be SR-neurons. Proprioceptors, i.e. internal sensors or "bias" neurons are configured as linear buffer neurons.

To get an impression of behaviors generated by networks of SR-neurons the following examples will all use the parameter values $\beta = 0.01$, $\gamma = 0.001$, $\delta = 0.002$. Reflex loops and walking behaviors are chosen because in these cases it is best observed how dynamic inputs are generating a desired behavior, and how resulting synaptic weights are varying around mean values.

3.1   Example 1: Reflex Loops

We will first demonstrate that a SR-neuron with self-connection, acting in a negative feedback loop, will lead to reflex oscillations of a pendulum. The pendulum is driven by a servo motor in the velocity mode. The following set-up (figure 3) is considered: The sensor neuron *AngleSensor* measures the joint angle of the pendulum with zero position in the vertical (down) position, and the motor neuron *Motor* drives the servo motor. The bias of the buffer neuron CC controls the center angle of the pendulum oscillations. Neuron $N1$ and the *Motor* neuron are SR-neurons. Starting the network will result in oscillations around a position given by θ_{CC} (compare figures 4 and 5).

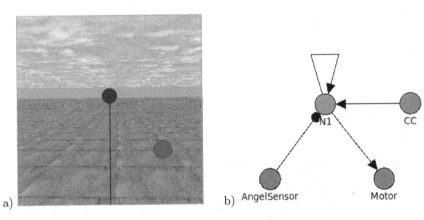

Fig. 3. a) The ODE pendulum simulator. b) A simple reflex loop with bias neuron CC defining the center of the oscillations; arrows are excitatory, dots are inhibitory synapses.

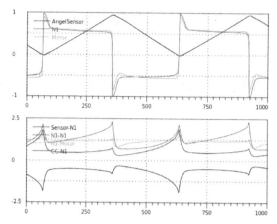

Fig. 4. Outputs of the sensor neuron, the hysteresis neuron N1, and the motor neuron (upper); and the weights of connections between them (lower) for $\theta_{CC} = 0.2$

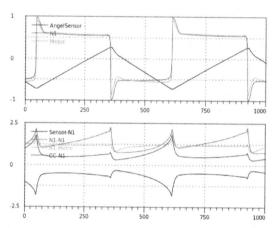

Fig. 5. Outputs of the sensor neuron, the hysteresis neuron N1, and the motor neuron (upper); and the weights of connections between them (lower) for $\theta_{CC} = -0.5$

3.2   Example 2: Single Leg Walking

In the second example we demonstrate that, based on simple reflex loops, the walking behavior of a single leg with three joints can be realized. The physical simulation is that of a leg belonging to the modular walking machine Octavio [9]. That simple sensorimotor reflex loops, instead of coupled oscillators, are able to produce interesting walking behaviors was already demonstrated for instance in [8]. The leg has three angle sensors and one foot contact sensor and three servo motors driving the joints in the angle control mode. The structure of the control network is displayed in figure 6b.

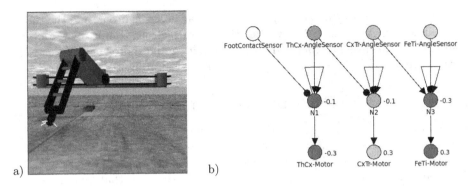

Fig. 6. a) The ODE simulator of a single leg, b) The structure of the neural network with SR-neurons

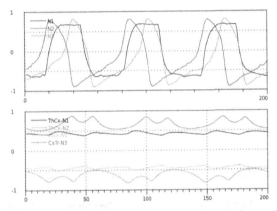

Fig. 7. The output signals of the hidden neurons N1, N2, N3 (upper), and the weights from the sensor neurons to these hidden neurons (lower)

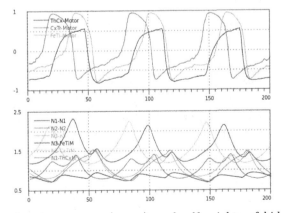

Fig. 8. Outputs of motor neurons (upper), and self-weights of hidden neurons and weights from hidden neurons to motor neurons(lower)

The sensor neurons (upper row) are again linear buffer neurons, and the three internal neuron $N1$, $N2$, $N3$ as well as the motor neurons (lower row) are SR-neurons. The resulting output signals of the hidden and motor neurons as well as the corresponding synaptic dynamics are displayed in figures 7 and 8.

3.3   Example 3: Structure Dependence of Behavior

To demonstrate the structural effects appearing with the application of SR-neurons we present a neural controller (figure 9a) for single leg walking which generates an equally good walking behavior as the controller of the last section. The remarkable observation for this structure is the fact that removing the marked synapse between neurons $N2$ and $N3$ will directly switch from forward movement to backward movement of the leg.

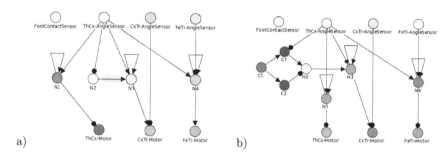

Fig. 9. Networks with SR-neurons for forward walking: a) Switching marked synapse from N2 to N3 on and off results in forward-backward walking (no bias terms used). b) A hybrid controller version to switch this synapse on and off by neural signals.

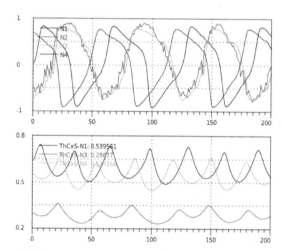

Fig. 10. Outputs of hidden neurons N1 to N4 (upper), and the four weights from sensor neurons to these hidden neurons (lower). Controller of figure 9a.

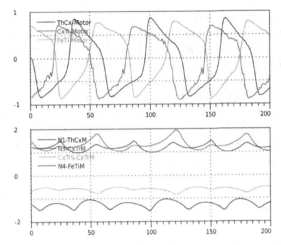

Fig. 11. Outputs of the motor neurons (upper), and the four weights to the motor neurons. Controller of figure 9a

One may switch the marked synapse in figure 9a on and off by using a neural structure with standard neurons CC, $C1$, $C2$ shown in figure 9b. The additional synapses have weights $|w| = 100$. For a bias value $\theta_{CC} = 0$ the leg moves forward, for $\theta_{CC} = 1$ it moves backward.

4   Discussion

We showed that SR-neurons can serve as homeostatic units able to regulate towards two desired output activities, which, if one likes, correspond to low and high firing rates. Having an excitatory self-connection they mimic a kind of binary neuron. In this mode they can display a hysteresis effect. The second operational mode of SR-neurons, due to an inhibitory self-coupling not shown here, is as a period-2 oscillation with varying amplitude and shift.

The approach to self-regulation developed here differs from others [6,11,3] by the fact that bias terms and scaling factors for the activation of the neurons are not directly involved in the homeostatic process.

The presented examples demonstrate that there is no special "learning rule" varying the synapses, but it is the structure of the network which allows a controller with SR-neurons to achieve the behavior by its "intrinsic" dynamics. Thus, the structure of a network is as essential for a resulting behavior as is the synaptic plasticity.

Observing the dynamics of the receptor strength ξ and transmitter strength η as in figure 7 one can conclude that, in general, these quantities vary around a definite mean value; as expected for instance in the single leg walking experiment. In fact, fixing the synapses at these mean values will keep the running behavior as before.

Thus, considering SRN-dynamics as a learning process one would need a mechanism to turn the SRN-dynamics on and off appropriately if the system meets a situation which demands learning or adaptation. Such a mechanism has to be based on proprioceptive signals and was developed, for example, in [5].

Another interesting property of a SR-neuron is that it can turn off its input by reducing its receptor strength to zero. In fact, receptor strength diverges if bias terms θ_i are outside the regulatory domain; i.e. $\theta_i \notin [a_+^*, a_-^*]$. This property may be useful in the context of evolutionary robotics by introducing a mechanism like "programmed death" of SR-neurons, for those which can not contribute to behavior relevant synaptic dynamics. This allows for an effective structure evolution of neurocontrollers with SR-neurons.

Although SR-neurons were used here as elements of reflex loops it can be shown that they can also be used to drive wheeled robots. This was presented in [12] where a different type of SRN-dynamics was successfully applied.

Acknowledgements. This research was funded in part by DFG-grant PA 480/7-1. Valuable discussions with Christian Rempis and Hazem Toutounji are gratefully acknowledged.

References

1. Davis, G.W., Bezprozvanny, I.: Maintaining the stability of neural function: a homeostatic hypothesis. Annual Review Physiology 63, 847–869 (2001)
2. Di Paolo, E.A.: Homeostatic adaptation to inversion of the visual field and other sensorimotor disruptions. In: Paris, J.-A., Meyer, A., Berthoz, D., Floreano, H. (eds.) From Animals to Animals, Proc. of the Sixth International Conference on the Simulation of Adaptive Behavior, SAB 2000, pp. 440–449. MIT Press (2000)
3. Hoinville, T., Tapia, C., Hénaff, P.: Flexible and Stable Pattern Generation by Evolving Constrained Plastic Neurocontrollers. Adaptive Behavior 19, 187–207 (2011)
4. Hülse, M., Pasemann, F.: Dynamical neural schmitt trigger for robot control. In: Dorronsoro, J.R. (ed.) ICANN 2002. LNCS, vol. 2415, pp. 783–788. Springer, Heidelberg (2002)
5. Rempis, C.W., Toutounji, H., Pasemann, F.: Controlling the Learning of Behaviors in the Sensorimotor Loop with Neuromodulators in Self-Monitoring Neural Networks. In: Proceedings of the IEEE International Conference on Robotics and Automation (ICRA) (to be published 2013)
6. Triesch, J.: Synergies between intrinsic and synaptic plasticity mechanisms. Neural Computation 19, 885–909 (2007)
7. Turrigiano, G.G.: The self-tuning neuron: synaptic scaling of excitatory synapses. Cell 135, 411–435 (2008)
8. von Twickel, A., Pasemann, F.: Reflex-oscillations in evolved single leg neurocontrollers for walking machines. Natural Computing 6, 311–337 (2007)
9. von Twickel, A., Hild, M., Siedel, T., Patel, V., Pasemann, F.: Neural control of a modular multi-legged walking machine: Simulation and hardware. Robotics and Autonomous Systems 60, 227–241 (2012)

10. Vargas, P.A., Moioli, R.C., de Castro, L.N., Timmis, J., Neal, M., Von Zuben, F.J.: Artificial homeostatic system: a novel approach. In: Capcarrère, M.S., Freitas, A.A., Bentley, P.J., Johnson, C.G., Timmis, J. (eds.) ECAL 2005. LNCS (LNAI), vol. 3630, pp. 754–764. Springer, Heidelberg (2005)
11. Williams, H., Noble, J.: Homeostatic plasticity improves signal propagation in continuous-time recurrent neural networks. Biosystems 87, 252–259 (2007)
12. Zahedi, K., Pasemann, F.: Adaptive behavior control with self-regulating neurons. In: Lungarella, M., Iida, F., Bongard, J.C., Pfeifer, R. (eds.) 50 Years of AI, Festschrift. LNCS (LNAI), vol. 4850, pp. 196–205. Springer, Heidelberg (2007)

Comparison of Two Memristor Based Neural Network Learning Schemes for Crossbar Architecture

Janusz A. Starzyk[1,2] and Basawaraj[1]

[1] School of Electrical Engineering and Computer Science, Ohio University, Athens, OH, USA
[2] University of Information Technology and Management, Rzeszow, Poland
{starzykj,basawaraj.basawaraj.1}@ohio.edu

Abstract. This paper compares two neural network learning schemes in crossbar architecture, using memristive elements. Novel memristive crossbar architecture with dense synaptic connections suitable for online training was developed. Training algorithms and simulations of the two proposed learning schemes, winner adjustment training (WAT) and multiple adjustments training (MAT) are presented. Tests performed using MNIST handwritten character recognition benchmark dataset confirmed the functionality of proposed learning schemes. Proposed learning schemes were compared accounting for noise, device variations and multipath effects. The proposed learning schemes improve available neural network learning schemes.

Keywords: neural network, memristive crossbar, self-organizing, learning, dense analog memories, character recognition.

1   Introduction

Memristors, first hypothesized by Chua 40 years ago [1], are elements relating charge and magnetic flux and change their resistance based upon the input current or voltage. Resistance switching in Titanium dioxide (TiO_2) was first recognized as related to the memristor characteristics in 2008 [2] and sparked considerable interest in their potential applications, including neural networks [3 – 6]. Later work has shown the existence of memristor characteristics in WO_3 [7] and other metal oxides like ZrO_2, NiO, Nb_2O_5, HfO_2, and CeO_x [8].

The ability to perform on-line training and scale the architecture to a large system are the major evaluation criterion of memristive architectures in neural network applications. The goal of this work is to make memristive on-line learning feasible in high density crossbar architectures. Specifically, we propose a neural network organization and two training scheme, suitable for high density architectures with interconnection weights that can be easily controlled during the network operation.

Neural networks using high-density crossbar architecture have been proposed in earlier work [9 – 11]. The architecture proposed in [9] is easily scalable as it is based on a simple crossbar structure with memristors formed at the junctions. But [9] lacks in details regarding mechanism for training the memristors in the crossbar architecture, probably due to its emphasis on fabrication of memristors. [10] proposes a neural

I. Rojas, G. Joya, and J. Cabestany (Eds.): IWANN 2013, Part I, LNCS 7902, pp. 492–499, 2013.
© Springer-Verlag Berlin Heidelberg 2013

crossbar architecture, provides a training scheme and shows its application to design of fault tolerant circuits using supervised learning. A related design in [11], shows application of a crossbar architecture using memristors for handwritten character recognition. Unlike [11] our approach uses neither lateral inhibition among the output neurons nor homoeostasis at the output neurons to dynamically adjust their thresholds. This decreases the circuitry needed and results in a more compact architecture.

2   Memristor Characteristics

Memristors change their resistance as a function of the flux applied to their terminals and remember their "state" when the flux is removed. Fig. 1 shows a physical model of the memristor as described in [2]. It consists of titanium dioxide (TiO_2) and an oxygen poor TiO_{2-x} layer, fabricated between platinum contacts. The titanium dioxide layer is considered the undoped region and has high resistance while the oxygen poor region is considered the doped region and has low resistance due to the oxygen vacancies acting as positive dopants. The effective resistance is the sum of the resistances of the doped and undoped regions:

$$M(x) = R_{ON}x + R_{OFF}(1 - x)$$
(1)

where the state variable

$$x = \frac{w}{D} \in (0,1)$$
(2)

And R_{ON} and R_{OFF} are the resistance of the doped region and undoped regions respectively. The memristor's current and voltage are related through Ohm's law, with memristance M(x) representing current resistance value.

Fig. 1. (a) Memristor model according to [2] (b) memristor symbols used in this paper left-to-right: traditional symbol, symbols used in this paper for p-type and n-type memristors.

Memristor is an asymmetrical device and its properties depend on the direction of the current that passes between its terminals. Conductivity increases when a positive voltage is applied across the memristor (plus at doped region and minus at undoped region), otherwise its conductivity goes down. Thus both the voltage value and duration of the applied voltage are important, so the resistance is in effect a function of total flux as well as initial state of the memristor.

Experiments and analysis performed following publication of [2] has furthered the understanding of the memristor characteristics. For example, [12] predicted that the speed of switching ON was faster than switching OFF; applying ON-switching for sufficiently long time would turn the device OFF i.e., the switching polarity can be reversed. These devices have ohmic i-v characteristics near the ON states and are non-linear towards the OFF states. Moreover, in [13] it was shown that increasing the applied current caused an exponential decrease in the required switching energy. The model from [2], used here, though limited is still a very good approximation of memristor characteristics and was used due to its simplicity and wide use in literature. Using a more detailed model, would not considerably affect either the algorithm or the results presented here.

3   Design of Neural Networks

To be compatible with neural network learning schemes the memristor conductance values are normalized to [0-1] interval and are used to represent weights of synaptic connections. For simplicity only excitatory neurons are considered here. A simple two layer neural network is shown in Fig. 2 (a). In this network neuron N is a postsynaptic neuron and it receives inputs from n presynaptic neurons $x_1,... x_n$. Here, for convenience, we assume that the output of an excitatory neuron is $V_{out} = 1$ V when activated and $V_{out} = 0$ V otherwise.

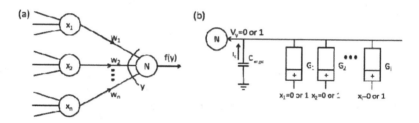

Fig. 2. (a) Two layer feed forward neural network; (b) Memristors connected to a neuron's input.

Fig. 2 (b) shows an input to a neuron N in a crossbar architecture with a number of memristor elements connected to its input line. When the neuron N's input in Fig. 2 (b) is charged to the voltage V_y above a threshold value V_{th}, the neuron fires and sets its feedback output voltage to V_{yf} for a short period of firing time T_f, regulated by adjusting load capacitance C_{large}. Thus, with 1 volt on the active input V_i the corresponding memristor increases its conductance (weight) whereas the inactive input ($V_i =0$ V) reduces its conductance. Here V_y can be obtained from

$$V_y = \sum \frac{V_i G_i}{\sum G_i} = \sum w_i V_i \tag{3}$$

where

$$w_i = \frac{G_i}{\sum G_i} \tag{4}$$

3.1 Neural Network Training

According to Hebbian learning, in a simple feedforward neural network, excitatory links increase their weights when both presynaptic and postsynaptic neurons fire. We adopt this rule to the memristor circuit with different schemas of weight adjustment and resulting training algorithms. Neural network training in crossbar architecture algorithm is as follows.

Neural Network Training in Crossbar Architecture

1. Select s random training data and use them to assign initial weights to the output neurons.
2. For all training samples repeat 3 - 6.
3. Compute similarity between training data and neuron weights using (3) and obtain input excitations of the output neurons V_y.
4. Sort all output neurons excitations V_y to determine dynamic threshold value V_{th} for the neurons' input excitation V_y.
5. Provide training feedback value $V_{yf} = 1$ V to all output neurons y with $V_y < V_{th}$ and provide training feedback value $V_{yf} = 0$ V to all output neurons y with $V_y > V_{th}$. That is, if the output neuron, N, fires, it uses feedback to reduce its input potential V_{yf} to zero. Thus with 1 volt on the active input V_i, the corresponding memristor increases its conductance (weight). However, if N does not fire, then it changes its input potential V_{yf} to 1 V for a period of firing time T_f. This produces a negative flux across memristor connected to the inactive input with $V_i = 0$ V reducing its conductance (weight). Neuron's firing time is regulated by adjusting its load capacitance C_{large}.
6. Adjust memristor weights using $\Delta w_i = (V_i - V_{yf})\Delta_i(G)$. $\Delta_i(G)$ is the increment of Δw_i per unit voltage applied across the memristor. V_{yf} is the provided feedback signal $V_{yf} = \bar{V}_{out}$. The resulting weight adjustments are a result of the memristor voltage polarization. For instance if the input voltage $V_i = 0\,V$ and $V_{yf} = 1\,V$ memristor weight is lowered and when $V_i = 1\,V$ and $V_{yf} = 0\,V$, it increases.

Two versions of the neural network training algorithm are used. They differ by the way a threshold value is established in step 4 of the algorithm. The first version known as **multiple adjustments training** (MAT) uses median value of V_y to establish the threshold and at every step of the training cycle up to 50% of memristor values are adjusted upwards and up to 50% downwards. This balances off total amount of flux that each memristor receives. According to Step 6 the adjustment upwards takes place only when $V_y = 0$ V and $V_i > 0$ V and the adjustment downwards takes place only when $V_y = 1$ V and $V_i < 1$ V.

MAT requires additional circuits to adjust threshold V_{th} dynamically. Its advantage is that training of memristors can be performed in dense crossbar architecture since no additional switches are required to select which memristors need to be adjusted upwards or downwards. Instead, memristors' control is done by adjusting threshold and feedback voltages only.

The second version known as **winner adjustment training** (WAT) selects a single winner that is most similar to the training sample. Thus V_{th} is equal to the winning neuron's V_y. Subsequently, all memristor connections are open except for the winning neuron, and the feedback voltage V_{yf} is set to 0.5 V. Thus, memristors of the winning neuron connected to inputs higher than 0.5 V are adjusted downwards and those lower that 0.5 V are adjusted downwards, moving neuron in the direction of the input data.

For compactness of the neural network design it is desirable that no elements other than memristors at the crossbar junctions are added to the crossbar architecture. This is the main advantage of the proposed approach. WAT requires only a winner-take-all circuit, thus it is easier to implement in hardware than MAT. However, since all memristors of losing neurons must be disconnected during training, WAT crossbar architecture must contain switching transistors in series with memristors to implement synaptic connections. All memristors in a single column (one output neuron) can be controlled by the same control signal, simplifying wiring of the control signals.

4   Benchmark Testing

In benchmark testing neural network has its memristive weights adjusted according to either WAT or MAT algorithm. We investigated test results for WAT and MAT training using MNIST [14] handwritten character database. Neural network with 784 inputs and 80 outputs was constructed and trained using 1280 data points. Once training was completed and memristor values were adjusted, we applied a new set of 10,000 test data from the MNIST database. The average test performance over 25 runs was 73.2% with standard deviation of 1.8% for WAT and 69.9% with standard deviation of 1.6% for MAT.

4.1   Analysis for Robustness

To test the robustness of the learning schemes we performed analysis to determine the effect of multipath, input noise, and memristor manufacturing tolerances. Memristors are not connected directly to ideal voltage sources but use drivers with nonzero resistance hence neural network training and testing will be subject to multipath effects. To estimate this we simplify our analysis to the one in which the input signals correspond to the input digit intensity and the feedback signal in the neural network training is obtained with V_y connected to a voltage source (0 V or 1 V) through a fixed conductance G_s. In such circuit, deviation of V_y from its ideal value (0 V or 1 V) will depend on the input signal as follows:

$$V_y = \frac{\sum_{k=1}^{m_c} V_i G_i + G_s V_s}{\sum_{k=1}^{m_c} G_i + G_s} \tag{5}$$

where m_c is the number of memristors in this category (either winners or losers). In case $V_s = 1$ we use (3) for memristors of losing neurons, and in this case deviation of the feedback voltage ΔV_y form a desired value equals

$$\Delta V_y = 1 - V_y = \frac{\sum_{k=1}^{m_l} G_i (1 - V_i)}{\sum_{k=1}^{m_l} G_i + G_s} \tag{6}$$

where m_l is the number of memristors that are in the losing neurons. For $V_s = 0$ we use (3) for memristors of winning neurons, and in this case ΔV_y equals

$$\Delta V_y = V_y = \frac{\sum_{k=1}^{m_w} V_i G_i}{\sum_{k=1}^{m_w} G_i + G_s} \tag{7}$$

where m_w is the number of memristors that are in the winning neurons. Due to similar deviations on the driver side memristor training will be subject to a significant noise. In the worst case this noise is uncorrelated, so we tested how such noise will affect testing performance. It was observed that, for driver conductance range of 10 μS to 100 mS, the performance in both tested methods (MAT and WAT training) did not suffer.

To analyze the robustness of learning schemes the test data was corrupted with uniform noise signal. The analysis was performed using a single layer neural network with 784 inputs and with 250 output neurons. Fig. 3 shows the test results and shows that the recognition rate is better than chance even for noise to signal ratios of 10 (for this dataset the chance level is 10%).

The last test of robustness performed used Monte-Carlo analysis to test the ability of the proposed neural network learning scheme to handle process variations leading to changes in nominal values of all memristors. While memristors fabricated on different dies will have significant deviations from their nominal values, crossbar memristors are fabricated on the same die and will have memristance parameters tracking each other.

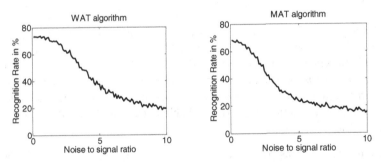

Fig. 3. Correct recognition performance on 10000 test data with various noise levels

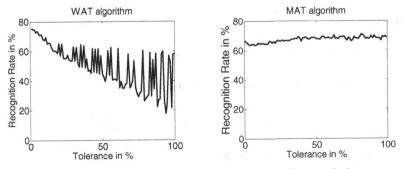

Fig. 4. Correct recognition performance in Monte Carlo analysis

Absolute changes in memristive conductivity are not important as long as the ratio in (4) remains unchanged. Critical for neural network performance are local variations within the same die which are small and correlated, thus their effect on the performance of a neural network will be also small.

We performed 100 tests, each starting from the same training data set (based on 2000 randomly selected digits) and the same 10,000 test data points but with memristor characteristics randomly varied within specified tolerance. The neural network had 300 output neurons. The tolerances on memristor parameters were increased from 0.5% until 100% of their nominal values. Fig. 4 shows the test results. Results show that recognition performance for WAT quickly goes down, so the network is more sensitive to variations in memristor parameters than it was to the input noise. However, since typically, within the same die, the memristances do not vary more than 5% (a reasonable assumption for medium size dies with present manufacturing capabilities [15]), the loss of the recognition performance is small. There is no observable loss in performance in MAT training even at larger tolerances.

5   Conclusion

In this paper, we presented two memristor training schemes in the crossbar organization such that neural network interconnection weights can be easily implemented and controlled during the network operation. We performed simulations to test the proposed neural network learning schemes considering multipath effects, device variations and noisy test data. The results showed that the proposed approach is compact, tolerant to noise and device variations, and it can be trained online. The results obtained verified correct adjustment of memristor values for the selected training data. In this work, the focus was on developing neural network learning schemes for crossbar architecture, rather than on specific data bases or advanced neural network application. The proposed solution improves available in the literature training methods for memristive neural networks.

Acknowledgements. This work was supported by the grant from the National Science Centre DEC-2011/03/B/ST7/02518.

References

1. Chua, L.: Memristor-the missing circuit element. IEEE Trans. Circuit Theory CT-18(5), 507–519 (1971)
2. Strukov, D.B., Snider, G.S., Stewart, D.R., Williams, R.S.: The missing memristor found. Nature 453, 80–83 (2008)
3. Pershin, Y.V., Di Ventra, M.: Experimental demonstration of associative memory with memristive neural networks. Neural Netw. 23(7), 881–886 (2010)
4. Pershin, Y.V., Di Ventra, M.: Solving mazes with memristors: A massively parallel approach. Phys. Rev. E 84(4), 046703 (2011)

5. Cantley, K.D., Subramaniam, A., Stiegler, H., Chapman, R., Vogel, E.: Neural learning circuits utilizing nano-crystalline silicon transistors and memristors. IEEE Trans. Neural Networks Learning Syst. 23(4), 565–573 (2012)

6. Kim, H., Sah, M.P., Yang, C., Roska, T., Chua, L.O.: Memristor Bridge Synapses. Proceedings of the IEEE 100(6), 2061–2070 (2012)

7. Chang, T., Jo, S.-H., Kim, K.-H., Sheridan, P., Gaba, S., Lu, W.: Synaptic behaviors and modeling of a metal oxide memristive device. Appl. Phys. A 102(4), 857–863 (2011)

8. Liu, L., Chen, B., Gao, B., Zhang, F., Chen, Y., Liu, X., Wang, Y., Han, R., Kang, J.: Engineering oxide resistive switching materials for memristivedevice application. Appl. Phys. A 102(4), 991–996 (2011)

9. Jo, S.H., Chang, T., Ebong, I., Bhadviya, B.B., Mazumder, P., Lu, W.: Nanoscale memristor device as synapse in neuromorphic systems. Nano Lett. 10(4), 1297–1301 (2010)

10. Chabi, D., Klein, J.-O.: Hight fault tolerance in neural crossbar. In: 2010 5th International Conference on Design and Technology of Integrated Systems in Nanoscale Era (DTIS), March 23-25, pp. 1–6 (2010)

11. Querlioz, D., Zhao, W.S., Dollfus, P., Klein, J.-O., Bichler, O., Gamrat, C.: Bioinspired networks with nanoscale memristive devices that combine the unsupervised and supervised learning approaches. In: 2012 IEEE/ACM International Symposium on Nanoscale Architectures (NANOARCH), July 4-6, pp. 203–210 (2012)

12. Strukov, D.B., Borghetti, J.L., Williams, R.S.: Coupled ionic and electronic transport model of thin-film semiconductor memristive behavior. Small 5(9), 1058–1063 (2009)

13. Pickett, M., Strukov, D.B., Borghetti, J., Yang, J., Snider, G., Stewart, D., Williams, R.S.: Switching dynamics in a titanium dioxide memristive device. J. of App. Phys. 106, art. 074508 (2009)

14. The MNIST database of handwritten digits,
http://yann.lecun.com/exdb/mnist/

15. Gray, D.T.: Optimization of the Process for Semiconductor Device Fabrication in the MicrON 636 Whittemore Cleanroom Facility. M.S. thesis, Materials Sc. Eng., Virginia Polytechnic Inst. and State Uni., Blacksburg, VA (2002)

Geometrical Complexity of Data Approximators

Evgeny M. Mirkes[1], Andrei Zinovyev[2,3,4], and Alexander N. Gorban[1]

[1] Department of Mathematics, University of Leicester, UK
{em322,ag153}@le.ac.uk
[2] Institut Curie, rue d'Ulm 26, Paris, 75005, France
[3] INSERM U900, Paris, France
[4] Mines ParisTech, Fontainebleau, France
andrei.zinovyev@curie.fr

Abstract. There are many methods developed to approximate a cloud of vectors embedded in high-dimensional space by simpler objects: starting from principal points and linear manifolds to self-organizing maps, neural gas, elastic maps, various types of principal curves and principal trees, and so on. For each type of approximators the measure of the approximator complexity was developed too. These measures are necessary to find the balance between accuracy and complexity and to define the optimal approximations of a given type. We propose a measure of complexity (geometrical complexity) which is applicable to approximators of several types and which allows comparing data approximations of different types.

Keywords: Data analysis, Approximation algorithms, Data structures, Data complexity, Model selection.

1 Introduction

1.1 Complexity of Data as Complexity of Approximator

It would be useful to measure the complexity of datasets which can be represented as a set of vectors in a potentially high-dimensional space. In this study, we analyse a measure of data complexity based on the analysis of the geometry of the data approximators. Therefore, we call it the geometrical measure of complexity.

There are many ways to define the complexity of data distribution. For example, Akaike information criterion (AIC) can be used to select models of data of minimal complexity, using information theory [1], or Structural risk minimization principle is applied in some areas [2].

Recently we proposed to measure data complexity through complexity of data approximators and developed three measures of data complexity: geometrical complexity, structural complexity and construction complexity [3-5].

Actually, each data approximation method is equipped with a measure of the approximator complexity. For example, in the classical data approximation methods, the number of centroids in k-means clustering, the number of principal components in Principal Component Analysis, and curvature or length of the principal curve serve as

I. Rojas, G. Joya, and J. Cabestany (Eds.): IWANN 2013, Part I, LNCS 7902, pp. 500–509, 2013.

measures of complexity. A close to optimal approximator must be able to catch the hypothetical intrinsic shape of the data distribution without trying to approximate the data's "noise" (though "one man's noise is another man's signal" [6]).

1.2 Complexity of Data Approximators

Measuring complexity of data approximators is important because of the following principle, which is an application of the famous Occam's razor: between two approximators of the same accuracy one should prefer the one with smaller complexity [7]. Thus, complexity measure becomes a computational tool for selecting the best approximator type and structure, a tool for model selection. The model selection problem is the classical problem of mathematical modelling and statistics. Its solution is always based on the complexity/accuracy trade-off [8]. Model selection should be based not solely on goodness-of-fit, but must also consider model complexity [9]. The positive effect of model complexity on estimation error for new data points from the same data domain can be directly demonstrated by cross-validation computational experiments in some settings [9].

Selecting a model of data with the least complexity can be even more crucial for *model generalizability* on new data domains not accessible through available sampling [10]. This model ability cannot be directly demonstrated by cross-validation. Using complexity criteria here gives us hope that the simpler model will better fit outside its definition domain, sometimes at the expense of worse accuracy for available data (accuracy/generalizability trade-off). However, this remains a strong hypothesis in statistics. In other fields of science (for example, theoretical physics) the internal simplicity ("beauty") of a model or a theory often guides theoretical constructions with many examples of success.

However, measuring and even defining complexity is a difficult task. Many methodological studies converge on that the notion of *complexity* should be distinguished from the notions of *size*, *order* and *variety*, and that any measure of complexity is dependent on the language of object's representation [11]. The most common attributes used to compare alternative models are the level of detail and the model complexity although these terms are used in many different ways [12]. There may be proposed many complexity criteria. Each of them represents an aspect of complexity and together they can be considered as a representation of the "objective complexity". At least, there is no other "objective complexity" besides a combination of various technical complexity measures.

Coming back to the problem of data approximation, we can distinguish two important characteristics of any approximator: 1) its "internal" structure, and 2) its "flexibility", i.e. its ability to be tuned in order to fit the data. For example, a polynomial approximator is characterized by its maximal degree (internal structure), and the constraints on the parameter values (flexibility). Therefore, talking about approximator's complexity, we can distinguish the complexity of the approximator's structure and the complexity of its configuration in the data space.

A large class of data approximators can be represented in the form of graphs, embedded into a data space [3], [13]. Here one can talk about the complexity of the graph structure itself and the complexity of the mapping of the graph's nodes into the data space. For example, a one-dimensional grid with k nodes can represent both linear and curvilinear approximations of data, with linear mapping being evidently less complex.

In this paper, we compare approximators represented by graphs of two different types: one-dimensional grids and trees. From what was said above, we have two options to compare them in terms of complexity: the complexity of their structures and the complexity of their mappings. Comparing structures of graph-based approximators is not always meaningful: they can be generated using different types of generating grammars (description languages). Here, the grammar generating trees has more types of elementary transformations (more variety) than the one generating one-dimensional grids.

In this study we focus on comparing approximators for their mapping complexity rather than comparing their structures. In this paper we present and test a concrete recipe for this, *the geometrical complexity*.

1.3   Three Measures of Approximators' Complexity

Three measures of the approximator's complexity were introduced in [3-5]: (1) Geometrical measure, (2) Structural measure, (3) Construction measure.

Structural and construction complexity measures introduced below estimate the complexity of the "internal" structure of the approximating graph, while *geometrical complexity* is an analogue of non-linearity for the graph-based approximators.

Geometrical Complexity. The geometrical measure of complexity estimates the deviation of the approximating object from some *"idealized" configuration*. The simplest such ideal configuration is linear: in this case the nodes of the approximator are located on some linear surface. Deviation from the linear configuration would mean some degree of non-linearity.

However, the notion of non-linearity is applicable only to relatively simple situations. For example, in the case of branching data distributions, non-linearity is not applicable as a good measure of geometrical complexity. In [13] it was suggested that a good generalization of linearity as "idealized" configuration can be the notion of harmonicity. An embedding of a graph into linear space is called *harmonic* if, in each star of the graph, the position of the central node of the star coincides with the mean of its leaf vectors [13]. For many applications it is convenient to consider not all the stars but a set of selected stars and to introduce *pluriharmonic* embedding [5,13]. For pluriharmonic embedding, the position of the central node of each selected star coincides with the mean of its leaf vectors.

In our estimations of the geometrical complexity of trees we will use the deviation from a harmonic embedment as analogue and generalization of the non-linearity.

In [4] geometrical complexity was used to optimize the balance between accuracy and complexity (see Fig. 1).

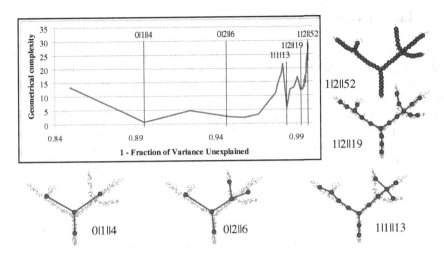

Fig. 1. The "accuracy–complexity" graph for a tree-like 2D data distribution. Several scales of the approximator's complexity are shown. Two of them, corresponding to the structural complexity barcodes 0|1||4 (1 3-star and 4 nodes) and 1|2||19 (1 4-star, 2 3-star and 19 nodes) are optimal and approximate the distribution structure at a certain "depth".

Structural Complexity. The structural complexity defines how complex is an approximator in terms of its structural elements (number of nodes, edges, stars of various degrees). In general, this index should be a non-decreasing function of these numbers. Contribution of some of the elements (for example, nodes and edges) might be not interesting for measuring the structural complexity and, hence, have zero weight (not present) in the resulting quantitative measure.

Construction Complexity. We derive our approximators by the systematic application of the graph grammar operations, in the way which is the most optimal in terms of the objective function. The construction complexity of an approximator can be defined as a minimum number of elementary graph transformations which were needed to produce it, using given grammar. This measure can be similar to the structural complexity in some implementations but is not equivalent to it. There is some similarity to the idea of the Kolmogorov complexity (see, for example, [14]) but the Kolmogorov complexity is defined on the base of the necessary length of the program and the construction complexity is the number of operation actually used in construction.

In this paper we show that the geometrical complexity can be used to compare approximators of different types. In particular, we compare Growing Self-Organising Maps and Growing Principal Trees in terms of this measure.

As for the other two measures, structural and construction complexities, it happens that for the two types of graph-based approximators considered in this paper, we can only compare them in terms of the number of nodes in the graph (see Table 1).

2   Materials and Methods

To check the ability of geometrical complexity (GC) to compare two different approximators, we designed several benchmarks datasets (see Fig. 2), which we approximated using two data approximation methods: Growing Self-Organizing Map (GSOM) [15] and Principal Tree (PT) [13].

First, for both methods we used the fraction of variance unexplained (FVU) as the measure of the approximation accuracy. In all tests we stopped the approximator's growth when the approximation accuracy was less or equal to a specified threshold value, which was the same for both GSOM and PT. For "*thin*" patterns the threshold value of FVU was equal to 0.1%, for "*scattered*" patterns the threshold value of FVU was equal to 0.2% and for "*scattered and noised*" patterns threshold value of FVU was equal to 1%.

The same parameters of approximation methods were used for all benchmarks.

For comparison of the constructed approximators we used three criteria: 1) number of nodes, 2) length of the constructed approximating graph, which, by definition, is the sum of all graph edge lengths, and 3) geometrical complexity.

2.1   Fraction of Variance Unexplained

The Fraction of Variance Unexplained (FVU) is the dimensionless least square evaluation of the approximation error. It is defined as the ratio between the sum of squared distances from data to the approximating line and the sum of squared distances from data to the mean point.

The distance from a point x_i to a line is the distance p_i between x_i and its orthogonal projection onto the line. The FVU of a line, used as an approximator, is simply

$$FVU = \sum_{i=1}^{n} p_i^2 \left/ \sum_{i=1}^{n} \|x_i - \bar{x}\|^2 \right.$$, where $\bar{x}$ is the mean point $\bar{x} = (1/n)\sum_{i=1}^{n} x_i$. For a po-

lygonal line (an ordered set of nodes $\{y_i\}$ $(i = 1,2,\cdots,k)$, connected by line fragments) FVU is defined in the following way. For a given set of nodes $\{y_i\}$ $(i = 1,2,\cdots,k)$ we compute the distance from each data point x_i to the polygonal line specified by a sequence of points $\{y_1, y_2, \cdots, y_k\}$. For this purpose, we calculate all the distances from x_i to each segment $[y_s, y_{s+1}], s \in [1..k-1]$ and find the minimal distance $d(x_i)$.

In this case we define $FVU = \sum_{i=1}^{n} d^2(x_i) \left/ \sum_{i=1}^{n} \|x_i - \bar{x}\|^2 \right.$.

2.2   Growing Self-Organizing Maps

Growing self-organizing maps is a well-known method to approximate sets of vectors embedded in high dimensional space by self-organizing maps with ability to increase the number of nodes [15]. For this study we use the line version of GSOM.

A detailed description of the algorithm used can be found in [16]: below we provide a brief description.

1. Initiate SOM by two nodes, connected by an edge, which is a linear fragment of the first principal component, centred at the mean point.
2. Optimize the node positions by using the Batch SOM learning algorithm [17].
3. Compute the current FVU (CFVU).
4. If CFVU is less than or equal to the specified threshold value then stop. If not then
5. Glue to the first node of the polygonal line a new edge, of the same length and direction as the first edge of the polygonal line, and calculate FVU for the new polygonal line (BFVU).
6. Extend the last node of the polygonal line with a new edge, of the same length and direction as the last edge of the polygonal line, and calculate FVU for new polygonal line (EFVU).
7. If BFVU<EFVU and BFVU<CFVU then glue to the first node of the polygonal line a new edge, of the same length and direction as the first edge of the polygonal line, and go to step 2.
8. If EFVU<BFVU and EFVU<CFVU then extend the last node of the polygonal line with a new edge, of the same length and direction as the last edge of the polygonal line, and go to step 2.

For each GSOM in this study we used the linear BSpline neighbourhood function with neighbourhood radius 3.

2.3   Principal Tree

Principal graphs form a rich family of tools for nonlinear dimensionality reduction for data of complex topology [18]. The method of topological grammars for construction of principal graphs and cubic complexes was proposed in [13]. The detailed description of principal tree (PT) construction algorithm can be found also in [4], [16]. The parameters which were used in this work are described below.

We used the graph growing grammar with two rules: "add new node to an existing node" and "bisect an existing edge". We used the graph shrinking grammar with two rules: "remove a leaf", "remove an edge, and glue its nodes". The grammar sequence "growing, growing, growing, shrink" was used for constructing the principal tree. In this study we used the elastic moduli 0.01 for stretching and 0.001 for bending.

1. Initiate PT by two nodes, connected by an edge, which is a linear fragment of the first principal component, centred at the mean point. Set stage = 1.
2. Optimize the energy functional of the graph
3. Calculate the current FVU. If FVU is less than or equal to the specified threshold value then stop. If not then
4. If stage < 4 then try all possible applications of growing grammar. Select the most optimal in terms of energy decrease operation. Apply the selected graph grammar operation. Add 1 to stage. Go to step 2.
5. If stage = 4 then try each possible application of shrinking grammar and save the best operation. Apply saved operation. Put stage = 1. Go to step 2.

2.4   Geometrical Complexity Used in the Study

A k-star is defined as a fragment of a graph, which is a node v_0 with all connected to it neighbour nodes $v_i : i = 1, \cdots, k$ (leaves). Let us denote by v_i both the tree (polygonal line) node and the embedment (position) of the corresponding vector in the data space. For each embedment of a k-star into the data space, the measure of non-harmonicity is the squared deviation of the central node v_0 from the mean point of

the leaf positions $GC(v_0) = \left\| v_0 - (1/k) \sum_{i=1}^{k} v_i \right\|^2$. This formula is analogous to the bend-

ing energy of a plate which is an integral of the squared first derivates of the deviation from a surface (k>2) or from a straight line (k=2). If we use an analogy with the method of finite elements then we have to multiply the sum of the bending energies in all k-stars by the squared number of nodes. Let us agree that for a 1-star the measure of non-harmonicity is equal to zero. Then the geometrical complexity of a tree (polygon-

al line) is $GC = n^2 \sum_{i=1}^{n} GC(v_i)$.

2.5   Available Implementations

A user-friendly graphical interface (Java-applet) for constructing principal trees and GSOM in 2D is available at [16].

3   Results and Discussion

The resulting maps and trees for each benchmark are shown in Fig. 2 and 3. The numerical values of the complexity measures for them are listed in Table 1. Table 1 allows us to compare four characteristics of constructed approximators: the number of nodes N, FVU, their lengths, and GC. We can see that for the approximators with the same FVU three complexity measures may be significantly different. The difference between GC values may be huge (one order of magnitude) while the length and N do not differ significantly. Thus, one can conclude the GC measure can be more sensitive in distinguishing the least complex approximator.

We can also compare two algorithms for approximation, GSOM and PT. For two patterns, "spiral thin" and "spiral scattered" they work similarly (the results are qualitatively the same, some measures are better for GSOM, and some for PT, without huge difference). The differences in the approximator lengths for the same patterns and given FVU are not big. The length of PT approximators is smaller in most cases. The number of nodes for the same FVU is always smaller for PT, and GS of the PT approximators is almost always much smaller than for GSOM.

Fig. 2. Approximating polygonal lines constructed by GSOM

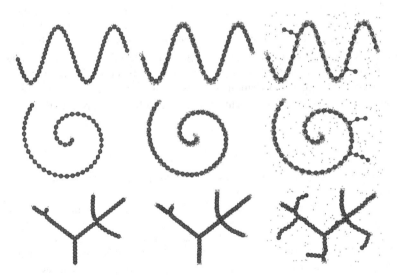

Fig. 3. Approximating trees constructed by the principal tree algorithm

Table 1. Complexity measures for the constructed approximators. N is number of nodes which reflects the construction complexity measure in this case. In the column "Pattern" S means "scattered" pattern, SN means "scattered and noised" pattern. The winner value of the complexity measures for each pattern is highlighted in bold and underlined.

Pattern	SOM				PT			
	N	FVU	GC	Length	N	FVU	GC	Length
Sinus	101	0.10%	3 895.64	1 671.30	**56**	0.10%	**263.89**	**1 236.06**
Sinus S	91	0.20%	1 043.05	1 454.64	**49**	0.20%	**238.59**	**1 232.75**
Sinus SN	90	0.98%	939.10	1 551.04	**41**	1.00%	**325.26**	**1 357.12**
Spiral	66	0.10%	**83.90**	**1 247.49**	**46**	0.08%	107.89	1 254.20
Spiral S	71	0.19%	**74.14**	1 253.22	**51**	0.19%	96.23	**1 251.32**
Spiral SN	158	0.99%	3 164.63	1 680.79	**46**	0.99%	**166.84**	**1 442.07**
Tree	82	1.05%	1 951.71	1 101.49	**59**	0.09%	**25.17**	**973.18**
Tree S	156	0.20%	6 143.35	1 434.03	**71**	0.19%	**29.93**	**995.80**
Tree SN	115	1.00%	4 699.45	1 822.96	**75**	0.99%	**319.64**	**1 392.00**

4   Conclusion

Accordingly to the general principles of model selection, among approximators having the same accuracy (FVU), the one with the smallest complexity should be preferred. Here we demonstrated that the geometrical complexity (GC) is applicable for comparing self-organising maps and principal tree approximators. Moreover, we show that the geometrical complexity investigated here is a more sensitive measure for model selection between these two types of graph-based approximators, compared to other measures such as the length and the number of nodes in the graph. Further investigations are required to demonstrate the utility of geometrical complexity for model selection among many different types of graph-based approximators.

References

1. Hirotugu, A.: A new look at the statistical model identification. IEEE Transactions on Automatic Control 19(6), 716–723 (1974)
2. Vapnik, V., Chervonenkis, A.: Ordered risk minimization I. Automation and Remote Control 35, 1226–1235 (1974)
3. Gorban, A.N., Zinovyev, A.: Principal graphs and manifolds. In: Olivas, E.S., Guererro, J.D.M., Sober, M.M., Benedito, J.R.M., Lopes, A. (eds.) Handbook of Research on Machine Learning Applications and Trends: Algorithms, Methods and Techniques, Information Science Reference, pp. 28–59. IGI Global, Hershey (2009)
4. Zinovyev, A., Mirkes, E.: Data complexity measured by principal graphs. Computers and Mathematics with Applications (2013) doi:10.1016/j.camwa.2012.12.009, arXiv:1212.5841

5. Gorban, A.N., Zinovyev, A.: Principal manifolds and graphs in practice: from molecular biology to dynamical systems. International Journal of Neural Systems 20(3), 219–232 (2010)
6. Blakeslee, S.: Lost on earth: wealth of data found in space, An Edward Ng's quote from the article in New York Times (March 1990)
7. Burnham, K.P., Anderson, D.R.: Model Selection and Multimodel Inference: A Practical Information-Theoretic Approach, 2nd edn. Springer (2002)
8. Akaike, H.: A new look at the statistical model identification. IEEE Transactions on Automatic Control 19(6), 716–723 (1974)
9. Myung, I.J.: The Importance of Complexity in Model Selection. Journal of Mathematical Psychology 44, 190–204 (2000)
10. Forster, M.R.: Key Concepts in Model Selection: Performance and Generalizability. Journal of Mathematical Psychology 44, 205–231 (2000)
11. Edmonds, B.: What is complexity? – The philosophy of complexity per se with application to some examples in evolution. In: Heylighen, F., Aerts, D. (eds.) The Evolution of Complexity. Kluwer, Dordrecht (1998)
12. Brooks, R.J., Tobias, A.M.: Choosing the best model: Level of detail, complexity, and model performance. Mathematical and Computer Modelling 24(4), 1–14 (1996)
13. Gorban, A.N., Sumner, N., Zinovyev, A.: Topological grammars for data approximation. Applied Mathematics Letters 20(4), 382–386 (2007)
14. Kolmogorov, A.N.: Three approaches to the quantitative definition of information. Problems of Information Transmission 1(1), 1–7 (1965)
15. Alahakoon, D., Halgamuge, S.K., Sirinivasan, B.: A self growing cluster development approach to data mining. In: Proceedings of IEEE International Conference on Systems, Man and Cybernetics, San Diego, USA, pp. 2901–2906 (1998)
16. PCA Master applet, Mirkes, E., University of Leicester (2011)
   http://bioinfo.curie.fr/projects/elmap
17. Kohonen, T.: The Self-Organizing Map (SOM).,
   http://www.cis.hut.fi/projects/somtoolbox/
   theory/somalgorithm.shtml
18. Gorban, A.N., Kégl, B., Wunch, D.C., Zinovyev, A. (eds.): Principal Manifolds for Data Visualisation and Dimension Reduction. LNSE, vol. 58. Springer, Heidelberg (2008)

Self-Organization Process in Large Spiking Neural Networks Leading to Formation of Working Memory Mechanism

Mikhail Kiselev

Megaputer Intelligence Ltd., Moscow
mkiselev@megaputer.ru

Abstract. The subject of this work is evolutionary process in initially chaotic and homogenous spiking neural networks leading to formation of the neuron groups with partially synchronized activity (so called polychronous groups) which are not only capable of recognizing input patterns but also can keep information about pattern presentation in form of their specific activity for a long time. This result is demonstrated for very simple neuron – coincidence detector and for standard synaptic plasticity model (STDP).

Keywords: spiking neural network, working memory, neural network self-organization, polychronization.

1   Introduction

A number of articles published in the last two decades were devoted to simulation of working memory mechanisms. For example, the realization of working memory in form of permanent activity of cyclic neuronal circuits was considered in [1]. Stability of this activity was studied in [2]. Working memory mechanisms were described in terms of attractors (in particular, in terms of continuous sets of attractors) in [3]. Possibility to stabilize these attractors using short-term potentiation of the involved synapses was also explored in this work. The same mechanism but with greater stress on physiological realism was considered in [4].

In these and many other papers working memory is simulated due to complex network structure, specially selected distribution of synaptic weights and delays etc. The question how all these structures are formed is usually left beyond the scope. Moreover, pattern memorization is not connected with their recognition – it is assumed that the network has learnt somehow to produce specific reaction to presented stimuli in form of activity of a certain neuron population. However, it is evident that in case of large network and complex system of stimuli detailed specification of individual synaptic weights and delays is impossible – the structures responsible for recognition and memorization should appear together as a result of network evolution. Evolutionary processes of this kind were considered in [5]. It was demonstrated in this work that evolution of neural network may lead under certain conditions to formation of structures recognizing different patterns and keeping information about their presentation.

I. Rojas, G. Joya, and J. Cabestany (Eds.): IWANN 2013, Part I, LNCS 7902, pp. 510–517, 2013.

Nevertheless, the model proposed in [5] did not achieve all desired goals. Although the initial distribution of synaptic weights was random the network even in its initial state had complex architecture consisting of set of layers and columns.

In order to make network informational capacity sufficiently high it was proposed in the article [6] to utilize so called neuron group polychronization. Neurons in a polychronous group (PNG) exhibit stereotypical time-locked spatiotemporal spike-timing patterns. There may be very many PNGs in a network, they may overlap extensively. Even if one PNG corresponds to one memorized pattern the whole network can store information about presentation of very great variety of different patterns. However, this paper proposing the approach to increase dramatically network informational capacity leaves unanswered many other questions. How does the network learn to recognize the input patterns? How is the necessary recognition selectivity level achieved? Although the proposed model seems to assume some network self-organization process based on the long-term synaptic plasticity it is not described.

In the present paper I would like to fill this gap and to show how self-organization process based on the standard synaptic plasticity model (STDP) [7] in a big originally chaotic and homogenous spiking neural network leads to development of the PNGs which learn to recognize and memorize (in form of their specific activity) spatiotemporal patterns. We consider the neuron model in Section 2. Structure of the simulated network is described in Section 3. Section 4 is devoted to synaptic plasticity rules. In Section 5 we discuss organization of the experiments and the input signal used. The obtained results are presented in Section 6.

2   Coincidence Detector Neuron Model

In order to make results of this research more general a simple neuron model reproducing only it's most significant and undoubted properties was selected. The most significant characteristic feature of neuron is that it behaves as a detector of coincidence of presynaptic spikes. It emits postsynaptic spike when several spikes arrive to its excitatory synapses inside a short time window (and if there are no sufficient number of spikes on its inhibitory synapses). For convenience, we assume that this time window is equal to 1 ms.

Thus, discrete time step in the described computational experiments equals to 1 ms. In each step t neuron may either stay silent or emit spike induced by presynaptic spikes obtained in this step if

$$\sum_{i \in \mathbf{A}_t} w_i - w^- n_t^- > T , \tag{1}$$

where w_i is weight of the i-th excitatory synapse, $\mathbf{A}_t$ – set of the excitatory synapses receiving spikes at the time t, w^- - weight of inhibitory synapses (it is the same for all inhibitory synapses and is constant in time), n_t^- - number of inhibitory synapses receiving spikes at the time t, T – membrane potential threshold.

Neurons are subdivided to excitatory and inhibitory. The excitatory neurons are connected only to excitatory synapses and vice versa. Value of T (= 8.5) and maximum values of weights are selected so that at least 5 spikes from other neurons or 3 spikes from the network input nodes (they are described in next section) are necessary for neuron to fire. It corresponds to weight values 2 and 4, respectively. Inhibitory synapse weights are always equal to 10 – the correct balance of excitation and inhibition is achieved by variation of percentage of inhibitory neurons in the network.

Time of spike propagation from neuron to neuron lays in the range 1 – 30 ms that is close to physiological values. Setting the spike propagation delays is considered in next section.

3   Network Topology

As it was mentioned, the considered neural network has homogeneous structure in the sense that all neurons and all connections of the same kind (excitatory and inhibitory) have originally the same distributions of weights, delays, connection probabilities etc. All neurons have the equal numbers of excitatory synapses and the equal numbers of inhibitory synapses. All neurons are connected via excitatory synapses to the equal number of network input nodes. Through these connections (we will call them afferent connections) neurons receive the external signals consisting of noise (random spikes with constant mean frequency) and patterns to be recognized. Neuron's axon can never be connected to a synapse of the same neuron. Provided that all these conditions are met, selection of sets of presynaptic neurons and input nodes was absolutely random for every neuron.

The part of inhibitory neurons was about 30% - it corresponded to necessary balance of excitation and inhibition. The number of input nodes was always 900 while the number of neurons and number of neuron synapses were variable.

Selection of synaptic delays was found to be very important for evolution of neural network in the desired direction. For example, it is essential that the propagation delay of inhibitory connections is substantively less than of excitatory connections – and it is in concordance with neurophysiological measurements. Inhibitory connection delays were selected randomly using the lognormal distribution with parameters ln(3) and 1. The experiments with other distributions showed no significant influence on results. On the contrary, it was discovered that selection of excitatory connection delays is important so that we consider it in more detail. In my model (like in [6]) the key role in realization of working memory is played by PNGs. But it is evident that potential number of PNGs in the network depends strongly on its topology and distribution of synaptic delays in it. Indeed, PNGs are characterized by great number of short paths between the same pair of neurons such that the total delay in every path is (almost) identical. It is intuitively clear (and it can be proven rigorously) that in case of completely random delays the average number of equal-delay paths connecting two neurons is significantly less than in case of neurons organized in some spatial order with connection delays proportional to the distance between the neurons connected. From this point of view it is efficient to place neurons on a sphere setting the synaptic delays equal to spherical distances between the neurons. It may be also an N-dimensional hypersphere. In the discussed experiments the best results were obtained for 4-dimensional hypersphere. Maximum delay (corresponding to link between diametrically opposite neurons on the sphere) was set to 30 ms. Afferent connection delays were random (lognormal distribution with parameters ln(15) and 3).

4   Synaptic Plasticity and Network Self-Organization

Similar to majority of other models, evolution of network is described in our case in terms of the synaptic weight changes. As it was said, only the excitatory synapses are plastic in the described model. The synaptic plasticity rules correspond to the standard STDP (spike timing dependent plasticity) [7], implying potentiation (LTP) of the synapses participating in the postsynaptic spike generation and depression (LTD) of the synapses receiving spikes short time after it. In my slightly simplified model, LTP affects only the synapses receiving spike in the time quant when the neuron fires. The LTD period is 3 ms after firing. It is shorter than the characteristic time of LTD effect adopted in the neurophysiological models (10 – 30 ms), however it is compensated by the fact that in our case LTD effect is constant during this whole period while in the standard model it decreases exponentially. This simplification is also aimed at reducing the computation complexity.

Since synaptic weight cannot grow infinitely as well as cannot become negative value of LTD and LTP should depend on the current weight value. In order to include this feature in the model it was proposed in [8] that object of LTP and LTD is the so called *synaptic resource* W rather than the synaptic weight w itself. These values are bound by the equation

$$w = w_{\min} + \frac{(w_{\max} - w_{\min})\max(W,0)}{w_{\max} - w_{\min} + \max(W,0)} \ . \tag{2}$$

It is evident, that $w_{\min} \leq w < w_{\max}$. For interneuron connections $w_{\min} = 0$. For afferent synapses $w_{\min} = 2$. This selection guarantees that the network will always keep at least potential ability to react to input signals.

In this approach LTP and LTD act by incrementing or decrementing W by a certain value. Therefore, the synaptic resource determines not only the value of the synaptic weight but also its **stability**. Indeed, if W is negative or significantly greater than $w_{\max}$ then w is close to $w_{\min}$ or $w_{\max}$ and value of W determines how many times LTP or LTD should be used in order to begin to change value of w significantly. In case of small positive W its variation causes approximately equal variation of w. The value of W modification for single LTP or LTD was set to 0.5.

At last, let us mention another factor influencing synaptic plasticity which was simulated in some of the experiments described. It is known that many afferent projections to cortex use glutamate as neurotransmitter. But serving as neurotransmitter glutamate is also capable of increasing the synaptic plasticity [9]. In order to simulate this influence the glutamate concentration γ was introduced as an additional neuron state component. The LTP/LTD effect depends linearly on it. Value of γ is increased by a certain constant whenever the neuron receives spike from a network input node until it reaches its maximum possible value (it corresponds to the plasticity value equal to 0.5 – see above) and decays exponentially with the time constant 10 ms. The maximum plasticity is reached when at least 8 afferent spikes reach neuron simultaneously. Although the results considered below were obtained even without this mechanism it makes the PNGs responsible for working memory much more stable.

5   Input Signal and Network Evolution Monitoring

In the described experiments the input signal consisted of random sequence of 3 different patterns presented with 1 sec interval but in such a way that pattern is never presented just after the same pattern. Every pattern was a sequence of spikes from 300 input nodes corresponding to this pattern which was repeated with 16 ms period during 30 ms time interval. Delays of first spikes from every node relatively to the pattern beginning were constant for all its presentations and randomly selected from the interval [0, 15] but every pattern presentation began with a randomly selected delay relatively to the pattern beginning. Therefore, the spike sequences in the different presentations of the same pattern were not identical. Fig. 1 illustrates why presentations of the same pattern may be different.

At the beginning of each experiment resources of all excitatory synapses were set to 0. Then prevailing effect of LTP caused systematic growth of synaptic weights until this process was balanced by LTD and synaptic weight saturation (when the weight stability grew instead of the weights themselves). Usually it required ~ 10^3 pattern presentations.

Formation of PNGs with activity specific for a single patter was monitored in process of network evolution. The stable neuron firing sequences were used for finding PNGs (in contrast with [7] where PNGs were determined from analysis of synaptic delays). In our approach, PNG is defined by a sequence <neuron id, firing time>. Since PNGs are interesting as a working memory mechanism it makes sense to

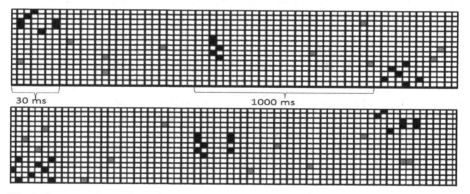

Fig. 1. Two sequences of input patterns – A, B, C and C, B, A. Here pattern period is 5 cells; pattern presentation time is 7 cells. Black cells denote patterns; grey cells denote noise.

analyze the network history fragments lasting from 100 ms to 300 ms from pattern presentation onset (while the pattern duration is 30 ms). We will call the PNGs found using these fragments the memory PNGs (MPNGs). Let us consider the pattern A. Let P_{Ai} be a set of pairs <neuron id, time after the beginning of i-th presentation of the pattern A> corresponding to all spikes emitted in the period 100-300 ms since the beginning of i-th presentation of this pattern. Set of all such sets corresponding to the pattern A will be denoted as $\vartheta_A = \{P_{Ai}\}$. Then the algorithm for finding in ϑ_A a MPNG with support n ($G(\vartheta_A, n)$) is the following:

1. Create the matrix C_{at}, initializing by 1s all its elements for which $<a, t> \in P_{A1}$, and by 0s – all the rest elements.
2. Iteratively for each $i, 2 \leq i \leq N_A$, find the shift s, for which the value of
$$\sum_{<a, t+s> \in P_{Ai}} C_{at}$$ is greatest, then increment by 1 those C_{at}, for which $<a, t+s> \in P_{Ai}$.
3. If $\forall t \forall a C_{at} < n$, then $G(\vartheta_A, n) = \varnothing$, else if $\bar{t}$ is the least t, for which $C_{at} \geq n$, then $<a, t-\bar{t}> \in G(\vartheta_A, n) \Leftrightarrow C_{at} \geq n$.

Using this algorithm the MPNGs keeping activity for a long time after pattern presentation are found. But only the MPNGs specifically reacting to only one pattern are interesting. Let us define activity of the MPNG G in the history fragment P_{Ai} as $A_{Ai}(G) = \max_s |\text{shift}(G, s) \cap P_{Ai}|$, where shift(G, s) is built from G by addition of s to second elements in all pairs in G. Then the strength of reaction of G to the pattern A can be defined as $R_A(G) = \min_i A_{Ai}(G)$, and the measure of selectivity of $G(\vartheta_A, n)$ - as

$$S(A, n) = \frac{N_A}{\left|\{<B, i>: A_{Bi}(G(\vartheta_A, n)) \geq R_A(G(\vartheta_A, n))\}\right|}.$$ It is evident that if reaction of

$G(\vartheta_A, n)$ to any pattern different from A is weaker than $R_A(G(\vartheta_A, n))$, then $S(A, n) = 1$.

The experiments were arranged in the following way. The network evolution started from the initial state described above. During this process the general network parameters were measured – such as mean firing frequency or sum of all synaptic weights. When these parameters reached their equilibrium values that was indicated by absence of significant long-term trend of their values the evolution process was stopped. In typical case the equilibrium was reached after thousands of pattern presentations. After that the algorithm described above was used to find $G(\vartheta_A, n)$ for various A и n on the basis of 30 last presentations of every pattern. For every $G(\vartheta_A, n)$ its reaction to the recognized pattern and to the other patterns was determined that was used to calculate their selectivity. Results obtained in these experiments are discussed in next section.

6   Results

The goal of the performed experiments was demonstration of network self-organization process leading to formation of MPNGs capable of recognizing input

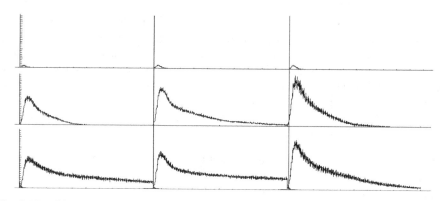

Fig. 2. The firing rates of MPNGs recognizing three different patterns after pattern presentation at 0 sec, 3000 sec and 10000 sec of network evolution. The firing rates are averaged over 100 most recent pattern presentations. Length of each period equals to 1 sec.

patterns and keeping information about pattern presentation for a time interval significantly greater than pattern duration. More precisely, it was required that for all patterns there would exist an MPNG with 100% selectivity and whose reaction strength for any presentation of the recognized pattern would equal to at least one tenth of the MPNG size. This goal was achieved – a region of network parameter values was found for which the network evolution always lead to formation of MPNGs with the desired properties. In typical case significant activity of MPNGs lasted until next pattern presentation (i.e. > 1 sec) – see Fig. 2, the bottom row.

Formation of these MPNGs in process of network self-organization is also illustrated on Fig. 2. We see that the rudiments of MPNGs existed even in the very beginning of evolution due to fluctuations of neuron interconnections although their response to the patterns was very weak and did not last after pattern presentation – so that their function was just recognition rather than memorization. Interestingly, in earlier evolution stage growth of the MPNG reaction strength was more significant; later stages were characterized by predominant growth of the reaction duration.

The following factors were discovered to be most important for successful evolution.

- Network size. The desired effect was only observed in sufficiently large networks – it was never observed if the number of neurons was less than 6000 while the networks including 10000 neurons produced the selective MPNGs under conditions of 10-30% variations of other network parameters.
- Excitation/inhibition balance. It is the most crucial factor requiring careful selection of the respective network parameters (part of inhibitory neurons, ratio of inhibitory/excitatory synapse weights etc.). In our case this balance was achieved by variation of percentage of inhibitory neurons in the network. For network size equal to 10000 neurons this parameter should lay in the range 31-34%. This range seems to be too narrow so that some homeostatic mechanism is necessary for balancing excitation and inhibition under various conditions. Future development of the discussed model should include an implementation of such mechanism.

- Synaptic delay distribution. Importance of this factor has been already mentioned. Inhibitory connections should be several times faster than excitatory ones - otherwise the undesired oscillatory activity involving the whole network develops and the network ceases to respond to input signals. Several distributions of excitatory delays were tested: random, corresponding to spherical region of N-dimensional Euclidean space and corresponding to N-dimensional hyperspheres. Stable successful results were obtained only for hyperspheres. In case of Euclidean delays the highly selective PNGs sometimes appeared but not for all patterns.
- Presence of fluctuations. In the best configuration number of presynaptic neurons for each neuron is about 20% of the total number of neurons. If this value was close to 100% then the initial fluctuations were insufficient for development of the rudimentary MPNGs (see the discussion above).

7   Conclusion

The present work demonstrates a self-organization process which leads to development in originally chaotic and homogenous network polychronous neuron groups able to recognize and memorize presented spatiotemporal patterns. Use of polychronous groups is essential since it is a natural approach to dramatic increase of network informational capacity. The development of the self-organization process in the desired direction is facilitated by combination of several features of the model such as synaptic weight stabilization effect, N-sphere geometry based values of synaptic delays and simulation of glutamate-modulated plasticity.

References

1. Zipser, D., Kehoe, B., Littlewort, G., Fuster, J.: A Spiking Network Model of Short-Term Active Memory. Journal of Neuroscience 13(8), 3406–3420 (1993)
2. Wang, X.-J.: Synaptic reverberation underlying mnemonic persistent activity. Trends in Neurosciences 24(8), 455–463 (2001)
3. Itskov, V., Hansel, D., Tsodyks, M.: Short-term facilitation stabilize parametric working memory trace. Frontiers in Computational Neuroscience 5, Article 40 (2011)
4. Mongillo, G., Barak, O., Tsodyks, M.: Synaptic Theory of Working Memory. Science 319, 1543–1546 (2008)
5. Kiselev, M.: Self-organized Short-Term Memory Mechanism in Spiking Neural Network. In: Dobnikar, A., Lotrič, U., Šter, B. (eds.) ICANNGA 2011, Part I. LNCS, vol. 6593, pp. 120–129. Springer, Heidelberg (2011)
6. Szatmary, B., Izhikevich, E.: Spike-Timing Theory of Working Memory. PLoS Comput Biol. 6(8), e1000879 (2010)
7. Gerstner, W., Kistler, W.: Spiking Neuron Models. Single Neurons, Populations, Plasticity. Cambridge University Press (2002)
8. Kiselev, M.: Self-organized Spiking Neural Network Recognizing Phase/Frequency Correlations. In: Proceedings of IJCNN 2009, Atlanta, Georgia, pp. 1633–1639 (2009)
9. Nolte, J.: The Human Brain: an introduction to its functional anatomy. Mosby Elsevier (2009)

Self-organized Learning by Self-Enforcing Networks

Christina Klüver and Jürgen Klüver

Institute for Computer Science and Business Information Systems, Essen, Germany
{c.stoica-kluever,juergen.kluever}@uni-due.de
www.cobasc.de

Abstract. We describe a new type of self-organized learning neural networks, namely the Self-Enforcing Network SEN. After introducing our theoretical and methodical frame, basically orientated to the theory of Piaget, we show the logical principles of SEN, including a new activation function, a new learning rule, and a specific visualization algorithm. The operations of SEN are demonstrated with the examples of assimilating new perceptions of animals, the application of a SEN to direct marketing, and the usage of a SEN as consulting system for pupils and beginners at the university. Apparently SEN can be used in many different contexts.

Keywords: self-organized learning, neural networks, Self-Enforcing Network, SEN, prototype, reference type.

1 Theoretical and Methodical Frame

Self-organized learning is usually defined as learning without an immediate feed back by the environment of the learning system, in contrast to, e.g. supervised learning or reinforcement learning. More precisely, in the process of self-organized learning the learning system changes its structure according to the learning task via an immanent logic, which is not directly influenced by environmental learning conditions. A self-organized learning process, hence, might be understood as the construction of an internal cognitive structure, which might be a conscious or an unconscious one. One might also say that by self-organized learning the learning system constructs an explicit order when before the learning process there was only an implicit order.

The most famous learning theoretical model of self-organized learning processes is probably the model of cognitive development of Piaget (e.g. [5]). As is well known, Piaget postulates that cognitive development is basically the construction of cognitive schemas by the permanent interplay of assimilation and accommodation. Assimilation means the process of integrating external perceptions into an already existing schema; accommodation is the construction of cognitive schemas or their corrections, i.e. the process of ordering perceptions according to the internal development logic. Both processes that have the goal of reaching equilibrium between the schema and the perceived environment are self-organized in the sense that it is the task of the learning system to build the schemas and to assimilate perceptions into the schemas without direct environmental feed back. To be sure, some times the constructed schemas are not correct and have to be changed according to negative feed back from the learning

I. Rojas, G. Joya, and J. Cabestany (Eds.): IWANN 2013, Part I, LNCS 7902, pp. 518–529, 2013.

system's environment. Yet the two basic processes of assimilation and accommodation operate according to their specific logic and are influenced by the environment only as regards content, i.e. which perceptions the environment presents to the learning system.

The model of Piaget model demonstrates the basic importance of self-organized learning processes not only for human learning systems. That is why we base our own development of a self-organized learning neural network, namely the Self-Enforcing Network (SEN), on the thought and results of Piaget.

One first step is the idea to define cognitive schemas as so-called semantical networks. Such networks are basically directed and frequently weighted graphs with semantical concepts as knots and "associative" connections between the concepts as edges. One can understand the weighted connections as "associative" links in the way that strong connections, i.e. high weight values, between two concepts A and B mean immediate associations of B if A is perceived and vice versa.

The next step is the characterization of the knots, i.e. the concepts. When using a SEN the concepts usually represent objects that are characterized by certain attributes. Each object, hence, is unambiguously defined by an attribute vector. One might interpret the strength of connections between them as a measure for the similarities of the according attribute vectors.

A third important methodical step is the introduction of so-called prototypes. The cognitive psychologist Rosch [6] discovered that human learners do not construct general categories according to logical principles but that they rather use prototypes to organize external perceptions. If a child, for example, learns that there are several objects that are all "dogs" the child does not define a category "dog" to which all according perceptions belong but selects a "prototypical" dog and clusters all objects around that prototype of "dog". The usage of prototypes can be very useful also in contexts outside cognitive problems.[1] Because the term "prototype" has a cognitive meaning we use the term "reference type" for problems that are not cognitive ones.

The last step is the introduction of a learning rule that is orientated to the famous principle of Hebb [1], namely the enforcing of the connections between the units of the artificial network. We use the Hebb principle for the construction of SEN in its simplest form.

2   The Self-Enforcing Network SEN

The chief function of a SEN is the ordering or classifying respectively of data sets, i.e. objects with certain attributes. Hence each SEN operates on a database consisting of such objects and attributes. Usually these data are represented in a "semantical matrix": The rows of the matrix are constituted by the objects and the columns by the according objects. The elements of the matrix are numerical values usually between zero and one, i.e. the affiliation degree of an attribute to an object. The values of the semantical matrix must, of course, be externally inserted by the user. The user hence

---

[1] Already Kohonen [4] proposed the usage of prototypes.

represents the environment of a self-organized learning system; self-organized learning is only possible by ordering external perceptions or inputs respectively.

The numerical values of the matrix are obtained by scaling techniques known for long in, e.g., quantitative social research. One might call these methods as quantifications of qualitatively perceived or observed proportional relations.

The network SEN consists of only one layer. Its activation function is usually the so-called *logarithmic linear* function

$$A_j = \sum_i w_{ij} \log_3 (A_i + 1)$$ (1)

where A_j is the activation value of the receiving neuron j, A_i the activation values of the sending neurons i, and w_{ij} as usual the according weight values; this function was developed by us.

The base 3 for the logarithm was chosen for the obvious reason that $\log A_i$ of course would become a negative number for all activation values between 0 and 1. Accordingly the logarithm is applied to $(A_i + 1)$. Base 3 was chosen for simple practical reasons: Base 2 would result in too large values, base 4 in too small ones. One can interpret the use of the logarithm as a dampening factor that is "internal" to the function in contrast to "external" factors like, e.g., scale or decay in interactive networks. According to our numerous experiences the logarithmic linear function operates very well in all cases.

The learning rule of SEN that transforms the values of the semantical matrix into the weight matrix of the network is:

$$w(t+1) = w(t) + \Delta w \text{ and}$$

$$\Delta w = c * v_{sm},$$ (2)

where c is a learning rate and v_{sm} is the according value in the semantical matrix.

In other words, the operations of a SEN start by analyzing the values of the semantical matrix v_{sm}. If an object o does not have the attribute a and the according semantical value $v_{oa} = 0$ then the weight value $w_{oa} = 0$ and remains so; in all other cases the weight value

$$w_{oa} = c * v_{oa}.$$

If more learning steps are necessary, i.e. if SEN has not reached an attractor, then

$$w(t + 1) = w(t) + c,$$

$$\text{if } w(t) \neq 0,$$ (3)

if $w(t) = 0$ then $w(t + 1) = 0$ for all learning steps.

In most cases, according to our numerous experiences, it is sufficient to use c = 0.1. To be sure, the weight values between two different objects a and b and two different attributes x and y always are

$$w_{ab} = w_{xy} = 0.$$ (4)

To put it into a nutshell, a SEN learning process consists of a) the transformation of the semantical matrix into a weight matrix according to equation (2), and b) the learning runs, i.e. the enforcing of the weight values, according to (3). The learning process, i.e. the accommodation, is finished when a point attractor has been reached. The result of this learning process is given by the end activation values of those neurons that represent the specific objects.

The assimilation of a new object is done by the comparison of the attribute values of the new object with the end activation vectors of the objects that are already part of the network.

An additional part of a SEN is a visualization algorithm. The classification of the objects, in particular by comparison with one or several reference types, is measured by computing the distance between the final activation values of the respective attributes that characterize the objects. In other words, the final activation values are ordered as a vector and the "distance" to other attribute vectors is measured by the Euclidean distance. A visualization of these distances on a two-dimensional grid can be done in two different ways, according to the problem the user investigates:

The first way is the so-called input centered modus. If a user wants the classification of a new object with respect to objects the SEN already contains he inserts the object and its attributes into the SEN and externally activates the attribute neurons of the new object, namely with those attribute values the user assigns to his object. In the case of an ideal reference type, for example, the attribute neurons are all activated with an activation value $av = 1$. The visualization algorithm of the SEN places the name of the new object in the center of the grid – hence the name of the modus – and places the other objects at random at the periphery of the grid. Then the objects will be "drawn" to the center, according to the size of the distances to the attribute vector of the new object. The nearer one of the drawn objects is to the center the more similar is this object to the new one and vice versa. The second way is the "reference type centered modus". If one wants to classify several objects according to their similarity to a particular reference type the reference type is placed into the center by externally activating the attributes of the reference type. Then by the same logic as in the first modus the objects are drawn to the center; their nearness to the center defines the respective similarity to the reference type.

3   Examples

3.1   Accommodation and Assimilation of Animals

In order to simulate such basic cognitive processes we constructed a semantical matrix as basis for a SEN with 12 animal names as objects and 12 attributes that characterize the objects "more or less", i.e. the values of the semantical matrix are coded as real numbers between 0 and 1. Speaking in the theoretical terms of Piaget the fundamental step of accommodation is the construction of the weight matrix via the learning rule (2) from the semantical matrix and the learning steps according to (3). The animals are among others "dog", "cat", "horse", "chicken", and "hare". The attributes were chosen according to these animals; an abbreviated list is "eats flesh", "domestic animal", "barks", "flying", "eats plants", "swims", "has fur", and "has fins".

To simulate an assimilating process we chose as new input the object "dachshund", characterized by an attribute vector consisting of, e.g., "fur", "four legs", "barking", and "small". This vector was inserted into the "animal SEN", namely placed in the center (input centered modus). The already classified animals serve as reference types and are placed by the visualization algorithm at the periphery of the visualization grid. When the new input was activated according to the procedure described above the different reference types were drawn into the center according to their similarity to the input vector. The result is shown in Fig. 1:

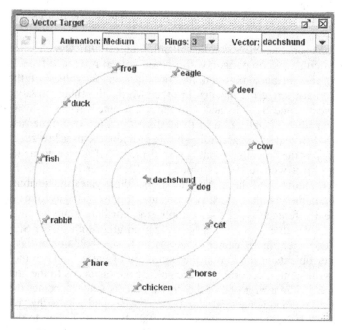

Fig. 1. Result of the assimilation process of "dachshund"

One sees immediately that indeed "dog" is placed very near to "dachshund"; "cat" is also rather near to the center but significantly farer away than "dog". Obviously the result is such, as one would expect from a correct assimilation because "dachshund" is correctly identified as a dog but also similar to a cat. Note that in this example the assimilation is the logical subordination of a single perception under a general category. By the way, in this example the semantical matrix was manually constructed. In case of large data sets the matrix will be automatically constructed from the according database (cf. [2]).

We show this rather simple example just to demonstrate how to simulate the fundamental principles of assimilation and accommodation as basis for self-organized learning. The following examples are rather concrete and show how and that a SEN can be applied to very practical problems.

3.2   Second Example: Selection of an Offshore Location for Winder Energy Plants

The second example deals with the problem of planning new plants for wind energy. Because of the climate crisis and the dangers of nuclear plants (for example the catastrophe at Fukushima in spring 2011) new plants for the generation of regenerative energy, in particular of wind energy, are planned. Germany has rather early developed techniques of generating electricity by wind energy; unfortunately many regions of Germany are not very suited for such plants because the mean wind force is not very high. When the German government decided in summer 2011 to switch off all nuclear power stations until 2022 the need for more and reliable wind energy plants became very urgently. Offshore plants seem to be a suited possibility because in the northern seas there is mostly wind enough; the coasts of Germany all are situated at the North Sea and the Baltic Sea. Yet the decision for the location of a new wind plant depends on many different factors.

One of our students in economy collected for his MSc-thesis real data concerning different possible offshore locations with respect to their (relative) suitability for wind energy stations. Accordingly the semantical matrix of his SEN contains the values of a) seven possible locations in the North Sea and the Baltic Sea near the German coasts, and b) the values for 10 attributes like distance to the coast, depth of water, distance to the nearest harbors, connection possibilities to cable networks, and so forth. Fig. 2 shows the semantical matrix:

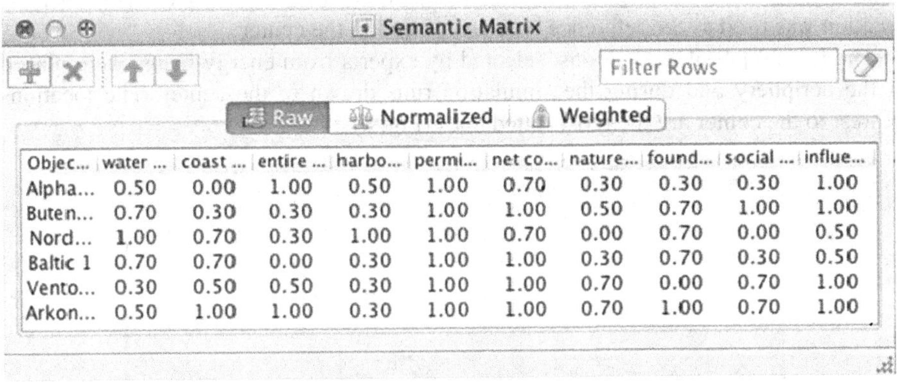

⊗ ○ ⊕				Semantic Matrix						
➕ ✖ ⬆ ⬇							Filter Rows			📝
			Raw	Normalized	Weighted					
Objec...	water ...	coast ...	entire ...	harbo...	permi...	net co....	nature...	found...	social ...	influe...
Alpha...	0.50	0.00	1.00	0.50	1.00	0.70	0.30	0.30	0.30	1.00
Buten...	0.70	0.30	0.30	0.30	1.00	1.00	0.50	0.70	1.00	1.00
Nord...	1.00	0.70	0.30	1.00	1.00	0.70	0.00	0.70	0.00	0.50
Baltic 1	0.70	0.70	0.00	0.30	1.00	1.00	0.30	0.70	0.30	0.50
Vento...	0.30	0.50	0.50	0.30	1.00	1.00	0.70	0.00	0.70	1.00
Arkon...	0.50	1.00	1.00	0.30	1.00	1.00	0.70	1.00	0.70	1.00

Fig. 2. The semantical matrix of the "wind energy SEN"

To impart impressions about the technical processes we show two weight matrices, namely of the first matrix after the transformation of the semantical matrix (Fig. 3) and of the final matrix after the network has reached an attractor (Fig. 4).

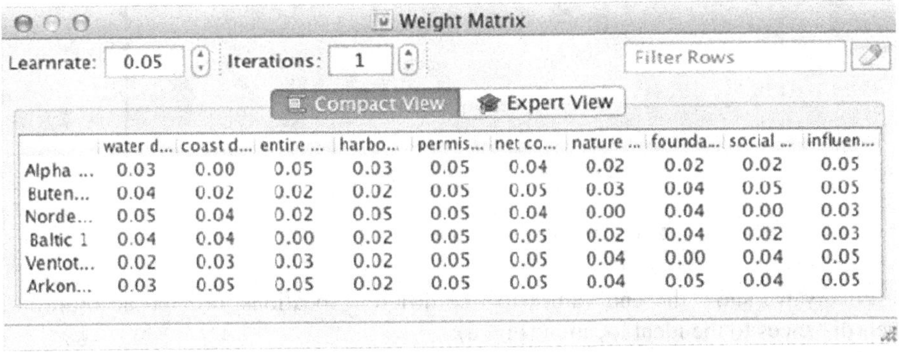

● ○ ○				Weight Matrix						
Learnrate:	0.05	Iterations:	1				Filter Rows			📝
			Compact View		Expert View					
	water d...	coast d...	entire ...	harbo...	permis...	net co...	nature ...	founda...	social ...	influen...
Alpha ...	0.03	0.00	0.05	0.03	0.05	0.04	0.02	0.02	0.02	0.05
Buten...	0.04	0.02	0.02	0.02	0.05	0.05	0.03	0.04	0.05	0.05
Norde...	0.05	0.04	0.02	0.05	0.05	0.04	0.00	0.04	0.00	0.03
Baltic 1	0.04	0.04	0.00	0.02	0.05	0.05	0.02	0.04	0.02	0.03
Ventot...	0.02	0.03	0.03	0.02	0.05	0.05	0.04	0.00	0.04	0.05
Arkon...	0.03	0.05	0.05	0.02	0.05	0.05	0.04	0.05	0.04	0.05

Fig. 3. The initial weight matrix

	water d...	coast d...	entire ...	harbo...	permis...	net co...	nature ...	founda...	social ...	influen...
Alpha ...	0.10	0.00	0.20	0.10	0.20	0.14	0.06	0.06	0.06	0.20
Buten...	0.14	0.06	0.06	0.06	0.20	0.20	0.10	0.14	0.20	0.20
Norde...	0.20	0.14	0.06	0.20	0.20	0.14	0.00	0.14	0.00	0.10
Baltic 1	0.14	0.14	0.00	0.06	0.20	0.20	0.06	0.14	0.06	0.10
Ventot...	0.06	0.10	0.10	0.06	0.20	0.20	0.14	0.00	0.14	0.20
Arkon...	0.10	0.20	0.20	0.06	0.20	0.20	0.14	0.20	0.14	0.20

Learnrate: 0.05   Iterations: 4   Filter Rows   Compact View   Expert View   Weight Matrix

Fig. 4. The final weight matrix

Afterwards the student constructed an "ideal location" that is characterized by best values for all attributes, regardless if such an ideal location factually exists. This ideal location was used as the reference type and placed in the center.

The factual possible locations, selected by experts from energy firms, were placed at the periphery and during the simulation runs drawn to the center. The locations nearest to the center are the most suited ones (Fig. 5).

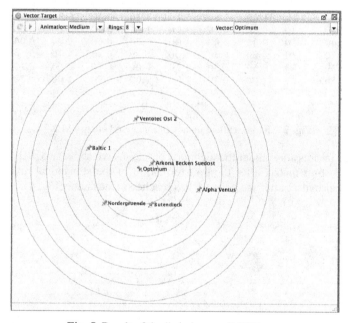

Fig. 5. Result of the "wind energy" SEN runs

Finally we show the end values for the different locations, ordered according to their distances to the ideal location (Fig. 6):

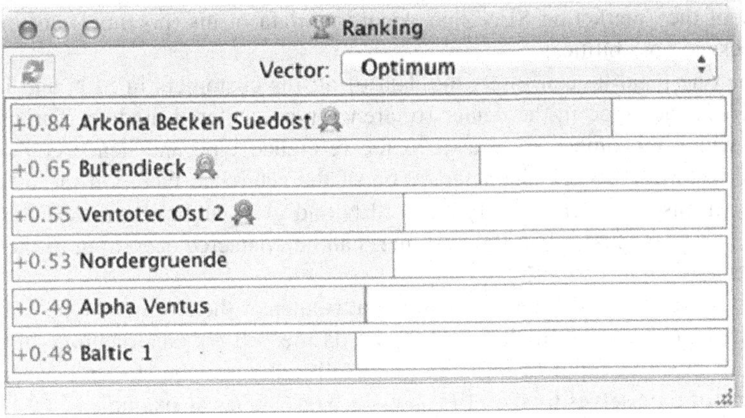

Fig. 6. End activation values of the location objects

The recommendations of the SEN were accepted by experts of energy firms as sound proposals. A comparison of this SEN with a standard software AHP (Analytic Hierarchy Process) for such classification problems showed that in most cases the proposals were equivalent; in cases where the two algorithms differed the experts mainly preferred the proposals of SEN. In addition, SEN has several practical advantages because more possible locations can be analyzed with a SEN at the same time than with AHP. A SEN usually is significantly faster.

3.3   Classifying Customers in Direct Marketing

Direct marketing means contacting potential customers with respect to their interests in buying certain products. The according marketing strategies, hence, must consider the buying behavior and habits of the persons one wishes to win as customers. There exist different techniques for the classification of customers in the field of so-called data mining; we applied a SEN to this problem with the following methodical procedure:

In the first step we constructed five reference types of customers that formed a reference frame for factual customers (in the sociological terminology introduced by Max Weber one could call these reference types *ideal types*). The attributes for the reference types we took from literature about typical customers behavior. We show an extract from the whole set of types and attributes to give an exemplary impression:

Type A: New products (innovations) will be immediately bought;
        direct advertising contacts by e-mails or post are successful.
Type C: New innovative products are only bought after their prize has been reduced;
        New products are only bought if they are factually needed
Type E: advertising does not influence the buying behavior at all;
        New products are only bought when they are needed;

If a user of the "marketing SEN" has according data of his (potential) customers he has two usage possibilities:

On the one hand he can insert the data of all his customers in SEN and inserts a specific reference type in the center (reference type centered modus). The user then can see which customers are drawn to the reference type and hence could be approached according to the characterization of the reference type. On the other hand the user can insert all reference types in SEN and gives as input a specific customer (input centered modus). Then the customer can be contacted according to the nearest reference type.

We constructed a questionnaire with the attributes of the reference types and asked twenty students of us to fill it out. Afterwards the self-characterizations of the students were given as input to SEN. In practically all cases the students accepted the classifying of themselves by the SEN as valid. Fig. 7 gives an example:

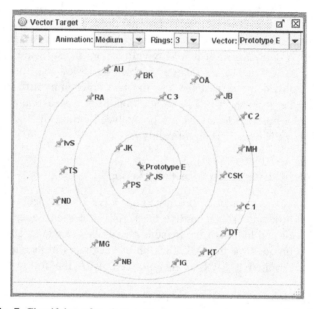

Fig. 7. Classifying of students with the reference type centered modus

Obviously only the students JK, PS and JS are customers of type E

In a former experiment we undertook the same categorizations with a specific Kohonen Feature Map (KFM). As this is a well-known algorithm we just show the result for the same prototype.

The results are apparently equivalent; according to several experts in direct marketing the SEN results are much easier to understand. In addition the SEN is much faster than the KFM, which needed several hundreds iterations in contrast to 3 – 4 learning steps of the SEN.

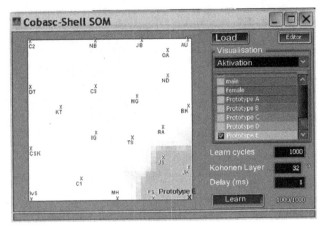

Fig. 8. Classifying of the same students by a KFM

3.4   The Consulting System OSWI

The faculty of economics and computer science at the university of Duisburg-Essen consists of four major disciplines and the according courses of study, namely economics, business administration, computer science, and business information technology. In commission of the faculty we developed a consulting system for pupils in the last school classes and for beginners at the university to advise potential students interested in the faculty, which courses of study they should (and could) select. The result is the consulting system OSWI, Orientating System for Economics and Computer Science, that is constructed as a twofold SEN, i.e. SEN based advising on two levels.

The first level gives advices for the choice of one of the four courses of study. We asked each of the forty professors of the faculty to name 5 – 7 abilities and fields of interest a student should have if he wishes to study the according course. We got 26 meaningful answers like "ability to think mathematically", "interest in economical problems", "ability to think in systemic models", "interest and experience in software development", and so forth. This way we characterized each course of study with the according attribute vector. The different attribute vectors are of course not disjoint because, e.g., the attribute "ability to think mathematically" was named by nearly all professors of the different disciplines. The four courses of study were inserted into the "first level SEN" as reference types.

A user of this system gets the list of 26 attributes and is asked to characterize himself by choosing 10 – 12 attributes according to his interests and his self-image with respect to his abilities. He shall order his selections according to the relevance for him: The first is most important for him or he judges himself to be very good in it respectively and so on. This "self estimation vector" is inserted into SEN (input centered modus) and externally activated, namely the first component with the activation value 1, the second with 0.9, and so on, i.e. only the first 10 elements are taken into regard. Fig. 9 shows the recommendation for one pupil:

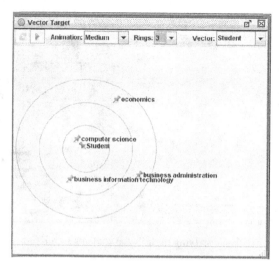

Fig. 9. Recommendation for one potential student, who should choose computer science

The second level enables the user to select certain sub disciplines of a course of study, for example "software development" and "system modeling" in the case of computer science or "marketing" and "commercial law" in the case of business administration. The construction process for this level was methodically the same as for level 1; to be sure, the answers of the professors had to be more in detail than those of the first level. The usage is the same as in the case of the first level.

Since finishing the construction of OSWI it has been used by several hundred students and pupils. In nearly all cases the users were rather content with the recommendations OSWI gave them. In particular OSWI confirmed many advanced students that their choice was right for them; in some cases OSWI on the contrary confirmed the doubts of the students if they had chosen the right course or sub disciplines in their course. Several students told us that they considered changing their main focus of study after using OSWI. Because OSWI is apparently very suited for consulting pupils and students we plan at present to construct other SEN based consulting systems for the other faculties of our university.

4   Conclusion

SEN based models have been proved to be very useful in a lot of other fields. Some additional examples can be found in [2] and [3], for example a SEN based medical diagnosis system, the forecasting of the probable selling success of cell phones, and the selection of the best methods of procedure in project planning. Apparently SEN is not only able to simulate processes of self-organized learning but can also successfully be used for the ordering of data sets according to the interests of users.

References

1. Hebb, D.O.: The Organization of Behavior. Wiley, New York (1949)
2. Klüver, C.: Solving problems of project management with a self enforcing Network (SEN). In: Klüver, C., Klüver, J. (eds.) Social-cognitive Complexity, Computational Models and Theoretical Frames. Special issue of CMOT. Dordrecht (NL), vol. 18(2), pp. 145–152. Springer Science+Business Media, Dordrecht, NL (2012)
3. Klüver, J., Klüver, C.: Social Understanding. On Hermeneutics, Geometrical Models, and Artificial Intelligence. Springer, Dordrecht, NL (2011)
4. Kohonen, T.: The »Neural« Phonetic Typewriter. IEEE Computer 21(3), 11–22 (1988)
5. Piaget, J.: The Principles of Genetic Epistemology. Routledge, London (1972)
6. Rosch, E.: Natural Categories. In: Cognitive Psychology, vol. 4, pp. 328–350 (1973)

Network Anomaly Detection with Bayesian Self-Organizing Maps

Emiro de la Hoz Franco[1,3], Andrés Ortiz García[2], Julio Ortega Lopera[1], Eduardo de la Hoz Correa[1,3], and Alberto Prieto Espinosa[1]

[1] Computer Architecture and Technology Department,
CITIC University of Granada, 18060 Granada, Spain
[2] Department of Communications Engineering
University of Málaga, 29071 Málaga, Spain
[3] Systems Engineering Program
Coast University, Barranquilla, Colombia

Abstract. The growth of the Internet and consequently, the number of interconnected computers through a shared medium, has exposed a lot of relevant information to intruders and attackers. Firewalls aim to detect violations to a predefined rule set and usually block potentially dangerous incoming traffic. However, with the evolution of the attack techniques, it is more difficult to distinguish anomalies from the normal traffic. Different intrusion detection approaches have been proposed, including the use of artificial intelligence techniques such as neural networks. In this paper, we present a network anomaly detection technique based on Probabilistic Self-Organizing Maps (PSOM) to differentiate between normal and anomalous traffic. The detection capabilities of the proposed system can be modified without retraining the map, but only modifying the activation probabilities of the units. This deals with fast implementations of Intrusion Detection Systems (IDS) necessary to cope with current link bandwidths.

1 Introduction

Nowadays, with the growth of Internet, not only the number of interconnected computers but also the relevance of network applications has increased considerably. At the same time, the trend to on-line services available through the Internet has exposed a lot of sensitive information to intruders and attackers [2]. This way, there are three main alternatives to protect information. The first consist in avoid sending data without any encryption. These systems encrypt the information before sending in order to preserve its privacy. The second consist on using a separate physical or logical channel to transfer the information as in Virtual Private Networks (VPN). However, the previous approaches do not react to attackers or intruders but they only suppose a passive protection to reduce the exposure of the information. On the other hand, the complexity of newer attacks makes necessary the use of elaborated techniques such as pattern classification or artificial intelligence techniques for successfully detecting an attack or

I. Rojas, G. Joya, and J. Cabestany (Eds.): IWANN 2013, Part I, LNCS 7902, pp. 530–537, 2013.
© Springer-Verlag Berlin Heidelberg 2013

just to differentiate among normal and anomalous traffic. There are two design approaches to IDS/IPS [4]. The first consists on looking for patterns that corresponds to known signatures of intrusions. The second one searches for abnormal patterns by using more complex features which allow discovering not only an intrusion but also a potential intrusion. This can be figured out, for instance, by discovering a misuse of the protocol flags or an abnormal number of certain events (such as the number of connection attempts in TCP). Nevertheless, discovering these attacks requires the calculation of more complex features.

The rest of the paper provides the description of the proposed procedure based on probabilistic SOM by using Gaussian Mixture Models (GMM) in Section 2. Then Section 3 describes the dataset used for the experiments and the feature selection accomplished to distinguish between anomalies and normal traffic in the experiments. Finally, Section 4 provides the experimental results obtained to analyze the performance of the proposed approach and Section 5 summarizes the conclusions of the paper.

2   Probabilistic SOM Using Gaussian Mixture Models

SOM [5] is one of the most used artificial neural network models for unsupervised learning. SOM groups similar data instances close in into a two or three dimensional lattice (output map). On the other hand, different data instances will be apart in the output map. Moreover, some important features of the input space can be inferred from the output space [3].

1. Input space modelling. The prototypes ω_{ij} (where (i,j) refers to the unit index in the map) computed during the SOM training provide an approximation to the input space.
2. Topological order. Units on the output map are arranged into a 2D or 3D lattice, and their position depends on the specific features of the input space.
3. Feature selection. SOM algorithm produces a number of prototypes from the input data space. Thus, the algorithm reduces the input space, as it is represented by the prototype vectors.

The SOM algorithm can be briefly explained as follows. Let $D \in \mathbb{R}^n$ be a n-dimensional data manifold. The SOM map is composed by d units, each represented by a n-dimensional model vector ω_i. For each input data instance, the Best Matching Unit (BMU) is defined as the unit with the closest model vector ω_i:

$$\|\omega_i - v\| \le \|\omega_j - v\| \forall v \in D, i \neq j \tag{1}$$

where $\|\cdot\|$ is the Euclidean distance. Once the BMU is determined for the current iteration, the model vectors are updated according to the rule

$$\omega_i(t+1) = \omega_i(t) + \alpha(t)h_i(t)(v - \omega_i(t)) \tag{2}$$

where $\alpha(t)$ is the learning rate and $h_i(t)$ is a function which defines the neighbourhood around the BMU ω_i Usually, $\alpha(t)$ diminishes following an exponential decay rule and h_i is a gaussian hat whose width shrinks in time (iterations). SOM has been linearly initialized in order to avoid random effects. Hence, SOM prototypes are initialized along the two first principal directions of the training data [13]. Once the map is already trained (i.e. the training data has been already presented to the map), each vector represents a set of input vectors. In other words, SOM quantizes the data manifold in d n-dimensional model vectors. This learning model activates a specific unit when a new data instance is presented to the SOM (i.e. the BMU). However, it is possible to measure the response of the map units instead of calculating the BMU as the unit which is closest to the input data. This probabilistic interpretation of the SOM is addressed by modelling the prototypes using a Gaussian Mixture Model (GMM), which fuzzifies the response of the SOM units [1]. The main goal is to train the map only once, while further tuning of the map response (i.e. the classification results) can be achieved by modifying the prior activation probabilities of the SOM units, as the activation level of the SOM units can be used to recognize patterns associated to normal connections and network anomalies. Thus, anomalies can be detected as they impose a deviation from the normal activation patterns [11]. Thus, the BMU is determined not only computing the minimum distance from an input vector but also taking into account the likelihood of an unit to be the BMU. In our experiments, the prior probability of each map unit i is computed taking into account the activation probability for the training set, in a similar way than in [1], as shown in equation 3

$$p(i) = \frac{\#\widetilde{X_i}}{\#\widetilde{X}} \tag{3}$$

where $\#\widetilde{X}$ is the total number of input vectors and $\#\widetilde{X_i}$ is the number of vectors whose closest prototype is ω_i is the number of sample vectors as defined on equation 4.

$$\widetilde{X_i} = \{x \in V \ / \ \|x - \omega_i\| \leq \|x - \omega_k\| \ k = 1, ...N\} \tag{4}$$

Thus, $\#\widetilde{X_i}$ can be defined as the set of data samples whose first BMU is the unit i (Voronoi set of unit i).

The GMM is built according to the equation 5 using d components, where the weights p_i for each gaussian component corresponds to the prior probabilities computed in equation 3.

$$P(x_1...x_n) = \sum_N p_i P_i(x_1..x_n) \tag{5}$$

In Equation 5, each individual Gaussian component P_i corresponds to the *n-dimensional* weights associated to each unit [1,11]. The mean of each individual gaussian component (kernel center) is the weight vector of the corresponding unit itself, while the covariance matrix for the i-component is given by the dispersion of the data samples around the model vector ω_i. Once the GMM model has been built, the response of the unit k can be computed as the posterior probability by using the Bayes theorem.

$$p(\omega_k|x) = \frac{p(x|\omega_k)P(\omega_k)}{p(x)} \tag{6}$$

In equation 6, $p(\omega_k|x)$ represents the probability that a sample vector x belongs to class ω_k, while $p(x|\omega_k)$ is the probability density function of the prototype ω_k computed from the GMM and $p(x)$ is a normalization constant.

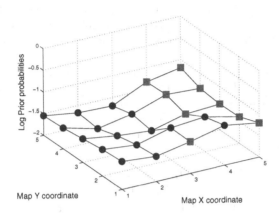

Fig. 1. SOM *a priori* activation probabilities for each unit. Circles and squares represent normal connections and network anomalies, respectively.

Figure 1 shows *a priori* activation probabilities for each map unit. In addition, normal connections are mapped in units represented as circles, while network anomalies are mapped in units represented as squares.

3 Experimental Setup: Benchmark and Feature Selection

3.1 Databases

The KDD'99 dataset [8] contains about 4GB of compressed data from captures of *tcpdump* [6] in the DARPA'98 IDS evaluation program [7], corresponding to about 7 weeks of network traffic. The extracted features contained on the KDD'99 dataset can be classified into three classes: basic features, traffic-based features and content-based features.

Nevertheless, the KDD'99 dataset has inherent problems due to the synthetic characteristic of the data [8,7], that may not be representative enough of real attacks. Thus, the Network Security Lab - Knowledge Discovery and Data Mining (NSL-KDD) dataset was proposed to overcome most of the deficiencies of KDD'99 stated in [7]. Moreover, in NSL-KDD redundant records were removed and the attacks were labeled and sorted by their level of detection difficulty. Then as in the most recent works, we also use the NSL-KDD dataset. However, feature description, indexing and attack classification described for KDD'99 database also applies to NSL-KDD as it can be considered as a subset of the KDD'99 dataset.

3.2   Feature Selection

NSL-KDD dataset provides 41 different features to identify a network anomaly (or normal connections). However, not all the features are discriminative enough for the classifier. Additionally, we noticed that many features are 0 or singular for almost all the data samples. These features could add noise to the classifier diminishing the classification performance. Unfortunately, removing singular or almost singular features does not guarantee a higher classification performance. This way, a feature selection stage is used to keep only the features which maximizes the separation ability. Moreover, this reduces the dimensionality of the input space and results in a lower computational burden. The feature selection has been accomplished by using the *Fisher Discriminant Ratio* which is characterized by its separation ability as shown in [9,12]. For the two-class separation case, FDR is defined as

$$FDR = \frac{(\mu_1 - \mu_2)^2}{\sigma_1^2 + \sigma_2^2} \tag{7}$$

where μ_i and σ_i are the mean and variance values of each input variable respectively. This way, the FDR value increases as the variable is more discriminant between the two classes and FDR is used along all the training set to compute the most discriminating features. Then, features whose FDR value is above a predefined threshold are selected.

4   Experimental Results

The proposed method to detect network anomalies has been evaluated by using cross-validation. Training and testing data is provided by the NSL-KDD as separate datasets. Thus, it is not necessary to extract subsets for cross-validation assessment from the database. Classification performance has been assessed by computing three statistical measures. Sensitivity can be defined as the ability of the classifier to detect positive results (i.e. anomalies) and it is described in equation 8. Specificity measures the ability to detect negative results (i.e. normal connections) as described in equation 9. In addition, Accuracy, as defined in equation 10, measures the percentage of samples correctly classified.

$$SENS = \frac{TP}{TP + FN} \tag{8}$$

$$SPEC = \frac{TN}{TN + FP} \tag{9}$$

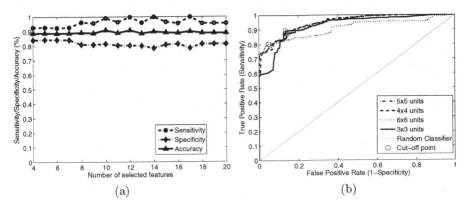

Fig. 2. Classification results as a function of the number of selected features (a) and ROC curves for different map sizes (b)

$$ACC = \frac{TP + TN}{TP + TN + FP + FN} \tag{10}$$

Two different experiments have been performed. The first one aims to determine the number of features which maximizes the classification performance. The results of these experiments are shown in Figure 2. As shown in Figure 2a, the classifier is able to discriminate two classes using 10 features. The specific selected features are *34, 29, 12, 38, 25, 39, 26, 35, 30, 40* as indexed in [8]. Although selecting more than 10 features provides higher accuracy values, it increases the computational burden.

The second one has been performed to find out the number of SOM units for the maximum performance. The number of SOM units plays an important role on the quantization auto-organization process. Furthermore, the distance between units belonging to different classes also depends on the map size. Thus, the *Receiver Operating Curves* (ROC) have been computed measuring the difference between the mean probability activation for units labeled as *normal* and *anomaly* while varying the map size. The ROC curves are shown in Figure 2b, where it is shown that maximum performance occurs for map size of 5x5 units (i.e. maximum area under ROC curve). In this case, the cut-off point provides sensitivity values of 0.9 and Specificity values of 0.89. Additionally, we show the activation of the map for both, normal and anomaly samples. Thus, Figure 3 shows the mean posterior activation probabilities of the SOM units for 100 normal and anomaly samples (Figures 3a and 3b, respectively). Thus, normal samples activate most of normal units, keeping anomaly units unactivated. On the other hand, anomaly entries activate anomaly units and the activation probability for normal units is lower in this case.

This way, the classification results with the SOM trained using the above indicated parameters, can be summarized in Table 1.

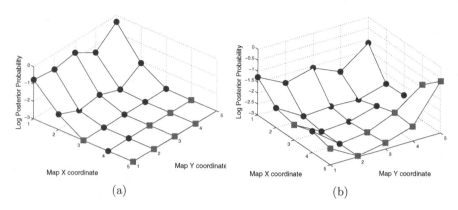

Fig. 3. Activation maps for (a) normal and (b) anomaly samples. SOM trained with 10 features as indicated in the text.

Table 1. Classification results for Different classification methods

Method	Number of features	Accuracy
Our Approach	**10**	**88.3%**
Naïve Bayes [10]	41	76.5%
Random Forest [10]	41	80.67%
Decision Trees [10]	41	81.05%

5   Conclusions

In this paper we present an unsupervised method for network anomaly detection based on the self-organizing map and gaussian mixture models. The prototypes generated by the auto-organization process are modelled by d gaussians where d is the number of SOM units. The proposed system is trained only once, and classification is performed by means of the mixture models. Additionally, a feature selection stage based on the *Fisher Discriminant Ratio* has been used to keep the most discriminative features. Moreover, parameters of the classifier such as the map size, have been optimized by computing the ROC curves. Using the activation probabilities computed during the training stage, we obtained sensitivity, specificity and accuracy values up to 93%, 81% and 88%, respectively. On the other hand, classification capabilities can be modified by varying the prior activation probabilities of the SOM units, avoiding to train the SOM. This way, the accuracy may be improved by tuning the detection threshold. Although these prior probabilities can be modified by the network administrator, it is also possible to adjust them automatically. Thus, as future work, we plan to tune the prior activation probabilities by means of multiobjective optimization. On the other hand, several SOMs could be combined in a SOM ensemble to build a hierarchical model in order to classify not only normal and abnormal connections but also the four types of attacks described in the dataset.

Acknowledgements. This work has been funded by the Ministerio de Ciencia e Innovación of the Spanish Government under Project No. TIN2012-32039.

References

1. Alhoniemi, E., Himberg, J., Vesanto, J.: Probabilistic measures for responses of self-organizing map units. In: Proc. of the International ICSC Congress on Computational Intelligence Methods and Applications (CIMA), vol. 1, pp. 286–290 (1999)
2. Ghosh, J., Wanken, J., Charron, F.: Detecting anomalous and unknown intrusions against programs. In: Proc. of the Annual Computer Security Applications Conference, vol. 1, pp. 259–267 (1998)
3. Haykin, S.: Neural Networks, 2nd edn. Prentice-Hall (1999)
4. Hoffman, A., Schimitz, C., Sick, B.: Intrussion detection in computer networks with neural and fuzzy classifiers. In: International Conference on Artificial Neural Networks (ICANN), vol. 1, pp. 316–324 (2003)
5. Kohonen, T.: Self-Organizing Maps. Springer (2001)
6. Lippmann, R.P., Fried, D.J., Graf, I., Haines, J.W., Kendball, K.R., McClung, D., Weber, D., Webster, S.E., Wyschgrod, D., Cuningham, R.K., Zissman, M.A.: Evaluating intrusion detection systems: the 1998 darpa off-line intrusion detection evaluation. Descex 2, 1012–1027 (2000)
7. McHugh, J.: Testing intrusion detection systems: a critique of the 1998 and 1999 darpa instrusion detection systems evaluation as performed by lyncoln laboratory. ACM Transactions on Information and Systems Security 3(4), 262–294 (2000)
8. Network Security Lab - Knowledge Discovery and Data MininG (NSL-KDD) (2007), http://kdd.ics.uci.edu/databases/kddcup99/kddcup99.html
9. Padilla, P., López, M., Górriz, J.M., Ramírez, J., Salas-González, D., Álvarez, I.: The Alzheimer's Disease Neuroimaging Initiative. NMF-SVM based CAD tool applied to functional brain images for the diagnosis of Alzheimer's disease. IEEE Transactions on Medical Imaging 2, 207–216 (2012)
10. Panda, M., Abraham, A., Patra, M.R.: Discriminative multinomial naïve bayes for network intrusion detection. In: Proc. of the 6th International Conference on Information Assurance and Security, IAS (2010)
11. Riveiro, M., Johansson, F., Falkman, G., Ziemke, T.: Supporting maritime situation awareness using self organizing maps and gaussian mixture models. In: Proceedings of the 2008 Conference on Tenth Scandinavian Conference on Artificial Intelligence (SCAI), vol. 1, pp. 84–91 (2008)
12. Theodoridis, S., Koutroumbas, K.: Pattern Recognition. Academic Press (2009)
13. Vesanto, J., Himberg, J., Alhoniemi, E., Parhankangas, J.: Som toolbox. Helsinki University of Technology (2000)

A Novel Neural Network Parallel Adder

Fangyue Chen[1,*], Guangyi Wang[2], Guanrong Chen[3], and Qinbin He[4]

[1] School of Sciences, Hangzhou Dianzi University
Hangzhou, Zhejiang, 310018, P.R. China
fychen@hdu.edu.cn
[2] School of Electronics Information, Hangzhou Dianzi University
Hangzhou, Zhejiang, 310018, P.R. China
[3] Department of Electronic Engineering,
City University of Hong Kong, Hong Kong SAR, P.R. China
[4] Department of Mathematics,
Taizhou University, Linhai 317000, China

Abstract. Addition is the most commonly used arithmetic operation and is the speed-limiting element in the core of arithmetic logic unit (ALU) in a microprocessor. Perceptron of feedforward neural networks, inspired by the threshold logic unit neuron model of McCulloch and Pitts, is one of the most important aspects of artificial neural networks (ANN). This paper proposes a design of neural network parallel adder (NNPA) under the framework of multi-layer perceptron (MLP) of binary feedforward neural networks (BFNN). The DNA-like learning algorithm proposed by the present authors is successfully used for training the weight-threshold values of NNPA. Moreover, the efficiency of NNPA is compared with that of the conventional adder such as carry-ripple adder and carry-look-ahead adder. It is shown that some advantages of ANN such as synchronous, parallel and fast speed in information processing are sufficiently taken by the current NNPA.

Keywords: Arithmetic logic unit, perceptron, binary feedforward neural networks, neural network parallel adder.

1 Introduction

It is well known that the design of microprocessors for many applications relies on three major criteria: low power consumption, reduced chip size, and especially high speed. Addition is the most commonly used arithmetic operation and is the speed-limiting element in the core of the arithmetic logic unit (ALU) in every microprocessor. Thus, optimizing the adder's structure and increasing its speed is always an issue to be studied in microprocessor design [1,2,3,4].

When two n-bit binary numbers, $A = (a_{n-1}, a_{n-2}, \cdots, a_0)$ and $B = (b_{n-1}, b_{n-2}, \cdots, b_0)$, are added to obtain the n-bit sum, $S = (s_{n-1}, s_{n-2}, \cdots, s_0) = A + B$, the following relations must are satisfied:

$$\begin{cases} s_i = a_i \oplus b_i \oplus c_i \\ c_{i+1} = a_i \cdot b_i + (a_i + b_i) \cdot c_i, \ (i = 0, 1, \cdots, n-1), \end{cases} \tag{1}$$

---

* Corresponding author.

I. Rojas, G. Joya, and J. Cabestany (Eds.): IWANN 2013, Part I, LNCS 7902, pp. 538–546, 2013.
© Springer-Verlag Berlin Heidelberg 2013

where "$\oplus$", "+", and "·" are the logic XOR, OR and AND operations respectively, and c_0 the initial carry and c_{i+1} is the carry of the i-th operation $(i = 0, 1, \cdots, n - 1)$.

A network with inputs a_i, b_i $(i = 0, 1, \cdots, n - 1)$ and c_0, which produces outputs s_i $(i = 0, 1, \cdots, n - 1)$, and the c_n specified above, is called an n-bit parallel adder. A logic network with inputs a_i, b_i, and c_i, which realizes s_i and c_{i+1} as two output functions for a particular i, is a one-bit full adder. An n-bit parallel adder constructed by cascading n stages of one-bit full adders is called a carry-ripple adder (CRA). Although CRAs are usually used where a high-speed adder is not required or the compactness of a network is most important, the CRA is faster than the carry-look-ahead adder (CLAA) [1] in some electronic implementation (e.g., the CRA was preferred for high speed in Intel's MOS microprocessor 8080 due to greater parasitic capacitance of the CLAA).

Consider a network of n-bit parallel adder as shown in **Fig. 1**, which defines $n + 1$ Boolean functions s_0, s_1, $\cdots$, s_{n-1} and c_n with $2n + 1$ input variables $(c_0, a_{n-1}, a_{n-2}, \cdots, a_0, b_{n-1}, b_{n-2}, \cdots, b_0)$. Here, it should be noted that the network may not be in the form of a cascade of identical one-bit full adder module.

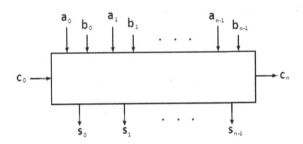

Fig. 1. Schematic diagram of n-bit parallel adder

In this paper, we propose a design of a group of multi-layer perceptron (MLP) of binary feedforward neural networks (BFNN) to realize the parallel adder. The MLPs obtained here is called a neural network parallel adder (NNPA).

2   Design of NNPA

2.1   Truth Table of NNPA

Based on the definition of the addition operation, the truth table of NNPA is shown in Table 1. In the table, $O = (0, 0, \cdots, 0)^T$ and $I = (1, 1, \cdots, 1)^T$ are $(2^n \times 1)$-order arrays respectively, and every $O^{(i)} = (0, 0, \cdots, 0, \underbrace{1, \cdots, 1}_{i})^T$ is also

Table 1. Truth table of NNPA

c_0	$a_{n-1}, \cdots, a_0$	$b_{n-1}, \cdots, b_0$	c_n	$s_{n-1}, \cdots, s_0$
O	$A^{(0)}$	B	$O^{(0)}$	$S^{(0)}$
O	$A^{(1)}$	B	$O^{(1)}$	$S^{(1)}$
$\vdots$	$\vdots$	$\vdots$	$\vdots$	$\vdots$
O	$A^{(2^n-2)}$	B	$O^{(2^n-2)}$	$S^{(2^n-2)}$
O	$A^{(2^n-1)}$	B	$O^{(2^n-1)}$	$S^{(2^n-1)}$
I	$A^{(0)}$	B	$O^{(1)}$	$S^{(1)}$
I	$A^{(1)}$	B	$O^{(2)}$	$S^{(2)}$
$\vdots$	$\vdots$	$\vdots$	$\vdots$	$\vdots$
I	$A^{(2^n-2)}$	B	$O^{(2^n-1)}$	$S^{(2^n-1)}$
I	$A^{(2^n-1)}$	B	I	$S^{(0)}$

a $(2^n \times 1)$-order array, whose last i elements are "1", $(i = 0, 1, \cdots, 2^n - 1)$. At the same tine, B is a $(2^n \times n)$-order array:

$$
B = \begin{pmatrix}
0 & 0 & \cdots & 0 & 0 \\
0 & 0 & \cdots & 0 & 1 \\
0 & 0 & \cdots & 1 & 0 \\
& & \vdots & & \\
1 & 1 & \cdots & 1 & 0 \\
1 & 1 & \cdots & 1 & 1
\end{pmatrix},
$$

where the decimal code of the i-th row $(b_{i,0}, b_{i,1}, \cdots, b_{i,n-1})$ of B, $\sum_{j=0}^{n-1} b_{i,j} \times 2^{n-1-j}$, is i $(i = 0, 1, \cdots, 2^n - 1)$. $A^{(0)}$ is a $(2^n \times n)$-order zero array, and $A^{(i)}$ is the array with all identical rows, and the decimal code of every row is i $(i = 1, 2, \cdots, 2^n - 1)$.

Moreover, $S^{(0)} = B$, and $S^{(i+1)}$ is the array generated by shifting the 1-th row to the last row in $S^{(i)}$ $(i = 0, 1, 2, \cdots, 2^n - 2)$. For example,

$$
S^{(1)} = \begin{pmatrix}
0 & 0 & \cdots & 0 & 1 \\
0 & 0 & \cdots & 1 & 0 \\
& & \vdots & & \\
1 & 1 & \cdots & 1 & 0 \\
1 & 1 & \cdots & 1 & 1 \\
0 & 0 & \cdots & 0 & 0
\end{pmatrix}, \quad
S^{(2)} = \begin{pmatrix}
0 & 0 & \cdots & 1 & 0 \\
0 & 0 & \cdots & 1 & 1 \\
& & \vdots & & \\
1 & 1 & \cdots & 1 & 1 \\
0 & 0 & \cdots & 0 & 0 \\
0 & 0 & \cdots & 0 & 1
\end{pmatrix}.
$$

Furthermore, the 2^{2n+1}-bit binary outputs of the $n+1$ Boolean functions implementing the n-bit parallel adder, say s_0, s_1, s_2, $s_3, \cdots,$ s_{n-1} and c_n, are the last $n + 1$ columns in Table 1. For example, the 2^{2n+1}-bit output of s_{n-1} is: $s_{n-1} = \left(s^{(0)}, s^{(1)}, \cdots, s^{(2^n-1)}, s^{(1)}, s^{(2)}, \cdots, s^{(2^n-1)}, s^{(0)}\right)^T$,

where $s^{(0)} = (\underbrace{0,0,\cdots,0}_{2^{n-1}\ 0s}, \underbrace{1,1,\cdots,1}_{2^{n-1}\ 1s})^T$, and $s^{(i+1)}$ is the 2^n-bit sequence generated by shifting the 1-th symbol of $s^{(i)}$ to its last location, i.e., the first $(2^n - 1)$
symbols of $s^{(i+1)}$ are the last $(2^n - 1)$ symbols of $s^{(i)}$, and the last symbol of $s^{(i+1)}$
is the 1-th symbol of $s^{(i)}$, $(i = 0, 1, \cdots, 2^n - 1)$. Hence, s_0, s_1, s_2, $s_3, \cdots$, s_{n-1}
and c_n are called parallel adder Boolean functions (PABFs).

The main result on PABFs, c_n and s_i $(i = 0, 1, \cdots, n - 1)$, is the following
theorem, which can be proved by using the criterion for LSBF and the DNA-like
learning algorithm proposed in [8,9,14].

Theorem 1. *For any $n \in Z^+$, among all PABFs, c_n is a linearly separable
Boolean function (LSBF), and each s_i is a linearly nonseparable Boolean function
(non-LSBF) but it can be decomposed as the logic XOR operations of 3 LSBFs
$(i = 0, 1, \cdots, n - 1)$.*

2.2   SLP and MLP of BFNN

Perceptron of BFNN, inspired by the threshold logic unit neuron model of Mc-
Culloch and Pitts, introduced by Rosenblatt, is one of the most important
and significant aspects of artificial neural networks (ANN) [5,6,7,8,9]. The SLP
(single-layer perceptron) and MLP (Multi-layer perceptron) are respectively:

(A) SLP with a hard-limiter activation function f:

$$y = f(\textstyle\sum_{i=1}^n w_i u_i - \theta), \tag{2}$$

where f is the first-order jump function defined by

$$f(x) = sign(x) = \begin{cases} 1 & if \ \ x > 0 \\ 0 & if \ \ x \le 0, \end{cases} \tag{3}$$

w_i $(i = 1, 2, \cdots, n)$ and θ are weight-threshold values, $u_i \in \{0, 1\}$ $(i = 1, 2, \cdots, n)$
are binary inputs, and y is the output.

(B) MLP consists of m SLPs connecting the input layer and the hidden layer
as well as one SLP connecting the hidden layer and the output layer, as follows:

$$\begin{cases} y_j = f(\sum_{i=1}^n w_{ij} u_i - \theta_j) \ (j = 1, 2, \cdots, m) \\ y = f(\sum_{j=1}^m \overline{w}_j y_j - \overline{\theta}), \end{cases} \tag{4}$$

where y_j is the output from the j-th neuron in the hidden layer, w_{ij} and θ_j are
the connection weight-threshold values between the i-th input and j-th neuron
in the hidden layer, and $\overline{w}_j$ and $\overline{\theta}$ are the weight-threshold values of the neuron
of the output layer.

It was known that LSBF can be realized by an SLP, and non-LSBF needs
MLP to realize it.

2.3 NNPA Design

For convenience, only consider the case of $n = 4$ in the following NNPA design. For the general case, the results are similar. Let $U = (u_1, u_2, \cdots, u_9) = (c_0, a_3, a_2, a_1, a_0, b_3, b_2, b_1, b_0) \in \{0, 1\}^9$ be the inputs of the five Boolean functions s_0, s_1, s_2, s_3 and c_4, so that their 2^9-bit binary outputs can be obtained from Table 1, as follows.

$s_0 =$
$(0101010101010101 \quad 1010101010101010 \quad 0101010101010101 \quad 1010101010101010$
$0101010101010101 \quad 1010101010101010 \quad 0101010101010101 \quad 1010101010101010$
$0101010101010101 \quad 1010101010101010 \quad 0101010101010101 \quad 1010101010101010$
$0101010101010101 \quad 1010101010101010 \quad 0101010101010101 \quad 1010101010101010$
$1010101010101010 \quad 0101010101010101 \quad 1010101010101010 \quad 0101010101010101$
$1010101010101010 \quad 0101010101010101 \quad 1010101010101010 \quad 0101010101010101$
$1010101010101010 \quad 0101010101010101 \quad 1010101010101010 \quad 0101010101010101$
$1010101010101010 \quad 0101010101010101 \quad 1010101010101010 \quad 0101010101010101)^T.$

$s_1 =$
$(0011001100110011 \quad 0110011001100110 \quad 1100110011001100 \quad 1001100110011001$
$0011001100110011 \quad 0110011001100110 \quad 1100110011001100 \quad 1001100110011001$
$0011001100110011 \quad 0110011001100110 \quad 1100110011001100 \quad 1001100110011001$
$0011001100110011 \quad 0110011001100110 \quad 1100110011001100 \quad 1001100110011001$
$0110011001100110 \quad 1100110011001100 \quad 1001100110011001 \quad 0011001100110011$
$0110011001100110 \quad 1100110011001100 \quad 1001100110011001 \quad 0011001100110011$
$0110011001100110 \quad 1100110011001100 \quad 1001100110011001 \quad 0011001100110011$
$0110011001100110 \quad 1100110011001100 \quad 1001100110011001 \quad 0011001100110011)^T.$

$s_2 =$
$(0000111100001111 \quad 0001111000011110 \quad 0011110000111100 \quad 0111100001111000$
$1111000011110000 \quad 1110000111100001 \quad 1100001111000011 \quad 1000011110000111$
$0000111100001111 \quad 0001111000011110 \quad 0011110000111100 \quad 0111100001111000$
$1111000011110000 \quad 1110000111100001 \quad 1100001111000011 \quad 1000011110000111$
$0001111000011110 \quad 0011110000111100 \quad 0111100001111000 \quad 1111000011110000$
$1110000111100001 \quad 1100001111000011 \quad 1000011110000111 \quad 0000111100001111$
$0001111000011110 \quad 0011110000111100 \quad 0111100001111000 \quad 1111000011110000$
$1110000111100001 \quad 1100001111000011 \quad 1000011110000111 \quad 0000111100001111)^T.$

$s_3 =$
$(0000000011111111 \quad 0000000111111110 \quad 0000001111111100 \quad 0000011111111000$
$0000111111110000 \quad 0001111111100000 \quad 0011111111000000 \quad 0111111110000000$
$1111111100000000 \quad 1111111000000001 \quad 1111110000000011 \quad 1111100000000111$
$1111000000001111 \quad 1110000000011111 \quad 1100000000111111 \quad 1000000001111111$
$0000000111111110 \quad 0000001111111100 \quad 0000011111111000 \quad 0000111111110000$
$0001111111100000 \quad 0011111111000000 \quad 0111111110000000 \quad 1111111100000000$
$1111111000000001 \quad 1111110000000011 \quad 1111100000000111 \quad 1111000000001111$
$1110000000011111 \quad 1100000000111111 \quad 1000000001111111 \quad 0000000011111111)^T.$

$c_4 =$

(0000000000000000   0000000000000001   0000000000000011   0000000000000111
0000000000001111   0000000000011111   0000000000111111   0000000001111111
0000000011111111   0000000111111111   0000001111111111   0000011111111111
0000111111111111   0001111111111111   0011111111111111   0111111111111111
0000000000000001   0000000000000011   0000000000000111   0000000000001111
0000000000011111   0000000000111111   0000000001111111   0000000011111111
0000000111111111   0000001111111111   0000011111111111   0000111111111111
0001111111111111   0011111111111111   0111111111111111   1111111111111111)^T.

Based on Theorem 1, c_4 is an LSBF, and s_0, s_1, s_2, s_3 are non-LSBFs, each of which can be decomposed as the logic $\oplus$ (XOR) operations of 3 LSBFs. If one adopts the decimal code of a Boolean function, then the LSBF decomposition expression of s_0 is

89384704253646338690737862001100389574698410373911265249572016271365425978903643969858446494182087942847195802169569459242311972750555508584002883319016 10
=134078079299425970995740249982058461274793658205923933777235614437217640300735469762129342630858744139340823690496223562829008549262201149695 62649886436010
$\oplus$134077397353454995689317945328453209999629629775885900984198388775043797274581546720819649310230756228913548398868527879162923407888525042576598507308318720
$\oplus$893840223076753633843155573474951382995343819438732413143202895845857584239836414159255969656309060092641856100358624688042537502373604784654361457537 84320.

Furthermore, by using the DNA-like learning algorithm [8,9,14], all SLP and MLP implementing c_4, s_0, s_1, s_2 and s_3 can be obtained as follows.

The SLP implementing c_4 is

$$y = f(2u_1 + 16u_2 + 8u_3 + 4u_4 + 2u_5 + 16u_6 + 8u_7 + 4u_8 + 2u_9 - 31). \qquad (5)$$

The MLP implementing s_0 is

$$\begin{cases} y_1^{(0)} = f(2u_1 + 2u_5 + 2u_9 - 1) \\ y_2^{(0)} = f(2u_1 + 2u_5 + 2u_9 - 3) \\ y_3^{(0)} = f(2u_1 + 2u_5 + 2u_9 - 5) \\ y \;\;= f(2y_1^{(0)} - 2y_2^{(0)} + 2y_3^{(0)} - 1). \end{cases} \qquad (6)$$

The MLP implementing s_1 is

$$\begin{cases} y_1^{(1)} = f(2u_1 + 4u_4 + 2u_5 + 4u_8 + 2u_9 - 3) \\ y_2^{(1)} = f(2u_1 + 4u_4 + 2u_5 + 4u_8 + 2u_9 - 7) \\ y_3^{(1)} = f(2u_1 + 4u_4 + 2u_5 + 4u_8 + 2u_9 - 11) \\ y \;\;= f(2y_1^{(1)} - 2y_2^{(1)} + 2y_3^{(1)} - 1). \end{cases} \qquad (7)$$

The MLP implementing s_2 is

$$\begin{cases} y_1^{(2)} = f(2u_1 + 8u_3 + 4u_4 + 2u_5 + 8u_7 + 4u_8 + 2u_9 - 7) \\ y_2^{(2)} = f(2u_1 + 8u_3 + 4u_4 + 2u_5 + 8u_7 + 4u_8 + 2u_9 - 15) \\ y_3^{(2)} = f(2u_1 + 8u_3 + 4u_4 + 2u_5 + 8u_7 + 4u_8 + 2u_9 - 23) \\ y \ = f(2y_1^{(2)} - 2y_2^{(2)} + 2y_3^{(2)} - 1). \end{cases} \quad (8)$$

The MLP implementing s_3 is

$$\begin{cases} y_1^{(3)} = f(2u_1 + 16u_2 + 8u_3 + 4u_4 + 2u_5 + 16u_6 + 8u_7 + 4u_8 + 2u_9 - 15) \\ y_2^{(3)} = f(2u_1 + 16u_2 + 8u_3 + 4u_4 + 2u_5 + 16u_6 + 8u_7 + 4u_8 + 2u_9 - 31) \\ y_3^{(3)} = f(2u_1 + 16u_2 + 8u_3 + 4u_4 + 2u_5 + 16u_6 + 8u_7 + 4u_8 + 2u_9 - 47) \\ y \ = f(2y_1^{(3)} - 2y_2^{(3)} + 2y_3^{(3)} - 1). \end{cases}$$

$$(9)$$

Summarizing (5) to (9), the structure of 4-bit NNPA is obtained (**Fig. 2**).

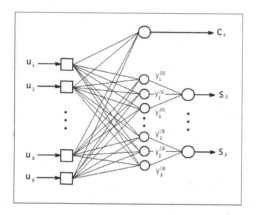

Fig. 2. MLPs implementing NNPA

3   Comparison of NNPA and Digital Circuits

The efficiency of NNPA and digital circuits can be evaluated by comparing the number of neurons in the hidden layers of the MLPs with the number of product terms required to implement the PABFs in Table 1.

First, it is easily known from formula (1) that $s_i = a_i \oplus b_i \oplus c_i = \bar{a}_i\bar{b}_ic_i + \bar{a}_ib_i\bar{c}_i + a_i\bar{b}_i\bar{c}_i + a_ib_ic_i$, and $c_{i+1} = a_ib_i + a_ic_i + b_ic_i$ $(i = 0, 1, 2, 3)$, thus, the number of product terms of 4-bit CRA cascaded by 4 identical one-bit full adder modules is $7 \times 4 = 28$.

On the other hand, in the carry-look-ahead circuits, the expressions of the carries c_1, c_2, c_3 and c_4 are, respectively,

$$c_1 = a_0 b_0 + (a_0 + b_0)c_0,$$
$$c_2 = a_1 b_1 + (a_1 + b_1)[a_0 b_0 + (a_0 + b_0)c_0],$$
$$c_3 = a_2 b_2 + (a_2 + b_2)\{a_1 b_1 + (a_1 + b_1)[a_0 b_0 + (a_0 + b_0)c_0]\},$$
$$c_4 = a_3 b_3 + (a_3 + b_3)\{a_2 b_2 + (a_2 + b_2)\{a_1 b_1 + (a_1 + b_1)[a_0 b_0 + (a_0 + b_0)c_0]\}\}.$$

The numbers of product terms are 3 for c_1, 7 for c_2, 15 for c_3, and 31 for c_4, respectively. In addition, the number of product terms for s_i is 4. So, the sum of the numbers of product terms of 4-bit CLAA is $3 + 7 + 15 + 31 + 4 \times 4 = 72$.

In comparison, the number of neurons in the hidden layers of the MLPs implementing 4-bit NNPA is only 12.

An alternative measure for the efficiency is the connectivity of an implementation, i.e., the number of nonzero weights of the neurons in the hidden layers of the MLP, which can be used to compare the total number of inputs of the AND-gates of the logic circuits implementing the same parallel adder [10]. Due to space limitation, this discussion is omitted here.

One of the most important advantages of ANN is its real-time operation, to be synchronously carried out in parallel. The run speed of NNPA thus is significantly faster than the traditional logic circuit adder such as CRA and CLAA based on the same hardware derive.

It is worth noting that cellular neural network (CNN) is a biologically inspired system in which computation emerges from the collective behavior of some locally coupled simple cells [11,12,13,14,15]. Similarly, other combinational logic circuits such as multiplier, selector, and so on [16,17], can be considered to be designed by using ANN. Thus, a combination of the advantages of both CNN and ANN (especially binary feedforward neural networks, BFNN) will be able to promote various real-world applications in the microprocessor technology.

Acknowledgment. This research was jointly supported by NSFC (Grants 11171084, 60872093 and 61271064) and the Hong Kong Research Grants Council (Grant No. CityU117/10).

References

1. Hung, C.L., Muroga, S.: Minimum parallel binary adders with NOR (NAND) gates. IEEE Trans. Computers 28, 648–659 (1979)
2. Sakurai, A., Muroga, S.: Parallel binary adders with a minimum number of connections. IEEE Trans. Computers 32, 969–976 (1983)
3. Jiang, Y.T., Al-Sheraidah, A., Wang, Y., Sha, E., Chung, J.G.: A novel multiplexer-based low-power full adder. IEEE Trans. Circ. Syst. (II) 51, 345–348 (2004)
4. Lin, S.-H., Sheu, M.-H.: VLSI design of diminished-one modulo $2^n + 1$ adder using circular carry selection. IEEE Trans. Circ. Syst. (II) 55, 897–901 (2008)
5. McCulloch, W.S., Pitts, W.: A logical calculus of the ideas immanent in nervous activity. Bulletin of Math. Biophysics 5, 115–133 (1943)
6. Rosenblatt, F.: The perceptron: A probabilistic model for information storage and organization in the brain. Cornell Aeronautical Laboratory, Psychological Review 65, 386–408 (1958)

7. Hopfield, J.J.: Neural networks and physical systems with emergent collective computational abilities. Proc. Natl. Acad. Sci. 79, 2554–2258 (1982); Neurons with graded response have collective computational properties like those of two-state neurons. Proc. Natl. Acad. Sci. 81, 3088–3092 (1984)

8. Chen, F.Y., Chen, G., He, G.L., Xu, X.B., He, Q.B.: Universal perceptron and DNA-like learning algorithm of binary neural networks: LSBF and PBF Implementation. IEEE Trans. Neural Netw. 20, 1645–1658 (2009)

9. Chen, F.Y., Chen, G., He, Q.B., He, G.L., Xu, X.B.: Universal perceptron and DNA-like learning algorithm of binary neural networks: Non-LSBF Implementation. IEEE Trans. Neural Netw. 20, 1293–1301 (2009)

10. Andree, H.M.A., Barkema, G.T., Lourens, W., Taal, A.: A comparison study of binary feedforward neural networks and digital circuits. Neural Network 6, 785–790 (1993)

11. Chua, L.O., Yang, L.: Cellular neural networks: theory. IEEE Trans. Circuit Syst. 35, 1257–1272 (1988)

12. Chua, L.O., Yang, L.: Cellular neural networks: application. IEEE Trans. Circuit Syst. 35, 1273–1290 (1988)

13. Roska, T.: Cellular wave computers for brain-like spatial-temporal sensory computing. IEEE Circuit Syst. Mag. 5, 5–19 (2005)

14. Chen, F.Y., He, G.L., Chen, G.: Realization of Boolean functions via CNN: mathematical theory, LSBF and template design. IEEE Trans. Circ. Syst. I 53, 2203–2213 (2006)

15. Chen, F.Y., Chen, G., He, Q.B.: DNA-like learning algorithm of CNN template implementing Boolean functions. In: Proc. 2009 IEEE Int. Symposium on Circuit and Syst., Taibei, Taiwan, pp. 2701–2704 (2009)

16. Franco, L., Cannas, S.A.: Solving arithmetic problems using feed-forward neural networks. Neurocomputing 18, 61–79 (1998)

17. Ken-Ichiro, S., Mititada, M., Hideyuki, K.: Simulation of neural network digital adder and multiplier. Electronics and Communications in Japan (Part III: Fundamental Electronic Science) 75, 47–58 (2007)

Improved Swap Heuristic for the Multiple Knapsack Problem

Yacine Laalaoui

IT Department, Taif University, Taif, Kingdom of Saudi Arabia
yacine.laalaoui@gmail.com

Abstract. In this paper, we describe two new improvements of the well known Martello and Toth Heuristic Method (MTHM). Our new improvements are very simple and at the same time they are very efficient since they yield to more than 15% over MTHM with an excellent execution time performance in relatively large problem instances. Further, the new improvements give a very close results to sophisticated meta-heuristics namely Genetic Algorithms with a gap less than 1% within a time slot less than a second.

Keywords: Multiple Knapsack Problem, Heuristics, Swap.

1 Introduction

Multiple Knapsack Problem (MKP) is an assignment problem[1] and it consists of packing n items to m available knapsacks. Each item i has a weight w_i and a profit p_i. Each knapsack i has a capacity c_i. The task is to assign items to bins under the constraint of respecting the capacity of each knapsack and each item can be assigned to one knapsack at most. The objective function is to maximize the total profit of all assigned items. The mathematical formulation of the problem is as follows :

$$maximize \sum_{i=1}^{m} \sum_{j=1}^{n} p_j y_{ij}, \tag{1}$$

$$subject\ to \sum_{j=1}^{n} w_j y_{ij} \le c_i,\ \forall i \in M = \{1, ...m\}, \tag{2}$$

$$\sum_{i=1}^{m} y_{ij} \le 1,\ \forall j \in N = \{1...n\}, \tag{3}$$

$$y_{ij} \in \{0, 1\}, \quad i \in M, j \in N. \tag{4}$$

Where y_{ij} takes 1 if the jth item is assigned to i^{th} knapsack and 0 otherwise.

---

[1] MKP is strongly NP-Complete problem.

I. Rojas, G. Joya, and J. Cabestany (Eds.): IWANN 2013, Part I, LNCS 7902, pp. 547–555, 2013.

It is assumed that items are sorted increasingly according to their ratio profit to weight: $p_1/w_1 \le p_2/w_2.... \le p_n/w_n$. It is also assumed that knapsack capacities are sorted in non-increasing order : $c_1 \ge c_2.... \ge c_m$.

The overall objective of the paper is to increase the ability of finding good solutions in MTHM heuristic. Since MTHM is a simple heuristic, we aim to design a simple, fast and efficient heuristics to solve hard MKP instances. A simple heuristic means a technique which is easy to understand and to implement. A fast heuristic means it has a polynomial time complexity while an efficient heuristic means a technique which is able to approach optimal algorithms.

The remainder of this paper includes MTHM brief description in section 2, the independent swaps in section 3 and the integration of the independent swaps in section 4. Section 5 describes some of our experimental study. The paper is ended with a conclusion in section 6.

2   MTHM Heuristic Technique

Martello and Toth Heuristic Method (MTHM) [1] is a famous heuristic to solve the MKP problem. MTHM is an approximate algorithm with a polynomial time complexity. In MTHM, it is assumed that initially all capacities are sorted increasingly and all items are sorted decreasingly according to their ratios profit to weight(p_i/w_i). In phase #1, a greedy heuristic is applied to generate an initial solution. Once the greedy heuristic terminates, items are removed from the solution to be rearranged (phase #2) and to reconsider them according to increasing profit per unit weight where each item would be assigned to the next knapsack in a cyclic manner. Phase #3 in MTHM considers all pairs of items assigned to different knapsacks. If the interchange of a pair of items would allow the insertion of one more item, then apply such interchange to increase the total profit. The last phase (phase #4) attempts to exclude each item currently in the solution and trying to replace it with one or more items not in the solution to increase the total profit. Notice that each phase in MTHM heuristic takes at most $O(n^2)$ time complexity. The last phase in MTHM (phase #4) attempts to replace each item from the solution with one or more non-assigned items. In this phase, there is an attempt to build combinations (called Y) of items with size above 1 item to be inserted into the solution. The size of each combination Y should not exceed the item size j (currently in the solution) plus the free space currently present in such knapsack. Each time an item i is appended to the combination Y, the size of the later would be decreased until no more items could be inserted into Y. The obvious remark is that there are dependency between items appended into the combination Y. For example, the second item $i2$ would not be appended unless the first item $i1$ has left some free space to hold the second one. If the profit of $i2$ is greater than the profit of $i1$, then it will be much better to insert $i1$ instead of $i2$. Further, if another item $i3$ has a better profit than $i1$ and $i2$, then it would be much better also to append $i3$ instead of $i2$.

3   Independent Swaps

To handle the problem of dependent insertions in MTHM, we propose and describe the following simple solutions.

1. Replace One Item by One Item: The first simple attempt to increase the total profit is to try replacing one item from the solution by another item currently not in the solution. In this matter, items should be checked one by one and to try all possible couples of items. The corresponding pseudo-code of this heuristic is shown in **Fig.** 1 where cr is an array of m elements to hold the free space present in each knapsack, x is an array of n elements to hold either the knapsack index where the current item is assigned or 0 otherwise and z is the current maximum profit obtained so far. For each item i in the solution and for each item j not in the solution, check whether the replacement of i by j would yield to the increase of the total profit. If it yields to some improvement, then keep such couple (i,j) as the maximum if it yields to more improvement than the previous maximum (steps 06,07 and 08 in **Fig.**1). Once all couples (i,j) for the same item i have been checked, then apply the replacement of the maximum which has been stored so far (steps 09,...,14 in **Fig.**1). This simple heuristic needs not more than $O(n^2)$ time complexity.

Algorithm *Replace-One-By-One*;
Inputs : problem instance, the array x and z;
Outputs : the array x and z ;
Begin

```
01.for(i = n; i ≥ 1; i − −)
02.    max = 0;
03.    if(x[i] ≤ m)
04.       for(j = 1; j ≤ n; j + +)
05.          if((i! = j)&&(x[j] == 0))
06.             if((cr[x[i]] + w[i] − w[j] ≥ 0)&&(max < p[j] − p[i]))
07.                index_max = i;
08.                max = p[j] − p[i];
09.          if(max > 0)// swap i and j
10.             j = index_max;
11.             cr[x[i]] = cr[x[i]] − w[j] + w[i];
12.             x[j] = x[i] ;
13.             x[i] = 0;
14.             z = z + p[j] − p[i];
```

End.

Fig. 1. *Replace-One-By-One* pseudo-code

2. Replace One Item by Two Items: The second simple attempt is to try replacing one item from the solution by two items which are currently not in the solution. For each item i in the solution, is there a combination of two items (j, k) such that the replacement of i by j and k would yield to the increase of the total profit ? If it yields to some improvement, then keep the couple which yields to the maximum improvement. After that, apply such swap of i and (j, k). The corresponding pseudo-code of this heuristic is shown in **Fig**. 2. In is worth to note that this simple process needs not more than $O(n^3)$ time complexity.

Algorithm *Replace-One-By-Two*;
Inputs : problem instance, the array x and z;
Outputs : the array x and z ;
Begin
01.**for** $(i = n; i \leq 1; i - -)$
02. $max = 0;$
03.   **if** $(x[i] \leq m)$
04.   **for** $(j = 1; j \leq n; j + +)$
05.   **if** $(x[i] == 0)$
06.   **for** $(k = 1; k \leq n; k + +)$
07.   **if** $((i! = j)\&\&(i! = k)\&\&(j! = k)\&\&(x[k] == 0))$
08.   **if** ($cr[x[i]] + w[i] - w[j] - w[k] \geq 0$ && $max < p[j] + p[k] - p[i]$)
09. $index\_max1 = j; index\_max2 = k;$
10. $max = p[j] + p[k] - p[i];$
11.   **if** $(max > 0)$// do swap if possible
12. $j = index\_max1; k = index\_max2;$
13. $cr[x[i]] = cr[x[i]] - w[j] - w[k] + w[i];$
14. $x[j] = x[k] = x[i];$
15. $x[i] = 0;$
16. $z = z + p[j] + p[k] - p[i];$
End.

Fig. 2. *Replace-One-By-Two* pseudo-code

3.1 Replace Two Items by One Item: For the furtherance of the global efficiency, the attempt to replace two items by one item could be added. For each couple of items (j, k) currently within the solution, there is a checking of whether it exists one item i not assigned yet. If the total profit increases when (j, k) are replaced by i, then save the couple (j, k) which would yield to the maximum improvement. The corresponding pseudo-code is shown in **Fig**.3. This process is too similar to the previous one and it needs not more than $O(n^3)$ time complexity since $O(n^2)$ is needed to generate all possible couples (j, k). Notice that this process doesn't exist in MTHM heuristic.

Property 1. each swap of items would never decrease the solution quality.

This property is applicable for the above simple heuristics (**Fig**.1, 2 and 3). Each heuristic receives an initial solution and it attempts to improve its quality without any decreases.

```
Algorithm Replace-Two-By-One;
Inputs : problem instance, the array x and z;
Outputs : the array x and z ;
Begin
01.for (i = 1; i ≤ n; i + +)
02.    max = 0;
03.    if(x[i] == 0)
04.        for(j = n; j ≥ 1; j − −)
05.            if(x[j] ≤ m)
06.                for(k = n; k ≥ 1; k − −)
07.                    if((i! = j)&&(i! = k)&&(j! = k)&&(x[k] == x[j]))
08.                        if ( cr[x[k]] − w[i] + w[j] + w[k] ≥0 && max < p[i] − p[j] − p[k] )
09.                            index_max1 = j;index_max2 = k;
10.                            max = p[i] − p[j] − p[k];
11.        if(max > 0 )// do swap if possible
12.            j = index_max1;k = index_max2;
13.            cr[x[k]] = cr[x[k]] − w[i] + w[k] + w[j];
14.            x[i] = x[k];
15.            x[k] = x[j] = 0;
16.            z = z + p[i] − p[j] − p[k];
End.
```

Fig. 3. *Replace-Two-By-One* pseudo-code

4   Integration of Independent Swaps

The solutions described above solve completely the problem of dependent swaps found in MTHM. The swap which yields to the maximum improvement is applied instead of the first encountered. Basically, there are three cases to use such algorithm s either 1) as a new heuristic which integrates the extra three simple heuristics (**Fig.**1, 2 and 3), 2) add the new heuristics in the tail of the MTHM technique or 3) add and iterate the new three heuristics at the end of the MTHM. Since our aim is to increase the efficiency of MTHM, we disregard the first case since its efficiency is unknown compared to MTHM.

1. Single Running Technique SRT): In this case, the MTHM heuristic would be used as an initial guess and the three extra steps would be placed after MTHM. It is clear that this new global heuristic would enhance the MTHM's efficiency since it contains more steps. Further, in this case all steps (MTHM plus the extra three steps) would be executed for only one time. As it is said above, the new heuristic would take a longer time than MTHM since three extra steps were added. One can skip one or two steps to handle this issue. The corresponding pseudo-code is shown in **Fig.**4(a).

2. Iterative Running Technique (IRT): In this case, the MTHM heuristic would be an initial guess to an iterative swaps technique. All steps (MTHM plus the extra three steps) would be executed along many iterations to improve the solution quality. The iterations would be stopped if no more improvements could be done. The corresponding pseudo-code is shown in **Fig.**4(b).

Lemma 1. *IRT technique terminates for finite problem instances.*

Proof. This lemma means that the *IRT* technique would never loop forever if the input problem instance is finite. the proof is easy and it is by converse. □

Algorithm *SRT*;	Algorithm *IRT*;
Inputs : pb instance,z and the array x;	**Inputs :** pb instance,z and the array x;
Outputs : z and the array x;	**Outputs :** z and the array x;
Begin	**Begin**
	01. $prev\_z = 0$;
01. MTHM(x,z);	02. MTHM(x,z);
02. *Replace-One-By-Two*(x,z);	03. *Replace-One-By-Two*(x,z);
03. *Replace-Two-By-One*(x,z);	04. *Replace-Two-By-One*(x,z);
04. *Replace-One-By-One*(x,z);	05. *Replace-One-By-One*(x,z);
	06. **if**($prev\_z < z$)
	07. $prev\_z = z$;
	08. **goto** step 03.
End.	**End.**
(a) single running	(b) iterative running

Fig. 4. Pseudo codes of the new heuristics

5   Experimental Results

To show the efficiency of our improvements, we have used the MTHM source code[2] to implement *SRT* and *IRT* heuristics. The original FORTRAN source code is converted first into C language using f2c converter. All source codes are compiled using gcc gnu compiler on Linux Ubuntu operating system. The used hardware platform is an HP machine with Intel Core 2 vPro processor (3.00 GHZ) and 4 GB of RAM space. It is worth to note that our hardware platform is similar to the one used by A. Fukunaga in [2,3] (Intel Core 2 Due processor with 2.4GHz). We have conducted an intensive experimental study on more than 480 random problem instances belonging to four different classes, uncorrelated, weakly correlated, strongly correlated and multiple subset-sum instances. Both cases of high and low ratio of items to bins n/m are considered. Due to the space limitation, we show only results on strongly correlated and multiple subset-sum instances in case of low ratio n/m. All random instances are generated according to the method described in [1] [2,3]. We have also conducted an experimental study on a data-set with 14 problem instances from literature [2,3] where each instance has 300 items and 100 knapsacks.

Results of our experimental study are shown in tables 1 and 2. Table. 1 contains a comparison to GA and MTHM techniques on a data-set from literature [2,3] while **Table.** 2 contains a comparison to MTHM on randomly generated data-sets. Notice that GA column shown in (Table.1) contains results of four

---
[2] http://www.or.deis.unibo.it/ staff_pages/martello/cvitae.html

different GA techniques : Hybrid Grouping GA (HGGA) [5], Weighted Coding GA (WCGA) [4], Undominated Grouping GA (UGGA)[2] and Representation-Switching GA (RSGA)[3].

The parameter %GAP in In **Table**. 1 means how far is the results of MTHM, SRT and IRT from GA's results. If %GAP is smaller by a technique A, then the technique A gives results very close to the GA's ones. In column "Best result" from In **Table**. 1, we report the best profit obtained by GA techniques in 14 problem instances. In columns "GAP%" from In **Table**. 1, we report the percentage of difference between GA meta-heuristic and the three heuristics MTHM, SRT and IRT. For example, if %GAP is 1% for IRT and GA result is 750000, then the obtained profit by IRT is 750000 - (1%)*750000 which is equal to 750000 - (0.01)*750000 = 750000 - 7500 = 742500. In column "%inc" from **Table**. 2, we report the percentage of increase of the obtained profit by our techniques SRT and IRT over MTHM results. For example, if SRT has 2% and the MTHM has 750000, then it means that SRT profit is 750000+2%*750000 which is equal to 750000+(0.02)*750000 = 765000. In **Table** 1, everything written in bold means the best GAP percentage between our new techniques and GA techniques. In **Table**. 2, everything written in bold means the percentage of increasing the solution quality over MTHM. Straightforwardly, the *IRT* GAP column's values are less than 1% in **Table**. 1 and the *SRT* GAP column's values are around than 1% which would mean that both techniques are very close to GA techniques.

Table 1. Results on strongly correlated and multiple subset-sum Instances compared to GA and MTHM techniques. GAP columns show the percentage of difference between GA and the other techniques. Time columns show the time in seconds.

Strongly correlated instances								
	GA		MTHM		SRT		IRT	
pb instance	Best result	time	GAP(%)	time	GAP(%)	time	GAP(%)	time
instance1	753370	180	7.116	0.001	**0.753**	0.090	**0.162**	0.444
instance2	769614	180	11.471	0.001	**1.645**	0.083	**0.424**	0.606
instance3	712937	180	11.738	0.001	**1.780**	0.080	**0.308**	0.411
instance4	728296	180	7.521	0.002	**1.372**	0.083	**0.238**	0.655
instance5	774614	180	8.163	0.001	**1.733**	0.083	**0.174**	0.595
instance6	740188	180	10.684	0.002	**1.041**	0.083	**0.236**	0.424
instance7	744783	180	10.167	0.002	**1.069**	0.082	**0.298**	0.503
instance8	757975	180	6.991	0.003	**1.021**	0.086	**0.205**	0.347
instance9	758685	180	9.276	0.002	**1.845**	0.080	**0.683**	0.494
instance10	786644	180	8.353	0.002	**0.439**	0.085	**0.160**	0.516
Multiple subset-sum instances								
	GA		MTHM		SRT		IRT	
pb instance	Best result	time	GAP(%)	time	GAP(%)	time	GAP(%)	time
instance1	751507	180	0.596	0.004	**0.189**	0.099	**0.181**	0.288
instance2	767773	180	0.646	0.005	**0.061**	0.096	**0.054**	0.277
instance3	711062	180	0.560	0.010	**0.255**	0.096	**0.238**	0.280
instance4	726409	180	0.536	0.013	**0.195**	0.100	**0.166**	0.371

Further, the time taken by both techniques *SRT* and *IRT* is less than 1 second for all problem instances which is an incomparable to the time taken by GA (180 seconds). Further, MTHM's GAP is not less than 6%. This would be mean that our techniques improve significantly MTHM and they approaches closely GA techniques within a time slot less than one second. Table.2 shows that *SRT* and *IRT* reach more than 15% of improvement over MTHM within a time slot not more than 15 second in relatively large problem instances (1000 items) which is a very suitable time for real-time decision making.

In fact, the efficiency of MKP solvers varies according to the ration n/m. So, if this ratio is high, then it would solved quickly using Mulknap solver [6]. In contrast, if this ratio is low, then the input problem instance is very hard to solve and to the best our knowledge, there is no optimal solver for this type of instances. In other words, the difficulty of an MKP problem instance is oriented according to the ratio n/m. A. Fukunaga in [7] showed the behavior of the Mulknap solver in different values of the ratio n/m and proposed a branch-and-bound solver to overcome Mulknap's shortcomings. A. Fukunaga's in [7] concluded that a samll change in the ration n/m would result to a huge search-space. A. Fukunaga's solver remains modest since the largest reported problem instance contains less than 200 items. In table 2, we can easily see that for the same ratios n/m, all obtained results are very close. For example, for ratios 200/40, 500/100 and 1000/200 SRT gives 0.813%, 0.947% and 0.952% which are almost the same results. The reader is invited to [7] for more details.

It is clear that the proposed methods could be used on-line since the their complexity is not more than $O(n^3)$. In one hand, an on-line solution to the Multiple Knapsack Problem is important in the sense that it can be easily integrated into a sophisticated branch-and-bound and meta-heuristic techniques as an initial guess. It is well know that the initial solution is very important to

Table 2. Results on strongly correlated and multiple subset-sum Instances with low ratios n/m. %inc is the percentage of increasing the solution quality. time columns show the time in seconds. Each result is the average of 20 problem instances.

| | | | Strongly correlated Instances | | | | Multiple subset-sum Instances | | | | |
| | | *MTHM* | *SRT* | | *IRT* | | *MTHM* | *SRT* | | *IRT* | |
n	m	time	%inc	time	%inc	time	time	%inc	time	%inc	time
	40	0.0005	**0.813**	0.0254	**1.436**	0.0826	0.0021	**1.157**	0.0383	**1.335**	0.1361
200	65	0.0008	**4.309**	0.0267	**6.030**	0.0824	0.0026	**1.879**	0.0374	**2.375**	0.1488
	80	0.0008	**5.667**	0.0275	**10.149**	0.1316	0.0037	**1.907**	0.0364	**3.084**	0.2213
	100	0.0003	**0.200**	0.0285	**0.242**	0.0602	0.0015	**0.497**	0.0407	**0.497**	0.0790
	100	0.0036	**0.947**	0.3528	**1.770**	1.3676	0.0132	**0.823**	0.5633	**0.894**	1.7874
500	160	0.0074	**4.275**	0.3794	**5.964**	1.2987	0.0237	**1.288**	0.5563	**1.568**	2.0857
	200	0.0084	**7.871**	0.3941	**15.023**	3.3082	0.0281	**1.511**	0.5370	**2.546**	4.2748
	250	0.0020	**0.083**	0.4024	**0.092**	0.8427	0.0108	**0.188**	0.6025	**0.188**	1.2055
	200	0.0159	**0.952**	2.8055	**1.836**	10.5777	0.0730	**0.681**	4.3884	**0.713**	14.4269
1000	330	0.0336	**5.030**	2.9585	**6.964**	10.2202	0.1405	**1.283**	4.2518	**1.607**	13.7301
	400	0.0505	**8.017**	3.1264	**14.375**	15.000	0.1641	**1.493**	4.1199	**2.181**	15.000
	500	0.0056	**0.035**	3.1861	**0.042**	6.8400	0.0637	**0.112**	4.7792	**0.113**	9.6387

converge quickly to the best solutions. If the initial guess is far from the best solution, then bigger computation times are needed to reach the best solutions. In the other hand, an on-line method could be used to solve very large problem instances which could not be solved using exact solvers.

6 Conclusion

In this paper, we described two new improvements of the well known Martello and Toth Heuristic Method (MTHM). Our new improvements are very simple and at the same time they are very efficient since they yield to more than 15% over MTHM with an excellent execution time performance in relatively large problem instances. Further, the new improvements give a very close results to sophisticated meta-heuristics namely Genetic Algorithms with a gap less than 1% within a time slot less than a second.

References

1. Martello, S., Toth, P.: Heuristic algorithms for the multiple knapsack problem. Computing 27, 93–112 (1981)
2. Fukunaga, A.: A new grouping genetic algorithm for the multiple knapsack problem. In: Proc. IEEE Congress on Evolutionary Computation, pp. 2225–2232 (2008)
3. Fukunaga, A., Tazoe, S.: Combining Multiple Representations in a Genetic Algorithm for the Multiple Knapsack Problem. In: Proc of the 11th IEEE Congress on Evolutionary Computation, pp. 2423–2430 (2009)
4. Raidl, R.: The multiple container packing problem: A genetic algorithm approach with weighted codings. ACM SIGAPP Applied Computing Review, 22–31 (1999)
5. Falkenauer, E.: A hybrid grouping genetic algorithm for bin packing. Journal of Heuristics 2, 5–30 (1996)
6. Pisinger, D.: An exact algorithm for large multiple knapsack problems. European Journal of Operational Research 114, 528–541 (1999)
7. Fukunaga, A.: A branch-and-bound algorithm for hard multiple knapsack problems. Annals of Operations Research 184(1), 97–119 (2011)

Maximum Margin Clustering for State Decomposition of Metastable Systems

Hao Wu

Department of Mathematics and Computer Science,
Free University of Berlin, Arnimallee 6, 14195 Berlin, Germany
`hwu@zedat.fu-berlin.de`

Abstract. When studying a metastable dynamical system, a prime concern is how to decompose the state space into a set of metastable states. However, the metastable state decomposition based on simulation or experimental data is still a challenge. The most popular and simplest approach is geometric clustering, which was developed based on the classical clustering technique but only works for simple diffusion processes. Recently, the kinetic clustering approach based on state space discretization and transition probability estimation has attracted many attentions for it is applicable to more general systems, but the choice of discretization policy is a difficult task. In this paper, a new decomposition method, called maximum margin metastable clustering, is proposed, which converts the problem of metastable state decomposition into a unsupervised learning problem use the large margin technique to search for the optimal decomposition without state space discretization. Some simulation examples illustrate the effectiveness of the proposed method.

Keywords: metastable states, large margin methods, clustering analysis, nonsmooth optimization.

1 Introduction

For many chemical, physical and biological systems, metastability analysis plays a very important role in studying and modeling the dynamical behavior. Examples include conformational transitions in macromolecules [1], autocatalytic chemical reactions [2] and climate changes [3]. Roughly speaking, the metastability of a dynamical system implies that the state space of the system can be decomposed into a set of macrostates, called metastable states, such that the local equilibrium of a metastable state can be reached on a fast timescale and transition rates between different metastable states are very small relative to lifetimes of metastable states. Therefore the dynamical behavior of a metastable system can be approximately described as a transition network of metastable states exhibiting the Markov property, and the approximate description is able to significantly simplify the system dynamics with preserving the essential dynamical properties. A large number of theoretical and experimental studies (see e.g., [4,5,6,7]) demonstrated the usefulness of the metastability analysis based

I. Rojas, G. Joya, and J. Cabestany (Eds.): IWANN 2013, Part I, LNCS 7902, pp. 556–565, 2013.

dynamical approximation. In the past decades, a lot of mathematical techniques for metastability analysis were developed, such as large deviations [8], spectral approach [9] and hitting time based method [10]. However, direct applications of these techniques to metastable state decompositions are generally infeasible because they are computationally expensive for complex systems. In most practical cases, the metastable state decomposition can only be performed through a statistical analysis of the simulation or experimental data.

The simplest approach for data based metastable state decomposition is geometric clustering, which assumes that the local maxima of the density function of the equilibrium state distribution are also centers of metastable states. Under this assumption, the metastable states can be identified through grouping data points in state space which are geometrically similar, and the decomposition can be implemented manually or by using some classical clustering algorithms [11,12,13]. It is clear that this approach can only be used for simple stochastic diffusion processes, where peaks of the state density function and centers of metastable states are both determined by positions potential wells.

A more general and accurate approach is kinetic clustering, which involves two steps: In the first step, we discretize the state space into small bins, and estimate the transition probabilities between bins from simulation or experimental trajectories. In the second step, the discrete bins are lumped to metastable states through minimizing the transition probabilities between different metastable states. In contrast to the geometric clustering approach, this approach can utilize the information of the system dynamics contained in data and be applied to more complex systems. One of the most popular methods belonging to this approach is PCCA+ [6], but it is only applicable to time reversible systems. In [14], an SVD decomposition based lumping algorithm for nonreversible systems was proposed. Furthermore, Jain and Stock [15] presented a "most probable path algorithm" for bin lumping, which avoids the computation of large matrix decompositions. The main difficulty of kinetic clustering comes from the choice of the bin size. Too small bin size may cause overfitting in the transition probability estimation, but if the size of bins is too large, the boundaries of metastable states are unable to be accurately described. In order to solve this problem, Chodera et al. [16] proposed an adaptive decomposition method which can iteratively modify metastable state boundaries.

The objective of this paper is to propose a new clustering method for metastable state decomposition based on the large margin principle. Recently, large margin techniques have received increased interest in the machine learning community [17]. For a given supervised or unsupervised classification problem, large margin techniques can improve the robustness and generalization capability of the classifier through maximizing the margin between training data and classification boundaries. In this paper, we derive an optimization model for metastable state decomposition by combining the large margin criterion and metastability criterion, and develop a nonsmooth programming based algorithm for searching for the optimal decomposition. Unlike the kinetic clustering approach, the presented method does not need any discretization or probability estimation.

2   Preliminaries

We first briefly review some of the necessary background of large margin learning. Given a set of training data $\{x_i\}_{i=1}^N$ and their labels $\{y_i\}$, where each x_i is a data point sampled from a domain $\mathcal{X} \subseteq \mathbb{R}^n$ and $y_i \in \{1, \ldots, \kappa\}$, the support vector machine (SVM) finds a linear decision rule $y = \text{argmax}_{1 \leq r \leq \kappa} \, w_r^\mathrm{T} x$, which can map most of training data to correct labels and maximize the margin between training data and decision boundaries, by solving the following optimization problem:

$$
\begin{aligned}
\min_{\mathbf{w},\xi} \ & \tfrac{1}{2}\beta \|\mathbf{w}\|^2 + \tfrac{1}{N}\mathbf{1}^\mathrm{T}\xi \\
\text{s.t.} \ & \forall i = 1,\ldots,N, \quad r = 1,\ldots,\kappa \\
& w_{y_i}^\mathrm{T} x_i + 1_{y_i=r} - w_r^\mathrm{T} x_i \geq 1 - \xi_i \\
& \xi_i \geq 0
\end{aligned}
\tag{1}
$$

where $\mathbf{w} = \left(w_1^\mathrm{T}, \ldots, w_\kappa^\mathrm{T}\right)^\mathrm{T}$, $\xi = (\xi_1, \ldots, \xi_N)^\mathrm{T}$ is the vector of slack variables, $\|\mathbf{w}\|$ is inversely proportional to the minimal classification margin, and the objective function balances the empirical misclassification loss versus the margin through selecting an appropriate $\beta > 0$.

Remark 1. It is worth pointing out that SVM was originally designed for two-class classification [18]. A lot of strategies have been proposed to extend the SVM for multi-class problems, such as one-against-all [19], one-against-one [20] and error-correcting output coding [21]. Here we select the multi-class SVM proposed in [22] which minimizes the the multi-class margin loss directly, because it is more "logical" than the other strategies from the perspective of large margin learning and can be easily applied to unsupervised learning problems (see below).

Maximum margin clustering (MMC) [23] extends the maximum margin principle to unsupervised learning. For a set of unlabeled data $\{x_i\}$, MMC targets to construct a maximum margin decision rule by optimizing both $\mathbf{w}$ and data labels $\mathbf{y} = (y_1, \ldots, y_N)$ as follows:

$$
\begin{aligned}
\min_{\mathbf{y}\in\{1,\ldots,\kappa\}^N,\mathbf{w},\xi} \ & \tfrac{1}{2}\beta \|\mathbf{w}\|^2 + \tfrac{1}{N}\mathbf{1}^\mathrm{T}\xi \\
\text{s.t.} \ & \forall i = 1,\ldots,N, \quad r = 1,\ldots,\kappa \\
& w_{y_i}^\mathrm{T} x_i + 1_{y_i=r} - w_r^\mathrm{T} x_i \geq 1 - \xi_i \\
& \xi_i \geq 0
\end{aligned}
\tag{2}
$$

Note that SVM and MMC can also get nonlinear decision rules by nonlinearly mapping the training data into high dimensional feature spaces, and the kernel trick can be used to compute the mapping implicitly. However, the kernel trick for MMC may lead to unacceptably large computational times even for small data sets [24]. Here we select the strategy in [25], which performs the feature mapping and compression by kernel principal component analysis (KPCA). In what follows, we let $\mathcal{X}$ be the KPCA feature space and $\{x_i\}$ denote coordinates of training data obtained by KPCA prepossessing.

3   Maximum Margin Metastable Clustering

In this section, we apply the framework of large margin learning to metastability analysis. Suppose that we have L trajectories $\{s_l\}_{l=1}^{L}$ with $s_l = x_{1:M_l}^l$ in a phase space $\mathcal{X}$ which are generated by a dynamical system, where the point x_t^l represents the system state at time t in the lth run. Furthermore, for convenience of notation, we let $\mathcal{S} = \{x_t^l | 1 \leq l \leq L, 1 \leq t \leq M_l\}$ be the set of all data points in the trajectories and $\left\{\left(x_i^s, x_i^f\right)\right\}_{i=1}^{N} = \{(x_t^l, x_{t+1}^l) \,|\, 1 \leq l \leq L, 1 \leq t \leq M_l - 1\}$ denote the state transition pair set. Then we can also construct a large margin decision rule for the metastable state decomposition based on the following criteria: (a) For most of state transition pairs $\left(x_i^s, x_i^f\right)$, x_i^s and x_i^f should be classified to the same metastable state, which implies that the boundaries between metastable states are rarely crossed in runs of the system. (b) The metastable state boundaries should be placed as far away from the trajectory data as possible. The two criteria leads to a maximum margin metastable clustering (M^{3}C) method that can be described by the following optimization problem:

$$
\begin{aligned}
&\min_{y \in \{1,\ldots,\kappa\}^N, w, \xi^s, \xi^f} \quad \tfrac{1}{2}\beta \|w\|^2 + \tfrac{1}{2N} \mathbf{1}^T \left(\xi^s + \xi^f\right) \\
&\text{s.t.} \quad \forall i = 1, \ldots, N, \quad r = 1, \ldots, \kappa \\
&\quad w_{y_i}^T x_i^s + 1_{y_i = r} - w_r^T x_i^s \geq 1 - \xi_i^s \\
&\quad w_{y_i}^T x_i^f + 1_{y_i = r} - w_r^T x_i^f \geq 1 - \xi_i^f \\
&\quad \xi_i^s \geq 0, \xi_i^f \geq 0
\end{aligned}
\tag{3}
$$

where $\xi^s = (\xi_1^s, \ldots, \xi_N^s)^T$ and $\xi^f = \left(\xi_1^f, \ldots, \xi_N^f\right)^T$ are slack variables, κ is the number of metastable states, and y_i denotes the metastable state label of x_i^s and x_i^f. Fig. 1 illustrates the difference between the ideas of MMC and M^{3}C. It can be observed that the M^{3}C method is able to exploit the information on both phase space distribution and metastable dynamics contained in the trajectory data through enforcing that x_i^s and x_i^f are assigned to the same metastable state for each state transition pair $\left(x_i^s, x_i^f\right)$.

Remark 2. It can be seen that the proposed M^{3}C can only identify metastable areas which are connected in the feature space. But for most practical purposes, this limitation is not important becuase the connectivity of metastable areas holds for most actual chemical and physical processes.

4   Cluster Balance Constraint

As MMC and some other clustering methods, the M^{3}C method also suffers from the imbalanced clustering problem, i.e., the solution of (3) may lead to some "empty metastable states". In order to avoid such physically meaningless solutions, we can consider the following cluster balance constraint for M^{3}C:

$$
\varrho \leq \frac{1}{|\mathcal{S}|} \sum_{x \in \mathcal{S}} 1_{\mathrm{argmax}_k\, w_k^T x = r} \leq \bar{\varrho}, \quad \forall r = 1, \ldots, \kappa
\tag{4}
$$

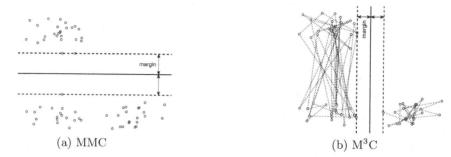

(a) MMC   (b) M³C

Fig. 1. Illustration of decision boundaries obtained by MMC and M³C, where dotted lines represent phase space trajectories, circles denote data points sampled from the trajectories

with $0 < \underline{\varrho} < 1/\kappa < \bar{\varrho} < 1$. However, the direct application of (4) is very difficult for the discontinuous constraint functions. In [26], Xu proposed a relaxation of (4), but the relaxation version is still not practical for large scale data set because it involves a matrix variable of size $|\mathcal{S}| \times |\mathcal{S}|$.

To simplify the optimization procedure, here we approximate the indicator function in (4) by a smooth function $m_r(x, \mathbf{w})$ with $m_r(x, \mathbf{w}) \propto \exp\left(\mu w_r^T x\right)$ and $\sum_{r=1}^{\kappa} m_r(x, \mathbf{w}) \equiv 1$, where $\mu > 0$ is the smoothing parameter and $\lim_{\mu \to \infty} m_r(x, \mathbf{w}) = 1_{\text{argmax}_k w_k^T x = r}$. This leads to a soft constraint:

$$\underline{\varrho} \leq \frac{1}{|\mathcal{S}|} \sum_{x \in \mathcal{S}} m_r(x, \mathbf{w}) \leq \bar{\varrho}, \quad \forall r = 1, \ldots, \kappa \tag{5}$$

Note that for a given $\mathbf{w}$, $m_1(\cdot, \mathbf{w}), \ldots, m_\kappa(\cdot, \mathbf{w})$ define a fuzzy clustering [27] of $\mathcal{S}$ with $m_r(x, \mathbf{w})$ representing the degree of x belonging to the rth cluster, and the decision rule $y = \text{argmax}_{1 \leq r \leq \kappa} w_r^T x$ is equivalent to the crisp result produced by the fuzzy clustering. Therefore (5) can be interpreted as a fuzzified cluster balance constraint.

Remark 3. Another simplified version of (4) is [25]

$$-\varrho \leq \frac{1}{|\mathcal{S}|} \sum_{x \in \mathcal{S}} \left(w_{r_1}^T x - w_{r_2}^T x\right) \leq \varrho, \quad \forall r_1, r_2 = 1, \ldots, N \tag{6}$$

where ϱ is generally set to be zero or a small positive number such that the decision boundaries are placed close to the centroid of $\mathcal{S}$. We do not adopt this constraint because it is actually redundant if the $\mathcal{S}$ is mean centered with $\sum_{x \in \mathcal{S}} x = 0$, and numerical experiments show that (6) often fails to prevent imbalanced clustering when $\kappa > 2$.

5   Optimization Method

We now investigate the optimization method for solving the M³C problem with the soft cluster balance constraint. It is easy to prove that if the value of $\mathbf{w}$

is given, the optimal $\mathbf{y}$, ξ^s and ξ^f should be $y_i = \check{y}_i(\mathbf{w})$, $\xi_i^s = \check{\xi}(\mathbf{w}, x_i^s, \check{y}_i(\mathbf{w}))$ and $\xi_i^f = \check{\xi}(\mathbf{w}, x_i^f, \check{y}_i(\mathbf{w}))$ with $\check{y}_i(\mathbf{w}) = \operatorname{argmin}_r \check{\xi}(\mathbf{w}, x_i^s, r) + \check{\xi}\left(\mathbf{w}, x_i^f, r\right)$ and $\check{\xi}(\mathbf{w}, x, r) = \max\{1 - w_r^T x + \max_{l \neq r} w_l^T x, 0\}$. Thus, we can eliminate the variables $\mathbf{y}$, ξ^s and ξ^f in the M^{3}C problem and get an equivalent formulation:

$$\min_{\mathbf{w}} f(\mathbf{w})$$
$$\text{s.t.} \ \ \forall r = 1, \ldots, \kappa \tag{7}$$
$$\varrho \leq h_r(\mathbf{w}) \leq \bar{\varrho}$$

where $f(\mathbf{w}) = \frac{1}{2}\beta \|\mathbf{w}\|^2 + \frac{1}{2N} \sum_{i=1}^{N}(\check{\xi}(\mathbf{w}, x_i^s, \check{y}_i(\mathbf{w})) + \check{\xi}(\mathbf{w}, x_i^f, \check{y}_i(\mathbf{w})))$ and $h_r(\mathbf{w}) = \frac{1}{|S|} \sum_{x \in S} m_r(x, \mathbf{w})$. It is obvious that we can adopt the penalty function method to solve (7) by sequentially solving $\min_{\mathbf{w}} f_p(\mathbf{w}, k) = f(\mathbf{w}) + \sigma^k g(\mathbf{w})$ for $k = 0, 1, 2, \ldots$ until a feasible $\mathbf{w}$ is found, where $g(\mathbf{w}) = \sum_{r=1}^{\kappa}(\max\{\varrho - h_r(\mathbf{w}), 0\} + \max\{h_r(\mathbf{w}) - \bar{\varrho}, 0\})$ is a nonsmooth exact penalty function and $\sigma > 1$ denotes the penalty coefficient. The main difficulty in the above algorithm arises from that $f(\mathbf{w})$ and $g(\mathbf{w})$ are both nonsmooth functions. Fortunately, the two functions are differential almost everywhere and the nonsmooth limited-memory BFGS algorithm [28] can be applied to minimizing $f_p(\mathbf{w}, k)$.

Remark 4. In contrast to the optimization methods in [26,25] for MMC, our method avoid introducing a large number of extra auxiliary variables or constraints.

6   Experiments

In this section, the proposed M^{3}C method will be applied to two examples of metastable dynamical systems:

$$dZ(t) = -\nabla V_1(Z(t))\, dt + \sqrt{2D}dW(t) \tag{Model I} \tag{8}$$
$$dZ(t) = -\nabla V_2(Z(t))\, dt + \sqrt{2D}dW(t) + c\left(Z(t^-)\right) dN(t) \quad \text{(Model II)} \tag{9}$$

where $Z = (z_1, z_2)^T$ denotes the system state, $V_1(Z)$ and $V_2(Z)$ are the potential functions, $W(t)$ is a Wiener process, $N(t)$ is a Poisson process with mean interval time of 0.04s, $D = 1.67$, $c(Z) = (0, -2z_2)^T$. $V_1(Z)$ and $V_2(Z)$ are shown in Figs. 2 and 3, and their detailed definitions are omitted here due to space limitations. The jumping term $c(Z(t^-))\, dN(t)$ in (9) means if a change of the value of $N(t)$ occurs the state of Model II jumps instantaneously to the symmetric point with respect to the axis $z_2 = 0$. Therefore, the three potential wells of Model I correspond to three metastable states, and Model II also has three metastable states where each one of them consists of two potential wells. For the metastability analysis, we generate 100 independent trajectories with time length 0.2s from (8) and 40 trajectories with time length 0.5s from (9) (sample interval $\Delta t = 0.02$s).

For the sake of comparison, we also implement two other competitive methods: MMC [25] and PCCA+ [6,7]. Note that the MMC method was not developed for the metastability analysis, but it can be viewed as an implementation of the geometric clustering approach. In order to evaluate the performance of these methods, we utilize the following quantities:

1. $Q = \sum_{r=1}^{\kappa} P_{rr}(\Delta t)$, where

$$P_{ij}(\tau) = \lim_{t \to \infty} \Pr(Z(t+\tau) \in \text{metastable state } i | Z(t) \in \text{metastable state } j)$$

   It is clear that Q can measure the metastability and a decomposition with strongly metastable states will result in a large Q close to κ [16]. (We run an independent simulation with time length 10^4s to estimate $P_{ij}(\tau)$ for each model. The detailed estimation algorithm is given in [6].)
2. Implied timescale $\text{ITS}_i(\tau) = -\tau / \ln \lambda_i(\tau)$ with $i > 1$, where $\lambda_i(\tau)$ denotes the ith largest eigenvalue of transition probability matrix $P(\tau) = [P_{ij}(\tau)]$. As mentioned in the introduction of this paper, one of the main aims of metastability analysis is to construct a simple Markov state model of complex systems. Thus, here we check the Markovity of metastable states through comparing the implied timescales with different τ. (According to [29], the value of $\text{ITS}_i(\tau)$ is independent of τ if the transition between metastable states is exactly Markovian.)

The simulation data, metastable state decomposition results and their Q values are displayed in Figs. 2 and 3, and the corresponding implied timescales are

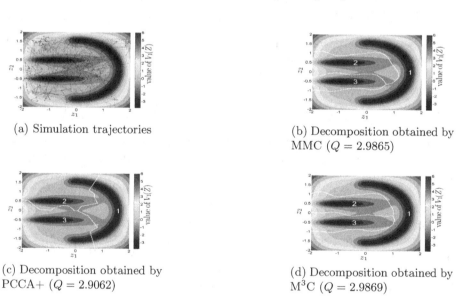

(a) Simulation trajectories

(b) Decomposition obtained by MMC ($Q = 2.9865$)

(c) Decomposition obtained by PCCA+ ($Q = 2.9062$)

(d) Decomposition obtained by M^{3}C ($Q = 2.9869$)

Fig. 2. Simulation and metastable state decomposition results of Model I

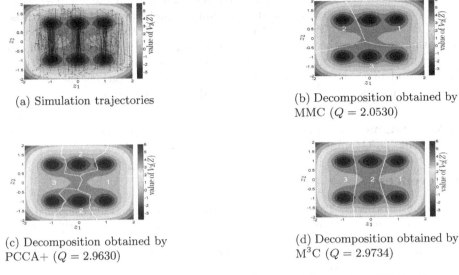

(a) Simulation trajectories

(b) Decomposition obtained by MMC ($Q = 2.0530$)

(c) Decomposition obtained by PCCA+ ($Q = 2.9630$)

(d) Decomposition obtained by M^3C ($Q = 2.9734$)

Fig. 3. Simulation and metastable state decomposition results of Model II

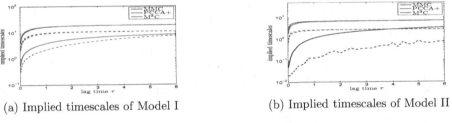

(a) Implied timescales of Model I

(b) Implied timescales of Model II

Fig. 4. Implied timescales, where solid lines indicate $ITS_2(\tau)$ and dashed lines represent $ITS_3(\tau)$

summarized in Fig. 4. It can be observed that our M^3C method gets the most "accurate" decomposition for both of models, which achieves the highest Q values and convergence rates of implied timescales. The experimental results clearly demonstrate the superior performance of the M^3C method. It is worth pointing out that the MMC method gives a "wrong" decomposition for Model II because the spatial structure of metastable states in Model II does not only depend on the shape of equilibrium state distribution function.

7 Conclusion

In this paper, we propose an effective and efficient method for metastable state decomposition based on the large margin principle. The key step is in (3), where each "label variable" y_i corresponds to a state transition pair instead of a single data. Then the error of metastable state decomposition can be expressed by the misclassification loss function in large margin learning, and the computational techniques from large margin learning including kernel tricks and balanced clustering can be applied. Moreover, we present a nonsmooth programming based algorithm to reduce the size of M^3C problem and search for the optimal decomposition. In the future, we will investigate the manner of parameter choice in M^3C and more efficient algorithms for real-world applications such as data analysis of molecular dynamics and autocatalytic chemical reactions.

Acknowledgments. This work is supported by the Deutsche Forschungsgemeinschaft (DFG) under grant Number WU 744/1-1.

References

1. Noé, F., Fischer, S.: Transition networks for modeling the kinetics of conformational change in macromolecules. Current Opinion in Structural Biology 18(2), 154–162 (2008)
2. Biancalani, T., Rogers, T., McKane, A.: Noise-induced metastability in biochemical networks. Physical Review E 86(1), 010106 (2012)
3. Berglund, N., Gentz, B.: Metastability in simple climate models: Pathwise analysis of slowly driven langevin equations. Stochastics and Dynamics 2(03), 327–356 (2002)
4. Aldhaheri, R., Khalil, H.: Aggregation and optimal control of nearly completely decomposable Markov chains. In: Proceedings of the 28th IEEE Conference on Decision and Control, pp. 1277–1282. IEEE (1989)
5. Chodera, J., Swope, W., Pitera, J., Dill, K.: Long-time protein folding dynamics from short-time molecular dynamics simulations. Multiscale Modeling & Simulation 5(4), 1214–1226 (2006)
6. Noé, F., Horenko, I., Schütte, C., Smith, J.: Hierarchical analysis of conformational dynamics in biomolecules: Transition networks of metastable states. Journal of Chemical Physics 126, 155102 (2007)
7. Prinz, J., Wu, H., Sarich, M., Keller, B., Senne, M., Held, M., Chodera, J., Schütte, C., Noé, F.: Markov models of molecular kinetics: Generation and validation. Journal of Chemical Physics 134, 174105 (2011)
8. Olivieri, E., Vares, M.: Large deviations and metastability. Cambridge University Press (2005)
9. Mathieu, P.: Spectra, exit times and long time asymptotics in the zero-white-noise limit. Stochastics: An International Journal of Probability and Stochastic Processes 55(1-2), 1–20 (1995)
10. Bovier, A.: Metastability: a potential theoretic approach. In: Proceedings of the International Congress of Mathematicians, Madrid, pp. 499–518 (August 2006)
11. Groningen, N.: Essential dynamics of reversible peptide folding: memory-free conformational dynamics governed by internal hydrogen bonds. Journal of Molecular Biology 309(1), 299–313 (2001)
12. Swope, W., Pitera, J., Suits, F., Pitman, M., Eleftheriou, M., Fitch, B., Germain, R., Rayshubski, A., Ward, T., Zhestkov, Y., Zhou, R.: Describing protein folding kinetics by molecular dynamics simulations. 2. example applications to alanine dipeptide and a β-hairpin peptide. Journal of Physical Chemistry B 108(21), 6582–6594 (2004)

13. Elmer, S., Park, S., Pande, V.: Foldamer dynamics expressed via markov state models. II. state space decomposition. Journal of Chemical Physics 123, 114903 (2005)
14. Mehrmann, V., Szyld, D., Virnik, E.: An SVD approach to identifying metastable states of Markov chains. Electronic Transactions on Numerical Analysis 29, 46–69 (2008)
15. Jain, A., Stock, G.: Identifying metastable states of folding proteins. Journal of Chemical Theory and Computation 8(10) (2012)
16. Chodera, J., Singhal, N., Pande, V., Dill, K., Swope, W.: Automatic discovery of metastable states for the construction of markov models of macromolecular conformational dynamics. Journal of Chemical Physics 126, 155101 (2007)
17. Xu, L.: Convex Large Margin Training Techniques: Unsupervised, Semi-supervised, and Robust Support Vector Machines. PhD thesis, University of Waterloo, Waterloo, Ontario, Canada (2007)
18. Vapnik, V.: Statistical Learning Theory. Wiley, New York (1998)
19. Bottou, L., Cortes, C., Denker, J.S., Drucker, H., Guyon, I., Jackel, L.D., LeCun, Y., Muller, U.A., Sackinger, E., Simard, P., Vapnik, V.: Comparison of classifier methods: a case study in handwriting digit recognition. In: Proceedings of the 12th International Conference on Pattern Recognition, vol. 2, IEEE, pp. 77–82. IEEE Computer Society Press (1994)
20. Friedman, J.: Another approach to polychotomous classifcation. Technical report, Department of Statistics, Stanford University (1996)
21. Allwein, E.L., Schapire, R.E., Singer, Y.: Reducing multiclass to binary: A unifying approach for margin classifiers. Journal of Machine Learning Research 1, 113–141 (2001)
22. Crammer, K., Singer, Y.: On the algorithmic implementation of multiclass kernel-based vector machines. Journal of Machine Learning Research 2, 265–292 (2001)
23. Xu, L., Neufeld, J., Larson, B., Schuurmans, D.: Maximum margin clustering. Advances in Neural Information Processing Systems 17, 1537–1544 (2004)
24. Zhang, K., Tsang, I., Kwok, J.: Maximum margin clustering made practical. IEEE Transactions on Neural Networks 20(4), 583–596 (2009)
25. Zhao, B., Wang, F., Zhang, C.: Efficient multiclass maximum margin clustering. In: Proceedings of the 25th International Conference on Machine learning, pp. 1248–1255. ACM (2008)
26. Xu, L., Schuurmans, D.: Unsupervised and semi-supervised multi-class support vector machines. In: Proceedings of the National Conference on Artificial Intelligence, vol. 20, p. 904. AAAI (2005)
27. Bezdek, J., Keller, J., Krisnapuram, R., Pal, N.: Fuzzy models and algorithms for pattern recognition and image processing, vol. 4. Springer (2005)
28. Skajaa, A.: Limited memory BFGS for nonsmooth optimization. Master's thesis, Courant Institute of Mathematical Science, New York University (2010)
29. Elmer, S., Park, S., Pandea, V.: Foldamer dynamics expressed via Markov state models. I. explicit solvent molecular-dynamics simulations in acetonitrile, chloroform, methanol, and water. Journal of Chemical Physics 123, 114902 (2005)

Hybrid Approach for 2D Strip Packing Problem Using Genetic Algorithm

Jaya Thomas and Narendra S. Chaudhari

Department of Computer science & Engineering, Indian Institute of Technology, Indore
{Jayat,narendra}@iiti.ac.in, nsc183@gmail.com

Abstract. In this paper we have studied the two-dimensional cutting stock problem, in which large number of small rectangles are to be placed in the big container such that the trim loss and height of the layout is minimized. We have proposed a placement approach along with a relevant fitness function to evaluate the overall goodness of the design layout. The computation results validate the solution and the effectiveness of the approach.

Keywords: Strip Packing, Genetic Algorithm, Biased Random Key.

1 Introduction

Manufacturing industries like glass, textile, wood, leather etc. faces a challenging problem of cutting material from the available stock. The problem is space minimization where placing smaller parts in the available large container such that the overall height of the layout is minimized. The problem is similar to many real world industrial problems which involve cutting of a number of rectangular shapes from the larger available so as to optimize the use of the material. Multiple objective functions [1] can be associated with this problem. In some cases, the material cost is negligible, or can be easily produced thus priority would be more on enhancing the packaging speed rather than the waste produced. That is the speed at which the shapes can be cut, in order to maximize efficient usage of the cutting equipment. Another objective is to minimize the setup cost. The setup cost can be minimized if we can reduce the number of cuts and prefer cuts that have single straight movements. The strip packing problem (SPP) can be considered both in two (2D-SPP) and in three dimensions. Together with the container loading problem and the bin packing problem, the SPP represents a further basic type of more-dimensional packing and cutting problems [2] [3] with considerable practical relevance. The 2D-SPP occurs, in the cutting of rolls of paper, metals, cloths in textile etc. The higher version of the problem even helps in decision making related to improving the design of the container, vehicle selection for transportation of goods etc. However in this paper, we have considered 2D aspect of the problem.

Gilmore and Gomory [4] where the first to work and propose solution to resolve these problems in a single dimension. Later, they widened the scope of their focus in 1965 to two-dimensional problems. Analysis of this kind of problem has since then spread rapidly. The SPP is NP-hard [5] [6]. The solution approaches are broadly classified as exact, approximation algorithm, heuristic and meta heuristic. In Exact very few approaches are used as their applicability is limited as the number of

I. Rojas, G. Joya, and J. Cabestany (Eds.): IWANN 2013, Part I, LNCS 7902, pp. 566–574, 2013.
© Springer-Verlag Berlin Heidelberg 2013

rectangles to be placed increases. A branch and bound approach [7] [17] are used for solving 2D-SPP. In [8] [9] gave a general framework based on exact for higher dimensional problem. The utility of exact algorithms for cutting and packing is limited, as the computational requirement grows exponentially with the number of shapes. Approximation algorithms commonly use heuristics to decide how to place the shapes. Bottom Left (BL) algorithm [10] was the very first documentation in which placement sequence was successive down placement and then movement of the piece to the left most corner of the strip. Another complex modified approach for Bottom-Left Fill (BLF) algorithm [11] fill holes in packing by maintaining a list of bottom-left location points. Another simple shelf-based algorithms with short running-time [12] using the concept of absolute ratio is also proposed. However these algorithms fail on complex strip placement.

The classical approaches are complex and results in huge computation. Moreover, they fail to find solution in some cases. To overcome these drawbacks we use a hybrid approach, heuristic with a genetic algorithm to handle 2D-SPP. GA are used widely in search, optimization [21] and also finds applicability in the design of models [13]. GA applied for 2D-SPP with additional constraints like orientation constraint and the guillotine constraint is discussed in [14]. However in this paper we have not considered these constraints at the preliminary stage. These are, to an increasing degree metaheuristics, mainly genetic algorithms (GA), but simulated annealing (SA), tabu search (TS) [18] and other types of metaheuristics are also applied. A common approach is to combine heuristics with a meta-search [15] [16]. One such is combining the genetic algorithm where each solution is represented as a sequence of shapes and is evaluated using BL. Some compare several meta-heuristics hybridized with BL and BLF: simulated annealing performed best. Literature reveals the progress made in solving combinatorial problems to get sub optimal and optimal solution by employing techniques like branch-and-bound, cutting plane, branch-cut-price, and approximation algorithm. It is evident from facts that many combinatorial optimization problems arising in practice benefit from heuristic methods that quickly produce good-quality solutions. To improve further the quality of solution meta-heuristic approaches are used. In this paper we also use a meta-heuristic approach to solve 2D-SPP and observe that the result obtained are quite promising. The paper is organized as follows in section 2 we discuss the 2D-SPP. Section 3 details the proposed approach along the various phases of genetic algorithm. Experimental results and discussions are briefed in section 4. The concluding remarks are presented in section 5.

2   Problem Description

The two-dimensional strip packing problem (2D-SPP) consists of orthogonally packing all the smaller pieces within the available large one with fixed width and variable height such that the overall height of the layout is minimized. The placement is subject to the following constraints

1. Placed rectangles cannot overlap each other and must be confined within the boundary of the strip.
2. Non-Guillotine cut (i.e. Rectangles are packed parallel to sides with no rotation).

The problem is addressed as follows: Given a large available rectangle (*W,* *,H*) where *W* and *H* represents the Width and Height respectively. Note that the height is large enough to accommodate all the rectangles to be placed. Let *m* small rectangles to be placed with specification w_i x h_i, where i=1,..., m. Each rectangle has a fixed orientation. Here, we have used non guillotine cut thus no rotation of the rectangle is allowed at placement. Each rectangle is placed parallel to the edge of the side. All the dimensions used are integer. However the problem can be easily extended to real numbers without much change. Each rectangle has an associated area and the objective is to minimize the height of the layout with all rectangles placed. The problem is elaborated with help of figure 1.

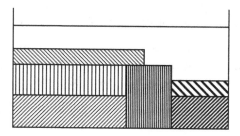

Fig. 1. Strip Packing

3   Proposed Approach

The approach combines the rectangle placement approach with the new evaluation fitness function for better convergence of the genetic algorithm. In this approach the initial population is generated using biased random key generator approach. The approach produces a feasible solution and their by the resulting crossover would lead to feasible solution. The algorithm works in three major stages. Initial population generation stage that helps in decoding of the rectangle. Later, referred as reproduction stage. Secondly the rectangle placements approach. Third, the population evaluation stage, where the strength of each population is evaluated based on the new fitness evaluation method. The following subsection details each stage.

3.1   Genetic Algorithm

GA is inspired by the mechanism of natural selection, a biological process in which stronger individuals are likely being the winners in a computing environment, Here, GA uses a direct analogy of such natural evolution. It presumes that the potential solution of a problem is an individual and can be represented by a set of parameters. These parameters are regarded as the genes of a chromosome and can be structured by a string of values in binary form. A positive value, generally known as fitness value, is used to reflect the degree of "goodness" of the chromosome for solving the problem, and this value is closely related to its objective value.

Proc Genetic Algorithm ()

Initialization;
Evaluation;
While (! termination criterion reached)

Selection and Reproduction;
Crossover;
Mutation;
Evaluation;
end
end *Proc*

Fig. 2. Overview Genetic Algorithm

3.2   Chromosome Representation

Each chromosome in the population represents a solution of the problem. The genetic algorithm described in this paper uses a randomly generated real number in the interval [0,1]. The evolutionary strategy used is similar to the one proposed by Bean [19] and Jose [20]. The random numbers are generated based on the number of strips to be packed. Further, these generated numbers are sorted to generate the placement sequence of the strip in the available block size. The placement sequence is illustrated in the figure.

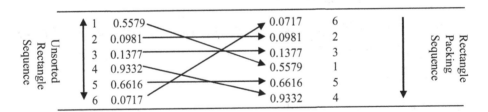

Fig. 3. Chromosome decoding

3.3   Population Evolution

This section details the evolution process of the genetic algorithm along with the placement approach used and fitness function evaluation.

3.3.1 Placement Approach

In this section we discuss the rectangle placement heuristic approach. In this approach the placement of the rectangle is done preferably at the bottom left of the container or in an appropriate formed residual space whichever preferable such that the overall objective of minimizing the height of all placed rectangles is achieved. The rectangle placement is checked against major governing factors. The factors like the rectangle is to be placed at the available bottom left coordinate. This placement is subject to the constraint that the preference will be given to available residual space such that with the placement of new block the maximum height remains unchanged. Second, the scenario being if more than one residual space is available then we select the residual space with least y-coordinate and more left oriented.

The procedure for rectangle placement is discussed below:

Repeat
                    //Finding Residual Space to accommodate Rectangle
         If((Rectangle Length <= Residual Space Length) and
                         (Rectangle Height <= Residual Space Height))

                         Flag found=true
         If (found)

         1.   Check Whether there exists any Residual Space
         Such that placing rectangle does not change the maximum height of layout
                    //Place the rectangles at bottom left
                    If more than one found then prefer one with least X-Coordinate &
                    Y- Coordinate

         2.   If condition 1 is false then find Residual Space with least Y-Coordinate

         3.   Create new Residual place after each rectangle placement as per figure 4

             end
           end
Until (all rectangles are placed)

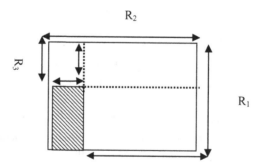

Fig. 4. Residual Space Creation

3.3.2 Crossover and Mutation

In order to facilitate the GA evolution cycle, two fundamental operators-crossover and mutation-are required. In the biased random key method both are integrated. In this approach mutants are introduced during the reproduction stage [20] in the non elite population. The parameterized uniform crossover is used for population evolution. The parent selection approach for mating is the major difference between a biased random key genetic algorithm RKGA [19] and BRKGA [20]. In the later both the parents are selected at random from the current population where as in BRKGA, one parent is selected at random from the elite and the other is selected from non-elite this ensures that the offspring is likely to have characteristic from elite parent. In this method, repetition of the elite parent is allowed thus an individual can produce more than one offspring for the same generation. A parameterized uniform crossover is used for the mating process of the chromosome. In this the fact that one parent is always selected from the elite set. Here, the probability of elite selection> 0.5 be a user-chosen parameter. This parameter is the probability that an offspring inherits the allele of its elite parent. In this way, the offspring is more likely to inherit characteristics of the elite parent than those of the non-elite parent. Since we assume that any random key vector can be decoded into a solution, then the offspring resulting from mating is always valid, i.e. can be decoded into a solution of the combinatorial optimization problem. When the next population is complete, i.e. when it has p individuals, fitness values are computed for all of the newly created random-key vectors and the population is partitioned into elite and non-elite individuals to start a new generation.

3.3.3 Fitness Function

The fitness evaluation function is the sole means of judging the quality of the evolved solutions. The fitness evaluation function is also necessary in the selection stage, where fitter individuals stand a good chance of being selected as parents and can pass their genetic material on to future generations. The main objective here is to minimize the overall height of the layout. To achieve this objective function we define fitness in terms of the height parameter.

In this approach we have considered two aspects first that minimum space must be left in packing between the placed rectangles towards the left of the container fig 5. Secondly, the placement should be such that the height is minimized. Thus the fitness function must hold on both the aspects. Trim loss factor cannot be the only criteria for fitness value assignment as it will evaluate to same value for all placement patterns. The trim loss is given as

$$Trim\ Loss\ (TL) = \frac{Left\ Residual\ Area}{Available\ Container\ Area} \tag{1}$$

The Fitness Function is evaluated as:

$$fitness = TL - \left(\frac{1}{RA_L}\right) * \left(\frac{1}{h_l}\right) \tag{2}$$

Where RA_L represents the unused residual area on the left side of the designed layout and h_l represents the maximum height of the designed layout. Here both the parameters are nonzero. In case of no left space the value is assumed to be 1.

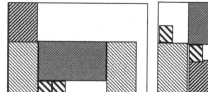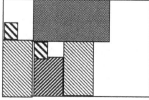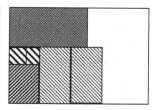

Fig. 5. Layout Structure for Rectangles in Container

Table 1. Initialization Parameters of Each Run

Population Size	500
Maximum Generations	30
Crossover Probability	0.7-0.85
Reproduction Probability	0.5
Selection Method	Tournament selection

Table 2 gives the average experimental results obtained by taking 5 instances from each category. The result reported are at the initial stage and would be revised by testing on large data sets. The fields in the table are indicating the dimensions, the number of small rectangles placed in the container. The Average trim loss and the optimum trim loss.

Table 2. Experimental Results

Sr. No.	Dimensions W x H	No. Of Rectangles	Minimum Average Loss	Optimum Trim loss
1.	20 x 20	16	0.00005610	0
2.	40 x 15	25	0.00017512	0
3.	60 x30	29	0.00592137	0
4.	80 x 120	97	0.00514932	0
5.	160 x 240	196	0.00158734	0

Table 3. Proposed Vs Existing Approach

Sr. No.	Existing Minimum Avg	Proposed Minimum Avg
1.	0.01099797	0.00005610
2.	0.00144487	0.00017512
3.	0.00738327	0.00592137
4.	0.00575832	0.00514932
5.	0.00378125	0.00158734

The approach is also compared with the existing approach. The results indicate that the performance of proposed is at par and in some case better than the existing approach. The observation is briefed in Table 3. The result indicates that consideration of minimizing height as one of the objectives along with the unused area helps to improve the results.

4   Conclusion

In this paper, we introduced a new genetic algorithm for the two-dimensional (2D) non-guillotine strip packing problem. Compared with previous GAs for the 2D-SPP, the proposed GA has the particularities of integrating a placement approach with proposed fitness function evaluation. The preliminary results show that the proposed GA competes very well with these state-of-the-art methods on small-sized instances. The results are reported in the initial results obtained which are at par and in some cases better than existing. Finally, we are studying the fitness function which may impact highly the performance of the genetic algorithm.

References

1. Coello Coello, C., Lamont, G., Van Veldhuizen, D.: Evolutionary Algorithms for Solving Multi-objective Problems. Springer, Berlin (2007)
2. Coffman Jr., E.G., Shor, P.W.: Average-case analysis of cutting and packing in two dimensions. European Journal of Operational Research 44, 134–144 (1990)
3. Bortfeldt, A.: A genetic algorithm for the two-dimensional strip packing problem with rectangular pieces. Discrete Optimization, European Journal of Operational Research 172, 814–837 (2006)
4. Gilmore, P., Gomory, R.: A linear programming approach to the cutting stock problem. Operations Research 9(6), 849–859 (1961)
5. Hopper, E., Turton, B.: A genetic algorithm for a 2D industrial packing problem. Computers & Industrial Engineering 37(1-2), 375–378 (1999)
6. Hopper, E., Turton, B.: An empirical investigation of meta-heuristic and heuristic algorithms for a 2D packing problem. EJOR 128(1), 34–57 (2000)
7. Martello, S., Monaci, M., Vigo, D.: An exact approach to the strip-packing problem. Informs J. Computing 15(3), 310–319 (2003)
8. Fekete, S., Schepers, J.: A new exact algorithm for general orthogonal d-dimensional knapsack problems. In: Burkard, R.E., Woeginger, G.J. (eds.) ESA 1997. LNCS, vol. 1284, pp. 144–156. Springer, Heidelberg (1997)
9. Fekete, S.P., Schepers, J.: A combinatorial characterization of higher-dimensional orthogonal packing. Mathematics of Operations Research, 353–368 (2004)
10. Baker, B., Coffman Jr, E., Rivest, R.: Orthogonal packings in two dimensions. SIAM J. Computing 9(4), 846–855 (1980)
11. Chazelle, B.: The bottom-left bin packing heuristic: an efficient implementation. IEEE Trans. Computers C-32(8), 697–707 (1983)
12. Bougeret, M., Dutot, P.F., Jansen, K., Otte, C., Trystram, D.: Approximation Algorithms for Multiple Strip Packing. In: Bampis, E., Jansen, K. (eds.) WAOA 2009. LNCS, vol. 5893, pp. 37–48. Springer, Heidelberg (2010)
13. Morgado-León, A., Escuín, A., Guerrero, E., Yáñez, A., Galindo, P.L., Sanchis, L.: Genetic Algorithms Applied to the Design of 3D Photonic Crystals. In: Cabestany, J., Rojas, I., Joya, G. (eds.) IWANN 2011, Part I. LNCS, vol. 6691, pp. 291–298. Springer, Heidelberg (2011)

14. Bortfeldt, A.: A genetic algorithm for the two-dimensional strip packing problem with rectangular pieces. Discrete Optimization, European Journal of Operational Research 172, 814–837 (2006)
15. Leung, S.C.H., Zhang, D.: A fast layer-based heuristic for non-guillotine strip packing. Expert Systems with Applications Elsivier 38, 13032–13042 (2011)
16. Zhang, D.: A Binary Search Heuristic Algorithm Based on Randomized Local Search for the Rectangular Strip-Packing Problem. Informs Journal on Computing (June 2012)
17. Kenmochi, M.: Exact algorithms for the two-dimensional strip packing problem with and without rotations, Discrete Optimization. European Journal of Operational Research 198, 73–83 (2009)
18. Alvarez-Valdes, R., et al.: A tabu search algorithm for a two-dimensional non-guillotine cutting problem. European Journal of Operational Research 183, 1167–1182 (2007)
19. Bean, J.C.: Genetic algorithms and random keys for sequencing and optimization. ORSA J. Comput. 6, 154–160 (1994)
20. Gonçalves, J.F., Resende, M.G.C.: Biased random-key genetic algorithms for combinatorial optimization. J. Heuristics 17, 487–525 (2011)
21. Qu, B.Y., Suganthan, P.N., Liang, J.J.: Differential Evolution With Neighborhood Mutation for Multimodal Optimization. IEEE Transactions on Evolutionary Computation 16(5), 601–614 (2012)
22. Pizzuti, C.: A Multiobjective Genetic Algorithm to Find Communities in Complex Networks. IEEE Transactions on Evolutionary Computation 16(3), 418–430 (2012)
23. Fernando, J.: A hybrid genetic algorithm-heuristic for a two-dimensional orthogonal packing problem. European Journal of Operational Research 183, 1212–1229 (2007)

Sea Clutter Neural Network Classifier: Feature Selection and MLP Design

Jose Luis Bárcena-Humanes, David Mata-Moya, María Pilar Jarabo-Amores,
Nerea del-Rey-Maestre, and Jaime Martín-de-Nicolás

Signal Theory and Communications Department,
Superior Polytechnic School, University of Alcala,
28805 Alcala de Henares, Madrid, Spain
{jose.barcena,david.mata,mpilar.jarabo,
nerea.rey,jaime.martinn}@uah.com

Abstract. The design of radar detectors in sea clutter environments is really a complex task. A neural network based automatic sea clutter classifier has been designed, as part of an adaptive detector capable of exploiting all the capabilities of detectors designed for specific clutter environments. The most extended sea clutter models have been considered (Gaussian, Weibull and K-distributed). Results show that an MLP with 3 inputs (the variance, the entropy of the modulus of the samples and the correlation coefficient), 6 hidden neurons and 4 outputs, is able to provide a performance similar to the $K - NN$ algorithm with $K = 10$ with a significant reduction in computational cost, a very important feature in real time applications.

Keywords: Sea clutter, radar detection, neural network, classifier, feature extraction, K-Nearest Neighbor.

1 Introduction

Marine radars are extensively used in maritime traffic monitoring and coastal surveillance tasks. One of the more important problems is related to the detection of small boats. Any object in the coverage area can intercept the transmitted signal and reradiate part of it towards the radar. This is the radar echo, that will be acquired by the receiver and applied to the detection stage. The objects to be detected are denoted as *targets*, while all the radar echoes from other non-desired objects are called *clutter*. In marine environments, the radar echo generated by the sea surface is of great interest. It can be significantly stronger than target radar echoes, and it is characterized by a high variability.

The objective is to decide between two hypotheses, target present (H_1) or target absent (H_0), fulfilling the Probability of Detection, P_D, and Probability of False Alarm, P_{FA}, requirements in the area of coverage. The P_D is the probability of detecting a target when it is present. The P_{FA} is the probability of deciding in favor of a target when there is no target. Under both hypotheses, the received signal has a noise component (receiver chain noise factor), and a

I. Rojas, G. Joya, and J. Cabestany (Eds.): IWANN 2013, Part I, LNCS 7902, pp. 575–583, 2013.

clutter component (radar echoes from the sea surface). The Neyman-Pearson, NP, detector is the most extended. It maximizes the P_D, maintaining the P_{FA} lower than or equal to a given value [1]. If $\tilde{\mathbf{z}}$ is the observation vector provided by the radar receiver, a possible implementation of the NP detector is the one based on the comparison of the likelihood ratio, $\Lambda(\tilde{\mathbf{z}})$, to a detection threshold selected according to the desired P_{FA} (1). This detector requires a complete statistical characterization of the observation vector under both hypotheses.

$$\Lambda(\tilde{\mathbf{z}}) = \frac{f(\tilde{\mathbf{z}}|H_1)}{f(\tilde{\mathbf{z}}|H_0)} \underset{H_0}{\overset{H_1}{\gtrless}} \eta_{P_{FA}} \tag{1}$$

In radar scenarios, target and clutter models are variable. Since the estimation of target parameters from observation vectors is difficult, a target model is usually assumed, and detection losses are suffered when actual target statistics differ from those assumed in the detector design [2,3]. Robust detectors with respect to target parameters have been analyzed and proposed, assuming constant and known clutter models [4,5,6]. Clutter parameters can be estimated from the environment.

In this paper, a sea clutter classifier is designed for allowing a dynamic selection of the best detection chain, among those designed assuming different clutter models. This classifier is composed of a feature extractor and a Multi-Layer Perceptron (MLP) designed for estimating the posterior probabilities of the classes, using the features provided by the feature extractor. The learning capabilities of the MLP are exploited for achieving performances similar to the ones provided by the K-Nearest Neighbor (KNN), with an important reduction in computational cost, a critical feature for real time implementations [7].

2   Characterization of the Observation Space

The general architecture of a coherent radar is presented in Fig. 1. After each antenna scan, at the output of the digital demodulator, a complex matrix is generated (Fig. 2). The observation vectors will be extracted from this matrix and applied to the detection stage. The anti-clutter processes are usually based on filtering approaches (MTI, Moving Target Indicator, or MTD, Moving Target Detector) and a detection threshold controlled by Constant False Alarm (CFAR) techniques, all of which assume a clutter model [8].

Sea clutter depends on multiple factors: the height of the waves, the wind speed, the transmitted signal frequency and polarization, the radar antenna pointing direction with respect to the direction of waves and the wind, the grazing angle (the complementary of the incident angle) and the size of the observation area (the radar resolution cell projected on the sea surface). In Fig. 3 the general geometry of a marine radar is shown.

The sea state is a term extensively used by mariners as a measure of the wave height, and it is commonly used to describe the roughness of the sea, although it doesn't provide a complete characterization of sea clutter [8].

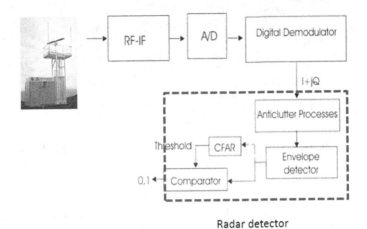

Radar detector

Fig. 1. General architecture of a coherent radar

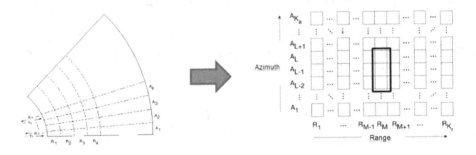

Fig. 2. Output of the digital demodulator for each radar scan

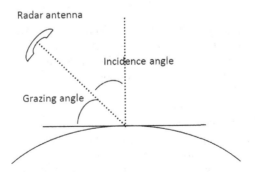

Fig. 3. Geometry of a marine radar

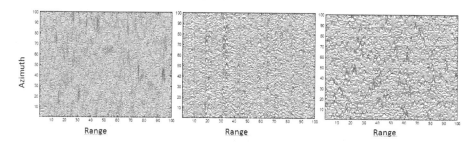

Fig. 4. Amplitude of clutter matrixes at the output of the digital demodulator: Correlated Gaussian (left), spiky K-distributed (center) and Weibull (right)

In low resolution radars (high resolution cells), sea clutter can be modeled as a Gaussian process. This model is also valid in systems with high grazing angles, and high resolution radars with low grazing angles and sea states 0 or 1 (wave height ranging from 0 to 0.1 meters). For higher sea states (wave height higher than 0.1 meters), the Weibull and the K-distributions have been proposed [9,10,11]. The probability density functions, pdfs, for K-distributed and Weibull processes are shown in (2) and (3), respectively. Their associated parameters are shape,ν, and scale, μ, for K, and shape,c, and scale, b, for Weibull. In Fig. 4, amplitude matrixes of clutter samples obtained at the output of the digital demodulator are presented.

$$f(r) = \frac{\sqrt{\frac{4\nu}{\mu}}}{2^{\nu-1}\Gamma(\nu)} \left(\sqrt{\frac{4\nu}{\mu}}r\right)^{\nu} K_{\nu-1}\left(\sqrt{\frac{4\nu}{\mu}}r\right)u(r) \qquad (2)$$

$$f(r) = \frac{c}{b}\left(\frac{r}{b}\right)^{c-1} \exp\left(-\left(\frac{r}{b}\right)^{c}\right)u(r) \qquad (3)$$

3   Proposed Scheme

Multiple detection strategies have been proposed for different clutter models. In order to exploit the potential of these detectors, a clutter automatic pre-classification stage is proposed, to select the more suitable detection strategy among those implemented (Fig. 5). Two operation modes are distinguished:

- *Off-line classifier design (dotted arrows).* A data base of pre-classified clutter matrixes is created. The most suitable sub-set of features is selected according to the estimated performance of classifiers trained in a supervised manner using different sub-sets of features.
- *On-line classifier operation (continuous arrows).* The selected sub-set of features is extracted from sub-matrixes of the actual scan to generate classifier inputs and select the best detector from the available bank.

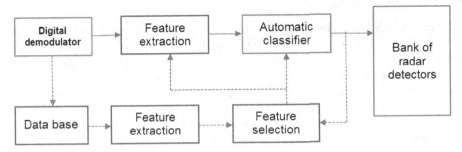

Fig. 5. Proposed detection scheme based on a bank of detectors and a clutter classifier capable of selecting the more suitable detector among those available

3.1   Training and Test Sets

A data base was generated synthetically for training and testing purposes, using the statistical models proposed in the literature and summarized in section 2. These models have been assessed using real data, because of that they can be used for generating synthetic data sets for design and testing purposes. As coherent models are considered, the generated samples are synthetic samples of the in phase and in quadrature components of the received radar echoes, which are provided by a synchronous detector in a real system. In Fig. 1, the digital demodulator encompasses the matched filter and the synchronous detector.

The data base was composed of clutter matrixes obtained at the output of the digital demodulator (Fig. 1). Assuming a radar system similar to SCANTER 2001 [12], the size of the matrixes was selected to guarantee a good estimation of the desired features, controlling the associated computational cost. Finally, matrixes of $M = 625 \times 1000$ complex elements were synthetically generated for covering an azimuth sector of 60 degrees and a range interval of $7.5km$. The data base was composed of four clutter classes:

- Correlated Gaussian: 3000 patterns with correlation coefficient uniformly distributed in $[0.9; 0.95]$.
- Uncorrelated Gaussian: 3000 patterns.
- K-distributed: 3000 patterns with a shape parameter, ν, uniformly distributed in $[0.4; 4.1]$.
- Weibull: 3000 patterns with shape parameter uniformly distributed in $[0.4; 2.1]$.

In all matrixes, the effects associated to the propagation of the electromagnetic wave and the size variation of the ground resolution cell of the radar were compensated. After that, all patterns were normalized to have unit power. Training, validation and testing sets were built from the database using $1,200$, 300, and $1,500$ patterns of each class, respectively.

3.2   Classifier Design

MLP *and K-NN based* classifiers have been analyzed. A study of clutter statistical features has been carried out, to identify the most suitable set to be used as classifier inputs. This is the common strategy for controlling the computational cost and increasing the generalization capabilities of classifiers. The features defined in Table 1 have been considered, together with the correlation coefficient, ρ. Symmetry and kurtosis are complex, and each one gives rise to 2 real values; the entropy is estimated from the modulus and the real and imaginary parts, generating 3 real values. A total of 9 features have been studied.

Table 1. Analyzed statistical features

Features	Definition	Features	Definition
variance	$\sigma^2(x) = \sum_M \dfrac{(x - \mu(x))^2}{M}$	Entropy	$E(x) = -\sum_m p(x) \cdot log_2(p(x))$
symmetry	$S(x) = \sum_M \dfrac{(x - \mu(x))^3}{\sigma^3}$	kurtosis	$K(x) = \sum_M \dfrac{(x - \mu(x))^4}{\sigma^4}$

Symmetry and kurtosis are several orders of magnitude higher than the other features. As the classification problem conveys the calculus of distances, these two features can mask the information provided by others. Because of that, feature vectors have been normalized for having zero mean and unity variance.

4   Results

MLPs with different number of inputs (ranging from 1 to 9), one hidden layer and four inputs have been trained considering the 512 possible feature combinations. The Levenberg-Marquardt algorithm, [13], has been used in combination with a cross-validation technique, and MLPs have been initialized using the Nguyen-Widrow method [14]. For each feature combination, the training process has been repeated twenty times. Only the cases where the performances of all the trained networks were similar in average, have been considered to extract conclusions.

In Tables 2 and 3, a summary of the results is presented:

- The best classification performance has been obtained using $[\rho; \sigma^2; E(|\tilde{\mathbf{z}}|)]$.
- Table 2 shows the improvement associated to the variance.
- Although the input clutter matrixes have been normalized to have unity variance, the statistical properties of the variance estimator are different (Fig. 6).

Table 2. Miss-classification matrix for the most relevant features (%). Input patterns (rows), classifier outputs (columns).

	$[\rho; \sigma^2; E(|\tilde{z}|)]$, L=6, NO=97				$[\rho; E(|\tilde{z}|)]$, L=5, NO=72			
	G_{nc}	G_c	Weibull	K	G_{nc}	G_c	Weibull	K
G_{nc}	**100**	0	6.21	0	**100**	0	5.23	0
G_c	0	**100**	0	0	0	**100**	0	0
Weibull	0	0	**85.26**	14.98	0	0	**63.72**	10.25
k	0	0	8.53	**85.02**	0	0	31.05	**89.75**

Table 3. Miss-classification matrix for other features (%). Input patterns (rows), classifier outputs (columns).

	$[\sigma^2; E(|\tilde{z}|)]$, L=5, NO=72				$[\rho; \sigma^2; S; K]$, L=9, NO=196			
	G_{nc}	G_c	Weibull	K	G_{nc}	G_c	Weibull	K
G_{nc}	**78.64**	43.64	1.43	0	**84.83**	0	57.69	11.29
G_c	21.36	**56.01**	3.58	0	0	**100**	0	0
Weibull	0	0.27	**87.50**	14.88	15.02	0	**23.64**	11.97
k	0	0.07	7.48	**85.12**	0.15	0	18.67	**76.74**

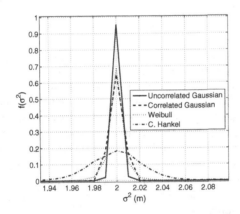

Fig. 6. Estimated fdp of the variance estimates

- Using $[\rho; \sigma^2; E(|\tilde{z}|)]$, the 6.21% of the Weibull patterns are classified as Gaussian. This can be explained considering that the Weibull distribution for a shape parameter equal to 2 is the same as the Rayleigh, and the entropy of the modulus of the complex clutter samples is used. An analysis of the test set has revealed that the shape parameter for Weibull patterns classified as Gaussian is very close to 2.

In the first row of each table, the input feature vector, the number of hidden neurons, L, and the number of required operations NO, are provided. If P is the number of real inputs, NO is calculated as $(NO = (2P + 1)L + (2L + 1)4 + 3)$.

Solutions based on the KNN algorithm (euclidean distance) have been designed. The results obtained for $K = 10$ and 25 are presented in Table 4. The performances for both K values are very similar, so $K = 10$ is selected because of its lower computational cost (NO). For $N = 1,500$ training patterns for each class, $NO = 4N(3P + K + 1) - 0.5K(K - 1) + 3$. The performance of this solution is very similar to that provided by the MLP for the same input features, but the associated computational cost is significantly higher.

Table 4. Miss-classification matrix (%) for the K-NN classifier. Input patterns (rows), classifier outputs (columns)

$[\rho; \sigma^2; E(|\tilde{z}|)]$, k=10, NO=107,958					$[\rho; \sigma^2; E(|\tilde{z}|)]$, k=25, NO=197,703			
	G_{nc}	G_c	Weibull	K	G_{nc}	G_c	Weibull	K
G_{nc}	100	0	3.27	0	100	0	5.33	0
G_c	0	100	0	0	0	100	0	0
Weibull	0	0	86.80	14.53	0	0	87.13	15.27
k	0	0	9.93	85.47	0	0	7.54	84.73

5 Conclusions

A neural network based pre-classification stage has been designed in order to improve the performance of radar detectors in sea clutter environments. The design of these detectors is really complex. If actual clutter statistics are different from those assumed in the design, performance losses can be very high. In order to design an adaptive detector capable of exploiting all the capabilities of solutions proposed in the bibliography, an automatic sea clutter classifier has been designed. The most extended sea clutter models have been used (Gaussian, Weibull and K-distributed), and the parameters of a commercial radar have been assumed for generating a synthetic data base from which training, validation and testing sets have been generated.

In order to reduce the complexity of the classifier and increase its generalization capabilities, a feature extraction stage has been designed for providing the classifier inputs. Variance, symmetry, entropy, kurtosis and correlation coefficient have been combined to determine the most suitable input vector. Results show that the best combination among those considered is the variance, the entropy of the modulus of the samples and the correlation coefficient. An MLP with 3 inputs, 6 hidden neurons and 4 outputs have been proposed. This classifier is able to provide a performance similar to the $K - NN$ algorithm with $K = 10$ with a significant reduction in computational cost, a very important feature in real time applications.

Acknowledgments. This work has been supported by the Spanish Ministerio de Economía y Competitividad, under Project TEC2012-38701

References

1. Neyman, Pearson: On the problem of the most efficient test of statistical hypotheses. Phil. Trans. Roy. Soc. A 231(9), 492–510 (1933)
2. Aloisio, V., di Vito, A., Galati, G.: Optimum detection of moderately fluctuating radar targets. IEE Proc. Radar, Sonar and Navigation 141(3), 164–170 (1994)
3. di Vito, A., Naldi, M.: Robustness of the likelihood ratio detector for moderately fluctuating radar targets. IEE Proceedings on Radar, Sonar and Navigation 146(2), 107–112 (1999)
4. Nayebi, M.M., Aref, M.R., Bastani, M.H.: Detection of coherent radar signals with unknown Doppler shift. IEE Proc. Radar, Sonar and Navigation. 143(2), 73–86 (1996)
5. Mata-Moya, D., Jarabo-Amores, P., Rosa-Zurera, M., Nieto-Borge, J.C., Lopez-Ferreras, F.: Combining MLPs and RBFNNs to Detect Signals With Unknown Parameters. IEEE Transactions on Instrumentation and Measurement 58(9), 2989–2995 (2009)
6. de la Mata-Moya, D., Jarabo-Amores, P., Vicen-Bueno, R., Rosa-Zurera, M., López-Ferreras, F.: Neural Network Detectors for Composite Hypothesis Test. In: Corchado, E., Yin, H., Botti, V., Fyfe, C. (eds.) IDEAL 2006. LNCS, vol. 4224, pp. 298–305. Springer, Heidelberg (2006)
7. Bishop, C.M.: Neural Networks for Pattern Recognition. Oxford Unviersity Press (1995)
8. Skolnik, M.: Radar Handbook, 3rd edn. Mc-Graw Hill (2008)
9. Conte, E., De Maio, A., Galdi, C.: Statistical analysis of real clutter at different range resolutions. IEEE Trans. Aerospace and Electronic Systems 40(3), 903–918 (2004)
10. Farina, A., et al.: High resolution sea clutter data: statistical analysis of recorded live data. IEE Proceedings Radar, Sonar and Navigation 144(3), 121–130 (1997)
11. Watts, S., Baker, C.J., Ward, K.D.: Maritime surveillance radar. Part 2: Detection performance prediction in sea clutter. IEE Proceedings Radar and Signal Processing 137(2), 63–72 (1990)
12. SCANTER 2001 Naval 2D Surface Surveillance, http://www.terma.com
13. Hagan, M.T., Menhaj, M.B.: Trainning feedfordward networks with the Marquardt alforithm. IEEE Trans. Neural Networks 6, 989–993 (1994)
14. Nguyen, D., Widrow, B.: Improving the Learning Speed of 2-Layer Neural Networks by Choosing Initial Values of the Adaptive Weights. In: Proc. of the Int. Conf. on Nerual Networks, IJCNN, pp. 21–26 (2009)

SONN and MLP Based Solutions for Detecting Fluctuating Targets with Unknown Doppler Shift in Gaussian Interference

David Mata-Moya, Pilar Jarabo-Amores, Nerea del-Rey-Maestre,
Jose Luis Bárcena-Humanes, and Jaime Martín-de-Nicolás

Signal Theory and Communications Department,
Superior Polytechnic School, University of Alcala,
28805 Alcala de Henares, Madrid, Spain
{david.mata,mpilar.jarabo,nerea.rey,jose.barcena,
jaime.martinn}@uah.es

Abstract. SONN and MLP based detection schemes are designed for approximating the Neyman-Pearson, NP, detector for detecting fluctuating targets with unknown Doppler shift in Gaussian interference. The optimum NP detector conveys a complex integral, so sub-optimum approaches based on the Constrained Generalized Likelihood Ratio, CGLR, are proposed as reference solutions. Detectors based on a single MLP, a single SONN, and mixtures of them are studied, and their detection capabilities and computational costs evaluated. Results show that the detector based on a mixture of SONNs is able to approximate the CGLR, outperforming the other proposed solutions, with lower computational cost.

Keywords: Average Likelihood Ratio, Multilayer Perceptrons, Second Order Neural Network, Mixture of Experts.

1 Introduction

Active pulse radars are extensively used for surveillance and monitoring tasks. Any object in the coverage area can intercept the radar transmitted signal and reradiate part of it towards the radar. This is the radar echo, that will be acquired by the receiver, which will generate the observation vector $\widetilde{\mathbf{z}}$, a complex vector composed of the in-phase and in-quadrature components of the digitized samples. The objects to be detected are denoted as *targets*, while all radar echoes from other non-desired objects are called *clutter*.

Radar detection can be formulated as a binary hypothesis test: H_0 is the null hypothesis (the received signal consists of interference, clutter plus noise), and H_1 is the alternative one (the received signal consists of target echo-plus-interference). The Neyman-Pearson (NP) detector, widely used in radar applications, maximizes the probability of detection (P_D), while maintaining the probability of false alarm (P_{FA}) lower than or equal to a given value [1]. If $f(\widetilde{\mathbf{z}}|H_0)$ and $f(\widetilde{\mathbf{z}}|H_1)$ are the likelihood functions, a possible implementation of the NP

I. Rojas, G. Joya, and J. Cabestany (Eds.): IWANN 2013, Part I, LNCS 7902, pp. 584–591, 2013.

detector consists in comparing the Likelihood Ratio (LR), $\Lambda(\tilde{\mathbf{z}})$, to a threshold selected according to P_{FA} requirements [1], and decide in favor of H_1 when the LR output is higher than the selected threshold. This approach requires a complete characterization of the likelihood functions, and detection losses are expected when the interference and/or target statistical properties vary from those assumed in the detector design [2,3].

In practical situations, clutter properties can be estimated, and the radar detector can be selected accordingly, but target ones are difficult to estimate. If θ is the target unknown parameter, defined in a space Θ, the detection problem can be formulated as a composite hypothesis test [1]. When the probability density function (pdf) of θ, $f(\theta)$, is known, the Average Likelihood Ratio (ALR) detector is optimum in the NP sense (1) [1]. This approach usually leads to intractable integrals without a closed solution, and sub-optimum ones are considered [4].

$$\Lambda(\tilde{\mathbf{z}}) = \frac{\int_{\Theta} f(\tilde{\mathbf{z}}|H_1, \theta) f(\theta) d\theta}{f(\tilde{\mathbf{z}}|H_0)} \underset{H_0}{\overset{H_1}{\gtrless}} \eta_{alr}(P_{FA}) \tag{1}$$

Learning machines trained in a supervised manner using a suitable error function have been proved to be able to approximate the NP detector [5]. Multi-Layer Perceptrons (MLP), [6], Radial Basis Function Neural Networks (RBFNN) [7], Second Order Neural Networks (SONN) [8] and Support Vector Machines (SVM) [9] have been applied to approximate the NP detector. In [10,11] committees of Neural Networks (NNs) were designed for detecting Gaussian targets with unknown correlation coefficient in Additive White Gaussian Noise (AWGN).

Due to the variety of targets and their possible dynamics, target Doppler frequency can vary in a wide range. As clutter usually presents zero, or very low, Doppler, many actual solutions apply clutter filtering techniques. These solutions are very far from those achievable with a good approximation to the ALR [4].

The present article tackles the design of NN based detectors for detecting targets with unknown Doppler shift in Gaussian interference. As a first step, a study of the CGLR is presented to define a good approximation to the ALR that can be used as reference. The objective is to exploit NN learning capabilities to obtain approximations capable of competing against the CGLR with a significant computational cost reduction, what will allow their real time implementation.

2   Case of Study

Target and clutter echoes are modeled as Gaussian complex vectors, $\tilde{\mathbf{z}} \in \mathcal{C}^P$, with the following autocorrelation matrixes:

- Target: $(\mathbf{\Sigma}_{\tilde{\mathbf{s}}\tilde{\mathbf{s}}})_{h,k} = p_s \cdot \exp(j(h-k)\Omega)$; $(h, k \in \{1, 2, ..., P\})$
- Interference: $(\mathbf{\Sigma}_{\tilde{\mathbf{c}}\tilde{\mathbf{c}}})_{h,k} = p_n \delta_{hk} + p_c \cdot \rho_c^{|h-k|^2}$; $(h, k \in \{1, 2, ..., P\})$

p_s and Ω are, respectively, the target power and mean Doppler shift. p_n is the noise power, δ_{hk} is the Kronecker delta, p_c and ρ_c are, respectively, the clutter power and one-lag correlation coefficient. The target one-lag correlation coefficient is 1 (Swerling I targets) [12]. Assuming these target and interference models, the NP detector for Ω varying uniformly in $[0; 2\pi]$ is formulated in (2), where T denotes the transpose operation, and $*$ denotes the complex conjugate. As this integral has no closed solution, the Constrained Generalized Likelihood Ratio $(CGLR_K)$ is considered [1]. In (3), K is the finite number of LR detectors designed for equispaced discrete values of Ω in $[0; 2\pi]$. This solution is used as reference to evaluate the NN detectors taking into consideration computational complexity and detection performance.

$$\int_0^{2\pi} \frac{\det(\mathbf{\Sigma}_{\widetilde{\mathbf{cc}}}) \exp\left(\widetilde{\mathbf{z}}^T[\mathbf{\Sigma}_{\widetilde{\mathbf{cc}}}^{-1} - (\mathbf{\Sigma}_{\widetilde{\mathbf{cc}}} + \mathbf{\Sigma}_{\widetilde{\mathbf{ss}}}(\Omega))^{-1}]\widetilde{\mathbf{z}}^*\right)}{\det(\mathbf{\Sigma}_{\widetilde{\mathbf{cc}}} + \mathbf{\Sigma}_{\widetilde{\mathbf{ss}}}(\Omega))} \cdot \frac{1}{2\pi} d\Omega \underset{H_0}{\overset{H_1}{\gtrless}} \eta_{alr}(P_{FA}) \tag{2}$$

$$\max_{\Omega_k} \frac{\det(\mathbf{\Sigma}_{\widetilde{\mathbf{cc}}}) \exp\left(\widetilde{\mathbf{z}}^T[\mathbf{\Sigma}_{\widetilde{\mathbf{cc}}}^{-1} - (\mathbf{\Sigma}_{\widetilde{\mathbf{cc}}} + \mathbf{\Sigma}_{\widetilde{\mathbf{ss}}}(\Omega_k))^{-1}]\widetilde{\mathbf{z}}^*\right)}{\det(\mathbf{\Sigma}_{\widetilde{\mathbf{cc}}} + \mathbf{\Sigma}_{\widetilde{\mathbf{ss}}}(\Omega_k))} \underset{H_0}{\overset{H_1}{\gtrless}} \eta_{cglr}(P_{FA})$$
$$k = 1, \ldots, K \tag{3}$$

The noise power is equal to 2, and the Clutter-to Noise (CNR) and the Signal-to-Interference ratios (SIR) are defined as $CNR = 10 \log_{10}(p_c/p_n)$ and $SIR = 10 \log_{10}(p_s/(p_n + p_c))$, respectively. The results obtained for $CNR = 20$ dB and $\rho_c = 0.7$ and 0.995 are presented in the following sections. These cases of study correspond to a non-dominant clutter (thermal noise can not be discarded in the study), and examples of low and high correlated clutter [2,3]. $P = 8$ has been assumed, without loss of generality.

3   CGLR Detectors

The final detection performance and the associated computational cost of the $CGLR_k$ detector are related to the parameter k . Receiver Operating Characteristic (ROC) curves for $K = 8, 16, 32$ have been estimated using Montecarlo simulations (maximum relative error in the estimation of P_{FA} equal to 10%). In Figure 1, SIR values capable of providing $P_D > 80\%$ for $P_{FA} > 10^{-6}$ have been selected. In section 5, the higher SIR values have been considered. ROC curves presented in Figure 1 allow us to conclude that a $CGLR_{16}$ can be selected as a compromise solution between detection performance and computational cost. The difference between $CGLR_8$ and $CGLR_{16}$ ROC curves decreases as ρ_c increases.

Considering each sum, product and exponential function as a simple operation on the processor, a $CGLR_K$ detector requires $(K) \times (8P^2 + 6P - 2) + K - 1$ operations. The number of operations required by a $CGLR_{16}$ is $8,943$.

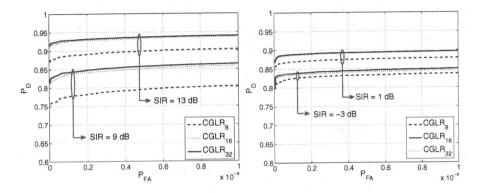

Fig. 1. Estimated ROC curves for $CGLR_K$ detectors for Swerling I targets with unknown Doppler shift and CNR=20dB: $\rho_c = 0.7$ (left), $\rho_c = 0.995$ (right)

4   NN Detectors

NNs are trained to minimize the mean squared error. This error function fulfills the sufficient condition that guarantees the theoretical capability of approximate the NP detector when the output of the neural detector is compared to a detection threshold fixed according to P_{FA} requirements [5]. The final approximation error depends on the NN architecture, the training set and the training algorithm. Detectors using a MLP or a SONN, and mixtures of experts are designed and tested.

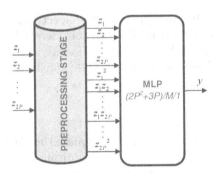

Fig. 2. SONN scheme: transformation from a $2P$-dimension input vector to a $(2P^2 + 3P)$-dimension one with first and second order terms, and a MLP with $(2P^2 + 3P)$ inputs and a hidden layer with $M = 1$ neuron

4.1   Single NNs Detectors

MLPs and SONNs with only one output are trained. As real arithmetic is used, the number of inputs is equal to $2P = 16$ (the real and imaginary parts of the complex samples). MLPs with a hidden layer with M neurons have been

trained ($MLP2P/M/1$) [13]. SONNs are expected to have a higher computational cost, but the second-order terms tend to increase the generalization
capabilities. Because of that, SONNs with only one quadratic neural unit are
considered ($SONN2P/2P^2 + 3P/1$) (Figure 2). For $P = 8$, the number of operations required by a single MLP is $35M + 2$, while a single SONN with one
quadratic neural unit needs 441 operations.

4.2 Mixture of Experts Based on NNs

According to the principle divide and conquer, committees of NNs are proposed
to improve the detection performance and/or reduce the computational cost.
Dynamic structures have been considered and mixtures of experts specialized
into input subspaces have been designed. The output of the mixture of NNs is
selected applying the maximum selection criterion following the philosophy of
the CGLR (Figure 3).

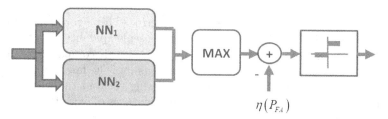

Fig. 3. Mixture of experts applying the maximum selection criterion

Taking into account the periodicity of the spectrum, subintervals of $\Omega \in$
$[0; \pi/4] \bigcup [3\pi/4; 2\pi]$ and $\Omega \in [\pi/4; 3\pi/4]$ are considered. NNs trained with patterns generated under H_1 with Ω varying uniformly in the first subinterval are
denoted as NN_1, while NNs specialized on $\Omega \in [\pi/4; 3\pi/4]$ are denoted as NN_2.

5 Results

NNs have been trained for minimizing the squared mean error, using a quasi-
Newton error minimization algorithm. This algorithm involves the estimation of
the Hessian matrix, which can be computationally prohibitive for large NNs. In
[14] a strategy is proposed for estimating the NN coefficients that reduce the
computational burden, retaining the fast convergence properties of the quasi-
Newton algorithm. A cross-validation technique has been used to avoid over-
fitting and all MLPs have been initialized using the Nguyen-Widrow method
[15]. For each case, the training process has been repeated ten times. Only the
cases where the performances of the ten trained networks were similar in average,
have been considered to extract conclusions.

P_{FA} and P_D values have been estimated using conventional Monte Carlo simulation with a relative error lower than 10% in the presented results.

According to the results presented in Section 3, the training parameters are the following: $CNR = 20$ dB, $SIR = 13$ dB for $\rho_c = 0.7$, and $SIR = 1$ dB for $\rho_c = 0.995$.

- In single NN based detectors, Swerling I targets have been generated with Ω uniformly distributed in $[0; 2\pi]$.
- In the mixtures of experts, two training sets have been generated: one with H_1 patterns with Ω uniformly distributed in $[0; \pi/4] \bigcup [3\pi/4; 2\pi]$, and other with H_1 patterns with Ω uniformly distributed in $[\pi/4; 3\pi/4]$.

Training and validating sets of $50,000$ patterns (half from H_0 and half from H_1) have been generated.

In Figure 4, ROC curves estimated for single MLPs with different number of hidden neurons are presented. MLPs with 17 and 14 hidden neurons, for $\rho_c = 0.7$ and 0.995 respectively, have been selected as compromise solutions between detection performance and computational cost. But the detection performance of single SONN solutions is better, as shown in Figure 5. In order to reduce the approximation error, mixtures of MLPs and SONNs have been also considered (Figure 5). Results show that the mixture of two SONNs, $SONN_1 + SONN_2$, is the best detector among those considered. On the other hand, the computational cost of this detector is really much lower than that associated to the proposed CGLRs, and the number of operations required by the single SONN and the mixture of two SONNs is lower than the one required by the single MLP and the mixture of MLPs, respectively (Table 1).

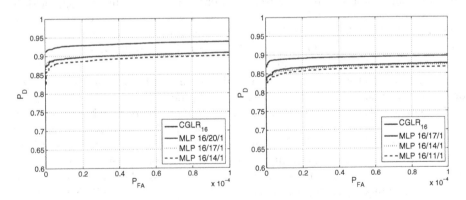

Fig. 4. ROC curves for single MLP detectors for Swerling I targets with unknown Doppler shift: CNR=20dB and $\rho_c = 0.7$ (left), CNR=20dB and $\rho_c = 0.995$ (right)

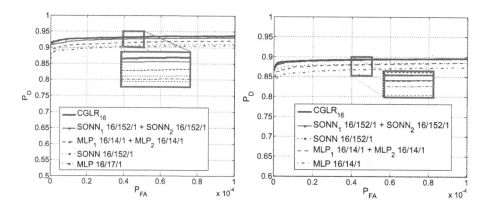

Fig. 5. ROC curves for the CGLR and the NN based detectors, for Swerling I targets with unknown Doppler shift: CNR=20dB, $\rho_c = 0.7$ and $SIR = 13$ (left), CNR=20dB, $\rho_c = 0.995$ and $SIR = 1$ (right)

Table 1. Number of operations, N_{op}, required by the considered detectors

Detectors	$CGLR_{16}$	$SONN_1 + SONN_2$	$MLP_1 + MLP_2$	SONN	MLP
N_{op}	8,943	882	1,194 $(\rho_c = 0.7)$ 984 $(\rho_c = 0.995)$	441	597 $(\rho_c = 0.7)$ 492 $(\rho_c = 0.995)$

6   Conclusions

In this paper, SONN and MLP based detection schemes have been designed and tested for approximating the NP detector in composite hypothesis-tests. The problem of detecting fluctuating targets (Swerling I) with unknown Doppler shift in Gaussian interference has been considered. Two clutter scenarios have been assumed, with correlation coefficients that define the expected variation interval in practical situations. In both cases, $CNR = 20$dB has been considered, which doesn't allow to discard the receiver thermal noise contribution.

The ALR detector has been formulated and sub-optimum approaches based on the CGLR analyzed, in order to define a reference detector for evaluating the performances of the NN based solutions. $CGLR_{2P}$ detectors have been proposed.

Detectors based on a single MLP and a single SONN have been designed and compared to mixtures of two MLPs and two SONNs, which output is selected applying the maximum selection criterion (following the philosophy of the CGLR). Each MLP and SONN of each mixture has been specialized on a subspace of the unknown parameter observation space. Results show that single SONNs trained with target Doppler shift varying uniformly in $[0; 2\pi]$, outperform single MLPs due to SONN inherent non-linearity. On the other hand, the computational cost associated to the SONN based solutions is lower than that of the MLP based ones. The mixture of two SONNs is a good compromise between detection capabilities and computational cost: $SONN_1$ has been trained with target Doppler shift

$\Omega \in [0; \pi/4] \bigcup [3\pi/4; 2\pi]$ and $SONN_2$ with $\Omega \in [\pi/4; 3\pi/4]$. The computational cost associated to the $SONN_1 + SONN_2$ detectors is really much lower than that associated to the proposed CGLRs, and lower than the one calculated for the proposed mixture of MLPs. Another interesting advantage of SONN based detectors against MLP ones is that, for the two clutter environments considered, the structure of the NN detector is the same. In the case of the single MLP and the mixture of MLPs, the selected number of hidden units is different for each scenario.

References

1. Van Trees, H.L.: Detection, estimation, and modulation theory, vol. 1. Wiley (1968)
2. Aloisio, V., di Vito, A., Galati, G.: Optimum detection of moderately fluctuating radar targets. IEE Proc. Radar, Sonar and Navigation 141(3), 164–170 (1994)
3. di Vito, A., Naldi, M.: Robustness of the likelihood ratio detector for moderately fluctuating radar targets. IEE Proceedings on Radar, Sonar and Navigation 146(2), 107–112 (1999)
4. Nayebi, M.M., Aref, M.R., Bastani, M.H.: Detection of coherent radar signals with unknown Doppler shift. IEE Proc. Radar, Sonar and Navigation 143(2), 73–86 (1996)
5. Jarabo-Amores, P., Rosa-Zurera, M., Gil-Pita, R., Lopez-Ferreras, F.: Study of Two Error Functions to Approximate the Neyman-Pearson Detector Using Supervised Learning Machines. IEEE Tran. Signal Processing 57(11), 4175–4181 (2009)
6. Gandhi, P., Ramamurti, V.: Neural networks for signal detection in non-Gaussian noise. IEEE Trans. Signal Process. 45(11), 2846–2851 (1997)
7. Casasent, D., Chen, X.: Radial Basis Function Neural Network for Nonlineal Fisher Discrimination and Neyman-Pearson Classification. IEEE Trans. Aerosp. Electron. Syst. 16(56), 529–535 (2003)
8. Mata-Moya, D., Jarabo-Amores, P., Martin de Nicolas-Presa, J.: High Order Neural Network Based Solution for Approximating the Average Likelihood Ratio. In: Proc. of IEEE Statistical Signal Processing Workshop (SSP), pp. 657–660 (2011)
9. Davenport, M.A., Baraniuk, R.G., Scott, C.D.: Tuning Support Vector Machines for Minimax and Neyman-Pearson Classification. IEEE Tran. Pattern Analysis and Machine Intelligence. 32(10), 1888–1898 (2010)
10. de la Mata-Moya, D., Jarabo-Amores, P., Vicen-Bueno, R., Rosa-Zurera, M., López-Ferreras, F.: Neural Network Detectors for Composite Hypothesis Tests. In: Corchado, E., Yin, H., Botti, V., Fyfe, C. (eds.) IDEAL 2006. LNCS, vol. 4224, pp. 298–305. Springer, Heidelberg (2006)
11. Mata-Moya, D., Jarabo-Amores, P., Rosa-Zurera, M., Nieto-Borge, J.C., Lopez-Ferreras, F.: Combining MLPs and RBFNNs to Detect Signals With Unknown Parameters. IEEE Transactions on Instrumentation and Measurement 58(9), 2989–2995 (2009)
12. Eaves, J.L., Reedy, E.K.: Principles of modern radar. Van Nostrand Reinhold (1987)
13. Cybenko, G.: Approximation by supeerpositions of a sigmoidal function. Mathematics of Control, Signals and Systems 2, 303–314 (1989)
14. El-Jaroudi, A., Makhoul, J.: A New Error Criterion for Posterior Probability Estimation With Neural Nets. In: Proc. Int. Conf. on Nerual Networks, IJCNN, pp. 185–192 (1990)
15. Nguyen, D., Widrow, B.: Improving the Learning Speed of 2-Layer Neural Networks by Choosing Initial Values of the Adaptive Weights. In: Proc. of the Int. Conf. on Nerual Networks, IJCNN, pp. 21–26 (2009)

An Ensemble of Computational Intelligence Models for Software Maintenance Effort Prediction

Hamoud Aljamaan, Mahmoud O. Elish, and Irfan Ahmad

Information & Computer Science Department
King Fahd University of Petroleum & Minerals
Dhahran, Saudi Arabia
{hjamaan,elish,irfanics}@kfupm.edu.sa

Abstract. More accurate prediction of software maintenance effort contributes to better management and control of software maintenance. Several research studies have recently investigated the use of computational intelligence models for software maintainability prediction. The performance of these models however may vary from dataset to dataset. Consequently, computational intelligence ensemble techniques have become increasingly popular as they take advantage of the capabilities of their constituent models toward a dataset to come up with more accurate or at least competitive prediction accuracy compared to individual models. This paper proposes and empirically evaluates an ensemble of computational intelligence models for predicting software maintenance effort. The results confirm that the proposed ensemble technique provides more accurate prediction compared to individual models, and thus it is more reliable.

Keywords: Computational intelligence, Ensemble techniques, Software maintenance, Prediction.

1 Introduction

Software maintenance is one of the most difficult and costly tasks in the software development lifecycle [24, 40]. Accurate prediction of software maintainability can be useful to support and guide [9]: software related decision making; maintenance process efficiency; comparing productivity and costs among different projects; resource and staff allocation, and so on. As a result, future maintenance effort can be kept under control.

Recent research studies have investigated the use of computational intelligence models for software maintainability prediction [10, 22, 40]. These models have different prediction capabilities and none of them has proved to be the best under all conditions. Performance of these models may vary from dataset to dataset. Computational intelligence ensemble techniques take advantage of the capabilities of their constituent models toward a dataset to come up with more accurate or at least competitive prediction accuracy compared to individual models. They have high potential in providing reliable predictions.

I. Rojas, G. Joya, and J. Cabestany (Eds.): IWANN 2013, Part I, LNCS 7902, pp. 592–603, 2013.
© Springer-Verlag Berlin Heidelberg 2013

This paper proposes and empirically evaluates an ensemble technique of computational intelligence models for predicting software maintenance effort. The rest of this paper is organized as follows. Section 2 reviews related work. In Section 3, we describe the proposed computational intelligence ensemble technique. In Section 4, we describe the ensemble constituent models. In Section 5, we present the discussions on the conducted empirical evaluation and its results. In Section 6, we present the conclusions and suggest directions for future work.

2 Related Work

Several research studies have investigated the relationship between object-oriented metrics and the maintainability of object-oriented software systems, and they found significant correlations between them [2, 6, 12, 24, 26]. These metrics can thus be used as good predictors of software maintainability. Furthermore, recent research studies have investigated the use of computational intelligence models for software maintainability prediction. These models were constructed using object-oriented metrics as input variables. Such models include TreeNet [11], multivariate adaptive regression splines [40], naïve bayes [22], artificial neural network [35, 40], regression tree [22, 40], and support vector regression [40].

Thwin and Quah [35] predicted the software maintainability as the number of lines changed per class. Their experimental results found that General Regression neural network predict maintainability more accurately than Ward network model. Koten and Gray [22] evaluated and compared the naïve bayes classifier with commonly used regression-based models. Their results suggest that the naïve bayes model can predict maintainability more accurately than the regression-based models for one system, and almost as accurately as the best regression-based model for the other system. Zhou and Leung [40] explored the employment of multiple adaptive regression splines (MARS) in building software maintainability prediction models. MARS was evaluated and compared against multivariate linear regression models, artificial neural network models, regression tree models, and support vector models. Their results suggest that, for one system, MARS can predict maintainability more accurately than the other four typical modeling techniques. Then, Elish and Elish [11] extended the work done by Zhou and Leung [40] to investigate the capability of TreeNet technique in software maintainability prediction. Their results indicate that TreeNet can yield improved, or at least competitive, prediction accuracy over previous maintainability prediction models.

Recently, computational intelligence ensemble models have received much attention and have demonstrated promising capabilities in improving the accuracy over single models [4, 34]. Ensemble models have been used in the area of software engineering prediction problems. They have been used in software reliability prediction [39], software project effort estimation [4], and software fault prediction [1, 19]. In addition, they have been used in many real applications such as face recognition [14, 18], OCR [25], seismic signal classification [33] and protein structural class prediction [3]. However, according to the best knowledge of the authors, none of the computational intelligence ensemble techniques have been used in the area of software maintenance effort prediction.

3   The Ensemble Technique

An ensemble of computational intelligence models uses the outputs of all its individual constituent prediction models, each being assigned a certain priority level, and provide the final output with the help of an arbitrator [29]. There are single-model ensembles and multi-model ensembles. In single-model ensembles, the individual constituent prediction models are of the same type (for example, all of them could be radial basis function network), but each with randomly generated training set. Examples of single-model ensembles include Bagging [5] and Boosting [13]. In multi-model ensembles, there are different individual constituent prediction models. This study focuses on multi-model ensembles.

The multi-model ensembles can be further classified, according to the design of the arbitrator, into linear ensembles and nonlinear ensembles [20]. In linear ensembles, the arbitrator combines the outputs of the constituent models in a linear fashion such as average, weighted average, etc. In nonlinear ensembles, no assumptions are made about the input that is given to the ensemble [20]. The output of the individual prediction models are fed into an arbitrator, which is a nonlinear prediction model such as neural network which when trained, assigns the weights accordingly. In this study, we propose a linear computational intelligence ensemble technique, which is described next.

The proposed ensemble takes the advantage of the fact that individual prediction models have different errors across the used dataset partitions. The idea behind this ensemble is that across the dataset partitions, take the best model in training based upon a certain criterion in that partition. In this study, the criterion is mean magnitude of relative error (MMRE). Fig. 1 provides a formal description of the ensemble.

```
Choose dataset with N observations
Choose M individual prediction models
Set K for K folds cross validation
For each k ∈ K fold
    For each m ∈ M model
        Apply model m on the training set for fold(k)
        Calculate training error E, based on a certain criterion
        Store error E
    End for
    Select the best model b ∈ M, based on training error E
    For each n ∈ N observation in the testing set for fold(k)
        EnsembleOutput = the result of applying model b on observation n
    End for
End for
```

Fig. 1. The ensemble technique

4   Ensemble Constituent Models

In this section we briefly describe the individual computational intelligence models that are used as base for the computational intelligence ensemble technique, i.e. the ensemble constituent models. These models were chosen because they are commonly and widely used in the literature of software quality and effort prediction. These models were built using WEKA machine learning toolkit [38], and their parameters were initialized using the default values.

4.1   Multilayer Perceptron

Multilayer Perceptron (MLP) [17] are feedforward networks that consist of an input layer, one or more hidden layers of nonlinearly activating nodes and an output layer. Each node in one layer connects with a certain weight to every other node in the following layer. MLP uses backpropagation algorithm as the standard learning algorithm for any supervised-learning.

The parameters of this model were initialized as follows. Backpropagation algorithm was used for training. Sigmoid was used as an activation function. Number of hidden layers was 5. Learning rate was 0.3 with momentum 0.2. Network was set to reset with a lower learning rate. Number of epochs to train through was 500. Validation threshold was 20.

4.2   Radial Basis Function Network

Radial Basis Function Network (RBF) [30] is an artificial neural network that uses radial basis functions as activation functions to provide a flexible way to generalize linear regression function. Commonly used types of radial basis functions include Gaussian, Multiquadric, and Polyharmonic spline. RBF models with Gaussian basis functions possess desirable mathematical properties of universal approximation and best approximation. A typical RBF model consists of three layers: an input layer, a hidden layer with a non-linear RBF activation function and a linear output layer.

The parameters of this model were initialized as follows. A normalized Gaussian radial basis function network was used. Random seed to pass on to K-means clustering algorithm was 1. Number of clusters for K-means clustering algorithm to generate was 2, with minimum standard deviation for clusters set to 0.1.

4.3   Support Vector Machines

Support Vector Machines (SVMs) was proposed by Vapnik [36] based on the structured risk minimization (SRM) principle. SVMs are a group of supervised learning methods that can be applied to classification or regression problems. SVMs aim to minimize the empirical error and maximize the geometric margin. SVM model is defined by these parameters: complexity parameter C, extent to which deviations are tolerated ε, and kernel.

The parameters of this model were initialized as follows. The cost parameter C was set to 1, with polynomial as SVMreg kernel. The most popular (RegSMOImproved) algorithm [32] was used for parameter learning.

4.4   M5 Model Tree

M5 Model tree (M5P) [31, 38] is an algorithm for generating M5 model trees that predicts numeric values for a given instance. To build a model tree, the M5 algorithm starts with a set of training instances. The tree is built using a divide-and-conquer method. At a node, starting with the root node, the instance set that reaches it is either associated with a leaf or a test condition is chosen that splits the instances into subsets based on the test outcome. In M5, the test that maximizes the error reduction is used. Once the tree has been built, a linear model is constructed at each node. The linear model is a regression equation.

The parameters of this model were initialized as follows. M5 algorithm was used for generating M5 model trees [31, 37]. Pruned M5 model trees were built, with 4 instances as the minimum number of instances allowed at a leaf node.

5   Empirical Evaluation

This empirical evaluation aims to determine the extent to which the proposed ensemble technique offers an increase in software maintenance effort prediction accuracy over individual models.

5.1   Datasets

We used two popular object-oriented software maintainability datasets published by Li and Henry [24]: UIMS and QUES datasets. These datasets are publicly available which makes our study verifiable, repeatable, and reputable [9]. The UIMS dataset contains class-level metrics data collected from 39 classes of a user interface management system, whereas the QUES dataset contains the same metrics collected from 71 classes of a quality evaluation system. Both systems were implemented in Ada. Both datasets consist of eleven class-level metrics: ten independent variables and one dependent variable.

The independent (input) variables are five Chidambar and Kemerer metrics [7]: WMC, DIT, NOC, RFC, and LCOM; four Li and Henry metrics [24]: MPC, DAC, NOM, SIZE2; and one traditional lines of code metric (SIZE1). Table 1 provides brief description for each metric.

The dependent (output) variable is a maintenance effort proxy measure, which is the actual number of lines in the code that were changed per class during a 3-year maintenance period. A line change could be an addition or a deletion. A change in the content of a line is counted as a deletion and an addition [24].

Table 1. Independent variables in the datasets

Metric	Description
WMC	Count of methods implemented within a class
DIT	Level for a class within its class hierarchy
NOC	Number of immediate subclasses of a class
RFC	Count of methods implemented within a class plus the number of methods accessible to an object class due to inheritance
LCOM	The average percentage of methods in a class using each data field in the class subtracted from 100%
MPC	The number of messages sent out from a class
DAC	The number of instances of another class declared within a class
NOM	The number of methods in a class
SIZE1	The number of lines of code excluding comments
SIZE2	The total count of the number of data attributes and the number of local methods in a class

Previous studies [10, 22, 40], on both datasets, indicate that both datasets have different characteristics, and therefore, considered heterogeneous and a separate maintenance effort prediction model is built for each dataset.

5.2 Accuracy Evaluation Measures

We used de facto standard and commonly used accuracy evaluation measures that are based on magnitude of relative error (MRE) [8]. These measures are mean magnitude of relative error (MMRE), standard deviation magnitude of relative error (StdMRE), and prediction at level q (Pred(q)). MMRE over a dataset of n observations is calculated as follows:

$$MMRE = \frac{1}{n}\sum_{i=1}^{n} MRE_i$$

where MREi is a normalized measure of the discrepancy between the actual value (x_i) and the predicated value ($\hat{x}_i$) of observation i. It is calculated as follows:

$$MRE_i = \frac{|x_i - \hat{x}_i|}{x_i}$$

In addition to MMRE, we used StdMRE since it is less sensitive to the extreme values compared to MMRE. Pred(q) is a measure of the percentage of observations whose MRE is less than or equal to q. It is calculated as follows:

$$\mathrm{Pr}ed(q) = \frac{k}{n}$$

where k is the number of observations whose MRE is less than or equal to a specified level q, and n is the total number of observations in the dataset. An acceptable value for level q is 0.3, as indicated in the literature [8, 22, 40]. We therefore adopted that value.

5.3   Results and Analysis

We used a 10-fold cross validation [21] (i.e. k-fold cross validation, with k set to 10), which is a common validation technique used to evaluate the performance of prediction models. In 10-fold cross validation; a dataset is randomly partitioned into 10 folds of equal size. For 10 times, 9 folds are picked to train the models and the remaining fold is used to test them, each time leaving out a different fold.

Table 2 provides the results obtained from applying the individual computational intelligence models on UIMS dataset, as well as the results achieved by the ensemble model. Among the individual models, the MLP model achieved the best result in general, whereas the RBF model was the worst. It can be observed that the ensemble model outperformed all the individual models.

Table 2. Prediction accuracy results: UIMS dataset

	Individual Models				Ensemble Model
	MLP	**RBF**	**SVM**	**M5P**	
MMRE	1.39	3.23	1.64	1.67	**0.97**
StdMRE	2.40	4.43	2.38	2.75	**1.61**
Pred(0.3)	23.33	15	20	23.33	**25**

Fig. 2 shows the box plot of MRE values for each model, where the middle of each box represents the MMRE for each model. As can be seen, the ensemble model has the narrowest box and the smallest whiskers (i.e. the lines above and below from the box). Moreover, its box and whiskers are lower than those of the individual models, which clearly indicate that the ensemble model outperforms the individual models. Fig. 3 shows a histogram of the achieved Pred(0.30) value by each model. From the figure it can be seen clearly that the ensemble model achieved the best Pred(0.30).

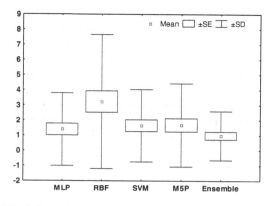

Fig. 2. Box plots of MRE for each model: UIMS dataset

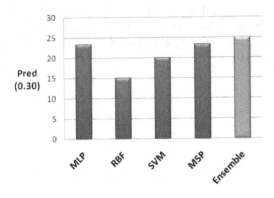

Fig. 3. Pred(0.30) for each model: UIMS dataset

Table 3 provides the results obtained from applying the individual computational intelligence models on QUES dataset, as well as the results achieved by the ensemble model. Among the individual models, the SVM model achieved the best result, whereas the RBF model was the worst. It also can be observed that the ensemble model outperformed all the individual models.

Table 3. Prediction accuracy results: QUES dataset

	Individual Models				**Ensemble Model**
	MLP	**RBF**	**SVM**	**M5P**	
MMRE	0.71	0.96	0.44	0.54	**0.41**
StdMRE	0.65	1.52	0.39	0.56	**0.32**
Pred(0.3)	40	36.66	56.66	51.66	**60**

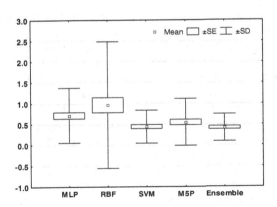

Fig. 4. Box plots of MRE for each model: QUES dataset

Fig. 4 shows the box plot of MRE values for each model, where the middle of each box represents the MMRE for each model. It can be observed that the ensemble model has the narrowest box and the smallest whiskers. Its box and whiskers are also lower than those of the individual models, which clearly indicate that the ensemble model outperforms the individual models in this dataset too. Fig. 5 shows a histogram of the achieved Pred(0.30) value by each model. The ensemble model achieved the highest Pred(0.30) value, i.e. 60%.

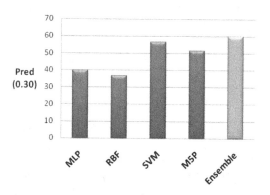

Fig. 5. Pred(0.30) for each model: QUES dataset

When considering the results from both datasets, there are two main interesting observations. First, the results support that the performance of the individual prediction models may vary from dataset to dataset; the MLP model was the best in the UIMS dataset while the SVM model was the best in the QUES dataset. Second, the ensemble model outperformed the individual models in both datasets.

6   Conclusions and Future Work

In this paper we presented an ensemble of computational intelligence models for predicting software maintenance effort. As ensemble constituent models, four popular prediction models (MLP, RBF, SVM, and M5P) were used. The prediction accuracy of the ensemble technique was empirically evaluated using two public object-oriented software maintainability datasets. The results indicate that ensemble technique provides more accurate prediction compared to individual models, and thus it is more reliable. It is worth noting that ensemble technique will be in general more complex and computationally expensive as compared to using a single prediction model but the benefits of using the ensemble in terms of prediction accuracy and robustness outweighs this penalty.

There are possible directions for future work, which include: investigating nonlinear ensemble models and comparing their performance with linear ensemble models; considering other ensemble constituent models; applying computational intelligence ensemble techniques to other software engineering prediction problems such as fault prediction. Both theoretical [15, 23] and empirical [16, 27, 28] research studies have demonstrated that a good ensemble is one where the individual prediction models in the ensemble are both accurate and make their errors on different parts of the input space. Therefore, one important direction of future work is to investigate different sets of ensemble constituent models.

Acknowledgements. The authors wish to acknowledge King Fahd University of Petroleum and Minerals (KFUPM) for utilizing the various facilities in carrying out this research.

References

1. Aljamaan, H., Elish, M.: An Empirical Study of Bagging and Boosting Ensembles for Identifying Faulty Classes in Object-Oriented Software. In: IEEE Symposium on Computational Intelligence and Data Mining, pp. 187–194 (2009)
2. Bandi, R.K., Vaishnavi, V.K., Turk, D.E.: Predicting Maintenance Performance Using Object-Oriented Design Complexity Metrics. IEEE Transactions on Software Engineering 29(1), 77–87 (2003)
3. Bittencourt, V.G., Abreu, M.C.C., Souto, M.C.P.D., Canuto, A.M.D.P.: An empirical comparison of individual machine learning techniques and ensemble approaches in protein structural class prediction. In: International Joint Conference on Neural Networks, pp. 527–531 (2005)
4. Braga, P.L., Oliveira, A.L.I., Ribeiro, G.H.T., Meira, S.R.L.: Bagging Predictors for Estimation of Software Project Effort. In: International Joint Conference on Neural Networks, pp. 1595–1600 (2007)
5. Breiman, L.: Bagging predictors. Machine Learning 24(2), 123–140 (1996)
6. Briand, L.C., Bunse, C., Daly, J.W.: A Controlled Experiment for Evaluating Quality Guidelines on the Maintainability of Object-Oriented Designs. IEEE Transactions on Software Engineering 27(6), 513–530 (2001)
7. Chidamber, S.R., Kemerer, C.F.: A Metrics Suite for Object Oriented Design. IEEE Transactions on Software Engineering 20(6), 476–493 (1994)
8. Conte, S., Dunsmore, H., Shen, V.: Software Engineering Metrics and Models. Benjamin/Cummings, Menlo Park (1986)
9. De Lucia, A., Pompella, E., Stefanucci, S.: Assessing effort estimation models for corrective maintenance through empirical studies. Information and Software Technology 47(1), 3–15 (2005)
10. Elish, M., Elish, K.: Application of TreeNet in Predicting Object-Oriented Software Maintainability: A Comparative Study. In: 13th European Conference on Software Maintenance and Reengineering (CSMR 2009), pp. 69–78 (2009)
11. Elish, M.O., Elish, K.O.: Application of TreeNet in Predicting Object-Oriented Software Maintainability: A Comparative Study. In: 13th European Conference on Software Maintenance and Reengineering (CSMR 2009), pp. 69–78 (2009)
12. Fioravanti, F., Nesi, P.: Estimation and prediction metrics for adaptive maintenance effort of object-oriented systems. IEEE Transactions on Software Engineering 27(12), 1062–1084 (2001)
13. Freund, Y.: Boosting a weak learning algorithm by majority. Information and Computation 121(2), 256–285 (1995)
14. Gutta, S., Wechsler, H.: Face Recognition Using Hybrid Classifier Systems. In: IEEE International Conference on Neural Networks, pp. 1017–1022 (1996)
15. Hansen, L., Salamon, P.: Neural Network Ensembles. IEEE Transactions on Pattern Analysis and Machine Intelligence 12(10), 993–1001 (1990)

16. Hashem, S., Schmeiser, B., Yih, Y.: Optimal linear combinations of neural networks. Neural Networks 3, 1507–1512 (1994)
17. Haykin, S.: Neural Networks: A Comprehensive Foundation New Jersey. Prentice Hall, New Jersey (1999)
18. Huang, F.J., Zhou, Z., Zhang, H.-J., Chen, T.: Pose invariant face recognition. In: Proc. 4th IEEE Int. Conf. on Automatic Face and Gesture Recognition, France, pp. 245–250 (2000)
19. Khoshgoftaar, T.M., Geleyn, E., Nguyen, L.: Empirical Case Studies of Combining Software Quality Classification Models. In: Third International Conference on Quality Software, p. 40 (2003)
20. Kiran, N., Ravi, V.: Software reliability prediction by soft computing techniques. Journal of Systems and Software 81(4), 576–583 (2008)
21. Kohavi, R.: A study of cross-validation and bootstrap for accuracy estimation and model selection. In: Proceedings of the 14th International Joint Conference on Artificial Intelligence (IJCAI), pp. 1137–1143 (1995)
22. Koten, C., Gray, A.: An application of Bayesian network for predicting object-oriented software maintainability. Information and Software Technology 48(1), 59–67 (2006)
23. Krogh, A., Vedelsby, J.: Neural Network Ensembles, Cross Validation, and Active Learning. In: Advances in Neural Information Processing Systems, vol. 7, pp. 231–238 (1995)
24. Li, W., Henry, S.: Object-Oriented Metrics that Predict Maintainability. Journal of Systems and Software 23(2), 111–122 (1993)
25. Mao, J.: A case study on bagging, boosting and basic ensembles of neural networks for OCR. In: Proc. IEEE Int. Joint Conf. on Neural Networks, pp. 1828–1833 (1998)
26. Misra, S.C.: Modeling Design/Coding Factors That Drive Maintainability of Software Systems. Software Quality Control 13(3), 297–320 (2005)
27. Opitz, D.W., Shavlik, J.W.: Actively searching for an effective neural-network ensemble. Connection Science 8(3/4), 337–353 (1996)
28. Opitz, D.W., Shavlik, J.W.: Generating Accurate and Diverse Members of a Neural-Network Ensemble. In: Advances in Neural Information Processing Systems, vol. 8, pp. 535–541 (1996)
29. Optiz, D., Maclin, R.: Popular Ensemble Methods: An Empirical Study. Journal of Artificial Intelligence Reseach 11, 169–198 (1999)
30. Poggio, T., Girosi, F.: Networks for approximation and learning. Proceedings of the IEEE 78(9), 1481–1497 (1990)
31. Quinlan, R.J.: Learning with Continuous Classes. In: 5th Australian Joint Conference on Artificial Intelligence, Singapore, pp. 343–348 (1992)
32. Shevade, S.K., Keerthi, S.S., Bhattacharyya, C., Murthy, K.R.K.: Improvements to the SMO Algorithm for SVM Regression. IEEE Transactions on Neural Networks 11(5), 1188–1193 (2000)
33. Shimshoni, Y., Intrator, N.: Classification of seismic signals by integrating ensembles of neural networks. IEEE Transactions on Signal Processing 46(5), 1194–1201 (1998)
34. Sollich, P.: Learning with Ensembles: How over-fitting can be useful. In: Advances in Neural Information Processing Systems, vol. 8, pp. 190–196 (1996)
35. Thwin, M.M.T., Quah, T.-S.: Application of Neural Networks for Software Quality Prediction Using Object-Oriented Metrics. Journal of Systems and Software 76(2), 147–156 (2005)
36. Vapnik, V.: The Nature of Statistical Learning Theory. Springer, New York (1995)

37. Wang, Y., Witten, I.H.: Induction of model trees for predicting continuous classes. In: Poster papers of the 9th European Conference on Machine Learning (1997)
38. Witten, I., Frank, E.: Data Mining: Practical Machine Learning Tools and Techniques, 2nd edn. Morgan Kaufmann, San Francisco (2005)
39. Zheng, J.: Predicting software reliability with neural network ensembles. Expert Systems with Applications (2007)
40. Zhou, Y., Leung, H.: Predicting object-oriented software maintainability using multivariate adaptive regression splines. Journal of Systems and Software 80(8), 1349–1361 (2007)

Sleep Stage Classification Using Advanced Intelligent Methods

José Manuel Sánchez Pascualvaca[1], Carlos Fernandes[1,2], Alberto Guillén[1],
Antonio M. Mora[1], Rogerio Largo[2], Agostinho C. Rosa[2], and Luis Javier Herrera[1,*]

[1] Department of Computer Architecture and Technology,
University of Granada, Granada, Spain
[2] Laseeb, ISR-IST, Av. Rovisco Pais,
1049-001 Lisbon, Portuga
jherrera@ugr.es

Abstract. Manual sleep stage classification is a tedious process that takes a lot of time to sleep experts performing data analysis or studies on this field. Moreover errors and inconsistencies between classifications of the same data are frequent. Due to this, there is a great need of automatic classification systems to support reliable classification. This work extends the work by Herrera et al. (*International Journal of Neural Systems* 10.1142/S0129065713500123), inspecting the use of two techniques to improve the accuracy of sleep stage classifiers based on support vector machines from electroencephalogram, electrooculogram and electromyogram signals. Moreover, three different support vector machine multi-classifiers have been tested to evaluate and compare their performance. To accomplish these tasks, three different feature extraction techniques are applied to the electroencephalogram signals. First, the joint use of these feature sets, together with the electrooculogram and electromyogram information, is inspected (and compared with the use of each feature extraction method separately). Second the possibility of using nearby stages information to predict the current stage is inspected. Results obtained show significant improvements in the classification rates achieved using the two proposed techniques.

Keywords: SVM Multi-classification; EEG Feature Extraction; Sleep stage classification.

1 Introduction

Most of the investigations that aim at automating sleep staging rely on polysomnography (PSGs), a multi-parameter test that registers data coming from many body functions, including electroencephalography (EEG), electromyography (EMG), electrooculography (EOG), heart activity and respiration [4][10][11]. All this information, measuring activity from the brain and other parts of the body allows to study the sleep phases of the patients and find out the existence of strange activities produced by a sleep disease. This process is performed with the aid of hypnograms, a graph

---

* Corresponding author.

I. Rojas, G. Joya, and J. Cabestany (Eds.): IWANN 2013, Part I, LNCS 7902, pp. 604–612, 2013.

representing the stages of sleep during the sleep time. However, obtaining the hypnograms is a tedious and difficult manual process, due to the amount of data that the specialists need to analyse, and the different criteria used in sleep stage classification processes [1][2].

The goal of this work is to provide an enhanced classification model to aid the experts in the acquisition of these hypnograms. This study is based a previous work for sleep stage classification [20] and extends it to consider EOG and EMG signals and to introduce a different times series prediction treatment of the problem. It first analyses the performance of different feature extraction methods from EEG signals and their joint use to sleep stages prediction, together with EOG and EMG information. In parallel, it is investigated the performance of three different multi-classification techniques, based on support vector machines (SVM), that have been used in the literature for other applications with satisfactory results. Finally, a hybrid methodological approach treating the classification problem also as a time series prediction problem (treating the patient signals as a data series) is attempted. This way, information from previous (and/or posterior) epochs is used for a given epoch[1] classification.

The rest of the paper is organized as follows: section 2 presents a brief summary of some of the methodologies used in the literature related with sleep studies. Section 3 introduces the feature extraction techniques used to process the EEG signals. Section 4 describes the SVM multi-classification methods used. Section 5 presents the time series approach to deal with the sleep data classification. In section 6 the most important results obtained in the experimental work experiments are shown. Section 7 concludes this work.

2   Background Review

Sleep is divided into a number of phases. This division includes stages with rapid eye movements (REM) and no rapid eye movement (NREM); the last one is usually divided into four phases (NREM1-4). Besides, sleep studies include the awaken stage. Therefore, there is a total of six stages to distinguish. This standard division of sleep states was carried out in 1968 by Allan Rechtschaffen and Anthony Kales [1].

Stage classification uses full night sleep records from the individuals. This registry is divided into epochs, and each one of them is assigned to one of six sleep phases. This mapping is performed by experts in a manual way. They use data obtained from EEGs, EOGs and EMGs for identifying relevant patterns and decide what stage to assign to each epoch. Several difficulties appear at this stage, mainly due to the large amount of data to analyse, the difficulties in classifying some stages, and the lack of consensus existing among experts from different institutions [2]. Even the same expert, when given the same data in different moments, might perform different sleep stage classifications for some epochs. Consequently, there have been many studies on automatic classifiers to aid the experts to perform this task [4][7][10].

---

[1] *Epoch*: time unit of measure of sleep (normally of 30 seconds). Hypnograms indicate the sleep stage the patient was in per each epoch period.

Information taken from the PSG has to be processed to extract the final features to be used in the classification process. This point is very important as it can have a large influence in the final performance of the classification system. There is a wide range of feature extraction methods from EEG. Some of the most well-known methods include Hjorth parameters [3], harmonic parameters or those related to energy bands [4], Fourier transforms [5] or Wavelet transforms [6].

There is also a wide range of classification paradigms applied to this problem. Some techniques are focused on imitating the manual classification process, identifying and measuring some typical signal patterns associated with each possible sleep stage, and then performing the classification. This type of techniques can employ fuzzy or neuro-fuzzy systems [7][8]. Moreover, other approaches use traditional classification techniques applied to numerical features extracted from the data, such as KNN clustering, artificial neural networks [10], or support vector machines [11].

The ultimate goal of all systems is to imitate or support the process of manual classification, besides trying to make a system as robust and widespread as possible. The latter is really difficult to achieve; although presenting a good accuracy rate, several research works found in the literature are not as generalized as possible (they do not test their methods on different datasets) or do not make difference among the six possible sleep R&K stages [7][9][12].

A previous work by Herrera et al. [20] introduced two techniques to improve sleep stage classification accuracy using SVMs. First the joint use of different feature extraction methods from EEG signals was presented showing an important improvement over the use of a single feature extraction method. Second the use of Stacked Sequential (SSL) learning [21], a meta-learning method in which the base classifier is augmented by making it aware of the labels of nearby patterns, was successfully used for improving the accuracy of the classifier. However this study considered 4 sleep stages (awake, REM, NREM1 and NREM2 as "light sleep" and NREM3 and NREM4 as "deep sleep"), and the single source of information treated was the EEG. Moreover, other alternatives to include nearby epochs' information need to be studied in the aim of approaching the experts' opinion.

3   Feature Extraction Methods

This paper tests and compares the following feature extraction methods (we refer to appropriate references for a detailed explanation of the respective methods [20]). Three main feature groups are evaluated separately (together with EOG and EMG) and jointly.

- Symbolic representation of the EEG signal by segmentation and extraction of frequency events [13].
- Features extracted from Wavelet transform of the EEG signal, showing activity or not in the different frequency bands considered [14]. We have 14 features in this group.

- Hjorth´s features. A set of three parameters that describe the EEG signal in the time domain, mean amplitude, mean frequency and wavelength signal, keep in mind their average values, their standard deviation and variance [3] [15]. In total, 9 features.
- EMG, average signal power.
- EOG, saccades density.

Therefore, each sleep epoch of 30 seconds (which is our data entry) has a total of 56 features describing it, from the four sets of features discussed above.

4   Multi-class SVM

The dominant approach to perform SVM multi-classification is to reduce the single multiclass problem into multiple binary classification problems [16]. This paper evaluates the performance of three different methods for multiclass SVM based on the construction and combination of several binary SVM models. The three methods are briefly described below:

One-vs-One. This algorithm builds $N(N-1)/2$ binary classifiers, which are all possible combinations of pair of classes (from all N classes). Each classifier is trained using examples of one class as positives and another as negatives. To find the class to be assigned, a voting scheme is taken. The class with the larger number of votes is selected.

One-vs-All. For a problem with N classes (with N>2), N binary classifiers are learned. Each classifier is trained with data belonging to class i labelled as positive and the rest of data as negative data. In the test phase, each data is tested on the N trained SVM models and is labelled according to the classifier providing the larger likelihood in its output.

SVM Binary Trees [17]. In this methodology the binary SVM models are organized in a binary tree structure. At each node of the tree, an SVM is trained using two sets of classes. Taking all data belonging to classes from one set as positive and all data belonging to classes from the other set as negative. All samples in the main node are assigned to one of two derived branch-nodes. This step is repeated at each branch-node until a node contains only samples from one class, which is the class assigned to the respective assigned data.

The goal of this analysis is to evaluate the performance of different types of SVM multi-classifiers, in order to identify whether any of the modalities of combination of binary classifiers provides a better operation than others due to the nature of the specific problem tackled. To carry out the implementation of the different alternatives we have relied on the LIBSVM library for Matlab [18]. This library is widely and successfully used in the scientific literature using SVM as classification paradigm.

The RBF kernel has been chosen for the experiments due to its good behaviour in a wide range of problems in the literature. Hyper-parameters C and γ were estimated using cross-validation and grid search, using for C a range between 2^{-5} and 2^{15}, and from 2^{-15} to 2^5 for γ [19].

5   Approach as Time Series Prediction Problem

Applying the principles of operation of time series prediction, the following reformulation of the problem can be applied: the class of each sleep epoch is defined as a function of its own features (the features from the same 30 seconds period coming from EEG, EOG and EMG), plus the features from nearby epochs. Therefore, the system will have more features to predict with. This can be understood as making the sleep stage of an epoch not only dependent on the behavior of the signals in the same epoch, but also dependent of the behavior of the signals in the previous and posterior epochs.

In a more formal way, it can be defined as:

$$y(t) \cong F(x(t), x(t-1), \dots, x(t-n), \ x(t+1), \dots, x(t+n)) \tag{1}$$

where, in this problem, $y(t)$ is the sleep stage epoch t belong to, $x(t)$ is the stage t input features vector, and each $x(t\pm n)$ are the input features vector from epoch $t\pm n$. Then, the hypothesis of this treatment is to assess whether sleep experts, in the manual sleep stages classification process, not only take into account information from the current stage, but directly or indirectly, also take into account information from previous and/or posterior epochs.

The main advantage of this approach over others including information from nearby epochs to perform the classification, such as the Stacked Sequential Learning approach [21], is that there is no need to perform a second classification process, and therefore the classification model remains simpler.

6   Results

This section presents the most relevant results obtained in the study performed after the application of each of the different alternatives and improvements proposed.

6.1   Data Description

This study used the all night recordings of 10 subjects from a normal population, from both genders (nine female, one mail) and between 18 and 31 years. The data was provided by Meditron Sleep Laboratory of the Neurology Department, State University of Sao Paulo, Botucatu, SP, Brazil [11][20]. Specifically, 8031 data (epochs) are available, each one covering a non-overlapping period of 30 seconds. Manual scoring of sleep stages according to Rechtschaffen & Kales criteria was performed by a very experienced sleep expert and provided for all subjects. The available data was randomly subdivided in 80% for training and 20% for testing. Results from 10 different executions with different training-test subdivisions are presented. Comparison among different alternatives was assessed using a paired t-test for the test results coming from the 10 different executions.

The numbering of the classes was performed this way: stage NREM 1→ S1, stage NREM 2→ S2, stage NREM 3→ S3, stage NREM 4→ S4, stage REM → S5 and stage Awake → S6.

6.2   First Results

First tests in this study aimed to compare the performance of the different SVM multi-classification alternatives over each of the EEG features sets available (plus EOG and EMG), separately and jointly. Table 1 shows the respective mean accuracy results for ten different executions with different training-test subdivisions. As it can be seen, the joint use of all features set provides a higher accuracy rate, with an improvement up to a 3% over any feature set alone (confirmed also by t-test with significance level $\alpha = 0.05$). Comparing the three feature extraction methods separately, the Wavelet transform method attained a better performance than the Segmentation method and the Hjorth method. Results using the different SVM multi-classification alternatives were very similar, being slightly superior for the OvsAll alternative for using the full feature set. This extreme was confirmed by a paired t-test with significance level (significance level $\alpha = 0.1$).

Table 1. Accuracy rate comparison of the different techniques multiclassification using each feature set separately (plus EOG and EMG) and the joint (full) set of features

Features group	OvsOne	OvsAll	B. Tree
Segmentation EEG + EOG + EMG	67.5 %	67.3%	66.5%
Horth´s parameters + EOG + EMG	65.2 %	63.6%	63.5%
Wavelet transf.+ EOG + EMG	69.1 %	68.8%	67.7%
Full features set	**72.1 %**	**72.4%**	**71.5%**

A drawback found in this problem is the unbalance in the data available for each different sleep stage. Thus, as an additional treatment, the idea proposed in [11] to handle this problem was used. It is based on a modification of the SVM formulation to take into account different C parameters for each class, in accordance with unbalance of the data. This parameter was then taken proportional to the data size for each stage. Table 2, shows the global and per-class performance for each multiclass alternative for the full feature set. When techniques based on decision trees are used, global performance ratios are slightly lower compared to other techniques.

Table 2. Global and by phase accuracy of each technique using the full features set

Technique	% S1	% S2	% S3	% S4	% S5	% S6	%Acc Total
One-vs-One	10%	83%	16%	79%	78%	85%	**72.1%**
One-vs-All	9%	82%	14%	81%	80%	86%	**72.4%**
Binary Tree	11%	81%	22%	78%	77%	84%	**71.5%**

6.3   Earlier and/or Later Epochs Information

Two different alternatives to use features from nearby epochs were evaluated. The first one adds the features from the two previous epochs and the two posterior epochs for the prediction of the current epoch sleep stage. The second uses of the features from the four previous epochs for the prediction of the current epoch sleep stage. For these tests, the full set of features was used. In both cases the number of features per epoch increases up to 280 features in total (corresponding to 5 epochs). Experimental results averaging ten executions with different training-test subdivisions are shown in Tables 3 and 4.

Table 3. Global and stage accuracy of each model using features from the two previous and the two posterior epochs

Technique	% S1	% S2	% S3	% S4	% S5	% S6	%Acc Total
One-vs-One	27%	85%	34%	79%	83%	83%	**76.2%**
One-vs-All	28%	84%	33%	81%	82%	85%	**76.5%**
Binary Tree	30%	85%	29%	79%	83%	83%	**75.5%**

Table 4. Global and stage accuracy of each model using features from the four previous epochs

Technique	% S1	% S2	% S3	% S4	% S5	% S6	%Acc Total
One-vs-One	25%	85%	29%	81%	82%	82%	**75.6%**
One-vs-All	25%	85%	29%	82%	82%	83%	**76.1%**
Binary Tree	26%	85%	26%	81%	78%	81%	**75.0%**

Results first show that using previous and posterior epochs' features largely enhances the classification accuracy of the models, with an improvement up to a 4% (confirmed also by t-test with significance level $\alpha = 0.05$). The use of previous and posterior epochs' features shows a slightly better performance in the three executions and for the three multiclass methods than using only previous epochs' features (however not confirmed by t-test). Results also show that the *"binary tree"* SVM multiclass method attains a lower performance than *"One-vs-One"* and *"One-vs-All"* alternatives. *"One-vs-All"* alternative again shows the best results for both cases, both considering previous and posterior epochs' features and previous epochs' features alone (confirmed in all cases by t-test with significance level $\alpha = 0.05$).

7   Conclusions

After analyzing each group of features and their joint use it is concluded that the combined use of these feature groups, obtained by heterogeneous extraction methods from EEG, and together with EOG and EMG, lead to the improvement the classification ratios (up to a 3%). These results confirm previous results obtained in [20] in which 4 sleep stages were considered using a patient-cross-validation framework. This points toward the relevancy of using different heterogeneous methods to extract information from EEG to perform sleep stage classification. This work used a sleep database of 10 normal subjects, and random training-test subdivisions of the available data.

Moreover, the additional treatment of the problem as time series prediction problem showed a significant improvement in the classification rates. This points out that the achievement of efficient automatic sleep classifiers have to take into account previous and posterior information from the respective epoch in the hypnogram constructions [20]. Accuracy rates increased up to a 4% using this problem modification.

Additionally, a parallel analysis of three SVM multiclass alternatives has shown that, for the given problem, *One-vs-All* technique has achieved a better accuracy than *One-vs-One*, and both obtain better classification rates than the binary tree option.

Acknowledgements. This work was partially supported by the Spanish Ministry of Science Project SAF-2010-20558 and by the FCT project [PEst-OE/EEI/LA0009/2011]. Fernandes wishes to thank FCT, Ministério da Ciência e Tecnologia, his Research Fellowship SFRH / BPD / 66876 / 2009. We would like to thank Meditron Electromedicine UNESP Sleep Lab for providing the data and manual scoring used in this work.

References

1. Rechtschaffen, A., Kales, A.: A manual of standardized terminology, techniques and scoring system for sleep stages of human subjects, Ser. National Institutes of Health (U.S.). Bethesda, Md: U.S. National Institute of Neurological Diseases and Blindness. Neurological Information Network 204 (1968)
2. Norman, R., Pal, I., Stewart, C., Walsleben, J., Rappaport, D.: Interobserver Agreement Among Sleep Scorers From Different Centers in a Large Dataset. Sleep 23, 901–908 (2000)
3. Hjorth, B.: EEG Analysis Based on Time Domain Properties. Electroencephalography and Clinical Neurophysiology 29, 306–310 (1970)
4. Van Hese, P., Philips, W., De Koninck, J., Van de Walle, R., Lemahieu, I.: Automatic Detection of Sleep Stages Using the EEG. In: 23th Annual IEEE/EMBS Conference (October 2001)
5. Görür, D., Halici, U., Aydin, H., Ongun, G., Ozgen, F., Leblebicioglu, K.: Sleep Spindles Detection Using Short Time Fourier Transform and Neural Networks. IEEE 2, 1631–1636 (2002)

6. Mora, A.M., Fernandes, C.M., Herrera, L.J., Castillo, P.A., Rosa, A.C.: Automatic Sleep Classification Procedure Using Wavelet Based Feature Extraction. In: Proceedings of the 2010 International Conference on Metaheuristics and Nature Inspired Computing, META-heuristics (2010)

7. Jo, H.G., Park, J.Y., Lee, C.K., An, S.K., Yoo, S.K.: Genetic Fuzzy Classifier for Sleep Stage Identification. Comp. in Biol. and Med. Elseiver 40, 629–634 (2010)

8. Heiss, J.E., Held, C.M., Estevez, P.A., Perez, C.A., Holzmann, C.A., Perez, J.P.: Classification of sleep stages in infants: a neuro fuzzy approach. Engineering in Medicine and Biology Magazine 21(5), 147–151 (2002)

9. Gunes, S., Polat, K., Yosunkaya, S.: Efficient Sleep Stage Recognition System Based on EEG Signal Using k-Means Clustering Based Feature Weighting. Expert Systems With Applications (2010)

10. Tagluk, M.E., Sezgin, N., Akin, M.: Estimation of Sleep Stages by an Artificial Neural Network Employing EEG, EMG and EOG. J. Med. Syst. (2009)

11. Herrera, L.J., Mora, A.M., Fernandes, C., Migotina, D., Guillén, A., Rosa, A.C.: Symbolic Representation of the EEG for sleep stage classification. In: 11th International Conference on Intelligent Systems Design and Applications (ISDA) (2011)

12. Huang, L., Sun, O., Cheng, J.: Novel Method of Fast Automated Discrimination of Seep Stages. In: Annual International Conference of the IEEE EMBS, Mexico, September 17-21 (2003)

13. Migotina, D., Rosa, A.: Segmentation of Sleep EEG Signal by Optimal Thresholds. In: Proc. Bio. Med. 2012, Austria, pp. 114–121 (February 2012)

14. Mallat, S.: A Wavelet Tour of Signal Processing, 2nd edn. Academic Press (1998)

15. Hjorth, B.: Time domain descriptors and their relation to a particular model for generation of eeg activity. CEAN -Computerized EEG Analysis, 3–8 (1975)

16. Hsu, C.W., Lin, C.J.: A comparison of methods for multiclass support vector machines. IEEE Transactions on Neural Networks 13(2), 415–425 (2002)

17. Takahashi, F., Abe, S.: Decision-tree-based multiclass support vector machines. In: Proceedings of the 9th International Conference on Neural Information Processing, ICONIP 2002, vol. 3, pp. 1418–1422 (2002)

18. Chang, C., Lin, C.: LIBSVM: a library for support vector machines. ACM Transactions on Intelligent Systems and Technology (2011),
    http://www.csie.ntu.edu.tw/~cjlin/libsvm/

19. Hsu, C.-W., Chang, C.-C., Lin, C.-J.: A Practical Guide to Support Vector Classification (2003-2010)

20. Herrera, L.J., Fernandes, C.M., Mora, A.M., Migotina, D., Largo, R., Guillen, A., Rosa, A.C.: Combination of Heterogeneous EEG Feature Extraction Methods and Stacked Sequential Learning for Sleep Stage Classification. International Journal of Neural Systems, doi:10.1142/S0129065713500123

21. Gatta, C., Puertas, E., Pujol, O.: Multi-scale stacked sequential learning. Pattern Recognition 44(10-11), 2414–2426 (2011)

An *n*-Spheres Based Synthetic Data Generator for Supervised Classification*

Javier Sánchez-Monedero, Pedro Antonio Gutiérrez, María Pérez-Ortiz,
and César Hervás-Martínez

University of Córdoba, Dept. of Computer Science and Numerical Analysis
Rabanales Campus, Albert Einstein building, 14071 - Córdoba, Spain

Abstract. Synthetic datasets can be useful in a variety of situations, specifically when new machine learning models and training algorithms are developed or when trying to seek the weaknesses of an specific method. In contrast to real-world data, synthetic datasets provide a controlled environment for analysing concrete critic points such as outlier tolerance, data dimensionality influence and class imbalance, among others. In this paper, a framework for synthetic data generation is developed with special attention to pattern order in the space, data dimensionality, class overlapping and data multimodality. Variables such as position, width and overlapping of data distributions in the n-dimensional space are controlled by considering them as *n*-spheres. The method is tested in the context of ordinal regression, a paradigm of classification where there is an order arrangement between categories. The contribution of the paper is the full control over data topology and over a set of relevant statistical properties of the data.

Keywords: synthetic data, data generator, data complexity, ordinal classification, ordinal regression, experimental design.

1 Introduction

Pattern recognition, machine learning and data mining communities compare and evaluate methods in terms of theoretical and experimental analysis. This experimental analysis typically consists on analysing the performance of a newly proposal against the performance of related state-of-the-art methods with a set on benchmark datasets (real-world or synthetic ones).

Ideally, dataset selection must test specific issues of a method identified after a complete theoretical analysis. However, this is not the general experimental tendency as Macià has pointed out in Chapter 2 of her PhD. Thesis, where a criticism of experimental assessment of learners based on arbitrary selection of datasets is done [1]. Macià et. al remark the necessity of a proper selection of experimental data guided by data complexity measures [2].

* This work was supported in part by the TIN2011-22794 project of the Spanish Ministerial Commision of Science and Technology (MICYT), FEDER funds and the P2011-TIC-7508 project of the "Junta de Andalucíaía" (Spain).

I. Rojas, G. Joya, and J. Cabestany (Eds.): IWANN 2013, Part I, LNCS 7902, pp. 613–621, 2013.

The motivations for selecting synthetic datasets are the following: a) the researcher wants to create a challenging problem to observe the behavior of a method (i.e. regarding data complexity issues [3,2]); b) the cost and difficulty to obtain samples from a specific domain are high. The reason for a) motivation is that it is difficult to characterize some of the inherent properties of real data sets, leading to the difficulty of selecting proper real-world problems for assessing a method. In the b) case, the motivation may be related to economical issues regarding the cost of sampling new data (for instance gene micro-array data) or domain specific reasons. For instance, ordinal classification (also known as ordinal regression) problems are based on the assumption that a given category ordering is somehow present in the input space. However, the presence of this order of the data may be difficult to verify. There, in comparison to nominal classification, the number of available benchmark datasets is low. This assertion can be easily checked in the literature [4]. The present work focuses on both motivations to synthetic data generation, and, without loss of generality, contextualize the problem in the field of ordinal classification.

Even though the relevance of proper synthetic data generation, scientific works on data generation methods based on controlled statistical data properties are scarce [5], as pointed by [6]: "Surprisingly, little work has been done on systematically generating artificial datasets for the analysis and evaluation of data analysis algorithms in data mining area.". The state-of-the-art of synthetic data generation can be divided in two categories [5]: specific approaches linked to an specific domain, for instance handwriting recognition, and general approaches focused on the properties of independent variables. For a brief review of related works please check the work of Pei and Zaïane [6] and Frasch et. al [5].

We focus on the general case dealing with challenging data complexity topics that are interesting in general, and more specifically, in the context of ordinal classification. Those topics are: class ordering in the input space, data dimensionality, class overlapping and data multimodality. Up to the authors knowledge there is no framework dealing with such a controlled data generator for covering these features, which is the main motivation for this article.

As a result of the present work, we have developed a synthetic data generator based on the aforementioned ideas. The program is written in Matlab language and it allows the generation of datasets with multidimensional isotropic Gaussian – also known as white Gaussian – distributions presenting the following parameters: number of classes, number of patterns per class, number of input dimensions, variance of the Gaussians and number of modes for each class. The framework allows data visualization which can be exported in PDF and SVG formats (via Matlab toolboxes), and datasets files can be exported to Weka and Matlab formats[1].

The rest of the paper is organized as follows. Section 2 identifies data complexity issues covered by the proposal. Section 3 explains the proposed framework.

---

[1] Source code of this framework is available at
http://www.uco.es/grupos/ayrna/iwann2013-syntheticdatagenerator

Experiments are covered at Section 4 and finally conclusions are depicted at the last section.

2   Ordinal Classification Peculiarities

As previously mentioned, ordinal classification deals with supervised classification problems in which there is an order within categories. This order is typically deduced from the problem nature by an expert or by a simple inference about the data. For instance, bond rating can be considered as an ordinal classification problem, where the purpose is to assign the right ordered category to bond, being the category labels $\{C_1 = \text{AAA}, C_2 = \text{AA}, C_3 = \text{A}, C_4 = \text{BBB}, C_5 = \text{BB}\}$, where labels represent the bond quality assigned by credit rating agencies. Here there is a natural order between classes $\{\text{AAA} \prec \text{AA} \prec \text{A} \prec \text{BBB} \prec \text{BB}\}$, AAA being the highest quality and BB the worst one. In this case, the ordinal nature of the problem can be deduced, not only from the dependent variables, i.e. the labels, but also for the independent variables, that reflect this order in the space of attributes $\mathcal{X}$. It is straightforward to observe that the public debt or the unemployment are variables that can grown or decrease in relation to class labelling. According to [4], an ordinal classifier must exploit this a priori knowledge about the input space and the topological distribution of the classes. To do so, datasets keeping this relative order among classes are needed. Conversely, the number of available benchmark datasets in this field is scarce. This is true up to the point that some of the most employed datasets are standard regression datasets that have been discretized [7].

In the literature, the most widely known dataset is the *toy* dataset, proposed by Herbrich et al. [8]. Figure shows an example of the *toy* dataset, reflecting five ordered classes non-linearly separable. Recently, Sánchez-Monedero et al. proposed the *spiral* dataset, which is generated by adding a Gaussian noise to points on a spiral [9]. Although these two datasets present interesting non-linear cases, it could be difficult to extend them to more dimensions or to clearly control the overlap between classes.

Formally, in an ordinal classification problem, the purpose is to learn a mapping ϕ from an input space $\mathcal{X}$ to a finite output set $\mathcal{C} = \{C_1, C_2, \ldots, C_Q\}$ containing Q labels, where the label set presents an order relation $C_1 \prec C_2 \prec \ldots \prec C_Q$. The symbol $\prec$ denotes this ordering between different ranks. A rank for an ordinal ranked category can be defined as $\mathcal{O}(C_q) = q$. Each pattern is represented by a n-dimensional feature vector $\mathbf{x} \in \mathcal{X} \subseteq \mathbb{R}^n$ and a class label $y \in \mathcal{C}$.

Starting from this bond example, some data characteristics widely recognized as challenging issues in machine learning can be pointed out. First, since class labels often represent ranks or degrees, class overlap can occur between all or a subset of classes. In the bond rate example, the top quality classes are highly overlapped and it is common that rating agencies move some company or country debt from one category to another when reevaluating the rating.

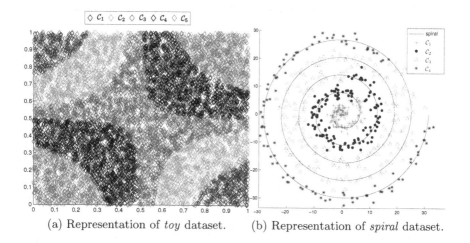

(a) Representation of *toy* dataset.   (b) Representation of *spiral* dataset.

Secondly, high dimensionality can be a problem in ordinal regression, specially for latent variable models, which are by far the most extended methods [10]. These methods assume that ordinal response is a coarsely measured latent continuous variable, and model it as real intervals in one dimension. These algorithms project samples onto a one-dimensional variable and fix or estimate thresholds dividing the latent space in consecutive intervals representing ordinal categories. In the presence of high dimensionality and non-linearity separable data, the mapping function ϕ can results into complex models that impose a highly non-linear transformation from the input space to the latent space. In the literature, many of the existing datasets present high dimensionality [4,10].

Thirdly, and related to the latent variable proposals, the influence of data multimodality, i.e. when different clusters of data in the same or different classes can be found, is suitable of analysis.

Finally, last requirement for the data generator is class imbalance, also present in ordinal classification problems [11,10]. This issue has been widely covered in the literature both for binary and multiclass problems.

3   Isotropic Gaussian Synthethic Data Generation

In order to simplify the requirements of the data generation, the patterns will be generated by random sampling from isotropic Gaussian distributions (i.e. distribution variance is the same through all the dimensions). Being, $\mathcal{N}(\mu, \sigma^2)$ a one-dimension Gaussian distribution, where μ is the mean and σ^2 is the variance. For higher dimensionality, the multivariate Gaussian distribution is defined as $\mathcal{N}(\boldsymbol{\mu}, \boldsymbol{\Sigma})$, where $\boldsymbol{\mu}$ is a n-dimensional mean vector, and $\boldsymbol{\Sigma}_{n \times n}$ is the covariance matrix. In the case of the multivariate isotropic Gaussian distribution, the distribution can be expressed as:

$$\mathbf{x} = \mathcal{N}(\boldsymbol{\mu}, \sigma^2 \mathbf{I}_{n \times n}), \tag{1}$$

where sample $\mathbf{x} \in \mathbb{X} \subseteq \mathbb{R}^n$ and $\mathbf{I}_{n \times n}$ is the identity matrix. Note this formulation reflects that the variance σ^2 is the same across all dimensions and that all the dimensions' variance are independent.

From now on, we will work with the multidimensional isotropic Gaussian distributions (Eq. (1)). With these premises, a hyper-sphere or n-sphere with center in $\mathbf{x}$ and radio r can be defined as follows:

$$S^n = \left\{ \mathbf{x} \in \mathbb{R}^{n+1} : \|\mathbf{x}\| = r \right\}, \tag{2}$$

in our case, $\mathbf{x} = \boldsymbol{\mu}$. Considering $r = 3\sigma$, nearly all (99.73%) of the population lies inside the n-sphere.

Then, we want to place the set of n-spheres with a separation between them of α in a Euclidean space, this is, the distance between one center $\boldsymbol{\mu}$ and another center $\boldsymbol{\mu}' = \boldsymbol{\mu} + \Delta_{\boldsymbol{\mu}}$ will be α:

$$d(\boldsymbol{\mu}, \boldsymbol{\mu}') = \sqrt{\sum_{i=1}^{n}(\boldsymbol{\mu}_i - \boldsymbol{\mu}_i')^2} = \sqrt{\sum_{i=1}^{n}(\boldsymbol{\mu}_i - \boldsymbol{\mu}_i + \Delta_{\boldsymbol{\mu}})^2} = \alpha, \tag{3}$$

$$\Delta_{\boldsymbol{\mu}} = \pm\alpha/\sqrt{n}. \tag{4}$$

Then, using an increment of $\Delta_{\boldsymbol{\mu}}$ guarantees a separation of α in the Euclidean space independently of the dimensionality. With this separation between the centers of the n-spheres the percentage of overlapped surface ($n = 2$) or volume ($n \geq 3$) respecting the n-sphere will be constant. Then, the overlap can be expressed in terms of σ. Note that working with anisotropic Gaussian would imply defining multiple hyper-ellipses, thus dealing with more complex calculations to effectively control class overlapping in multiple dimensions.

Regarding multimodality of the data, we proceed in the following way: the previously defined Gaussian distribution is considered as the *main* distribution of each class, and additional distributions are added to each class in order force multimodality. For each class q, the additional Gaussian distributions are centered in a random location within the surface of the n-sphere with center $\boldsymbol{\mu}_q$ and radius Δ_r. In order to sample points only on the n-sphere surface, the norm of the samples is used as the n-space position (see Eq. (2)), then, the center of each additional mode is obtained as:

$$\boldsymbol{\mu}_q^i = \Delta_r \times \|\mathbf{x}\|, \ \mathbf{x} \in C_q, \tag{5}$$

where $\boldsymbol{\mu}_q^i$ is the i-th mode of the q-th class, $\mathbf{x}$ is sampled from Eq. (1) and $\Delta_r = \pm\lambda/\sqrt{n}$, being $\lambda < \alpha$ the desired separation between each mode of the class. In this way, each class can be composed of different modes, however the overlap of the additional distributions can not be controlled. We denote the number of modes per class with m. Figure 1 shows an example of two generated datasets and theoretical n-spheres.

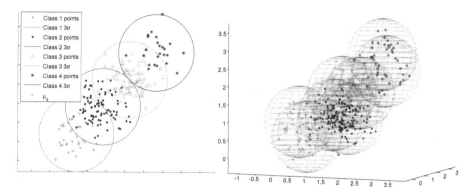

Fig. 1. Example of two synthetic datasets. The left dataset has $n = 2$, $\sigma^2 = 0.33$ and $m = 1$. The right dataset is generated with $n = 3$, $\sigma^2 = 0.33$ and $m = 2$.

4   Experimental Section

This section presents some experiments with a set of datasets generated with the proposed methodology, an analyse the performance of classifiers with these data. Evaluating the performance of ordinal classifiers is different from nominal classification evaluation, since corresponding performance metrics must consider this ordering relation, in such a way that misclassification between adjacent classes is considered less important than the one between non-adjacent classes, which are more separated in rank. Given the bond rating example, it is direct to conclude that predicting class BB when the real class is AA represents a more severe error than that associated with a AAA prediction. Therefore, specialized measures are needed for evaluating ordinal classifiers performance [11]. In this work, due to space restrictions, the MAE measure will be used since this is the more extended metric for ordinal classification.

4.1   Datasets and Experimental Design

In order to test the framework, we have conducted three types of experiments to observe the performance of ordinal classification methods regarding data

Table 1. Characteristics of the generated datasets

Statistical property tested values		
Experiment #1 variance (σ^2)	Experiment #2 dimensions (n)	Experiment #3 modes (m)
0.25	2	1
0.50	10	2
0.75	50	3
1.00	100	4
Constant parameters		
$n = 10$	$\sigma^2 = 0.33$	$\sigma^2 = 0.33$
$m = 1$	$m = 1$	$n = 10$

overlapping (variance, σ^2), data dimensionality (n) and data multimodality (m). The parameters for generating the datasets are detailed at Table 1.

For the sake of simplicity, the number of classes was set to four for all the experiments conducted, and the parameters for controlling distance between Gaussians were $\alpha = 1.0$ and $\lambda = 0.75$. The number of patterns generated for each mode was 36, 90, 90 and 24 for class 1, 2, 3 and 4, respectively, (in order to assess also the problem of class imbalance) using the same number for all the modes of a class. Then, in the bimodal case the corresponding total number of pattern per class was 72, 180, 180 and 48. The experimental procedure was a stratified 30-holdout (30 random repetitions) using 75% of the instances for training and 25% for testing.

In these experiments we included three ordinal regression methods. The Support Vector Ordinal Regression with implicit constraints (SVORIM) [12]; the Ordinal Extreme Learning Machine (ELMOR) [13] and the Proportional Odds Model (POM) [14]. For each data partition, a grid search with a 5-fold nested cross-validation was performed for algorithms' parameters optimization. For SVORIM, the following ranges of hyper-parameters were tested for the cost parameter and the width of the Gaussian kernel: $C \in \{10^3, 10^2, \ldots, 10^{-3}\}$ and $\gamma \in \{10^3, 10^2, \ldots, 10^{-3}\}$. For ELMOR, the sigmoid basis function was used and the number of neurons tested in the hidden layer was $\{5, 10, 20, \ldots, 100\}$. The POM method does not have hyper-parameters.

4.2   Experimental Results

In order to graphically compare the methods, three figures have been included showing the MAE performance of the methods for the three studied situations.

From these results, several conclusions can be drawn. Firstly, it can be seen that for the methods analysed, there are several data features that have a similar impact on classification, which is the case of the class overlapping (variance) and the number of modes. In the case of the variance, the three selected algorithms present very similar performance, although this performance is extremely damaged when increasing the variance. For the number of modes, there is not such a clear connection, although the methods also present a similar trend. Nevertheless, for the case of the number of dimensions, each algorithm present a differentiated tendency. More specifically, the ELMOR method seems to be very dependent of data dimensionality, probably due to the initial random choice characteristic of Extreme Learning Machines (note that setting two random parameters is not the same than 100). On the other side, the SVORIM, which is a extension of the Support Vector Machine paradigm to ordinal regression, seems to be totally indifferent to the increase of the dimensionality, perhaps because of the use of the kernel trick, which leads the algorithm to not work in the input space, but rather in the feature one.

These experiments are not really aimed to compare ordinal classifiers, but to emphasize the need to compare new proposals in a more controlled environment, in order to properly analyse their performances. Otherwise, conclusions may be biased towards a wrong direction. For example, in the case of two-dimensions

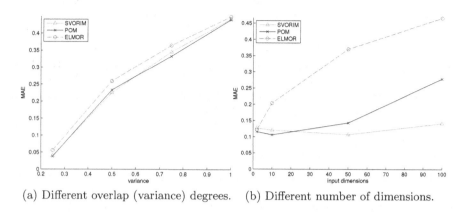

(a) Different overlap (variance) degrees.   (b) Different number of dimensions.

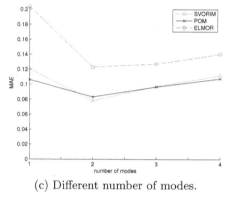

(c) Different number of modes.

Fig. 2. *MAE* performance of the classifiers for the different datasets (see Table 1) for variance, dimensionality and multimodality values)

the ELMOR, POM and SVORIM behave almost the same, despite the fact that the ELMOR is very influenced by dimensionality increase and that the POM is a linear model.

5   Conclusions and Future Work

In this work, we justify the study of methods for proper synthetic data generation. We have proposed a n-spheres based approach for controlling statistical properties of the data, and we have tested the generated datasets with different classifiers. Results prove the correctness of the proposal since controlled experiments match weak points of the classifiers used in the study. Nevertheless the simplification that the use of isotropic Gaussian carries out can involve a lack of representativity of synthetic data with respect to real data. Finally, in order to complete the current proposal, the challenge of defining a set of problem families and relating it to algorithm characteristics and limitations could be considered.

In this way, a preliminary line could be to connect our work to the work of Macià et. al [2] and the data complexity measures collected in the work of [3].

References

1. Macià, N.: Data complexity in supervised learning: A far-reaching implication. PhD thesis, La Salle – Universitat Ramon Llull (October 2011)
2. Macià, N., Bernadó-Mansilla, E., Orriols-Puig, A., Ho, T.K.: Learner excellence biased by data set selection: A case for data characterisation and artificial data sets. Pattern Recognition 46(3), 1054–1066 (2013)
3. Ho, T.K., Basu, M.: Complexity measures of supervised classification problems. IEEE Transactions on Pattern Analysis and Machine Intelligence 24(3), 289–300 (2002)
4. Hühn, J.C., Hüllermeier, E.: Is an ordinal class structure useful in classifier learning? Int. J. of Data Mining, Modelling and Management 1(1), 45–67 (2008)
5. Frasch, J.V., Lodwich, A., Shafait, F., Breuel, T.M.: A bayes-true data generator for evaluation of supervised and unsupervised learning methods. Pattern Recogn. Lett. 32(11), 1523–1531 (2011)
6. Pei, Y., Zaïane, O.: A synthetic data generator for clustering and outlier analysis. Technical report, Computing Science Department. University of Alberta (2006)
7. Chu, W., Ghahramani, Z.: Gaussian processes for ordinal regression. Journal of Machine Learning Research 6, 1019–1041 (2005)
8. Herbrich, R., Graepel, T., Obermayer, K.: Support vector learning for ordinal regression. In: Ninth International Conference on Artificial Neural Networks, ICANN 1999, vol. 1, pp. 97–102 (1999) (Conf. Publ. No. 470)
9. Sánchez-Monedero, J., Gutiérrez, P.A., Tiño, P., Hervás-Martínez, C.: Exploitation of pairwise class distances for ordinal classification. Neural Computation (accepted, 2013)
10. Gutiérrez, P.A., Pérez-Ortiz, M., Fernández-Navarro, F., Sánchez-Monedero, J., Hervás-Martínez, C.: An Experimental Study of Different Ordinal Regression Methods and Measures. In: Corchado, E., Snášel, V., Abraham, A., Woźniak, M., Graña, M., Cho, S.-B. (eds.) HAIS 2012, Part II. LNCS, vol. 7209, pp. 296–307. Springer, Heidelberg (2012)
11. Cruz-Ramírez, M., Hervás-Martínez, C., Sánchez-Monedero, J., Gutiérrez, P.A.: A Preliminary Study of Ordinal Metrics to Guide a Multi-Objective Evolutionary Algorithm. In: Proceedings of the 11th International Conference on Intelligent Systems Design and Applications, ISDA 2011, Cordoba, Spain, pp. 1176–1181 (2011)
12. Chu, W., Keerthi, S.S.: Support Vector Ordinal Regression. Neural Computation 19(3), 792–815 (2007)
13. Deng, W.Y., Zheng, Q.H., Lian, S., Chen, L., Wang, X.: Ordinal extreme learning machine. Neurocomputing 74(1-3), 447–456 (2010)
14. McCullagh, P.: Regression models for ordinal data. Journal of the Royal Statistical Society. Series B (Methodological) 42(2), 109–142 (1980)

Improving the Classification Performance of Optimal Linear Associative Memory in the Presence of Outliers

Ana Luiza Bessa de Paula Barros[1,2] and Guilherme A. Barreto[2]

[1] Department of Computer Science, State University of Ceará
Campus of Itaperi, Fortaleza, Ceará, Brazil
`analuiza@larces.uece.br`
[2] Department of Teleinformatics Engineering, Federal University of Ceará
Center of Technology, Campus of Pici, Fortaleza, Ceará, Brazil
`guilherme@deti.ufc.br`

Abstract. The optimal linear associative memory (OLAM) proposed by Kohonen and Ruohonen [16] is a classic neural network model widely used as a standalone pattern classifier or as a fundamental component of multilayer nonlinear classification approaches, such as the extreme learning machine (ELM) [10] and the echo-state network (ESN) [6]. In this paper, we develop an extension of OLAM which is robust to labeling errors (outliers) in the data set. The proposed model is robust to label noise not only near the class boundaries, but also far from the class boundaries which can result from mistakes in labelling or gross errors in measuring the input features. To deal with this problem, we propose the use of M-estimators, a parameter estimation framework widely used in robust regression, to compute the weight matrix operator, instead of using the ordinary least squares solution. We show the usefulness of the proposed classification approach through simulation results using synthetic and real-world data.

Keywords: Linear Associative Memory, Moore-Penrose Generalized Inverse, Pattern Classification, Outliers, M-Estimation.

1 Introduction

The OLAM model, as proposed by Kohonen and Ruohonen [16], is a well known computational paradigm of associative memory. As such, information in OLAM is stored distributively in a matrix operator, so that it can recall a stored data by specifying all or portion of a key (degraded key). The OLAM has the property of providing rapid recall of information, and it can tolerate local damage without a great degradation in performance.

Previous studies have evaluated empirically and/or theoretically the robustness of OLAM to noisy input patterns [3,19,20]. The main conclusion that these works have pointed out is that when input key vectors are degraded (noisy), the

I. Rojas, G. Joya, and J. Cabestany (Eds.): IWANN 2013, Part I, LNCS 7902, pp. 622–632, 2013.

model becomes extremely sensitive (unstable) and its association error becomes unacceptably large. Authors have tackled this limitation of the OLAM by including nonlinear features into the associative memory model [13] or by taking into consideration the properties of the noise directly into the developing of the model [1].

In this paper, we are interested in OLAM not for associative memory, but rather for pattern classification. In this context, the OLAM is theoretically equivalent to the least-squares classifier [4,21] and has been used either as a standalone classifier [2,5,15] or as a fundamental building block of multilayer nonlinear classification approaches, such as the radial basis functions network [18], the extreme learning machine (ELM) [10] and the echo-state network (ESN) [6].

In many real-world classification problems the labels provided for the data are noisy. There are typically two kinds of noise in labels. Noise near the class boundaries often occurs because it is hard to consistently label ambiguous data points. Labelling errors far from the class boundaries can occur because of mistakes in labelling or gross errors in measuring the input features. Labelling errors far from the boundary comprises a particular category of *outliers* [14].

In order to allow the OLAM classifier to handle outliers efficiently, in this paper we propose the use of M-estimators [12], a broad framework widely used for parameter estimation in robust regression problems, to compute the weight matrix operator instead of using the ordinary least squares solution. We show through simulations on synthetic and real-world data that the resulting OLAM classifier is very robust to outliers.

Despite the fact that M-estimation has been widely used in regression problems (see, e.g. references [9,17]), its application to supervised pattern classification problems is much less studied. In reality, we were not able to find a single paper on the combined use of M-estimation and neural network classifiers. Furthermore, to the best of our knowledge, this is the first time the performance of the OLAM model as a classifier is evaluated under the presence of outliers.

The remainder of the paper is organized as follows. In Section 2, we briefly review the fundamentals of OLAM in the context of pattern classification. Then, in Section 3 we describe the basic ideas and concepts behind the M-estimation framework and introduce our approach to robust supervised pattern classification using OLAM. In Section 4 we present the computer experiments we carried out using synthetic and real-world datasets and also discuss the achieved results. The paper is concluded in Section 5.

2   Fundamentals of OLAM

Let us assume that N data pairs $\{(\mathbf{x}_\mu, \mathbf{d}_\mu)\}_{\mu=1}^N$ are available for building and evaluating the model, where $\mathbf{x}_\mu \in \mathbb{R}^{p+1}$ is the μ-th input pattern[1] and $\mathbf{d}_\mu \in \mathbb{R}^K$ is the corresponding target class label, with K denoting the number of classes. For the labels, we assume an 1-of-K encoding scheme, i.e. for each label vector

---

[1] First component of $\mathbf{x}_\mu$ is equal to 1 in order to include the bias.

$\mathbf{d}_\mu$, the component whose index corresponds to the class of pattern $\mathbf{x}_\mu$ is set to "+1", while the other $K - 1$ components are set to "-1".

Then, let us randomly select N_1 ($N_1 < N$) data pairs from the available data pool and arrange them along the columns of the matrices $\mathbf{D}$ and $\mathbf{X}$ as follows:

$$\mathbf{X} = [\mathbf{x}_1 \mid \mathbf{x}_2 \mid \cdots \mid \mathbf{x}_{N_1}] \quad \text{and} \quad \mathbf{D} = [\mathbf{d}_1 \mid \mathbf{d}_2 \mid \cdots \mid \mathbf{d}_{N_1}]. \tag{1}$$

where $\dim(\mathbf{X}) = (p+1) \times N_1$ and $\dim(\mathbf{D}) = m \times N_1$. Our goal is to use the matrices $\mathbf{X}$ and $\mathbf{D}$ to build the following linear mapping:

$$\mathbf{D} = \beta\mathbf{X} \quad \text{(batch recall)} \quad \text{or} \quad \mathbf{d}_\mu = \beta\mathbf{x}_\mu \quad \text{(pattern-by-pattern recall)}, \tag{2}$$

for $\mu = 1, \ldots, N_1$. For both recall modes, the dimension of β is $K \times (p+1)$.

The ordinary least squares (OLS) solution of the linear system in Eq. (2) is given by the Moore-Penrose generalized inverse as follows:

$$\hat{\beta} = \mathbf{D}\mathbf{X}^T \left(\mathbf{X}\mathbf{X}^T\right)^{-1}, \tag{3}$$

where the hat symbol ($\wedge$) indicates an estimate of the matrix operator β. A minimum-norm solution for Eq. (2) is given by the regularized version of Eq. (3):

$$\hat{\beta} = \mathbf{D}\mathbf{X}^T \left(\mathbf{X}\mathbf{X}^T + \lambda\mathbf{I}\right)^{-1}, \tag{4}$$

where $\mathbf{I}$ is the identity matrix of dimension $(p+1) \times (p+1)$ and λ is a very small positive regularization parameter.

Once we have computed $\hat{\beta}$, the remaining $N_2 = N - N_1$ data pairs are used to validate the model. In this regard, for the pattern-by-pattern recall mode, the output of the OLAM is given by

$$\mathbf{y}_\mu = \hat{\beta}\mathbf{x}_\mu, \tag{5}$$

while for the batch recall mode we have $\tilde{\mathbf{Y}} = \hat{\beta}\tilde{\mathbf{X}}$.

The predicted class index i_μ^* for the μ-th testing input pattern is then given by the following decision rule:

$$i_\mu^* = \arg\max_{i=1,\ldots,K}\{y_{i\mu}\} = \arg\max_{i=1,\ldots,K}\{\hat{\beta}_i^T \mathbf{x}_\mu\}, \tag{6}$$

where $y_{i\mu}$ is the i-th component of vector $\mathbf{y}_\mu$ computed as in Eq. (5), with the vector β_i^T being the i-th row of the matrix $\hat{\beta}$.

It is worth noting that the parameter vector $\beta_i \in \mathbb{R}^m$, $i = 1, \ldots, m$, can be computed individually by means of the following equation:

$$\hat{\beta}_i = \left(\mathbf{X}\mathbf{X}^T\right)^{-1}\mathbf{X}\mathbf{D}_i^T, \tag{7}$$

where the vector $\mathbf{D}_i$ corresponds to the i-th row of the matrix $\mathbf{D}$.

3   Basics of M-Estimation

An important feature of OLS is that it assigns the same importance to all error samples, i.e. all errors contribute the same way to the final solution. A common approach to handle this problem consists in removing outliers from data and then try the usual least-squares fit. A more principled approach, known as *robust regression*, uses estimation methods not as sensitive to outliers as the OLS.

Huber [11] introduced the concept of M-estimation, where M stands for "maximum likelihood" type, where robustness is achieved by minimizing another function than the sum of the squared errors. Based on Huber theory, a general M-estimator applied to the i-th output neuron of the OLAM classifier minimizes the following objective function:

$$J(\boldsymbol{\beta}_i) = \sum_{\mu=1}^{N} \rho(e_{i\mu}) = \sum_{\mu=1}^{N} \rho(d_{i\mu} - y_{i\mu}) = \sum_{\mu=1}^{N} \rho(d_{i\mu} - \boldsymbol{\beta}_i^T \mathbf{x}_\mu), \tag{8}$$

where the function $\rho(\cdot)$ computes the contribution of each error $e_{i\mu} = d_{i\mu} - y_{i\mu}$ to the objective function, $d_{i\mu}$ is the target value of the i-th output neuron for the μ-th input pattern $\mathbf{x}_\mu$, and $\boldsymbol{\beta}_i$ is the weight vector of the i-th output neuron. The OLS is a particular M-estimator, achieved when $\rho(e_{i\mu}) = e_{i\mu}^2$. It is desirable that the function ρ possesses the following properties:

Property 1: $\rho(e_{i\mu}) \geq 0$.
Property 2: $\rho(0) = 0$.
Property 3: $\rho(e_{i\mu}) = \rho(-e_{i\mu})$.
Property 4: $\rho(e_{i\mu}) \geq \rho(e_{i'\mu})$, for $|e_{i\mu}| > |e_{i'\mu}|$.

Parameter estimation is defined by the estimating equation which is a weighted function of the objective function derivative. Let $\psi = \rho'$ to be the derivative of ρ. Differentiating ρ with respect to the estimated weight vector $\hat{\boldsymbol{\beta}}_i$, we have

$$\sum_{\mu=1}^{N} \psi(y_{i\mu} - \hat{\boldsymbol{\beta}}_i^T \mathbf{x}_\mu)\mathbf{x}_\mu^T = \mathbf{0}, \tag{9}$$

where $\mathbf{0}$ is a $(p+1)$-dimensional row vector of zeros. Then, defining the weight function $w(e_{i\mu}) = \psi(e_{i\mu})/e_{i\mu}$, and let $w_{i\mu} = w(e_{i\mu})$, the estimating equations are given by

$$\sum_{\mu=1}^{n} w_{i\mu}(y_{i\mu} - \hat{\boldsymbol{\beta}}_i^T \mathbf{x}_\mu)\mathbf{x}_\mu^T = \mathbf{0}. \tag{10}$$

Thus, solving the estimating equations corresponds to solving a weighted least-squares problem, minimizing $\sum_\mu w_{i\mu}^2 e_{i\mu}^2$.

It is worth noting, however, that the weights depend on the residuals (i.e. estimated errors), the residuals depend upon the estimated coefficients, and the estimated coefficients depend upon the weights. As a consequence, an iterative estimation method called *iteratively reweighted least-squares* (IRLS) [7] is commonly used. The steps of the IRLS algorithm in the context of training the OLAM classifier using Eq. (7) as reference are described next.

IRLS Algorithm for OLAM Training

Step 1 - Provide an initial estimate $\hat{\beta}_i(0)$ using the regularized least-squares solution in Eq. (7).

Step 2 - At each iteration t, compute the residuals from the previous iteration $e_{i\mu}(t-1)$, $\mu = 1, \ldots, N$, associated with the i-th output neuron, and then compute the corresponding weights $w_{i\mu}(t-1) = w[e_{i\mu}(t-1)]$.

Step 3 - Solve for new weighted-least-squares estimate of $\beta_i(t)$:

$$\hat{\beta}_i(t) = \left[\mathbf{X}\mathbf{W}(t-1)\mathbf{X}^T\right]^{-1}\mathbf{X}\mathbf{W}(t-1)\mathbf{D}_i^T, \tag{11}$$

where $\mathbf{W}(t-1) = \text{diag}\{w_{i\mu}(t-1)\}$ is an $N \times N$ weight matrix. Repeat Steps 2 and 3 until the convergence of the estimated coefficient vector $\hat{\beta}_i(t)$.

Several weighting functions for the M-estimators can be chosen, such as the Huber's weighting function:

$$w(e_{i\mu}) = \begin{cases} \frac{k}{|e_{i\mu}|}, & \text{if } |e_{i\mu}| > k \\ 1, & \text{otherwise.} \end{cases} \tag{12}$$

where the parameter k is a tuning constant. Smaller values of k produce more resistance to outliers, but at the expense of lower efficiency when the errors are normally distributed. In particular, $k = 1.345\sigma$ for the Huber function, where σ is a robust estimate of the standard deviation of the errors[2].

In a sum, the basic idea of the proposed approach is very simple: replace the OLS estimation of the weight matrix $\hat{\beta}$ described in Eq. (3) with the one provided by the combined use of the M-estimation framework and the IRLS algorithm. From now on, we refer to the proposed approach by *Robust OLAM* classifier (or ROLAM, for short). In the next section we present and discuss the results achieved by the ROLAM classifier on synthetic and real-world datasets.

4   Simulations and Discussion

As a proof of concept, in the first experiment we aim at showing the influence of outliers in the final position the decision line between two linear separable data classes. For this purpose, we created a synthetic two-dimensional dataset consisting of $N = 120$ samples plus N_{out} outliers. The OLAM and the ROLAM classifiers are trained twice. The first time they are trained with the outlier-free dataset. The second time, they are trained with the outliers added to the original

---

[2] A usual approach is to take $\sigma = \text{MAR}/0.6745$, where MAR is the median absolute residual.

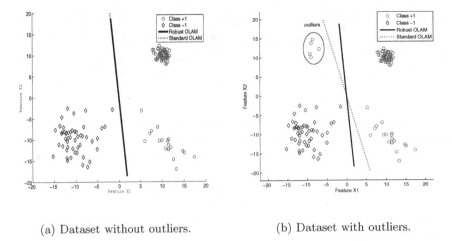

(a) Dataset without outliers.   (b) Dataset with outliers.

Fig. 1. Decision lines of the standard OLAM and the proposed robust OLAM classifiers. (a) Dataset without outliers. (b) Dataset with outliers.

dataset. It is worth mentioning that all data samples are used for training the classifiers, since the goal is to visualize the final position of the decision line and not to compute recognition rates.

For this experiment, the Huber weighting function was used for implementing the ROLAM classifier and the regularization constant required for implementing the standard OLAM classifier was set to $\lambda = 10^{-2}$. The default tuning parameter k of Matlab's `robustfit` function was used. In order to evaluate the final decision lines of the OLAM and the ROLAM classifiers in the presence of outliers, we added $N_{out} = 10$ outliers to the dataset and labelled them as beloging to class $+1$. The outliers were located purposefully far from the class boundary found for the outlier-free case; more specifically, at the decision region of class -1.

The results for the training without outliers are shown in Fig. 1a, where as expected the decision lines of both classifiers coincide. The results for the training with outliers are shown in Fig. 1b, where this time the decision line of the OLAM classifier moved towards the outliers, while the decision line of the ROLAM classifier remained at the same position, thus revealing the robustness of the proposed approach to the presence of outliers. The dataset (with and without outliers) used in the first experiment can be made available by the authors upon request.

In the second and third experiments we aim at evaluating the robustness of the ROLAM classifier using real-world datasets. In these experiments, four weighting functions were tested for implementing the ROLAM classifier and the regularization constant required for implementing the standard OLAM classifier

Table 1. Performance comparison of OLAM and ROLAM classifiers (Iris dataset)

Classifier	$N_{out} = 0\%$	$N_{out} = 5\%$	$N_{out} = 10\%$	$N_{out} = 20\%$	$N_{out} = 30\%$
OLAM	93.05 ± 4.76	91.75 ± 5.52	85.60 ± 9.03	66.60 ± 11.78	56.25 ± 10.65
ROLAM (Bisquare)	93.35 ± 4.55	$\mathbf{94.05 \pm 4.36}$	$\mathbf{93.40 \pm 5.02}$	69.10 ± 11.90	$\mathbf{57.80 \pm 11.20}$
ROLAM (Fair)	93.25 ± 5.52	93.35 ± 5.13	92.05 ± 5.73	$\mathbf{70.20 \pm 11.28}$	53.15 ± 12.30
ROLAM (Huber)	$\mathbf{94.20 \pm 4.75}$	93.10 ± 5.31	93.10 ± 5.49	67.00 ± 10.96	53.35 ± 11.55
ROLAM (Logistic)	93.55 ± 4.68	93.50 ± 5.29	93.30 ± 5.23	68.95 ± 9.33	56.80 ± 11.77

was set to $\lambda = 10^{-2}$. The default tuning parameter k of Matlab's `robustfit` function was adopted for all weighting functions.

In order to evaluate the classifier's robustness to outliers we follow the methodology introduced by Kim and Ghahramani [14]. Thus, the original labels of some data samples of a given class are deliberately changed to the label of the other class. Two datasets were selected, Iris and Vertebral Column (VC), which are publicly available for download from the UCI Machine Learning Repository website [8].

For the Iris dataset, we labelled the samples of classes Virginica ($N_{vir} = 50$ samples) and Versicolor ($N_{ver} = 50$ samples) as $+1$ and -1, respectively. Data from category Setosa ($N_{set} = 50$ samples) will be labelled as belonging to class $+1$, i.e. will be treated as outliers of class $+1$. A certain number N_{out} of outliers are randomly selected and added to the training set of the classifiers. Thus, the total number of training samples is given by $N = P_{train} \times (N_{ver} + N_{vir}) + N_{out}$, where P_{train} is a percentage of samples randomly selected from classes Versicolor and Virginica. We evaluate the performance of the OLAM and ROLAM classifiers for increasing values of N_{out}.

The results are given in Table 1 for different values of N_{out} and for different weighting functions. In this table, we show the values of the classification rates and the corresponding standard deviations averaged over 100 training/testing runs. By analyzing the results, we verify that the ROLAM classifier always performed better than the OLAM, even for the case without outliers ($N_{out} = 0\%$). The ROLAM classifier using the bisquare function achieved the best overall performance. For $N_{out} = 20\%$, the performances of the ROLAM classifier using the bisquare and the fair functions are statistically equivalent. It is worth noting that the standard deviation of the classification rate increases with the increase in N_{out}. Also interesting is the fact that, for $N_{out} > 30\%$, the ROLAM classifier performs as badly as the OLAM classifier (results are not shown here for lack of space). However, when the number of outliers is too high, perhaps they should not be considered as outliers anymore, but as usual data samples of the class. In this situation, we recommend a more powerful classifier (e.g. the ELM) to be used, since it can produce a curved decision surface.

For the VC dataset, we labelled the samples of classes Normal ($N_{nor} = 100$ samples) and Spondylolisthesis[3] ($N_{spl} = 150$ samples) as $+1$ and -1, respectively. Samples from category Spondylolisthesis will be labelled as belonging to

---

[3] Spondylolisthesis is the displacement of a vertebra or the vertebral column in relation to the vertebrae below.

Table 2. Performance comparison of OLAM and ROLAM classifiers (VC dataset)

Classifier	$N_{out} = 0\%$	$N_{out} = 5\%$	$N_{out} = 10\%$	$N_{out} = 20\%$	$N_{out} = 30\%$
OLAM	90.70 ± 4.05	85.34 ± 5.18	79.80 ± 6.12	63.66 ± 7.79	53.66 ± 7.30
ROLAM (Bisquare)	$\mathbf{91.98 \pm 3.54}$	$\mathbf{94.32 \pm 2.95}$	$\mathbf{87.92 \pm 5.21}$	$\mathbf{67.00 \pm 6.11}$	$\mathbf{54.74 \pm 7.79}$
ROLAM (Fair)	91.54 ± 3.58	89.16 ± 4.43	82.36 ± 5.47	63.52 ± 7.16	52.76 ± 7.47
ROLAM (Huber)	91.58 ± 3.61	90.14 ± 4.29	84.26 ± 5.88	64.42 ± 6.77	53.88 ± 8.82
ROLAM (Logistic)	91.76 ± 3.58	90.10 ± 4.25	84.62 ± 5.35	64.14 ± 7.53	51.84 ± 8.83

class $+1$, i.e. will be treated as outliers of class $+1$. For generating the training set, we first compute the centroids of both classes. Then, we select randomly a certain number of samples from the total available (e.g. $0.8 \times (N_{norm} + N_{spl})$). Finally, among the selected samples of class Spondylolisthesis, we select a certain quantity (N_{out}) of the most distant ones to the centroid of the class Normal to have their labels changed to $+1$.

The results are given in Table 2 for different values of N_{out} and for different weighting functions. In this table, we show the values of the classification rates and the corresponding standard deviations averaged over 100 training/testing runs. One can easily note that the ROLAM classifier using the bisquare function performed much better than the standard OLAM classifier, even for the case without outliers. Again, when the percentage of outliers reaches high values (e.g. $N_{out} \geq 20\%$) the performances of both classifiers begin to deteriorate considerably, with the performance of the ROLAM classifier degrading at a lower rate.

In the previous experiments we used the default value of the tuning parameter k. Since this is a free parameter, we show in a final experiment that the performance of the ROLAM classifier can be improved considerably if an optimal tuning parameter is searched during the training phase. In this experiment we used the Wisconsin Breast Cancer (Diagnostic) dataset, which is also publicly available for download from the UCI Machine Learning Repository website [8]. The range of the search for the optimal value of the tuning parameter for each weighting function covered the interval from 0.1 to 10.

The results are shown in Table 3, where k_{def} and k_{opt} denote the default and the optimal values of the tuning parameter, respectively. In this table, we show the values of the classification rates and the corresponding standard deviations averaged over 100 training/testing runs. Below each value of the pair (classification rate, standard deviation) we show the associated value of k_{def} or k_{opt} for a specific percentage of outliers. We labelled the samples of class Malignant as -1 and of class Benign as -1. During training a certain number of randomly selected samples from the category Benign (class -1) will be labelled as belonging to class malignant (class $+1$), i.e. will be treated as outliers of class -1. For generating the outliers we followed the same procedure used in the second and third experiments. The regularization constant required for implementing the standard OLAM classifier was again set to $\lambda = 10^{-2}$.

The results in Table 3 emphasize the power of the M-estimation method in providing a principled approach for robust pattern classification. It is easy to

Table 3. Performance comparison of OLAM and ROLAM classifiers (Breast Cancer dataset)

Classifier	0%	5%	10%	20%	30%
OLAM	95.39±1.87	92.47±2.51	85.76±3.31	74.58±4.45	62.12±4.50
ROLAM (**Andrews**)	94.33±1.96	95.24±1.83	86.50±3.27	75.02±3.86	62.94±4.48
	$(k_{def}=1.339)$	$(k_{def}=1.339)$	$(k_{def}=1.339)$	$(k_{def}=1.339)$	$(k_{def}=1.339)$
	95.89±1.80	95.40±2.00	89.75±5.09	80.46±4.04	66.85±4.60
	$(k_{opt}=3.5)$	$(k_{opt}=1.5)$	$(k_{opt}=1)$	$(k_{opt}=0.5)$	$(k_{opt}=0.5)$
ROLAM (**Bisquare**)					
	94.78±1.96	95.17±1.88	85.69±3.38	75.29±4.15	62.76±4.22
	$(k_{def}=4.685)$	$(k_{def}=4.685)$	$(k_{def}=4.685)$	$(k_{def}=4.685)$	$(k_{def}=4.685)$
	95.57±1.77	95.31±1.85	93.61±2.78	81.44±4.79	70.37±5.30
	$(k_{opt}=9.5)$	$(k_{opt}=4.5)$	$(k_{opt}=3)$	$(k_{opt}=1.5)$	$(k_{opt}=1)$
ROLAM (**Cauchy**)					
	94.89±1.69	93.59±2.66	86.12±3.51	75.47±3.93	62.44±5.00
	$(k_{def}=2.385)$	$(k_{def}=2.385)$	$(k_{def}=2.385)$	$(k_{def}=2.385)$	$(k_{def}=2.385)$
	95.82±1.91	94.09±2.23	90.72±2.72	77.53±4.14	69.82±5.02
	$(k_{opt}=8.5)$	$(k_{opt}=2)$	$(k_{opt}=0.5)$	$(k_{opt}=0.1)$	$(k_{opt}=0.1)$
ROLAM (**Fair**)					
	94.80±1.94	92.55±2.43	85.88±3.19	75.96±3.80	62.60±5.06
	$(k_{def}=1.400)$	$(k_{def}=1.400)$	$(k_{def}=1.400)$	$(k_{def}=1.400)$	$(k_{def}=1.400)$
	95.61±1.69	93.40±2.11	87.17±2.65	76.03±3.89	64.94±4.46
	$(k_{opt}=9)$	$(k_{opt}=0.1)$	$(k_{opt}=0.1)$	$(k_{opt}=4.5)$	$(k_{opt}=0.1)$
ROLAM (**Huber**)					
	94.07±2.00	93.94±2.27	86.21±3.07	74.61±3.57	62.82±4.68
	$(k_{def}=1.345)$	$(k_{def}=1.345)$	$(k_{def}=1.345)$	$(k_{def}=1.345)$	$(k_{def}=1.345)$
	95.87±1.78	93.71±2.33	87.91±3.04	76.61±3.61	65.65±5.07
	$(k_{opt}=7.5)$	$(k_{opt}=1.5)$	$(k_{opt}=0.1)$	$(k_{opt}=0.5)$	$(k_{opt}=0.1)$
ROLAM (**Logistic**)					
	94.68±2.12	92.97±2.43	86.23±3.14	75.79±4.10	64.29±5.16
	$(k_{def}=1.205)$	$(k_{def}=1.205)$	$(k_{def}=1.205)$	$(k_{def}=1.205)$	$(k_{def}=1.205)$
	95.86±1.60	93.29±2.55	86.46±3.39	76.16±3.90	65.20±6.10
	$(k_{opt}=6.5)$	$(k_{opt}=1)$	$(k_{opt}=1.5)$	$(k_{opt}=2)$	$(k_{opt}=0.1)$
ROLAM (**Talwar**)					
	95.54±1.86	95.02±2.15	85.82±3.31	74.76±3.77	62.16±5.33
	$(k_{def}=2.795)$	$(k_{def}=2.795)$	$(k_{def}=2.795)$	$(k_{def}=2.795)$	$(k_{def}=2.795)$
	95.99±1.80	95.44±1.90	92.19±2.79	78.75±3.69	68.01±8.55
	$(k_{opt}=5)$	$(k_{opt}=2.5)$	$(k_{opt}=1.5)$	$(k_{opt}=1)$	$(k_{opt}=0.5)$
ROLAM (**Welsch**)					
	94.61±1.93	94.89±1.98	85.63±3.25	74.97±3.87	62.19±5.46
	$(k_{def}=2.985)$	$(k_{def}=2.985)$	$(k_{def}=2.985)$	$(k_{def}=2.985)$	$(k_{def}=2.985)$
	95.70±1.72	94.94±1.89	91.81±3.05	80.45±4.62	69.73±4.91
	$(k_{opt}=8)$	$(k_{opt}=2.5)$	$(k_{opt}=1.5)$	$(k_{opt}=1)$	$(k_{opt}=0.5)$

note that the performances of the ROLAM classifier are much better than those achieved by the OLAM classifier and by the ROLAM classifier using default values of the tuning parameter, specially in the presence of a high number of outliers (10%, 20% and 30%). The improvement is particularly sharp for the following weighting functions: Andrews, Bisquare, Cauchy, Talwar and Welsch.

5   Conclusion

In this paper we introduced a robust OLAM classifier for supervised pattern classification in the presence of labeling errors (outliers) in the data set. The robust OLAM classifier is designed by means of M-estimation methods which are used to compute the weight matrix operator instead of using the ordinary least squares solution. We have shown that the resulting classifier is robust to label noise not only near the class boundaries, but also far from the class boundaries which can result from mistakes in labelling or gross errors in measuring the input features.

Currently we are evaluating the use of M-estimation techniques in the design of robust ELM-based classifiers. The results we obtained so far suggests that this is a promising approach, since the ELM is in fact using the OLAM classifier in the output layer.

References

1. Baek, D., Oh, S.Y.: Improving optimal linear associative memory using data partitioning. In: Proceedings of the 2006 IEEE International Conference on Systems, Man, and Cybernetics (SMC 2006), vol. 3, pp. 2251–2256 (2006)
2. Barreto, G.A., Frota, R.A.: A unifying methodology for the evaluation of neural network models on novelty detection tasks. Pattern Analysis and Applications 16(1), 83–972 (2013)
3. Cherkassky, V., Fassett, K., Vassilas, N.: Linear algebra approach to neural associative memories and noise performance of neural classifiers. IEEE Transactions on Computers 40(12), 1429–1435 (1991)
4. Duda, R.O., Hart, P.E., Stork, D.G.: Pattern Classification, 2nd edn. John Wiley & Sons (2006)
5. Eichmann, G., Kasparis, T.: Pattern classification using a linear associative memory. Pattern Recognition 22(6), 733–740 (1989)
6. Emmerich, C., Reinhart, R.F., Steil, J.J.: Recurrence enhances the spatial encoding of static inputs in reservoir networks. In: Diamantaras, K., Duch, W., Iliadis, L.S. (eds.) ICANN 2010, Part II. LNCS, vol. 6353, pp. 148–153. Springer, Heidelberg (2010)
7. Fox, J.: Applied Regression Analysis, Linear Models, and Related Methods. Sage Publications (1997)
8. Frank, A., Asuncion, A.: UCI machine learning repository (2010), http://archive.ics.uci.edu/ml
9. Horata, P., Chiewchanwattana, S., Sunat, K.: Robust extreme learning machine. Neurocomputing 102, 31–44 (2012)
10. Huang, G.B., Wang, D.H., Lan, Y.: Extreme learning machines: a survey. International Journal of Machine Learning and Cybernetics 2, 107–122 (2011)
11. Huber, P.J.: Robust estimation of a location parameter. Annals of Mathematical Statistics 35(1), 73–101 (1964)
12. Huber, P.J., Ronchetti, E.M.: Robust Statistics. John Wiley & Sons, LTD. (2009)
13. Hunt, B., Nadar, M., Keller, P., VonColln, E., Goyal, A.: Synthesis of a nonrecurrent associative memory model based on a nonlinear transformation in the spectral domain. IEEE Transactions on Neural Networks 4(5), 873–878 (1993)
14. Kim, H.-C., Ghahramani, Z.: Outlier robust gaussian process classification. In: da Vitoria Lobo, N., Kasparis, T., Roli, F., Kwok, J.T., Georgiopoulos, M., Anagnostopoulos, G.C., Loog, M. (eds.) SSPR&SPR 2008. LNCS, vol. 5342, pp. 896–905. Springer, Heidelberg (2008)
15. Kohonen, T., Oja, E.: Fast adaptive formation of orthogonalizing filters and associative memory in recurrent networks of neuron-like elements. Biological Cybernetics 25, 85–95 (1976)

16. Kohonen, T., Ruohonen, M.: Representation of associated data by matrix operators. IEEE Transactions on Computers 22(7), 701–702 (1973)
17. Li, D., Han, M., Wang, J.: Chaotic time series prediction based on a novel robust echo state network. IEEE Transactions on Neural Networks and Learning Systems 23(5), 787–799 (2012)
18. Poggio, T., Girosi, F.: Networks for approximation and learning. Proceedings of the IEEE 78(9), 1481–1497 (1990)
19. Stiles, G.S., Denq, D.: On the effect of noise on the Moore-Penrose generalized inverse associative memory. IEEE Transactions on Pattern Analysis and Machine Intelligence 7(3), 358–360 (1985)
20. Stiles, G., Denq, D.L.: A quantitative comparison of the performance of three discrete distributed associative memory models. IEEE Transactions on Computers 36(3), 257–263 (1987)
21. Webb, A.: Statistical Pattern Recognition, 2nd edn. John Wiley & Sons, LTD. (2002)

SMBSRP: A Search Mechanism Based on Interest Similarity, Query Relevance and Distance Prediction[*]

Fen Wang, Changsheng Xie, Hong Liang, and Xiaotao Huang

School of Computer Science, HuaZhong University of Science & Technology,
Wuhan, 430074, China

Abstract. In the study of unstructured peer-to-peer networks, due to the lack of network structure, efficient search for resource discovery remains a fundamental challenge. With a combination of the Vector Space Model, this paper presents SMBSRP, *a Search Mechanism Based on interest Similarity, query Relevance and distance Prediction* to improve search performance. SMBSRP groups nodes with similar interests together to build the overlay network. By evaluating the query "condition", SMBSRP decides whether to choose part of neighbors to forward query messages. Besides that, this paper proposes a forwarding factor, which is used to estimate which neighbors to be selected to forward query. The forwarding factor is constructed by neighbors interest similarity, connectivity degree and network distance. The experiment results show that, compared with flooding and random walk, without losing the query hit rate, SMBSRP can reduce the redundant information efficiently.

Keywords: Unstructured Peer-to-Peer, Vector Space Model, Interest Similarity, Query Relevance, Distance Prediction.

1 Introduction

In the 21th Century, peer-to-peer (P2P) network applications are more and more popular. Compared with traditional C/S model, P2P network has better fault tolerance and higher reliability. In unstructured P2P networks, each node does not have global information about the whole topology and the location of queried resources. Due to the lack of network structure, efficient search for resource discovery remains a fundamental challenge in the study of unstructured P2P networks.

Typically, there are two types search technologies for unstructured P2P networks: flooding and random walk. Unstructured P2P networks like Gnutella[1] use flooding. Using flooding, node will forward search messages to its all neighbors [2]. With the expansion of network scale and the increase of network nodes, the redundant search messages will be growth in exponents. Random walk[3], [4] chooses a certain number of neighbors to forward messages. With the expenses of the increase of response time, it overcomes the weakness of flooding.

---

[*] This work was supported by the National Science Foundation of Hubei Province under Granted No. 2011CDB048, and by the Fundamental Research Funds for the Central Universities.

I. Rojas, G. Joya, and J. Cabestany (Eds.): IWANN 2013, Part I, LNCS 7902, pp. 633–646, 2013.

In this paper, we propose a *Search Mechanism Based on interest Similarity, query Relevance and distance Prediction* (SMBSRP), which build neighbor set based on the interest similarity, forward query message based on query relevance, and choose forwarding nodes by a forwarding factor.

1.1   Overview of SMBSRP

To evaluate a search mechanism, usually we should consider three aspects: the successful search ratio, the redundant message number in p2p network, and the result response time. According these three aspects, to design a search mechanism, there are three problems which should be resolved. The first is sending query messages to what peers to get results. The corresponding design scheme will mainly determine the successful search ratio. The second problem is choosing how many peers to forward the query message. The selected peer numbers will have great influence on the number of redundant messages. The third problem is employing what nodes to forward the query message. The message transfer time between the sending query node and the employed node will determine the result response time to some extent. Below, we discuss our solution schemes one by one for these three problems.

Studies show that when the query message been forwarded, there's spatial locality. It means that nodes which have the same or similar shared network resources are more likely to deal with each other queries issued [5],[6],[7],[8]. According to it, SMBSRP evaluates the similarity between nodes, group nodes with similar interest resource together to construct one node's neighbor set, and forward query messages to nodes in neighbor set. Taking into account of unstructured P2P networks highly dynamic, the neighbor node set will be updated regularly. By *the neighbor set building and maintenance algorithm*, we realize this scheme.

To measure nodes interest similarity, we use vector space model (VSM) model. For each node, according to the documents on the node, based on VSM, we can build a node vector named the node interest vector. For two nodes, the interest similarity between them can be computed by vectors relevance computing method. The details of this method and VSM will be introduced in section 3.

Studies show that user's queries has temporal locality. It means that user's queries having great relevance with nodes' historic queries [9],[10]. According to it, SMBSRP use node's search history to measure the query "condition". To measure the query "condition", it also uses VSM and vectors relevance computing method. A *new query* should be forwarded by all nodes in the neighbor set. By contrast, an *older query* can be forwarded by part of neighbor nodes. Compared with flooding, this scheme can reduce the forwarding message's redundancy. It's achieved by *the message query and forwarding algorithm* of SMBSRP.

The response time of the query information is one of the important indicators to measure a search mechanism. A typical method is using the average number of hops as the evaluation of the response time. For the number of hops can only show the length of the search path. It cannot reflect the real spending time of messages physical transmission. As we know that the real spending time can be technically measured by the nodes round trip time (RTT). Usually, when nodes have shorter physical distance, there is smaller RTT. So, in SMBSRP, while selecting partial neighbors to forward message, it chooses nodes with smaller RTT, which means the message transfer time should be short. This network distance prediction method will more conducive to a rapid return of the search results. The specific implementation of this scheme is realized in *the message query and forwarding algorithm* of SMBSRP.

As mentioned above, SMBSRP will group the high interest similarity nodes into a neighbor set, select neighbors with small RTT to forward query messages, and consider query "condition" to decide choosing how many neighbors forwarding messages. For "old" query, it only chooses part of neighbor nodes to forward messages. The core implementation of the design for SMBSRP is *the neighbor set building and maintenance algorithm* and *the message query and forwarding algorithm*.

1.2   Structure of the Rest of the Paper

The remainder of this paper is structured as follows: Section 2 introduces the related work. Section 3 provides basic background on vector space model (VSM) and vectors relevance computing. In Section 4, we describe the system design of SMBSRP. In Section 5, we present our experimental results and performance evaluation. We finally conclude the paper in Section 6.

2   Related Work

A number of searches have been proposed to improve search performance on peer-to-peer systems.

Reference [11] proposed a clustering in demand pattern. It models clustering in file popularity distributions and the consequent non-uniform distribution of file replicas, and achieves optimal search time and optimal search cost. Reference [12] proposes a modified random BFS technique where each peer randomly selects a subset of its peers to propagate a request. It efficiently minimizes the number of messages. In [13], it proposes three technologies. The *iterative deepening* uses a changing TTL to control the iterative search depth; the *Directed BFS* technique only sends query messages to a subset of neighbors; and the *local indices* needs each node n maintains an index over the data of all nodes within a certain hops, then the query message can be searched by the index and only send to nodes within limited nodes in the index table.

Reference [14] describes an Adaptive Probabilistic Search method (APS). The scheme utilizes feedback from previous searches to probabilistically guide future ones. It performs efficient object discovery while inducing zero overhead over dynamic network operations. When the hot of network resources is change, APS cannot work well. In [15], according to the characteristics of the power-law link distribution of the connectivity of the network nodes, Adamic LA etc. propose to choose the highest degree neighbor to forward messages.

Reference [16] proposes GES: *Gnutella with Efficient Search*. GES uses distributed topology adaptation algorithm to restructure and initial P2P overlay. The overlay network is constructed by semantic neighbors and random neighbors. To construct and maintain the overlay network, each node need to periodically send two types query messages, and nodes received it need to response. This paper does not consider the expenses caused by the reconstruction of overlay network.

Reference [17] introduces *Product Search*. For every step the message is forwarded to the neighbor with the minimum product of neighbor's degree and common neighbors' size. In [18], Changze Wu proposes a controllable cost search algorithms, which would ensure stable number of search nodes and achieve different search branches using different search depth. In [19], it presents *Plexus*, a peer-to-peer search protocol that provides an efficient mechanism for advertising a bit sequence (pattern), and discovering it using any subset of its 1-bits in a partial decentralized P2P system. In [20], Tsungnal Lin etc. propose the dynamic search (DS) algorithm, which is a generalization of flooding and random work. It resembles flooding for short-term search and RW for long-term search. In [21], it introduces a search protocol named assisted search with partial indexing, which builds a partial index of shared data to enhance search in unstructured P2P overlay networks. In [22], it introduces a full-text retrieval scheme named BloomCast. It hybridizes a lightweight DHT with an unstructured P2P overlay, and replicates the items uniformly at random across the P2P networks to support random node sampling and network size estimation. Furthermore, it uses Bloom Filter encoding instead of replicating the raw data to reduce the communication and storage costs.

In this paper, we propose SMBSRP, which builds overlay network based on node's interest similarity. When forwarding query messages, it will judge query condition to decide how many neighbors be chosen to forward query. Without losing the query hit rate, SMBSRP effectively reduce the redundant message number.

3   Background: VSM and Vectors Relevance Computing

In VSM[23], documents and queries are represented as vectors. Each dimension in vectors corresponds to a separate term which occurs within the document/query. Each term is assigned a weight by a term weighting scheme such as tf-idf weighting scheme. Usually, the weight can present the frequency of term occurring in the document. To evaluate whether a document is relevant to a query, or whether a document is relevant to another document, VSM uses vectors relevance computing method to measure the relevance between the query vector and the document vector, or the two document-vectors. To describe how to compute the similarity between two vectors, we choose a document vector D and a query vector Q as example.

According to the document similarities theory, the vectors similarity between the query vector Q and the document vector D can be calculated by comparing the deviation of angles between these two vectors. In practice, we calculate the cosine of the angle between the vectors, instead of the angle itself. The similarity value is computed as (1).

$$Sim(D, Q) = \frac{D \bullet Q}{\|D\| * \|Q\|} = \frac{\sum_{i=1}^{n} W_{id} * W_{iq}}{\sqrt{\sum_{i=1}^{n} W_{id}^2} * \sqrt{\sum_{i=1}^{n} W_{iq}^2}} \tag{1}$$

In (1), $\|D\|$ is the norm of vector D, $\|Q\|$ is the norm or vector Q, W_{iq} is the weight of term iq in query Q, W_{id} is the weight of term id in document vector D. The value of Sim(D,Q) reflects the relevance between vectors. If the value of Sim(D,Q) is zero, it means that the query and document vectors are orthogonal and have no match. Along with the increase of the value, the degree of the vectors relevance is rising. When the value exceeds a threshold, we can say that the document matches the query. This value can also be used in document relevance ranking.

4   System Design

In this section, we detail the design of SMBSRP. We in turn describe node profile, discuss neighbor set building and maintenance algorithm, present message query and forwarding algorithm.

4.1   Node Profile

Node profile is represented as node's data structure. To support *the neighbor set building and maintenance algorithm* and *the message query and forwarding algorithm*, the data structure of a node contains five areas: *node's IP address, a document vector list Doc_Vec_List, an interest vector I, a historical query List Que_Vec_List and a neighbor list Nei_List.*

The document vector list Doc_Vec_List consists of all document vectors Di in the node. Each document vectors Di is represented by VSM. Each dimension in document vectors Di corresponds to a separate term j which occurs within the document. Each term j is assigned a weight D_{ji} to present the frequency of term j occurring in the document i.

The *interest vector I* will summarize all documents information in the node. For interest vector I, each *term j* component has a weight $W_j = \sum_{j=1}^{m} D_{ji}$, where *m* is the number of documents on the node. For each *term j*, if using the "dampened" tf scheme, we replace its weight W_j as $1 + \log W_j$. Then, we construct the *interest vector I.*

The *historical query List Que_Vec_List* is constituted of the historical queries which have been searched in this node. Each query entry consists of two parts. One is a *query vector Q_j* that using VSM to present the historical query j. The other is a flag to mark whether the query is *successful*. The *successful query* means node has documents to match this query. The *neighbor list Nei_List* is comprised of each neighbor node's IP address, messages round-trip time RTT, and the neighbor node's connectivity. The node structure is shown in Figure 1.

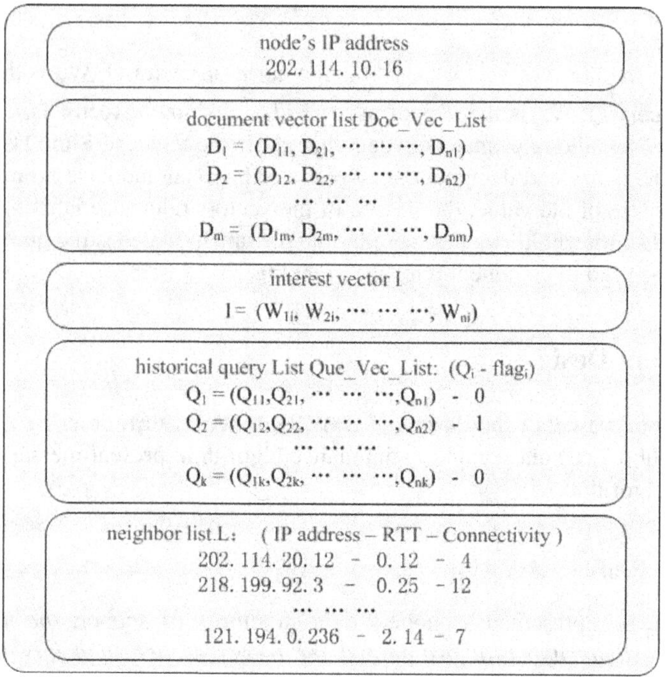

Fig. 1. Node's data structure

We use the vector space model VSM to construct *document vector D_i, interest vector I* and *query vector Q_j*.

4.2   Neighbor Set Building and Maintenance

The neighbor set building and maintenance is an important component in SMBSRP. The task of this scheme is to construct and periodically refresh the logical neighbor set for each node in P2P network. When one node joins into network, SMBSRP will select nodes in P2P network as its logical neighbors, and thereby establish the logical neighbor set for it. We call this node as *source node* in this paper. Considering the dynamics of unstructured P2P network, we need to periodically update the logical neighbor set. To implement the neighbor set building and maintenance scheme, we should follow four steps as describes below:

Step1. Source node broadcasts messages. When source node joins in P2P network, it will broadcast a Ping message. The message contains source node's *IP address, interest vector I_s*, and package *TTL* (time-to-live).

Step2. Target node replies a response message. Node in P2P network who receive the Ping message will reply a Pong message to the source node. This node called as *target node*. The *target node* will compute vectors similarity $Sim(I_s,I_t)$ between itself interest vector I_t and source node interest vector I_s. The response message consists of *target node's IP address, similarity value, connectivity,* and *RTT* (Round-trip Time).

Step3. Building logical neighbor list *Nei_List*. While receiving the response message, source node will store the target node's information in local neighbor candidate cache. The candidate cache is maintained throughout the lifetime of the node. By using *building neighbor set algorithm*, which is outlined by *build_neigh_set()*, *source node* constructs its neighbor list *Nei_List*. This algorithm is the core of neighbor set building and maintenance scheme.

The *building neighbor set algorithm* consists of two steps. The first step is to construct a response queue *res_que* and a slow response queue *slow_res_que*. While RTT exceeds a value, the response should be seemed as slow response. We predefine a RTT threshold constant *RTT_THRESHOLD*. While target node's RTT exceeds *RTT_THRESHOLD*, it will be put into *slow_res_que*; else it will be put into *res_que*. To sort nodes in these two queues, we follow two principles: 1) nodes with smaller vectors similarity value have higher priority and 2) while nodes are in the same similarity, nodes with less RTT have higher priority. According to these two rules, target nodes are sorted in the two queues.

The second step is choosing neighbor candidates into neighbor list *Nei_List*. Source node will get target nodes from *res_que* one by one into *Nei_List*, until the total target nodes' number reaches the maximum number *MAX_NEI_NUM* of neighbor list. If the number of nodes in the *res_que* is less than *MAX_NEI_NUM*, after getting empty the *res_que*, it will get nodes from *slow_res_que* one by one into *Nei_List*, until the total nodes number reaches the *MAX_NEI_NUM*, or the *slow_res_que* is empty. Target node in neighbor list is called neighbor node or neighbor in this paper.

Step4. Update and maintain the logical neighbor list *Nei_List*. After building logical neighbor set, source node will send renew message to neighbors periodically (such as 10s). If there's no renew in the neighbor node, it will send a specific response message to source node, else it will recompute $Sim(I_s,I_t)$, and send a response message to source node. The response message consists of its IP address similarity value, nodes connectivity and RTT. While receiving response message, source node will update its neighbor candidate cache. If a neighbor leaves P2P networks, the source node will not get any response message from this neighbor, and then this neighbor's information will be deleted from source node neighbor candidate cache.

Due to the dynamic of unstructured P2P network, after building the logical neighbor set, source node will periodically random select a certain number nodes to send Ping message. After receiving these nodes' response messages, source node will insert their information into its neighbor candidate cache.

Then, according to the *building neighbor set algorithm*, source node renew response queue *res_que* and slow response queue *slow_res_que*, restructure the logical neighbor list *Nei_List*.

Building neighbor set algorithm:
//   The targ_node[] is an array constituted by nodes in
//source node X's neighbor candidate cache.

```
X.build_neigh_set(targ_node[]){
```

//   Classify target nodes into two queues

```
[1]  i <- 1
[2]  while targ_node[i]!= NULL {
[3]    if targ_node[i].rtt > RTT_THRESHOLD
         then
[4]        slow_res_que <- targ_node[i]
[5]      else
[6]        res_que  <- targ_node[i]
[7]    end if
[8]    i<- i+1
[9]  }
```

//   Using sort method to rank target nodes in two queues

```
[10] call  Sort(res_que)
[11] call  Sort(slow_res_que)
```

//   Building neighbor list nei_list from two queues

```
[12] i <- j <- 1
[13] while i<=MAX_NEI_NUM  and
res_que.node[j]!= NULL  {
[14] nei_list.node[i] <- res_que.node[j]
[15]   i <- i+1
[16]   j <- j+1
[17] }
[18] j <- 1
[19] while i<=MAX_NEI_NUM  and
slow_res_que.node[j]!= NULL {
[20]   nei_list.node[i] <-
slow_res_que.node[j]
[21]   i <- i+1
[22]   j <- j+1
[23]   }
[24] }
```

4.3   Message Query and Forwarding

The message query and forwarding is an important component in SMBSRP. The task of it is to find the matched documents in P2P network for a query, and to choose neighbors forwarding query message.

In SMBSRP, a query message consists of *source query node's IP address, query vector Q, timestamp,* and *TTL of query message.* The node who creates and initializes a query message called as *source query node. Timestamp* records the occurrence time of the query. When one node receives a query, it will compare the *query vector Q* to each *document vector* D_i in its *document vector list Doc_Vec_List.* As discussed in section 3, if the vectors similarity *Sim(Q, D_i)* exceeds a threshold value, it means that document i matches query. Then document i will be returned as a result. All return documents should be sorted by the similarity value.

While the TTL is greater than zero, to get more matched documents, the query message should be forwarded. Here, we call node that sends query as *query node.* The *query node* will choose its neighbor nodes to forward query message. The *query node* will firstly compute the vectors similarity Sim(Q,Q_j) between the *query vector Q* and each *successful query*'s vector Q_j in *the historical query List Que_Vec_List.* It will record the largest similarity value as Max_Sim_His_Query. The system should predefine a similarity threshold SIM_THRESHOLD. If the Max_Sim_His_Query is smaller than SIM_THRESHOLD, it means that the query is a *new query.* The query message should forwarded by all neighbors. Else means that the query is an *older query,* then the *query node* needs to choose parts of neighbors to forward query message. While the query message is send by the *query node* to its neighbors, the TTL will be decreased by 1. When the TTL is zero, the query message should be stopped to forwarding.

To choose part of neighbors to forward query message, here we comprehensive consider neighbor node's interest vector, connectivity, and its physical distance to determine which neighbors should be chosen. We use a *forwarding factor E* to evaluate neighbors. Below we discuss the construction of *forwarding factor E* in detail.

1) The similarity between neighbor interest vector $I_{neighbor}$ and the query vector Q, presented as SQI.

$$SQI = Sim(Q, I_{neighbor}) \qquad (2)$$

The compute of SQI is described in section 3. According to (2), neighbors with higher similarity will have larger SQI. Obviously, the value of SQI is between [0,1].

2) The relative value of node's connectivity degree, which is presented as RCD.

The *query node* has a neighbor list *Nei_List.* Each neighbor has a connectivity value $N_{neighbor}$ in *query node's* neighbor list. RCD presents one neighbor's relative value of its connectivity in all neighbors of the neighbor list.

$$RCD = \frac{N_{neighbor} - N_{min}}{N_{max} - N_{min}} \qquad (3)$$

Here, N_{max} and N_{min} separately present the maximum connectivity degree and the minimum connectivity degree of neighbors in the neighbor list Nei_List. So,

$$0 \leq N_{neighbor} - N_{min} \leq N_{max} - N_{min}$$

Through (3), we can know that the value of RCD is between [0,1]. Obviously neighbors with larger connectivity will have larger RCD.

3) The relative value of the neighbor node's round trip time RTT, presented as RVR.

In *query node's* neighbor list, each neighbor has a round trip time $RTT_{neighbor}$. Here using RVR to show the relative RTT value of one neighbor node. We need to define RVR as that neighbors that have larger RVR will have shorter physical distance. For nodes with shorter physical distance have smaller RTT, so the calculation method of RVR is shown as (4):

$$RVR = \frac{RTT_{max} - RTT_{neighbor}}{RTT_{max} - RTT_{min}} \tag{4}$$

In (4), RTT_{max} and RTT_{min} mean the maximum round trip time and the minimum round trip time of neighbors in one *query node's* neighbor list. So,

$$0 \le RTT_{max} - RTT_{neighbor} \le RTT_{max} - RTT_{min}$$

Calculated by (4), we can draw a conclusion that the value of RVR is between [0,1].

As we know that, if *query node* chooses neighbor nodes with higher interest similarity, larger connectivity degree, and shorter physical distance to forward query message, the P2P system should have higher query hit rate, smaller redundant message number, and shorter result response time. Combination of these three conditions, we define a *forwarding factor E*, to sort neighbors in the neighbor list. The *forwarding factor E* is defined as follows:

$$E = W_\alpha \times SQI + W_\beta \times RCD + W_\gamma \times RVR \tag{5}$$

The $W_\alpha, W_\beta, W_\gamma$ separately denote the weight coefficient of one neighbor node's interest similarity SQI, relative connectivity RCD, and relative round trip time RVR. The value of them must meet the following (6):

$$W_\alpha + W_\beta + W_\gamma = 1 \tag{6}$$

and,

$$W_\alpha \in [0,1], \quad W_\beta \in [0,1], \quad W_\gamma \in [0,1]$$

Since the value of SQI, RCD and RVR are all in [0, 1], according to (5), we know that $0 \le E \le 1$.

While selecting neighbors to forward query message, query node will prior chooses nodes with high *forwarding factor E*. We can use genetic algorithm to determine the weight coefficient W_α, W_β, and W_γ. To simplify calculation, the value of W_α, W_β, and W_γ are all defined as 0.3333 by default.

5   Experimental Results and Performance Evaluation

In this section, we introduce our simulate system and evaluate the performance of SMBSRP. We simulate the proposed mechanism and algorithms. Programming by java language, we use Gnutella Graph class to generate an undirected graph, which is used to simulate a fully distributed unstructured P2P network topology. In simulation system, there are 300 nodes, 2000 files. We assume nodes join and leave the P2P system at a constant rate. Requests arrival follows Zipf distribution of θ=0.25. The experiments evaluate system performance while the TTL of query message changes from 0 to 12. Here, we focus on three main parameters value: the query message number, average number of hops, and the success rate of search. In order to evaluate the performance of SMBSRP, we compare it with flooding and random walk. Here, we choose walker equal to 4 in random walk.

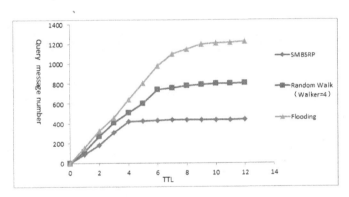

Fig. 2. The query message number (300 nodes in P2P system)

Fig.2 shows under the three search mechanisms, the query message number changes with the increase of the TTL. It shows that, when TTL increases to 4, the message number of SMBSRP mechanism tends to be a fixed value, and there is no longer a substantial increase. While TTL is from 4 to 12, compared to flooding, the query message number of SMBSRP reduces nearly 67%, and compared to random walk, it reduces almost one half. When the TTL value is the same, using SMBSRP, the query message number is always the minimum number.

Fig.3 shows the average number of hops changes with the increase of TTL. We can see that, with the TTL value increasing from the initial zero, average number of hops increases rapidly. When TTL increases to a certain value, the average number of hops tends to converge. In SMBSRP, it convergence first, followed by flooding mechanism, the random walk convergence slowest. After TTL greater than 5, the average number of hops for SMBSRP is the minimum.

Fig.4 shows the success rate of search for three mechanisms. As can be seen, when the TTL is less than 5, the search success rates of all three search mechanism are all not high. In this condition, SMBSRP has significant advantage. When the TTL value is increased, the number of query message arriving at nodes is more and more, and the success rate of search is also rising. Theoretically, when the TTL value reaches a certain level (in most condition, it should be greater than 7), query messages of the flooding mechanism can be forwarded to all nodes in the network. Fig.4 shows that when TTL greater than 7, the success rate of search for flooding mechanism is the maximum, SMBSRP follows it, and the rate for random walk is the minimum.

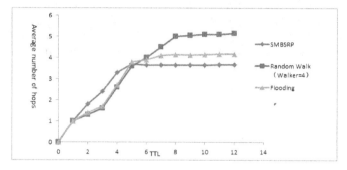

Fig. 3. The average number of hops (300 nodes in P2P system)

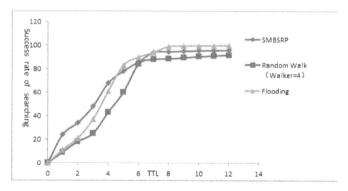

Fig. 4. The success rate of search (300 nodes in P2P system)

6   Conclusions and Future Work

In this paper, we proposed an unstructured P2P network search mechanism named SMBSRP, which is based on interest similarity, query relevance and network distance prediction. Experiments show that it can efficiently reduce the redundant messages, along with a short average search length. Future work involves implementing the genetic algorithm in message forwarding algorithm on unstructured P2P network that contains super nodes.

References

[1] RFC-Gnutella 0.6 (2008),
http://rfc-gnutella.sourceforge.net/
developer/testing/index.html

[2] Clip2.com, The gnutella protocol speci_cation v0.4. (2000),
http://www9.limewire.com/developer/gnutellaprotocol0.4.pdf

[3] Gkantsidis, C., Mihail, M., Saberi, A.: Random Walks in Peer-to-Peer Networks. In: INFOCOM 2004, vol. 63(3), pp. 241–263 (2004)

[4] Lv, Q., Cao, P., Cohen, E., Li, K., Shenker, S.: Search and replication in unstructured Peer-to-Peer networks. In: Proceedings of the 16th Annual International Conference on Super Computing, pp. 84–95 (2002)

[5] Yang, Y., Li, Y., Sun, H.: A new keyword-based similarity measuring model on peer groups in peer-to-peer networks. In: CECNet 2012, pp. 50–52 (2012)

[6] Le Fessant, F., Handurukande, S.B., Kermarrec, A.-M., Massoulié, L.: Clustering in Peer-to-Peer File Sharing Workloads. In: Voelker, G.M., Shenker, S. (eds.) IPTPS 2004. LNCS, vol. 3279, pp. 217–226. Springer, Heidelberg (2005)

[7] Iamnitchi, A., Ripeanu, M., Santos-Neto, E., Foster, I.: The Small World of File Sharing. IEEE Trans. Parallel and Distributed Systems 22(7), 1120–1134 (2011)

[8] Sripanidkulchai, K., Maggs, B., Zhang, H.: Efficient Content Location Using Interest-Based Locality in Peer-to-Peer Systems. In: Proc. IEEE INFOCOM 2003, pp. 2166–2176 (2003)

[9] Yeferny, T., Arour, K.: LearningPeerSelection: A Query Routing Approach for Information Retrieval in P2P Systems. In: ICIW 2010, pp. 235–241 (2010)

[10] Ishak, I., Salim, N.: A Similarity and Feedback Based Query Routing Across Unstructured Peer-to-Peer Networks. In: International Conference on Advanced Computer Theory and Engineering (ICACTE 2008), pp. 600–604 (2008)

[11] Tewari, S., Kleinrock, L.: Optimal Search Performance in Unstructured Peer-to-Peer Networks With Clustered Demands. IEEE Journal on Selected Areas in Communications 25(1), 84–95 (2007)

[12] Kalogeraki, V., Gunopulos, D., Zeinalipour-Yazti, D.: A Local Search Mechanism for Peer-to-Peer Networks. In: Proceedings of the Eleventh International Conference on Information and Knowledge Management (CIKM 2002), pp. 300–307 (2002)

[13] Yang, B., Garcia-Molina, H.: Improving Search in Peer-to-Peer Networks. In: Proc. 22th IEEE Int'l Conf. Distributed Computing Systems (ICDCS 2002), pp. 12–19 (2002)

[14] Tsoumakos, D., Roussopoulos, N.: Adaptive Probabilistic Search for Peer-to-Peer Networks. In: IEEE International Conference on P2P Computing, vol. 53(3), pp. 46–51 (2003)

[15] Adamic, L.A., Lukose, R.M., Puniyani, A.R., Huberman, B.A.: Search in Power-law Networks. Physical Review E 64, 046135, 1–8 (2001)

[16] Zhu, Y., Hu, Y.: Enhancing search performance on Gnutella-like P2P Systems. IEEE Trans. on Parallel and Distributed System 17(12), 1482–1485 (2006)

[17] Ke, B.W.Q., Dong, Y.: Degree and similarity based search in networks. In: FSKD, pp. 1267–1270 (2011)

[18] Wu, C., Wu, K., Zeng, J., Chen, X.: Controllable Cost Search Strategy in Unstructured P2P. In: Sixth Annual ChinaGrid Conference, pp. 180–187 (2011)

[19] Ahmed, R., Boutaba, R.: Plexus: a scalable peer-to-peer protocol enabling efficient subset search. IEEE/ACM Transactions on Networking 17(1), 130–143 (2009)

[20] Lin, T., Lin, P., Wang, H., Chen, C.: Dynamic Search Algorithm in Unstructured Peer-to-Peer Networks. IEEE Transactions on Parallel and Distributed Systems 20(5), 654–666 (2009)

[21] Zhang, R., Charlie Hu, Y.: Assisted Peer-to-Peer Search with Partial Indexing. IEEE Transactions on Parallel and Distributed Systems 18(8), 1146–1158 (2007)

[22] Chen, H., Jin, H., Luo, X., Liu, Y., Gu, T., Chen, K., Ni, L.M.: BloomCast: Efficient and Effective Full-Text Retrieval in Unstructured P2P Networks. IEEE Transactions on Parallel and Distributed Systems 23(2), 232–241 (2012)

[23] Berry, M.W., Drmac, Z., Jessup, E.R.: Matrices, Vector Spaces, and Information Retrieval. SIAM Rev. 41(2), 335–362 (1999)

An Unfolding-Based Preprocess
for Reinforcing Thresholds in Fuzzy Tabulation*

Pascual Julián-Iranzo[1], Jesús Medina-Moreno[2,**], Pedro J. Morcillo[3],
Ginés Moreno[3], and Manuel Ojeda-Aciego[4]

[1] Dept. of Information Technologies and Systems, University of Castilla-La Mancha
pascual.julian@uclm.es
[2] Department of Mathematics, University of Cadiz
jesus.medina@uca.es
[3] Department of Computing Systems, University of Castilla-La Mancha
gines.moreno@uclm.es, pmorcillo@dsi.uclm.es
[4] Department of Applied Mathematics, University of Málaga
aciego@uma.es

Abstract. We have recently proposed a technique for generating thresholds (filters) useful for avoiding useless computations when executing fuzzy logic programs in a tabulated way. The method was conceived as a static preprocess practicable on program rules before being executed with our *fuzzy thresholded tabulation* principle, thus increasing the opportunities of prematurely disregarding those computation steps which are redundant (tabulation) or directly lead to non-significant solutions (thresholding). In this paper we reinforce the power of such static preprocess—which does not require the consumption of extra computational resources at execution time—by re-formulating it in terms of the fuzzy unfolding technique initially designed in our group for transforming and optimizing fuzzy logic programs.

Keywords: Fuzzy Logic Programming, Tabulation, Thresholding, Unfolding.

1 Introduction

The fields of logic programming and fuzzy logic have shown its complementarity for more than two decades [2, 4, 8, 10]. In this work we continue our efforts to provide a refined and improved fuzzy query answering procedure for multi-adjoint logic programming MALP [9, 10]. Roughly speaking, the general idea is that, when trying to perform a computation step by using a given program rule $\mathcal{R}$, we firstly analyze if such step might contribute to reach further significant solutions (not yet tabulated, saved or stored). After recalling the static approach

---

* Work supported by the Spanish MICINN projects TIN2009-14562-C05-01, TIN2009-14562-C05-03, TIN2012-39353-C04-01, TIN2012-39353-C04-04, and TIN2011-25846, and by Junta de Andalucía project P09-FQM-5233.
** Corresponding author.

I. Rojas, G. Joya, and J. Cabestany (Eds.): IWANN 2013, Part I, LNCS 7902, pp. 647–655, 2013.
© Springer-Verlag Berlin Heidelberg 2013

introduced in [5], the main contribution starts in Section 3, where we focus on a new refinement in order to improve the existing static preprocessing step.

The MALP approach[1] considers a language, $\mathcal{L}$, containing propositional variables, constants, and a set of logical connectives. In our fuzzy setting, we use implication connectives $(\leftarrow_1, \leftarrow_2, \ldots, \leftarrow_m)$ together with a number of aggregators. They will be used to combine/propagate truth values through the rules. The general definition of aggregation operators subsumes conjunctive operators (denoted by $\&_1, \&_2, \ldots, \&_k$), disjunctive operators $(\vee_1, \vee_2, \ldots, \vee_l)$, and average and hybrid operators (usually denoted by $@_1, @_2, \ldots, @_n$). Aggregators are useful to describe/specify user preferences. The language $\mathcal{L}$ will be interpreted on a *multi-adjoint lattice*, $\langle L, \preceq, \leftarrow_1, \&_1, \ldots, \leftarrow_n, \&_n \rangle$, which is a complete lattice equipped with a collection of adjoint pairs $\langle \leftarrow_i, \&_i \rangle$, where each $\&_i$ is a conjunctor[2] intended to provide a *modus ponens*-rule w.r.t. $\leftarrow_i$. In general, the set of truth values L may be the carrier of any complete bounded lattice but, for simplicity, in the examples of this work we shall select L as the set of real numbers in the interval $[0,1]$.

A *rule* is a formula $A \leftarrow_i \mathcal{B}$, where the *head* A is a propositional symbol and the *body* $\mathcal{B}$ is a formula built from propositional symbols $B_1, \ldots, B_n$ $(n \geq 0)$, truth values of L and conjunctions, disjunctions and aggregations. Rules with an empty body are called *facts*. A *goal* is a body submitted as a query to the system. Roughly speaking, a MALP program is a set of pairs $\langle R; \alpha \rangle$, where R is a rule and α is a value of L, which might express the confidence which the user of the system has in the truth of the rule R (note that the truth degrees in a given program are expected to be assigned by an expert).

The standard procedural semantics of the multi–adjoint logic language $\mathcal{L}$ is based on the notion of *admissible step* (which can be seen as a fuzzy extension of classical SLD-resolution [3]) whose definition is also crucial for describing the unfolding transformation [7]. Hereafter, $\mathcal{C}[A]$ denotes a formula where A is a subexpression (usually a propositional symbol) which occurs in the (possibly empty) context $\mathcal{C}[]$, whereas $\mathcal{C}[A/A']$ means the replacement of A by A' in context $\mathcal{C}[]$. In the following definition, we always consider that A is the selected propositional symbol in goal $\mathcal{Q}$.

Definition 1 (Admissible Steps). *Let $\mathcal{Q}$ be a goal, which is considered as a state, and let $\mathcal{G}$ be the set of goals. Given a program $\mathbb{P}$, an admissible computation is formalized as a state transition system, whose transition relation $\to_{AS} \subseteq (\mathcal{G} \times \mathcal{G})$ is the smallest relation satisfying the following admissible rules:*

1. *$\mathcal{Q}[A] \to_{AS} \mathcal{Q}[A/v\&_i\mathcal{B}]$ if there is a rule $\langle A \leftarrow_i \mathcal{B}; v \rangle$ in $\mathbb{P}$ and $\mathcal{B}$ is not empty.*
2. *$\mathcal{Q}[A] \to_{AS} \mathcal{Q}[A/v]$ if there is a fact $\langle A \leftarrow_i; v \rangle$ in $\mathbb{P}$.*

---

[1] Visit `http://dectau.uclm.es/floper/` and `http://dectau.uclm.es/fuzzyXPath/` for downloading related tools [1, 11].

[2] An increasing operator satisfying boundary conditions with the top element.

[3] The acronym **SLD** stands for "**S**election-function driven **L**inear resolution for **D**efinite clauses".

It is obvious that if we exploit all propositional symbols of a goal, by applying admissible steps as much as needed, then the goal becomes a formula (with no propositional symbols) which can then be directly interpreted in the multi–adjoint lattice L and thus obtaining the desired *fuzzy computed answer* (or f.c.a., in brief) for that goal.

Although the procedural principle we have just seen suffices for executing MALP programs, there exists a much more efficient mechanism for solving queries as occurs with the thresholded tabulation procedure proposed in [5, 6] that we are going to summarize in the following section.

2   Fuzzy Thresholded Tabulation with Static Preprocess

Tabulation arises as a technique to solve two important problems in deductive databases and logic programming: termination and efficiency. The datatype we will use for the description of the proposed method is that of a *forest*, that is, a finite set of trees. Each one of these trees has a root labeled with a propositional symbol together with a truth-value from the underlying lattice (called the *current value* for the *tabulated* symbol); the rest of the nodes of each of these trees are labeled with an "extended" formula in which some of the propositional symbols have been substituted by its corresponding value.

Thresholded Tabulation uses threshold filters in order to prune some useless branches or, more exactly, for avoiding the use (at execution time) of those program rules whose weights do not surpass a given "threshold" value. The following notations are considered in the description of filters.

- Let $\mathcal{R} = \langle A \leftarrow_i \mathcal{B}; \vartheta \rangle$ be a program rule.
- Let $\mathcal{B}'$ be an expression with no atoms, obtained from the body $\mathcal{B}$ by replacing each occurrence of a propositional symbol by $\top$.
- Let $v \in L$ be the result of interpreting $\mathcal{B}'$ under a given lattice.
- Then, $Up\_body(\mathcal{R}) = v$.

Apart from the truth degree ϑ of a program rule $\mathcal{R} = \langle A \leftarrow_i \mathcal{B}; \vartheta \rangle$ and the maximum truth degree of its body $Up\_body(\mathcal{R})$, in the multi-adjoint logic setting, we can consider a third kind of filter for reinforcing thresholding. The idea is to combine the two previous measures by means of the adjoint conjunction $\&_i$ of the implication $\leftarrow_i$ in rule $\mathcal{R}$. Now, we define the *maximum truth degree of a program rule*, symbolized by function Up_rule, as:

$$Up\_rule(\mathcal{R}) = \vartheta \&_i (Up\_body(\mathcal{R}))$$

As shown in [5], such filters can be safely compiled on program rules after applying an easy static preprocess whose benefits will be largely redeemed on further executions of the program. So, for any MALP program $\mathbb{P}$, we can obtain its extended version $\mathbb{P}+$ (for being used during the "query answering" process) by adding to its program rules their *proper threshold* $Up\_rule(\mathcal{R})$ as follows:

$$\mathbb{P}+ = \{\langle A \leftarrow_i \mathcal{B}; \vartheta; Up\_rule(\mathcal{R}) \rangle \mid \mathcal{R} = \langle A \leftarrow_i \mathcal{B}; \vartheta \rangle \in \mathbb{P}\}.$$

Example 1. In the following extended program $\mathbb{P}+$ (with mutual recursion) all rules include the appropriate threshold $Up\_rule(\mathcal{R})$ in their third component. For instance, this value for Rule $\mathcal{R}_4$ is lower than the weight of the rule, which is on account of the presence of a truth degree in its body.

$$
\begin{array}{ll}
\mathcal{R}_1 : \langle\, p \;\leftarrow_P\; q \,\&_G\, s \,;\, 0.6 \,;\, 0.6 \,\rangle & \mathcal{R}_4 : \langle\, r \;\leftarrow_L\; 0.9 \,\&_P\, s \,;\, 0.8 \,;\, 0.7 \,\rangle \\
\mathcal{R}_2 : \langle\, p \;\leftarrow_P\; r \quad\quad\;\; ;\, 0.7 \,;\, 0.7 \,\rangle & \mathcal{R}_5 : \langle\, q \;\leftarrow \quad\quad\quad\quad\;\; ;\, 0.8 \,;\, 0.8 \,\rangle \\
\mathcal{R}_3 : \langle\, q \;\leftarrow_L\; p \,\&_G\, s \,;\, 0.9 \,;\, 0.9 \,\rangle & \mathcal{R}_6 : \langle\, r \;\leftarrow \quad\quad\quad\quad\;\; ;\, 0.8 \,;\, 0.8 \,\rangle
\end{array}
$$

Operations for Tabulation with Thresholding Using Extended Programs

The tabulation procedure requires four basic operations: Root Expansion, New Subgoal/Tree, Value Update, and Answer Return. In the first case we take profit of the *filters* for thresholding compiled on extended programs, whose further use will drastically diminish the number of nodes in trees (note that by avoiding the generation of a single node, the method implicitly avoids the generation of all its possible descendants as well). New Subgoal is applied whenever a propositional variable is found without a corresponding tree in the forest. Value update is used to propagate the truth-values of answers to the root of the corresponding tree. Finally, answer return substitutes a propositional variable by the current truth-value in the corresponding tree. Let us formally describe such operations:

Rule 1: Root Expansion. Given a tree with root $A\colon r$ in the forest, if there is a program rule $\mathcal{R} = \langle A \leftarrow_i \mathcal{B}; \vartheta; Up\_rule(\mathcal{R})\rangle \in \mathbb{P}+$ not consumed before and verifying $Up\_rule(\mathcal{R}) \not\leq r$, append the new child $\vartheta \&_i \mathcal{B}$ to the root of the tree.

Rule 2: New Subgoal/Tree. Select a non-tabulated propositional symbol C occurring in a leaf of some tree (this means that there is no tree in the forest with the root node labeled with C), then create a new tree with a single node, the root $C\colon \bot$, and append it to the forest.

Rule 3: Value Update. If a tree, rooted at $C\colon r$, has a leaf $\mathcal{B}$ with no propositional symbols, and $\mathcal{B} \rightarrow_{IS}^* s$, where $s \in L$, then update the current value of the propositional symbol C by the value of $\sup_L\{r, s\}$.

Furthermore, once the tabulated truth-value of the tree rooted by C has been modified, for all the occurrences of C in a non-leaf node $\mathcal{B}[C]$, such as the one in the left of the following figure, then update the whole branch substituting the constant u by $\sup_L\{u, t\}$ (where t is the last tabulated truth-value for C, i.e. $\sup_L\{r, s\}$) as in the right of the figure.

$$
\begin{array}{ccc}
\vdots & \qquad\qquad & \vdots \\
\mid & & \mid \\
\mathcal{B}[C] & & \mathcal{B}[C] \\
\mid & & \mid \\
\mathcal{B}[C/u] & & \mathcal{B}[C/\sup_L\{u, t\}] \\
\vdots & & \vdots
\end{array}
$$

Rule 4: Answer Return. Select in any leaf a propositional symbol C which is tabulated, and assume that its current value is r; then add a new successor node as the figure shows:

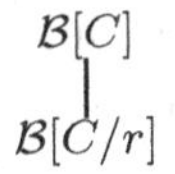

$$\mathcal{B}[C]$$
$$|$$
$$\mathcal{B}[C/r]$$

Example 2. Let us see now how our method proceeds when solving query $?p$ w.r.t. the extended program $\mathbb{P}+$ of Example 1.

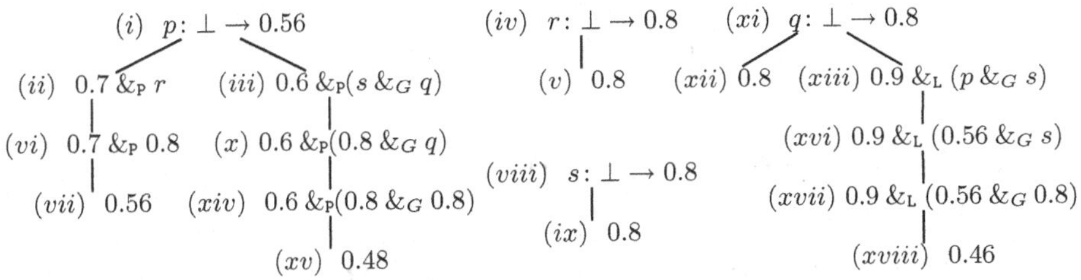

$$(i)\ \ p: \bot \rightarrow 0.56$$

$(ii)\ \ 0.7\ \&_{\text{P}}\ r \qquad (iii)\ 0.6\ \&_{\text{P}}(s\ \&_G\ q)$

$(vi)\ \ 0.7\ \&_{\text{P}}\ 0.8 \qquad (x)\ 0.6\ \&_{\text{P}}(0.8\ \&_G\ q)$

$(vii)\ \ 0.56 \qquad (xiv)\ 0.6\ \&_{\text{P}}(0.8\ \&_G\ 0.8)$

$$(xv)\ \ 0.48$$

$$(iv)\ \ r: \bot \rightarrow 0.8 \qquad (xi)\ \ q: \bot \rightarrow 0.8$$

$(v)\ \ 0.8 \qquad (xii)\ 0.8 \quad (xiii)\ 0.9\ \&_{\text{L}}\ (p\ \&_G\ s)$

$(viii)\ \ s: \bot \rightarrow 0.8 \qquad (xvi)\ 0.9\ \&_{\text{L}}\ (0.56\ \&_G\ s)$

$(ix)\ \ 0.8 \qquad (xvii)\ 0.9\ \&_{\text{L}}\ (0.56\ \&_G\ 0.8)$

$$(xviii)\ \ 0.46$$

Fig. 1. Example forest for query $?p$ w.r.t. the extended program $\mathbb{P}+$

First of all, node (i) is introduced. Now, *Root Expansion* is applied and nodes $(ii), (iii)$ are generated, see Figure 1. Then *New Subgoal* is applied on r and a new tree is generated with node (iv). Node (v) is obtained on r, from a new application of *Root Expansion* and Rule $\mathcal{R}_6$; and, thru *Value Update*, node (iv) current value is directly updated to 0.8. However, Rule $\mathcal{R}_4$ does not generate another node since the inequality $Up\_rule(\mathcal{R}_4) \leq 0.8$ holds and so this rule does not verify the filter in *Root Expansion*. By using the value of r, *Answer Return* extends the initial tree with node (vi). Now *Value Update* generates node (vii) and updates the current value of p to 0.56. Then, we continue in the other branch of the first tree and *New Subgoal* and *Root Expansion* are applied on s, and a new tree is generated with nodes $(viii), (ix)$. *Value Update* increases the current value to 0.8. By using this value, *Answer Return* extends the initial tree with node (x) and q open a new tree in node (xi). After applying *Root Expansion*, nodes (xii) and $(xiii)$ are introduced from Rules $\mathcal{R}_3$, $\mathcal{R}_5$ and the value of q is updated. *Answer Return* extends the initial tree with node (xiv). A further application of *Value Update* generates node (xv), however, the value of p is not updated, because $0.48 \leq 0.56$. The forest is not terminated since one branch in the forth tree is open.

This tree is continued from node $(xiii)$ and, using the values of p and s, nodes (xvi) and $(xvii)$. *Value Update* generates node $(xvii)$, however this value does not increase the value of p and the forest is terminated.

The advantage of the improved method has been shown when $\mathcal{R}_4$ was not used on node (iv). One more interesting application is given, if we replaced the first program rule by: $\mathcal{R}'_1 : \langle p \leftarrow_P (0.9 \&_G q \&_G s); \ 0.6; \ 0.54 \rangle$. It is important to note now that even when the truth degree of the rule is 0.6, its threshold decreases to $Up\_rule(\mathcal{R}'_1) = 0.6 * 0.9 = 0.54 < 0.56$, which reduce considerably the number of steps in the tabulation procedure, as Figure 2 shows.

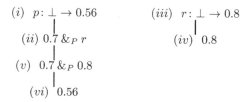

$$(i) \ \ p : \perp \rightarrow 0.56 \qquad\qquad (iii) \ \ r : \perp \rightarrow 0.8$$
$$(ii) \ 0.7 \&_P r \qquad\qquad\qquad (iv) \ \ 0.8$$
$$(v) \ \ 0.7 \&_P 0.8$$
$$(vi) \ \ 0.56$$

Fig. 2. Example threshold forest for query $?p$ w.r.t. the extended program $\mathbb{P}'+$

3   Improving the Static Preprocess with Fuzzy Unfolding

Program transformation is an optimization technique for computer programs that starting with an initial program $\mathbb{P}_0$ derives a sequence $\mathbb{P}_1, \ldots, \mathbb{P}_n$ of transformed programs by applying *elementary transformation rules* (fold/unfold) which improve the original program. The basic idea is to divide the program development activity, starting with a (possibly naive) problem specification written in a programming language, into a sequence of small transformation steps. *Unfolding* [3, 12] is a well-known, widely used, semantics-preserving program transformation rule. In essence, it is usually based on the application of operational steps on the body of program rules. The unfolding transformation is able to improve programs, generating more efficient code. Unfolding is the basis for developing sophisticated and powerful programming tools, such as fold/unfold transformation systems or partial evaluators, etc.

On the other hand, as revealed in the examples of the previous sections, the presence of truth degrees in the body of fuzzy program rules is always desirable for optimizing the power of thresholding at tabulation time. In [7], we show that it is possible to transform a program rule into a semantically equivalent set of rules with the intended shape. The following definition is recalled from [7], but we have slightly simplified it in the sense that here we deal with propositional (instead of first order) MALP programs:

Definition 2 (Fuzzy Unfolding). *Let $\mathbb{P}$ be a program and let $\mathcal{R} : \langle A \leftarrow_i B; \alpha \rangle \in \mathbb{P}$ be a program rule which is not a fact. Then, the fuzzy unfolding of program $\mathbb{P}$ with respect to rule $\mathcal{R}$ is the new program: $\mathbb{P}' = (\mathbb{P} - \{\mathcal{R}\}) \cup \mathcal{U}$ where $\mathcal{U} = \{\langle A \leftarrow_i B'; \alpha \rangle \mid B \rightarrow_{AS} B'\}$.*

It is important to note in the previous definition that the set of *unfolded rules* $\mathcal{U}$ is not a singleton in general. For instance, assume a program $\mathbb{P}$ with the following two rules $\mathcal{R}_1 : \langle p \leftarrow; 0.5 \rangle$ and $\mathcal{R}_2 : \langle p \leftarrow_\mathsf{P} p; 1 \rangle$. Note that the first rule does not admit unfolding since there are no propositional symbols on its body (it is a "fact"), but $\mathcal{R}_2$ can be unfolded by using both rules for obtaining $\mathcal{U} = \{\mathcal{R}_{2-1} : \langle p \leftarrow_\mathsf{P} 0.5 ; 1 \rangle, \quad \mathcal{R}_{2-2} : \langle p \leftarrow_\mathsf{P} 1 \&_\mathsf{P} p ; 1 \rangle \}$.[4]

Now the unfolded program $\mathbb{P}'$ is composed by the following set of rules defining p (note that we have replaced the original rule $\mathcal{R}_2$ by the transformed rules $\mathcal{R}_{2-1}$ and $\mathcal{R}_{2-2}$):

$$
\begin{array}{llll}
\mathcal{R}_1 : & \langle\, p & \leftarrow & ; 0.5 \,\rangle \\
\mathcal{R}_{2-1} : & \langle\, p & \leftarrow_\mathsf{P} 0.5 & ; 1 \,\rangle \\
\mathcal{R}_{2-2} : & \langle\, p & \leftarrow_\mathsf{P} 1 \&_\mathsf{P} p ; 1 & \,\rangle
\end{array}
$$

Assume now that we add more rules to the previous program $\mathbb{P}$ (note that the tabulation procedure would never end if the program is infinite) as follows:

$$
\begin{array}{llllll}
\mathcal{R}_1 : & \langle\, p & \leftarrow & ; 0.5 \,\rangle & \qquad \mathcal{R}_5 : \langle\, q_2 & \leftarrow_\mathsf{P} q_3 ; 0.7 \,\rangle \\
\mathcal{R}_2 : & \langle\, p & \leftarrow_\mathsf{P} p & ; 1 \,\rangle & \qquad \mathcal{R}_6 : \langle\, q_3 & \leftarrow_\mathsf{P} q_4 ; 0.7 \,\rangle \\
\mathcal{R}_3 : & \langle\, p & \leftarrow_\mathsf{P} q_1 & ; 0.7 \,\rangle & \qquad \vdots \\
\mathcal{R}_4 : & \langle\, q_1 & \leftarrow_\mathsf{P} q_2 & ; 0.7 \,\rangle
\end{array}
$$

Now, by unfolding rule $\mathcal{R}_2$ we obtain three transformed rules, the two ones seen before (i.e., $\mathcal{R}_{2-1}$ and $\mathcal{R}_{2-2}$) as well as $\mathcal{R}_{2-3} : \langle p \leftarrow_\mathsf{P} 0.7 \&_\mathsf{P} q_1 ; 1 \rangle$. Two more examples: the unfolding of this last rule (using $\mathcal{R}_4$) replaces $\mathcal{R}_{2-3}$ by $\mathcal{R}_{2-3-4} : \langle p \leftarrow_\mathsf{P} 0.7 \&_\mathsf{P} 0.7 \&_\mathsf{P} q_2 ; 1 \rangle$, whereas the unfolding of $\mathcal{R}_3$ (using again $\mathcal{R}_4$) replaces such rule by $\mathcal{R}_{3-4} : \langle p \leftarrow_\mathsf{P} 0.7 \&_\mathsf{P} q_2 ; 0.7 \rangle$.

On the other hand, let us remember that the static preprocess described in [5] and summarized at the beginning of the previous section, simply consists in generating extended programs where each program rule $\mathcal{R}$ is annotated with its proper $Up\_rule(\mathcal{R})$ value in order to increase the opportunities of bounding branches when applying thresholded tabulation. Now, our new proposal reinforces the static preprocess (since the new annotated $Up\_rule(\mathcal{R})$ values tends to be lower and thus, more powerful) by applying several unfolding steps (at least one) on program rules before generating the extended program $\mathbb{P}+$. The question now is when to stop the application of unfolding steps.

To answer this question, let us first introduce a new concept. Assume that, for a given propositional symbol p, we define the truth degree $\tau_p \in L$ as the infimum one among the weights of the rules with head p in the original program $\mathbb{P}$. For instance, in our previous example we have that $\tau_p = \inf\{0.5, 1, 0.7\} = 0.5$, $\tau_{q_1} = 0.7$, $\tau_{q_2} = 0.7$ and so on. Then, after applying the first unfolding step on each program rule $\mathcal{R} \in \mathbb{P}$ we obtain a new set of transformed rules, say $\mathcal{R}'_i$, $1 \leq i \leq n$, such that no more unfolding steps are applied on each rule $\mathcal{R}'_i$ if obviously it has no propositional symbols on its body (as occurs with $\mathcal{R}_{2-1}$) or one of the following conditions hold:

---

[4] Note that in this step the body of $\mathcal{R}_{2-2}$ is not computed, although the interpretation of $1 \&_\mathsf{P} p$ is clearly p, since this part is made in the tabulation procedure.

- $Up\_rule(\mathcal{R}'_i) \leq Up\_rule(\mathcal{R})$. This situation is represented by rule $\mathcal{R}_{2-2}$ in our example (since $Up\_rule(\mathcal{R}_{2-2}) = Up\_rule(\mathcal{R}_2) = 1$), and its is useful for preventing an infinite unfolding loop on rules whose Up_rule values cannot be improved by unfolding.

- $Up\_rule(\mathcal{R}'_i) \leq \tau_p$. This case is really the more interesting one in our technique, as revealed in rules $\mathcal{R}_{2-3-4}$ and $\mathcal{R}_{3-4}$ since:
  - $Up\_rule(\mathcal{R}_{2-3-4}) = 1 * 0.7 * 0.7 * 1 = 0.49 < 0.5 = \tau_p$.
  - $Up\_rule(\mathcal{R}_{3-4}) = 0.7 * 0.7 * 1 = 0.49 < 0.5 = \tau_p$.

We have just seen how the unfolding process finishes in our example, which let us now to generate the following extended program $\mathbb{P}+$:

$$
\begin{array}{llllll}
\mathcal{R}_1: & \langle & p & \leftarrow & & ; 0.5 & ; 0.5 \rangle \\
\mathcal{R}_{2-1}: & \langle & p & \leftarrow & 0.5 & ; 1 & ; 0.5 \rangle \\
\mathcal{R}_{2-2}: & \langle & p & \leftarrow_\mathsf{p} & 1 \&_\mathsf{p} p & ; 1 & ; 1 \rangle \\
\mathcal{R}_{2-3-4}: & \langle & p & \leftarrow_\mathsf{p} & 0.7 \&_\mathsf{p} 0.7 \&_\mathsf{p} q_2 & ; 1 & ; 0.49 \rangle \\
\mathcal{R}_{3-4}: & \langle & p & \leftarrow_\mathsf{p} & 0.7 \&_\mathsf{p} q_2 & ; 0.7 & ; 0.49 \rangle \\
\mathcal{R}_{4-5}: & \langle & q_1 & \leftarrow_\mathsf{p} & 0.7 \&_\mathsf{p} q_3 & ; 0.7 & ; 0.49 \rangle \\
\mathcal{R}_{5-6}: & \langle & q_2 & \leftarrow_\mathsf{p} & 0.7 \&_\mathsf{p} q_4 & ; 0.7 & ; 0.49 \rangle \\
& \vdots
\end{array}
$$

It is not difficult to check that, by following our improved thresholded tabulation technique, the rule $\mathcal{R}_{3-4}$ is not considered in the computation of p and no loops arise in this program. Therefore, the unique solution 0.5 for our initial query $?p$, which is obtained in a small number of computation steps.

4   Conclusions and Future Work

In this paper we have taken profit of our previous advances on fuzzy unfolding in order to reinforce the power of the static preprocess we described in [5], with the new aim of increasing the performance of the fuzzy thresholded tabulation procedure conceived in [6] for the fast execution of MALP programs. We are nowadays lifting our results to the first order case and implementing the method drawn in this paper into the FLOPER platform [11].

References

1. Almendros-Jiménez, J.M., Luna Tedesqui, A., Moreno, G.: A Flexible XPath-based Query Language Implemented with Fuzzy Logic Programming. In: Bassiliades, N., Governatori, G., Paschke, A. (eds.) RuleML 2011 - Europe. LNCS, vol. 6826, pp. 186–193. Springer, Heidelberg (2011)
2. Baldwin, J.F., Martin, T.P., Pilsworth, B.W.: Fril- Fuzzy and Evidential Reasoning in Artificial Intelligence. John Wiley & Sons, Inc. (1995)
3. Burstall, R.M., Darlington, J.: A Transformation System for Developing Recursive Programs. Journal of the ACM 24(1), 44–67 (1977)

4. Ishizuka, M., Kanai, N.: Prolog-ELF Incorporating Fuzzy Logic. In: Aravind, K. (ed.) Proceedings of the 9th International Joint Conference on Artificial Intelligence (IJCAI 1985), pp. 701–703. Morgan Kaufmann (1985)
5. Julián, P., Medina, J., Morcillo, P.J., Moreno, G., Ojeda-Aciego, M.: A static preprocess for improving fuzzy thresholded tabulation. In: Cabestany, J., Rojas, I., Joya, G. (eds.) IWANN 2011, Part II. LNCS, vol. 6692, pp. 429–436. Springer, Heidelberg (2011)
6. Julián, P., Medina, J., Moreno, G., Ojeda, M.: Efficient thresholded tabulation for fuzzy query answering. Studies in Fuzziness and Soft Computing (Foundations of Reasoning under Uncertainty) 249, 125–141 (2010)
7. Julián, P., Moreno, G., Penabad, J.: On Fuzzy Unfolding. A Multi-adjoint Approach. Fuzzy Sets and Systems 154, 16–33 (2005)
8. Kifer, M., Subrahmanian, V.S.: Theory of generalized annotated logic programming and its applications. Journal of Logic Programming 12, 335–367 (1992)
9. Medina, J., Ojeda-Aciego, M., Vojtáš, P.: Multi-adjoint logic programming with continuous semantics. In: Eiter, T., Faber, W., Truszczyński, M. (eds.) LPNMR 2001. LNCS (LNAI), vol. 2173, pp. 351–364. Springer, Heidelberg (2001)
10. Medina, J., Ojeda-Aciego, M., Vojtáš, P.: Similarity-based Unification: a multi-adjoint approach. Fuzzy Sets and Systems 146, 43–62 (2004)
11. Morcillo, P.J., Moreno, G.: Programming with fuzzy logic rules by using the FLOPER tool. In: Bassiliades, N., Governatori, G., Paschke, A. (eds.) RuleML 2008. LNCS, vol. 5321, pp. 119–126. Springer, Heidelberg (2008)
12. Tamaki, H., Sato, T.: Unfold/Fold Transformations of Logic Programs. In: Tärnlund, S. (ed.) Proc. of Second Int'l Conf. on Logic Programming, pp. 127–139 (1984)

Author Index